Big Data and Artificial Intelligence—Volume III

Big Data and Artificial Intelligence—Volume III

Topic Editors

Miltiadis D. Lytras
Andreea Claudia Serban

MDPI

Basel • Beijing • Wuhan • Barcelona • Belgrade • Novi Sad • Cluj • Manchester

Topic Editors
Miltiadis D. Lytras
Computer Science
Effat University
Jeddah
Saudi Arabia

Andreea Claudia Serban
Faculty of Theoretical and
Applied Economics
The Bucharest University of
Economic Studies
Bucharest
Romania

Editorial Office
MDPI AG
Grosspeteranlage 5
4052 Basel, Switzerland

This is a reprint of the Topic, published open access by the journals *Big Data and Cognitive Computing* (2504-2289), *Future Internet* (ISSN 1999-5903), *Information* (ISSN 2078-2489), *Remote Sensing* (2072-4292) and *Sustainability* (ISSN 2071-1050), freely accessible at: https://www.mdpi.com/topics/Big_Data_Artificial_Intelligence.

For citation purposes, cite each article independently as indicated on the article page online and as indicated below:

Lastname, A.A.; Lastname, B.B. Article Title. *Journal Name* **Year**, *Volume Number*, Page Range.

Volume III
ISBN 978-3-7258-3975-9 (Hbk)
ISBN 978-3-7258-3976-6 (PDF)
https://doi.org/10.3390/books978-3-7258-3976-6

Set
ISBN 978-3-7258-3969-8 (Hbk)
ISBN 978-3-7258-3970-4 (PDF)

Contents

About the Editors ix

Preface xi

Youmeng Luo, Wei Li, Xiaoyu Ma and Kaiqiang Zhang
Image Retrieval Algorithm Based on Locality-Sensitive Hash Using Convolutional Neural Network and Attention Mechanism
Reprinted from: *Information* **2022**, *13*, 446, https://doi.org/10.3390/info13100446 1

Ziyang Bian, Tingfa Xu, Junjie Chen, Liang Ma, Wenjing Cai and Jianan Li
Auto-Learning Correlation-Filter-Based Target State Estimation for Real-Time UAV Tracking
Reprinted from: *Remote Sens.* **2022**, *14*, 5299, https://doi.org/10.3390/rs14215299 13

Meng Zhang, Jiatong Ling, Buyun Tang, Hang Zhang and Laibin Zhang
A Data-Driven Based Method for Pipeline Additional Stress Prediction Subject to Landslide Geohazards
Reprinted from: *Sustainability* **2022**, *14*, 11999, https://doi.org/10.3390/su141911999 35

Mateus Habermann, Elcio Hideiti Shiguemori and Vincent Frémont
Unsupervised Cluster-Wise Hyperspectral Band Selection for Classification
Reprinted from: *Remote Sens.* **2022**, *14*, 5374, https://doi.org/10.3390/rs14215374 51

Wenchao Ban and Liangduo Shen
PM2.5 Prediction Based on the CEEMDAN Algorithm and a Machine Learning Hybrid Model
Reprinted from: *Sustainability* **2022**, *14*, 16128, https://doi.org/10.3390/su142316128 70

Yang Zhang, Hongpo Zhang and Bo Zhang
An Effective Ensemble Automatic Feature Selection Method for Network Intrusion Detection
Reprinted from: *Information* **2022**, *13*, 314, https://doi.org/10.3390/info13070314 85

Xingyu Chai, Jincai Li, Jun Zhao, Wuxin Wang and Xiaofeng Zhao
LGB-PHY: An Evaporation Duct Height Prediction Model Based on Physically Constrained LightGBM Algorithm
Reprinted from: *Remote Sens.* **2022**, *14*, 3448, https://doi.org/10.3390/rs14143448 100

Jinghui Feng, Haopeng Kuang and Lihua Zhang
EBBA: An Enhanced Binary Bat Algorithm Integrated with Chaos Theory and Lévy Flight for Feature Selection
Reprinted from: *Future Internet* **2022**, *14*, 178, https://doi.org/10.3390/fi14060178 119

Jianhua Guo, Qingsong Xu, Yue Zeng, Zhiheng Liu and Xiaoxiang Zhu
Semi-Supervised Cloud Detection in Satellite Images by Considering the Domain Shift Problem
Reprinted from: *Remote Sens.* **2022**, *14*, 2641, https://doi.org/10.3390/rs14112641 135

Yong Zhou, Lingyu Wang and Junhao Qian
Application of Combined Models Based on Empirical Mode Decomposition, Deep Learning, and Autoregressive Integrated Moving Average Model for Short-Term Heating Load Predictions
Reprinted from: *Sustainability* **2022**, *14*, 7349, https://doi.org/10.3390/su14127349 158

Mohamed Soudy, Yasmine M. Afify and Nagwa Badr
GenericConv: A Generic Model for Image Scene Classification Using Few-Shot Learning
Reprinted from: *Information* **2022**, *13*, 315, https://doi.org/10.3390/info13070315 178

Siyun Hou, Wengen Li, Tianying Liu, Shuigeng Zhou, Jihong Guan, Rufu Qin and Zhenfeng Wang
MIMO: A Unified Spatio-Temporal Model for Multi-Scale Sea Surface Temperature Prediction
Reprinted from: *Remote Sens.* **2022**, *14*, 2371, https://doi.org/10.3390/rs14102371 . . . . **191**

Huanxin Zou, Shitian He, Xu Cao, Li Sun, Juan Wei, Shuo Liu and Jian Liu
Rescaling-Assisted Super-Resolution for Medium-Low Resolution Remote Sensing Ship Detection
Reprinted from: *Remote Sens.* **2022**, *14*, 2566, https://doi.org/10.3390/rs14112566 . . . . **207**

Jiahao Guo, Xiaohuo Yu and Lu Wang
Unsupervised Anomaly Detection and Segmentation on Dirty Datasets
Reprinted from: *Future Internet* **2022**, *14*, 86, https://doi.org/10.3390/fi14030086 . . . . **224**

Leyuan Liu, Jian He, Keyan Ren, Zhonghua Xiao and Yibin Hou
A LiDAR–Camera Fusion 3D Object Detection Algorithm
Reprinted from: *Information* **2022**, *13*, 169, https://doi.org/10.3390/info13040169 . . . . **241**

Qian Wang, Chunshan Wang, Huarui Wu, Chunjiang Zhao, Guifa Teng, Yajie Yu and Huaji Zhu
A Two-Stage Low-Altitude Remote Sensing Papaver Somniferum Image Detection System Based on YOLOv5s+DenseNet121
Reprinted from: *Remote Sens.* **2022**, *14*, 1834, https://doi.org/10.3390/rs14081834 . . . . **252**

Khalid Alfalqi and Martine Bellaiche
An Emergency Event Detection Ensemble Model Based on Big Data
Reprinted from: *Big Data Cogn. Comput.* **2022**, *6*, 42, https://doi.org/10.3390/bdcc6020042 . . . . **270**

Junjie Chen, Bo Huang, Jianan Li, Ying Wang, Moxuan Ren and Tingfa Xu
Learning Spatio-Temporal Attention Based Siamese Network for Tracking UAVs in the Wild
Reprinted from: *Remote Sens.* **2022**, *14*, 1797, https://doi.org/10.3390/rs14081797 . . . . **288**

Zhengwu Yuan, Wendong Huang, Aixia Yang, Chan Tang and Xiaobo Luo
Graph-Based Embedding Smoothing Network for Few-Shot Scene Classification of Remote Sensing Images
Reprinted from: *Remote Sens.* **2022**, *14*, 1161, https://doi.org/10.3390/rs14051161 . . . . **308**

Xinwei Ren, Wei Yang, Xianliang Jiang, Guang Jin and Yan Yu
A Deep Learning Framework for Multimodal Course Recommendation Based on LSTM+Attention
Reprinted from: *Sustainability* **2022**, *14*, 2907, https://doi.org/10.3390/su14052907 . . . . **327**

Fucong Liu, Hui Xu, Miao Qi, Di Liu, Jianzhong Wang and Jun Kong
Depth-Wise Separable Convolution Attention Module for Garbage Image Classification
Reprinted from: *Sustainability* **2022**, *14*, 3099, https://doi.org/10.3390/su14053099 . . . . **341**

Andrea Gasparetto, Matteo Marcuzzo, Alessandro Zangari and Andrea Albarelli
A Survey on Text Classification Algorithms: From Text to Predictions
Reprinted from: *Information* **2022**, *13*, 83, https://doi.org/10.3390/info13020083 . . . . **359**

Fanfan Liu, Wenzhe Zhao, Guangyao Zhou, Liangjin Zhao and Haoran Wei
SR-Net: Saliency Region Representation Network for Vehicle Detection in Remote Sensing Images
Reprinted from: *Remote Sens.* **2022**, *14*, 1313, https://doi.org/10.3390/rs14061313 . . . . **398**

Huaji Zhou, Jing Bai, Linchun Niu, Zhu Xiao, Jie Xu, Shilian Zheng, et al.
Electromagnetic Signal Classification Based on Class Exemplar Selection and Multi-Objective Linear Programming
Reprinted from: *Remote Sens.* **2022**, *14*, 1177, https://doi.org/10.3390/rs14051177 . . . . **419**

Jialin Shi, Chenyi Guo and Ji Wu
A Hybrid Robust-Learning Architecture for Medical Image Segmentation with Noisy Labels
Reprinted from: *Future Internet* **2022**, *14*, 41, https://doi.org/10.3390/fi14020041 . . . . **438**

Jianing Wang, Runhu Huang, Siying Guo, Linhao Li, Zhao Pei and Bo Liu
HyperLiteNet: Extremely Lightweight Non-Deep Parallel Network for Hyperspectral Image Classification
Reprinted from: *Remote Sens.* **2022**, *14*, 866, https://doi.org/10.3390/rs14040866 . . . . **451**

Beáta Novotná, Ľuboš Jurík, Ján Čimo, Jozef Palkovič, Branislav Chvíla and Vladimír Kišš
Machine Learning for Pan Evaporation Modeling in Different Agroclimatic Zones of the Slovak Republic (Macro-Regions)
Reprinted from: *Sustainability* **2022**, *14*, 3475, https://doi.org/10.3390/su14063475 . . . . **470**

Jan Pawel Musial and Jedrzej Stanislaw Bojanowski
Comparison of the Novel Probabilistic Self-Optimizing Vectorized Earth Observation Retrieval Classifier with Common Machine Learning Algorithms
Reprinted from: *Remote Sens.* **2022**, *14*, 378, https://doi.org/10.3390/rs14020378 . . . . **492**

Wendong Huang, Zhengwu Yuan, Aixia Yang, Chan Tang and Xiaobo Luo
TAE-Net: Task-Adaptive Embedding Network for Few-Shot Remote Sensing Scene Classification
Reprinted from: *Remote Sens.* **2022**, *14*, 111, https://doi.org/10.3390/rs14010111 . . . . **528**

Johann Desloires, Dino Ienco, Antoine Botrel and Nicolas Ranc
Positive Unlabelled Learning for Satellite Images' Time Series Analysis: An Application to Cereal and Forest Mapping
Reprinted from: *Remote Sens.* **2022**, *14*, 140, https://doi.org/10.3390/rs14010140 . . . . **547**

About the Editors

Miltiadis D. Lytras

Dr. Miltiadis D. Lytras is a faculty member at The American College of Greece, Athens, Greece, and a visiting researcher at Effat University. A world-renowned expert in cognitive computing, information systems, technology-enabled innovation, social networks, and knowledge management, Dr. Lytras has extensive experience bridging academia and industry across Europe and Asia. His expertise spans advanced computer science, business management, and digital transformation, with a focus on semantic technologies, AI-driven analytics, and smart systems.

Dr. Lytras has an impressive editorial and publication record, having co-edited over 110 high-impact special issues in leading ISI/Scopus-indexed journals and co-authored more than 80 books with top international publishers such as Elsevier, Emerald, IGI Global, and Springer. His research contributions include over 120 high-impact papers published in Q1 and Q2 Web of Science and Scopus journals, including *IEEE Transactions on Knowledge and Data Engineering, Future Generation Computer Systems, Journal of Business Research, International Journal of Information Management*, and *Technological Forecasting and Social Change.*

With 25 years of experience in research and development (R&D) projects, Dr. Lytras has played a key role in securing and executing over 70 R&D initiatives worldwide. His work has contributed significantly to digital transformation, AI ethics, smart cities, and higher education innovation. He has also held senior editorial positions in several prestigious journals, serving as the Founding Editor-in-Chief of the *International Journal on Semantic Web and Information Systems* (IJSWIS) and the *International Journal of Technology Enhanced Learning* (IJTEL).

Dr. Lytras continues to shape the academic and research landscape through his contributions to AI-driven education, knowledge-based decision making, and sustainable development, fostering collaboration between researchers, industry leaders, and policymakers.

Andreea Claudia Serban

Dr. Andreea Claudia Serban is a Professor at the Department of Economics and Economic Policies and Director of the Doctoral School of Economics I at Bucharest University of Economic Studies. Holding a PhD in Economics from the same institution and a postdoctoral degree from the Romanian Academy of Sciences, Dr. Șerban has established herself as a leading expert in sustainable development, knowledge economy, smart development, labor markets, education, and demographic challenges.

Her research portfolio includes the publication of seven books and numerous peer-reviewed articles in high-ranking international journals. She has actively participated in international conferences and contributed to key discussions on economic policies, digital transformation, and innovation-driven development. Dr. Șerban also serves as an Associate Editor for the *International Journal on Semantic Web and Information Systems* (IJSWIS) and as the Editor-in-Chief of the *Global Journal of Business, Economics, and Management: Current Issues.*

In addition to her editorial roles, Dr. Șerban is a recognized expert for the Lithuanian Research Council and a dedicated reviewer for prestigious international journals. She has been a member of editorial boards and organizing committees for numerous academic conferences, contributing to shaping discourse in economics, AI applications, and digital policies.

Through her work, Dr. Șerban continues to bridge economic theory and practical policymaking, advancing research in AI-powered economic modeling, digital transformation, and innovation-driven business strategies. Her contributions have been instrumental in promoting sustainable and inclusive growth in the era of artificial intelligence and Big Data.

Preface

The rapid evolution of Big Data and artificial intelligence (AI) has transformed industries, research paradigms, and everyday life. These technologies serve as catalysts for innovation, enabling data-driven decision-making, automation, and enhanced predictive capabilities. The second volume of this Topic continues to explore the theoretical and applied dimensions of Big Data and AI, highlighting their impact across domains ranging from computational science to business and public policy. By fostering interdisciplinary research and promoting the exchange of knowledge, this reprint presents a diverse range of studies that reflect the latest advancements in these fields.

This volume synthesizes research contributions spanning multiple dimensions of AI and Big Data, encompassing both theoretical approaches and applied contributions. The studies included in this collection address key areas such as machine learning, neural networks, natural language processing, image recognition, and data analytics. Researchers explore novel architectures for deep learning, innovative frameworks for managing large-scale datasets, and advanced methodologies for processing textual and visual data. Computational optimization and AI-driven decision-making models further expand the theoretical foundations of the field. Alongside these advancements, the volume presents applied contributions that demonstrate the practical implementation of AI and Big Data across multiple disciplines. Key applications include machine learning techniques for climate and environmental research, AI-driven healthcare solutions for disease detection and personalized medicine, business intelligence models for predictive analytics and fraud detection, and AI-enabled smart city initiatives that optimize urban planning, traffic management, and autonomous transportation systems. These studies not only introduce new methodologies but also provide real-world examples of AI and Big Data applications, emphasizing their transformative role in contemporary society.

By integrating cutting-edge research on methodologies, frameworks, and real-world case studies, this reprint offers valuable insights into the evolving landscape of AI and Big Data. The contributions highlight advancements in predictive AI models for weather forecasting, demand prediction, and anomaly detection, as well as innovations in high-dimensional data processing, data fusion, and automated feature selection. Ethical and societal considerations are also addressed, focusing on privacy, security, and governance challenges in AI applications. Additionally, the volume underscores the importance of interdisciplinary collaboration, showcasing industry–government–academic partnerships that drive innovation and technological progress. These findings serve as a foundation for future research, providing researchers, policymakers, and industry professionals with essential knowledge to harness the power of AI and Big Data effectively.

As technological breakthroughs continue to shape the future of AI and Big Data, new areas of development are emerging. Explainable AI and ethical AI are becoming increasingly critical, ensuring transparency and fairness in AI-driven decision-making. The integration of AI with quantum computing presents exciting possibilities for solving complex data problems, while AI for sustainability is gaining momentum as a tool to support the achievement of the United Nations Sustainable Development Goals. The convergence of AI and the Internet of Things (IoT) is also expected to enable real-time, autonomous decision-making in smart environments, further expanding the impact of intelligent systems. Given the ongoing advancements in AI and data science, we anticipate further volumes of this Topic, providing an ever-expanding repository of knowledge that reflects the latest innovations and research trends.

We extend our sincere gratitude to all contributors, peer reviewers, and the editorial team for their invaluable efforts in making this Topic a success. The MDPI team has played a crucial role in facilitating the publication process and ensuring high-quality research dissemination. We also acknowledge the scientific community's ongoing commitment to AI and Big Data research, which continues to drive innovation and foster knowledge sharing. We hope this reprint serves as a valuable resource for researchers, practitioners, and industry leaders, inspiring future research endeavors in Big Data and Artificial Intelligence.

Miltiadis D. Lytras and Andreea Claudia Serban
Topic Editors

information

MDPI

Article

Image Retrieval Algorithm Based on Locality-Sensitive Hash Using Convolutional Neural Network and Attention Mechanism

Youmeng Luo, Wei Li *, Xiaoyu Ma and Kaiqiang Zhang

School of Computer and Information Engineering, Xiamen University of Technology, Xiamen 361024, China
* Correspondence: drweili@hotmail.com

Abstract: With the continuous progress of image retrieval technology, in the field of image retrieval, the speed of a search for a desired image from a great deal of image data becomes a hot issue. Convolutional Neural Networks (CNN) have been used in the field of image retrieval. However, many image retrieval systems based on CNN have a poor ability to express image features, resulting in a series of problems such as low retrieval accuracy and robustness. When the target image is retrieved from a large amount of image data, the vector dimension after image coding is high and the retrieval efficiency is low. Locality-sensitive hash is a method to find similar data from massive high latitude data. It reduces the data dimension of the original spatial data through hash coding and conversion, and can also maintain the similarity between the data. The retrieval time and space complexity are low. Therefore, this paper proposes a locality-sensitive hash image retrieval method based on CNN and the attention mechanism. The steps of the method are as follows: using the ResNet50 network as the feature extractor of the image, adding the attention module after the convolution layer of the model, and using the output of the network full connection layer to retrieve the features of the image database, then using the local-sensitive hash algorithm to hash code the image features of the database to reduce the dimension and establish the index, and finally measuring the features of the image to be retrieved and the image database to get the most similar image, completing the content-based image retrieval task. The method in this paper is compared with other image retrieval methods on corel1k and corel5k datasets. The experimental results show that this method can effectively improve the accuracy of image retrieval, and the retrieval efficiency is significantly improved. It also has higher robustness in different scenarios.

Keywords: image retrieval; convolutional neural networks; attention mechanism; locality sensitive hash

Citation: Luo, Y.; Li, W.; Ma, X.; Zhang, K. Image Retrieval Algorithm Based on Locality-Sensitive Hash Using Convolutional Neural Network and Attention Mechanism. *Information* **2022**, *13*, 446. https://doi.org/10.3390/info13100446

Academic Editors: Miltiadis D. Lytras and Andreea Claudia Serban

Received: 1 September 2022
Accepted: 20 September 2022
Published: 24 September 2022

1. Introduction

With the advent of the information age, the number of digital images on the internet is growing exponentially. How to search for desired images from many image databases has become a hot issue that needs to be solved urgently [1], and content-based image retrieval (CBIR) can solve this problem. CBIR is a technology based on computer vision, which is used to search similar images from databases. Images are represented by high-dimensional features, which use different distance metric measures to calculate the similarity between images [2]. The retrieval performance is based on the shape, structure, color, or other features extracted from the image and the quality of matching.

Generally, content-based image retrieval includes two parts, feature extraction and similar image search. In the aspect of feature extraction, the Convolutional Neural Network can be used to extract features. However, the training samples of the actual data set are limited, which will lead to low robustness of the model, and extracting effective features is the key to improving the accuracy of image retrieval. However, most of the current image retrieval technologies only focus on the whole image feature extraction, and cannot identify the salient parts. Therefore, researchers have added an attention mechanism to the neural network. Attention mechanism is a method of imitating the human brain. It tends to selectively acquire some important parts of the observed things according to needs, while ignoring the unimportant parts [3], so that the model can make more accurate judgments. In the aspect of similar image searches, compared with the methods of manually selecting image features to express image content, such as SIFT, HOG, etc., the high-dimensional image features learned by Convolutional Neural Network can better express image information, but its computational complexity will also increase. In the past, researchers compared the feature vectors of high-dimensional query images with those of images in the database one by one. This method consumes feature storage space and matching time. The hash-based image retrieval technology can map high dimensional feature vectors to the same space while maintaining similarity through the hash algorithm to obtain binary hash codes that are much lower than the original dimensions. At the same time, the generated hash codes can be used to build efficient indexes, which not only realize data dimension reduction, but also improve the efficiency of feature matching.

Given the above problems, this paper designs a locality-sensitive hash image retrieval method based on CNN and attention mechanism. The main contributions of this paper are:

1. Preprocessing before CNN training. Adopting standard data enhancement methods for existing training data, such as random scaling, rotation, clipping, noise addition and color contrast change, to increase the sample size, avoiding overfitting and promoting the robustness of the model.
2. By adding a simple and effective attention module (CBAM) to the Convolutional Neural Network, it can be extensively adopted to promote the representation ability of CNN and improve image retrieval accuracy to a certain extent.
3. According to the features extracted by CNN, a locality-sensitive hashing dimension reduction method is designed to build a hash index, which solves the problem of large-scale and high-dimensional images and greatly shortens the retrieval time.

2. Related Work

In the early CBIR system, people used manually extracted features to represent the content of images. In recent years, with the appearance of large-scale labeled datasets such as ImageNet, CNN have been widely used for deep learning. People have found that deep convolution neural network has a strong feature extraction ability in image processing, and the use of Convolution Neural Network has become the mainstream feature extraction method [4]. As various CNN have been put forward, such as AlexNet, VGGNet, ResNet, and GoogleNet, they have achieved good results in many fields such as image classification [5], object detection, and image retrieval.

In the field of image retrieval, the attention mechanism also has relevant applications. Noh et al. [6] posed a spatial domain attention model, which was added to the convolution layer of the CNN to select the key points in the image and raise the accuracy of image retrieval. Li ZongMin et al. [7] employed an attention mechanism to the field of hand-drawn image retrieval, added an attention module to the VGG16 network, and used the extracted 4068-dimensional feature vector for hand-drawn image retrieval, achieving excellent retrieval results. Ng et al. [8] combined the second-order loss and the second-order spatial attention to re-weight the features to perform image matching. However, the feature dimension extracted by these methods is too high, which is not conducive to calculation.

In 2012, Krizhevsky et al. [9] used the full connection layer output of the AlexNet network proposed by him as the feature vector of the image for image retrieval. However, its output feature vector is up to 4096 dimensions, which has the disadvantages of too high dimensions and large computation. Babenko et al. [10] used Principal Component Analysis (PCA) to compress the dimension of the feature vector to lower the dimension of the data, which significantly accelerated the retrieval speed. However, comparing the similarity of the two matrices is still not an efficient operation. The feature representation of CNN learning is inefficient for image retrieval applications because of its high dimensional features. To adapt the feature representation learned from CNN to the image retrieval task, hash technology has been successfully applied, because it has the characteristics of fast query processing speed and low storage cost [11]. In 2014, Xia et al. [12] proposed the CNNH method, which is divided into two steps: the first step is to learn the hash code, the second step is to train the neural network, and learn the image features and hash function simultaneously. After that, Lai et al. [13] improved the CNNH method and proposed the DNNH model, so that the features learned by the neural network can be fed back to the hash code in time. The literature [14,15] proposes to use the approximate nearest neighbor (ANN) algorithm to speed up the operation, such as local sensitive hash, map the high dimensional feature data into the low-dimensional binary space, form the binary code, and compare their Hamming distance to further improve the retrieval speed.

To sum up, the existing large-scale image retrieval methods have their advantages and disadvantages, especially in terms of retrieval accuracy and retrieval time. These problems will be effectively addressed in the methods proposed in this study.

3. Image Retrieval Framework Based on Locality-Sensitive Hash Using CNN and Attention Mechanism

This study is the image retrieval based on the descending dimension of locality-sensitive hash. The process consists of feature extraction, hash coding, index construction and similarity calculation. Firstly, we obtain the feature vector from the image database through the feature extraction model, then save the binarized feature encoding of the image after the feature vector is processed by locality-sensitive hashing, and build an index to record in the image feature library. Finally, we input the image to be retrieved into the network model to obtain the feature code of the image and the feature database in the database for similarity calculation and comparison, and the closest n images are returned. The retrieval is completed. The flow diagram is shown in Figure 1.

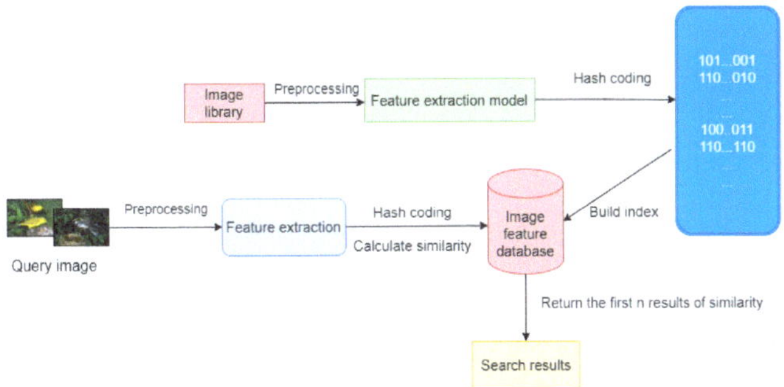

Figure 1. Flow chart of the image retrieval algorithm.

3.1. Feature Extraction

The framework feature extraction model is based on CNN and attention mechanism. The feature extraction model is composed of image preprocessing, convolution layer, attention module and full connection layer, as shown in Figure 2. ResNet50 is selected as the feature extraction network. The user first enhances the training set in the image library with standard data to increase the sample size, and then performs preprocessing operations. The preprocessed image is input into the model for training, and an attention module is added after the convolution layer of the network. By training the attention module, the important features in the retrieval task are given higher weights, so that the model can extract more effective features. Lastly, the output feature map is inputted to the full connection layer.

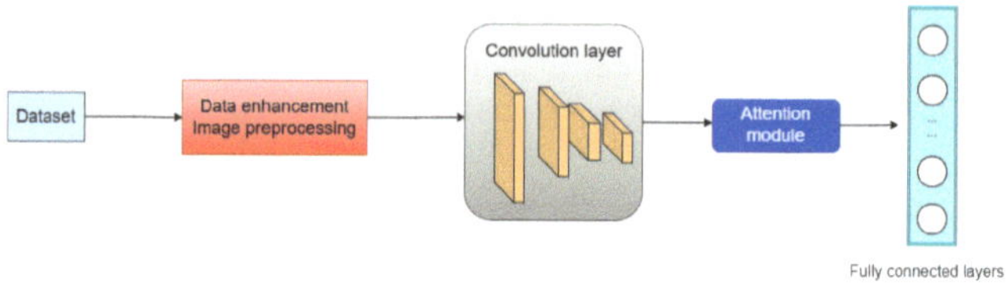

Figure 2. Feature extraction model.

Image Preprocessing

Deep convolution neural networks need a large number of training data to obtain good results, prevent overfitting, and improve robustness. However, it is often difficult to obtain enough training samples, so this paper adopts standard image enhancement methods before convolution training, image enhancement is a process of creating new training data from existing training data. Figure 3 shows the original image conversion examples of several image enhancement methods adopted in this paper. This paper applies these transformations to the original image data set and adds some training data. Image enhancement greatly improves the quality and performance of the deep neural network.

Figure 3. Several data enhancement examples.

After image enhancement of the training set, the training set is preprocessed to make the network model in this paper converge faster during training and feature extraction, and increase the accuracy of the retrieval results.

(1) Image size processing. The image dimension of the image database used in this study is 192 × 128 and 128 × 192, and the size is inconsistent. The size is processed as 192 × 192 to maintain the original characteristics.
(2) Mean and normalization. In the mean removal process, the mean value is subtracted from the RGB 3D, and the image data is centered to 0 to prevent overfitting, see Formula (1). For normalization processing, calculate the RGB maximum value, and compress the image data between 0–1. After normalization processing, the data can better respond to the activation function and improve its expressiveness of the data. The conversion function of the data is shown in Formula (2).

$$Y = Y_{\mathrm{n}} - \frac{1}{m}\sum_{n=1}^{m} Y_n \tag{1}$$

$$Y = \frac{Y_n - Y_{\min}}{Y_{\max} - Y_{\min}} \tag{2}$$

3.2. Using Attention Mechanism (CBAM) to Improve Retrieval Accuracy

The attention mechanism used in this paper is the CBAM module, and CBAM can be added to any position of the convolution layer. Spatial attention and channel attention have different functions. CBAM combines the two to better weigh the prominent parts in the image. In this study, the attention mechanism is applied to the locality-sensitive hashing image retrieval method. This study inputs an image of size 192 × 192 into the network model. A 7 × 7 × 2048 feature map is obtained after the image passes through the convolution layer of the model. The attention module will perform weighting operations on the channel domain and the spatial domain of the feature map, respectively, to generate a new feature map. The size of the feature map does not change before and after being input into the attention module. Lastly, the new feature map is flattened and input to the full connection layer of the model.

CBAM [16] is composed of a one-dimensional channel attention module $F_c \in R^{C\times 1\times 1}$ and a two-dimensional spatial attention module $F_s \in R^{1\times H\times W}$, which are arranged in series, as shown in Figure 4. The image features are first input into the channel attention module, and the channel features are generated using the channel relations of the features β'. Then, the generated features β' go through the spatial attention module to obtain the final features β''. The channel attention module retains more image texture information and detailed semantic information, while the spatial attention module extracts effective spatial features by focusing on the contour and spatial structure of the image. The calculation formula of β'' is as follows:

$$\beta' = F_c(\beta) \otimes \beta \tag{3}$$

$$\beta'' = F_s(\beta') \otimes \beta' \tag{4}$$

where: β' is the resulting output of the channel attention module and $\otimes$ is the array element multiplied in sequence.

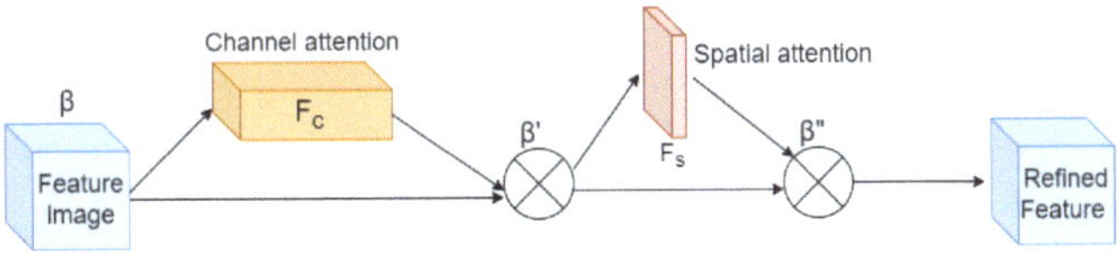

Figure 4. CBAM module (channel and spatial).

3.2.1. Channel Attention Module

The channel attention module can be expressed as:

$$F_c(\beta) = \sigma(MLP(Avg\,\mathrm{Pool}(\beta)) + MLP(\mathrm{MaxPool}(\beta))) \tag{5}$$

Firstly, the input feature map is entered into the channel attention, and then the max pooling and average pooling are carried out. Then, the output of the features by the MLP are added through the shared network, and then the RELU function is used to activate to obtain the final channel attention feature map. The flow is shown in Figure 5. Finally, we multiply this channel attention feature map by the input feature map to generate the input features required by the spatial attention module. Among β expressed as input characteristic map, $F_c(\beta)$ shows an output characteristic diagram. In this paper, the size of the input feature map of the channel attention module is $7 \times 7 \times 2048$, and the weight of each channel can be obtained after the operation of the pooling and perceptron. The weight is $1 \times 1 \times 2048$.

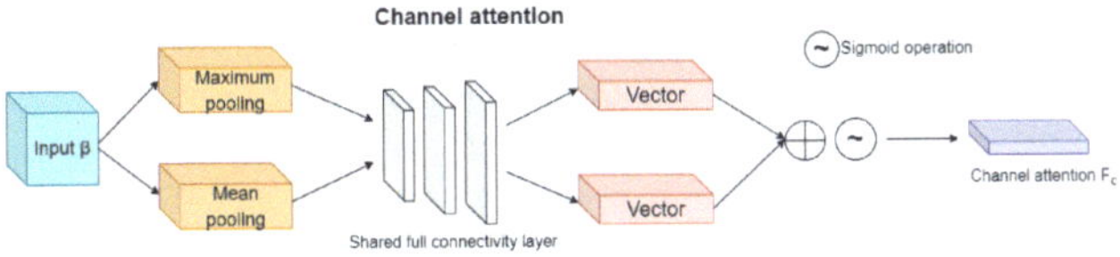

Figure 5. Channel attention structure.

3.2.2. Spatial Attention Module

The spatial attention module can be expressed as:

$$F_s(\beta') = \sigma(\mathrm{Conv}([\mathrm{mean}(\beta'); \mathrm{max}(\beta')])) \tag{6}$$

The input of this module is the feature map β' output by the channel attention module, We subject the feature map to maximum pooling along the channel dimension and mean pooling for dimensionality reduction. Then, the two feature maps with the number of channels of 1 are synthesized into one, and the convoluted attention feature map is obtained after Conv operation. Finally, the final spatial attention feature map $F_s(\beta)$ is obtained after RELU function activation, and the flow is shown in Figure 6. Finally, we multiply the spatial attention feature map with the input feature map β' to obtain the ultimate features.

From the above, it can be seen that the feature map output by the convolution module is weighted by the attention module to gain a new feature map of the same size. It focuses on the target area of the input image, filters out some invalid background information, and effectively represents the characteristics of the input image.

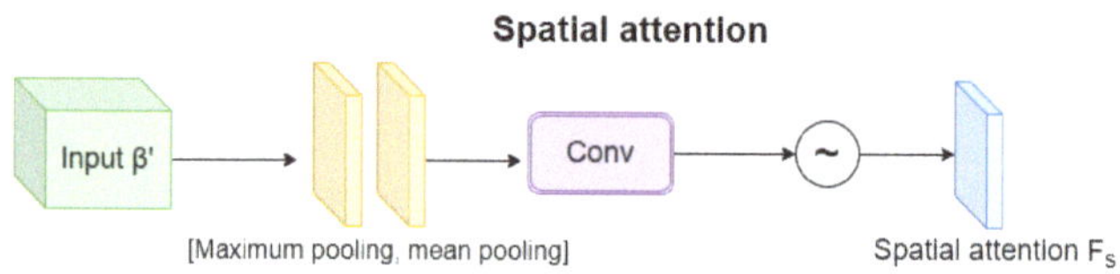

Figure 6. Spatial attention structure.

3.3. Using Local-Sensitive Hash Algorithm to Improve Retrieval Speed

There are many common similarity search algorithms in the image search part. The first thought is linear search. Firstly, the similarity measure is defined, and then the similarity is calculated in pairs, and the top-n is calculated for filtering. However, under the background demand of large-scale image retrieval, this time complexity is too large. At the same time, for high-dimensional sparse data, the calculation of similarity is very time-consuming. At this time, we need some approximation algorithms to improve the retrieval speed. In this study, the LSH algorithm is used to represent the original image with a low latitude binary hash code to avoid directly storing the high latitude image features. Moreover, the generated hash code is used to build an index to search for the nearest neighbor of the image. By calculating the distance between the image to be retrieved and each image in the database, the most similar result to the retrieved image is obtained. In the actual calculation, since the hash code is composed of 0 and 1, their hash distance can be directly performed by the bit operation in the computer, so that the operation speed is greatly improved. Compared with using image features in the original feature space for retrieval, using hash coding for nearest neighbor search reduces the cost of feature storage, improves the efficiency of feature matching, and significantly speeds up the retrieval speed.

LSH constructs a hash function so that the closer the feature points in the original feature space are to be mapped by this hash function, the more likely they will fall into the same hash bucket, and vice versa. Taking Figure 7 as an example, there are three images in total. The last two images have high similarities. From the perspective of a semantic level, the latter two images look more similar. Then, after mapping into the hash code, the distance should be smaller than that of the first image. Then, when querying the image, it is only necessary to obtain the bucket number, and then take out all the data in the bucket from the corresponding bucket, perform linear matching on them, and finally find the similar images that meet the query requirements. In other words, a super large original data set is divided into several subsets after the mapping transformation of the hash function, so as to reduce the query scope to improve retrieval performance.

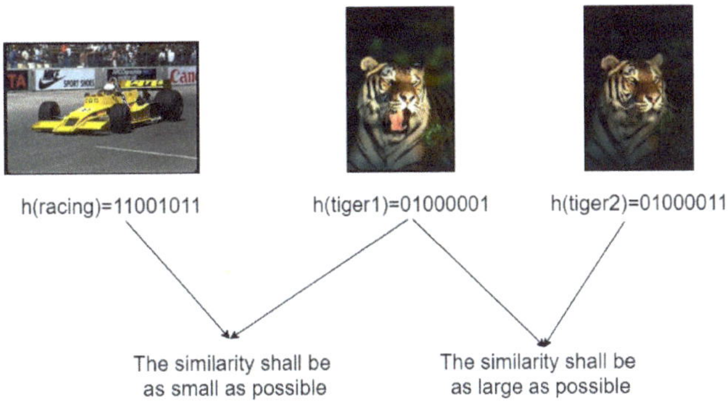

Figure 7. Hash code and image similarity.

(1) Similarity measure and LSH hash function

(a) LSH hash function family:

It is known that $d(x,y)$ is the distance between two points x, y, h is the hash function, $h(x)$ and $h(y)$ are the hash transformation of points x, y, and $P_1 > P_2$. For any two points x, y in the high dimensional space:

If $d(x,y) \leq h_1$, there must be $Pr[h(x) = h(y)] \geq P_1$.

If $d(x.y) \geq h_2$, there must be $Pr[h(x) = h(y)] \leq P_2$.

In similarity calculation, different methods can be adopted to measure the similarity between two points. For dissimilar measurement methods, the hash function used in the local hash is different. There are many methods to measure the similarity of two vectors. This study uses the LSH function based on cosine distance.

(b) LSH hash function of the cosine of the included angle of vector:

Almost similar eigenvectors can be found by representing them in k-dimensional space based on cosine distance. We use the angle between two vectors to measure whether two vectors are similar. The smaller the angle between two vectors, the more similar they are. The calculation formula of vector included angle is as follows:

$$\cos(\theta) = \frac{x \cdot y}{||x|| \cdot ||y||} \tag{7}$$

$$\cos(\theta) = \frac{\sum_{i=1}^{n} x_i y_i}{\sqrt{\sum_{i=1}^{n} x_i^2} \sqrt{\sum_{i=1}^{n} y_i^2}} \tag{8}$$

where x and y represent two sample data points, $\theta \in [0, \pi]$.

(2) Building Index

(a) The step to building index:

- Calculate a hash function $h(x)$ to store similar points in the same bucket.
- For a new query point $x_n \rightarrow h(x_n) \rightarrow \text{Bucket}_n$, calculate that x_n should belong to a certain slot.

Since the image features extracted in this paper are a large number of high latitude features, this paper projects the high latitude features into the multi-dimensional space, and each dimension represents a basic feature. However, it is very complicated to project K feature vectors in the high dimensional space with n features. Therefore, the feature vectors are projected into the m-dimensional space using the random projection method, where $m \ll n$, while maintaining the same cosine distance. Firstly, this paper divides the original data space with a random hyperplane. After each data is projected, it will fall into one side of the hyperplane. After multiple random hyperplane partitions, the original space is divided into many cells, and the data in each cell is considered likely to be adjacent, then, hashes the data in each cell into the corresponding slot through the hash function $h(x)$ to form a hash table. In a hash table, different numbers of buckets and their corresponding feature vectors may be created. Figure 8 shows LSH using the random projection method to create a set of hash tables to build an index. For example, x_1, x_5, x_8, x_{13} are similar feature vectors that are divided into the same slot.

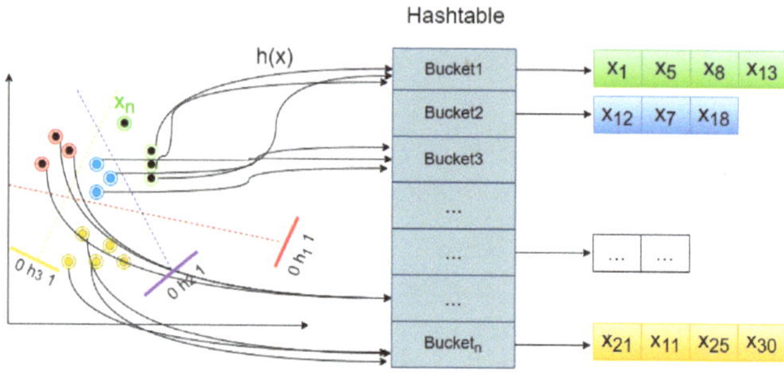

Figure 8. LSH builds index.

(3) Online Searching

The query image is first input into the network, and the feature vector is produced after the feature extraction model, and then the feature vector is encoded and hashed into the homologous hash bucket of the hash table employing the LSH algorithm random projection, and the corresponding bucket number is obtained. Then, the corresponding data in the bucket number is taken out, the distance between the query image to be retrieved and these data is calculated, and the most similar n data are returned. The search process is shown in Figure 9.

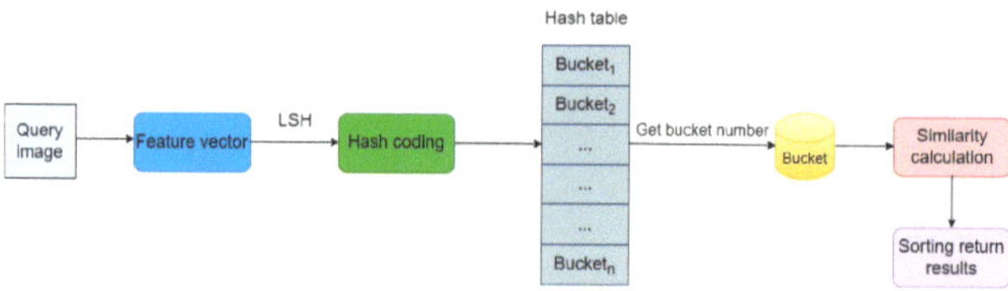

Figure 9. LSH lookup process.

4. Experimental Results and Analysis

4.1. Experimental Environment

The operating system of this experiment is ubuntu 20.04, the CPU is 2.5 GHz 6-core Intel Core i5, the GPU is NVIDIA GeForce RTX 3080, the memory is 12 GB, the programming language of the program is Python 3.6, and the framework based on deep learning is PyTorch.

4.2. Data Sets

This paper selects two data sets that are extensively adopted in the field of image retrieval: corel1k and corel5k. The corel1k dataset contains 10 categories of images such as buses and dinosaurs, with 100 images in each category, totaling 1000 images. Corel5k contains 50 categories of images, including racing cars, beaches, cats, airplanes, etc. Similar to the corel1k dataset, each category contains 100 images and a total of 5000 images. In the experiment, we will randomly extract 10 images from each type of image from corel1k and corel5k as query images, and the rest of corel5k as training sets.

4.3. Evaluating Indicator

The performance of content-based image retrieval is generally evaluated using recall, average retrieval accuracy (MAP) and retrieval time. Therefore, this paper uses these three evaluation indicators. The calculation formula of recall is:

$$B_{Recall} = \frac{1}{m} \sum_{i=1}^{n} f(i) \tag{9}$$

where m is the total number of images in the image library that are similar to the image to be retrieved, $f(i)$ is the similarity relationship between the ith image of the top n images in the search results and the image to be retrieved. If the image to be retrieved is similar to the retrieved image, the value of $f(i)$ is 1, otherwise it is 0.

For k searches, the average accuracy is calculated as follows:

$$B_{MAP} = \frac{1}{x} \sum_{b=1}^{x} \left[\frac{1}{n} \sum_{i=1}^{n} f(i) \right] \tag{10}$$

4.4. Performance Evaluation

4.4.1. Performance Comparison

We compare the proposed method with four different existing image retrieval methods to demonstrate the effectiveness of our method. The five comparison algorithms are SIFT, SVM active learning [17], LeNet-L [18], ref. [19] and ResNet50+CBAM+LSH (ours). Among the five models compared in this paper, the SIFT algorithm combined with the k-means clustering algorithm can achieve fast retrieval of small-scale simple images, which belong to the traditional image retrieval algorithm. The SVM active learning algorithm uses a classifier to find the optimal retrieval target by maximizing the interval. LeNet-L is an improvement of the LeNet-5 model in ref. [18]. Its structure is relatively simple and clear, and its parameters are medium. It is the best choice to compare other networks. Ref. [19] uses CNN to extract features, and further uses the LSH algorithm to build an index for searching. For fairness, we reproduce the method of ref. [19] on the same dataset. ResNet50+CBAM+LSH is the method suggedted in this study. ResNet50, as the feature extractor, adds the CBAM module after the convolution layer to increase the representation ability of CNN. Then, the LSH algorithm is used for hash coding to drop the dimension of high latitude features in the image library before searching. Table 1 shows the changing trend of the MAP of the five models on the dataset corel1k with the number of returned images and the comparison results of the recall rate.

Table 1. MAP, recall and Top-n of different methods on the dataset corel1k.

Method	MAP	Recall	Top-5	Top-10	Top-20	Top-30
Traditional SIFT	0.7748	0.7024	0.8543	0.7933	0.7893	0.7693
SVM Active Learning [17]	0.7241	0.6512	0.8403	0.7806	0.7425	0.6916
VGG-N [18]	0.8951	0.7858	0.9304	0.9125	0.8927	0.8841
Literature [19]	0.9186	0.8192	1	0.9932	0.9842	0.9646
Ours	0.9578	0.842	1	1	1	0.9994

According to Table 1, for the same Top-n, the method in this paper is superior to SIFT and other CNN models. At the same time, when the number of returned images grows, the method in this paper can still maintain a high MAP. In the dataset corel1k, the MAP of this paper is about 18.3% higher than that of the traditional SIFT method, about 23.4% higher than that of the SVM active learning method, and about 6.3% higher than that of the VGG-N method, which shows that the model in this study is better than other models in MAP. From Table 1 we know that compared with the literature [19], which does not use an attention mechanism, the MAP of the model in this paper using an attention mechanism is increased by about 3.9% on average. This result proves that under the same other conditions, the neural network with the CBAM module can enhance the representation ability of CNN, thus improving the accuracy of image retrieval. In addition, the recall rate of this method is also significantly better than the other four existing image retrieval methods, which proves the utility of the algorithm model proposed in this study.

In order to further examine the effectiveness of the algorithm model suggested in this study, we tested the MAP under this method and other methods on the corel5K dataset. The experimental results can be seen from Table 2. For the MAP on the dataset corel5K, this paper is about 17.6% higher than SIFT, 22% higher than SVM active learning, 5.5% higher than VGG-N, and 4.5% higher than literature [19]. This shows that the model suggested in this study has an effective role in improving MAP on different data sets, and further verifies that the CBAM module added to the neural network in this study can effectively heighten the accuracy of image retrieval.

Table 2. The MAP of different methods on data set corel5k.

Method	Traditional SIFT	SVM Active Learning [17]	VGG-N [18]	Literature [19]	Ours
MAP	0.7528	0.7091	0.8733	0.8841	0.9286

4.4.2. Comparison of Retrieval Time of Different Methods

This paper compares the retrieval time of this method with that of other algorithms on corel5k and corel1k through experiments, as shown in Table 3.

Table 3. Retrieval time of different methods on data sets corel1k and corel5k (time/s).

Method	corel1k	corel5k
Traditional SIFT	18.81	76.91
SVM Active Learning [17]	19.56	78.57
VGG-N [18]	12.42	43.81
Literature [20]	2.33	11.56
Ours	0.13	0.64

Table 3 illustrates the retrieval time of five different methods on the corel1k and corel5k datasets. Based on the traditional SIFT and SVM, the retrieval time is the longest, and the retrieval time of the method in this paper is significantly lower than that of VGG-N and literature [20]. Compared with VGG-N and literature [20] that do not use the LSH algorithm, the retrieval speed of this method is significantly improved. Although the ResNet50 network model is used for feature extraction in this paper, with high feature dimensions, the local-sensitive hash algorithm is adopted to reduce the feature dimensions, and there is no need to compare and calculate the image features in the image library one by one, which greatly reduces the retrieval time. Thus, the feasibility of the algorithm model in this study to improve the efficiency of retrieval time is verified.

4.4.3. Model Robustness

In this experiment, the target retrieval image is preprocessed by various methods such as rotation, clipping and brightness change, and compare it with the primitive image input to test the robustness of this model. The experimental results are displayed in Figure 10. It is manifest from Figure 10, the input image can still maintain good search results after being rotated, cropped, and brightness changed. It shows that the model in this paper uses the standard data enhancement method to augment the training amount of samples in the training. In the process of image retrieval, under the influence of factors such as rotation, clipping, scaling and brightness change in the image, the model has certain robustness.

Figure 10. For example, the top 8 return results of image retrieval under different processing.

5. Conclusions

This study puts forward a local sensitive hash image retrieval algorithm based on CNN and attention mechanism. The feature is extracted by Convolutional Neural Network, and the channel and spatial attention module are embedded in the network so that the attention module can effectively identify the importance of different regions in the feature map, and give more weight to the key regions in a weighted way, to raise the feature expression ability of hash coding. The experiments were carried out on corel1k and corel5k datasets, respectively. It can be known from the experimental results that the proposed method has an advantage to improve the retrieval accuracy and recall rate after adding the attention mechanism module, and can still maintain a good accuracy rate when top-n increases. The search part uses a locally sensitive hash algorithm to hash code the high-dimensional features of the image and map them to a low latitude hash space to establish an image index and retrieve the image. The algorithm not only ensures a certain accuracy, but also greatly reduces the retrieval time, which fully demonstrates the effectiveness of the algorithm and good retrieval performance. At last, the experiment fully shows that the algorithm has certain robustness under the influence of rotation, cropping, brightness change, and other factors by using data enhancement methods. Because this paper uses the features of the full connection layer of the network to retrieve, high-level features are easy to lose a lot of detailed information; therefore, it will influence the capability of image retrieval in part. So, in future research, we will not only use the semantic information of high-level features, but also consider the details of texture information of low-level features, and improve the attention module to integrate the high-level semantic features and low-level features of images, to make image retrieval more accurate.

Author Contributions: Conceptualization, W.L.; Data curation, Y.L., X.M. and K.Z.; Formal analysis, Y.L.; Funding acquisition, W.L.; Investigation, K.Z.; Methodology, W.L.; Project administration, W.L.; Resources, Y.L., X.M. and K.Z.; Software, Y.L.; Supervision, W.L.; Validation, X.M.; Visualization, X.M.; Writing—original draft, Y.L.; Writing—review & editing, W.L. All authors have read and agreed to the published version of the manuscript.

Funding: This research was funded by Fujian Natural Science Foundation of China (2022J011233) and Xiamen University of Technology (XPDKT20027).

Data Availability Statement: Not applicable.

Conflicts of Interest: The authors declare no conflict of interest.

References

1. Wang, H.; Qu, H.; Xu, J.; Wang, J. Texture image retrieval based on fusion of local and global features. *Multimed. Tools Appl.* **2022**, *81*, 14081–14104. [CrossRef]
2. Mohite, N.B.; Gonde, A.B. Deep features based medical image retrieval. *Multimed. Tools Appl.* **2022**, *81*, 11379–11392. [CrossRef]
3. Xiong, B.; Lou, L.; Meng, X.; Ma, H.; Wang, Z. Short-term wind power forecasting based on Attention Mechanism and Deep Learning. *Electr. Power Syst. Res.* **2022**, *206*, 107776. [CrossRef]
4. Zhu, Y.; Liu, R.; Huang, Q. Weakly supervised information fine-grained image recognition based on deep neural network. *J. Electron. Meas. Instrum.* **2020**, *32*, 8.
5. Qin, J.; Pan, W.; Xiang, X.; Tan, Y.; Hou, G. A biological image classification method based on improved CNN. *Ecol. Inform.* **2020**, *58*, 101093. [CrossRef]

6. Noh, H.; Araujo, A.; Sim, J.; Weyand, T.; Han, B. Large-Scale Image Retrieval with Attentive Deep Local Features. In Proceedings of the 2017 IEEE International Conference on Computer Vision (ICCV), Venice, Italy, 22–29 October 2017.
7. Li, Z.; Li, S.; Liu, Y.; Li, H. Hand-drawn image retrieval method based on attention model. *Comput. Sci.* **2020**, *47*, 6.
8. Ng, T.; Balntas, V.; Tian, Y.; Mikolajczyk, K. SOLAR: Second-Order Loss and Attention for Image Retrieval. *arXiv* **2020**, arXiv:2001.08972.
9. Sutskever, I.; Hinton, G.E.; Krizhevsky, A. Imagenet classification with deep convolutional neural networks. In *Advances in Neural Information Processing Systems 25*; Neural Information Processing Systems Foundation, Inc.: San Diego, CA, USA, 2012; pp. 1097–1105.
10. Babenko, A.; Slesarev, A.; Chigorin, A.; Lempitsky, V. *Neural Codes for Image Retrieval*; Springer International Publishing: Berlin, Germany, 2014.
11. Wang, W.; Jiao, P.; Liu, H.; Ma, X.; Shang, Z. Two-stage content based image retrieval using sparse representation and feature fusion. *Multimed. Tools Appl.* **2022**, *81*, 16621–16644. [CrossRef]
12. Xia, R.; Pan, Y.; Lai, H.; Liu, C.; Yan, S. Supervised Hashing for Image Retrieval via Image Representation Learning. In Proceedings of the Twenty-Eighth AAAI Conference on Artificial Intelligence, Québec City, QC, Canada, 27–31 July 2014.
13. Lai, H.; Pan, Y.; Ye, L.; Yan, S. Simultaneous Feature Learning and Hash Coding with Deep Neural Networks. In Proceedings of the IEEE International Conference on Pattern Recognition and Computer Vision, Boston, MA, USA, 7–12 June 2015; pp. 3270–3278.
14. Shan, T. Research on the Nearest Neighbor Search Algorithm Based on Image Features. Master's Thesis, University of Science and Technology of China, Hefei, China, 2017.
15. Gao, X. Research on Large-scale Image Nearest Neighbor Retrieval Algorithm Based on Hash Algorithm. Master's Thesis, University of Electronic Science and Technology of China, Chengdu, China, 2018.
16. Woo, S.; Park, J.; Lee, J.Y.; Kweon, I.S. *CBAM: Convolutional Block Attention Module*; Springer: Cham, Switzerland, 2018.
17. Wang, X.; Luo, G.; Qin, K.; Zhang, Y. An Image Retrieval Method Based on SVM and Active Learning. *Comput. Appl. Res.* **2016**, *33*, 3836–3846.
18. Wei, Y.; Yan, Z. Research on Image Retrieval Technology Combined with Attention Convolutional Neural Network. *Small Microcomput. Syst.* **2021**, *42*, 2368–2374.
19. Balasundaram, P.; Muralidharan, S.; Bijoy, S. An mproved Content Based Image Retrieval System using Unsupervised Deep Neural Network and Locality Sensitive Hashing. In Proceedings of the 2021 5th International Conference on Computer, Communication, and Signal Processing, ICCCSP 2021, Chennai, India, 24–25 May 2021; pp. 65–71.
20. Qin, J.; Huang, J.; Xiang, X.; Tan, Y. Image retrieval based on convolutional neural network and attention mechanism. *Telecommun. Technol.* **2021**, *61*, 304–310.

remote sensing

MDPI

Article

Auto-Learning Correlation-Filter-Based Target State Estimation for Real-Time UAV Tracking

Ziyang Bian [1,2], Tingfa Xu [1,3], Junjie Chen [1], Liang Ma [2], Wenjing Cai [2] and Jianan Li [1,*]

[1] Key Laboratory of Photoelectronic Imaging Technology and System, Beijing Institute of Technology, Beijing 100081, China
[2] North China Research Institute of Electro-Optics, Beijing 100015, China
[3] Beijing Institute of Technology Chongqing Innovation Center, Chongqing 401120, China
* Correspondence: lijianan@bit.edu.cn

Abstract: Most existing tracking methods based on discriminative correlation filters (DCFs) update the tracker every frame with a fixed learning rate. However, constantly adjusting the tracker can hardly handle the fickle target appearance in UAV tracking (e.g., undergoing partial occlusion, illumination variation, or deformation). To mitigate this, we propose a novel auto-learning correlation filter for UAV tracking, which fully exploits valuable information behind response maps for adaptive feedback updating. Concretely, we first introduce a principled target state estimation (TSE) criterion to reveal the confidence level of the tracking results. We suggest an auto-learning strategy with the TSE metric to update the tracker with adaptive learning rates. Based on the target state estimation, we further developed an innovative lost-and-found strategy to recognize and handle temporal target missing. Finally, we incorporated the TSE regularization term into the DCF objective function, which by alternating optimization iterations can efficiently solve without much computational cost. Extensive experiments on four widely-used UAV benchmarks have demonstrated the superiority of the proposed method compared to both DCF and deep-based trackers. Notably, ALCF achieved state-of-the-art performance on several benchmarks while running over 50 FPS on a single CPU. Code will be released soon.

Keywords: single object tracking; UAV tracking; correlation filters; target state estimation; auto-learning

Citation: Bian, Z.; Xu, T.; Chen, J.; Ma, L.; Cai, W.; Li, J. Auto-Learning Correlation-Filter-Based Target State Estimation for Real-Time UAV Tracking. *Remote Sens.* **2022**, *14*, 5299. https://doi.org/10.3390/rs14215299

Academic Editors: Miltiadis D. Lytras and Andreea Claudia Serban

Received: 24 August 2022
Accepted: 17 October 2022
Published: 23 October 2022

1. Introduction

Given a target object specified by a bounding box in the first frame, visual object tracking aims to determine the target's exact location sequentially in a video, which serves as a fundamental task in the computer vision community. Recently, increasing attention has been paid to unmanned aerial vehicle (UAV) tracking [1–5] due to its promising development in the field of geomatics [6], agroforestry [7], transportation [8] and mid-air aircraft tracking [9].

Existing object tracking methods can mainly be divided into two types: discriminative correlation filter (DCF)-based [10–14] and deep-based [15–18]. Since UAVs only provide limited power capacity and computational resources, deep-based trackers usually suffer low inference speeds and high computational overheads. In contrast, DCF-based trackers achieve an ideal trade-off between accuracy and speed by efficiently computing the Fourier frequency domain, providing good solutions to real-time UAV tracking.

DCF-based trackers [3,11–13] detect the target by generating a response map and assigning the peak position as the target position. Then, the model will be updated with a fixed learning rate after every frame to adapt to the target's appearance change. However, due to the common challenges of fast motion and out-of-view in UAV tracking, target appearance could change drastically, rending the ineffectiveness of a stable update strategy. An appropriate update strategy is a goal to pursue. Recently, some trackers [2,19–21] have tried to mine information from response maps, such as fluctuation degree and peak height, as updated guidance. Intuitively, the peak height can provide a rough indication of the similarity between the target and the tracker, and the fluctuation degree indicates background clutter to some extent. LMCF [19] proposed an criterion called average peak-to-correlation energy (APCE), beyond which a model update is required. We regard it as a good measure of the fluctuation degree, but this single criterion seems not robust enough to cope with the appearance change, as shown in Figure 1. It is not sensitive to slight target state changes (e.g., in frames 1300, 1400), and there is an unnecessarily sharp drop when the model state is credible (in frame 1268). In addition, although LMCF stops the model from updating in some frames, which can prevent model degradation to a certain extent, it still relies on the pre-set single learning rate to cope with the model updating.

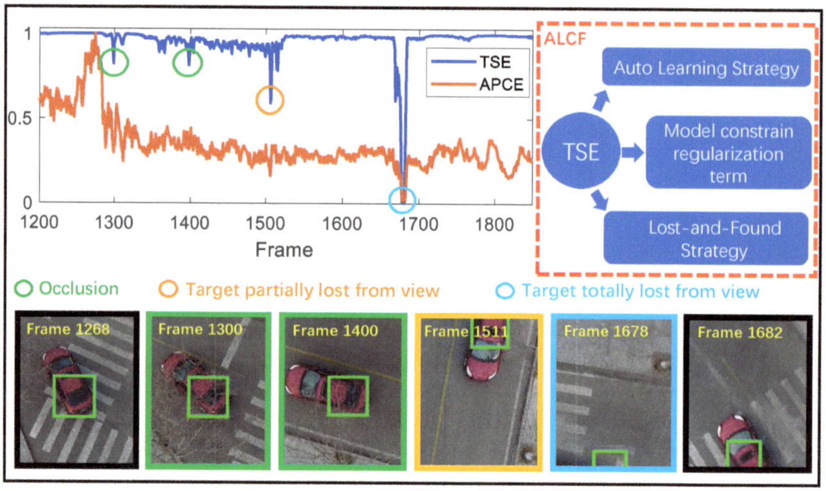

Figure 1. Comparison of the proposed TSE with APCE. The green box shows the tracking result.

1.1. Related Works

1.1.1. Discriminative Correlation Filter

Discriminative correlation filter (DCF)-based methods have been widely applied to visual tracking due to their high computational efficiency. Bolme et al. [22] proposed a minimum output sum of squared error (MOSSE) filter and were the first to apply a correlation filter to object tracking. Henriques et al. [23] introduced a circulant matrix to produce sufficient samples for training and detection while maintaining a fast tracking speed. However, the resulting periodic repetitions at boundary positions limit the discrimination of the tracker. To mitigate this issue, various regularization terms have been introduced [2,12,24]. Danelljan et al. [12] adopted a spatial regularization term that penalizes the filter coefficients for background regions to force the tracker to focus more on the target center. Huang et al. [3] used response maps as regularization terms for model training and repressed the aberrance of response maps to improve tracking performance. Chen et al. [4] designed a Gaussian-like response map as expected output to enhance the representation learning ability of the correlation filter. Xu et al. [25] proposed to gauge the relevance of multi-channel features for the purpose of channel selection to improve the performance of DCF in accuracy and provide a parsimonious model from the attribute perspective. Wang et al. [26] proposed a context and saliency-aware CF to strengthen the ability to extract targets of interest from complex background. Fu et al. [27] introduced an object-saliency-aware, dual-regularized correlation filter to suppress the boundary effect and consequently enhance the performance of the tracker. Li et al. [28] introduced a temporal regularization term, which uses the model trained in the previous frame as the reference model to constrain the model training in the current frame, thereby preventing model degradation. However, it uses a fixed constant as the regularization coefficient and adopts the model in the previous frame as the reference model regardless of whether the target state in the previous frame is credible.

1.1.2. Prior Knowledge to Model Update

Recently, increasing attention has been paid to the model updating. Wang et al. [19] introduced a criterion called APCE to measure response map quality and guide model updates. Li et al. [2] used the variations of response maps between consecutive frames to estimate the credibility of the target state and restricted model update when needed. Huang et al. [29] updated the Siamese tracker by transferring the previous target knowledge with attention. Dai et al. [30] used an offline-trained meta-updater to determine whether or not to update the tracker with two fixed learning rates (0 or 1), which fails to adapt to fickle target appearances varying at different rates in UAV tracking scenarios. Though adopting adaptive learning rates, Yang et al. [31] highly relies on LSTM [32], causing a high computational cost. Chen et al. [33] constructed a Gaussian-like (GL) function label for correlation filter training and proposed an accurate incremental update to mitigate model degradation by combining target samples with adaptive aspect ratios. In this paper, we fully exploit the information of the response map and directly obtain the learning rate from it, which is computationally efficient and suitable for real-time tracking. To the best of our knowledge, no previous works have exploited the quality of response maps to evaluate the target state and adjust the model's learning rate adaptively.

1.1.3. Redetect the Lost Target

Due to the complex situations during tracking, targets may be missed in some frames. To address this problem, previous works [34–36] often designed an additional detector to search for the target on the entire screen after target loss. Some trackers [37,38] even introduce global detection into tracking and locate the target through spatio-temporal constraints. However, these methods are challenging to run in real-time on GPU, which limits their applications in UAV tracking. Huang et al. [9] exploited motion features of a moving target to redetect the target from clutter background after loss, but this rough approach cannot handle static targets or visible videos. Zhu et al. [21] expanded the search area to increase the possibility of recapturing the target. Recently, Alan Lukezic et al. [36] verified the possibility of correlation filters as a detector using multiple correlation filters updated at different time scales. However, the large number of filters used in [36] slows the tracking speed, which limits its application in the field of UAV. Therefore, exploring an efficient re-detection method for UAV tracking is very meaningful. Unlike previous works, we introduce an iterative search method with only one correlation filter, which tracks the target and searches for the target after it is lost, achieving tracking performance improvement at the expense of only a slight decrease in tracking speed.

1.2. Contributions

We present a novel auto-learning correlation filter (short for ALCF) tracking algorithm for efficient UAV tracking. We first introduce an innovative criterion, target state estimation (TSE), to comprehensively get aware of the target state. Considering the peak height and fluctuation degree, both informative features of response maps, simultaneously, the resulting TSE value can represent the uncertainty in the target state of each frame reliably, thereby having a keener sense of slight target state changes and being more resistant to noise. As shown in Figure 1, TSE remains steady when the target moves smoothly and fluctuates significantly when occlusion or target lost occurs (as denoted by the circles). Our TSE successfully indicates whether the target is currently in the presence of occlusions (green circle), partially lost (yellow circle), or out-of-view (blue circle). It is noted that the comparison of APCE and TSE in Figure 1 was performed on the same tracker (ALCF).

Based on the TSE, we propose a novel auto-learning strategy that automatically derives a rapid and optimal learning rate from TSE adaptively depending on the estimated target states. Specifically, when the TSE remains large, indicating a transparent and credible target state, no update for the tracker is required, since it nicely fits the target. Moreover, when the TSE decreases slightly, indicating the target undergoes deformation, a significant learning rate is produced to adapt to rapid appearance changes. Furthermore, when the TSE drops sharply and the target state is uncertain, the target may suffer severe occlusion or be lost from view such that a small learning rate is generated to retain prior knowledge of the target and avoid model degradation. This way, ALCF can adjust its tracker properly under complex and changeable situations.

Moreover, the target in UAV tracking usually suffers large positional displacement between successive frames or even temporal disappearance from view, which quickly results in tracking failures. The challenges lie in recapturing the target when it appears in subsequent frames, as large positional displacement may exist. Thanks to the TSE criterion, which would decrease sharply when the target is lost (as exampled by the blue circle in Figure 1), we present a new lost-and-found strategy that perceives such a TSE decrease and enlarging search areas centered at the location, where the target last appears immediately. ALCF can quickly recapture the lost target with a new alternate search strategy with negligible extra computational cost.

The proposed TSE can also contribute to model optimization. We introduce a novel regularization term by using the trusted model trained with previous frames as a reference model to prevent model degradation. Instead of manually setting the coefficient for the regularization term, ALCF can automatically adjust the coefficient according to TSE. When the target state is credible, the coefficients are small to relax the model variation, and vice versa—the coefficients are significant in constraining the variation of the model in the current frame. We also incorporate two additional regularization terms to help suppress the aberration of computed responses for consecutive frames [3] and focus more on the central portion of targets [12].

We extensively evaluated our ALCF on several commonly used benchmark datasets, including UAVDT [39], UAV123@10fps [40], DTB70 [41], and VisDrone2018-test-dev [42], which contain large-scale, challenging aerial video sequences captured by UAVs. We provide two version trackers of ALCF, one based on a simple scale regression strategy (proposed by DSST [43]), which far exceeds real-time on a single CPU and performs favorably against other state-of-the-art trackers with hand-drafted features. Another version of ALCF exploits deep features for scale regression (proposed by Alpha-Refine [44]), which shows competitive performance against recent deep trackers.

To sum up, this work makes the following contributions:

- We present a novel TSE metric to the community which allows accurate estimation of the target state.
- We introduce a new ALCF tracker upon TSE, which consists of a novel auto-learning strategy, a fast lost-and-found strategy, and an effective regularization term for efficient and robust UAV tracking.
- We propose a deep version model which outperforms numerous recently popular deep framework-based trackers and established the new state-of-the-art on several UAV datasets.
- We provide the community with an optimal hand-drafted feature-correlation filter at over 50 FPS on a single CPU.

2. Method

In this section, we first describe the tracking mechanism of the correlation filter and the overall framework of the ALCF tracker, followed by a detailed description of the construction process of target state estimation (TSE), and finally describe the three strategies based on TSE.

2.1. Overview

The overall framework of our method is shown in Figure 2. The DCF-based tracker consists of three stages: the detection stage for target localization, the online model update stage for learning target appearance information, and the training stage for computing correlation filters. Based on the TSE value computed from the response map, we propose a lost-and-found strategy in the detection stage to cope with target loss. Then, an auto-learning strategy to learn new information from the target sample is presented, which learns adaptively according to the target state. Finally, a model that constrains regularization is introduced into model training to obtain a robust correlation filter model for tracking.

In the detection stage, the correlation filter f_{t-1} from the previous frame is used to search for the target in the current frame. The correlation response map is used to locate the target position, and the peak position is considered the target position of the current frame. Besides the peak position of the response map, we also exploit other information (i.e., peak value, fluctuation degree) to calculate TSE to evaluate the target state. When there is a sudden drop in the TSE value, we judge that the target has been lost and start using the lost-and-found strategy to re-detect the target, skipping the online update stage and the training stage if the target is lost.

For the online update stage, we propose an auto-learning strategy to learn new information from the current image. Unlike traditional DCF-based trackers that use a pre-set constant as the learning rate, ALCF uses TSE to compute an adaptively changing learning rate η, which allows the appearance model to learn more from the current image when the target state is credible and less when the target state is in doubt, thus maintaining robust tracking.

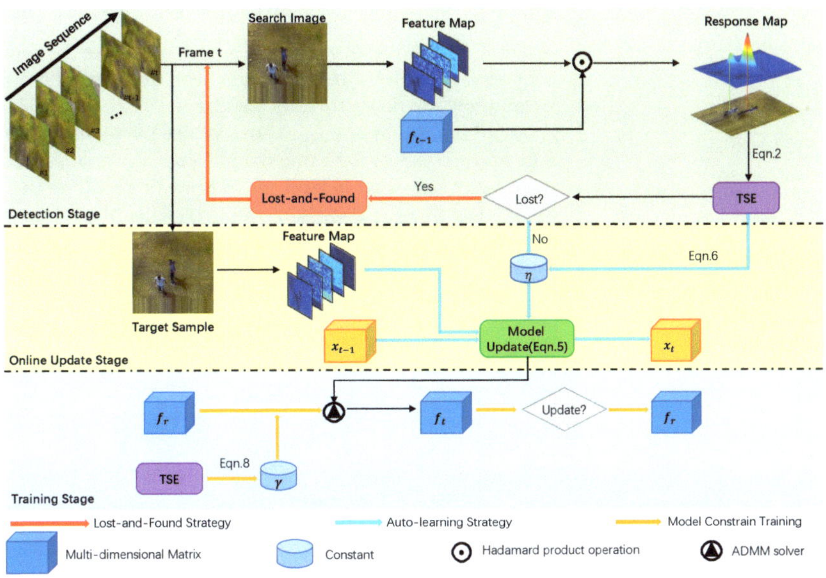

Figure 2. The overall framework of ALCF comprises three stages, i.e., detection, online updating, and training.

For the training phase, we present the model constrain regularization strategy to train the model. Specifically, we use the reference model f_r preserved from historical frames to prevent model degradation and use TSE to guide the update of the reference model and its weight coefficient γ during model training. When the target state is credible, we reduce the weight coefficient of the reference model and relax the restriction on model learning from the current frame, in the meanwhile updating the reference model to use the newly computed correlation filter f_t as the reference model for model training in subsequent frames. Conversely, we increase the weight coefficient of the reference model and restrict the model changes to prevent model degradation. Then, the model can be solved efficiently via the alternating direction method of multipliers (ADMM) algorithm.

2.2. *Target State Estimation (TSE)*

We developed TSE by comprehensively evaluating the peak value F_{max} and fluctuation degree (short for FD) of response maps, which can indicate the uncertainty in the target state and thus reflect the target state.

Denote F as the response map of dimensions $W \times H$ (width×height in pixels). We first compute mean-square error (MSE) to measure the difference between F and the ideal Gaussian response:

$$MSE = \frac{1}{WH} \cdot \sum_{i=1}^{W} \sum_{j=1}^{H} (\boldsymbol{F}(i,j) - \boldsymbol{y}(i,j))^2, \tag{1}$$

where $F(i,j)$ represents the response value on pixel (i,j), y is the ideal response map (Gaussian distribution), and F is expected to be similar to y when the target state is credible, leading to a low MSE value.

Given MSE, the TSE is computed as:

$$TSE = [1 + e^{-(T-\mu)}]^{-1}, \tag{2}$$

$$T = \alpha_1 F_{max} + \alpha_2 FD, \tag{3}$$

$$FD = \frac{{F_{max}}^2}{MSE}, \tag{4}$$

where F_{max} is the peak value of F, which measures the similarity score for the tracker and the target; FD reflects the fluctuation degree of F and the degree of background disturbance. When there are no distractors (i.e., frames 7, 21, and 307 in Figure 3), the value of FD is about 20 times that of F_{max}. Thus, we use two constant coefficients, α_1 and α_2, to ensure F_{max} and FD are of the same magnitude. μ is an offset constant. When the target state is credible, F_{max} is large while MSE is small, resulting in a considerable TSE value (close to 1). In contrast, TSE would drop to 0 when the target is lost. When the target suffers deformation (in frame 21, the target suffers out-of-plane rotation), the peak value drops slightly. When there exist distractors (in frame 228), sub-peaks lead to a significant drop to FD, but F_{max} is still high, keeping the TSE value enormous. Thus, with the combined effect of peak and fluctuation degree, our TSE is more robust and better guides the target state changes. More details about the construction of TSE can be found in Section 3.3.5.

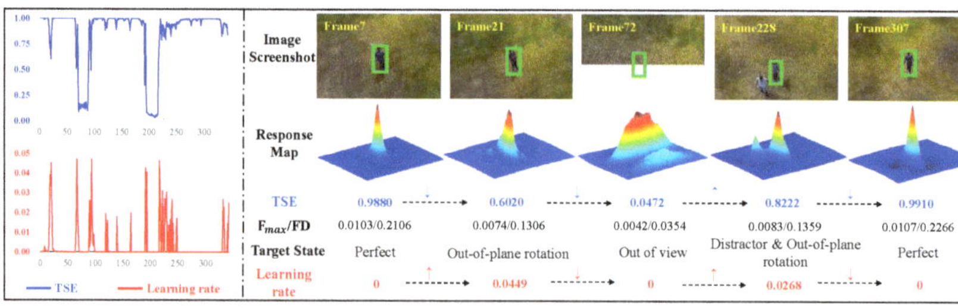

Figure 3. Illustration of the auto-learning strategy. The green box shows the tracking result.

2.3. *Auto-Learning Correlation Filter*

Based on TSE, we developed a new ALCF tracker comprised of an auto-learning strategy, a lost-and-found strategy, and an effective regularization term.

2.3.1. Auto-Learning Strategy

Existing methods update the tracking model with a single learning rate as follows:

$$\hat{x}_k^{model} = (1-\eta)\hat{x}_{k-1}^{model} + \eta\hat{x}_k, \tag{5}$$

where ˆ denotes the discrete fourier transform (DFT) of a signal, $\hat{x}_k$ is the vectorized image of frame k, and $\hat{x}_{k-1}^{model}$ is the tracking model learned from frame $k-1$. η is a fixed learning rate that is hard to set manually for UAV tracking, as constant and uncertain appearance changes exist. For example, when the correlation filter model can describe the target well, the model does not need to learn new information from the target sample image. When the target suffers severe occlusion or is lost from the screen, the model should stop updating to prevent contamination by untrustworthy samples. On the other hand, when there exists a difference between the model and the current target appearance, the model should accelerate learning new information from the target appearance. When there exists a huge difference between the target appearance and the model, we argue that maybe the target is lost, and the learning rate should reduce to keep the prior knowledge of the target.

Given this, we developed an auto-learning strategy that updates the tracking model with varying learning rates depending on the target state indicated by TSE, as shown in Figure 3. We automatically adjust the learning rate depending on different target states revealed by TSE computed from the distribution of response maps. When the target moves smoothly (frame 12), TSE is close to one, and the learning rate drops to zero, since updating the tracker is not necessary. However, when the target suffers occlusion or deformation (frames 32, 48, and 156), TSE decreases while the learning rate increases to adapt to rapid changes in target appearance. Moreover, when the target is lost from view (frame 20), TSE drops sharply, and the learning rate becomes zero to stop the model updating and retain the prior knowledge of the target. Concretely, we use TSE to compute the learning rate of frame k from a Gaussian distribution:

$$\eta_k = \frac{1}{\sqrt{2\pi\sigma}} e^{-\frac{(TSE-\beta)^2}{2\sigma^2}} - \varphi, \tag{6}$$

where β is set as 0.5, since TSE is bounded between 0 and 1, the standard deviations σ is set as 1, and φ is a constant to make η_k fall in a reasonable value range. When the value of TSE is close to 1, the target state is trusted, and the learning rate is low; and as the value of TSE drops, the learning rate takes the lead and then decreases. When TSE decreases to 0, we think the target may have been lost and stopped updating.

2.3.2. Lost-and-Found Strategy

We designed a novel lost-and-found strategy based on TSE. When the target is lost from view, TSE will drop suddenly. For example, it decreases by over seventy percent within ten frames, as illustrated in Figure 4. ALCF keeps a close eye on the trend of TSE and initiates the search for the lost target around the location where it last appeared (green rectangle) upon a sharp drop in the TSE. TSE drops rapidly when the target suffers complete occlusion, as shown by the TSE curve. ALCF initiates four additional search windows (yellow dashed box) and iterates their space intervals over three increased distances frame by frame to enlarge search areas. Until any one of the search windows regains a large TSE (more extensive than the sum value of the rest windows, as shown in frame 49), the tracker successfully recaptures the lost target and terminates the re-found process.

Since large positional displacement may exist when the lost target reappears in subsequent frames, previous works have introduced additional detectors to search for the target in the entire frame, leading to heavy memory and computation cost. Instead, we increase the number of search windows to cover a large search area and thus avoid using a detector. However, simply increasing search windows, though increasing the chances of recapturing the target, would hurt tracking speed at the same time. Instead of augmenting massive search windows in every single frame, we explored an efficient alternating search scheme that gradually enlarges the distance of four additional search windows in consecutive frames to cover large search areas sequentially, as shown by the yellow dashed rectangle in Figure 4. We set three distance intervals based on the width w and height h of the target and iterate over them every three consecutive frames until the target is re-found.

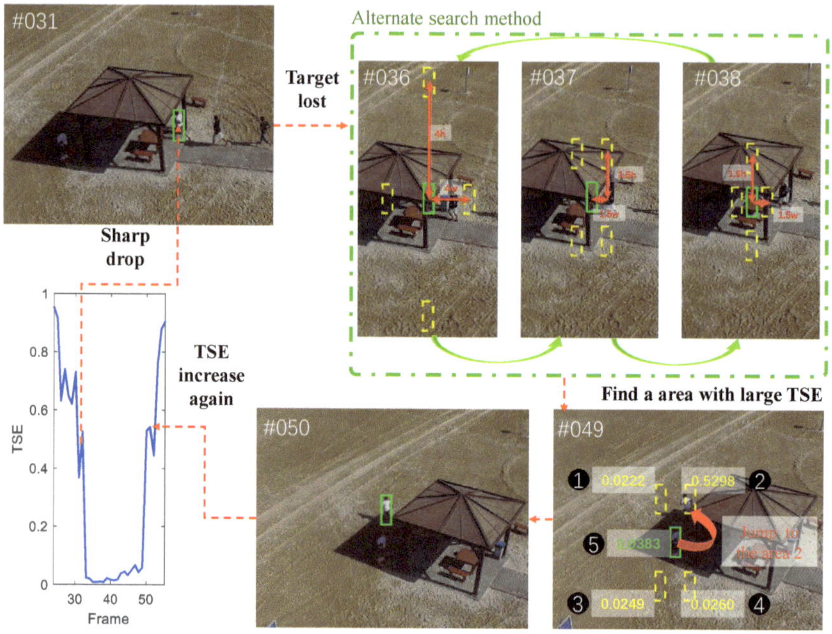

Figure 4. Diagram of lost-and-found strategy.

Our lost-and-found strategy exploits the correlation filter as the detector, which has two advantages: (i) The re-detection process is efficiently computed in the Fourier domain. The alternate search mechanism does not introduce a significant additional computational burden, which is friendly to the limited computational resources of UAVs. (ii) The correlation filter can be used not only as a tracking model but also as a detector when the target is lost, which brings no additional memory pressure and has low memory requirements for the UAVs.

2.3.3. Model Constrain Regularization

We further introduced a new regularization term by employing the prior knowledge learned from previous frames to constrain the filter learning of the current frame:

$$R_1 = \frac{\gamma_k}{2} \| f_k - f_r \|_2^2, \quad (7)$$

where $f_k \in \mathbb{R}^T$ denotes the filter of frame k, and $f_r \in \mathbb{R}^T$ represents the reference filter learned from previous frames, which encodes informative prior knowledge about the target. Instead of updating f_r after each frame [28], we only use the filter learned from the frame with favorable target states (e.g., no severe occlusion and deformation) as a reliable reference filter. To this end, we use TSE to estimate the target state of every frame and only use the filter from the frame whose computed TSE is larger than a pre-set threshold τ to refresh f_r.

In addition, we use an adaptive coefficient γ_k to adjust the weight of the regularization term depending on the current target state:

$$\gamma_k = \frac{1}{2}(1 - TSE^2), \tag{8}$$

when the current target state is trustworthy, i.e., TSE is large, γ_k is small such that f_k learns less from f_r to adapt to the latest appearance change of the target. Otherwise, f_k borrows more knowledge from f_r to prevent model degradation.

2.4. Model Optimization

The objective function can be formulated as:

$$E(f_k) = \frac{1}{2}\left\| \sum_{d=1}^{D} (\boldsymbol{x}_k^d * \boldsymbol{f}_k^d) - \boldsymbol{y} \right\|_2^2 + \sum_{i=[1,2,3]} R_i, \tag{9}$$

where $x_k^d \in \mathbb{R}^T$ denotes the d-th channel of the vectorized image of frame k, and D is the total channel number. $y \in R^T$ is the expected response (Gaussian distribution). $*$ indicates the convolution operator. $R_{i\in\{1,2,3\}}$ are three additional regularization terms to improve the discrimination of the model, which are detailed below.

R_1 is the regularization term defined in Equation (7). We further employ the spatial regularization term [12] to eliminate boundary effects caused by circulant shifted samples:

$$R_2 = \frac{1}{2}\sum_{d=1}^{D}\left\| w \odot f_k^d \right\|_2^2, \tag{10}$$

where w is a negative Gaussian-shaped spatial weight vector to make the tracking model emphasize more on the target center. $\odot$ indicates the Hadamard product operation.

In addition, a regularization term for response maps [3] is also used to exploit informative features of response distribution:

$$R_3 = \frac{\theta}{2}\left\| \sum_{d=1}^{D} (\boldsymbol{x}_k^d * \boldsymbol{f}_k^d) - \sum_{d=1}^{D} (\boldsymbol{x}_{k-1}^d * \boldsymbol{f}_{k-1}^d)[\Psi_{p,q}] \right\|_2^2, \tag{11}$$

where θ is a regularization coefficient; $x_{k-1}, f_{k-1} \in \mathbb{R}^T$ denote the vectorized image and the filter of frame $k-1$, respectively; $\sum_{d=1}^{D}(\boldsymbol{x}_{k-1}^d * \boldsymbol{f}_{k-1}^d)$ is the computed response map. p and q denote the location differences of two peaks in both response maps in two-dimensional space, and $[\Psi_{p,q}]$ is used to align the peak position of two consecutive response maps.

2.5. Optimization Solutions

For the ease of optimization, we introduce an auxiliary variable $\hat{g}_k = \sqrt{N}\boldsymbol{F} f_k$, where F is the orthonormal $N \times N$ matrix of complex basis vectors for mapping any N dimensional vectorized signal to Fourier frequency domain. Thus, we transformed the objective function into the frequency domain:

$$\begin{aligned} E(f_k, \hat{g}_k^d) = {} & \frac{1}{2}\left\| \sum_{d=1}^{D} (\hat{x}_k^d * \hat{g}_k^d) - \hat{y} \right\|_2^2 + \frac{1}{2}\sum_{d=1}^{D}\left\| w \odot f_k^d \right\|_2^2 \\ & + \frac{\theta}{2}\left\| \sum_{d=1}^{D} (\hat{x}_k^d * \hat{g}_k^d) - \boldsymbol{R}_{pre} \right\|_2^2 + \frac{\gamma_k}{2}\sum_{d=1}^{D}\left\| \hat{g}_k^d - \hat{g}_r^d \right\|_2^2, \end{aligned} \tag{12}$$

where R_{pre} represents the discrete Fourier transformation of the previous frame's response map, which is a constant for the current frame. We used the alternating direction method of multipliers (ADMM) algorithm to minimize Equation (12) to achieve a locally optimal solution. The augmented Lagrangian form of the equation can be formulated as:

$$L_k(f_k, \hat{g}_k^d, \hat{\xi}^T) = E(f_k, \hat{g}_k^d) + \hat{\xi}^T(\hat{g}_k - \sqrt{N}\boldsymbol{F}f_k) + \frac{\rho}{2}\left\|\hat{g}_k - \sqrt{N}\boldsymbol{F}f_k\right\|_2^2, \tag{13}$$

where $\hat{\xi}^T = [\hat{\xi}_1^T, \hat{\xi}_2^T, \ldots, \hat{\xi}_D^T]$ is the $1 \times DN$ Lagrangian vector in the Fourier domain, ρ is the penalty factor, and superscript T indicates the conjugate transpose operation. Then, ADMM is applied by alternately solving the following two sub-problems:

2.5.1. The Solution to Sub-Problem $\hat{g}_{k+1}$

The sub-problem $\hat{g}_{k+1}$ can be formulated as:

$$\begin{aligned}\hat{g}_{k+1} = argmin \frac{1}{2}\left\|\sum_{d=1}^{D}(\hat{x}_k^d \odot \hat{g}_k^d) - \hat{y}\right\|_2^2 \\ + \frac{\gamma_k}{2}\sum_{d=1}^{D}\left\|\hat{g}_k^d - \hat{g}_r^d\right\|_2^2 + \frac{\theta}{2}\left\|\sum_{d=1}^{D}(\hat{x}_k^d \odot \hat{g}_k^d) - \boldsymbol{R}_{pre}\right\|_2^2 \\ + \hat{\xi}^T(\hat{g}_k - \sqrt{N}\boldsymbol{F}f_k) + \frac{\rho}{2}\left\|\hat{g}_k - \sqrt{N}\boldsymbol{F}f_k\right\|_2^2,\end{aligned} \tag{14}$$

which is too complex to solve directly. Given the fact that x_k is sparse-banded, and thus each element of $\hat{y}(\hat{y}(t), t = 1, 2, \ldots, N)$ is dependent only on each $\boldsymbol{x}_k(t) = [x_k^1(t), x_k^2(t), \ldots, x_k^D(t)]^T$ and $\hat{g}_k(t) = [conj(\hat{g}_k^1(t)), conj(\hat{g}_k^2(t)), \ldots, conj(\hat{g}_k^D)(t)]^T$ [45], $conj(.)$ denotes the complex conjugate operation. Thus, we divided the vectorized image $\boldsymbol{x}_k$ into N elements and solved the problem by N smaller and independent problems, over $t = [1, 2, \ldots, N]$:

$$\begin{aligned}\hat{g}_{k+1}(t) = argmin \frac{1}{2}\left\|\hat{x}_k^T(t) \odot \hat{g}_k(t) - \hat{y}(t)\right\|_2^2 \\ + \frac{\gamma_k}{2}\|\hat{g}_k(t) - \hat{g}_r\|_2^2 + \frac{\theta}{2}\left\|(\hat{x}_k^T(t) \odot \hat{g}_k(t)) - \boldsymbol{R}_{pre}\right\|_2^2 \\ + \hat{\xi}^T(\hat{g}_k(t) - \sqrt{N}\boldsymbol{F}f_k(t)) + \frac{\rho}{2}\left\|\hat{g}_k(t) - \sqrt{N}\boldsymbol{F}f_k(t)\right\|_2^2,\end{aligned} \tag{15}$$

where $f_k(t) = [f_k^1(t), f_k^2(t), \ldots f_k^D(t)]$, $\hat{x}_k^T(t)$ is the discrete Fourier transformation of $x_k^T(t)$. Each smaller problem can be calculated efficiently to achieve the solution of Equation (15):

$$\hat{g}_{k+1}(t) = ((1 + N\theta)\hat{x}_k(t)\hat{x}_k^T(t) + (\rho + \gamma_k)N)^{-1}\boldsymbol{H}, \tag{16}$$

$$\boldsymbol{H} = \hat{x}_k^T(t)\hat{y}(t) + N\theta\boldsymbol{R}_{pre}\hat{x}_k^T(t) + N\hat{g}_r - N\xi^T + \rho N\hat{f}_k. \tag{17}$$

We use Sherman–Morrison formula [46] to accelerate the inverse operation:

$$(M + uv^T)^{-1} = M^{-1} - M^{-1}uv^TM^{-1}(1 + v^TM^{-1}u). \tag{18}$$

Hence, Equation (16) can be reformulated as:

$$\hat{g}_{k+1}(t) = \frac{1}{(\rho + \gamma_k)N}(I - \frac{\hat{x}_k(t)\hat{x}_k^T(t)}{\frac{(\rho+\gamma_k)N}{1+\theta N} + \hat{x}_k(t)\hat{x}_k^T(t)})\boldsymbol{H}. \tag{19}$$

2.5.2. The Solution to Sub-Problem f_k

The subproblem f_k can be written as:

$$f_{k+1} = argmin \frac{1}{2} \sum_{d=1}^{D} \left\| w \odot f_k^d \right\|_2^2 + \hat{\xi}^T (\hat{g}_k - \sqrt{N} \mathbf{F} f_k) + \frac{\rho}{2} \left\| \hat{g}_k - \sqrt{N} \mathbf{F} f_k \right\|_2^2, \tag{20}$$

which can be easily solved as follows:

$$f_{k+1} = \frac{\xi^T N + \rho N g_{k+1}}{w \odot w + \rho N}. \tag{21}$$

2.5.3. Update of Lagrangian Parameter

We update Lagrangian multipliers as:

$$\xi^{i+1} = \xi^i + \rho(g_{k+1}^{i+1} - f_{k+1}^{i+1}), \tag{22}$$

where i denotes the iteration number. g_{k+1}^{i+1} and f_{k+1}^{i+1} indicate the solution of subproblem g_{k+1} and f_{k+1}, respectively. The coefficient ρ (initially equals 1) is computed as:

$$\rho^{i+1} = min(\rho_{max}, \beta\rho^i), \tag{23}$$

where β and ρ_{max} are set as 10 and 10,000, respectively.

3. Results and Discussion

In this section, we first compare our ALCF with state-of-the-art trackers with four challenging UAV datasets, then provide experimental analysis to validate our design choices. Finally, a deep version of ALCF is proposed to compare with recent deep trackers to validate the superior performance of ALCF.

3.1. Implementation Details

All experiments were performed using MATLAB R2019a on a computer with Intel i7-9700 CPU@3.0 GHz, 16 GB RAM, and NVIDIA GTX 1080. We set $\theta = 2.2, \tau = 0.9$, $\varphi = 0.35$, $\mu = 6$. Considering the numerical magnitude difference between F_{max} and FD, we set $\alpha_2 = 20 \times \alpha_1 = 500$. The iteration number of ADMM was 2. We extracted histogram of oriented gradients (HOG) [47] and color name (CN) [48] features for input images. A separate filter in [43] was used in our ALCF for scale estimation, and the scale refine module proposed in [44] was used in DeepALCF for scale estimation. Note that the parameters of ALCF remained fixed on all image sequences on all benchmarks. ALCF and DeepALCF had the same parameters except for the size estimation method.

Two widely-used metrics, the precision plot and success plot, describe the performance of all trackers through one-pass evaluation (OPE). Concretely, a precision plot computed the percentages of frames in which the estimated target location is within a given distance threshold from the ground truth. We used the score at the threshold of 20 pixels for ranking the trackers. The success plot measures the fractions of successful frames in which the intersection over union (IoU) between the predicted bounding box and the ground truth is more significant than a certain threshold varied from 0 to 1. We use the area under curve (AUC) of the success plots to rank all trackers.

3.2. Comparison with Hand-Crafted Trackers

We compared our ALCF with thirteen state-of-the-art hand-crafted trackers, including AutoTrack [2], STRCF [28], ECO [49], CSR-DCF [10], MCCT-H [50], ARCF [3], BACF [13], DSST [43], fDSST [11], SRDCF [12], Staple [51], SAMF [52], and KCF [53]. The detailed comparisons on different benchmarks are listed below. Note that the publicly available codes and default parameters offered by the authors were utilized for impartial comparison. In Figures 5–8, the numbers in the legend indicate the average precision scores for the precision plot and average AUC scores for the success plot.

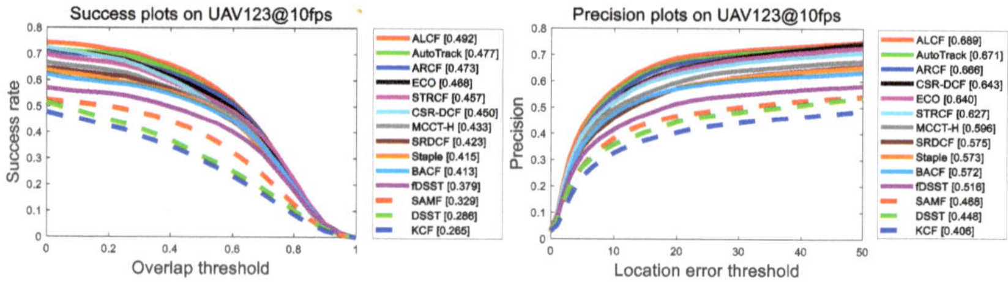

Figure 5. Overall performance of ALCF and other state-of-the-art trackers on UAV123@10fps.

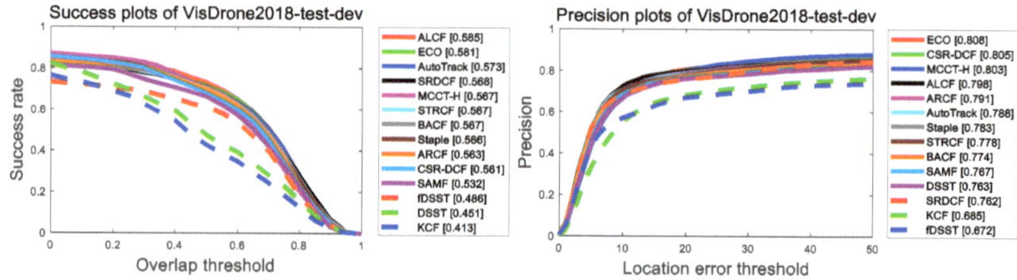

Figure 6. Overall performance of ALCF and other state-of-the-art trackers on VisDrone2018-test-dev.

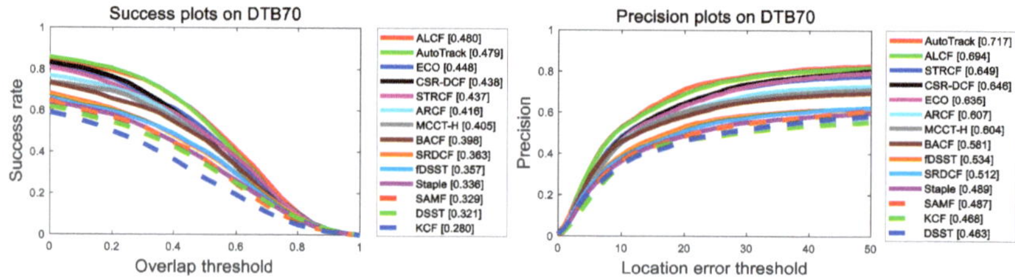

Figure 7. Overall performance of ALCF and other state-of-the-art trackers on DTB70.

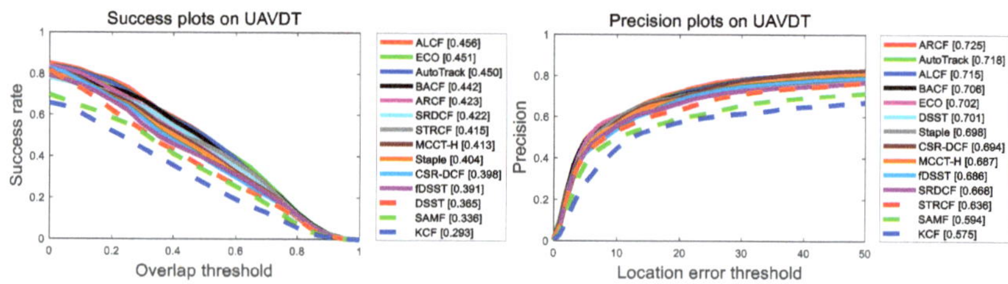

Figure 8. Overall performance of ALCF and other state-of-the-art trackers on UAVDT.

3.2.1. Results on UAV123@10fps

UAV123@10fps [40] consists of 123 challenging sequences with 12 different attributes. All sequences were temporally down-sampled to 10 FPS to simulate challenging large target displacements between consecutive frames. Figure 5 provides comparisons with other preeminent trackers. ALCF achieved the best performance in both precision and success. Notably, ALCF outperformed the previous state-of-the-art tracker, AutoTrack, by a large margin, i.e., 1.8% and 1.5%, in precision and success, respectively. This evidences that ALCF can cope well with sudden changes in target appearance and large position shifts between adjacent frames, which commonly appear in UAV tracking.

3.2.2. Results on VisDrone2018-Test-Dev

VisDrone2018-test-dev [42] contains 35 sequences captured by various UAV platforms in over 14 different cites in China, featuring diverse real-world scenarios. Figure 6 shows that our ALCF obtained the best success score of 0.585 and a competitive precision score of 0.798, which is on par with the best-performing tracker ECO, mainly due to the auto-learning and the lost-and-found strategy in ALCF. The auto-learning strategy can adjust the learning rate according to the target state to prevent model degradation. The lost-and-found strategy can help the tracker recapture target after the target is lost, enhancing the performance in coping with long-term tracking (the average length of VisDrone2018-test-dev is around 940 frames; the most extended sequence can reach a length of 2569 frames).

3.2.3. Results on DTB70

DTB70 [41] comprises 70 sequences covering different types of UAV movements, including rapid translation and rotation. Targets include humans, animals, and rigid objects. In Figure 7, the numbers in the legend indicate the average precision scores for the precision plot and average AUC scores for the success plot. Figure 7 depicts that our ALCF performed the best in terms of success rate and second-best in precision. We attribute this to the auto-learning update strategy in ALCF, since there are many viewpoint changes in DTB70, where auto-learning plays an important role.

3.2.4. Results on UAVDT

UAVDT [39] was selected from over 10 h of videos taken by a UAVs at different locations in urban areas. The targets are mainly vehicles with additional attributions, i.e., weather conditions, flying altitude, and camera view. As we can see in Figure 8, our tracker is the best in terms of success. It outperformed the previous best tracker, ECO [49], and second-best tracker, AutoTrack [2], by 0.5% and 0.6%, respectively.

3.2.5. Average Performance Results

Figure 9 reports the average precision and success rate of the 13 trackers and ALCF on the four benchmarks. ALCF yielded the best average success rate score and precision score. It is noted that ALCF was the only tracker with a success rate score (0.503) above 0.5, surpassing the second-best tracker AutoTracker (0.494) by 0.9%. As far as we know, ALCF is the best tracker with hand-drafted features on the four benchmarks.

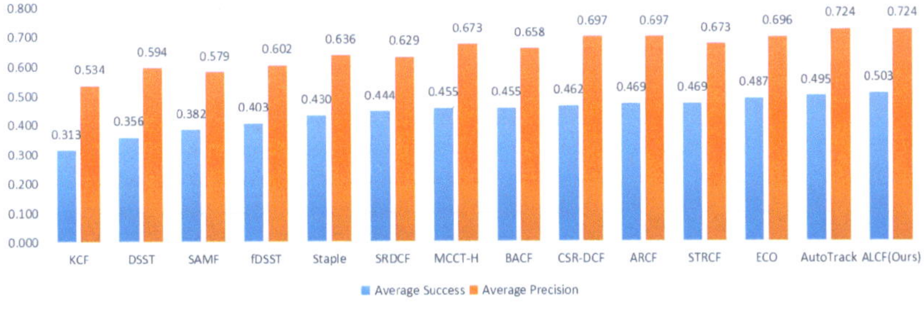

Figure 9. Overall evaluation of ALCF and other trackers regarding average precision and success rate.

3.2.6. Per-Attribute Evaluation

Figure 10 provides the precision comparison of twelve attributes defined on UAV123@10fps. The number in parentheses is the precision of ALCF. Detailed definitions of these attributions can be found on UAV123@10fps. Our method performed the best against state-of-the-art trackers (i.e., AutoTrack [2], ARCF [3], BACF [13], ECO [49], and STRCF [28]) in most attributes. It is worth noting that our ALCF performed extremely well not only in coping with rapid changes (e.g., viewpoint or illumination variation, and camera motion), but also in preventing model degradation (e.g., out-of-view, partial and full occlusion).

These challenges abound in real-world UAV object tracking, which is ample evidence that our strategies are quite effective for UAV tracking.

3.2.7. Visualization

Figure 11 illustrates some qualitative comparisons of ALCF with other top-performing trackers on some challenging sequences (from top down they are *ChasingDrones (DTB70), StreetBasketball3 (DTB70), bike2 (UAV123@10fps), person10 (UAV123@10fps),* and *S0602 (UAVDT)).* Our ALCF showed superiority over other trackers in handling challenging scenarios, including fast motion (the first sequence), heavy occlusion (the second sequence), low resolution (the third sequence), loss from view (the fourth sequence), and viewpoint change (the last sequence). Of course, our algorithm still has limitations. In the case of long-term occlusion and superimposition of target deformation, tracking loss will occur.

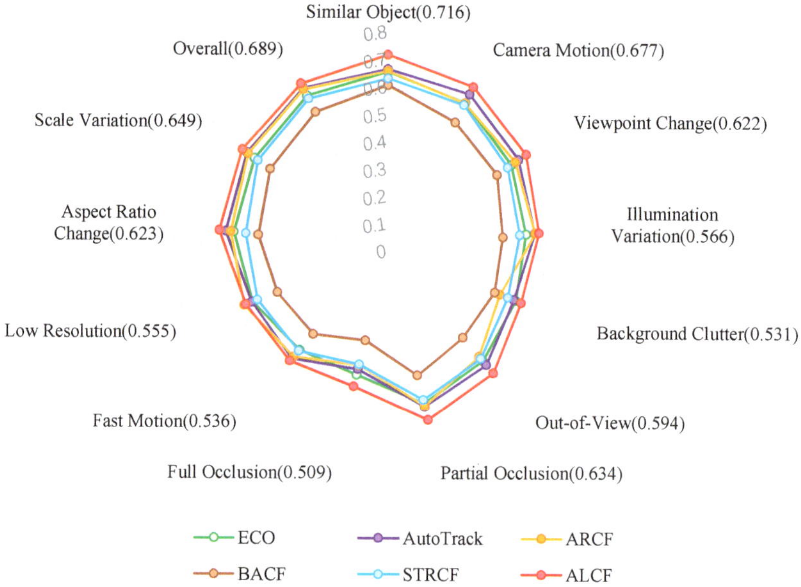

Figure 10. Attribute-based evaluation of ALCF and some other trackers in precision.

3.2.8. Speed

Table 1 compares the tracking speeds of ALCF and other DCF-based trackers (top-5 trackers with hand-drafted features in terms of success over four benchmarks). All the trackers can run on a single CPU. ALCF reached a real-time speed of 51.3 FPS, comparable to that of ECO and AutoTrack, with better predictive performance.

Table 1. Average speed (FPS) of top five trackers on [39–42], ranked by average success (top 3 in **bold**).

	ALCF	AutoTrack	ECO-HC	STRCF-HC	ARCF
FPS	**51.3**	**58.3**	**69.4**	23.2	23.3
Success	**0.503**	**0.494**	**0.487**	0.469	0.469
Precision	**0.724**	**0.723**	0.696	0.672	**0.697**

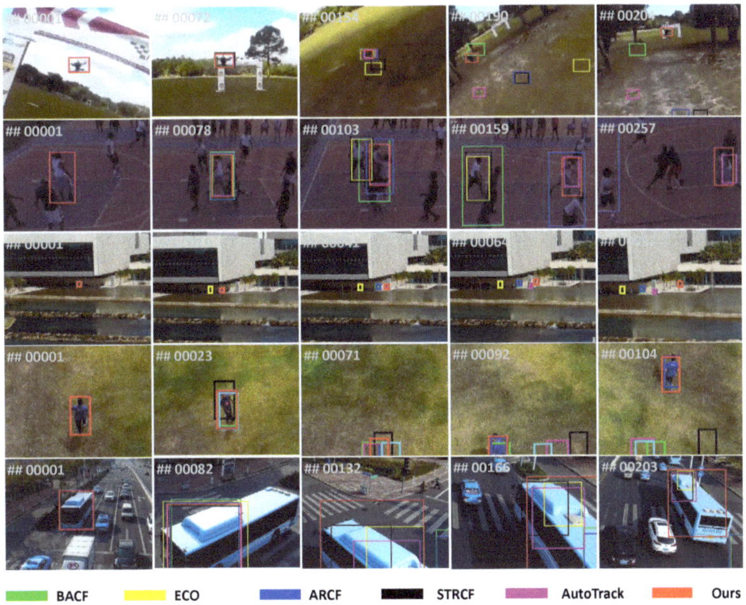

Figure 11. Qualitative comparison of ALCF with other top-performing trackers on some challenging sequences.

3.3. Ablation Studies

We investigated the effectiveness of different components of ALCF on UAV123@10fps. As presented in Table 2, we used the model variant, which employs the same objective function and fixes the learning rate of 0.025 as in SRDCF [12] as a baseline. For a fair comparison, the baseline model adopted the same features extraction and scale estimation scheme as ALCF (original SRDCF utilizes only HOG features, so the performance in Figure 5 is not as good as that in Table 2). AL, MC, and LF denote the auto-learning strategy, model constraint regularization term, and lost-and-found strategy, respectively. Different components are incrementally added to the baseline model individually or in combination to verify their effectiveness.

Table 2. Ablation analysis. R represents the response map regularization term. AUC represents the success rate (top 1 in **bold.**)

	AL	MC	LF	R	Precision	AUC	FPS
Baseline					0.600	0.432	36.94
	✓				0.637	0.460	45.64
		✓			0.649	0.463	34.03
			✓		0.648	0.465	32.38
		✓	✓		0.654	0.468	32.36
	✓	✓	✓		0.667	0.473	47.89
	✓			✓	0.654	0.469	42.44
		✓		✓	0.650	0.464	33.70
			✓	✓	0.652	0.474	36.52
		✓	✓	✓	0.662	0.475	29.91
	✓	✓	✓	✓	**0.689**	**0.492**	**41.31**
Baseline+STRCF					0.638	0.457	31.87
Baseline+MC+Fix					0.639	0.459	33.40

3.3.1. Effect of Auto-Learning Strategy

As shown in Table 2, Baseline+AL outperformed the baseline in both precision and success rate by 3.7% and 2.8%, respectively. In addition, incorporating an auto-learning strategy improves tracking speed by 23.6%, as it helps reduce meaningless or detrimental training of individual samples. The tracking speed in Table 2 is lower than the average speed in Table 1 because the targets in UAV123@10fps vary drastically between consecutive frames compared to those in the other three datasets, resulting in the frequent model updating in most frames. Nevertheless, our auto-learning strategy still improved the tracking speed significantly.

3.3.2. Effect of Regularization Term

Baseline+MC denotes the model variant that incorporated the proposed regularization term into the baseline model, Baseline+MC+Fix replaced the adaptive regularization coefficient γ calculated from Equation (8) with a fixed constant, and Baseline+STRCF updated the reference model in Equation (7) after each frame without considering target states [28]. As is shown in Table 2, using a credible model as a reference and an adaptive regularization coefficient contributes more to the performance boost.

3.3.3. Effect of the Lost-and-Found Strategy

We present a model variant (Baseline+LF) with a lost-and-found strategy added to the baseline model. Notably, employing the lost-and-found strategy boosted the performance significantly by 4.8% and 3.3% in precision and success rate, respectively, along with a negligible increase in tracking speed.

3.3.4. Effect of the Objective Function

We also found that all three components can be further improved to different degrees with the help of the additional response map regularization term used in [3], mainly attributed to its aberration suppression of the response maps between two adjacent frames, which helps the response maps to be more stable. The TSE proved to be more accurate (the performance of Baseline+AL+MC+LF+R (precision: 0.689, AUC: 0.492, FPS: 41.31) is higher than that of Baseline+AL+MC+LF (precision: 0.667, AUC: 0.473, FPS: 47.89) with a slight sacrifice in speed). We also verified the generality of our three innovations, which can improve the tracking performance for trackers with different objective functions.

3.3.5. Construction of Target State Estimation

In our experiment, we found that T (calculated from Equation (3)) has an extensive variation range, which exceeded 10 when the target state was trustworthy and dropped below 2 when the target was lost. We should focus on the variation when T falls between 2 to 10 to perceive the transition of target states. Hence, we introduce the sigmoid activation function (with μ as 6) upon T to magnify its variation in the interval [2,10] while suppressing noisy fluctuations when the target state is entirely credible ($T > 10$) or completely untrustworthy ($T < 2$).

We also experimented on UAV123@10fps to validate our parameter settings. Figure 12 shows the performance peaks at $\mu = 6$. When μ becomes small, the value of TSE becomes large (close to 1), which leads to a gradual reduction of the learning rate to 0. When $\mu = 0$, it is equivalent to not updating the model, so the model's performance was the worst. Moreover, when μ becomes large, the value of TSE will become smaller, which caused the model to use a large learning rate to update when it did not need to be updated and zero when it needed to be updated. However, the performance was slightly better compared to the case of not updating the model ($\mu = 0$), which verified the importance of the model update.

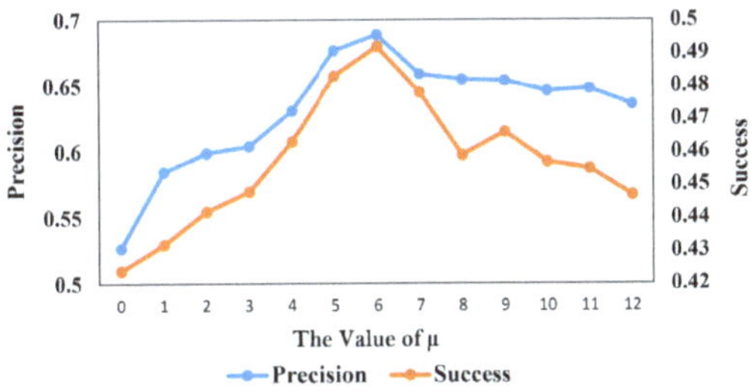

Figure 12. Effect of μ on the performance of ALCF.

3.3.6. Sensitivity Analysis of Model Constraint Regularization Term

Based on STRCF [28], we further designed an adaptive model constraint regularization term for UAV tracking, which can update the reference model adaptively and adjust the effect of the reference model on the correlation filter learning. Moreover, we validated our parameter settings on UAV123@10fps. We set a threshold τ to refresh f_r. When the value of TSE is larger than τ, the reference model f_r will be updated. To find the optimal value of τ, we refined the search within the range of values of τ, taking values at 0.1 intervals. The result is shown in Figure 13. We can see that when $\tau = 0$, the reference filter f_r is updated every frame with an adaptive parameter γ. When $\tau = 1$, the reference filter f_r is never updated during tracking. By comprehensively considering the precision and success rate, we set $\tau = 0.9$.

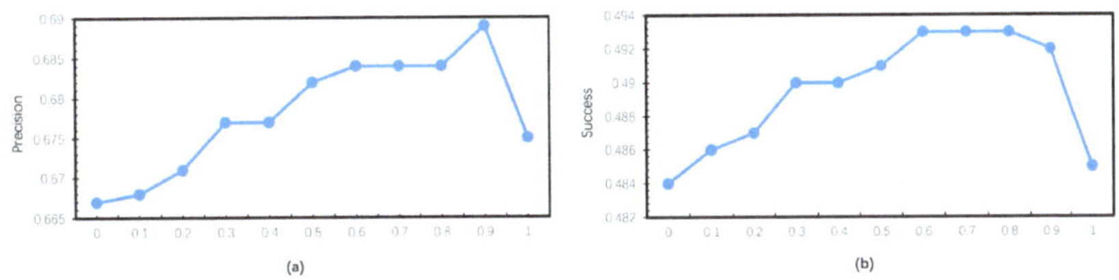

Figure 13. Effect of τ on the performance of ALCF. (**a**) shows the precision while (**b**) shows the success rate.

3.4. Extension to Deep Trackers

We further evaluated the tracking performance of ALCF in comparison with other state-of-the-art deep-based trackers on the four benchmarks. Inspired by [44], we proposed a deep version of ALCF called DeepALCF, which exploits Alpha-Refine as our scale estimation. Note that DeepALCF still exploits HOG and CN as feature presentations for object location, but it exploits ResNet-34 to extract deep features for scale estimation, which is more accurate. More detail about the scale estimation can be found in Alpha-Refine [44].

3.4.1. Results on VisDrone2018-Test-Dev

We selected 10 new state-of-the-art deep trackers for comparison. The results are shown in Table 3. DeepALCF yielded the best precision score (0.816), surpassing the second-best tracker KYS [54] and third-best tracker DiMP [55] by 0.6% and 1%. DeepALCF also obtained the best success score (0.619), outperforming the second-best tracker, Super_DiMP [56] (0.610), and the third-best tracker, DiMP [55] (0.605) by 0.9% and 1.4%. This proves that the auto-learning strategy can cope well with object tracking in natural scenes.

Table 3. Performance comparison of ALCF with other ten state-of-the-art deep trackers (top 3 in **bold**).

	MDNet [16]	UDT+ [15]	PrDiMP [57]	ATOM [20]	DiMP [55]	SiamRPN [17]	CFNet [58]	Super_DiMP [56]	KYS [54]	HiFT [59]	ALCF Ours	DeepALCF Ours
Precision	0.790	0.800	0.797	0.751	**0.806**	0.781	0.778	0.800	**0.810**	0.721	0.798	**0.816**
Success	0.579	0.584	0.600	0.563	**0.605**	0.573	0.568	**0.610**	0.600	0.527	0.585	**0.619**

3.4.2. Results on UAVDT

Table 4 shows the precision and success scores of ALCF and other 12 deep trackers on UAVDT. ALCF was very competitive in comparison with PrDiMP [57] in both precision and success and outperformed third-best tracker MDNet [16] by 2.3% and 7% in terms of precision and success, respectively.

3.4.3. Results on DTB70

We compared DeepALCF with 13 popular deep trackers. Table 5 shows the comparison results. DeepALCF outperformed the third-best tracker DaSiamRPN [21] (0.512) by 1.8% in terms of success, and DeepALCF (0.727) is very close to DaSiamRPN (0.735) in terms of precision. Compared with ARTracker [33], ALCF reached a real-time speed of 51.3 FPS, which is far more than ARTracker (FPS: 13.28).

3.4.4. Results on UAV123@10fps

We conducted a comprehensive experiment in Table 6, which compares DeepALCF and ALCF with numerous recent deep trackers. DeepALCF achieved the best precision (0.712) and success rate (0.516) scores. It can be found that ALCF with hand-drafted features obtained a third-best precision score, which outperformed a lot of deep trackers while running at over 50 FPS on a single CPU.

Table 4. Overall evaluation of ALCF, DeepALCF, and some other state-of-the-art trackers on UAVDT [39] (top 3 in **bold**).

Trackers	Venue	Precision	Success
MDNet [16]	CVPR2016	**0.725**	0.464
MCCT [50]	CVPR2018	0.691	0.448
ASRCF [24]	CVPR2019	0.720	0.449
TADT [60]	CVPR2019	0.700	0.441
UDT [15]	CVPR2019	0.674	0.442
UDT+ [15]	CVPR2020	0.696	0.415
PrDiMP [57]	CVPR2020	**0.757**	**0.559**
fECO [61]	TIP2020	0.699	0.415
fDSTRCF [61]	TIP2020	0.677	0.454
HiFT [59]	ICCV2021	0.652	**0.474**
LUDT [62]	IJCV2021	0.631	0.418
LUDT+ [62]	IJCV2021	0.701	0.406
ALCF	Ours	**0.725**	0.456
DeepALCF	Ours	**0.748**	**0.534**

Table 5. Overall evaluation of ALCF and some other state-of-the-art trackers on DTB70 [41] (top 3 in **bold**).

Trackers	Venue	Precision	Success
HCF [63]	ICCV2015	0.616	0.415
MDNet [16]	CVPR2016	0.703	0.466
CFNet [58]	CVPR2017	0.587	0.398
IBCCF [64]	ICCVW2017	0.669	0.460
CREST [65]	ICCV2017	0.650	0.452
ADNet [66]	CVPR2017	0.637	0.422
DaSiamRPN [21]	ECCV2018	**0.735**	**0.512**
UDT+ [15]	CVPR2019	0.650	0.457
MCCT [50]	CVPR2019	0.725	0.484
TADT [60]	CVPR2019	0.690	0.460
ASRCF [24]	CVPR2019	0.696	0.468
KAOT [67]	TMM2020	0.692	0.469
ARTracker [33]	GRSL2022	**0.752**	**0.588**
ALCF	Ours	0.694	0.480
DeepALCF	Ours	**0.727**	**0.530**

Table 6. Overall evaluation of ALCF and some other state-of-the-art trackers on UAV123@10fps [40] (top 3 in **bold**).

Trackers	Precision	Success	Trackers	Precision	Success
HCF	0.601	0.426	MDNet	0.664	0.477
CFNet	0.568	0.422	IBCCF	0.651	0.481
SiamFC [68]	0.678	0.472	CREST	0.600	0.445
C-COT [69]	**0.704**	**0.502**	DeepSTRCF	0.680	0.499
DaSiamRPN	**0.689**	0.481	UDT+	0.674	0.470
MCCT	**0.689**	0.496	ADNet	0.647	0.422
TADT	0.685	**0.507**	ASRCF	0.685	0.477
ALCF	**0.689**	0.492	DeepALCF	**0.712**	**0.516**

4. Conclusions

In this work, we proposed a novel tracking method called ALCF to deal with challenging situations (occlusion, deformation, etc.) in UAV tracking. Specifically, we introduced a novel criterion named target state estimation (TSE) to identify exact target states and adjust the learning rate for model update adaptively. In addition, a new model constraining regularization terms and an efficient lost-and-found strategy were further developed to enhance the robustness of the tracker to target variation and loss. Comprehensive experiments on four widely-used UAV datasets demonstrated that ALCF performs better than most state-of-the-art trackers in both precision and success scores while running over 50 FPS on a single CPU. We also provided a deep version tracker called DeepALCF, which is competitive against recently popularized deep trackers. We believe our work would contribute to the field of UAV tracking. We will focus on improving the real-time performance of the model in edge computing and running our algorithm on a hardware system on a UAV to evaluate its performance.

Author Contributions: Conceptualization, T.X.; methodology, Z.B. and J.L.; software, Z.B.; validation, L.M.; formal analysis, J.C.; investigation, J.C. and W.C.; resources, T.X.; data curation, J.C.; writing—original draft preparation, Z.B.; writing—review and editing, J.C. and J.L.; visualization, Z.B.; supervision, T.X.; project administration, T.X.; funding acquisition, T.X. All authors have read and agreed to the published version of the manuscript.

Funding: This research was funded by the National Key Laboratory Foundation of China, grant number TCGZ2020C004 and 202020429036.

Institutional Review Board Statement: Not applicable.

Informed Consent Statement: Not applicable.

Data Availability Statement: The data presented in this study are available on request from the corresponding author.

Acknowledgments: This work was supported by the Key Laboratory Foundation under grant TCGZ2020C004 and grant 202020429036.

Conflicts of Interest: The authors declare no conflict of interest.

References

1. Lin, B.; Bai, Y.; Bai, B.; Li, Y. Robust Correlation Tracking for UAV with Feature Integration and Response Map Enhancement. *Remote Sens.* **2022**, *14*, 4073. [CrossRef]
2. Li, Y.; Fu, C.; Ding, F.; Huang, Z.; Lu, G. AutoTrack: Towards High-Performance Visual Tracking for UAV with Automatic Spatio-Temporal Regularization. In Proceedings of the IEEE Conference on Computer Vision and Pattern Recognition, Seattle, WA, USA, 13–19 June 2020; pp. 11923–11932.
3. Huang, Z.; Fu, C.; Li, Y.; Lin, F.; Lu, P. Learning Aberrance Repressed Correlation Filters for Real-Time UAV Tracking. In Proceedings of the IEEE International Conference on Computer Vision, Seoul, Korea, 27 October–2 November 2019; pp. 2891–2900.
4. Chen, J.; Xu, T.; Li, J.; Wang, L.; Wang, Y.; Li, X. Adaptive Gaussian-Like Response Correlation Filter for UAV Tracking. In Proceedings of the ICIG, Haikou, China, 6–8 August 2021; pp. 596–609.
5. Wang, L.; Li, J.; Huang, B.; Chen, J.; Li, X.; Wang, J.; Xu, T. Auto-Perceiving Correlation Filter for UAV Tracking. *IEEE Trans. Circuits Syst. Video Technol.* **2022**, *32*, 5748–5761. [CrossRef]
6. Nex, F.; Remondino, F. UAV for 3D mapping applications: A review. *Appl. Geomat.* **2014**, *6*, 1–15. [CrossRef]
7. Torresan, C.; Berton, A.; Carotenuto, F.; Gennaro, S.F.D.; Gioli, B.; Matese, A.; Miglietta, F.; Vagnoli, C.; Zaldei, A.; Wallace, L. Forestry applications of UAVs in Europe: A review. *Int. J. Remote Sens.* **2017**, *38*, 2427–2447. [CrossRef]
8. Karaduman, M.; Cınar, A.; Eren, H. UAV Traffic Patrolling via Road Detection and Tracking in Anonymous Aerial Video Frames. *J. Intell. Robot. Syst.* **2019**, *95*, 675–690. [CrossRef]
9. Huang, B.; Chen, J.; Xu, T.; Wang, Y.; Jiang, S.; Wang, Y.; Wang, L.; Li, J. SiamSTA: Spatio-Temporal Attention based Siamese Tracker for Tracking UAVs. In Proceedings of the 2021 IEEE/CVF International Conference on Computer Vision Workshops (ICCVW), Montreal, BC, Canada, 11–17 October 2021; pp. 1204–1212.
10. Lukežič, A.; Vojíř, T.; Čehovin Zajc, L.; Matas, J.; Kristan, M. Discriminative Correlation Filter with Channel and Spatial Reliability. In Proceedings of the IEEE Conference on Computer Vision and Pattern Recognition, Honolulu, HI, USA, 21–26 July 2017; pp. 6309–6318.
11. Danelljan, M.; Häger, G.; Khan, F.S.; Felsberg, M. Discriminative Scale Space Tracker. *IEEE Trans. Pattern Anal. Mach. Intell.* **2017**, *39*, 1561–1575. [CrossRef] [PubMed]
12. Danelljan, M.; Häger, G.; Khan, F.; Felsberg, M. Learning Spatially Regularized Correlation Filters for Visual Tracking. In Proceedings of the IEEE International Conference on Computer Vision, Santiago, Chile, 7–13 December 2015; pp. 4310–4318.
13. Galoogahi, H.K.; Fagg, A.; Lucey, S. Learning Background-Aware Correlation Filters for Visual Tracking. In Proceedings of the IEEE International Conference on Computer Vision, Venice, Italy, 22–29 October2017; pp. 1144–1152.
14. Huang, B.; Xu, T.; Jiang, S.; Chen, Y.; Bai, Y. Robust visual tracking via constrained multi-kernel correlation filters. *IEEE Trans. Multimed.* **2020**, *22*, 2820–2832. [CrossRef]
15. Wang, N.; Song, Y.; Ma, C.; Zhou, W.; Liu, W.; Li, H. Unsupervised Deep Tracking. In Proceedings of the IEEE Conference on Computer Vision and Pattern Recognition, Long Beach, CA, USA, 15–20 June 2019; pp. 1308–1317.
16. Nam, H.; Han, B. Learning Multi-Domain Convolutional Neural Networks for Visual Tracking. In Proceedings of the IEEE Conference on Computer Vision and Pattern Recognition, Las Vegas, NV, USA, 27–30 June 2016; pp. 4293–4302.
17. Li, B.; Yan, J.; Wu, W.; Zhu, Z.; Hu, X. High Performance Visual Tracking With Siamese Region Proposal Network. In Proceedings of the IEEE Conference on Computer Vision and Pattern Recognition, Salt Lake City, UT, USA, 18–23 June 2018; pp. 8971–8980.
18. Wang, Y.; Xu, T.; Jiang, S.; Chen, J.; Li, J. Pyramid Correlation based Deep Hough Voting for Visual Object Tracking. In Proceedings of the Asian Conference on Machine Learning, PMLR, Virtual Event, 17–19 November 2021; pp. 610–625.
19. Wang, M.; Liu, Y.; Huang, Z. Large margin object tracking with circulant feature maps. In Proceedings of the IEEE Conference on Computer Vision and Pattern Recognition, Honolulu, HI, USA, 21–26 July 2017; pp. 4021–4029.
20. Danelljan, M.; Bhat, G.; Khan, F.S.; Felsberg, M. ATOM: Accurate Tracking by Overlap Maximization. In Proceedings of the IEEE Conference on Computer Vision and Pattern Recognition, Long Beach, CA, USA, 16–20 June 2019; pp. 4660–4669.
21. Zhu, Z.; Wang, Q.; Li, B.; Wu, W.; Yan, J.; Hu, W. Distractor-Aware Siamese Networks for Visual Object Tracking. In Proceedings of the European Conference on Computer Vision, Munich, Germany, 8–14 September 2018; pp. 101–117.
22. Bolme, D.; Beveridge, J.; Draper, B.; Lui, Y.M. Visual object tracking using adaptive correlation filters. In Proceedings of the IEEE Conference on Computer Vision and Pattern Recognition, San Francisco, CA, USA, 13–18 June 2010; pp. 2544–2550.
23. Henriques, J.F.; Caseiro, R.; Martins, P.; Batista, J.P. Exploiting the Circulant Structure of Tracking-by-Detection with Kernels. In Proceedings of the European Conference on Computer Vision, Florence, Italy, 7–13 October 2012; Springer: Berlin/Heidelberg, Germany, 2012; pp. 702–715.

24. Dai, K.; Wang, D.; Lu, H.; Sun, C.; Li, J. Visual Tracking via Adaptive Spatially-Regularized Correlation Filters. In Proceedings of the IEEE Conference on Computer Vision and Pattern Recognition, Long Beach, CA, USA, 16–20 June 2019; pp. 4670–4679.
25. Xu, T.; Feng, Z.; Wu, X.J.; Kittler, J. Adaptive Channel Selection for Robust Visual Object Tracking with Discriminative Correlation Filters. *Int. J. Comput. Vis.* **2021**, *129*, 1359–1375. [CrossRef]
26. Wang, F.; Yin, S.; Mbelwa, J.T.; Sun, F. Context and saliency aware correlation filter for visual tracking. *Multimed. Tools Appl.* **2022**, *81*, 27879–27893. [CrossRef]
27. Fu, C.; Xu, J.; Lin, F.; Guo, F.; Liu, T.; Zhang, Z. Object Saliency-Aware Dual Regularized Correlation Filter for Real-Time Aerial Tracking. *IEEE Trans. Geosci. Remote Sens.* **2020**, *58*, 8940–8951. [CrossRef]
28. Li, F.; Tian, C.; Zuo, W.; Zhang, L.; Yang, M.H. Learning Spatial-Temporal Regularized Correlation Filters for Visual Tracking. In Proceedings of the IEEE Conference on Computer Vision and Pattern Recognition, Salt Lake City, UT, USA, 18–22 June 2018; pp. 4904–4913.
29. Huang, B.; Xu, T.; Shen, Z.; Jiang, S.; Zhao, B.; Bian, Z. SiamATL: Online Update of Siamese Tracking Network via Attentional Transfer Learning. *IEEE Trans. Cybern.* **2021**, *52*, 7527–7540. [CrossRef] [PubMed]
30. Dai, K.; Zhang, Y.; Wang, D.; Li, J.; Lu, H.; Yang, X. High-performance long-term tracking with meta-updater. In Proceedings of the IEEE/CVF Conference on Computer Vision and Pattern Recognition, Seattle, WA, USA, 13–19 June 2020; pp. 6298–6307.
31. Yang, T.; Xu, P.; Hu, R.; Chai, H.; Chan, A.B. ROAM: Recurrently optimizing tracking model. In Proceedings of the IEEE/CVF Conference on Computer Vision and Pattern Recognition, Seattle, WA, USA, 13–19 June 2020; pp. 6718–6727.
32. Xingjian, S.; Chen, Z.; Wang, H.; Yeung, D.Y.; Wong, W.K.; Woo, W.c. Convolutional LSTM network: A machine learning approach for precipitation nowcasting. In Proceedings of the Advances in Neural Information Processing Systems, Montreal, QC, Canada, 7–12 December 2015; pp. 802–810.
33. Chen, J.; Xu, T.; Huang, B.; Wang, Y.; Li, J. ARTracker: Compute a More Accurate and Robust Correlation Filter for UAV Tracking. *IEEE Geosci. Remote Sens. Lett.* **2022**, *19*, 1–5. [CrossRef]
34. Kalal, Z.; Mikolajczyk, K.; Matas, J. Tracking-Learning-Detection. *IEEE Trans. Softw. Eng.* **2011**, *34*, 1409–1422. [CrossRef]
35. Fan, H.; Ling, H. Parallel Tracking and Verifying: A Framework for Real-Time and High Accuracy Visual Tracking. In Proceedings of the IEEE International Conference on Computer Vision, Venice, Italy, 22–29 October 2017; pp. 5487–5495.
36. Lukezic, A.; Zajc, L.C.; Vojír, T.; Matas, J.; Kristan, M. FCLT-A Fully-Correlational Long-Term Tracker. *arXiv* **2018**, arXiv:1804.07056.
37. Voigtlaender, P.; Luiten, J.; Torr, P.H.; Leibe, B. Siam R-CNN: Visual Tracking by Re-Detection. In Proceedings of the IEEE Conference on Computer Vision and Pattern Recognition, Seattle, WA, USA, 13–19 June 2020; pp. 6578–6588.
38. Huang, L.; Zhao, X.; Huang, K. GlobalTrack: A Simple and Strong Baseline for Long-Term Tracking. In Proceedings of the AAAI Conference on Artificial Intelligence, New York, NY, USA, 7–12 February 2020; Volume 34, pp. 11037–11044.
39. Du, D.; Qi, Y.; Yu, H.; Yang, Y.; Duan, K.; Li, G.; Zhang, W.; Huang, Q.; Tian, Q. The Unmanned Aerial Vehicle Benchmark: Object Detection and Tracking. In Proceedings of the European conference on computer vision (ECCV), Munich, Germany, 8–14 September 2018; pp. 370–386.
40. Müller, M.; Smith, N.; Ghanem, B. A Benchmark and Simulator for UAV Tracking. In Proceedings of the European Conference on Computer Vision, Amsterdam, The Netherlands, 11–14 October 2016; Springer: Berlin/Heidelberg, Germany, 2016; pp. 445–461.
41. Li, S.; Yeung, D.Y. Visual object tracking for unmanned aerial vehicles: A benchmark and new motion models. In Proceedings of the AAAI Conference on Artificial Intelligence, San Francisco, CA, USA, 4–9 February 2017; Volume 31.
42. Zhu, P.; Wen, L.; Bian, X.; Ling, H.; Hu, Q. Vision Meets Drones: A Challenge. In Proceedings of the European Conference on Computer Vision, Munich, Germany, 8–14 September 2018; Springer: Berlin/Heidelberg, Germany, 2018.
43. Danelljan, M.; Häger, G.; Khan, F.; Felsberg, M. Accurate scale estimation for robust visual tracking. In Proceedings of the British Machine Vision Conference, Nottingham, UK, 1–5 September 2014; Bmva Press: Newcastle, UK, 2014.
44. Yan, B.; Zhang, X.; Wang, D.; Lu, H.; Yang, X. Alpha-refine: Boosting tracking performance by precise bounding box estimation. In Proceedings of the CVPR, Virtual, 19–25 June 2021; pp. 5289–5298.
45. Galoogahi, H.K.; Sim, T.; Lucey, S. Multi-channel Correlation Filters. In Proceedings of the IEEE International Conference on Computer Vision, Sydney, Australia, 1–8 December 2013; pp. 3072–3079.
46. Sherman, J.; Morrison, W.J. Adjustment of an Inverse Matrix Corresponding to a Change in One Element of a Given Matrix. *Ann. Math. Stat.* **1950**, *21*, 124–127. [CrossRef]
47. Dalal, N.; Triggs, B. Histograms of oriented gradients for human detection. In Proceedings of the 2005 IEEE Computer Society Conference on Computer Vision and Pattern Recognition (CVPR'05), San Diego, CA, USA, 20–26 June 2005; Volume 1, pp. 886–893.
48. Danelljan, M.; Khan, F.; Felsberg, M.; van de Weijer, J. Adaptive Color Attributes for Real-Time Visual Tracking. In Proceedings of the IEEE Conference on Computer Vision and Pattern Recognition, Columbus, OH, USA, 23–28 June 2014; pp. 1090–1097.
49. Danelljan, M.; Bhat, G.; Shahbaz Khan, F.; Felsberg, M. ECO: Efficient Convolution Operators for Tracking. In Proceedings of the IEEE Conference on Computer Vision and Pattern Recognition, Honolulu, HI, USA, 21–26 July 2017; pp. 6638–6646.
50. Wang, N.; Zhou, W.; Tian, Q.; Hong, R.; Wang, M.; Li, H. Multi-cue Correlation Filters for Robust Visual Tracking. In Proceedings of the IEEE Conference on Computer Vision and Pattern Recognition, Salt Lake City, UT, USA, 18–22 June 2018; pp. 4844–4853.
51. Bertinetto, L.; Valmadre, J.; Golodetz, S.; Miksik, O.; Torr, P.H.S. Staple: Complementary Learners for Real-Time Tracking. In Proceedings of the IEEE Conference on Computer Vision and Pattern Recognition, Las Vegas, NV, USA, 27–30 June 2016; pp. 1401–1409.

52. Li, Y.; Zhu, J. A Scale Adaptive Kernel Correlation Filter Tracker with Feature Integration. In Proceedings of the European Conference on Computer Vision, Zurich, Switzerland, 5–12 September 2014; pp. 254–265.
53. Henriques, J.F.; Caseiro, R.; Martins, P.; Batista, J. High-Speed Tracking with Kernelized Correlation Filters. *IEEE Trans. Pattern Anal. Mach. Intell.* **2015**, *37*, 583–596. [CrossRef]
54. Bhat, G.; Danelljan, M.; Gool, L.V.; Timofte, R. Know your surroundings: Exploiting scene information for object tracking. In *European Conference on Computer Vision*; Springer: Cham, Switzerland, 2020; pp. 205–221.
55. Bhat, G.; Danelljan, M.; Gool, L.V.; Timofte, R. Learning Discriminative Model Prediction for Tracking. In Proceedings of the IEEE International Conference on Computer Vision, Seoul, Korea, 27 October–2 November 2019; pp. 6182–6191.
56. Danelljan, M.; Bhat, G. PyTracking: Visual Tracking Library Based on PyTorch. Available online: https://github.com/visionml/pytracking (accessed on 8 January 2020).
57. Danelljan, M.; Gool, L.V.; Timofte, R. Probabilistic Regression for Visual Tracking. In Proceedings of the IEEE Conference on Computer Vision and Pattern Recognition, Seattle, WA, USA, 13–19 June 2020; pp. 7183–7192.
58. Valmadre, J.; Bertinetto, L.; Henriques, J.; Vedaldi, A.; Torr, P.H.S. End-To-End Representation Learning for Correlation Filter Based Tracking. In Proceedings of the IEEE Conference on Computer Vision and Pattern Recognition, Honolulu, HI, USA, 21–26 July 2017; pp. 2805–2813.
59. Cao, Z.; Fu, C.; Ye, J.; Li, B.; Li, Y. HiFT: Hierarchical Feature Transformer for Aerial Tracking. In Proceedings of the IEEE/CVF International Conference on Computer Vision, Montreal, QC, Canada, 10–17 October 2021; pp. 15457–15466.
60. Li, X.; Ma, C.; Wu, B.; He, Z.; Yang, M.H. Target-Aware Deep Tracking. In Proceedings of the IEEE Conference on Computer Vision and Pattern Recognition, Long Beach, CA, USA, 16–20 June 2019; pp. 1369–1378.
61. Wang, N.; Zhou, W.; Song, Y.; Ma, C.; Li, H. Real-time correlation tracking via joint model compression and transfer. *IEEE Trans. Image Process.* **2020**, *29*, 6123–6135. [CrossRef]
62. Wang, N.; Zhou, W.; Song, Y.; Ma, C.; Liu, W.; Li, H. Unsupervised deep representation learning for real-time tracking. *Int. J. Comput. Vis.* **2021**, *129*, 400–418. [CrossRef]
63. Ma, C.; Huang, J.B.; Yang, X.; Yang, M.H. Hierarchical Convolutional Features for Visual Tracking. In Proceedings of the IEEE International Conference on Computer Vision, Santiago, Chile, 7–13 December 2015; pp. 3074–3082.
64. Li, F.; Yao, Y.; Li, P.; Zhang, D.; Zuo, W.; Yang, M.H. Integrating Boundary and Center Correlation Filters for Visual Tracking with Aspect Ratio Variation. In Proceedings of the IEEE International Conference on Computer Orkshops (ICCVW), Venice, Italy, 22–29 October 2017; pp. 2001–2009.
65. Song, Y.; Ma, C.; Gong, L.; Zhang, J.; Lau, R.; Yang, M.H. CREST: Convolutional Residual Learning for Visual Tracking. In Proceedings of the IEEE International Conference on Computer Vision, Venice, Italy, 22–29 October 2017; pp. 2555–2564.
66. Yun, S.; Choi, J.; Yoo, Y.; Yun, K.; Young Choi, J. Action-decision networks for visual tracking with deep reinforcement learning. In Proceedings of the CVPR, Honolulu, HI, USA, 21–28 July 2017; pp. 2711–2720.
67. Li, Y.; Fu, C.; Huang, Z.; Zhang, Y.; Pan, J. Intermittent contextual learning for keyfilter-aware uav object tracking using deep convolutional feature. *IEEE Trans. Multimed.* **2020**, *23*, 810–822. [CrossRef]
68. Bertinetto, L.; Valmadre, J.; Henriques, J.F.; Vedaldi, A.; Torr, P.H.S. Fully-Convolutional Siamese Networks for Object Tracking. In *European Conference on Computer Vision*; Springer: Cham, Switzerland, 2016; pp. 850–865.
69. Danelljan, M.; Robinson, A.; Khan, F.; Felsberg, M. Beyond Correlation Filters: Learning Continuous Convolution Operators for Visual Tracking. In *European Conference on Computer Vision*; Springer: Cham, Switzerland, 2016; pp. 472–488.

sustainability

Article

A Data-Driven Based Method for Pipeline Additional Stress Prediction Subject to Landslide Geohazards

Meng Zhang [1,2,3], Jiatong Ling [4,*], Buyun Tang [2,5], Shaohua Dong [1,2] and Laibin Zhang [1,2]

1 College of Safety and Ocean Engineering, China University of Petroleum in Beijing, Beijing 102249, China
2 Key Laboratory of Oil and Gas Safety and Emergency Technology, Ministry of Emergency Management, Beijing 102249, China
3 CNPC International Pipeline Company, Beijing 102206, China
4 School of Engineering, The University of British Columbia, Kelowna, BC V1V 1V7, Canada
5 College of Mechanical and Transportation Engineering, China University of Petroleum in Beijing, Beijing 102249, China
* Correspondence: jiatong.ling@ubc.ca

Abstract: Pipelines that cross complex geological terrains are inevitably threatened by natural hazards, among which landslide attracts extensive attention when pipelines cross mountainous areas. The landslides are typically associated with ground movements that would induce additional stress on the pipeline. Such stress state of pipelines under landslide interference seriously damage structural integrity of the pipeline. Up to the date, limited research has been done on the combined landslide hazard and pipeline stress state analysis. In this paper, a multi-parameter integrated monitoring system was developed for the pipeline stress-strain state and landslide deformation monitoring. Also, data-driven models for the pipeline additional stress prediction was established. The developed predictive models include individual and ensemble-based machine learning approaches. The implementation procedure of the predictive models integrates the field data measured by the monitoring system, with k-fold cross validation used for the generalization performance evaluation. The obtained results indicate that the XGBoost model has the highest performance in the prediction of the additional stress. Besides, the significance of the input variables is determined through sensitivity analyses by using feature importance criteria. Thus, the integrated monitoring system together with the XGBoost prediction method is beneficial to modeling the additional stress in oil and gas pipelines, which will further contribute to pipeline geohazards monitoring management.

Keywords: oil and gas pipeline; landside monitoring; data-driven; pipeline additional stress; XGBoost

Citation: Zhang, M.; Ling, J.; Tang, B.; Dong, S.; Zhang, L. A Data-Driven Based Method for Pipeline Additional Stress Prediction Subject to Landslide Geohazards. *Sustainability* **2022**, *14*, 11999. https://doi.org/10.3390/su141911999

Academic Editors: Miltiadis D. Lytras and Andreea Claudia Serban

Received: 21 August 2022
Accepted: 20 September 2022
Published: 22 September 2022

1. Introduction

The pipeline is an important means for oil and gas transportation. The safe operation of the pipeline is threatened by various factors including corrosion, third-party damage, natural disaster, etc. For pipelines in areas with complex geological conditions such as mountainous regions, landslides have always been considered a dominant challenge for continuous pipeline service [1–3] . In this respect, landslides always impose additional stress on the pipeline, which may further lead to local deformation and displacement [4,5]. Once the additional stress generated exceeds the material strength or there are defects such as cracks and metal loss in the deformation area, it will seriously damage structural integrity of the pipeline [6], resulting in catastrophic accidents such as rupture and leakage. For example, in 2008–2009, two consecutive pipeline landslide accidents appeared in Zhejiang, China [7]. The earlier one was dealt with in time, whereas the latter one caused an explosion. On 5 March 2019, a pipeline rupture and subsequent explosion occurred in Alborz in Iran, leading to the suspension of gas supply to more than 12,000 customers [8]. With the random characteristic of natural disasters, it is difficult to predict and analyze their impact on pipelines. An accurate monitoring method is of great significance for protective measures to be taken at an early stage [9]. Therefore, it is fundamental to develop advanced methods for pipeline state monitoring under complicated geological circumstances.

It is known from the field that changed position of the pipeline under landslides imposes additional stresses on the pipeline, which put the pipeline in an unstable stress state [10,11]. To assess the resulting additional stress, the stress-strain monitoring for the pipelines across complex geological terrains is employed as an effective way for pipe condition monitoring. The widely used stress monitoring methods are categorized as long distance-based and point-based measurement. For long-distance real-time pipeline monitoring, distributed optical fiber sensors installed parallel to the pipe body are commonly utilized, which combines data acquisition capability including strss-strain behavior of soil, temperature, and vibration measurement [12,13]. It should be noted that this technology suffers equipment stability problems over the course of a long lifespan. In addition, point-based measurement adopts strain gauge sensors to stick around the pipe wall for strain monitoring, and the additional stress was calculated based on the measured strain value [14]. It directly reflects the pipeline behavior when compared with long-distance pipeline monitoring. However, the accuracy of the strain measurement sensor is susceptible to the disturbance of the external environment, resulting in the introduction of unreliable data.

Pipeline strain and stress are typically regarded as indicators that reflect the impact of landsliding on the pipeline [15], but an individual strain monitoring system cannot clearly explain the pipeline and its surrounding environmental conditions simultaneously. To improve the reliability of stress evaluation, landslide-related monitoring is complementary to comprehensive pipeline stress analysis. The landslide-induced failure of oil and gas pipelines is a complex process involving many parameters. Separately from strain monitoring, surface deformation monitoring and geohazard inducing monitoring are considered for landslide management [16,17]. For example, disaster reduction stick equipment has been designed to simultaneously monitor the landslide on precipitation, soil water content, pore water pressure, slope displacement, and slope inclination [18]. In most application cases, the monitoring of pipeline landslides is limited to the monitoring of the landslide hazard without the involvement of pipe interaction. There is still a lack of research on the stress state of pipelines under landslide interference. As a result, the relationship between landslide deformation data and pipeline safety conditions has not been effectively established.

Although pipeline landslide research in terms of soil-pipe interaction analysis [19], landslide behavior [20], and pipeline failure probability prediction [21] have been widely investigated. There are still some research gaps in this field. On the one hand, limited research has been done on the combined landslide hazard and pipeline stress state analysis. On the other hand, the interference of invalid measurement data of pipeline stress-strain monitoring has not been fully considered. To fill this gap, this paper focuses on establishing the quantitative relationship between landslide movement and pipeline mechanical state based on multi-parameter monitoring data. Advanced data-driven technologies are adopted for the prediction of additional stress. With joint monitoring tools and predictive models, effective prevention measures can be taken to reduce the probability of pipeline failure caused by landslides, which further mitigates the hazard of landslides posed on the pipeline. It should be noted that the model is applicable to slope soil layers, that is, slope sediment or residual soil or sedimentary soil. The contributions of this paper can be concluded as:

1. A multi-parameter integrated monitoring system was developed to monitor the pipeline and landslide conditions in a complex geologic environment.
2. The data-driven based predictive model was proposed for additional stress evaluation under landslide movement.
3. Field sites were selected for demonstration of the geohazard monitoring system, and the additional stress model was verified based on the on-site data.

The rest of this paper is arranged as follows: Section 2 presents the pipeline geohazards monitoring implementation process; Section 3 describes the proposed methodology for accurate prediction of the additional stress. Section 4 performs a case study to illustrate the feasibility and effectiveness of the methodology. Section 5 presents the conclusions of this paper.

2. Pipeline Geohazards Monitoring Implementation

2.1. Selection of Monitoring Sites

In mountainous regions, it is impossible to avoid every potential landslide crossing. Pipelines passing through potentially active landslides may be subjected to forces such as traction, compression, and shear [22]. Due to the instability of the disaster-causing body, the stress concentration of the pipeline and the deformation of the pipe body lead to the occurrence of pipeline geological disasters. When the strain of the pipe body exceeds the allowable range, the pipe will bend and deform or even fail. The stability and deformation of landslides are affected by internal factors (soil composition, slope structure, geological structure, etc.) and external factors (water factors, weathering, earthquakes, human activities, etc.). Internal factors determine the scale and form of landslide deformation and failure, and external factors promote the occurrence and development of landslide deformation.

The landsliding process is a very complex process in which multiple factors are involved. For the pipeline landslide program, in addition to considering the scale of the landslide, it is also necessary to consider the impact of the landslide body on the pipeline structure. The evolution process of the pipeline landslide geological disaster was shown in Figure 1. The pipeline traversing landslide terrain is the initial inducement. In the early stage of landslide deformation, a certain part of the landslide body is deformed because the shear stress is greater than the shear strength, resulting in small creeping. During this period, the deformation rate of the landslide is characterized by low speed and stability, and the pipeline is slightly deformed. After that, with the gradual rupture of the slip surface, penetrated cracks begin to appear on the slope surface. With the increase of cracks, the slope body is in an unstable state, and the deformation of the pipeline increases. Finally, the slope sliding occurs and causes the failure of the pipeline.

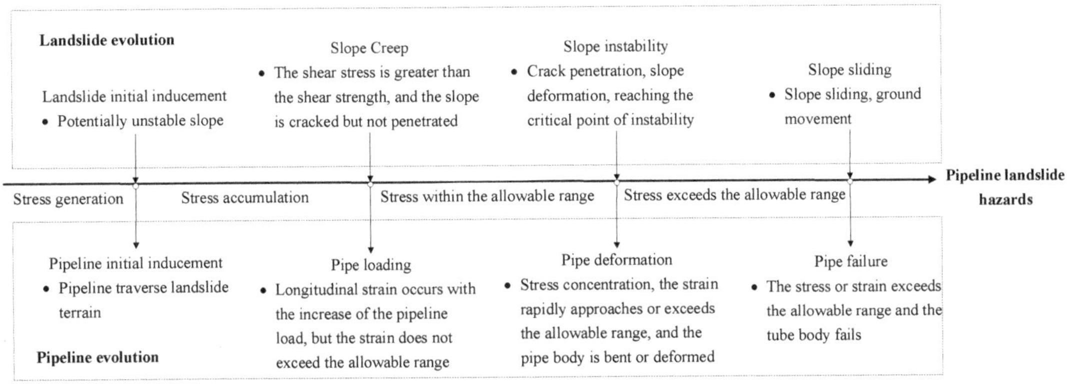

Figure 1. Schematic diagram of the evolution of pipeline geological hazards.

Mitigation measures are usually implemented to eliminate or reduce the severity of landslide effects on the pipeline. It is economically feasible to install modern sensors in specific locations susceptible to ground movement. Pipeline monitoring points and landslide monitoring points are two important components of the integrated system. For the pipeline monitoring site, the selected monitoring points should reflect the force law of the critical pipeline segment, and the load distribution of the pipeline should also be considered. For the landslide deformation monitoring site, the monitoring data should comprehensively reflect the development of the landslide and its impact on the pipeline.

Some preliminary analysis is required for the selection of monitoring points before sensor installation. The steps are as follows. Firstly, the geohazard-sensitive areas of the pipeline are determined according to the engineering geological disaster report, geological survey report, and other relevant assessment documents. During this procedure, necessary investigations of slope maps, structural geology, morphology, surface mechanisms, and environmental conditions help to explain the mechanisms of landslide deformation and failure in specific situations. Secondly, site investigations in the selected geohazard-sensitive area will be conducted. The experts are also invited to further verify the key areas for controlling slope stability, and analyze the feasibility of equipment installation. Then, finite element simulations of the pipeline are carried out to determine the stress concentration point of the pipeline, and it was selected as the pipeline monitoring point.

2.2. Selection of Monitoring Elements

Generally, the content of pipeline landslide monitoring includes pipeline monitoring and landslide deformation monitoring. The monitoring of pipeline stress and strain can directly reflect the pipeline condition subject to geological hazards. In landslide monitoring, the monitoring contents are different with various requirements, among which surface displacement monitoring and inner displacement monitoring are the most important objects. For the inducing factors of geological disasters, the monitoring elements mainly include meteorological conditions such as rainfall, snowmelt, temperature, and evaporation, as well as hydrological conditions such as surface water, river, ditch water level, pore water pressure, soil moisture content, and groundwater level [23]. In this paper, parameters including pipeline additional stress, landslide surface displacement, landslide inner displacement, anti-sliding pile inclination, landslide traction stress and compressive stress, and soil pressure are selected as the monitoring indicators to form a joint monitoring system for pipeline landslide events.

2.2.1. Pipeline Monitoring

Additional stress. For landslides that have undergone large deformation, the pipelines laid in the slope body have been subjected to additional stress [24]. The strain gauge sensor attached to the surface of the pipeline is used to determine the additional stress for pipeline safety state evaluation. The strain gauges can easily capture the response of the pipeline to ground movement. This type of sensor uses elastic sensitive elements to measure the strain generated by the deformation of the object under force, and converts the external stress on the pipeline into the change of strength circuit. Therefore, by calculating the change of the current or voltage in the circuit, the external stress of the pipeline can be determined. Pipeline stress and strain monitoring are the core technologies of pipeline integrity management. It can intuitively and quantitatively obtain the real-time stress and strain data of the pipe body under buried working conditions and issue deformation warnings in time. Compared with land displacement monitoring, it reflects pipe deformation quickly and accurately. In order to obtain the accurate axial strain of the pipeline, at least 3 strain gauges should be arranged in each pipeline section.

2.2.2. Landslide Deformation Monitoring

Surface land displacement. The sliding of the soil mass on the surface of the landslide will have a direct impact on the pipelines buried in it. As the pipelines are mostly buried in trenches, floods and debris flows can easily wash away the upper soil of the pipelines, resulting in exposed pipelines. Moreover, soil accumulated around the pipelines will lead to excessive external stress, resulting in pipeline bending with serious consequences. Therefore, it is necessary to monitor the surface displacement to prevent the damage caused by excessive displacement. The occurrence of a landslide may result in differential movement of the soil in all three directions. The z-axis is perpendicular to the ground, and the x-axis and y-axis are in the ground plane. To grasp the dynamic changes of each part of the landslide surface, it is necessary to monitor its displacement.

Inner land displacement. Deep soil displacement induced by landslides results in different stresses and deformation rates at different land depths. Accordingly, the force on the pipeline will also change as the buried depth increases. Inner land displacement leads to unequal stress on the pipeline, which leads to buckling and bending of the pipeline.

Anti-slide pile inclination. Pipelines crossing the mountainous area are prone to be affected by excessive stress concentration, and the establishment of anti-sliding piles in the landslide area is a commonly used prevention and control method. Anti-slide piles are piles that penetrate the landslide body to offset the sliding force and buffer the stress brought by the landslide, mainly used for landslide treatment in shallow and medium-thick layers. The inclination of the anti-slide pile can approximately reflect the degree of landslide hazard and is one of the main external monitoring factors.

Landslide stress. The soil movement caused by the landslide will impose external loads on the pipeline. Anti-slide piles have been widely used in landslide control. It is a kind of geotechnical mitigation alternative to reduce the severity of the landslide by reducing the driving force of the slide and providing additional resistance to the ground movement. The monitoring sensors are installed on anti-slide piles to reflect the external load distribution of the landslide.

Soil pressure. Soil pressure refers to the force of the surrounding soil acting on the pipeline. The soil pressure above the pipe increases as the stiffness of the fill increases, which will inevitably lead to the expansion of the pipeline defects. In most cases, the soil pressure will cause the pipeline to be unevenly stressed, which is prone to causing the pipeline to be damaged. By fixing the pressure fiber grating sensor on the pipeline, the soil pressure can be monitored.

2.3. *The Implementation of Monitoring System*

The monitoring system was installed at the selected monitoring sites, which consists of the field monitoring part and the data acquisition part. The on-site monitoring section consists of various monitoring instruments installed on the landslide and the pipe body, while the data acquisition section is able to automatically collect monitoring data and transmits it to the user terminal in real-time to reflect the pipeline and surrounding condition. In detail, the arrangement of the sensors with different functions is shown in Figure 2. Additional stress is measured by sensor 2, which is located in a different direction on pipelines. Soil pressure information can be obtained from sensor 3. Landslide traction and compressive stress, anti-slide pile inclination along with inner land displacement can be obtained from sensor 4, while land surface displacement is collected from sensor 5.

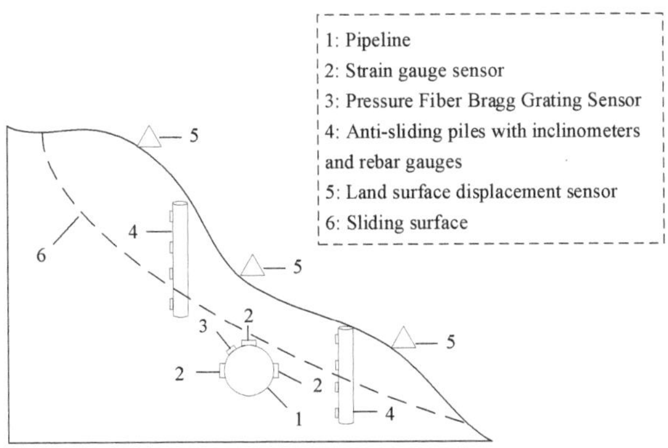

Figure 2. Schematic of the landslide monitoring system.

3. Proposed Methodology

3.1. The Framework of Proposed Method

The pipeline landslide hazard is a very complex process in which multiple factors are involved. It is difficult to determine the additional stress by traditional methods when landslide movement factors need to be considered. In recent years, it is known that data-driven models are usually combined with machine learning techniques to explore the relationship and patterns from a data perspective [25]. The machine learning approaches are capable of solving nonlinear regression problems with high performance. In this regard, the commonly used machine learning methods are adopted in this paper to tackle the relationship between additional stress and landslide movement parameters, which further realize the additional stress prediction. As pipeline additional stress is an important factor in failure analysis, the proposed predictive model could be used as an alternative source of additional stress indicator in cases of failures of strain monitoring sensors. Figure 3 presents the flowchart of the proposed approach for pipe additional stress prediction. Data preprocessing, model training, and performance evaluation comprises the main steps. Symbol definitions in Figure 3 can be found in Table 1.

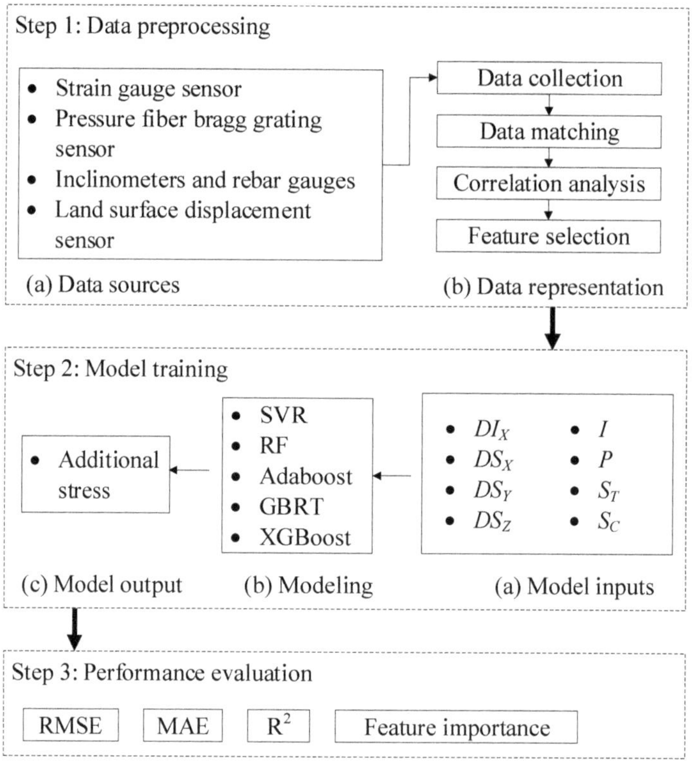

Figure 3. Flowchart of the proposed approach.

Table 1. Statistical description of the monitoring database.

Monitoring Type	Variables	Symbol	X_{min}	X_{max}	X_{mean}	X_{std}
Landslide monitoring	Inner x-axis displacement, mm	DI_X	0	7	2.31	1.8
	Inner y-axis displacement, mm	DI_Y	0	8	3.66	2.17
	Surface x-axis displacement, mm	DS_X	0.05	0.25	0.16	0.03
	Surface y-axis displacement, mm	DS_Y	0	0.25	0.16	0.03
	Surface z-axis displacement, mm	DS_Z	−0.08	−0.03	−0.04	0.01
	Anti-slide pile inclination, °	I	0	1.2	0.41	0.21
	Soil pressure, kPa	P	100	150	135.05	16.34
	Traction stress, kN	S_T	60	80	72.25	4.03
	Compressive stress, kN	S_C	50	70	61.92	3.52
Pipeline monitoring	Pipe additional stress, MPa	S	4	7	5.98	0.28

3.2. Data Preprocessing

In the data preprocessing processes, all the monitored elements are involved in the dataset for the additional stress prediction. Different types of monitored data are measured by multi-sensors, and data matching needs to be done to align them with the same time series. Then, a correlation analysis between every two features needs to be calculated to remove any redundant features with repeated information. The correlation coefficient (CC) ranging from −1 to 1 is calculated to check a linear correlation between two variables. The absolute value of CC represents the strength of the correlation between the two features. Generally, due to redundant features having no contribution to a predictive model, the features that have a strong positive correlation need to be removed. The standardization also needs to be done to eliminate the difference between the different dimensions.

3.3. Model Training

The processed data then can be used for the training model. The additional stress is used as the model output, whereas other features such as inner displacement and surface displacement are fed as model inputs. In this paper, individual and ensemble-based machine learning approaches are developed for the predictive model.

3.3.1. Support Vector Regression

Inspired by the Support vector machine (SVM), Support vector regression (SVR) was proposed to solve nonlinear regression estimation problems by including the new loss function [26]. The goal of the method is to find the optimized hyperplane for regression, which is defined as Equation (1).

$$f(x) = < w, x > +b \tag{1}$$

where w is the regression coefficient vector, and b is the bias. As the ε-insensitive loss function was introduced into the SVR model, where ε is a precision parameter that presents the radius of the internal tube region, the loss of the predicted sample in the ε-insensitive zone is zero [27]. Therefore, the estimation of w and b was based on the optimization problem in Equation (2), which is subject to the constraints in Equations (3)–(5).

$$min\frac{1}{2}||w||^2 + C\Sigma_{i=1}^{n}|\xi_i^* + \xi_i| \tag{2}$$

$$s.t. f(x_i) - y_i \leq \varepsilon + \xi_i \tag{3}$$

$$s.t. y_i - f(x_i) \leq \varepsilon + \xi_i^* \tag{4}$$

$$s.t. \varepsilon \geq 0, \xi_i^* \geq 0, \xi_i \geq 0 \tag{5}$$

where C is the constant coefficient, ξ_i^* and ξ_i are the distance between the sample data and the decision boundary. With the involvement of Lagrange multiplier and kernel function K, Equation (1) can be expressed as Equation (6):

$$f(x) = \sum_{i=1}^{n} (\alpha_i^* - \alpha) K(x_i, x_i') + b \tag{6}$$

where α_i^* and α_i are Lagrange multiplier. For the pipeline additional stress prediction, it can be regarded as a multi-regression analysis problem, of which the objective function is the minimization of MAE shown in Equation (7)

$$\text{MAE} = \frac{1}{N} \sum_{i=1}^{N} |f(x_i) - y_i| \tag{7}$$

where N is the number of data.

3.3.2. Random Forest

Random forest (RF) is a representative bagging ensemble learning (EI) algorithm, of which the base estimators are decision trees [28]. The bagging method relies on the average principle to decide the ensembled results for the regression task; it is expected that the accuracy of the random forest increase with higher individual decision tree accuracy. Since the base estimator will randomly select features for branching, each base estimator is independent of the other. In addition, the bootstrap technique is employed for data sampling in the training process to construct different datasets. The randomness of data selection and feature selection enables the random forests robust to noise [29].

3.3.3. Adaptive Boosting

Adaptive boosting (AdaBoost) is a representative boosting ensemble learning algorithm, which adaptively influences the subsequent modeling process according to the results of the previous weak evaluators [30]. For boosting technology, the output $H(x)$ integrates multiple weak learning models following Equation (8) .

$$H(x) = \sum_{t=1}^{T} \phi_t f_t(x_i) \tag{8}$$

where T is the total number of weak estimator, ϕ_t is the weight of t-th weak estimator, $f_t(x_i)$ is the result of x_i on the t-th weak estimator. AdaBoost adopts different sample weights according to the predicted sample results [31]. The weight of the sample with larger prediction error will increase, and the weight of the sample with smaller prediction error will decrease. Each iteration aims to minimize the total training error.

3.3.4. Gradient Boosting Regression Tree

Gradient boosting regression tree (GBRT) is also a sequential EL-model using boosting strategy [32]. Compared with AdaBoost, GBRT utilizes the negative gradient of the loss function as an approximation of the residual in the boosted tree algorithm to build a subsequent weak evaluator. The prediction steps based on GBRT are as follows [33]:

1. Weak learner initialization

$$f_0(x) = argmin \sum_{i=1}^{N} L(y_i, c) \tag{9}$$

   In Equation (9), N is the number of samples, c is the constant value with the smallest loss function; y_i is the actual target value.
2. Iteratively build M boosted trees

The negative gradient for samples $i = 1, 2, \ldots N$ is expressed as Equation (10).

$$r_{im} = \left[\frac{\partial L(y_i, f(x_i))}{\partial f(x_i)}\right]_{f(x)=f_{m-1}(x)} \quad (10)$$

r_{im} is the residual, (x_i, r_{im}) are used as the training data of the next tree, the corresponding node area of the newly established regression tree is $R_{jm}, j = 1, 2, \ldots J$. The expression of a new learner can be obtained as Equation (11).

$$f_m(x) = f_{m-1}(x) + \sum_{j=1}^{J} r_{jm} I(x \in R_{jm}) \quad (11)$$

$$r_{jm} = argmin \sum_{x \in R_{jm}} L(y_i, f_{m-1}(x_i) + c) \quad (12)$$

r_{jm} is the minimum value of the loss function for the m-th tree at the j-th iteration, which is shown in Equation (12). $I(x \in R_{jm})$ is the characteristic function, c is the constant value with the smallest loss function.

3. Final output
The predictive results are based on the ensemble predictions of the weak learner models as Equation (13).

$$f(x) = f_M(x) = f_0(x) + \sum_{m=1}^{M} \sum_{j=1}^{J} r_{jm} I(x \in R_{jm}) \quad (13)$$

where M is the maximum number of iterations.

3.3.5. Extreme Gradient Boosting

Extreme gradient boosting (XGBoost) is optimized on the basis of the GBRT algorithm, both the training speed and model accuracy are improved [34]. An objective function was designed as Equation (14) for model optimization.

$$Obj = \sum_{i=1}^{n} L(y_i, \hat{y}_i) + \sum_{k=1}^{K} \Omega f(k) \quad (14)$$

where Obj represents the objective function, which consists of the loss function and regularization term. Loss function $\sum_{i=1}^{n} L(y_i, \hat{y}_i)$ was designed to evaluate the loss between predicted and true values, whereas the regularization term $\sum_{k=1}^{K} \Omega f(k)$ was used to control the model complexity and avoid overfitting. Due to the advantages of parallel computing and allowing row and column sampling, XGBoost has attracted a lot of attention because of its high efficiency.

3.4. *Performance Evaluation*

The obtained results from the implementation of the regression models will be evaluated to determine the best performing model among the five approaches aforementioned. The whole dataset was split for model training (80% of data) and testing (20% of data). However, the division of the training set and the testing set will interfere with the model results. In this regard, the k-fold cross-validation is used to observe the generalization performance of the model. In this process, the data is divided into n subsets, one of which is used sequentially as the test set and the other n − 1 folds as the training set, in which $k = 5$ is taken. Finally, the accuracy of the model is evaluated by considering the mean of n cross-validation results. Several performance metrics including Root Mean Square Error

(RMSE) in Equation (15), Mean Absolute Error (MAE) in Equation (16), and coefficient of determination R^2 in Equation (17) are computed for model performance evaluation.

$$\text{RMSE} = \sqrt{\frac{\sum_{i=1}^{m} (y_i - \widehat{y_i})^2}{m}} \tag{15}$$

$$\text{MAE} = \frac{\sum_{i=1}^{m} |y_i - \widehat{y_i}|}{m} \tag{16}$$

$$R^2 = 1 - \frac{\sum_{i=1}^{m} (y_i - \widehat{y_i})^2}{\sum_{i=1}^{m} (y_i - \bar{y_i})^2} \tag{17}$$

where m is the total number of test sample, y_i is the true value, $\widehat{y_i}$ is the prediction value, $\bar{y_i}$ is the mean value

4. Case Study

4.1. Monitoring Sites Description

Pipeline segments crossing the mountain areas in southwest China were illustrated in the case study. The pipe material is X52 and the pipe diameter is 323.9 mm, with wall thickness of 8.7 mm and designed pressure of 10 MPa. In-line inspections were conducted in 2008, 2014, and 2018, respectively to acquire the integrity condition of the pipeline. In addition, the geohazard investigation along the pipeline has been completed. According to the above information, four potential landslide disaster sites were chosen for the implementation of the multi-parameter integrated monitoring system. The detailed description of the four monitoring sites is as follows. (1) The first monitoring site is located on the left bank of the river, in the middle and upper part of the area. The landslide elevation ranges from 2170 m to 2263 m, and the height difference is about 93 m. (2) In 2014, the landslide where the second monitoring site is located showed obvious signs of deformation, threatening the safety of the pipeline. Subsequently, the pipeline owner took timely control measures for the landslide by setting up the anti-slide piles. However, there are cracks in the lower part of the landslide with the rainfall in recent years. (3) The slope section where the third monitoring site is located is about 420 m long, with a height difference of about 290 m and a slope gradient of 45–50°. (4) The landslide zone where the fourth monitoring site is located is 280 m long, with a width of which is 320 m and a thickness is 15 m, which is an area with frequent natural disasters such as debris flow.

The integrated monitoring system described in Section 2 was implemented in selected monitoring sites. For the monitoring of surface displacement, the monitoring points are arranged in the sliding-sensitive area of the slope. Benefiting from the Global Satellite Navigation and Positioning System, the horizontal and vertical displacements of the monitoring points are automatically measured. In detail, the original observation data is automatically calculated and processed by professional deformation monitoring software, and the real-time millimeter-level coordinate value of the monitoring point is obtained. For the monitoring of inner displacement, anti-sliding piles with inclinometers and rebar gauges are arranged in the areas where the inner landslide soil shows a likely deformation trend. The inclinometer is used to collect the inner displacement and the inclination, whereas the rebar gauges are utilized for traction and compressive stress measurement. The pressure fiber bragg grating sensor is installed on the pipeline through the bracket for the measurement of soil pressure. In addition, the pipeline stress-strain monitoring is realized by attaching strain gauges at the stress concentration points indicated by the simulation analysis. In this paper, a total of 400 sets of data all come from these on-site sensors. Table 1 shows the statistical characteristics of the monitoring variables.

The correlation matrix between features is shown in Figure 4. Since a redundant feature provides no contribution but long training time to a predictive model, the Anti-slide pile inclination factor, which has a strong positive correlation (CC = 1) with the Inner y-axis displacement, needs to be removed.

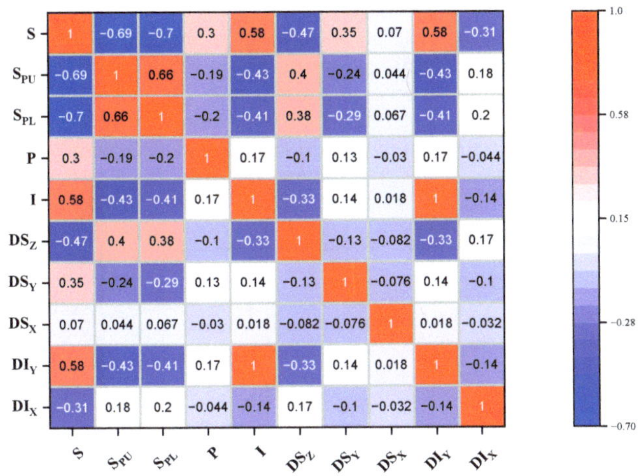

Figure 4. Correlation matrix of every two attributes.

4.2. Comparative Analysis

In the model training process, SVR, along with the other four ensemble-based methods (RF, AdaBoost, GBRT, XGBoost) are validated with the field monitoring data, the results with three performance metrics were presented in Table 2. The results are calculated using the test dataset (20% of the whole sample set). Remarkably, all the proposed ensemble models produced satisfactory predictions, showing that XGBoost had the best performance, with RMSE of 0.0154, MAE of 0.0118, and the highest R^2 of 0.9893. The predicted results of the RF model were not as accurate as the outputs from other ensemble models, with RMSE of 0.0288 and MAE of 0.0193. In addition, AdaBoost and GBRT models obtained comparable performance, whereas GBRT shows slightly better results in RMSE and R^2 metrics. On the contrary, the individual model SVR obtain unacceptable prediction results.

Table 2. Performance comparison of five regression models.

Model Type	Data-Driven Models	RMSE	MAE	R^2
Individual	SVR	0.0910	0.0666	0.6233
Ensemble	RF	0.0288	0.0193	0.9623
	AdaBoost	0.0217	0.0119	0.9758
	GBRT	0.0198	0.0147	0.9821
	XGBoost	0.0154	0.0118	0.9893

The additional stress prediction details for each model is visualized in Figures 5–9. The abscissa of the scatter plot represents the actual monitored value and the ordinate represents the model predicted value. To quantify the fitted relationship between the measured and predicted value, a linear fit is shown in each figure. The most perfect fit is that the measured value is equal to the predicted value, expressed as y = x. That is, the closer the black dots to line y = x, the better the prediction effect. It can be seen from Figure 5a that the results from SVR are quite different from the true data. Figure 6a shows the ensembled model outperforms the individual model, but the RF model for additional stress prediction is not accurate enough. From Figures 7a and 8a, satisfied prediction data are obtained, and most of the data are close to the monitored value. As shown in Figure 9a, XGBoost performs the best agreement between the predicted results and the measured data, with a linear fit equation of y = 0.991x + 0.055. The standard deviation of the predicted points is 0.142.

As the black dots in the scatter plots combined the true value and predicted value, each test sample and its corresponding predicted data are also compared. Figure 5b presents the predicted additional stress of the SVR model differs a lot from the true value, showing that the prediction accuracy is low. As observed in Figures 6b, 7b and 8b, the prediction accuracy of the ensembled model is improved. Although there is a noticeable error between the actual and predicted results at some testing sample points. Regarding the best results, Figure 9b shows that almost all the prediction values of additional stress are closer to the actual value.

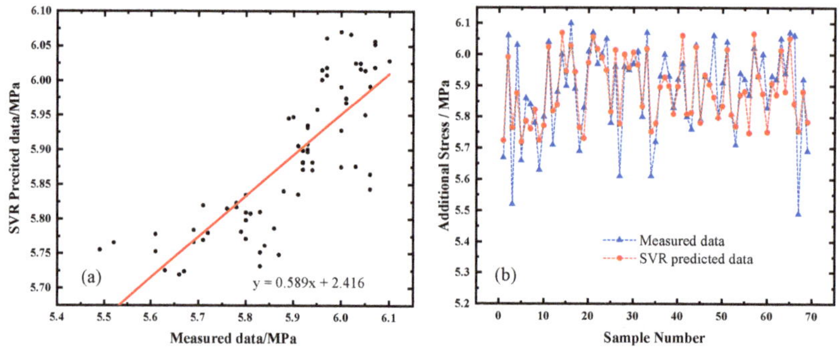

Figure 5. Predicted results of SVR. (**a**) Scatter plots of the predicted versus measured data. (**b**) Comparison of predicted and measured data.

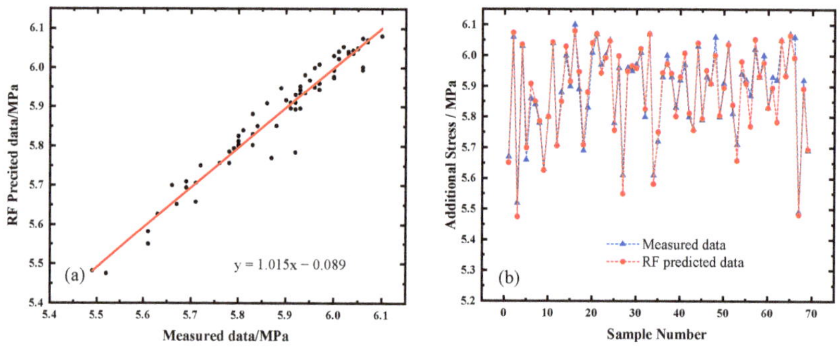

Figure 6. Predicted results of RF. (**a**) Scatter plots of the predicted versus measured data. (**b**) Comparison of predicted and measured data.

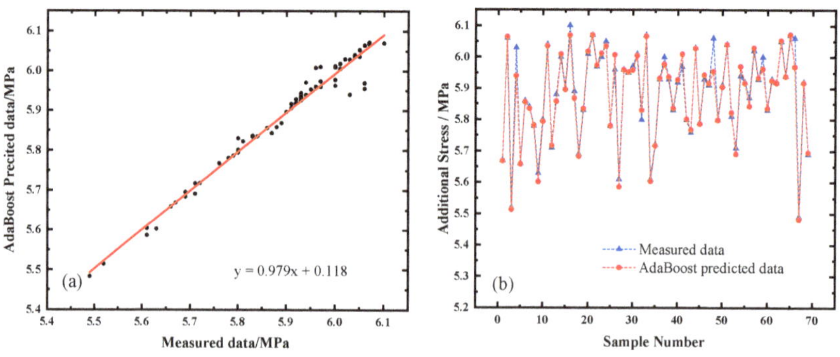

Figure 7. Predicted results of AdaBoost. (**a**) Scatter plots of the predicted versus measured data. (**b**) Comparison of predicted and measured data.

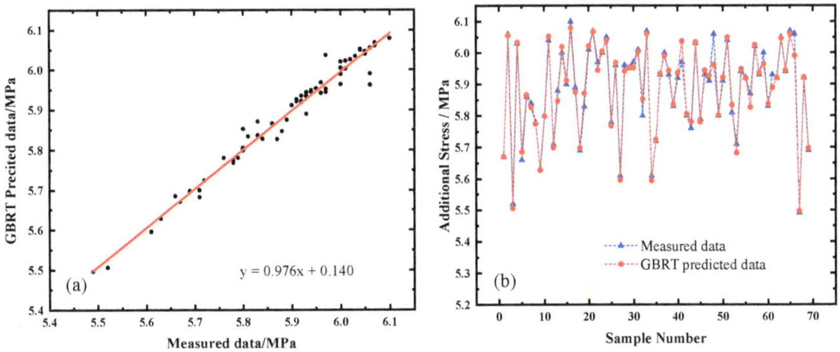

Figure 8. Predicted results of GBRT. (**a**) Scatter plots of the predicted versus measured data. (**b**) Comparison of predicted and measured data.

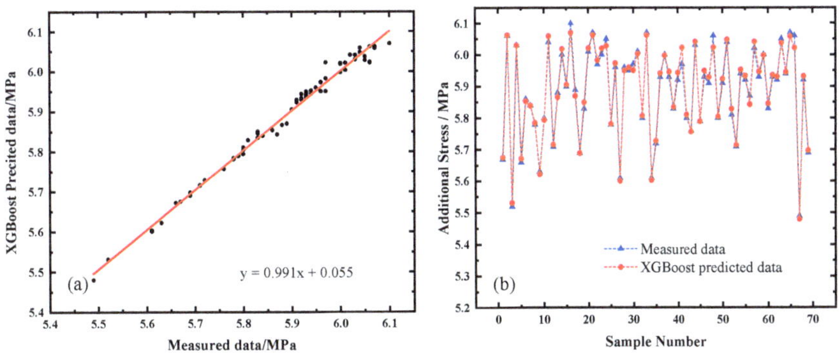

Figure 9. Predicted results of XGBoost. (**a**) Scatter plots of the predicted versus measured data. (**b**) Comparison of predicted and measured data.

4.3. Sensitivity Analysis

In the prediction models, the relationship between the input variables and the target output has been quantified. To illustrate the importance of the landslide features on the additional stress determination, the importance analysis is conducted for the explanation. The influence of each characteristic on the final predicted additional stress was calculated based on the XGBoost model due to its high performance.

The results were ranked in Table 3. The greater the influence on the model prediction, the more significant the feature is. According to the obtained results, compressive stress is listed as the most important feature with feature importance of 0.5070, followed by traction stress with 0.2406. The surface y-axis displacement and soil pressure are the third and fourth most important features, with relative relevance of 0.0910 and 0.0844, respectively. The top four parameters account for more than 90% importance of all of the input variables. It can be seen that the surface z-axis displacement showed 0.045 importance. Moreover, little difference in importance can be found in the inclination and surface x-axis displacement, weighted for 0.0162 and 0.0119, while the inner x-axis displacement indicated the lowest importance. It can be seen from the feature importance of the RF model that the top three factors are the same with XGBoost. Although the importance values vary, rankings other than the inclination factor remain constant for both models.

According to the obtained importance results, the stress plays a more significant role than the displacement, whereas the landslide surface displacement matters a lot more than the inner displacement. The results indicated that the pipe behavior was affected by soil-pipe interaction, and ground movements are consistent with the findings in [35].

Mitigation alternatives can be determined according to the importance of different external influencing factors. However, as the results of machine learning models are closely related to the training data [36], the obtained results are subject to change if a larger database that includes more information and data variability is used since the current database is limited.

Table 3. Feature importance analysis.

Rank (XGBoost)	Monitoring Parameter	Feature Importance (XGBoost)	Rank (RF)	Monitoring Parameter	Feature Importance (RF)
1	Compressive stress	0.5070	1	Compressive stress	0.4109
2	Traction stress	0.2406	2	Traction stress	0.3867
3	Surface y-axis displacement	0.0910	3	Surface y-axis displacement	0.1361
4	Soil pressure	0.0844	4	Inclination	0.0266
5	Surface z-axis displacement	0.0450	5	Soil pressure	0.0249
6	Inclination	0.0162	6	Surface z-axis displacement	0.0081
7	Surface x-axis displacement	0.0119	7	Surface x-axis displacement	0.0055
8	Inner x-axis displacement	0.0039	8	Inner x-axis displacement	0.0010

5. Conclusions

For the safe service of oil and gas pipelines that cross mountainous areas subject to landslide geohazard, a multi-parameter integrated monitoring system was developed. The joint monitoring for pipeline landslide hazards not only obtains the stress state information of the pipeline, but also the landslide displacement information. To tackle with the interference of invalid measurement data for pipeline stress-strain monitoring, advanced data-driven technologies including SVR, RF, AdaBoost, GBRT, and XGBoost are employed for the prediction of additional stress. Based on the field monitoring data, the relation between landslide movement and pipeline additional stress has been reasonably quantified. From this work, some conclusions are obtained:

- The results indicate that the XGBoost model has the highest performance in the prediction of the additional stress, with RMSE of 0.0154 MPa, MAE of 0.0118 MPa, and R^2 value of 0.9893.
- The top five factors contributing to the additional stress for the applied dataset are landslide compressive stress, landslide traction stress, landslide surface y-axis displacement, soil pressure, and landslide surface x-axis displacement.

For further development, the accuracy of the additional stress prediction from the landslide monitoring system can be improved with more monitoring data collected. Overall, the developed technique has allowed the implementation of preventive geotechnical works before a geohazard develops dramatically. In this sense, the model can be seen as a valuable tool that complements existing methodologies and can provide useful information to support the decision-making process.

Author Contributions: Conceptualization, methodology, writing—original draft preparation, M.Z. and J.L.; investigation, B.T.; writing—review and editing, supervision, S.D. and L.Z. All authors have read and agreed to the published version of the manuscript.

Funding: This research was funded by the Strategic Cooperation Technology Projects of CNPC and CUPB ZLZX2020-05.

Institutional Review Board Statement: Not applicable.

Informed Consent Statement: Not applicable.

Data Availability Statement: Not applicable.

Conflicts of Interest: The authors declare no conflict of interest.

References

1. Badida, P.; Balasubramaniam, Y.; Jayaprakash, J. Risk evaluation of oil and natural gas pipelines due to natural hazards using fuzzy fault tree analysis. *J. Nat. Gas Sci. Eng.* **2019**, *66*, 284–292. [CrossRef]
2. Zhang, S.-Z.; Li, S.-Y.; Chen, S.-N.; Wu, Z.-Z.; Wang, R.-J.; Duo, Y.-Q. Stress analysis on large-diameter buried gas pipelines under catastrophic landslides. *Pet. Sci.* **2017**, *14*, 579–585. [CrossRef]
3. Pu, H.; Xie, J.; Schonfeld, P.; Song, T.; Li, W.; Wang, J.; Hu, J. Railway alignment optimization in mountainous regions considering spatial geological hazards: A sustainable safety perspective. *Sustainability* **2021**, *13*, 1661. [CrossRef]
4. Teng, M.-C.; Ke, S.-S. Disaster impact assessment of the underground hazardous materials pipeline. *J. Loss Prev. Process. Ind.* **2021**, *71*, 104486. [CrossRef]
5. Zahid, U.; Godio, A.; Mauro, S. An analytical procedure for modelling pipeline-landslide interaction in gas pipelines. *J. Nat. Gas Sci. Eng.* **2020**, *81*, 103474. [CrossRef]
6. Yavorskyi, A.; Karpash, M.; Zhovtulia, L.; Poberezhny, L.Y.; Maruschak, P. Safe operation of engineering structures in the oil and gas industry. *J. Nat. Gas Sci. Eng.* **2017**, *46*, 289–295. [CrossRef]
7. Zheng, J.; Zhang, B.; Liu, P.; Wu, L. Failure analysis and safety evaluation of buried pipeline due to deflection of landslide process. *Eng. Fail. Anal.* **2012**, *25*, 156–168. [CrossRef]
8. Vasseghi, A.; Haghshenas, E.; Soroushian, A.; Rakhshandeh, M. Failure analysis of a natural gas pipeline subjected to landslide. *Eng. Fail. Anal.* **2021**, *119*, 105009. [CrossRef]
9. Ali, H.; Choi, J.-H. A review of underground pipeline leakage and sinkhole monitoring methods based on wireless sensor networking. *Sustainability* **2019**, *11*, 4007. [CrossRef]
10. Guo, W.; Liu, Y.; Li, C. Stress analysis and mitigation measures for floating pipeline. *IOP Conf. Ser. Earth Environ. Sci.* **2017**, *59*, 012014. [CrossRef]
11. Xu, P.; Zhang, M.; Lin, Z.; Cao, Z.; Chang, X. Additional stress on a buried pipeline under the influence of coal mining subsidence. *Adv. Civ. Eng.* **2018**, *2018*, 3245624. [CrossRef]
12. Zhang, S.; Liu, B.; He, J. Pipeline deformation monitoring using distributed fiber optical sensor. *Measurement* **2019**, *133*, 208–213. [CrossRef]
13. Yin, J.; Li, Z.-W.; Liu, Y.; Liu, K.; Chen, J.-S.; Xie, T.; Zhang, S.-S.; Wang, Z.; Jia, L.-X.; Zhang, C.-C.; et al. Toward establishing a multiparameter approach for monitoring pipeline geohazards via accompanying telecommunications dark fiber. *Opt. Fiber Technol.* **2022**, *68*, 102765. [CrossRef]
14. Li, X.; Wu, Q.; Jin, H.; Kan, W. A new stress monitoring method for mechanical state of buried steel pipelines under geological hazards. *Adv. Mater. Sci. Eng.* **2022**, *2022*, 4498458. [CrossRef]
15. Sarvanis, G.C.; Karamanos, S.A. Analytical model for the strain analysis of continuous buried pipelines in geohazard areas. *Eng. Struct.* **2017**, *152*, 57–69. [CrossRef]
16. Xin, L.; Mingzhou, B.; Bohu, H.; Pengxiang, L.; Lebin, Z.; Yun, C.; Hai, S. Safety analysis of landslide in pipeline area through field monitoring. *J. Test. Eval.* **2021**, *51*, e200751. [CrossRef]
17. Xu, W.; Xu, H.; Chen, J.; Kang, Y.; Pu, Y.; Ye, Y.; Tong, J. Combining numerical simulation and deep learning for landslide displacement prediction: An attempt to expand the deep learning dataset. *Sustainability* **2022**, *14*, 6908. [CrossRef]
18. Yan, Y.; Yang, D.-S.; Geng, D.-X.; Hu, S.; Wang, Z.-A.; Hu, W.; Yin, S.-Y. Disaster reduction stick equipment: A method for monitoring and early warning of pipeline-landslide hazards. *J. Mt. Sci.* **2019**, *16*, 2687–2700. [CrossRef]
19. Kunert, H.G.; Otegui, J.; Marquez, A. Nonlinear fem strategies for modeling pipe–soil interaction. *Eng. Fail. Anal.* **2012**, *24*, 46–56. [CrossRef]
20. Wei, A.; Yu, K.; Dai, F.; Gu, F.; Zhang, W.; Liu, Y. Application of tree-based ensemble models to landslide susceptibility mapping: A comparative study. *Sustainability* **2022**, *14*, 6330. [CrossRef]
21. Alvarado-Franco, J.P.; Castro, D.; Estrada, N.; Caicedo, B.; Sánchez-Silva, M.; Camacho, L.A.; Muñoz, F. Quantitative-mechanistic model for assessing landslide probability and pipeline failure probability due to landslides. *Eng. Geol.* **2017**, *222*, 212–224. [CrossRef]
22. Liu, S.; Wang, H.; Li, R.; Ji, B. A novel feature identification method of pipeline in-line inspected bending strain based on optimized deep belief network model. *Energies* **2022**, *15*, 1586. [CrossRef]
23. Yan, Y.; Xiong, G.; Zhou, J.; Wang, R.; Huang, W.; Yang, M.; Wang, R.; Geng, D. A whole process risk management system for the monitoring and early warning of slope hazards affecting gas and oil pipelines. *Front. Earth Sci.* **2022**, *9*, 812527. [CrossRef]
24. Ding, Y.; Yang, H.; Xu, P.; Zhang, M.; Hou, Z. Coupling interaction of surrounding soil-buried pipeline and additional stress in subsidence soil. *Geofluids* **2021**, *2021*, 7941989. [CrossRef]
25. Radwan, A.E.; Wood, D.A.; Radwan, A.A. Machine learning and data-driven prediction of pore pressure from geophysical logs: A case study for the mangahewa gas field, new zealand. *J. Rock Mech. Geotech. Eng.* **2022**. [CrossRef]
26. Awad, M.; Khanna, R. Support vector regression. In *Efficient Learning Machines*; Springer: Berlin/Heidelberg, Germany, 2015; pp. 67–80.
27. Jia, Z.; Ho, S.-C.; Li, Y.; Kong, B.; Hou, Q. Multipoint hoop strain measurement based pipeline leakage localization with an optimized support vector regression approach. *J. Loss Prev. Process. Ind.* **2019**, *62*, 103926. [CrossRef]

28. Rodriguez-Galiano, V.; Sanchez-Castillo, M.; Chica-Olmo, M.; Chica-Rivas, M. Machine learning predictive models for mineral prospectivity: An evaluation of neural networks, random forest, regression trees and support vector machines. *Ore Geol. Rev.* **2015**, *71*, 804–818. [CrossRef]
29. Roy, M.-H.; Larocque, D. Robustness of random forests for regression. *J. Nonparametr. Stat.* **2012**, *24*, 993–1006. [CrossRef]
30. Ying, C.; Qi-Guang, M.; Jia-Chen, L.; Lin, G. Advance and prospects of adaboost algorithm. *Acta Autom. Sin.* **2013**, *39*, 745–758.
31. Riccardi, A.; Fernández-Navarro, F.; Carloni, S. Cost-sensitive adaboost algorithm for ordinal regression based on extreme learning machine. *IEEE Trans. Cybern.* **2014**, *44*, 1898–1909. [CrossRef]
32. Huang, Y.; Liu, Y.; Li, C.; Wang, C. Gbrtvis: Online analysis of gradient boosting regression tree. *J. Vis.* **2019**, *22*, 125–140. [CrossRef]
33. Nie, P.; Roccotelli, M.; Fanti, M.P.; Ming, Z.; Li, Z. Prediction of home energy consumption based on gradient boosting regression tree. *Energy Rep.* **2021**, *7*, 1246–1255. [CrossRef]
34. Chen, T.; Guestrin, C. Xgboost: A scalable tree boosting system. In Proceedings of the 22nd ACM SIGKDD International Conference on Knowledge Discovery and Data Mining, San Francisco, CA, USA, 13–17 August 2016; pp. 785–794.
35. Ferreira, N.J.; Blatz, J.A. Measured pipe stresses on gas pipelines in landslide areas. *Can. Geotech. J.* **2021**, *99*, 1855–1869. [CrossRef]
36. Liu, Y.; Bao, Y. Review on automated condition assessment of pipelines with machine learning. *Adv. Eng. Inform.* **2022**, *53*, 101687. [CrossRef]

remote sensing

MDPI

Article

Unsupervised Cluster-Wise Hyperspectral Band Selection for Classification

Mateus Habermann [1,*], Elcio Hideiti Shiguemori [1] and Vincent Frémont [2]

1 Institute for Advanced Studies, Sao Jose dos Campos 12228-001, Brazil
2 Department of Automatics and Robotics, Nantes Université, École Centrale de Nantes, CNRS, LS2N, UMR 6004, F-44000 Nantes, France
* Correspondence: habermannmh@fab.mil.br

Abstract: A hyperspectral image provides fine details about the scene under analysis, due to its multiple bands. However, the resulting high dimensionality in the feature space may render a classification task unreliable, mainly due to overfitting and the Hughes phenomenon. In order to attenuate such problems, one can resort to dimensionality reduction (DR). Thus, this paper proposes a new DR algorithm, which performs an unsupervised band selection technique following a clustering approach. More specifically, the data set was split into a predefined number of clusters, after which the bands were iteratively selected based on the parameters of a separating hyperplane, which provided the best separation in the feature space, in a one-versus-all scenario. Then, a fine-tuning of the initially selected bands took place based on the separability of clusters. A comparison with five other state-of-the-art frameworks shows that the proposed method achieved the best classification results in 60% of the experiments.

Keywords: band selection; unsupervised; feature engineering

Citation: Habermann, M.; Shiguemori, E.H.; Frémont, V. Unsupervised Cluster-Wise Hyperspectral Band Selection for Classification. *Remote Sens.* **2022**, *14*, 5374. https://doi.org/10.3390/rs14215374

Academic Editor: Edoardo Pasolli

Received: 17 August 2022
Accepted: 18 October 2022
Published: 27 October 2022

1. Introduction

In pattern recognition problems, the separation among classes in the feature space is of great importance for the success of the classifier [1]. An appropriate separation may be achieved by means of effective data representation [2,3]. When it comes to hyperspectral image (HSI) classification, by selecting the right bands, one can provide a wider class separation [4], as well as attenuate the negative effects of the Hughes phenomenon [5] and avoid the overfitting of the classifier [6–8].

In such a scenario, feature extraction (FE), i.e., a combination of the original spectral bands, is capable of tackling the aforementioned problems, but it is not a recommended approach for dimensionality reduction of hyperspectral data, because the resulting features do not carry the physical information any longer [9], impairing, consequently, a proper understanding of the model [10,11]. Band selection (BS), on the other hand, is as good as FE in terms of providing class separability; moreover, it keeps the original information about the spectral bands [9,12]. Since a BS method provides suitable bands for a given task, it is possible to design tailored sensors to perform that application, consequently avoiding redundant and irrelevant bands [13].

BS methods can be grouped into one out of three major categories [14]: *wrapper methods*, when the selection of bands occurred during the training phase of the classifier; in this case, the classifier must be trained from scratch every time a band subset is assessed; *embedded methods*, when the classifier selects the bands by itself, for example, Lasso [15]; and *filter methods*, when the band selection process takes place before the classifier training phase; it has no relation to the classifier to be used [2].

In terms of the available data set to train the algorithm, a band selection framework can be considered supervised [16–19], semi-supervised [20–23], or unsupervised [24–26]. The latter ends up being the most feasible in real applications due to the difficulty in acquiring labeled samples [27].

It is known that unsupervised state-of-the-art BS frameworks follow either a ranking-based [28] or a clustering-based approach [29]. Ranking-based methods sort the spectral bands in relation to a specific criterion. However, they fail in terms of avoiding correlated bands [30]. Clustering-based BS frameworks, on the other hand, aim to find the most representative bands of each cluster of the data set, decreasing the correlation amongst bands [31]. Thus, clustering in the BS literature is normally used to form clusters of spectral bands. For instance, in [32], the authors propose a BS method that uses dynamic programming to cluster the spectral bands, which are considered continuous. In [33], the density peak is used for the clustering of the bands, due to its capability of tackling non-spherical data. The authors propose the weight between the normalized local density and cluster distance. In [34], the BS algorithm is performed based on correntropy-based clustering of the spectral bands. In [29], the authors propose a BS algorithm that initially clusters the spectral bands based on Euclidean distance. Then one band per cluster is selected, resulting in a band subset whose bands are ranked in a fine-tuning step. In [35], the proposed BS uses a self-tuning algorithm to cluster the spectral bands. In [13], a kernel-based probabilistic clustering of spectral bands is proposed, based on the assumption that there is a smooth transition between two adjacent clusters. Finally, in [36] the most representative bands for each pixel are selected by means of an attention mask. Then, an autoencoder reconstructs the original image using the selected bands. In the end, the final bands are selected by a clustering method.

As those cited papers performed the clustering operation on the spectral bands, structural information on the data set was not taken into account. Moreover, since the final objective of band selection frequently lies in a better classification of the data instances, an analysis based solely on the best representative bands (without looking at class separation on the feature space) ends up being of secondary importance. Furthermore, in an unsupervised BS framework, normally the filter approach is used. Thus, the band selection takes place in a preprocessing step, i.e., before the use of the classifier itself [37]. Thus, *a priori* one does not know beforehand which classifier will be used. For this reason, this paper presents a BS framework that seeks to maximize the distance among classes in the feature space. Therefore, our main purpose is not the representation of the data set by a few bands [38], but rather the selection of bands that best separate the classes. This class separability during the selection of bands is the gap we propose to fill in relation to other approaches. Since this framework is set to work in an unsupervised environment, the actual classes are represented by clusters.

Thus, in the proposed approach, the bands were iteratively selected based on data set portions, which, in turn, were defined by clustering algorithms. Eleven clustering methods were evaluated in order to provide the best match between the resulting clustering and the actual data classes. Thus, the clusters formed may be deemed as representatives of the actual classes, which enables an analysis based on the separability of the classes in the feature space; consequently, structural information was taken into account. Once the clusters were formed, a one-versus-all approach was adopted. In this way, the selected bands were those that provided the best separability between the cluster and the rest of the data set. Then, those bands were subjected to a fine-tuning procedure, which consisted of placing these bands into some clusters in order to select a combination of those that provided the biggest cluster separability in the feature space. The proposed method bears the acronym CW due to its cluster-wise approach.

The contributions of this paper are as follows:

- The use of a cluster-wise approach to solving the unsupervised band selection problem;
- Once two clusters were formed, the selection of bands was based on the parameters of a hyperplane defined by a single-layer neural network;
- Fine-tuning of the selected bands based on cluster separability in the feature space.

In Section 2, the proposed method is presented. In Section 3, the results of the proposed method are compared to five competitors by using three classifiers and three hyperspectral images commonly used in BS literature. Finally, in Section 5, we offer the conclusion of this work.

2. Method

Every BS algorithm is supposed to select relevant features—refer to [39] for a thorough definition of feature relevance. In short, a relevant spectral band (*i*) should provide useful information [40]; and (*ii*) should not be redundant [14]. Since the proposed band selection framework is designed for classification purposes, the bands considered to provide useful information are those that provide maximum separation between clusters in the feature space. When it comes to redundancy between spectral bands, in this work, it is measured by correlation.

Therefore, following this reasoning, the proposed method is composed of three parts: Data clustering; Selection of bands of interest; and Redundancy reduction.

2.1. Data Clustering

Regarding unsupervised problems—data reconstruction [2] and data structure analysis (DSA), for instance, are approaches that render feature selection feasible.

Data entry clustering can find natural groupings in data sets, and, for this reason, it is considered a DSA-based band selection approach, when used for this purpose.

Inspired by [41], the proposed method also performs clustering of the data entries. However, here, we adopted a partitional clustering instead of a hierarchical one, as illustrated in Figure 1a. With partitional clustering, each resulting cluster $\mathbf{C}_i$, $i \in \{1, 2, \ldots, k\}$, could be taken as a representative of the real class if (*i*) k equals the number of classes present in the data set, and (*ii*), the clustering algorithm, is appropriate for the data set at hand.

One generally wants to classify objects present in a known scene, supposing one knows beforehand the number k of classes is plausible.

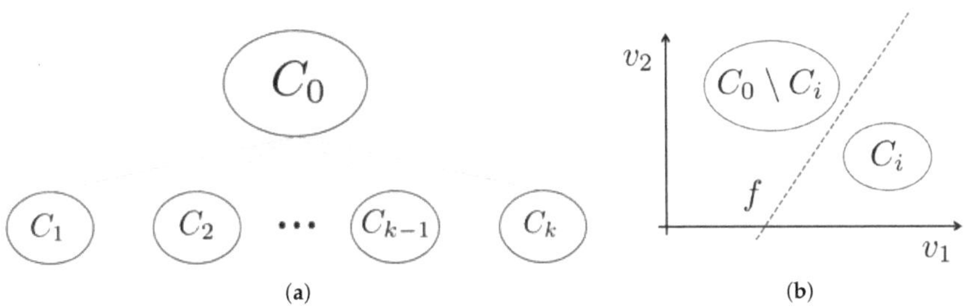

Figure 1. (**a**) Illustration of k clusters after the partitional clustering. (**b**) A representation of one-versus-all binary classification by means of a line segment f, where v_1 and v_2 are variables that enable the 2D representation.

Choice of the Clustering Algorithm

As for the fitness of a clustering algorithm to hyperspectral data, 11 methods were evaluated. It is worth noting that our focus was not on the best clustering algorithm available in the literature, but rather just to use some well-established clustering algorithms to show the efficiency of the proposed method.

The input data are the Salinas hyperspectral image [42], with 224 bands and 16 classes.

So, each clustering algorithm was set to find $k = 16$ clusters—as we will see later in this paper, the proposed method sets k equal to the number of classes in the image. It is important to clarify that, at this point, the focus is on the comparison of clustering methods, so the data labels will be used.

The measure of agreement between the two data sets—the clustering result and Salinas ground truth—was computed by means of the adjusted Rand index (r) [1]. In short, let κ_1 and κ_2 be two different clustering types of a given data set; where κ_1 is the real class of the Salinas image and κ_2 is the result of a clustering algorithm.

Considering all pairs of vectors $\mathbf{x}_j$ and $\mathbf{x}_l$, with $j \neq l$, let α_1 be the number of times that both vectors belong to the same clusters in the clustering types κ_1 and κ_2. Moreover, let α_2 be the number of times the vectors belong to different clusters in κ_1 and different clusters in κ_2.

Finally, the adjusted Rand index between clusters κ_1 and κ_2 is given by

$$r = (\alpha_1 + \alpha_2)/m, \tag{1}$$

where m is the number of possible vector pairs in the data set.

For each clustering algorithm, r was calculated 10 times. Table 1 shows the mean values for k-means and k-medoids algorithms with different distance metrics. K-means using the *cosine* similarity measure has the best outcome. For the sake of clarity, the bigger the values of r, the more similar the clustering types of κ_1 and κ_2.

Table 1. Adjusted Rand index ($0 \leq r \leq 1$) *(mean values out of 10 runs for Salinas HSI)*.

Clustering Algorithm	r
K-means (Euclidean)	0.6997
K-means (cityblock)	0.7382
K-means (cosine)	0.7941
K-means (correlation)	0.7170
K-medoids (Euclidean)	0.7062
K-medoids (Mahalanobis)	0.7685
K-medoids (cityblock)	0.7402
K-medoids (Minkowski)	0.7396
K-medoids (Chebychev)	0.7269
K-medoids (Spearman)	0.7674
K-medoids (Jaccard)	0.6146

Consequently, all of the clusters throughout this paper were obtained by k-means using the *cosine* similarity measure.

It is worth mentioning that an appropriate partitional clustering is able to turn supervised band selection algorithms into unsupervised ones, by taking the resulting clusters as class representatives, and the degree of success depends on r values. This paper follows that approach, by considering [4] as a reference. Since $0 \leq r \leq 1$, where 1 means the two clustering outcomes match identically, $r = 0.7941$ indicates a good match between the clusters and the real classes of Salinas HSI. At this point, we opted to analyze a hyperspectral image not used in Section 3 in order to maintain the unsupervised nature of the proposed approach.

2.2. *Selection of Bands of Interest*

Once the initial data set is split into k clusters, it is time to present the proposed band selection algorithm, which has k iterations. At each iteration, two steps take place: (*i*) the selection of candidate bands and (*ii*) fine-tuning.

2.2.1. Selection of Candidate Bands

Let $\mathbf{C}_0 = [\mathbf{b}_1, \mathbf{b}_2, \ldots, \mathbf{b}_d] \in \mathbb{R}^{n \times d}$ be the initial cluster, i.e., the HSI, where b_j is the jth band vector whose norm l_2 is scaled to 1, n is the number of pixels and d is the dimensionality of the data set.

Let $\mathbf{C}_i$, $\forall i \in \{1, 2, \ldots, k\}$, be the k clusters after the partitional clustering of $\mathbf{C}_0$, where k is the number of classes in the data set.

The following properties hold for the clusters:

- $\mathbf{C}_i \neq \emptyset$, for $i \in \{1, 2, \ldots, k\}$;
- $\cup_{i=1}^{k} \mathbf{C}_i = \mathbf{C}_0$;
- $\mathbf{C}_i \cap \mathbf{C}_l = \emptyset$, with $i \neq l$ and $i, l \in \{1, 2, \ldots, k\}$.

For each cluster, a one-versus-all binary classification was performed between $\mathbf{C}_i$ and $\mathbf{C}_0 \setminus \mathbf{C}_i$.

As in [4], we used a single-layer neural network to generate the separating hyperplane f. As an illustration, both the one-versus-all classification and the hyperplane f are shown in Figure 1b.

The cross-entropy loss function of the neural network is given by

$$L_f = -\frac{1}{\eta} \sum_{j=1}^{\eta} [y_j log(\hat{y}_j) + (1 - y_j) log(1 - \hat{y}_j)], \tag{2}$$

where η is the cardinality of the set containing the data points —since we make $|\mathbf{C}_0 \setminus \mathbf{C}_i| \approx |\mathbf{C}_i|$ in order to balance the two clusters, $\eta \leq n$—; $y_j \in \{0, 1\}$ is the expected output to the input vector $\mathbf{x}_j \in \mathbb{R}^{d \times 1}$, where label 1 corresponds to cluster $\mathbf{C}_i$; and $\hat{y}_j$ is the calculated output given by

$$\hat{y}_j = \frac{1}{1 + e^{z_j}}, \tag{3}$$

which is the sigmoid activation function. where e is the Euler's number, and z_j is the hyperplane equation

$$z_j = x_j^{(1)} w^{(1)} + x_j^{(2)} w^{(2)} + \ldots + x_j^{(d)} w^{(d)} + \beta, \tag{4}$$

where $\mathbf{w} \in \mathbb{R}^{d \times 1}$ and $\beta \in \mathbb{R}$—both calculated by a single-layer neural network—are the hyperplane f parameters.

The training phase of the network consists of 2000 training epochs, using the back-propagation algorithm, with 70% of the data set for training and the remaining 30% for the test.

After the neural network's training, a given input vector $\mathbf{x}_j$ will cause either $z_j \geq 0$ or $z_j < 0$. As z_j is the argument of a sigmoid function, if

- $z_j \geq 0$, then $\hat{y}_j \leftarrow round(\hat{y}_j) = 1$,
- $z_j < 0$, then $\hat{y}_j \leftarrow round(\hat{y}_j) = 0$,

where $round(\hat{y}_j \geq 0.5) = 1$, and $round(\hat{y}_j < 0.5) = 0$.

The band selection is based on the magnitude of weight vector components $w^{(l)}$, $l \in \{1, \ldots, d\}$. Indeed, according to (4), the biggest weights in magnitude, $|w^{(l)}|$, will strongly determine the signal of z_j. Therefore, the bands $x_j^{(l)}$—related to the biggest $|w^{(l)}|$—are the most relevant for the binary one-versus-all classification, and are, consequently, initially selected.

In order to provide an illustrative view on this matter, Figure 2 depicts a 2D situation in which a linear classifier, represented by a line segment, separates two different clusters, in red and blue colors, composed of synthetic data of variables v_1 and v_2. In Figure 2a, the clusters are linearly separable, and it is easy to perceive that this separation is provided by variable v_2, whereas variable v_1 bears similar values for both clusters. It is worth noting that the green line's parameters $w^{(1)}$ and $w^{(2)}$, calculated by a single-layer neural network (β is omitted), indicate the relative importance of variables in this binary classification. That is, $|w^{(2)}| = 4.8126 > |w^{(1)}| = 0.5782$ indicates a higher relevance of variable v_2 in relation to v_1. A similar situation occurs in Figure 2b, but this time $|w^{(1)}| > |w^{(2)}|$, indicating that variable v_1 provides better separability between the clusters. Figure 2c,d shows that the same analysis is valid even when the clusters overlap.

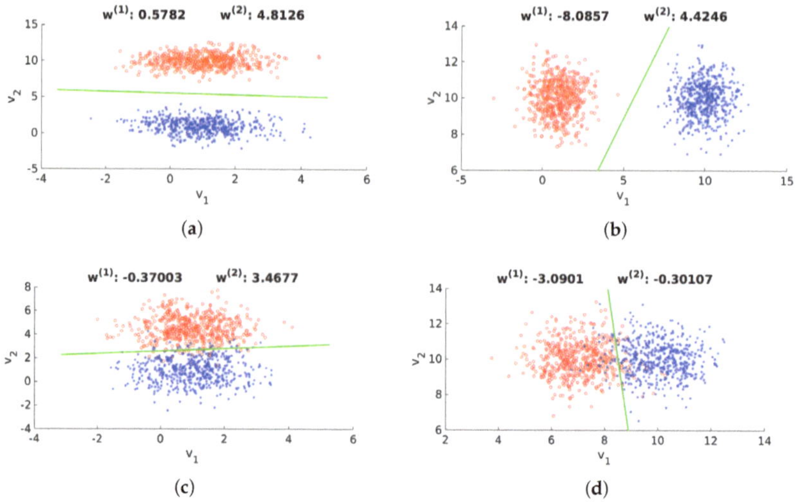

Figure 2. A 2D binary classification illustration using synthetic data, in variables v_1 and v_2. The hyperplane parameters $w^{(1)}$, $w^{(2)}$, and β (the latter not shown here) are calculated by a single-layer neural network, whose result is depicted by a green line segment. The magnitude of the neural networks (or the hyperplane parameters) indicate the relevance of its correspondent feature. (**a**) Two linearly separable classes. Clearly, attribute v_2 provides good separation between the clusters, which is corroborated by $w^{(2)} > w^{(1)}$. (**b**) A similar situation as in (**a**) occurs here, but this time v_1 provides the separation between the clusters, and $w^{(1)} > w^{(2)}$. In (**c**,**d**) it is possible to draw the same conclusion, even when the clusters overlap.

According to the proposed method, the number s of selected bands is defined by the user.

Since the method has k iterations, the selection of $(s/k) \in \mathbb{N}$ bands per iteration would be sufficient. However, at each iteration, the method selects $4(s/k)$ bands, from which only s/k are kept after the fine-tuning step. It is worth noting that, except for $4(s/k)$, other numbers have not been tested.

2.2.2. Fine-Tuning

At each iteration $i \in \{1, 2, \ldots, k\}$, $4(s/k)$ bands are selected based on the biggest weights $|w^{(l)}|$, according to (4).

Those bands are then placed in s/k clusters—by means of k-means (Euclidean)—q_l, $l \in \{1, 2, \ldots, (s/k)\}$, and from each cluster 1 band $\mathbf{b}$ will be initially selected.

By picking 1 band from each cluster q, several tuples t are formed. An example of it is shown in Figure 3. The exact number of tuples is $|q_1| \times |q_2| \times \ldots \times |q_{s/k}|$. Formally, at iteration i the set containing all tuples is given by

$$Q = \{(\mathbf{b}_i^{(1)}, \ldots, \mathbf{b}_j^{(s/k)}), \mathbf{b}_j^{(l)} \in q_l, l \in \{1, \ldots, (s/k)\}\}. \tag{5}$$

Note that this approach for refining the band selection is based on [43]; however, here we adopted a different criterion to assess the importance of each tuple of bands.

For each tuple $t \in Q$, its bands were evaluated according to the class separability they provided between $\mathbf{C}_i$ and $\mathbf{C}_0 \setminus C_i$. At this point, the data sets of both clusters contain only the bands in t.

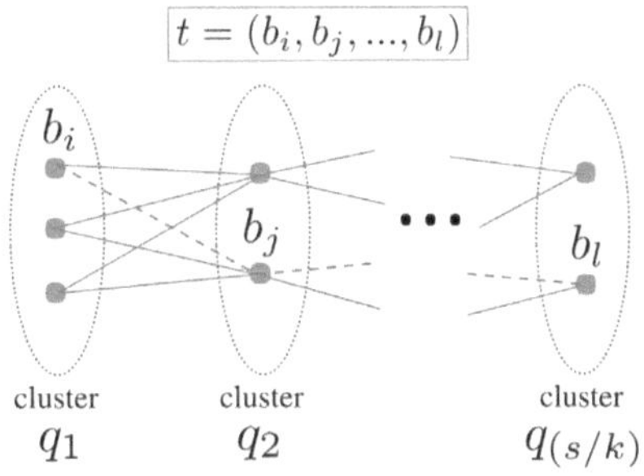

Figure 3. Each cluster provides one band for the composition of tuple t. In this example, the band $\mathbf{b}_i \in q_1$, the band $\mathbf{b}_j \in q_2$, and the band $\mathbf{b}_l \in q_{s/k}$, among others connected by the dashed line, form the tuple t. Each possible combination of bands in different clusters gives rise to all $t \in Q$.

The class separability is measured by the $\rho \in \mathbb{R}$ index,

$$\rho = \frac{trace(\mathbf{S}_w + \mathbf{S}_b)}{trace(\mathbf{S}_w)}, \tag{6}$$

where

$$\mathbf{S}_w = p_{\mathbf{C}_i}\Sigma_{\mathbf{C}_i} + p_{(\mathbf{C}_0 \setminus \mathbf{C}_i)}\Sigma_{(\mathbf{C}_0 \setminus \mathbf{C}_i)},$$

and

$$\mathbf{S}_b = p_{\mathbf{C}_i}(\mu_{\mathbf{C}_i} - \mu_0)(\mu_{(\mathbf{C}_0 \setminus \mathbf{C}_i)} - \mu_0)p_{(\mathbf{C}_0 \setminus \mathbf{C}_i)},$$

where $\mu_{\mathbf{C}_i}$ is the mean of cluster $\mathbf{C}_i$, μ_0 is the global mean, Σ is the covariance matrix, and p is the *a priori* probability. Since the clusters $\mathbf{C}_i$ and $\mathbf{C}_0 \setminus \mathbf{C}_i$ are balanced, i.e., $|\mathbf{C}_i| = |\mathbf{C}_0 \setminus \mathbf{C}_i|$, then $p_{\mathbf{C}_i} = p_{(\mathbf{C}_0 \setminus \mathbf{C}_i)} = 0.5$.

According to (6), the bigger the ρ, the more compact the clusters are, and the more distant they are from each other.

Finally, for each tuple $t \in Q$ there is a corresponding ρ value, and only the bands in t^{max}, whose ρ is the biggest, are selected at iteration i.

2.3. Redundancy Reduction

Let $\Upsilon \in \mathbb{R}^{d \times d}$ be the correlation matrix of the data set $\mathbf{C}_0$, calculated according to Pearson's correlation coefficient.

Let ψ be a set composed of the bands the most correlated to the band represented by ψ indices. Set ψ is calculated before the iterations start, according to Algorithm 1. For the sake of clarity, $\psi_j = \mathbf{b}_l$, for instance, means that $\mathbf{b}_l$ is the band the most correlated to band $\mathbf{b}_j$.

Algorithm 1 starts by creating a correlation matrix Υ, according to Pearson's correlation. After the initialization of matrix $\mathbf{I}$ and vector ψ, we sort all the columns of Υ in a descend fashion, and store the indices idx, which corresponds to the band indices. To the first position of ψ is assigned the band $\mathbf{b}_{\mathbf{I}(1,1)}$. Then the remaining positions of ψ receive the bands the most correlated to the band corresponding to that position in a way no same band is assigned to more than one position. The output is the vector ψ.

Given t^{max} and ψ, the subset $\delta_{t^{max}}$ of the bands the most correlated to those in t^{max} is given by Algorithm 2.

In Algorithm 2 we have t^{max} and ψ as input. Given the bands in t^{max}, the algorithm finds the most correlated bands to those in t^{max} and insert them in subset $\delta_{t^{max}}$, which is the output of the algorithm.

Algorithm 1 Most correlated bands

$\Psi = corr(\mathbf{C}_0)$ ▷ Pearson's correlation
Initialize matrix $\mathbf{I}$
$\psi = \emptyset$
for all the columns $\mathbf{c} \in \Psi$ **do**
    $[values, idx] = sort(\mathbf{c}, \text{"}descend\text{"})$ ▷ $idx \in \mathbb{N}^{1\times d}$
    $\mathbf{I} = [\mathbf{I}; idx]$
$\psi_1 \leftarrow \mathbf{b}_{\mathbf{I}(1,1)}$
for i = 2 : d **do**
    for j = 1 : d **do**
        if $\mathbf{b}_{\mathbf{I}(i,j)} \notin \psi$ **then**
            $\psi \leftarrow \psi \cup \mathbf{b}_{\mathbf{I}(i,j)}$
            Break
Return: ψ

Algorithm 2 The most correlated bands to a given subset

Input: t^{max}, ψ
$\delta_{t^{max}} = \emptyset$
for j = 1 : $|t^{max}|$ **do** ▷ Vector cardinality
    for l = 1 : $|\psi|$ **do**
        if $\psi_l = t_j^{max}$ **then**
            $\delta_{t^{max}} \leftarrow \delta_{t^{max}} \cup b_l$
Return: $\delta_{t^{max}}$

Finally, at each iteration i, the bands in t^{max} are selected and inserted in $\mathbf{S}$, which is the final subset of selected bands. So,

$$\mathbf{S} \leftarrow \mathbf{S} \cup t^{max}. \tag{7}$$

Once we have t^{max}, its most correlated bands $\delta_{t^{max}}$ are inserted in subset $\mathbf{D}$, which are the bands to be discarded. That is,

$$\mathbf{D} \leftarrow \mathbf{D} \cup \delta_{t^{max}}. \tag{8}$$

Then, for the next iteration $i+1$ the data set is updated according to

$$\mathbf{C}_0 \leftarrow \mathbf{C}_0 \setminus (\mathbf{S} \cup \mathbf{D}). \tag{9}$$

The method iterates until $i = k$. Then, the final output is the subset $\mathbf{S}$ of selected bands.

2.4. *Proposed Method's Overview*

Algorithm 3 presents the overview of the proposed method.

Algorithm 3 Proposed band selection algorithm

Input: Data set $\mathbf{C}_0$, number k of classes
$\mathbf{S} = \emptyset$ ▷ Subset of selected bands
$\mathbf{D} = \emptyset$ ▷ Subset of bands to be discarded
Proceed to k-means clustering (*cosine* distance) of $\mathbf{C}_0$ into k clusters $\mathbf{C}_i$
for i = 1 : k **do**
    Proceed to a binary classification between clusters $\mathbf{C}_i$ and $\mathbf{C}_0 \setminus \mathbf{C}_i$ (*one-versus-all*) using a single-layer neural net
    Select the $4(s/k) \in \mathbb{N}$ bands related to the biggest separating hyperplane parameters $|w|$, according to (4)
    Proceed to the band selection fine-tuning, according to Section 2.2.2
    Update subset of selected bands **S** according to (7)
    Update subset **D** according to (8)
    Update data set according to (9)
Return: S

3. Results

Normally the quality of the subset of selected bands is assessed by the performance of the classifiers. So, this approach is adopted here, with support vector machine (SVM), K-nearest neighbor (KNN), and classification and regression tree (CART) classifiers [1], via three hyperspectral data sets used in several BS papers: Botswana, Indian Pines, and Pavia University. All of them can be downloaded at [42].

3.1. Competitors

The proposed method is compared to five state-of-the-art BS methods: ASPS, MPWR, ONR, UBS, and VGBS.

3.1.1. ASPS

ASPS [44] is the acronym for hyperspectral band selection via adaptive subspace partition strategy. Its framework begins with a two-step partition of the data cube, starting with the coarse partition of the image cube into a predetermined number of parts, and then there is a fine subspace partition, from which bands with low noise levels are finally selected.

3.1.2. MPWR

In [45], the authors proposed a manifold-preserving and weakly redundant (MPWR) unsupervised band selection method. A manifold-preserving band-importance metric was used to measure the band-wise essentiality. Concerning the redundancy caused by the correlated bands, this paper establishes a constrained band-weight optimization model. Thus, both band-wise manifold-preserving capability and intraband correlation are integrated into the BS method.

3.1.3. ONR

The approach called 'hyperspectral band selection via optimal neighborhood reconstruction' is based on optimal neighborhood reconstruction (ONR) [46]. It proceeds to different band combinations in order to reconstruct the original data set, and also applies a noise reducer to minimize the influence of noisy bands.

3.1.4. UBS

UBS is used as the acronym for the approach presented in [47]. This method is based on the spectral decomposition of a matrix, then the loading-factors matrix can be constructed for band prioritization, according to the corresponding eigenvalues and eigenvectors.

3.1.5. VGBS

This paper states that there is a relation between the volume of a sub-simplex and the volume gradient of a simplex. Based on this, they proposed a BS method called VGBS [48]. It is unsupervised and seeks to remove the most redundant band based only on the gradient of volume instead of calculating the volumes of all sub-simplexes.

3.2. *Experimental Results*

In order to compare the outcome of the proposed method, five different bands were selected from scratch: 10, 20, 30, 40, and 50. A set with 50 bands does not necessarily contain the 10-band set, for example, due to the nature of the neural networks.

We compare the results to other BS methods for each hyperspectral image separately. All of the classifier accuracies exhibited here are the mean values of ten runs.

Our approach is dubbed CW—the unsupervised cluster-wise method.

3.2.1. (Case 1) Botswana HSI

The Botswana image is composed of 242 spectral bands and has 14 classes. See further details about this image at [42].

Table 2 shows the results of the BS methods using Botswana HSI. Bold values represent the highest scores attained. Figure 4 presents the same results in an illustrative way. In the figure, different marks are used in order to identify the competitors. The line connecting the marks does not mean interpolation, but rather helps the reader.

Table 2. Botswana image results (*overall accuracy and standard deviation in %*).

Method	10 Bands	20 Bands	30 Bands	40 Bands	50 Bands
			KNN classifier		
CW	**90.86** ± 0.69	88.81 ± 0.27	90.86 ± 0.91	**91.48** ± 0.97	**93.22** ± 0.69
ASPS	85.63 ± 0.75	89.73 ± 0.94	**91.07** ± 0.76	89.01 ± 0.63	91.58 ± 0.72
MPWR	81.52 ± 0.99	86.45 ± 1.43	90.76 ± 0.56	90.35 ± 1.05	90.25 ± 0.89
ONR	90.66 ± 0.91	**92.61** ± 0.53	89.94 ± 0.64	91.38 ± 0.89	91.99 ± 0.86
UBS	88.50 ± 1.07	89.53 ± 1.05	87.99 ± 0.86	89.94 ± 0.66	90.04 ± 0.96
VGBS	88.09 ± 0.79	90.55 ± 1.00	88.50 ± 1.30	87.17 ± 1.29	88.09 ± 1.10
			CART classifier		
CW	**84.80** ± 1.23	**86.55** ± 1.27	84.91 ± 1.49	**85.93** ± 1.03	86.04 ± 1.13
ASPS	81.72 ± 1.04	85.32 ± 1.27	83.78 ± 1.15	83.98 ± 1.38	84.50 ± 0.96
MPWR	72.59 ± 1.27	81.31 ± 1.13	84.29 ± 1.02	85.52 ± 1.13	85.01 ± 1.47
ONR	83.37 ± 0.74	84.80 ± 1.35	**84.91** ± 1.06	84.60 ± 1.01	84.70 ± 1.46
UBS	80.39 ± 1.14	83.26 ± 1.07	83.68 ± 1.36	85.32 ± 0.84	85.01 ± 0.98
VGBS	83.98 ± 0.81	85.22 ± 0.95	82.24 ± 1.35	83.47 ± 0.65	**86.24** ± 1.44
			SVM classifier		
CW	89.73 ± 0.72	**94.76** ± 0.91	**94.97** ± 0.69	93.94 ± 0.61	**94.15** ± 0.56
ASPS	87.78 ± 0.67	91.27 ± 0.67	93.84 ± 0.66	92.09 ± 0.69	94.05 ± 0.55
MPWR	87.06 ± 1.06	90.86 ± 0.97	93.73 ± 0.67	94.76 ± 0.75	94.15 ± 0.61
ONR	**92.91** ± 0.36	94.25 ± 0.48	93.83 ± 0.62	**94.76** ± 0.46	94.14 ± 0.77
UBS	89.42 ± 0.82	92.50 ± 0.98	92.71 ± 0.67	93.42 ± 0.86	92.91 ± 0.89
VGBS	90.04 ± 0.97	92.81 ± 0.58	93.63 ± 1.06	93.01 ± 0.68	93.53 ± 0.65

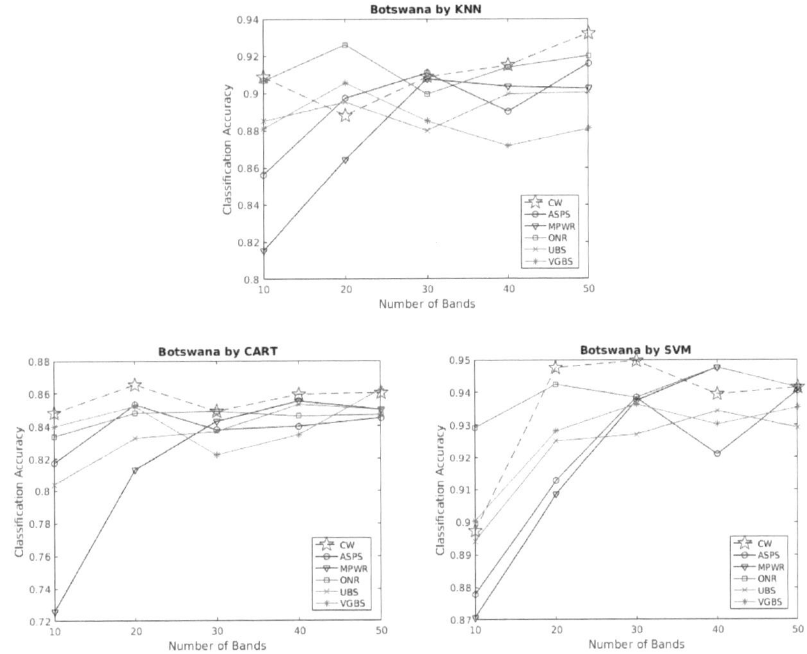

Figure 4. Overall Accuracies for Botswana Image according to KNN, CART, and SVM classifiers.

Since the Botswana image has 14 classes, the proposed method has the same number of iterations—see Algorithm 3. For each one-versus-all case, a single-layer neural net is run, and the error (of the test set) versus the epoch curves are shown in Figure 5. Clearly, there is convergence in all cases, meaning that it is possible to find a hyperplane to separate the clusters $\mathbf{C}_i$ and $\mathbf{C}_0 \setminus \mathbf{C}_i$. The different curve shapes indicate how fast the algorithm converged. For instance, in Figure 5, *Cluster 1 versus All* converged faster than *Cluster 6 versus All*.

Concerning the results, there were five different sets of selected bands, and each set was subjected to three classifiers. Thus, in total, we had 15 different experiments, from which the proposed CW framework surpassed its competitors in 9 out of 15 cases.

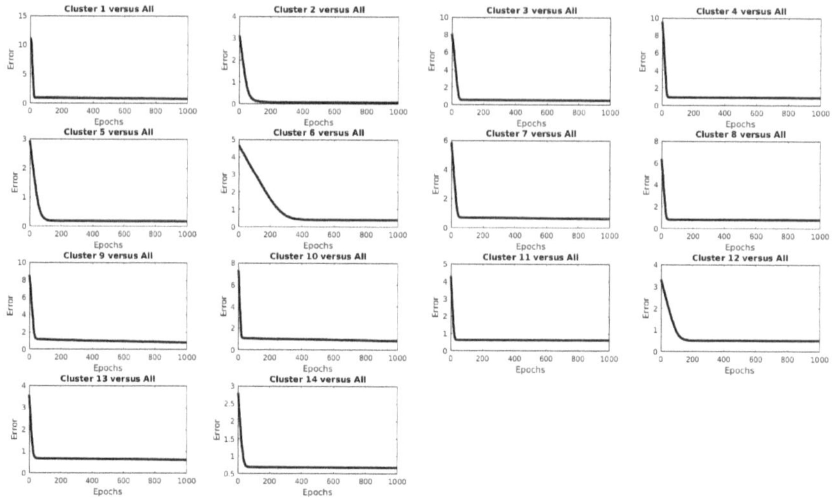

Figure 5. Error versus epoch curve of each one-versus-all case for the Botswana image.

3.2.2. (Case 2) Indian Pines HSI

This image has 224 bands and 16 classes. Further details about this image can be found at [42].

Table 3 shows the accuracies and standard deviations of the results.

Out of 15 experiments, the proposed CW method achieved the best results 10 times. In Figure 6, it is possible to have a visual idea of the performance of all BS methods.

According to Figure 7, for all iterations, the single-layer neural network converged. This convergence is seen by the fact that the error was lower as the number of epochs increased.

Table 3. Indian Pines image results *(overall accuracy and standard deviation in %)*.

Method	10 Bands	20 Bands	30 Bands	40 Bands	50 Bands
			KNN classifier		
CW	**76.81** ± 1.13	**80.00** ± 0.54	**78.14** ± 0.31	**81.13** ± 0.59	**79.45** ± 0.70
ASPS	68.61 ± 0.68	66.50 ± 0.93	66.41 ± 0.90	63.19 ± 0.33	63.32 ± 0.27
MPWR	69.66 ± 0.89	70.63 ± 0.59	72.52 ± 0.43	73.59 ± 0.49	70.63 ± 1.12
ONR	71.35 ± 0.35	70.99 ± 0.61	67.41 ± 1.62	71.45 ± 0.57	72.03 ± 0.96
UBS	64.62 ± 0.52	63.15 ± 0.69	64.78 ± 1.09	64.98 ± 0.82	63.22 ± 0.72
VGBS	69.14 ± 0.80	70.21 ± 0.90	67.94 ± 0.74	70.41 ± 0.55	69.98 ± 0.61
			CART classifier		
CW	70.82 ± 0.78	74.21 ± 0.67	72.55 ± 0.86	73.50 ± 0.70	72.75 ± 0.81
ASPS	69.49 ± 0.41	69.20 ± 0.33	71.94 ± 0.46	73.20 ± 0.77	73.20 ± 0.58
MPWR	63.45 ± 0.66	67.32 ± 0.73	70.73 ± 0.70	71.58 ± 0.65	71.28 ± 0.88
ONR	71.28 ± 0.99	**75.41** ± 0.76	73.30 ± 1.40	**74.57** ± 1.22	73.33 ± 0.82
UBS	**71.71** ± 0.96	73.66 ± 0.78	**74.67** ± 0.83	74.24 ± 1.12	**73.69** ± 0.86
VGBS	70.44 ± 0.30	70.60 ± 0.57	71.22 ± 1.28	70.83 ± 0.75	71.32 ± 0.78
			SVM classifier		
CW	**84.58** ± 0.80	**86.70** ± 0.16	**84.29** ± 5.14	**87.97** ± 4.48	**87.61** ± 0.78
ASPS	81.39 ± 0.58	80.36 ± 0.65	82.60 ± 0.58	80.72 ± 0.52	83.68 ± 0.20
MPWR	72.88 ± 0.95	78.82 ± 0.64	81.07 ± 0.82	84.39 ± 0.50	83.77 ± 0.50
ONR	82.70 ± 0.31	84.75 ± 0.70	84.10 ± 0.60	86.93 ± 0.82	84.88 ± 4.00
UBS	79.61 ± 0.51	82.86 ± 0.23	76.91 ± 0.61	82.08 ± 0.37	79.94 ± 3.60
VGBS	76.59 ± 0.70	79.97 ± 0.79	79.06 ± 0.52	80.68 ± 0.75	80.07 ± 1.04

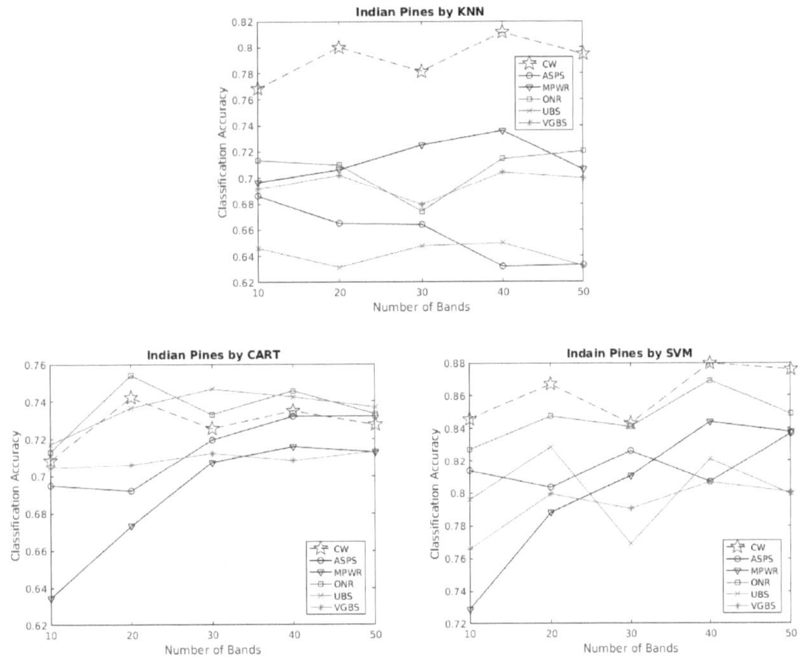

Figure 6. Overall accuracies for Indian pines image according to KNN, CART, and SVM classifiers.

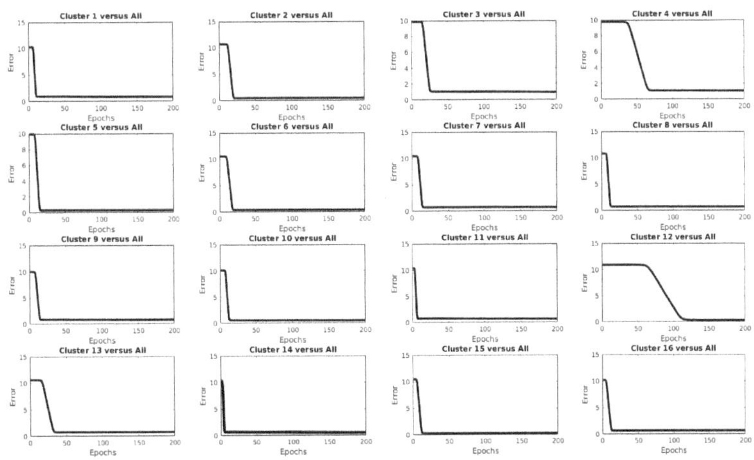

Figure 7. Error versus epoch curve of each one-versus-all case for the Indian pines image. Only the first 200 training epochs are shown here.

3.2.3. (Case 3) Pavia University HSI

This image has 102 spectral bands and 9 classes. See more information about this image at [42].

In Table 4, we see the results, in terms of overall accuracy and standard deviation, of all methods. The proposed CW algorithm has the best accuracies in 8 out of 15 experiments. It is also shown in Figure 8.

Table 4. Pavia University image results (*overall accuracy and standard deviation in %*).

Method	10 Bands	20 Bands	30 Bands	40 Bands	50 Bands
			KNN classifier		
CW	**92.19** ± 0.26	92.12 ± 0.48	91.73 ± 0.15	90.37 ± 0.24	90.99 ± 0.15
ASPS	87.67 ± 0.11	90.24 ± 0.39	89.90 ± 0.39	89.33 ± 0.29	90.54 ± 0.40
MPWR	91.80 ± 0.39	90.99 ± 0.14	**92.37** ± 0.93	90.45 ± 0.41	91.19 ± 1.19
ONR	88.89 ± 0.19	**92.30** ± 0.15	91.82 ± 0.29	**91.02** ± 0.16	**91.60** ± 0.11
UBS	86.00 ± 0.42	88.15 ± 0.44	87.84 ± 0.22	88.10 ± 0.22	88.17 ± 0.10
VGBS	84.26 ± 0.39	87.77 ± 0.34	86.42 ± 0.28	87.24 ± 0.39	88.30 ± 0.60
			CART classifier		
CW	**89.54** ± 0.12	89.18 ± 0.35	89.31 ± 0.29	**89.04** ± 0.39	**89.04** ± 0.37
ASPS	83.23 ± 0.35	86.69 ± 0.26	86.74 ± 0.48	87.06 ± 0.37	86.47 ± 0.10
MPWR	89.29 ± 0.67	88.82 ± 0.83	**89.67** ± 0.71	89.03 ± 0.95	88.81 ± 0.81
ONR	85.37 ± 0.13	**89.63** ± 0.23	89.03 ± 0.26	88.69 ± 0.16	89.00 ± 0.25
UBS	85.02 ± 0.49	87.27 ± 0.43	86.26 ± 0.31	86.50 ± 0.23	87.42 ± 0.36
VGBS	85.79 ± 0.23	88.26 ± 0.31	88.12 ± 0.32	88.49 ± 0.20	88.04 ± 0.31
			SVM classifier		
CW	**94.97** ± 2.38	90.77 ± 12.59	**75.56** ± 16.17	**68.26** ± 16.51	**67.01** ± 15.36
ASPS	89.58 ± 9.48	88.67 ± 9.41	40.19 ± 12.39	48.71 ± 15.96	40.56 ± 10.81
MPWR	94.93 ± 3.95	**91.85** ± 8.64	70.66 ± 13.81	39.13 ± 17.51	54.86 ± 15.50
ONR	91.31 ± 12.39	52.29 ± 24.56	46.62 ± 10.86	43.36 ± 21.41	43.48 ± 12.96
UBS	89.32 ± 2.49	54.75 ± 18.87	53.18 ± 13.99	38.06 ± 16.05	45.21 ± 13.29
VGBS	91.29 ± 13.59	89.74 ± 14.54	74.77 ± 17.88	50.58 ± 8.86	51.20 ± 12.23

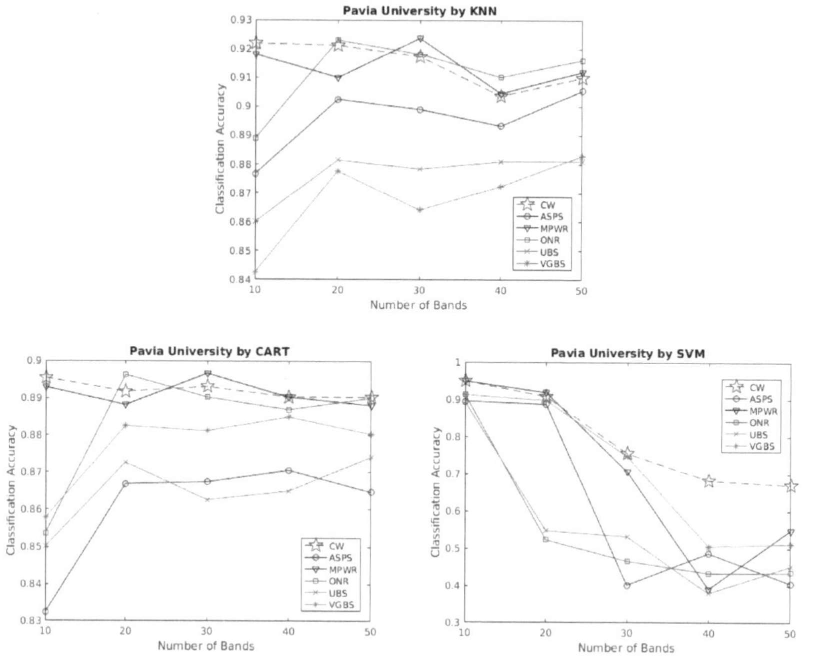

Figure 8. Overall accuracies for the Pavia University image according to KNN, CART, and SVM classifiers.

Figure 9 indicates that all the one-versus-all separating hyperplanes solutions converged.

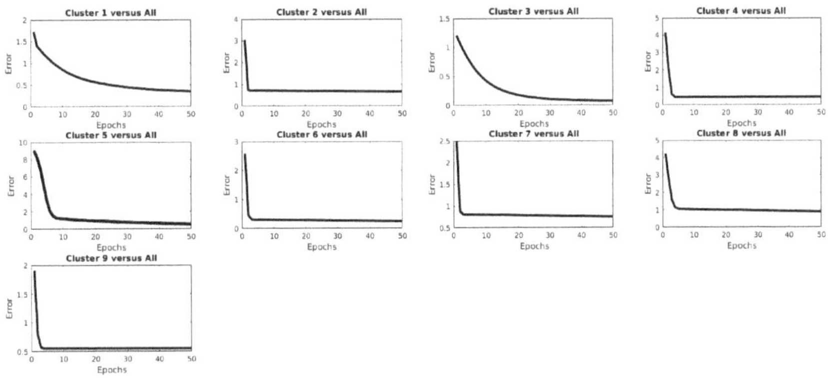

Figure 9. Error versus epoch curve of each one-versus-all case for the Pavia University image. Only the first 50 training epochs are shown here.

3.3. *Remark*

Finally, since this is a paper about band selection, Table 5 presents all of the bands selected by the proposed method CW.

Table 5. All the bands selected by the proposed CW method.

Number of Bands	Selected Bands
	Botswana image
10	4 11 17 21 24 41 69 101 105 120
20	7 10 12 28 32 33 37 39 40 49 55 58 59 65 67 73 75 76 113 124
30	2 4 5 7 21 27 29 30 32 33 34 35 37 43 44 47 54 56 57 58 61 62 71 74 78 80 82 118 120 124
40	2 3 4 5 6 8 13 14 16 27 28 31 32 33 34 35 36 39 41 42 52 54 58 60 63 65 69 70 72 74 78 88 89 92 97 100 101 105 135 142
50	1 2 4 5 6 8 10 16 19 21 22 23 24 25 26 27 30 31 32 33 34 36 41 42 47 57 58 59 66 67 69 72 74 75 77 78 87 89 94 96 98 100 101 102 104 109 110 113 130 144
	Indian Pines image
10	16 20 21 33 34 39 92 97 119 128
20	8 10 15 16 17 19 26 27 30 33 36 43 46 47 64 78 97 98 117 133
30	5 6 7 8 9 15 27 30 35 37 39 40 46 56 57 62 63 64 71 73 74 75 76 78 82 92 98 168 173 174
40	4 6 7 9 10 16 17 19 27 30 32 33 34 35 36 40 46 50 52 53 57 63 69 72 74 84 92 93 97 99 100 117 121 122 126 137 139 140 142 199
50	6 9 11 12 15 20 22 23 25 26 29 30 31 32 33 36 41 42 43 44 45 46 49 50 51 55 56 59 65 71 73 74 75 76 77 84 92 95 98 102 114 117 119 121 122 130 138 168 172 199
	Pavia University image
10	21 42 55 70 72 73 75 83 85 98
20	15 18 28 46 49 55 56 60 61 63 65 71 83 85 88 89 91 95 99 103
30	10 16 20 22 31 36 38 40 50 54 59 61 62 64 65 67 70 72 74 77 80 83 85 91 92 94 96 98 100 102
40	3 10 11 14 16 17 18 20 23 27 38 39 44 46 51 56 58 59 61 62 63 65 67 69 71 72 74 75 76 78 80 83 84 85 91 94 96 98 100 103
50	9 10 11 13 15 17 18 20 23 25 26 28 29 31 33 35 37 40 41 44 56 57 59 61 62 63 65 66 67 69 71 72 73 75 77 79 81 83 84 85 88 90 92 94 95 96 98 100 102 103

4. Discussion

The proposed method was introduced in Section 2.2 and its performance was reviewed in Section 3.2.

It is still important to emphasize the advantages of using the proposed method CW. On the other hand, the deficiencies of the algorithm will also be highlighted.

4.1. *Pros*

As shown in Tables 2–4, with their corresponding Figures 4, 6 and 8, each classifier, due to its intrinsic characteristics, performed differently when compared to the others. SVM classifiers are well known for their effectiveness in high-dimensional spaces, such as the feature space of a hyperspectral image. Thus, for this reason, we see SVM outperforming the other classifiers. On this topic, we see an exception in Table 4 and Figure 8, where SVM is less accurate than the other two classifiers. Given that KNN and CART exhibit stable performances as the band numbers increase, the reason for the bad performance of SVM, in this case, as the dimensionality became higher, may lie in the fact that the Pavia University data were not well discriminative for the SVM algorithm (this phenomenon happened with all competitors). When it comes to the KNN classifier, we set the number of neighbors equal to 7, i.e., $K = 7$ in all experiments; the objective here was not to find the best settings for the KNN but to provide equal conditions for the comparison of the band selection methods. As KNN classifies an input pattern based on its neighbors in the feature space, it performed better than the CART classifier, whose decision trees rely on binary rules, which become more complex as dimensionality increases.

In general, the proposed CW method outperformed its competitors in lower dimensions, due to the fact that the CW algorithm selected its bands based on the class separations in the feature space. Thus, even in lower dimensions, we saw a good performance of the proposed method, which was designed to be used as a filter method, i.e., a preprocessing step of hyperspectral data classification tasks. As the dimension increased, the CW method maintained good results when compared to its competitors. In fact, considering all of the results, we see that the proposed method achieved the best results in $\frac{9+10+8}{45} = 60\%$ in the experiments. This is likely due to the fact that the CW method is capable of selecting the best spectral bands for each individual cluster (or class) in a one-versus-all fashion, even in an unsupervised case. Moreover, the cluster separability criterion used during the band selection process makes the job of the classifiers easier.

In terms of processing times, the proposed method does not appear amongst the fastest ones, as Figure 10 indicates. However, its outstanding accuracy mean compensates for this fact. Moreover, the mean processing time of the CW method was less than 50 s, which caused no problems in offline applications.

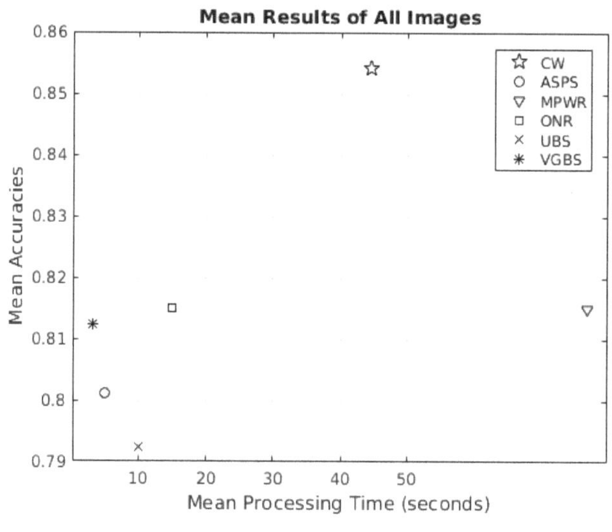

Figure 10. Mean processing time of all images and all classifiers together.

4.2. Cons

The proposed CW algorithm is not capable of addressing all the issues concerning a band selection application, such as the optimal number of bands to be selected. In fact, we do not address this topic in this paper. Here, the number s of bands to be selected is a user-defined parameter.

Moreover, it is necessary to know the number k of classes in the scene depicted by the image. Even though a remote sensing expert may easily infer the number of classes in a given scene, this topic remains unsolved.

5. Conclusions

The high dimensionality of a hyperspectral image can be useful in terms of good discrimination amongst objects and classes. On the other hand, it can also be a source of problems, such as the curse of dimensionality and overfitting of the classifier.

In order to alleviate such issues, this paper proposes a novel unsupervised band selection framework based on partitional clustering, in which each cluster stands for a real class of the data set. A hyperplane was used to separate all clusters in a one-versus-all fashion. After this, we proceeded to fine-tuning the initially selected bands based on the cluster separability in the feature space.

The proposed method achieved the best classification results in 60% of the experiments.

In future works, it is advisable to verify the performance of support vector machines to find a separating hyperplane between clusters. Furthermore, other numbers of initially chosen bands should be tested, rather than only $4(s/k)$. Moreover, some more recent clustering algorithms could be tested in order to check their effects on the final results. Finally, one could use optimization algorithms to find a suitable subset of bands during the fine-tuning process.

Author Contributions: Conceptualization, M.H. and V.F.; methodology, M.H.; software, M.H.; validation, M.H., V.F. and E.H.S.; formal analysis, M.H.; writing—original draft preparation, M.H.; writing—review and editing, M.H.; supervision, V.F. and E.H.S; funding acquisition, V.F. All authors have read and agreed to the published version of the manuscript.

Funding: This work was carried out in the framework of the NExT Senior Talent Chair DeepCoSLAM, which were funded by the French Government, through the program Investments for the Future managed by the National Agency for Research ANR-16-IDEX-0007, and with the support of Région Pays de la Loire and Nantes Métropole.

Data Availability Statement: Publicly available datasets were analyzed in this study. This data can be found here: http://www.ehu.eus/ccwintco/index.php/Hyperspectral_Remote_Sensing_Scenes, accessed on 16 August 2022.

Acknowledgments: The authors thank the reviewers for their critiques and comments.

Conflicts of Interest: The authors declare no conflict of interest.

References

1. Theodoridis, S.; Koutroumbas, K. *Pattern Recognition*, 4th ed.; Elsevier: Amsterdam, The Netherlands, 2009.
2. Guyon, I.; Elisseeff, A. An Introduction to Variable and Feature Selection. *J. Mach. Learn. Res.* **2003**, *3*, 1157–1182.
3. Chen, C.H. *Statistical Pattern Recognition*, 1st ed.; Spartan Books: Washington, DC, USA, 1973.
4. Habermann, M.; Frémont, V.; Shiguemori, E.H. Supervised band selection in hyperspectral images using single-layer neural networks. *Int. J. Remote Sens.* **2019**, *40*, 3900–3926. [CrossRef]
5. Houle, M.E.; Kriegel, H.P.; Kröger, P.; Schubert, E.; Zimek, A. Can Shared-Neighbor Distances Defeat the Curse of Dimensionality? In *Scientific and Statistical Database Management*; Gertz, M., Ludäscher, B., Eds.; Springer: Berlin/Heidelberg, Germany, 2010; pp. 482–500.
6. Duda, R.O.; Hart, P.E.; Stork, D.G. *Pattern Classification*, 2nd ed.; Wiley: New York, NY, USA, 2001.
7. Haykin, S.S. *Neural Networks and Learning Machines*, 3rd ed.; Pearson Education: Upper Saddle River, NJ, USA, 2009.
8. Kuncheva, L.I.; Matthews, C.E.; Arnaiz-González, Á.; Rodríguez, J.J. Feature Selection from High-Dimensional Data with Very Low Sample Size: A Cautionary Tale. *arXiv* **2020**, arXiv:abs/2008.12025.
9. Shang, X.; Song, M.; Wang, Y.; Yu, C.; Yu, H.; Li, F.; Chang, C. Target-Constrained Interference-Minimized Band Selection for Hyperspectral Target Detection. *IEEE Trans. Geosci. Remote Sens.* **2020**, *59*, 6044–6064. [CrossRef]
10. Zeng, M.; Cai, Y.; Cai, Z.; Liu, X.; Hu, P.; Ku, J. Unsupervised Hyperspectral Image Band Selection Based on Deep Subspace Clustering. *IEEE Geosci. Remote Sens. Lett.* **2019**, *16*, 1889–1893. [CrossRef]
11. Feng, J.; Chen, J.; Sun, Q.; Shang, R.; Cao, X.; Zhang, X.; Jiao, L. Convolutional Neural Network Based on Bandwise-Independent Convolution and Hard Thresholding for Hyperspectral Band Selection. *IEEE Trans. Cybern.* **2021**, *51*, 4414–4428. [CrossRef] [PubMed]
12. Ding, X.; Li, H.; Yang, J.; Dale, P.; Chen, X.; Jiang, C.; Zhang, S. An Improved Ant Colony Algorithm for Optimized Band Selection of Hyperspectral Remotely Sensed Imagery. *IEEE Access* **2020**, *8*, 25789–25799. [CrossRef]
13. Bevilacqua, M.; Berthoumieu, Y. Multiple-Feature Kernel-Based Probabilistic Clustering for Unsupervised Band Selection. *IEEE Trans. Geosci. Remote Sens.* **2019**, *57*, 6675–6689. [CrossRef]

14. Tang, J.; Alelyani, S.; Liu, H. Feature selection for classification: A review. In *Data Classification*; CRC Press: Boca Raton, FL, USA, 2014; pp. 37–64. [CrossRef]
15. Liu, T.; Xiao, J.; Huang, Z.; Kong, E.; Liang, Y. BP Neural Network Feature Selection Based on Group Lasso Regularization. In Proceedings of the 2019 Chinese Automation Congress (CAC), Hangzhou, China, 22–24 November 2019; pp. 2786–2790. [CrossRef]
16. Cao, X.; Xiong, T.; Jiao, L. Supervised Band Selection Using Local Spatial Information for Hyperspectral Image. *IEEE Geosci. Remote Sens. Lett.* **2016**, *13*, 329–333. [CrossRef]
17. Gao, J.; Du, Q.; Gao, L.; Sun, X.; Wu, Y.; Zhang, B. Ant colony optimization for supervised and unsupervised hyperspectral band selection. In Proceedings of the 2013 5th Workshop on Hyperspectral Image and Signal Processing: Evolution in Remote Sensing (WHISPERS), Gainesville, FL, USA, 25–28 June 2013; pp. 1–4. [CrossRef]
18. Habermann, M.; Frémont, V.; Shiguemori, E.H. Problem-based band selection for hyperspectral images. In Proceedings of the 2017 IEEE International Geoscience and Remote Sensing Symposium (IGARSS), Fort Worth, TX, USA, 23–28 July 2017; pp. 1800–1803. [CrossRef]
19. Wei, X.; Cai, L.; Liao, B.; Lu, T. Local-View-Assisted Discriminative Band Selection with Hypergraph Autolearning for Hyperspectral Image Classification. *IEEE Trans. Geosci. Remote Sens.* **2020**, *58*, 2042–2055. [CrossRef]
20. Feng, J.; Jiao, L.; Liu, F.; Sun, T.; Zhang, X. Mutual-Information-Based Semi-Supervised Hyperspectral Band Selection with High Discrimination, High Information, and Low Redundancy. *IEEE Trans. Geosci. Remote Sens.* **2015**, *53*, 2956–2969. [CrossRef]
21. Guo, Z.; Yang, H.; Bai, X.; Zhang, Z.; Zhou, J. Semi-supervised hyperspectral band selection via sparse linear regression and hypergraph models. In Proceedings of the 2013 IEEE International Geoscience and Remote Sensing Symposium—IGARSS, Melbourne, Australia, 21–26 July 2013; pp. 1474–1477. [CrossRef]
22. Cao, X.; Wei, C.; Ge, Y.; Feng, J.; Zhao, J.; Jiao, L. Semi-Supervised Hyperspectral Band Selection Based on Dynamic Classifier Selection. *IEEE J. Sel. Top. Appl. Earth Obs. Remote Sens.* **2019**, *12*, 1289–1298. [CrossRef]
23. Bai, J.; Xiang, S.; Shi, L.; Pan, C. Semisupervised Pair-Wise Band Selection for Hyperspectral Images. *IEEE J. Sel. Top. Appl. Earth Obs. Remote Sens.* **2015**, *8*, 2798–2813. [CrossRef]
24. Karoui, M.S.; Djerriri, K.; Boukerch, I. Unsupervised Hyperspectral Band Selection by Sequentially Clustering A Mahalanobis-Based Dissimilarity of Spectrally Variable Endmembers. In Proceedings of the 2020 Mediterranean and Middle-East Geoscience and Remote Sensing Symposium (M2GARSS), Tunis, Tunisia, 9–11 March 2020; pp. 33–36. [CrossRef]
25. Sui, C.; Tian, Y.; Xu, Y.; Xie, Y. Unsupervised Band Selection by Integrating the Overall Accuracy and Redundancy. *IEEE Geosci. Remote Sens. Lett.* **2015**, *12*, 185–189. [CrossRef]
26. Yang, C.; Bruzzone, L.; Zhao, H.; Tan, Y.; Guan, R. Superpixel-Based Unsupervised Band Selection for Classification of Hyperspectral Images. *IEEE Trans. Geosci. Remote Sens.* **2018**, *56*, 7230–7245. [CrossRef]
27. Wang, Q.; Zhang, F.; Li, X. Optimal Clustering Framework for Hyperspectral Band Selection. *IEEE Trans. Geosci. Remote Sens.* **2018**, *56*, 5910–5922. [CrossRef]
28. Xu, B.; Li, X.; Hou, W.; Wang, Y.; Wei, Y. A Similarity-Based Ranking Method for Hyperspectral Band Selection. *IEEE Trans. Geosci. Remote Sens.* **2021**, *59*, 9585–9599. [CrossRef]
29. Datta, A.; Ghosh, S.; Ghosh, A. Combination of Clustering and Ranking Techniques for Unsupervised Band Selection of Hyperspectral Images. *IEEE J. Sel. Top. Appl. Earth Obs. Remote Sens.* **2015**, *8*, 2814–2823. [CrossRef]
30. Jia, S.; Tang, G.; Zhu, J.; Li, Q. A Novel Ranking-Based Clustering Approach for Hyperspectral Band Selection. *IEEE Trans. Geosci. Remote Sens.* **2016**, *54*, 88–102. [CrossRef]
31. Xie, W.; Lei, J.; Yang, J.; Li, Y.; Du, Q.; Li, Z. Deep Latent Spectral Representation Learning-Based Hyperspectral Band Selection for Target Detection. *IEEE Trans. Geosci. Remote Sens.* **2020**, *58*, 2015–2026. [CrossRef]
32. Zhang, F.; Wang, Q.; Li, X. Hyperspectral image band selection via global optimal clustering. In Proceedings of the 2017 IEEE International Geoscience and Remote Sensing Symposium (IGARSS), Fort Worth, TX, USA, 23–28 July 2017; pp. 1–4. [CrossRef]
33. Tang, G.; Jia, S.; Li, J. An enhanced density peak-based clustering approach for hyperspectral band selection. In Proceedings of the 2015 IEEE International Geoscience and Remote Sensing Symposium (IGARSS), Milan, Italy, 26–31 July 2015; pp. 1116–1119. [CrossRef]
34. Sun, W.; Peng, J.; Yang, G.; Du, Q. Correntropy-Based Sparse Spectral Clustering for Hyperspectral Band Selection. *IEEE Geosci. Remote Sens. Lett.* **2020**, *17*, 484–488. [CrossRef]
35. Kumar, V.; Hahn, J.; Zoubir, A.M. Band selection for hyperspectral images based on self-tuning spectral clustering. In Proceedings of the 21st European Signal Processing Conference (EUSIPCO 2013), Marrakech, Morocco, 9–13 September 2013; pp. 1–5.
36. Dou, Z.; Gao, K.; Zhang, X.; Wang, H.; Han, L. Band Selection of Hyperspectral Images Using Attention-Based Autoencoders. *IEEE Geosci. Remote Sens. Lett.* **2021**, *18*, 147–151. [CrossRef]
37. Damodaran, B.B.; Courty, N.; Lefèvre, S. Sparse Hilbert Schmidt Independence Criterion and Surrogate-Kernel-Based Feature Selection for Hyperspectral Image Classification. *IEEE Trans. Geosci. Remote Sens.* **2017**, *55*, 2385–2398. [CrossRef]
38. Sun, W.; Peng, J.; Yang, G.; Du, Q. Fast and Latent Low-Rank Subspace Clustering for Hyperspectral Band Selection. *IEEE Trans. Geosci. Remote Sens.* **2020**, *58*, 3906–3915. [CrossRef]
39. Kohavi, R.; John, G.H. Wrappers for Feature Subset Selection. *Artif. Intell.* **1997**, *97*, 273–324. [CrossRef]
40. Cai, Y.; Liu, X.; Cai, Z. BS-Nets: An End-to-End Framework for Band Selection of Hyperspectral Image. *IEEE Trans. Geosci. Remote Sens.* **2020**, *58*, 1969–1984. [CrossRef]

41. Habermann, M.; Frémont, V.; Shiguemori, E.H. Unsupervised Hyperspectral Band Selection Using Clustering and Single-Layer Neural Network. *Revue Française de Photogrammétrie et de Télédétection* **2018**, *217–218*, 33–42. [CrossRef]
42. Graña, M.; Veganzons, M.A.; Ayerdi, B. Hyperspectral Remote Sensing Scenes. Available online: http://www.ehu.eus/ccwintco/index.php/Hyperspectral_Remote_Sensing_Scenes (accessed on 16 August 2022).
43. Fu, X.; Shang, X.; Sun, X.; Yu, H.; Song, M.; Chang, C.I. Underwater Hyperspectral Target Detection with Band Selection. *Remote Sens.* **2020**, *12*, 1056. [CrossRef]
44. Wang, Q.; Li, Q.; Li, X. Hyperspectral Band Selection via Adaptive Subspace Partition Strategy. *IEEE J. Sel. Top. Appl. Earth Obs. Remote Sens.* **2019**, *12*, 4940–4950. [CrossRef]
45. Sui, C.; Li, C.; Feng, J.; Mei, X. Unsupervised Manifold-Preserving and Weakly Redundant Band Selection Method for Hyperspectral Imagery. *IEEE Trans. Geosci. Remote Sens.* **2020**, *58*, 1156–1170. [CrossRef]
46. Wang, Q.; Zhang, F.; Li, X. Hyperspectral Band Selection via Optimal Neighborhood Reconstruction. *IEEE Trans. Geosci. Remote Sens.* **2020**, *58*, 8465–8476. [CrossRef]
47. Chang, C.-I.; Du, Q.; Sun, T.-L.; Althouse, M.L.G. A joint band prioritization and band-decorrelation approach to band selection for hyperspectral image classification. *IEEE Trans. Geosci. Remote Sens.* **1999**, *37*, 2631–2641. [CrossRef]
48. Geng, X.; Sun, K.; Ji, L.; Zhao, Y. A Fast Volume-Gradient-Based Band Selection Method for Hyperspectral Image. *IEEE Trans. Geosci. Remote Sens.* **2014**, *52*, 7111–7119. [CrossRef]

sustainability

Case Report

PM2.5 Prediction Based on the CEEMDAN Algorithm and a Machine Learning Hybrid Model

Wenchao Ban and Liangduo Shen *

School of Ocean Engineering Equipment, Zhejiang Ocean University, Zhoushan 316000, China
* Correspondence: slduo@163.com

Abstract: The current serious air pollution problem has become a closely investigated topic in people's daily lives. If we want to provide a reasonable basis for haze prevention, then the prediction of PM2.5 concentrations becomes a crucial task. However, it is difficult to complete the task of PM2.5 concentration prediction using a single model; therefore, to address this problem, this paper proposes a fully adaptive noise ensemble empirical modal decomposition (CEEMDAN) algorithm combined with deep learning hybrid models. Firstly, the CEEMDAN algorithm was used to decompose the PM2.5 timeseries data into different modal components. Then long short-term memory (LSTM), a backpropagation (BP) neural network, a differential integrated moving average autoregressive model (ARIMA), and a support vector machine (SVM) were applied to each modal component. Lastly, the best prediction results of each component were superimposed and summed to obtain the final prediction results. The PM2.5 data of Hangzhou in recent years were substituted into the model for testing, which was compared with eight models, namely, LSTM, ARIMA, BP, SVM, CEEMDAN–ARIMA, CEEMDAN–LSTM, CEEMDAN–SVM, and CEEMDAN–BP. The results show that for the coupled CEEMDAN–LSTM–BP–ARIMA model, the prediction ability was better than all the other models, and the timeseries decomposition data of PM2.5 had their own characteristics. The data with different characteristics were predicted separately using appropriate models and the final combined model results obtained were the most satisfactory.

Keywords: timeseries data; PM2.5 concentration prediction; CEEMDAN–LSTM–BP–ARIMA coupling model

Citation: Ban, W.; Shen, L. PM2.5 Prediction Based on the CEEMDAN Algorithm and a Machine Learning Hybrid Model. *Sustainability* **2022**, *14*, 16128. https://doi.org/10.3390/su142316128

Academic Editors: Gwanggil Jeon and Hong Tang

Received: 20 July 2022
Accepted: 30 November 2022
Published: 2 December 2022

1. Introduction

Atmospheric pollution [1] is closely related to agricultural production and human health. In recent years, with the expansion of industrial production, the problem of atmospheric pollution has become increasingly serious. Thus, we must pay attention to this problem [2]. PM2.5 refers to particles in the air with a particle size less than or equal to 2.5 microns, which can float in the outdoor air for a long time. Its content is a key factor in determining the degree of air pollution. A greater content of PM2.5 in the outdoor air indicates increased pollution. PM2.5 particles are also an important element in the formation of haze, as they are finer particles, are richer in harmful substances, and have a longer transmission distance and life span. Therefore, PM2.5 particles are harmful to human health, the quality of the atmospheric environment, and they have a more direct impact on air quality and visibility. A person's travel and health conditions in haze environments [3] are greatly affected. Therefore, it is necessary to establish accurate, reliable, and effective models to make predictions of atmospheric pollutant concentrations over a long period of time. The prediction results [4–7] can provide guidance for decision-making behavior, in addition to being important for the protection and management of ambient air.

Accurate long-term forecasting of PM2.5 concentration is more important than short-term forecasting, as it allows us more time to deal with the impact of air pollution. At present, the main timeseries forecasting methods are divided into two categories. One is the traditional probabilistic method [8], which determines the parameters of the model according to theoretical assumptions and prior knowledge of the data. If our actual data and theoretical models do not match, then this method cannot give satisfactory long-term prediction results. There is also the machine learning method [9]. The biggest difference with the traditional probability method is that the machine learning method does not need to determine the model parameters through theoretical assumptions and prior knowledge of the data. The algorithm is used for learning to obtain the law between the model parameters and the data to make predictions. Therefore, deep learning networks have actually surpassed traditional probabilistic prediction methods in many nonlinear modeling fields.

Traditional methods are mostly used in simple applications of the environment, such as threshold autoregressive (TAR) models [10] and hidden Markov models (HMM) [11] since these models are determined on the basis of theoretical assumptions and prior knowledge of the data parameters. Many times we cannot know the previous parameters, resulting in relatively large limitations, which seriously affect the accuracy of prediction.

Machine learning models are generally based on basic algorithms and historical data to build predictive models that can adaptively learn model parameters, obtain laws and relationships between complex data, and conduct simulation training through part of the data to obtain models to predict future development trends. Basic parameter identification algorithms include iterative algorithms [12], particle-based algorithms [13], and recursive algorithms [14].

Machine learning models are divided into shallow networks and deep networks [15,16]. Shallow networks include the short-term prediction method based on the back-propagation (BP) neural network proposed by Ni et al. [17], the improved grey neural network model [18], the radial basis function (RBF) neural network [19], etc. These methods have been used for PM2.5 concentration data, daily average temperature, and other pollutant concentration data.

However, due to the simple structure of the shallow network, it can only achieve short-term prediction performance, and long-term accurate prediction results must be captured using deep neural networks.

Deep machine learning networks show strong learning ability in complex time-series. They have good performance in capturing the high nonlinearity of timeseries data. They can highly abstract data analysis through the multi-nonlinear transformation of complex structures. Timeseries problem-solving functions include long short-term memory (LSTM) [20,21], differential integrated moving average autoregressive model (ARIMA) [22,23], and support vector machine (SVM) [24,25].

Although deep machine learning models have the ability to extract accurate information in complex environments, PM2.5 concentration sequences are datasets with strong randomness and nonlinearity, and the accuracy of long-term prediction needs further development. In recent years, the application of deep learning methods in the field of air pollution has also attracted the attention of researchers, combinatorial approaches to data decomposition have been shown to be effective ways to improve forecasting performance, and various hybrid models have been introduced to forecast nonlinear timeseries data.

For example, Huang and Kuo [26] used a hybrid model based on convolutional neural networks (CNNs) and LSTM to predict PM2.5 concentrations. Rojo used a loess-based seasonal decomposition program [27,28] to decompose the seasonal components of the data for the short-term prediction of air pollen, and Zhang and Li set the wavelet basis function through wavelet decomposition [29,30] to obtain the predicted decomposition information.

The fully adaptive noise ensemble empirical modal decomposition (CEEMDAN) algorithm [31,32] can completely decompose timeseries data into intrinsic mode function (IMF) components with different frequency characteristics [33] and sort them from high frequency to low frequency, which can greatly reduce the complexity of the original data. We believe that the predictive power of a single model is ultimately limited; therefore, we focused on developing a combined model based on data decomposition.

The focus of this study is to improve the long-term prediction accuracy of PM2.5 based on deep learning networks, combined with the CEEMDAN decomposition algorithm, the complexity of the original PM2.5 data is effectively reduced, and to predict all IMF components separately through four different timeseries machine learning models, the BP model, ARIMA model, LSTM model, and SVM model. Different characteristics were used to adapt to different models, and then the optimal prediction results corresponding to each IMF component were superimposed and summed to restore the optimal solution so as to obtain the combination model based on the above models that was most suitable for Hangzhou PM2.5 data. Our innovation priorities were as follows:

(1) The CEEMDAN decomposition algorithm was introduced for the long-term prediction of PM2.5 timeseries data.
(2) After the IMF component was obtained using the CEEMDAN decomposition algorithm, an adaptive model was further established according to the characteristics of the IMF component, so as to improve the accuracy of long-term prediction.
(3) This method of further predictive analysis of the IMF components can become a new framework, which can be applied for the prediction of more similar data, thus obtaining accurate long-term prediction results.

2. Materials and Methods

The air quality data used in this paper came from the air quality historical data query website (https://www.aqistudy.cn/historydata/, accessed on 19 July 2022), and the daily average PM2.5 concentration data records from December 2013 to December 2021 were selected from Hangzhou, Zhejiang Province, and Kunming, Yunnan Province. The samples were both 2952. The sampling interval was 1 day (the daily average data for each day are obtained from the arithmetic mean of the 24-h effective hourly concentration data of all single stations in the city where they are located). The PM2.5 unit was ug/m^3. The first 90% of the sample data was selected for simulation training, and the last 10% of the sample data was used to test the prediction results; i.e., we used the PM2.5 concentration data for 2952 days (December 2013–February 2021) to predict the trend of PM2.5 concentration data in the next 295 days (March 2021–December 2021). Lastly, in the 295-day forecast results, we intercepted the 50-day model forecast results with more obvious comparisons for graphical display and comparative analysis. The timeseries data of PM2.5 concentration in Hangzhou are shown in Figure 1, which shows that the PM2.5 concentration in Hangzhou had obvious changes and fluctuations.

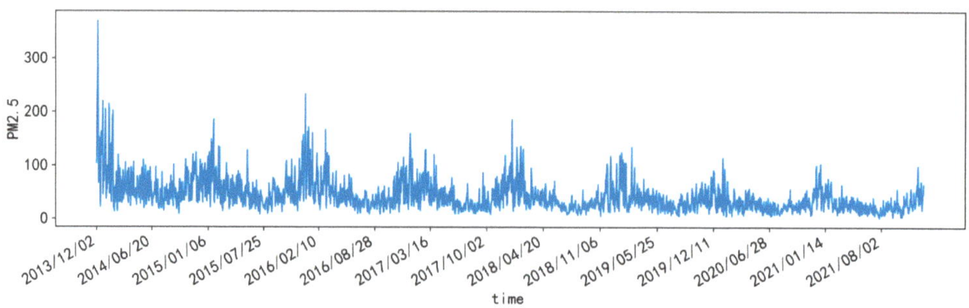

Figure 1. Hangzhou PM2.5 data sequence diagram.

2.1. Introduction to the Model

2.1.1. Fully Adaptive Noise Ensemble Empirical Modal Decomposition (CEEMDAN) Algorithm

The CEEMDAN algorithm is mostly used in the field of forecasting. It can completely decompose the original data, with strong volatility, into several intrinsic mode function (IMF) components with different frequency characteristics, thereby reducing the volatility of the data and improving the prediction accuracy. The CEEMDAN algorithm has three major advantages. The first is its completeness. In other words, the original data can be obtained by adding and summing the components decomposed by the algorithm, which is beyond the reach of many decomposition algorithms. Secondly, the CEEMDAN algorithm has a faster calculation speed, which can effectively improve the operation speed of the program. Lastly, the CEEMDAN algorithm has a better modal decomposition effect, preventing the occurrence of multiple low-frequency components with small amplitudes, which are of little significance for data analysis.

Let $E_i(\cdot)$ be the ith eigenmode component obtained through empirical mode decomposition (EMD). Then, the ith eigenmodal component obtained through CEEMDAN decomposition is $\overline{C_i(t)}$. V^j denotes the Gaussian white noise signal conforming to the standard normal distribution. $j = 1, 2, \ldots, N$ is the frequency at which white noise is added, ε is the standard table of white noise and $y(t)$ is the data to be decomposed. The CEEMDAN decomposition steps are described below.

(1) The new signal $y(t) + (-1)^q \varepsilon^j(t)$ is obtained by adding Gaussian white noise to the signal to be decomposed $y(t)$, where q = 1.2. The EMD decomposition of the new signal yields the first order eigenmode components.

$$E(y(t) + (-1)^q \varepsilon v^j(t)) = C_1^j(t) + r^j \tag{1}$$

(2) The first eigenmodal component of the CEEMDAN decomposition is derived by taking the resulting Nth modal component and balancing it overall.

$$\overline{C_1(t)} = \frac{1}{N}\sum_{j=1}^{N} C_1^j(t) \tag{2}$$

(3) The residuals are calculated after subtracting the first modal component.

$$r_1(t) = y(t) - \overline{C_i(t)} \tag{3}$$

(4) By adding a new signal of positive and negative paired Gaussian white noise to $r_1(t)$ and using the new signal as a medium to start the EMD decomposition, the first-order modal component, D_1, can be derived, which generates the second eigenmodal component of the CEEMDAN decomposition as follows:

$$\overline{C_2(t)} = \frac{1}{N}\sum_{j=1}^{N} D_1^j(t). \tag{4}$$

(5) The residuals are calculated after subtracting the second modal component from the results.

$$r_2(t) = r_1(t) - \overline{C_2(t)} \tag{5}$$

(6) The above process is repeated until the obtained residual signal is always a monotonic function and cannot be further decomposed. Then, the calculation is completed. The number of eigenmodal components obtained at this point is K. The original signal is then decomposed as

$$y(t) = \sum_{K=1}^{K} \overline{C_K(t)} + r_K(t). \tag{6}$$

An exploded view of the CEEMDAN algorithm is shown in Figure 2. The original PM2.5 concentration timeseries data in Hangzhou were completely decomposed into nine intrinsic mode function (IMF) components and residuals and then displayed in order from top to bottom according to high and low frequency. Then, the difference between each IMF component and residual was substituted into the machine learning timeseries model for prediction. Lastly, the final prediction result was obtained by superimposing and summing the predicted IMF components. The flowchart is shown in Figure 3.

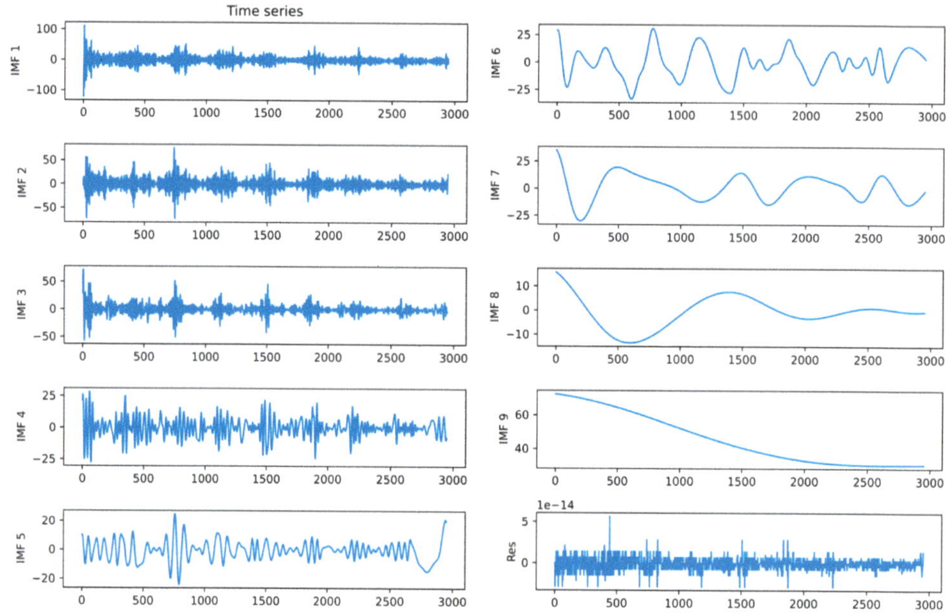

Figure 2. Splitting result diagram based on a fully adaptive noise ensemble empirical modal decomposition (CEEMDAN) algorithm.

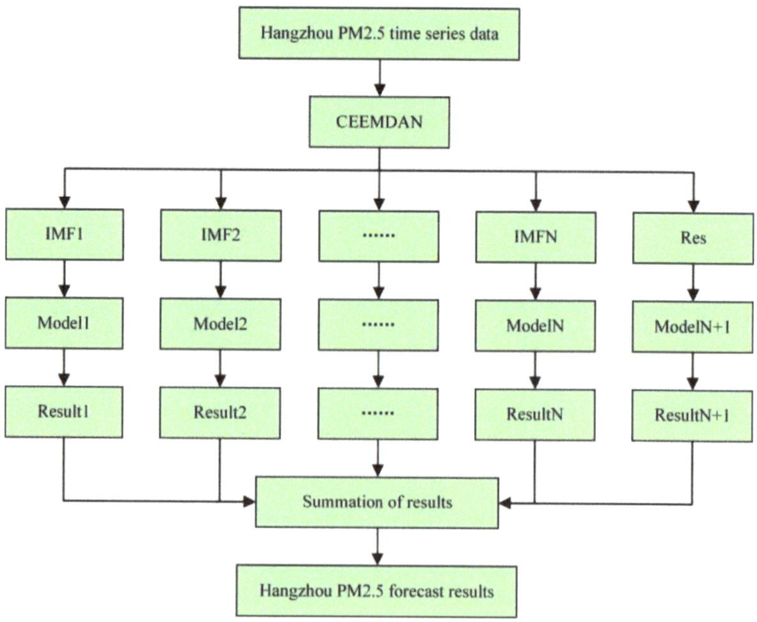

Figure 3. CEEMDAN model flowchart.

2.1.2. Long Short-Term Memory (LSTM) Model

LSTM has a strong ability to remember data in a delayed manner; when it grasps the current content, it acquires the relationships existing in the data over a longer time span to achieve long-term memory, enabling more accurate predictions of their development. An LSTM cell has three gates, called the forget gate, the input gate, the and output gate. Figure 4 shows the cell body of the neuronal structure.

$$f_t = \sigma(W_f * [h_{t-1}, x_t] + b_f) \tag{7}$$

Here, f_t is the forget gate, σ is the standard sigmoid activation function, h_{t-1} indicates the unit value at the previous time, W_f is the weight value, x_t is the input value at time t, and b_f is the vector of bias value.

$$i_t = \sigma(W_i * [h_{t-1}, x_t] + b_i) \tag{8}$$

$$\tilde{C}_t = \tan\text{h}(W_c * [h_{t-1}, x_t] + b_c) \tag{9}$$

Here, i_t is the input gate, $\tilde{C}_t$ is the current input state unit, and tanh $(\cdot)$ is the hyperbolic tangent activation function.

$$C_t = f_t * C_{t-1} + i_t * \tilde{C}_t \tag{10}$$

Here, C_t is the state cell of the hidden layer at moment t, f_t is the degree of the last forget function about information C_{t-1}, i_t is the degree to which $\tilde{C}_t$ is to be added, and C_t is the state of the cell obtained.

$$O_t = \sigma(W_O[h_{t-1}, x_t] + b_O) \tag{11}$$

$$\text{h}_t = \text{o}_t * \tanh(C_t) \tag{12}$$

Here, O_t is the current output gate.

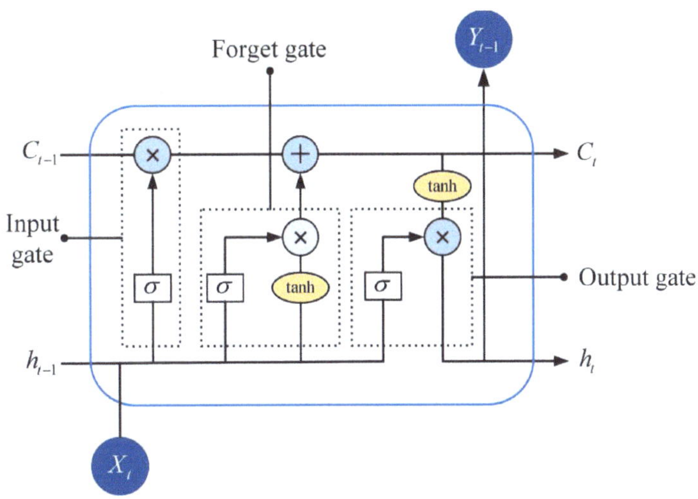

Figure 4. Long short-term memory (LSTM) neuronal structure.

2.1.3. Differential Integrated Moving Average Autoregressive Model (ARIMA)

The ARIMA model, known as the differential integrated moving average autoregressive model, is the most commonly used demand forecasting model in all industries. It treats the original timeseries data to be forecasted as a random dataset of equal length, uses mathematical methods to identify the characteristics of this dataset, and describes them by means of a numerical model. After modeling, it is possible to predict future values from known values. The flowchart of the ARIMA model is shown in Figure 5.

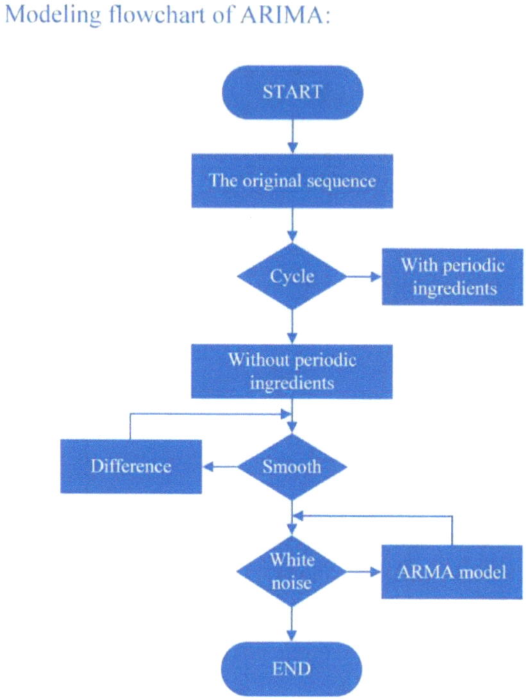

Figure 5. Differential integrated moving average autoregressive model (ARIMA) flowchart.

The ARIMA (p, d, q) model yields the following expression:

$$\left(1-\sum_{i=1}^{p}\phi_i L^i\right)(1-L^d)X_t=\left(1+\sum_{i=1}^{q}\theta_i L^i\right)\varepsilon_t, \tag{13}$$

where L is the lag operator, and p, d, and q represent the three basic parameters in the model; p is the number of lags of the original data itself in the ARIMA model, d is a positive integer parameter describing the number of orders of differentiation needed to stabilize the original data, and q is the number of lags of the prediction error used in the prediction model.

2.1.4. Backpropagation (BP) Neural Network Model

BP neural networks are a class of multilayer feedforward neural networks trained by error backpropagation, and they are one of the most popular neural network architectures in use today. Their network structure can be divided into an input layer, an implicit layer, and an output layer. The essence of this model is to convert the sample data from an input–output problem into a nonlinear optimization problem and use certain methods to change the weights along the negative trend of the error function. Figure 6 shows the structure of the BP neural network model.

Its mathematical expression is as follows:

$$\begin{aligned} y &= w_{11}^{(2,3)} * \tan sig(w_{11}^{(1,2)} * x_1 + w_{21}^{(1,2)} * x_2 + b_1^{(2)}) \\ &+ w_{21}^{(2,3)} * \tan sig(w_{12}^{(1,2)} * x_1 + w_{22}^{(1,2)} * x_2 + b_2^{(2)}) \\ &+ w_{31}^{(2,3)} * \tan sig(w_{13}^{(1,2)} * x_1 + w_{23}^{(1,2)} * x_2 + b_3^{(2)}) + b_1^{(3)} \end{aligned} \tag{14}$$

where w is the weight value, b is the threshold value, $w_{11}^{(2,3)}$ represents the weight from the first node of the second layer to the first node of the third layer, and $b_1^{(2)}$ represents the threshold of the first node of the second layer.

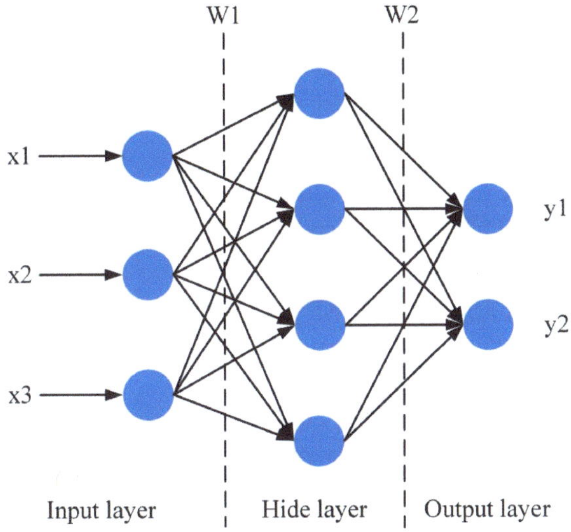

Figure 6. Backpropagation (BP) neural network structure diagram.

2.1.5. Support Vector Machine (SVM) Model

The SVM is essentially a small sample classification method. It maximizes the spatial distance of the samples by finding an optimal partitioning "plane", and it is superior in solving nonlinear high-dimensional spatial problems. It is often used in pattern recognition, function approximation, etc. The principle of the SVM method is shown in Figure 7.

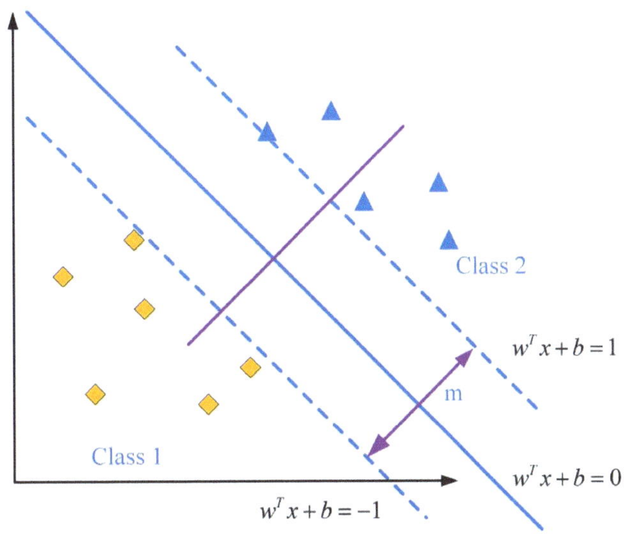

Figure 7. Support vector machine (SVM) schematic.

The expression of the optimal partition 'plane' is as follows:

$$w^T x + b = 0. \tag{15}$$

The two plane expressions that are parallel and equal to the optimal dividing 'plane' are expressed as follows:

$$w^T x + b = 1, \tag{16}$$

$$w^T x + b = -1. \tag{17}$$

2.1.6. Reference for Setting Important Parameters of the Model

In this model, the input dataset was the original PM2.5 timeseries data, the training set is the first 90% of the original data, and a select 10% of the training set data for 10-fold cross testing, reduce the chance caused by a random division, improve its generalization ability, and improve the efficiency of data use. the LSTM model was implemented using the keras package in Python 3.6. We set the most important parameter reference values to include a hidden layer with 300 dimensions, the 'relu' activation function, look back of 10, 'Adam' optimizer, batch size of 64, and 300 epochs. For the ARIMA model, we chose to use the Akaike information criterion to formulate the reasonable order of p and q, only limiting the parameter p. The maximum value of q was 3, with a scrolling step of 30. The model automatically determined the best order within three operations. The reference order of the BP neural network model function prediction order was 30, and the hidden layer reference was three layers. In the model prediction of SVM, a scrolling step of 30, the KernelFunction was set to 'polynomial', the KernelScale was set to 10, and the BoxConstraint was set to 5.

2.1.7. Model Evaluation Indices

In order to effectively compare the parameters of the models, we chose the root-mean-square error (RMSE), the mean absolute error (MAE), and the goodness of fit (R-squared, coefficient of determination) to judge the prediction accuracy of the models. The units of RMSE and MSE were the same as that of the PM2.5 concentration data, i.e., ug/m^3. Smaller values indicate a greater measurement accuracy. A closer value of R^2 to 1 indicates a better fit. The expressions are shown in Equations (18)–(20):

$$\text{RMSE} = \sqrt{\frac{1}{m}\sum_{i=1}^{m} (y_i - \hat{y}_i)^2}, \tag{18}$$

$$\text{MAE} = \frac{1}{m}\sum_{i=1}^{m} |y_i - \hat{y}_i|, \tag{19}$$

$$R^2 = 1 - \frac{\sum_{i=1}^{m} (y_i - \hat{y}_i)^2}{\sum_{i=1}^{m} (y_i - \overline{y})^2}, \tag{20}$$

where m is the number of test samples, and y_i and $\hat{y}_i$ are the original and predicted values of the data, respectively.

3. Results and Analysis

In this study, we first compared the prediction effectiveness of individual models for PM2.5.

The PM2.5 prediction results of the four single models are shown in Figure 8, and the error results are shown in Table 1. It can be concluded that the basic trend of the future PM2.5 concentration could be predicted simply by using single models, but there were clearly large errors, especially at the peaks and troughs. The ARIMA model was the best model for predicting PM2.5 in Hangzhou among the four single models, with an RMSE of 10.09, an MAE of 7.51, and an R^2 of 0.46.

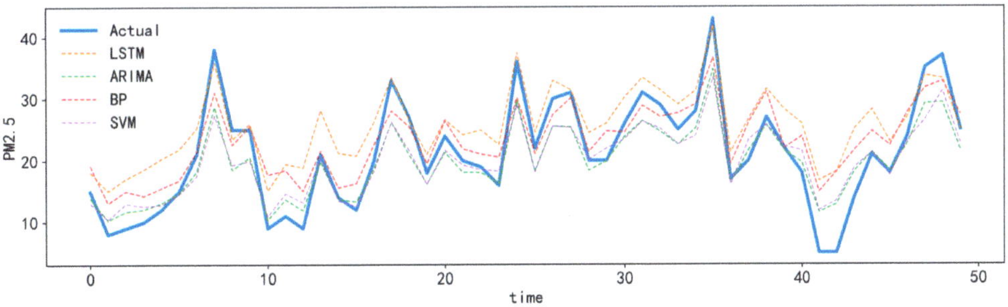

Figure 8. Comparison of single model prediction results.

Table 1. Single model error comparison.

	LSTM	ARIMA	BP	SVM
RMSE	10.27	10.09	10.55	10.61
MAE	7.88	7.51	8.21	7.93
R^2	0.42	0.46	0.37	0.43

We then optimized the single models using the CEEMDAN algorithm due to its predictive power. Figure 9 and Table 2 show the prediction and error comparison results for the CEEMDAN–LSTM, CEEMDAN–ARIMA, CEEMDAN–BP, and CEEMDAN–SVM models.

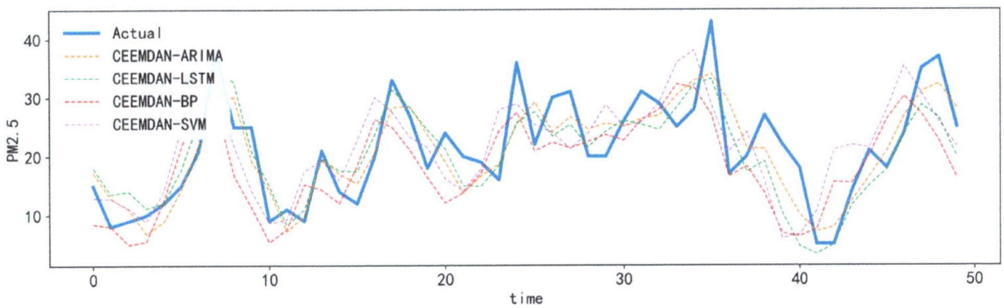

Figure 9. Comparison of prediction results using coupled models based on the CEEMDAN algorithm.

Table 2. Comparison of coupled model errors based on the CEEMDAN algorithm.

	CEEMDAN–LSTM	CEEMDAN–ARIMA	CEEMDAN–BP	CEEMDAN–SVM
RMSE	7.28	6.27	6.90	6.27
MAE	5.71	4.95	5.48	4.88
R^2	0.70	0.74	0.67	0.77

It can be concluded that the prediction ability of the coupled models processed using the CEEMDAN algorithm was greatly improved, and the trend of PM2.5 concentration in the future could be better predicted. Every error metric in the coupled model was better than that of the single models in Table 1. The fitting result of the CEEMDAN–SVM model was best, with an RMSE value of 6.27, an MAE value of 4.88, and an R^2 of 0.77. The comparison between the coupled models, combined with the CEEMDAN algorithm, and the simple models without the CEEMDAN algorithm fully reflect the superior performance of the CEEMDAN algorithm in improving the prediction accuracy of PM2.5.

Without the CEEMDAN algorithm, the predictive ability of the ARIMA model was better than that of the SVM model; whereas, with the CEEMDAN algorithm, the prediction ability of the CEEMDAN–SVM model was better than that of the CEEMDAN–ARIMA model. Their RMSE values were the same, but the MAE values of the CEEMDAN–SVM model were better. Therefore, we chose to compare the prediction results of each modal component after decomposition using the CEEMDAN algorithm in detail.

We used the CEEMDAN algorithm to completely decompose the original PM2.5 concentration data in Hangzhou into nine IMF components and residuals according to their frequency from high to low, and then substituted each IMF component and residual into the LSTM, ARIMA, SVM, and BP neural network models for prediction. The final prediction result was obtained by stacking and summing. The comparison results are shown in Tables 3–5. The LSTM network performed best in IMF1 component prediction. The BP neural network performed best in predicting the IMF2 components. The prediction performance of the ARIMA model was excellent for the IMF3–9 components. Since the residual value was minuscule, the value of all models when predicting the residual was infinitely close to 0. It can be concluded that the model corresponding to each data type with the best performance was different, and no model could be used for all components. Therefore, data types with different frequencies should be matched with different prediction models, and the optimal solution for each IMF component prediction was obtained by constructing the CEEMDAN–LSTM–BP–ARIMA model. The prediction results obtained from the experiment are shown in Figure 10, and the error results are shown in Table 6. The fitting degree in the figure is high, and the various error indicators are significantly better than all the previous models.

Table 3. Comparison of RMSE values of split result graphs based on the CEEMDAN algorithm.

	LSTM (RMSE)	ARIMA (RMSE)	SVM (RMSE)	BP (RMSE)
IMF1	5.84	6.16	6.12	6.90
IMF2	1.76	1.53	1.80	1.52
IMF3	0.27	0.25	0.71	0.29
IMF4	0.15	0.04	0.43	0.05
IMF5	0.10	1.57×10^{-3}	0.20	1.9×10^{-3}
IMF6	0.04	6.66×10^{-5}	0.46	1.1×10^{-3}
IMF7	0.06	3.72×10^{-6}	0.77	1.2×10^{-3}
IMF8	0.01	3.81×10^{-8}	0.14	3.21×10^{-4}
IMF9	0.13	1.38×10^{-8}	1.34	0.09
Residual	0.00	0	0	0

Table 4. Comparison of MAE values of split result graphs based on the CEEMDAN algorithm.

	LSTM (MAE)	ARIMA (MAE)	SVM (MAE)	BP (MAE)
IMF1	4.70	4.97	4.75	5.51
IMF2	1.37	1.20	1.31	1.17
IMF3	0.19	0.17	0.53	0.21
IMF4	0.13	0.02	0.35	0.03
IMF5	0.05	1×10^{-3}	0.14	1×10^{-3}
IMF6	0.03	2.17×10^{-5}	0.38	7.83×10^{-4}
IMF7	0.05	7.36×10^{-7}	0.68	8.78×10^{-4}
IMF8	0.01	7.88×10^{-9}	0.10	3.10×10^{-4}
IMF9	0.09	9.35×10^{-9}	1.34	0.074
Residual	0.00	0.00	0.00	0.00

Table 5. Comparison of R^2 values of split result graphs based on CEEMDAN algorithm.

	LSTM (R^2)	ARIMA (R^2)	SVM (R^2)	BP (R^2)
IMF1	0.20	0.16	0.17	0
IMF2	0.88	0.91	0.88	0.91
IMF3	0.99	1.00	0.98	1.00
IMF4	1.00	1.00	0.99	1.00
IMF5	1.00	1.00	1.00	1.00
IMF6	1.00	1.00	1.00	1.00
IMF7	1.00	1.00	1.00	1.00
IMF8	1.00	1.00	1.00	1.00
IMF9	1.00	1.00	1.00	1.00
Residual	0.00	0.00	0.00	0.00

Table 6. CEEMDAN–LSTM–BP–ARIMA model error comparison (Hangzhou).

	CEEMDAN–LSTM–BP–ARIMA
RMSE	5.9
MAE	4.63
R^2	0.79

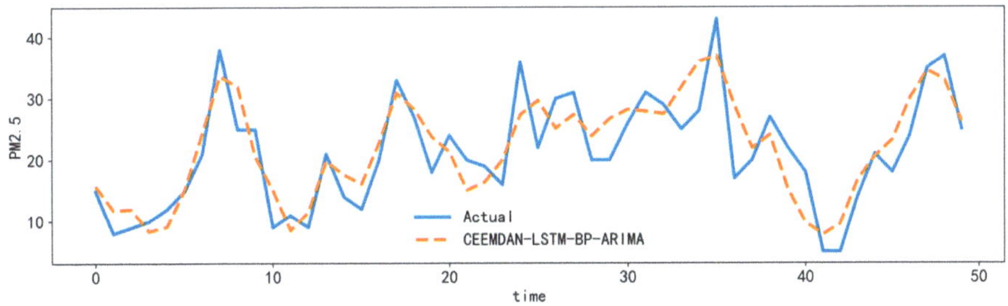

Figure 10. CEEMDAN–LSTM–BP–ARIMA prediction comparison (Hangzhou).

In order to better reflect the applicability of the model, and to prevent the occurrence of chance, we substitute the data from Kunming, Yunnan Province, into the optimal model for validation. The prediction results are shown in Figure 11, and the error results are shown in Table 7. We can obviously conclude that the optimal model applicable to Hangzhou PM2.5 prediction is also applicable to Kunming, and the prediction results fit well with the actual values. Therefore, in this experiment, it can be said that the model combining the machine learning time series model with the CEEMDAN algorithm is the ideal model for predicting PM2.5 concentration.

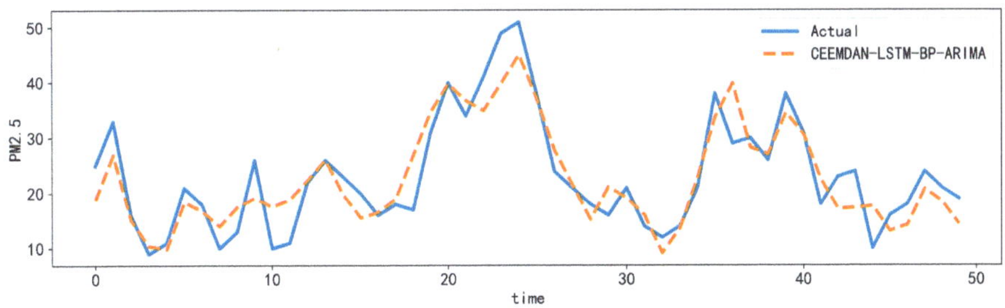

Figure 11. CEEMDAN–LSTM–BP–ARIMA prediction comparison (Kunming).

Table 7. CEEMDAN–LSTM–BP–ARIMA model error comparison (Kunming).

	CEEMDAN–LSTM–BP–ARIMA
RMSE	4.55
MAE	3.66
R^2	0.79

Comparing various methods, Qian [8] et al. used the traditional probabilistic method for PM2.5 concentration prediction, which was able to provide highly time-resolved particle concentrations, but required high parameters of the individual data itself, and required the combination of meteorological variables, land use terms, and spatial and temporal lag terms. In contrast, the learning ability of simple time series models such as LSTM, ARIMA, SVM, and BP is limited, and there is an upper limit to their ability to handle anomalous data. Zhao [20] et al. used LSTM models to model the local variation of PM2.5, Chen [18] et al. used BP neural networks for a 3-h short-term prediction of PM2.5 concentrations, and the experimental results proved that the single machine learning time series model has certain PM2.5 concentration prediction ability, which is mainly superior in the short term or locally, and the long-term results are not satisfactory. The CEEMDAN algorithm has been well applied in the hands of Rongbin [32] et al. The algorithm has good decomposition integrity by decomposing the original signal with complexity and nonsmoothness into eigenmodal components (IMF), and the decomposition completes with significantly lower data values, so the decomposed training data can better improve the prediction accuracy when applied to a single neural network model.

4. Conclusions

In recent years, air quality problems have had a serious impact on people's normal life. Environmental problems such as PM2.5 have received more and more attention and PM2.5 is characterized by a strong multilateral and a strong randomness. Thus, accurate long-term PM2.5 concentration prediction remains a formidable challenge for us.

In this study, we proposed a way to combine the CEEMDAN algorithm with the LSTM model, ARIMA model, BP neural network, and SVM model to predict the PM2.5 concentration. The results of various evaluation indicators showed that all models based on the CEEMDAN algorithm improved the prediction accuracy to varying degrees compared with the original simple models. The introduction of the CEEMDAN algorithm can provide new inspiration for a PM2.5 prediction, and the CEEMDAN algorithm can perhaps be combined with additional timeseries machine learning models. In this experiment, the predictive performance of the coupled model was higher than that of the single model.

Secondly, we discovered a new application of the IMF components obtained using the CEEMDAN algorithm. We carried out LSTM, ARIMA, BP neural network, and SVM modeling and prediction for each IMF component of the PM2.5 concentration timeseries data. The optimal prediction results of the components were added and summed. In this paper, the CEEMDAN-LSTM-BP-ARIMA model obtained the ideal results for a PM2.5 concentration prediction in Hangzhou and Kunming. Compared with the other models, the long-term prediction accuracy was significantly improved. Applying a single model is not optimal. We found that the best models differed according to IMF components, whereby a combination of timeseries machine learning models obtained the best prediction accuracy. We believe that this method also has good generalizability and can be used to predict additional characteristics such as wind speed and other pollutant concentrations.

Author Contributions: Conceptualization, W.B.; methodology, W.B.; software, W.B.; validation, W.B.; formal analysis, W.B.; investigation, W.B.; resources, W.B.; data curation, W.B.; writing—original draft preparation, W.B.; writing—review and editing, W.B.; visualization, W.B.; supervision, L.S.; project administration, L.S.; funding acquisition, L.S. All authors have read and agreed to the published version of the manuscript.

Funding: This research was financially supported by Provincial scientific research fund for basic research (2021JZ008) and General Projects of Zhoushan Science and Technology.

Institutional Review Board Statement: Not applicable.

Informed Consent Statement: Not applicable.

Data Availability Statement: Not applicable.

Conflicts of Interest: The authors declare no conflict of interest.

Abbreviations

Abbreviation	Full Name
CEEMDAN	Complete EEMD with adaptive noise
LSTM	Long short-term memory
BP	Backpropagation
ARIMA	Differential integrated moving average autoregressive
SVM	Support vector machine
IMF	Intrinsic mode function

References

1. Kampa, M.; Castanas, E. Human health effects of air pollution. *Environ. Pollut.* **2008**, *151*, 362–367. [CrossRef]
2. Xing, Y.-F.; Xu, Y.-H.; Shi, M.-H.; Lian, Y.-X. The impact of PM2. 5 on the human respiratory system. *J. Thorac. Dis.* **2016**, *8*, E69–E74.
3. Ding, S.; Xu, N.; Ye, J.; Zhou, W.; Zhang, X. Estimating Chinese energy-related CO2 emissions by employing a novel discrete grey prediction model. *J. Clean. Prod.* **2020**, *259*, 120793. [CrossRef]
4. Wang, J.; Tang, J.; Guo, K. Green Bond Index Prediction Based on CEEMDAN-LSTM. *Front. Energy Res.* **2022**, *9*, 793413. [CrossRef]
5. Aravazhi, A. Hybrid Machine Learning Models for Forecasting Surgical Case Volumes at a Hospital. *AI* **2021**, *2*, 512–526. [CrossRef]
6. Kim, T.Y.; Cho, S.B. Predicting residential energy consumption using CNN-LSTM neural networks. *Energy* **2019**, *182*, 72–81. [CrossRef]
7. Shahid, F.; Zameer, A.; Muneeb, M. Predictions for COVID-19 with deep learning models of LSTM, GRU and Bi-LSTM. *Chaos Solitons Fractals* **2020**, *140*, 110212. [CrossRef]
8. Di, Q.; Koutrakis, P.; Schwartz, J. A hybrid prediction model for PM2.5 mass and components using a chemical transport model and land use regression. *Atmos. Environ.* **2016**, *131*, 390–399. [CrossRef]
9. Bai, Y.; Wang, W.; Jin, X.; Su, T.; Kong, J.; Zhang, B. Adaptive filtering for MEMS gyroscope with dynamic noise model. *ISA Trans.* **2020**, *101*, 430–441. [CrossRef]
10. Aero, O.; Ogundipe, A. Fiscal Deficit and Economic Growth in Nigeria: Ascertaining a Feasible Threshold. *Int. J. Econ. Financ. Issue* **2018**, *8*, 296–306. Available online: https://ssrn.com/abstract=2861505 (accessed on 31 December 2016). [CrossRef]
11. Guo, H.; Pedrycz, W.; Liu, X. Hidden Markov Models-Based Approaches to Long-term Prediction for Granular Time Series. *IEEE Trans. Fuzzy Syst.* **2018**, *26*, 2807–2817. [CrossRef]
12. Xu, L.; Ding, F. Iterative parameter estimation for signal models based on measured data. *Circuits Syst. Signal Process.* **2018**, *37*, 3046–3069. [CrossRef]
13. Ding, J.; Chen, J.; Lin, J.X.; Wan, L.J. Particle filtering based parameter estimation for systems with output-error type model structures. *J. Frankl. Inst.* **2019**, *356*, 5521–5540. [CrossRef]
14. Ding, F.; Xu, L.; Meng, D.; Jin, X.-B.; Alsaedi, A.; Hayat, T. Gradient estimation algorithms for the parameter identification of bilinear systems using the auxiliary model. *J. Comput. Appl. Math.* **2020**, *369*, 112575. [CrossRef]
15. LeCun, Y.; Bengio, Y.; Hinton, G. Deep learning. *Nature* **2015**, *521*, 436–444. [CrossRef]
16. Deng, L.; Yu, D. Deep learning: Methods and applications. *Found. Trends Signal Process.* **2014**, *7*, 197–387. [CrossRef]
17. Ni, X.Y.; Huang, H.; Du, W.P. Relevance analysis and short-term prediction of PM 2.5 concentrations in Beijing based on multi-source data. *Atmos. Environ.* **2017**, *150*, 146–161. [CrossRef]
18. Chen, Y. Prediction algorithm of PM2.5 mass concentration based on adaptive BP neural network. *Computing* **2018**, *100*, 825–838. [CrossRef]
19. Liu, C.; Shu, T.; Chen, S.; Wang, S.; Lai, K.K.; Gan, L. An improved grey neural network model for predicting transportation disruptions. *Expert Syst. Appl.* **2016**, *45*, 331–340. [CrossRef]
20. Zhao, J.; Deng, F.; Cai, Y.; Chen, J. Long short-term memory-fully connected (LSTM-FC) neural network for PM2.5 concentration prediction. *Chemosphere* **2019**, *220*, 486–492. [CrossRef]
21. Huang, Z.; Xu, W.; Yu, K. Bidirectional LSTM-CRF Models for Sequence Tagging. *arXiv* **2015**, arXiv:1508.01991.
22. Duan, Y.; Wang, H.; Wei, M.; Tan, L.; Yue, T. Application of ARIMA-RTS optimal smoothing algorithm in gas well production prediction. *Petroleum* **2022**, *8*, 270–277. [CrossRef]

23. Xu, D.; Zhang, Q.; Ding, Y.; Zhang, D. Application of a hybrid ARIMA-LSTM model based on the SPEI for drought forecasting. *Environ. Sci. Pollut. Res.* **2022**, *29*, 4128–4144. [CrossRef] [PubMed]
24. Chauhan, V.K.; Dahiya, K.; Sharma, A. Problem formulations and solvers in linear SVM: A review. *Artif. Intell. Rev.* **2019**, *52*, 803–855. [CrossRef]
25. Huang, W.; Liu, H.; Zhang, Y.; Mi, R.; Tong, C.; Xiao, W.; Shuai, B. Railway dangerous goods transportation system risk identification: Comparisons among SVM, PSO-SVM, GA-SVM and GS-SVM. *Appl. Soft Comput.* **2021**, *109*, 107541. [CrossRef]
26. Huang, C.J.; Kuo, P.H. A deep cnn-lstm model for particulate matter (PM2.5) forecasting in smart cities. *Sensors* **2018**, *18*, 2220. [CrossRef]
27. Rojo, J.; Rivero, R.; Romero-Morte, J.; Fernandez-González, F.; Perez-Badia, R. Modeling pollen time series using seasonal-trend decomposition procedure based on LOESS smoothing. *Int. J. Biometeorol.* **2017**, *61*, 335–348. [CrossRef]
28. Xiong, T.; Li, C.; Bao, Y. Seasonal forecasting of agricultural commodity price using a hybrid STL and ELM method: Evidence from the vegetable market in China. *Neurocomputing* **2018**, *275*, 2831–2844. [CrossRef]
29. Cheng, Y.; Zhang, H.; Liu, Z.; Chen, L.; Wang, P. Hybrid algorithm for short-term forecasting of PM2.5 in China. *Atmos. Environ.* **2019**, *200*, 264–279. [CrossRef]
30. Qiao, W.; Tian, W.; Tian, Y.; Yang, Q.; Wang, Y.; Zhang, J. The forecasting of PM2.5 using a hybrid model based on wavelet transform and an improved deep learning algorithm. *IEEE Access* **2019**, *7*, 142814–142825. [CrossRef]
31. Zhou, Y.; Li, T.; Shi, J.; Qian, Z. A CEEMDAN and XGBOOST-based approach to forecast crude oil prices. *Complexity* **2019**, *2019*, 4392785. [CrossRef]
32. Rongbin, C.; Sanming, L. Research on Wind Power Prediction Method Based on CEEMDAN-SSA-GRU. In Proceedings of the 2021 IEEE International Conference on Electrical Engineering and Mechatronics Technology (ICEEMT), Qingdao, China, 2–4 July 2021; pp. 597–601.
33. Wang, Z.-Y.; Qiu, J.; Li, F.-F. Hybrid models combining EMD/EEMD and ARIMA for Long-term streamflow forecasting. *Water* **2018**, *10*, 853. [CrossRef]

information

MDPI

Article

An Effective Ensemble Automatic Feature Selection Method for Network Intrusion Detection

Yang Zhang, Hongpo Zhang * and Bo Zhang

School of Cyber Science and Engineering, Zhengzhou University, Zhengzhou 450001, China; yangzhang@gs.zzu.edu.cn (Y.Z.); bzhang@gs.zzu.edu.cn (B.Z.)
* Correspondence: zhp@zzu.edu.cn

Abstract: The mass of redundant and irrelevant data in network traffic brings serious challenges to intrusion detection, and feature selection can effectively remove meaningless information from the data. Most current filtered and embedded feature selection methods use a fixed threshold or ratio to determine the number of features in a subset, which requires a priori knowledge. In contrast, wrapped feature selection methods are computationally complex and time-consuming; meanwhile, individual feature selection methods have a bias in evaluating features. This work designs an ensemble-based automatic feature selection method called EAFS. Firstly, we calculate the feature importance or ranks based on individual methods, then add features to subsets sequentially by importance and evaluate subset performance comprehensively by designing an NSOM to obtain the subset with the largest NSOM value. When searching for a subset, the subset with higher accuracy is retained to lower the computational complexity by calculating the accuracy when the full set of features is used. Finally, the obtained subsets are ensembled, and by comparing the experimental results on three large-scale public datasets, the method described in this study can help in the classification, and also compared with other methods, we discover that our method outperforms other recent methods in terms of performance.

Keywords: cyber security; intrusion detection system (IDS); automatic feature selection; normalized score of mixed (NSOM); ensemble method

Citation: Zhang, Y.; Zhang, H.; Zhang, B. An Effective Ensemble Automatic Feature Selection Method for Network Intrusion Detection. *Information* **2022**, *13*, 314. https://doi.org/10.3390/info13070314

Academic Editor: Miltiadis D. Lytras

Received: 14 May 2022
Accepted: 24 June 2022
Published: 27 June 2022

1. Introduction

As technology advances, there are even more devices connected to the Internet and data traffic on the network increases. This results in the Internet facing an increased attack surface, increasing the possibility that the network will be destroyed and that this threat will always be present [1]. The resulting risk of property damage and information leakage caused by cyberattacks is becoming an important issue in need of a solution [2]. Intrusion detection systems (IDSs) can identify network attacks, including unauthorized access, denial of service attacks, etc. [3].

Anomaly-based detection, misuse-based detection, and hybrid detection are the most common types of intrusion detection systems [4,5]. Traditionally, anomaly detection and misuse detection have been studied from two perspectives [6]. Anomaly-based detection first establishes a normal network access pattern and then accesses that are found to be different from the normal pattern are judged to be anomalous. An anomaly-based IDS has the advantage of being able to detect known attacks and 0-day attacks, and the disadvantage is the high false alarm rate [7]. The misuse-based intrusion detection technique compares the behavior of an attacker trying to compromise the system with the activity behavior of a known user. It compiles a library of known malicious characteristics, which may then be compared to detect known assaults [8]. An IDS that is based on misuse does not produce a huge false-positive rate, but it does not discover new attacks and the feature database needs to be continuously updated [4].

As the network traffic continues to increase, the amount of data generated is getting enormous, which increases the computational cost to some extent [9]. Furthermore, the data contain redundant and irrelevant information, resulting in lower classification accuracies and more false positives [10]. The process of selecting the most useful features for a model is known as feature selection. It reduces the dimensionality and computational cost of the data and improves the classifier's performance, and it is more useful for visualization and human understanding [10,11].

Currently, the main feature selection methods are based on filtered, wrapped, and embedded methods [19]. The filtered method calculates the correlations between both features and labels, as well as between features, before ranking the features and selecting the top-ranked features as feature subsets [10]. Meanwhile, the wrapped method is a model-based feature selection, which optimizes the search subset by evaluating the objective function of the model to obtain the best subset [12]. The advantage of the filtered approach is that it is computationally efficient while not relying on a specific classification algorithm; the disadvantage is that the performance is slightly worse [13]. Wrapped methods have the advantage of good performance but require a search of the feature space and face the problem of a long search time [14]. Embedded methods combine the classification process with the feature selection process [15] and perform feature selection along with classification, depending on the specific classification algorithm. Moreover, ensemble feature selection models can address the bias of individual methods for feature evaluation and enhance the performance of feature subsets [16–21].

This paper designs a novel intrusion detection framework that removes irrelevant and redundant features to improve the model detection accuracy and reduce the training time. The following is a list of our contributions.

1. This paper proposes an automatic feature selection method and designs a hybrid normalized score of mixed (NSOM) for the comprehensive evaluation of a subset performance. This method is able to select feature subsets dynamically and overcome the drawback of fixed thresholds.
2. In this paper, an ensemble feature selection method is designed to improve the performance of feature subsets in classification by the intersection and union of different subsets.
3. The methodologies designed in this paper are validated on the UNSW-NB15, CIC-IDS2017, and CSE-CIC-IDS2018 datasets and compared with the latest results. The outcomes of the experiments suggest that our framework can produce better results.

The remainder of this work is constructed as follows. The development of the research on feature selection-based intrusion detection systems is presented in the Section 2. The method proposed in this research is detailed in full in the Section 3. The Section 4 contains the experimental results and analyses. In the Section 5, a summary of the entire article is presented.

2. Related Work

In intrusion detection systems, as network attacks become more complex and network traffic features become more numerous, it is easy to fall into dimensional disasters [10,22]. Selection is the method of determining a subset of features from the initial feature set that can be classified correctly [23] by minimizing the number of features, decreasing the computation time, and increasing model accuracy.

Bansal et al. [24] used XGBoost for data classification, and the experimental results on the CIC-IDS2017 dataset showed that XGBoost was more accurate than other methods. Fitni et al. [25] used the Spearman correlation coefficient to select important features from the CSE-CIC-IDS2018 dataset and achieved an accuracy of 98.8% on the integrated model. Lin et al. [26] utilized a Long Short Term Memory Network (LSTM) with the attention mechanism incorporated to identify attacks and imbalance processing using SMOTE and achieved 96.2% accuracy on the CSE-CIC-IDS2018 dataset.

A wrapped method based on a swarm intelligence search was also applied to choose the greatest subset. Chaouki et al. [27] applied a genetic algorithm to determine the optimal subset, and experiments on the KDD CUP99 and UNSW-NB15 datasets showed that the performance after feature reduction remained essentially the same as the original features. Zhou et al. [28] combined correlation and bat optimization to classify using an ensemble classifier with multi-classification accuracies of 99.81% on the NSL-KDD dataset, 99.52% on the AWID dataset, and 99.89% on the CIC-IDS2017 dataset. Nazir et al. [29] proposed a taboo search–random forest (TS-RF)-based wrapped feature selection method, which can minimize the number of features and the number of false alarms while improving the classification accuracy. On multiple datasets, Farahani et al. [30] suggested a Cross-Correlation (CCFS)-based feature selection method, which was compared to the Cuttlefish optimization method and mutual information, and the method achieved better results in several metrics, such as accuracy and precision. Benmessahel et al. [31] combined the locust swarm optimization (LSO) algorithm with a feedforward neural network. This contributed to improving the algorithm convergence speed and reducing the possibility of falling into local optima, achieving detection rates of 99.33% and 89.83% on the NSL-KDD and UNSW-NB15 datasets, respectively. Vijayanand et al. [32] proposed a genetic algorithm-based selection method that achieves better results for the evaluation of the CIC-IDS2017 dataset by selecting the important features of each class of attacks.

For filtered and embedded methods, there are no clear criteria for determining the subset size or threshold. On the KDD CUP99 dataset, Akashdeep et al. [33] selected features based on information gain and relevancy. Then, the features were divided into a group of 10 by importance and taken as merged and intersected sets, respectively. Finally, they achieved better results in the attack classes. Selvakumar et al. [34] combined filtered and wrapped methods on the KDD Cup99 dataset and finally used the selected 10 features to improve the detection performance while reducing the computation time. Kshirsagar et al. [35] combined information gain and correlation by selecting 0, 0.25, and 0.5 thresholds, union features greater than 0.5, intersecting features between 0.25 and 0.5, removing features less than 0.25, and finally merging to obtain the selected subset. Krishnaveni et al. [21] used a univariate integrated feature selection technique combined with the majority voting for validation on the NSL-KDD dataset, based on 10–100% of the different proportions selected for the experiment, and the accuracy and stability were improved. Osanaiye et al. [20] used a multi-filtered approach to obtain different feature sequences and then selected the features in the top third of the different approaches. Finally, based on the number of occurrences of the features, a subset of features with at least three occurrences was obtained. Muthamil et al. [36] first used multiple methods for selection and then selected the features that were selected no less than three times as the input to the later classifier. Bhatia et al. [37] selected the top one-third of features from different methods on the NSL-KDD dataset and then combined them. Finally, the number of features was then reduced to 10 by determining the best features subset according to the ant colony algorithm. Although the above studies clarify the criteria for selecting subsets, they are not yet uniform, so metrics are needed to measure subset sizes comprehensively.

The ensemble approach achieves better performance by integrating the results of multiple feature selection methods. Binbusayyis et al. [21] first selected the respective subset of features using four filtered methods and then combined the features that were selected more than three times to form a new subset, achieving an experimental accuracy of 95.87% on the UNSW-NB15 dataset and 99.88% on the CIC-IDS2017 dataset. Femi et al. [38] combined the CfsSubsetEval correlation model, the genetic algorithm-based wrapper model, and the rule-based model serially. The final selected subset was input to an artificial neural network (ANN) for classification. On the UNSW-NB15 dataset, the accuracy was 98.8%. Karna et al. [39] used mutual information, Chi-square, and Pearson correlation coefficients to combine a more stable subset of features. It achieved 99.16% accuracy on the CIC-IDS2017 dataset using 25 features. Shubhra et al. [40] used the maximum correlation–minimum redundancy algorithm (mRMR), joint mutual information (JMI), and maximization of mutual information (CMIM) to select the top features and then combined them into a new feature subset using frequency voting on the CIC-IDS2017 dataset with an accuracy of 99.25%. Leevy et al. [41] evaluated seven methods and finally selected features that appeared more than twice in these seven subsets, achieving a classification F1 score of 95.68% using LightGBM on the CSE-CIC-IDS2018 dataset. Ensemble feature selection methods are all used to achieve stronger and more stable classification results by integrating various feature selection approaches' outcomes, where different integration strategies can also have an impact on the classification results.

3. Materials and Methods

To increase the intrusion detection system's identification capability and efficiency, an ensemble automatic feature selection method (EAFS) is designed in this paper. The importance or ranking of each feature is initially calculated using various base approaches in feature selection, and then the features are added to the subset sequentially according to their importance. An NSOM is designed to evaluate the subset performance comprehensively. After obtaining a subset for each method, the final feature subset is obtained by ensembling the methods. Figure 1 illustrates the designed intrusion detection framework, which consists of four main components.

- Data pre-processing: converting the raw data into the data format required by the classifier, as well as performing a series of operations such as numerical and normalization of the data;
- Feature selection: to remove irrelevant and redundant features and improve model performance, we implement an ensemble automatic feature selection method;
- Classifier training: to test the efficiency of the proposed strategy, RF and DT classifiers are trained using selected subsets;
- Attack identification: to verify the effectiveness and generalization of the method, we test it on multiple intrusion detection public datasets using a 5-fold cross-validation method.

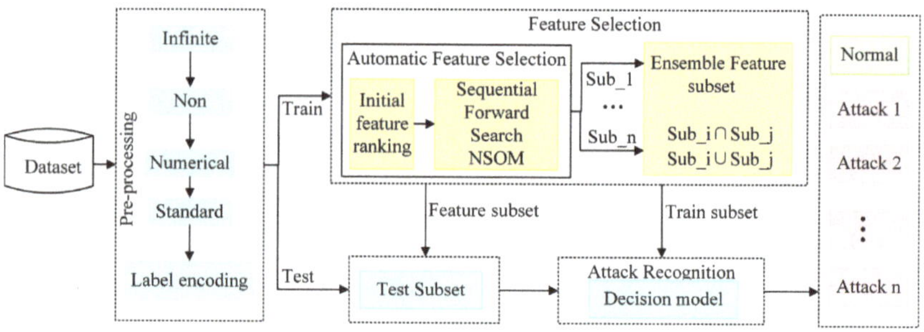

Figure 1. Feature selection framework diagram.

3.1. Benchmark Dataset

The UNSW-NB15 dataset [42] is a collection of network traffic from the University of New South Wales in Australia, and it contains a relatively complete set of normal activities and nine types of attack activities. The dataset has a total of 47 feature variables, and the last two columns are binary and multi-categorical labels, respectively.

The CIC-IDS2017 dataset [43] was collected by the Canadian Institute for Cybersecurity Research at the end of 2017. The dataset includes the most recent cyberattacks, close to real-world data. A total of seven types of attacks are implemented. The dataset has a total of 2,830,743 records, each containing 78 different features.

The CSE-CIC-IDS2018 dataset [43] is collected on a larger network than the IDS2017 dataset, with a sample size of 10 million, and contains a more comprehensive set of attack types. It contains seven attack scenarios. A total of 83 statistical features are generated, such as duration, the number of packets, bytes, packet length, etc. Table 1 shows the distribution of the three datasets mentioned above.

Table 1. Statistics of UNSW-NB15, CIC-IDS2017, and CSE-CIC-IDS-2018 datasets.

UNSW-NB15		CIC-IDS2017		CSE-CIC-IDS-2018	
Class	**Numbers**	**Class**	**Numbers**	**Class**	**Numbers**
Benign	2,218,764	Benign	2,273,097	Benign	13,484,708
Exploits	44,525	DDOS	128,027	DDOS	1,263,933
Reconnaissance	13,987	DOS	252,661	DOS	654,300
DoS	16,353	Bot	1966	Bot	286,191
Generic	215,481	Patator	13,835	Brute Force	380,949
Shellcode	1511	Heartbleed	11	Infiltration	161,934
Fuzzers	24,246	Infiltration	36	Web Attack	928
Worms	174	PortScan	158,930		
Backdoors	2329	Web Attack	2180		
Analysis	2677				
Total	2,540,047	Total	2,830,743	Total	16,232,943

3.2. Data Pre-Processing

Numeric and character-based features are included in the intrusion detection dataset, and some features contain infinite and missing values. Because most machine learning and deep learning methods can only work with numerical data, it is necessary to convert non-numerical features in the dataset to numerical features so that they can be analyzed and processed efficiently using methods such as machine learning [44].

For the three features of "proto", "state", and "service" in the UNSW-NB15 dataset, the category features are sequential integers. The timestamp data in the CSE-CIC-IDS2018 dataset is removed.

Meanwhile, noise and duplicate data exist in the dataset. In the CIC-IDS-2017 dataset, there are two identical columns, Fwd Header Length and Fwd Header Length.1, and this paper removes the column feature of Header Length.1. For the CSE-CIC-IDS-2018 dataset, there are 4 features in the file "Thuesday-20-02-2018_TrafficForML_CICFlowMeter.csv", which are Flow ID, Src IP, Src Port, and Dst IP, but there are no such features in other files, so we will delete them for consistency.

Observing the dataset reveals that there are missing values and infinity values in the file, which will directly affect the subsequent data processing. In this paper, missing values are filled with 0, while infinity values are substituted with the column's maximum value plus one.

It also uses standardization techniques to standardize the data to a uniform range and accelerate the data convergence [45]. As shown in Equation (1), the feature values are scaled to a distribution with a mean of 0 and a standard deviation of 1 using this method.

$$x' = \frac{x - \mu}{\sigma}, \quad (1)$$

where x is the initial value, x' is the standardized value, and μ and σ are the mean and standard deviation, respectively.

3.3. Feature Selection

In this paper, we compare the Pearson correlation coefficient (PCC), mutual information (MI), mRMR, ET, and XGBoost algorithms, respectively. Among them, Pearson correlation coefficient, MI, and mRMR are filtered methods, which can quickly obtain the importance of features by calculating the relationship between features and labels; ET and XGBoost methods are embedded methods, which depend on classification algorithms to quickly evaluate the performance of subsets. Filtered and embedded methods can reduce computational costs when performing feature selection and are more efficient when extended to high-dimensional datasets [16–21].

In the designed model, firstly, it calculates the variance of each feature, removes those with zero variance as they are not beneficial for classification [27,46], and keeps those with relatively high variance for the next step of feature selection. Then, it calculates the feature importance for each method and saves them in the list in descending order of importance. Thirdly, the prediction accuracy ACC for all features is calculated, which is used to limit the size of the subset; this is the reason that if the accuracy of a subset is worse than the accuracy using all features, then its performance is unlikely to be optimal. In the fourth step, based on the ranked list, features are added to the subset sequentially using the forward floating search method, starting with the most important features, and the performance of each subset is evaluated until ACC is reached. The NSOM is meant to evaluate subset performance by integrating three independent sub-objectives: subset accuracy, subset number of features, and subset training time. Finally, the target subset is determined by having the highest NSMO. Algorithm 1 uses a single feature selection approach to calculate each characteristic's importance and rank them in descending order. The automatic feature selection method's pseudo-code is Algorithm 2.

Algorithm 1 Feature ranking algorithm

Input:The training set $F = \{f_i, i = 1,2,\ldots,I\}$, i is the number of features;Feature selection method FS
Output:The ranked subset of features F_{rank}
for $i = 1$ to I **do**
    Var=VarianceThreshold(F)
    if $F_i = 0$ **then**
        F' = Remove F_i from F
    end if
end for
Initialize FS Method
Fit $FS(F')$
Get F_{imp} # Get the important of each feature
$F_{rank} = sorted(F_{imp})$
return F_{rank}

Algorithm 2 Automatic feature selection algorithm

Input:F_{rank},Feature ranked list;N is total numbers of features;ACC is original accuracy.
Output:Feature subset F
for F_i in F_rank **do**
    if fs_{acc}<ACC **then**
        add F_i into $F_{selected}$
        evaluate $F_{selected}$ get acc_i,$time_i$,num_i
        add acc_i to fs_{acc}
        add $time_i$ to fs_{time}
        add num_i to fs_{num}
    end if
end for
output fs_{acc},fs_{time} and fs_{num}
acc_N=normalization(fs_{acc})
$time_N$=normalization(fs_{time})
num_N=normalization(fs_{num})
$score_{temp}$=0
for $i = 1$ in range(len(fs_{acc})) **do**
    according Equation (2), calculate $score_i$
    if $score_i > score_{temp}$ **then**
        $score_{temp} = score_i$
        F=$F_{rank}[: i]$
    end if
end for
return F

In evaluating the subset performance, this paper does not use the commonly used evaluation functions because they tend to measure only a single metric. To consider the performance of subsets comprehensively, this paper designs a normalized score of mixed, which contains the accuracy of a subset's classification, the number of features in a subset, and the training time of a subset. Firstly, classification accuracy is a metric that is available for every evaluation function. In addition, feature selection's objective is to use as few features as feasible to improve classification performance, so the number of features is included in the evaluation function in this study, hoping to reduce the number of features in the subset. For classifiers, facing a large amount of data often consumes a lot of time to train, so while reducing the number of features, this paper considers the training time of the classifier to reduce the time cost. The NSOM designed in this paper is shown in Equation (2).

$$\text{NSOM} = FS\_acc_N - \alpha \cdot FS\_num_N - \beta \cdot FS\_time_N, \tag{2}$$

where α and β are the weight coefficients between [0,1], respectively; FS_acc_N, FS_num_N, and FS_time_N are the normalized subset accuracy, the number of features contained in the subset, and the training time, respectively. After normalizing NSOM, it takes values between [0,1]. The closer the NSOM is to 1, the greater the subset's performance.

The optimal subset of features for each method is obtained after processing by Algorithm 2. However, there is a certain bias in the importance of individual methods for different features, which affects the final classification effect. The subsets derived from multiple feature selection approaches are combined in this article using an ensemble strategy to eliminate this bias. Given F as a dataset, each sample in F has m features and a category label. Two feature selection meta-approaches are chosen after assessing the performance of several methods, and let the sequence obtained by the ith method for the features in order of importance be $F_i = (a_{i1}, a_{i2}, \ldots, a_{im})$, then for the intersection of two feature subsets, defined as follows: $F(inter) = F1 \cap F2$, for the union of two feature subsets is defined as $F(union) = F1 \cup F2$. By combining the feature subsets, *F*(*inter*) contains some important features commonly recognized by both methods, and *F*(*union*) contains features that can be more effective by combining, and the effectiveness of the methods is verified on decision trees and random forests classification algorithms.

4. Results and Analysis

4.1. Experimental Environment and Setup

The algorithm process designed in this paper is based on a Python implementation using the scikit-learn toolkit to simulate experiments on a machine with an Intel(R) Core™ i9-9900KF CPU at 3.6GHz. The UNSW-NB15, CIC-IDS2017, and CSE-CIC-IDS2018 datasets were used to test the model's performance. The number of samples in the dataset used in the experiments ranges from millions to tens of millions, covering different kinds of attacks, which are in line with modern real-network scenarios and can be a good evaluation of an IDS performance. All experimental results in this work are acquired using 5-fold cross-validation to avoid the impact of randomness on the results.

4.2. Evaluation Metrics

The evaluation metrics used in this paper include accuracy (Acc), detection rate (DR), F1 score, and false alarm rate (FAR). The metrics are calculated as shown in Equations (3)–(6).

$$\text{Acc} = \frac{TP + TN}{TP + TN + FP + FN}, \tag{3}$$

$$\text{Precision} = \frac{TP}{TP + FP}, \tag{4}$$

$$\text{DR} = Recall = \frac{TP}{TP + FN}, \tag{5}$$

$$\text{F1} = 2 * \frac{\text{Precision} * \text{Recall}}{\text{Precision} + \text{Recall}'} \tag{6}$$

$$\text{FAR} = \frac{FP}{FP + TN'} \tag{7}$$

4.3. Automatic Feature Selection Performance Analysis

This study compares the suggested feature selection method against one that does not include feature selection to assess its performance. Table 2 shows the performance comparison on the UNSW-NB15, CIC-IDS2017, and CSE-CICIDS2018 datasets. On the UNSW-NB15 dataset, the proposed method achieves 98.37% accuracy and a 98.31% F1 score, and the training time is reduced by more than half. When compared to no feature selection on the CIC-IDS2017 dataset, the proposed feature selection strategy will perform better, while the time after feature selection is greatly reduced, especially on the DT classifier, which is more than three times. The false alarm rate on the RF classifier is reduced to 0.085%. Moreover, based on the CSE-CIC-IDS2018 dataset, the accuracy after feature selection reaches 99.04%, especially in terms of training time, which is greatly reduced compared to using all features. It is also evident that the designed feature selection approach is critical for increasing the efficiency of large data detection.

Table 2. Classification performance based on the different dataset.

Classifier	Acc(%)	DR(%)	F1(%)	FAR(%)	Training Time(s)	Predicted Time(s)
Comparison of classification performance on the UNSW-NB15 dataset						
RF	98.31	98.31	98.24	1.70	42	1.78
DT	98.19	98.19	98.14	1.83	14	0.09
RF	98.37	98.37	98.31	1.65	19	1.55
DT	98.30	98.30	98.25	1.72	6	0.09
Comparison of classification performance on the CIC-IDS2017 dataset						
RF	99.90	99.90	99.90	0.10	139	2.21
DT	99.87	99.87	99.87	0.13	107	0.13
RF	99.92	99.92	99.91	0.09	79	1.80
DT	99.88	99.88	99.88	0.12	24	0.07
Comparison of classification performance on the CSE-CIC-IDS2018 dataset						
RF	98.93	98.93	98.77	1.07	1596	21
DT	98.73	98.73	98.69	1.27	1127	0.83
RF	99.04	99.04	98.83	0.96	243	12
DT	99.03	99.03	98.83	0.97	40	0.30

In previous problems of selecting the number of subsets for features, in the face of filtered and embedded methods, many used artificially set ratios to select features or those based on fixed thresholds [18,20,33–37]. Such methods are significantly subjective and require empirical knowledge, and they do not adequately consider the applicability to different scenarios. The top 10% and 30% of features were chosen for comparison with the regularly used selection ratios to demonstrate the efficiency of the method presented in this research. Figure 2 compares the experimental results of the three datasets under different proportions.

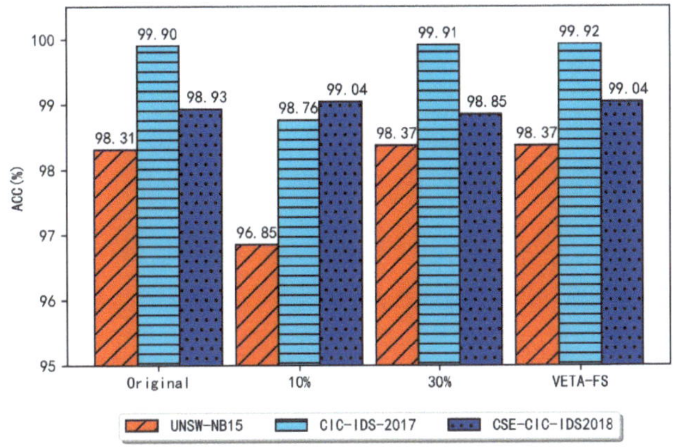

Figure 2. Performance comparison with other selection strategies.

As can be seen from Figure 2, for the UNSW-NB15 and CIC-IDS2017 datasets, the performance of the classifier is poor when the proportion reaches 10%, and the accuracy is even lower compared to the original features, only 96.85% and 98.76%, respectively. This is because when only the top 10% of the features are selected, it is not enough to retain the most critical features in the original dataset, and a part of the useful information is lost, leading to the degradation of the classifier performance. In contrast, for the CSE-CIC-IDS2018 dataset, the best performance has been achieved when 10% of the features are selected, which is consistent with the feature subset selected by the method designed in this paper, and both achieved the best performance with 99.04% accuracy. When the first 30% of the subset is selected, the UNSW-NB15 and CIC-IDS2017 datasets are the same as the features selected by the method designed in this paper, and the performance also remains consistent with an accuracy of 98.37% and 99.91%, respectively. For the CSE-CIC-IDS2018 dataset, the effect is significantly lower when the first 30% of the features are selected, which is because, for the CSE- CIC-IDS2018 dataset, the best feature set can be selected quickly using the method designed in this paper, and when further features are added, it is equivalent to adding redundant information, resulting in a lower classification performance.

4.4. Ensemble Feature Selection Performance Comparison

In this section, to select the method with a better performance for the later ensemble feature selection method, several commonly used feature selection methods are compared, including five methods such as the Pearson correlation coefficient, mutual information, mRMR, ET method, and XGBoost. Figures 3 and 4 reveal the obtained experimental results.

The above comparison reveals that for the UNSW-NB15 dataset, the best two of these methods are the ET algorithm and the mRMR method, respectively. For the CIC-IDS2017 dataset, the XGBoost and ET algorithms achieve better results and lower false alarm rates, so these two methods are chosen as the meta-methods for feature subset integration on the CIC-IDS2017 dataset later in this paper. For the CSE-CIC-IDS2018 dataset, it is clear that the subsets obtained using the two methods, the ET algorithm and mRMR algorithm, respectively, achieved the best classification performance and maintained a low false alarm rate. After identifying the two methods with the best performance on each dataset, this paper combines the subsets obtained by these two methods and compares the performance when taking the intersection and the union, respectively.

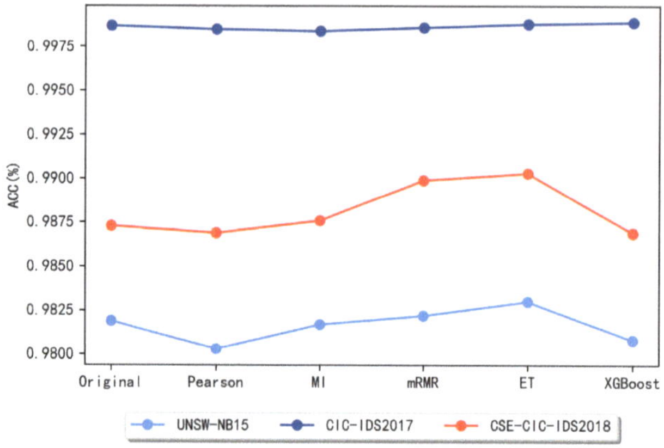

Figure 3. Comparison of accuracy of different methods.

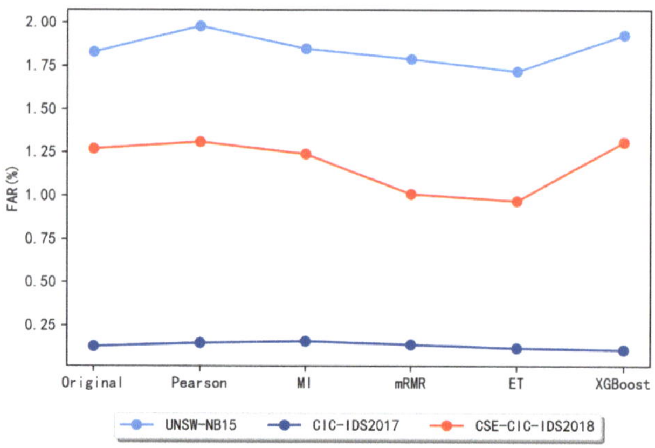

Figure 4. Comparison of false alarm rate of different methods.

Figures 5 and 6 compare the performance of the different combined approaches for each dataset on the RF and DT classifiers, respectively. In Figure 5, the UNSW-NB15 dataset achieves a classification accuracy of 98.41% and all other metrics are better than the original dataset. The CIC-IDS2017 dataset shows a slightly higher accuracy, detection rate, and F1 score when compared to the dataset without feature selection, and the false alarm rate decreases to 0.09%. On the CSE-CIC-IDS2018 dataset, the ensemble approach achieved better results, especially the false alarm rate was reduced to less than 1%, reaching 0.96%. In Figure 6, for UNSW-NB15, the highest accuracy of 99.34% was achieved using the ensemble feature subset and the false alarm rate was reduced to 1.69% compared to the original dataset. Compared with these two separate methods, the accuracy was improved by 0.12%. On the CIC-IDS2017 dataset, the ensemble method maintained a similar classification performance compared to the original dataset. On the CSE-CIC-IDS2018 dataset, the accuracy of the ensemble feature subset reached 99.03%, and the detection rate was 99.03%, F1 score was 98.82%, and the false alarm rate was 0.97%, respectively, compared to the accuracy of 98.73% on the original dataset.

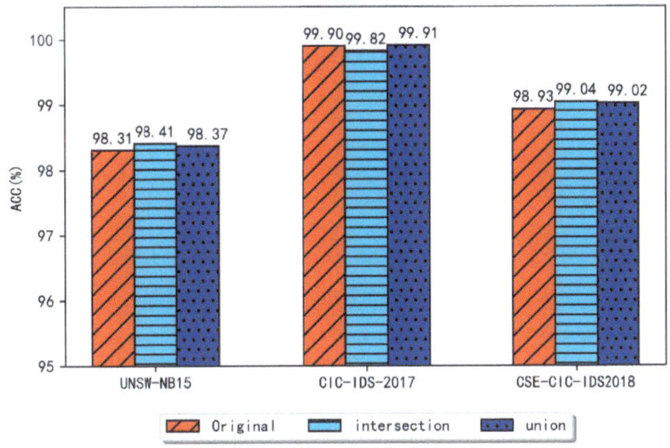

Figure 5. Using an RF classifier to compare the accuracy of several datasets.

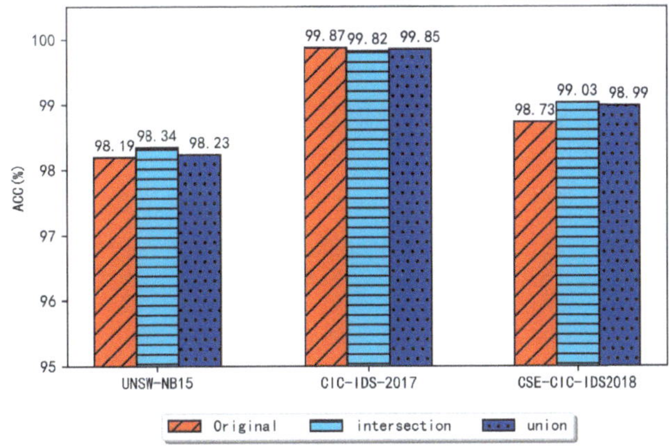

Figure 6. Using a DT classifier to compare the accuracy of several datasets.

5. Discussion

We compare the effectiveness of the approach described in this research to earlier studies to demonstrate its efficacy on the UNSW-NB15, CIC-IDS2017, and CSE-CIC-IDS2018 datasets. As the results show, we further demonstrate the effectiveness of the approach in selecting the most useful information for intrusion detection in this paper. In Table 3, this paper is compared with the latest research on the UNSW-NB15, CIC-IDS2017, and CSE-CIC-IDS2018 datasets, and the comparison items shown include the feature selection and classification techniques used, classification accuracy, and false alarm rate. For the UNSW-NB15 dataset, this paper is compared to swarm intelligence search-based methods, including GA [27], TS [29], and LSO [31], respectively, and the method designed in this paper achieves the highest accuracy and the lowest rate of false alarms. In comparison with ensemble strategies such as SCM3 [21] and RHF [38], the method in this paper also achieves the highest accuracy, and although the false alarm rate of the RHF method is reduced to 1.3%, our method works better in terms of accuracy. Overall, the method in this paper has more advantages.

Table 3. Comparison of results on multiple datasets with the latest(%).

Dataset	Reference	Technology	Acc	DR	F1	FAR
UNSW-NB15	Khammassi [27]	GA-LR	81.42	-	-	6.39
	Nazir [29]	TS-RF	83.12	-	-	3.70
	Benmessahel [31]	LSO-FNN	95.42	99.33	95.98	9.40
	Binbusayyis [21]	SCM3-RF	95.87	97.40	-	7.40
	Femi [38]	RHF-ANN	97.60	-	97.70	1.30
	This work	EAFS-RF	98.36	98.36	98.31	1.65
CIC-IDS2017	Bansal [24]	XGBoost	99.54	-	-	0.237
	Vijayanand [32]	GA-SVM	99.85	-	-	0.09
	Zhou [28]	CFS-BA	99.89	99.90	-	0.12
	Binbusayyis [21]	SCM3-RF	99.88	99.90	-	0.20
	Karna [39]	CPM-voting	99.16	-	99.17	-
	Shubhra [40]	EFW-SVM	99.25	99.52	-	0.007
	This work	EAFS- RF	99.92	99.92	99.92	0.08
CSE-CIC-IDS2018	Fitni [25]	spearman	98.80	97.10	97.90	-
	Lin [26]	LSTM-Attention	96.20	96.00	93.00	-
	Leevy [41]	vote-LightGBM	-	-	95.88	-
	This work	EAFS- RF	99.04	99.04	98.83	0.96

For the CIC-IDS2017 dataset, the method in this paper achieved 99.92%, 99.92%, and 0.08% for Acc, F1, and FAR, respectively. When compared to the classification results obtained without the use of feature selection approaches, the accuracy of the classification using the XGBoost [24] classifier is 99.54%, but the false alarm rate is 0.237%, which is nearly three times higher than the false alarm rate of this paper. By comparing the swarm intelligence search algorithms such as GA [32] and CFS [28], the method in this paper achieves a higher accuracy and higher false alarm rate than these two methods. In comparison with the ensemble feature selection strategies such as SCM3 [21], CPM [39], and EFW [40], these three methods achieve 99.88%, 99.16%, and 99.25% accuracy, respectively, while the accuracy of this paper reaches 99.92%, relative to the F1 value of 99.17% achieved by the CPM method, which achieves an F1 value of 99.92%.

For the CSE-CIC-IDS2018 dataset, the method in this paper achieved 99.14%, 99.14%, and 98.85% for Acc, DR, and F1, respectively. In the comparison with the classification results without feature selection, the accuracy of the classification using LSTM [26] reached 96.2% and the detection rate was 96%. The classification accuracy, detection rate, and F1 values were 98.8%, 97.1%, and 97.9%, respectively, when using the feature selection method with the Spearman correlation coefficient [25]. The classification F1 value was 95.68% when the feature selection results were ensembled using the voting [41] integration strategy. As seen in the table, the method proposed in this paper was able to achieve better results than previous studies in these metrics that were compared.

Our proposed ensemble-based automatic feature selection method improves the classification performance while minimizing the training and prediction time, according to the experimental data. The main reason for the better results of the method designed in this paper is that, firstly, when selecting the best features for each dataset, the search range of the subsets is reduced by the accuracy based on the full set, avoiding a large amount of time consumption for a large-scale search. Secondly, the automatic feature selection method designed in this paper can measure the performance advantages and disadvantages of different subsets and integrate multiple metrics to evaluate the performance of subsets. Third, the ensemble feature selection strategy takes into account the bias of different feature selection methods for the same feature and obtains a subset with more general significance. The above three reasons enable the method designed in this paper to select the best subset and perform the classification prediction quickly and accurately in large-scale data scenarios.

6. Conclusions

For intrusion detection systems where the number of acquired traffic features is large and redundant and irrelevant features tend to have a serious impact on the results, this paper designs an ensemble-based automatic feature selection method, EAFS. Firstly, features with zero variance are removed using a variance threshold. Next, the importance or ranking of each feature is obtained using a separate feature selection method, and the ranked features are gradually added to the subset selected. Then, the subset performance is evaluated using the designed NSOM scores and the subset with the highest NSOM value is taken as the final subset. Finally, the two methods with the best performance on each dataset are obtained and the results are ensembled to reduce the bias of different methods. Experiments are implemented on the commonly used intrusion detection datasets, UNSW-NB15, CIC-IDS2017, and CSE-CIC-IDS2018, and after feature selection, the final classification performance is improved compared to the original feature set. On the UNSW-NB15 dataset, the classification accuracy and false alarm rate are 98.36% and 1.65%, respectively. On the CIC-IDS2017 dataset, the classification accuracy, F1 value, and false alarm rate are 99.92%, 99.92%, and 0.08%, respectively. On the CSE-CIC-IDS2018 dataset, the classification accuracy, detection rate, and F1 value are 99.04%, 99.04%, and 98.83%. The performance of our method also achieved good results in comparison with other works of literature. In future research, we will continue to study automatic feature selection and extend it to more selection methods, and we will also investigate the effectiveness of lightweight classification models in time.

Author Contributions: Conceptualization, H.Z.; investigation, Y.Z. and H.Z.; methodology, Y.Z.; resources, H.Z.; supervision, H.Z.; validation, Y.Z. and B.Z.; writing—original draft, Y.Z.; writing—review and editing, H.Z. and B.Z. All authors have read and agreed to the published version of the manuscript.

Funding: This research was partly supported by the Key R&D and promotion projects of Henan Province (Technological research) (Grant No. 212102210143).

Institutional Review Board Statement: Ethical review and approval are not required because all datasets used in this study are in the public domain.

Informed Consent Statement: Not applicable.

Data Availability Statement: The UNSW-NB15 dataset used to support the findings of this study is available at https://www.unsw.adfa.edu.au/unsw-canberra-cyber/cybersecurity/ADFA-NB15-Datasets/ (accessed on 26 June 2022). The CICIDS2017 dataset used to support the findings of this study is available at https://www.unb.ca/cic/datasets/ids-2017.html (accessed on 26 June 2022). The CSE-CIC-IDS2018 dataset used to support the findings of this study is available at https://www.unb.ca/cic/datasets/ids-2018.html (accessed on 26 June 2022).

Conflicts of Interest: The authors declare no conflict of interest.

References

1. Verkerken, M.; D'hooge, L.; Wauters, T.; Volckaert, B.; Turck, D. Towards model generalization for intrusion detection: Unsupervised machine learning techniques. *J. Netw. Syst. Manag.* **2021**, *30*, 12. [CrossRef]
2. Habeeb, R.A.A.; Nasaruddin, F.; Gani, A.; Hashem, I.A.T.; Ahmed, E.; Imran, M. Real-time big data processing for anomaly detection: A Survey. *Int. J. Inf. Manag.* **2019**, *45*, 289–307. [CrossRef]
3. Kabir, M.; Hartmann, S. Cyber security challenges: An efficient intrusion detection system design. In Proceedings of the 2018 International Young Engineers Forum (YEF-ECE), Costa da Caparica, Portugal, 4 May 2018.
4. Buczak, A.; Guven, E. A survey of data mining and machine learning methods for cyber security intrusion detection. *IEEE Commun. Surv. Tutor.* **2015**, *18*, 1153–1176. [CrossRef]
5. Aljawarneh, S.; Aldwairi, M.; Yassein, M.B. Anomaly-based intrusion detection system through feature selection analysis and building hybrid efficient model. *J. Comput. Sci.* **2018**, *25*, 152–160. [CrossRef]
6. Maldonado, J.; Riff, M.C.; Neveu, B. A review of recent approaches on wrapper feature selection for intrusion detection. *Expert Syst. Appl.* **2022**, *198*, 116822. [CrossRef]
7. Fernandes, G.; Rodrigues, J.; Carvalho, L.; Al-Muhtadi, J.; Proença, M. A comprehensive survey on network anomaly detection. *Telecommun. Syst.* **2019**, *70*. 447–489. [CrossRef]

8. Thakkar, A.; Lohiya, R. A survey on intrusion detection system: Feature selection, model, performance measures, application perspective, challenges, and future research directions. *Artif. Intell. Rev.* **2021**, *55*, 453–563. [CrossRef]
9. Wei, G.; Zhao, J.; Feng, Y.; He, A.; Yu, J. A novel hybrid feature selection method based on dynamic feature importance. *Appl. Soft. Comput.* **2020**, *93*, 106337. [CrossRef]
10. Di Mauro, M.; Galatro, G.; Fortino, G.; Liotta, A. Supervised feature selection techniques in network intrusion detection: A critical review. *Eng. Appl. Artif. Intell.* **2021**, 101, 104216. [CrossRef]
11. Solorio-Fernández, S.; Carrasco-Ochoa, J.A.; Martínez-Trinidad, J.F. A review of unsupervised feature selection methods. *Artif. Intell. Rev.* **2019**, *53*, 907–948. [CrossRef]
12. Kasongo, S.M.; Sun, Y. A deep learning method with wrapper based feature extraction for wireless intrusion detection system. *Comput. Secur* **2020**, *92*, 101752. [CrossRef]
13. Alaiz-Rodriguez, R.; Parnell, A.C. An information theoretic approach to quantify the stability of feature selection and ranking algorithms. *Knowl.-Based Syst* **2020**, *195*, 105745. [CrossRef]
14. Thakkar, A.; Lohiya, R. Attack classification using feature selection techniques: A comparative study. *J. Amb. Intell. Hum. Comp.* **2020**, *12*, 1249–1266. [CrossRef]
15. Lu, M. Embedded feature selection accounting for unknown data heterogeneity. *Expert Syst. Appl.* **2019**, *119*, 350–361. [CrossRef]
16. Tsai, C.F.; Sung, Y.T. Ensemble feature selection in high dimension, low sample size datasets: Parallel and serial combination approaches. *Knowl.-Based Syst.* **2020**, *203*, 106097. [CrossRef]
17. Shukla, A.K.; Singh, P. Building an effective approach toward intrusion detection using ensemble feature selection. *Int. Inf. Secur. Priv.* **2019**, *13*, 31–47. [CrossRef]
18. Krishnaveni, S.; Sivamohan, S.; Sridhar, S.S.; Prabakaran, S. Efficient feature selection and classification through ensemble method for network intrusion detection on cloud computing. *Clust. Comput.* **2021**, *24*, 1761–1779. [CrossRef]
19. Bolón-Canedo, V.; Alonso-Betanzos, A. Ensembles for feature selection: A review and future trends. *Inf. Fusion* **2019**, *52*, 1–12. [CrossRef]
20. Osanaiye, O.; Cai, H.; Choo, K.R.; Dehghantanha, A.; Xu, Z.; Dlodlo, M. Ensemble-based multi-filter feature selection method for DDoS detection in cloud computing. *EURASIP J. Wirel. Commun. Netw.* **2016**, *2016*, 130. [CrossRef]
21. Binbusayyis, A.; Vaiyapuri, T. Identifying and benchmarking key features for cyber intrusion detection: An ensemble approach. *IEEE Access* **2019**, *7*, 106495–106513. [CrossRef]
22. Odhiambo Omuya, E.; Onyango Okeyo, G.; Waema Kimwele, M. Feature selection for classification using principal component analysis and information gain. *Expert Syst. Appl.* **2021**, *174*, 114765. [CrossRef]
23. Guyon, I.; Elisseeff, A. An introduction to variable and feature selection. *J. Mach. Learn. Res.* **2003**, *3*, 1157–1182.
24. Bansal, A.; Kaur, S. Extreme gradient boosting based tuning for classification in intrusion detection systems. In *Advances in Computing and Data Sciences*; Singh, M., Gupta, P., Tyagi, V., Flusser, J., Ören, T., Eds.; Springer: Singapore, 2018; pp. 372–380.
25. Fitni, Q.R.S.; Ramli, K. Implementation of ensemble learning and feature selection for performance improvements in anomaly-based intrusion detection systems. In Proceedings of the 2020 IEEE International Conference on Industry 4.0, Artificial Intelligence, and Communications Technology (IAICT), Bali, Indonesia, 7–8 July 2020.
26. Bansal, A.; Kaur, S. Dynamic betwork anomaly detection system by using deep learning techniques. In *Advances in Computing and Data Sciences*; Da Silva, D., Wang, Q., Zhang, LJ., Eds.; Springer: Cham, Switzerland, 2019; pp. 161–176.
27. Khammassi, C.; Krichen, S. A GA-LR wrapper approach for feature selection in network intrusion detection. *Comput. Secur.* **2017**, *70*, 255–277. [CrossRef]
28. Zhou, Y.; Cheng, G.; Jiang, S.; Dai, M. Building an efficient intrusion detection system based on feature selection and ensemble classifier. *Comput. Netw.* **2020**, *174*, 107247. [CrossRef]
29. Nazir, A.; Khan, R.A. A novel combinatorial optimization based feature selection method for network intrusion detection. *Comput. Secur.* **2021**, *102*, 102164. [CrossRef]
30. Farahani, G. Feature selection based on cross-correlation for the intrusion detection system. *Secur. Commun. Netw.* **2020**, *2020*, 8875404. [CrossRef]
31. Benmessahel, I.; Xie, K.; Chellal, M.; Semong, T. A new evolutionary neural networks based on intrusion detection systems using locust swarm optimization. *Evol. Intell.* **2019**, *12*, 131–146. [CrossRef]
32. Vijayanand, R.; Devaraj, D.; Kannapiran, B. Intrusion detection system for wireless mesh network using multiple support vector machine classifiers with genetic-algorithm-based feature selection. *Comput. Secur.* **2018**, *77*, 304–314. [CrossRef]
33. Akashdeep; Manzoor, I.; Kumar, N. A feature reduced intrusion detection system using ANN classifier. *Expert Syst. Appl.* **2017**, *88*, 249–257. [CrossRef]
34. Selvakumar, B.; Muneeswaran, K. Firefly algorithm based feature selection for network intrusion detection. *Comput. Secur.* **2019**, *81*, 148–155.
35. Kshirsagar, D.; Kumar, S. A feature reduction based reflected and exploited DDoS attacks detection system. *J. Ambient Intell. Hum. Comput.* **2022**, *13*, 393–405. [CrossRef]
36. Muthamil Sudar, K.; Deepalakshmi, P. An intelligent flow-based and signature-based IDS for SDNs using ensemble feature selection and a multi-layer machine learning-based classifier. *J. Intell. Fuzzy Syst.* **2021**, *40*, 4237–4256. [CrossRef]
37. Bhatia, M.P.S.; Sangwan, S.R. Soft computing for anomaly detection and prediction to mitigate IoT-based real-time abuse. *Pers. Ubiquit. Comput.* **2021**, 1–11. [CrossRef]

38. Ayo, F.; Folorunso, S.; Adebayo, A.-A.; Adekunle, A.; Bamidele, A. Network intrusion detection based on deep learning model optimized with rule-based hybrid feature selection. *Inf. Secur. J. Glob. Perspect.* **2021**, *29*, 267–283. [CrossRef]
39. Karna, I.; Madam, A.; Deokule, C.; Adhao, R.; Pachghare, V. Ensemble-based filter feature selection technique for building flow-based IDS. In Proceedings of the 2021 2nd International Conference on Advances in Computing, Communication, Embedded and Secure Systems (ACCESS), Ernakulam, India, 2–4 September 2021.
40. Dwivedi, S.; Vardhan, M.; Tripathi, S. An effect of chaos grasshopper optimization algorithm for protection of network infrastructure. *Comput. Netw.* **2020**, *176*. 107251. [CrossRef]
41. Leevy, J.L.; Hancock, J.; Zuech, R.; Khoshgoftaar, T.M. Detecting cybersecurity attacks across different network features and learners. *J. Big Data* **2021**, *8*, 38. [CrossRef]
42. Moustafa, N.; Slay, J. UNSW-NB15: A comprehensive data set for network intrusion detection systems. In Proceedings of the 2015 Military Communications and Information Systems Conference (MilCIS), Canberra, ACT, Australia, 10–12 November 2015.
43. Ghorbani, A.A.; Habibi Lashkari, A.; Sharafaldin, I. Toward generating a new intrusion detection dataset and intrusion traffic characterization. In Proceedings of the 4th International Conference on Information Systems Security and Privacy (ICISSP), Madeira, Portugal, 22–24 January 2018.
44. Salo, F.; Nassif, A.B.; Essex, A. Dimensionality reduction with IG-PCA and ensemble classifier for network intrusion detection. *Comput. Netw.* **2019**, *148*, 164–175. [CrossRef]
45. Çavuşoğlu, Ü. A new hybrid approach for intrusion detection using machine learning methods. *Appl. Intell.* **2019**, *49*, 2735–2761. [CrossRef]
46. Vaca, F.D.; Niyaz, Q. An Ensemble Learning Based Wi-Fi Network Intrusion Detection System (WNIDS). In Proceedings of the 2018 IEEE 17th International Symposium on Network Computing and Applications (NCA), Cambridge, MA, USA, 1–3 November 2018.

remote sensing

Article

LGB-PHY: An Evaporation Duct Height Prediction Model Based on Physically Constrained LightGBM Algorithm

Xingyu Chai [1], Jincai Li [1,*], Jun Zhao [1], Wuxin Wang [2] and Xiaofeng Zhao [1]

1 College of Meteorology and Oceanography, National University of Defense Technology, Changsha 410073, China; chaixingyu1997@nudt.edu.cn (X.C.); zhaojun@nudt.edu.cn (J.Z.); zhaoxiaofeng@nudt.edu.cn (X.Z.)

2 College of Computer Science, National University of Defense Technology, Changsha 410073, China; wuxinwang@nudt.edu.cn

* Correspondence: lijincai@nudt.edu.cn

Abstract: The evaporation duct is a special atmospheric stratification that significantly influences the propagation path of electromagnetic waves at sea, and hence, it is crucial for the stability of the radio communication systems. Affected by physical parameters that are not universal, traditional evaporation duct theoretical models often have limited accuracy and poor generalization ability, e.g., the remote sensing method is limited by the inversion algorithm. The accuracy, generalization ability and scientific interpretability of the existing pure data-driven evaporation duct height prediction models still need to be improved. To address these issues, in this paper, we use the voyage observation data and propose the physically constrained LightGBM evaporation duct height prediction model (LGB-PHY). The proposed model integrates the Babin–Young–Carton (BYC) physical model into a custom loss function. Compared with the eXtreme Gradient Boosting (XGB) model, the LGB-PHY based on a 5-day voyage data set of the South China Sea provides significant improvement where the RMSE index is reduced by 68%, while the SCC index is improved by 6.5%. We further carried out a cross-comparison experiment of regional generalization and show that in the sea area with high latitude and strong adaptability of the BYC model, the LGB-PHY model has a stronger regional generalization performance than that of the XGB model.

Keywords: evaporation duct prediction model; Babin–Young–Carton (BYC) model; Paulus–Jeske model; eXtreme Gradient Boosting (XGB) model; LightGBM algorithm; machine learning with physical constraints

Citation: Chai, X.; Li, J.; Zhao, J.; Wang, W.; Zhao, X. LGB-PHY: An Evaporation Duct Height Prediction Model Based on Physically Constrained LightGBM Algorithm. *Remote Sens.* **2022**, *14*, 3448. https://doi.org/10.3390/rs14143448

Academic Editor: Gabriel Senay

Received: 12 May 2022
Accepted: 15 July 2022
Published: 18 July 2022

1. Introduction

The atmospheric duct is a stratification in the atmospheric boundary layer which significantly affects the propagation of electromagnetic waves. This is because the refractive index in the atmospheric boundary layer is sharply decreased with the increase in altitude. Therefore, the electromagnetic waves may abnormally bend downward, resulting in the bending radius of the waves becoming smaller than that of the earth. For some frequencies and angle conditions, the energy of radio waves will be refracted repeatedly in the atmospheric stratification. Hence, the atmospheric stratification acts as a conductor that enables low attenuation over-the-horizon propagation of electromagnetic waves. Nonetheless, this limits the transmission of electromagnetic waves only within the atmospheric duct. Figure 1 shows the radar blind zone caused by the atmosphere duct. Efficient use of the atmospheric duct not only improves radio communications between the ships at sea but also shields the communication equipment in the air from radio interference.

The atmospheric duct includes the evaporation duct, surface duct, and suspended duct. The evaporation duct is a special type of surface duct that is created by the sharp humidity reduction by increasing altitude. The evaporation duct frequently occurs in marine environments with an average occurrence probability higher than 60% in the global sea area. In the South China Sea, the probability of the available evaporation duct is as high as 80%, and in some areas, even permanent evaporation ducts exist [1]. Generally, the evaporation duct appears lower than 40 m above the sea surface, hence significantly affecting the efficiency of radio communication.

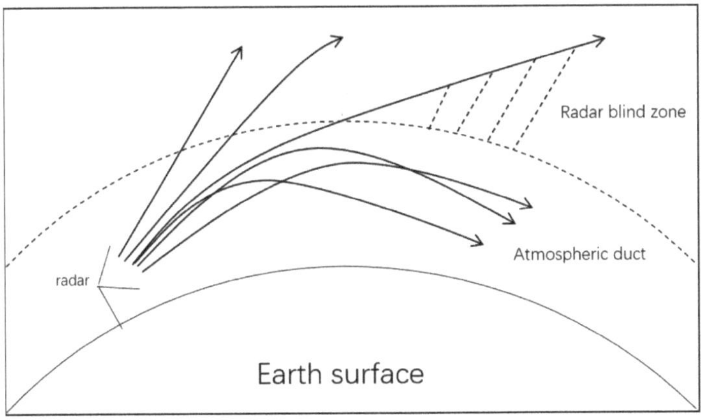

Figure 1. A radar blind zone caused by the atmospheric duct.

Evaporation duct height (EDH) directly reflects the lamination height of the evaporation duct and plays a crucial role in the strength of the evaporation duct [2]. Therefore, investigating the evaporation duct has attracted lots of attention in the research community.

The atmospheric refractive index, N, is an important physical quantity involved in the generation process of the atmospheric duct and determined by conventional meteorological elements as follows:

$$N = \frac{77.6}{T} \times \left(p + \frac{4810e}{T}\right) \tag{1}$$

Where p is atmospheric pressure in hPa, T is the atmospheric thermodynamic temperature in K, and e is the partial pressure of water vapor in hPa, i.e.,

$$e = \frac{6.105R_h e^x}{100} \tag{2}$$

$$x = 25.22 \times \frac{T - 273.2}{T} - 5.31 \times \ln\left(\frac{T}{273.2}\right) \tag{3}$$

and R_h is the atmospheric relative humidity.

The atmospheric modified refractive index, M, is also an essential index for incorporating the earth's curvature and determining the height and strength of the evaporation duct. The vertical and horizontal gradients of M determine the height and strength of the evaporation duct, respectively. The expression of M is:

$$M = N + \frac{h}{r_e} \times 10^6 = N + 0.157 \times h \tag{4}$$

where h is the target height for calculating the M.

The height of the evaporation duct is determined by either direct detection or indirect measurement. Direct detection methods use various tools such as meteorological gradiometer, microwave refractometer, low-altitude tethered balloon, or other sounding equipment to obtain the atmospheric refractive index profile structure. Using the atmospheric refractive index profile structure, the characteristic values of the evaporation duct, such as the height of the evaporation duct, are then determined. Nevertheless, direct detection methods are often highly costly and require complex operations.

Indirect measurement methods include using theoretical models, remote sensing inversion, and machine learning. The theoretical method is the most commonly used technique to obtain the height of the evaporation duct. The calculated value of the height of the evaporation duct is obtained based on known meteorological factors. For instance, in 1971, Jeske of the Hamburg University in Germany proposed the Jeske model [2]. In 1985, Paulus further improved the Jeske model by studying more accurate data resulting in the Paulus–Jeske (P-J) model [3]. In the 1980s, A. S. Gavrilov and other scholars of the Russian National University of Meteorology and Hydrology also proposed the RSHMU evaporation duct height prediction model [4], which was mainly used in the former Soviet Union. In 1992, Musson-Genon, Gauthier, and Bruth jointly proposed the Musson–Gauthier–Bruth (MGB) [5] model based on the Monin–Obukhov similarity theory. Later, Babin proposed the BYC model in 1997 [6], which applied the advanced COARE sea–air flux algorithm.

More recently, in 2000, Frederickson et al. of the U.S. Naval Graduate School put forward the Naval Postgraduate School (NPS) model [7], and in 2001, Liu Chengguo et al. put forward the pseudo-refractive index model by adding the concept of pseudo refractive index to the P-J model [8]. In 2002, Dai Fushan et al. replaced the M-O similarity theory with the local similarity theory based on the BYC model and came up with the Local model [9]. In 2015, Ding Juli et al. also proposed the Universal Evaporation Duct (UED) model [10]. Although the theoretical method is convenient, its accuracy and generalization are still weak.

The uncertainty of the theoretical model is from the empirical parameters of the model, these parameters are derived from local observations. If the natural conditions of the applied model area and the natural conditions of the summed empirical parameter area are very different, the theoretical models may have large deviations. In [6], the Babin–Young–Carton (BYC) model has a better performance than other models and can better fit the true value; the root mean square error (RMSE) of the model is 3.0, which results from an experiment based on the data from Wallops Island. However, in [11], the performance of the BYC model (the model in reference) in the Yellow Sea, China, is not very good in most cases and worse than the performance of other models, and the error is more than 5.0 in some cases. The above examples show that the theoretical model has uncertainties caused by different regional natural conditions.

The remote sensing inversion method is also one of the most important techniques to obtain the height of the evaporation duct. Different versions of the remote sensing inversion method include laser radar inversion, radar clutter inversion, and GNSS occultation signal inversion. In 2003, Peter Gerstoft et al. conducted sea clutter inversion of the refractive index of low-altitude atmosphere [12], compared the replica field with sea clutter observation by using the square error objective function, and conducted a global search for 11 environmental parameters in total by using a genetic algorithm. The inversion algorithm was based on S-band radar sea clutter data in Vopulos Island, Virginia. It was concluded that under the unconstrained environment model, the propagation loss and single detection loss are close to the actual duct. Yardim used the root mean square error considering the statistical characteristics of the duct region to evaluate the performance of radar sea clutter in retrieving the atmospheric refractive index [13]. Tian studied the antenna frame height of maritime search radar under evaporation duct [14] in 2010. The experimental results show that the appropriate antenna frame height is closely related to the radar working frequency and the meteorological environment in the sea area. After that, Zhang proposed a method to improve the inversion performance of duct RFC by using different antenna height combinations [15]. GNSS occultation inversion [16] method has also been widely used in the fields of geographic survey, meteorology and hydrology, etc. GNSS refers to the global navigation satellite system, which uses the ground-based GNSS receiver to receive satellite signals, calculates the time delay caused by the influence of atmospheric refraction, etc., and introduces the time delay into the calculation model as a parameter to be determined. Considering the sources of errors generated by the calculation model and the methods of eliminating errors, the delay parameters generated by atmospheric refraction and positioning parameters can be calculated together. Based on this, the relationship between the delay parameters and the profile of atmospheric elements can be constructed to retrieve the profile structure of atmospheric elements, and then evaporation duct height can be calculated. Wang used the forward propagation model and genetic algorithm to estimate the evaporation duct height by using the signal power information received by the ground-based GPS receiver [17]. Although the experimental results show that the effect is excellent, it is an experiment under ideal conditions: the forward propagation model ignores the refraction effect from the maximum parabolic equation area height to the GPS satellite height.

Nevertheless, these remote sensing inversion techniques have their own pros and cons, which need to be further investigated. Table 1 shows the advantages and disadvantages of remote sensing inversion methods.

Table 1. The advantages and disadvantages of remote sensing inversion methods.

Method	Advantages	Disadvantages
Laser radar inversion	Continuous and immediate security	The technology is not yet mature, and the scheme still needs to be improved.
Radar sea clutter inversion	The ability of regional generalization is strong, and the estimation effect is better by using suitable inversion algorithm	Due to the limitation of radiation silence, it is impossible to continuously monitor and diagnose, and it cannot realize large-scale duct monitoring.
GNSS occultation signal inversion	High accuracy under ideal conditions	There is a great error at a specific angle, which needs to be improved.

As Table 1 shows, the remote sensing inversion methods generally need to be improved; they are limited by the performance of the inversion algorithm or even the structure of scheme.

Using machine learning methods facilitates efficient exploration of the hidden patterns in the data and improves the traditional models. Such techniques may also be used to develop purely data-driven models that replace the traditional model with more efficient machine learning methods [18].

In 2007, S. A. Isaakidis et al. used a three-layer artificial neural network (ANN) to build a prediction model for the height of the evaporation duct and achieved efficient prediction results [19]. In 2009, Rémi Douvenot's team also proposed a technique based on LS-SVM to invert the height of the evaporation duct [20] and compared their proposed method with the common quadratic regression inversion method. Their experiments showed that compared with the common quadratic regression inversion method, LS-SVM avoids many significant errors and is superior in terms of stability. In 2013, Yang Chao compared the performance of three machine learning models (RVM, LS-SVM, and RBFNN) applied to the sea clutter inversion of evaporation duct height [21]. It was then concluded in [21] that the LS-SVM sea clutter inversion method provides the highest level of accuracy. Further on, in 2018, Zhu Xiaoyu and others tried to combine the traditional evaporation duct theoretical model with machine learning and optimized the P-J model by adopting the support vector regression (SVR) method [22]. Compared with the traditional P-J model, the obtained SVR_PJ model provides higher accuracy. In 2018, using the data collected in the South China Sea, Zhu Xiaoyu et al. built the SCS-MLP [23], a multilayer perceptron (MLP) [24] model for predicting the height of evaporation duct in the South China Sea. They then showed that their approach significantly improved the P-J model. In 2019, Zhao Wenpeng et al. also applied BP neural network to the prediction of evaporation duct height and achieved good results [25]. They further proposed a pure data-driven gradient lifting tree (GBDT) evaporation duct height prediction model, PDD_GBR, which significantly improved both the accuracy and regional generalization ability of their original model [26].

In 2020, Zhao Wenpeng used the method of Extreme Gradient Boosting (XGBoost) [27]. As an improved version of the GBDT algorithm [28], XGBoost is highly efficient in estimating the evaporation duct height. The experimental results also showed that compared with the traditional P-J theoretical model the accuracy is significantly improved. Furthermore, compared with the SCS-MLP and PDD_GBR models, the accuracy and generalization performance are significantly improved. In 2021, Han Jie used deep learning method to forecast the proximity of evaporation duct in the Yellow Sea area of China [29]. The characteristic of the research is that it uses sufficient data obtained from the Yellow Sea exploration. Using artificial neural network (ANN) and support vector machine (SVM) as benchmark comparison models, it is found that deep learning method has obvious advantages in accuracy.

Affected by the physical empirical parameters that are not universal, the traditional theoretical models of the evaporation duct result in limited accuracy and poor generalization ability. The accuracy, generalization ability and scientific interpretability of the existing purely data-driven height prediction model require further improvements.

In this paper, we propose the LightGBM evaporation duct height prediction model LGB-PHY. As the training set, we use the evaporation duct height data of the South China Sea throughout a five-day route. We further compare our proposed model with the XGB evaporation duct height prediction model. The generalization ability of the LGB-PHY model is also verified by the cross-validation method using data from four different sea areas. The experimental results confirm that given sufficient training data, the LGB-PHY model has higher accuracy and better regional generalization ability than that of the XGB model in most sea areas.

2. Brief Introduction of Existing Evaporation Duct Height Model

2.1. The Paulus–Jeske Model

The Paulus–Jeske (P-J) model is the most widely used theoretical model of the evaporation duct. In 1985, Paulus improved the Jeske model [3] with more accurate data and proposed the P-J model [2]. The P-J model is a classic and effective evaporation duct theoretical model which was also adopted in IREPS (Integrated Reflective Effects Prediction System). IREPS was a software package used by the US navy for evaluating electromagnetic wave propagation in 1978. This model was also applied to the IREPS upgraded version, AREPS (Advanced Reflective Effects Prediction System) and other systems. The input parameters of the model are sea surface temperature, air temperature, relative humidity, and wind speed, where the sea surface atmospheric pressure is assumed to be 1000 hPa.

The P-J model uses the potential refractive index, N_p, instead of atmospheric refractive index, N. The potential temperature also replaces ordinary temperature, and the potential water vapor pressure replaces the ordinary water vapor pressure. Assuming that the bit refractive index is equal to the refractive index, the calculation formula of the bit refractive index is as follows:

$$N_p = \frac{77.6}{\theta} \times \left(p + \frac{4810 \times e_p}{T}\right) \tag{5}$$

where the water vapor pressure e_p is:

$$e_p = e^{1+\frac{1000}{p}} \tag{6}$$

The p in the formula means air pressure.

The critical gradient is also defined as the potential refractive index gradient $\partial N_p / \partial z$. Differentiating Equation (5), we then obtain

$$\frac{\partial N_p}{\partial z} = \frac{\partial N}{\partial z} - \frac{\partial N}{\partial P} \times \frac{\partial P}{\partial z} \tag{7}$$

Assuming the standard temperature of 15 °C, the atmospheric pressure of 1000 hPa, and the acceleration of gravity g = 9.8 m/s^2, the critical refractive index gradient of the atmospheric duct is obtained as:

$$\frac{\partial N_p}{\partial z} = -0.125 \tag{8}$$

The potential refractive index is a similar variable and satisfies the following conditions:

$$\frac{\partial N_p}{\partial z} = \frac{S_{N_p}\varphi}{\rho\kappa u_*(z+z_0)} \tag{9}$$

where S_{N_p} is the vertical flux of potential refractive index, Z denotes the measuring height of more than 6 m, z_0 is dynamic roughness, φ denotes a universal function, k is the Kalman constant which is 0.4, and u_* is the friction speed. Universal function φ is a function of Z/L and represents the atmospheric stability near the surface. In stable atmospheric conditions, the universal function is

$$\varphi = 1 + \alpha_1 \times \frac{Z}{L} \tag{10}$$

where α_1 is a constant with a value of 5.2, and L is the Monin–Obukhov length.

Under unstable atmospheric conditions:

$$\varphi^4 - 4\alpha_2\frac{Z}{L}\varphi^3 = 1 \tag{11}$$

where α_2 is a constant with a value of 4.5.

Paulus concluded the empirical relationship from a large number of experiments. Combined with these empirical relations, the predicted value of the height of the evaporation duct is then obtained using Equation (11). Further details about the P-J model can be found in [2].

2.2. The BYC Model

The widely used P-J model does not consider the extension of similarity theory under low wind speed conditions and further ignores the correction of seawater salinity. These factors result in a deviation of prediction accuracy in some cases. Babin et al. obtained the BYC model by introducing the toga COARE air–sea flux algorithm. In the BYC model, the Buck equation [30] is used to improve the accuracy of water vapor partial pressure calculation, and the salinity is corrected. The M-O similarity theory is further extended to low wind speed [31]. In theory, the M-O is a more advanced traditional evaporation duct height prediction model than that of the P-J model. In the BYC model, the atmospheric refractive index gradient is written as

$$\frac{\partial N}{\partial z} = A + B\frac{\partial \theta}{\partial z} + C\frac{\partial q}{\partial z} \tag{12}$$

Using the Monin–Obukhov similarity theory, $\frac{\partial \theta}{\partial z}$ and $\frac{\partial q}{\partial z}$ are also obtained as

$$\frac{\partial \theta}{\partial z} = \frac{\theta_*\varphi_h}{\kappa z} \tag{13}$$

$$\frac{\partial q}{\partial z} = \frac{q_*\varphi_q}{\kappa z} \tag{14}$$

For the refractive index gradient $\frac{\partial N}{\partial z}$ = -0.157, the height of the evaporation duct can be determined based on the above formula. Furthermore, the height of the evaporation duct under stable or neutral atmospheric conditions is

$$z_{EDH} = \frac{-(B\theta_* + Cq_*)}{\kappa(A+0.157) + \frac{5}{L}(B\theta_* + Cq_*)} \tag{15}$$

Under unstable conditions [26], the height of the evaporation duct is

$$z_{EDH} = \frac{-(B\theta_* + Cq_*)\varphi_h}{\kappa(A + 0.157)} \tag{16}$$

where θ_*, q_* and φ_h are derived from the COARE air–sea flux algorithm. See [6] for further details.

In theory, the BYC model is more comprehensive than that of the P-J model. Furthermore, using the air–sea flux algorithm, COARE, enables a more accurate description of the physical phenomena, hence a more efficient extension of the M-O similarity theory [32]. Nevertheless, the BYC model has higher complexity, more physical constraints, and more empirical parameters, hence greater sensitivity than the P-J model. This may result in model deviations in the unsuitable air–sea environment.

2.3. The XGB Model

XGBoost (eXtreme Gradient Boosting) is a machine learning algorithm that is an optimization model based on gradient boosting. Gradient boosting decision tree (GBDT) [33,34] is an integrated model of the ensemble learning algorithm. Using GBDT for regression tasks is referred to as the Gradient Boosting Regression (GBR). The gradient boosting algorithm is a tree class integration method, through the integrated training of multiple weak learners. The overfitting problem is also addressed effectively by improving accuracy. In general, it is a model that combines several single weak models to improve the training effect. In this setting, each model learns from the mistakes of the previous weak learner model. For m weak learners, it needs m iterations, each iteration produces a new model, and in each iteration, the value of the loss function moves to its negative gradient. This results in a smaller loss function. Finally, the weighted sum of the models in each stage is used to obtain the final result. The basic formula is as the following:

$$F_m(x) = \sum_{i=1}^{m} \beta f_i(x) \tag{17}$$

where m is the number of weak learners, β is the weight coefficient, f is a weak learner, and F is the whole model. The objective here is to obtain a general model with the minimum loss function L:

$$\begin{aligned} &F_m \text{argmin} \sum_{i=1}^{n} L(y_i, F_m(x_i)) \\ &= \beta_m \text{argmin} \sum_{i=1}^{n} L(y_i, F_{m-1}(x_i) + \beta_m f_m(x_i)) \end{aligned} \tag{18}$$

Gradient boosting is based on the Greedy algorithm [35,36]. Therefore, it only solves the next weak learner as a new term at a time. *In* (18) $\beta_m f_m(x_i)$ is to obtain the fastest reduction of the loss function, and hence, each time, the new term should be equal to the negative gradient of the loss function, i.e.,

$$\beta_m f_m(x_i) = \gamma \frac{\partial L(y_i, F_{m-1}(x_i))}{\partial F_{m-1}(x_i)} \tag{19}$$

where γ is the step size, which sets the weight coefficients β and a negative sign representing the negative gradient. Nevertheless, $\beta_m f_m(x_i)$ is generally expressed as pseudo residual R, i.e.,

$$R_{im} = \gamma \frac{\partial L(y_i, F_{m-1}(x_i))}{\partial F_{m-1}(x_i)} \tag{20}$$

Equation (20) indicates that the pseudo residual error of F is obtained according to the previous model. For the known weak learner, a new weak learner, f_m, can be obtained through X and Y training in the known training samples. The step size γ can be further obtained by substituting f_m into the formula.

XGBoost generally continues the use of gradient boosting and is characterized by the second-order Taylor expansion of the loss function where the regularization term is also added to the loss function [37]. The loss function L is therefore defined as

$$L = \sum_i l(y\_hat_i, y_i) + \sum_k \Omega(f_k) \tag{21}$$

The formula in (21) is composed of two terms. The first term represents the loss of the gradient boosting algorithm, where i is the number of training samples and L is the loss of a single sample. The second term includes the regularization part, where

$$\Omega(f) = \gamma T + \frac{1}{2}\lambda||w||^2 \tag{22}$$

In 2020, Zhao Wenpeng proposed a pure data-driven evaporation duct prediction model based on the XGBoost algorithm, which is called the XGB model [27]. It was further shown through experimental studies that this model overperforms the accuracy of the traditional theoretical model. Its generalization performance is also significantly higher than the MLP model based on multilayer perceptron and the GBR model based on the traditional GBDT algorithm. In this paper, we consider the XGB model, where according to the observed data the parameters of the models are fine-tuned, and the XGB model obtained is compared with the LGB-PHY model.

3. LightGBM Evaporation Duct Height Prediction Model with Physical Information (LGB-PHY Model)

3.1. Introduction to the LightGBM Algorithm

LightGBM is a boosting regression algorithm proposed by Microsoft [38]. It is one of the latest machine learning algorithms also based on decision tree, and the logic about the contributions of each predictor to the final model is the same with the XGB model. It uses the idea of an algorithm based on a histogram to discretize continuous features into several discrete bins. The algorithm based on the histogram avoids calculating all continuous features and takes discrete bins as the unit. This method improves the training efficiency and reduces the memory overhead. In particular, the histogram algorithm also has a subtraction feature, where a target leaf can subtract its neighbor nodes from its parent node. This further accelerates the convergence speed of the algorithm.

In addition to the histogram algorithm (Figure 2), LightGBM adopts the leaf-wise tree growth (Figure 3) strategy and selects the leaf with the largest split gain to grow. Unlike many boosting algorithms that use a level-wise tree growth (Figure 4) strategy, it is a depth-first tree growth strategy. The leaf-wise selects the leaf with the largest split gain to grow, and hence, the depths of different leaf nodes are not the same. Experiments show that the leaf-wise tree growth strategy results in a lower loss. To avoid the over-fitting phenomenon caused by excessive longitudinal growth, LightGBM further sets the maximum depth of the tree.

As an optimized GBDT algorithm, LightGBM has relatively high complexity and is capable of solving nonlinear problems. LightGBM is both efficient and accurate with histogram algorithm, leaf-wise tree growth strategy and other characteristics. The tree model is efficient on the tabular data type used for height prediction of evaporation duct.

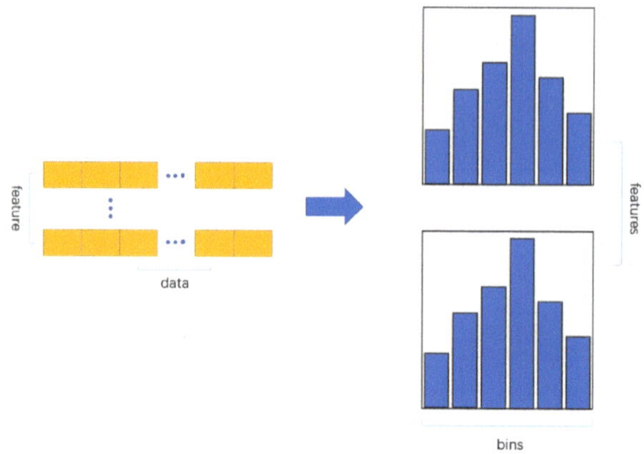

Figure 2. The histogram algorithm.

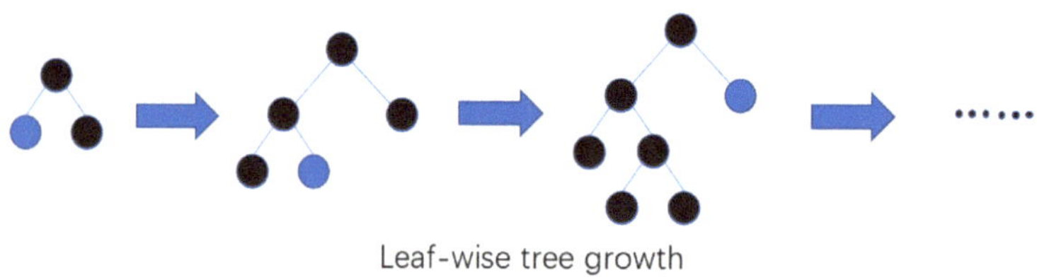

Figure 3. The leaf-wise algorithm.

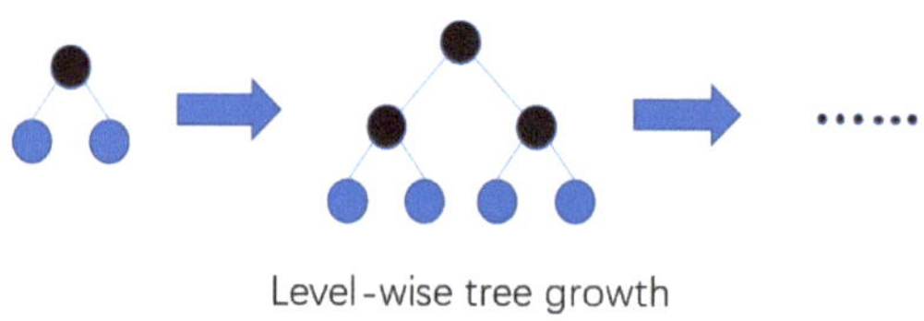

Figure 4. The level-wise algorithm.

3.2. Construction of LGB-PHY Model

The traditional theoretical model of evaporation duct height incorporates nonlinear factors, such as sea surface temperature, air temperature, relative humidity, wind speed, and atmospheric pressure. For example, the P-J model is a function of sea surface temperature (*SST*), air temperature (T_a), relative humidity (*RH*), and wind speed (*U*), where the function value is the evaporation duct height (*EDH*):

$$EDH = f(SST, T_a, RH, U, P) \tag{23}$$

The BYC model is a semi-empirical physical model based on a large number of regional observation experiences and the Monin–Obukhov similarity theory [3]. This model describes the complex physical relationship between the basic meteorological elements and evaporation duct height. This model combines the sea–air flux algorithm of COARE, which has complex physical constraints. After the calculation, we have obtained the contributions of each predictor, 26% from wind speed, 22.3% from air temperature, 21.3% from sea surface temperature, 18.9% from air pressure and 11.5% from relative humidity.

The LightGBM is efficient at dealing with the regression problem of low-dimensional data and can be applied to the low-dimensional observation data sets. The evaporation duct height prediction model based on the LightGBM replaces the traditional theoretical model (which has a large number of physical process descriptions and empirical parameter assumptions) with a machine learning algorithm. This relies on the capability of machine learning algorithms to capture the correlation between data, hence addressing the bottleneck of traditional theoretical models which are limited by physical assumptions and empirical parameters. At the same time, combining the physical formula of the BYC model with the loss function of the LightGBM model imposes certain physical constraints on the process of training and extracting data association. Such constraints lead to a more accurate and scientific machine learning model. Figure 5 shows the way to combine the physical constraints.

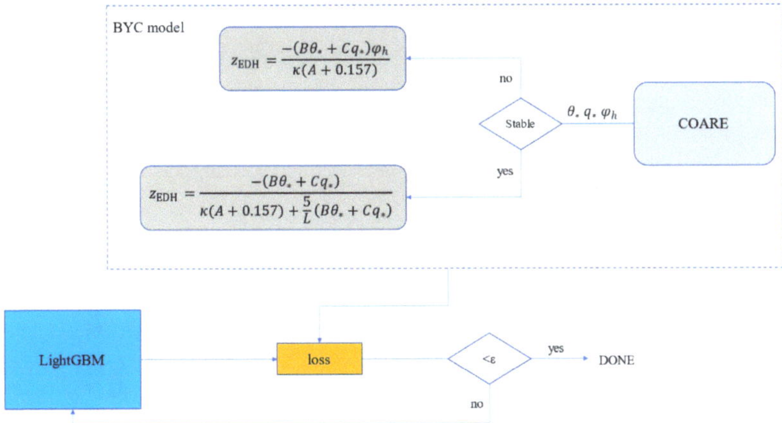

Figure 5. Combining the physical constraints.

The LightGBM evaporation duct height prediction model combining physical information takes the actual observation data as the starting point. Therefore, for the construction of the model the observation data set should be preprocessed first. The sample pair should be set as $(x_i, y_i)^n{}_{i=1}$, and vector x_i expressed as $x_i = (SST_i, T_i, RH_i, U_i, P_i)$ is a five-dimensional vector in which the eigenvalues represent the sea surface temperature, air temperature, relative humidity, wind speed and atmospheric pressure, respectively. Set the measured data as a label $y_i = EDH\_Obs_i$. Here, N is number of samples of which 80% is selected as the training set and the remaining 20% as the test set. Different application areas and data attributes will have different key parameters configuration. According to the specific conditions of the experimental data and some prior knowledge, the parameters are finally determined as follows through a large number of computer experiments: learning rate (learning_rate) = 0.1, number of weak learners (n_estimators) = 4000, number of leaves of the model (num_leaves) = 28, minimum leaf node sample weight (min_child_weight) = 0.001, and minimum leaf node samples (min_child_samples) = 20. The loss function is custom loss function, for training the function name it is custom_normal_train, and for validating the function name it is custom_normal_valid. The key parameters will be fine-tuned according to the amount of data. The flowchart of the model is depicted in Figure 6.

The custom loss function of the model is defined as:

$$Loss = MSE + \alpha MSE\_BYC \tag{24}$$

As shown in the above formula, the loss function is divided into two parts. The first part is *MSE*, calculated by the predicted value of evaporation duct height and the tag value, and the second part is *MSE_BYC*, calculated by the BYC model; the tag value, α, is a weight parameter to ensure that the magnitude balance of the *MSE* in the two parts. In our experiments, we tried different values of α and found α = 0.01 is an efficient choice.

$$MSE = \frac{1}{n}\sum_{i=1}^{n}(y\_pred - y\_true)^2 \quad (25)$$

The above formula is the RMSE obtained from the predicted value and label value of evaporation duct height, where *y_pred* denotes the predicted value of the model, and *y_true* represents the true value label of the height of the evaporation duct

$$MSE\_BYC = \frac{1}{n}\sum_{i=1}^{n}(y\_BYC - y\_true)^2 \quad (26)$$

The above formula, i.e., *MSE_BYC* is obtained from the height value and label value of the evaporation duct calculated by the BYC model, where *y_BYC* refers to the height value of the evaporation duct calculated by the BYC model.

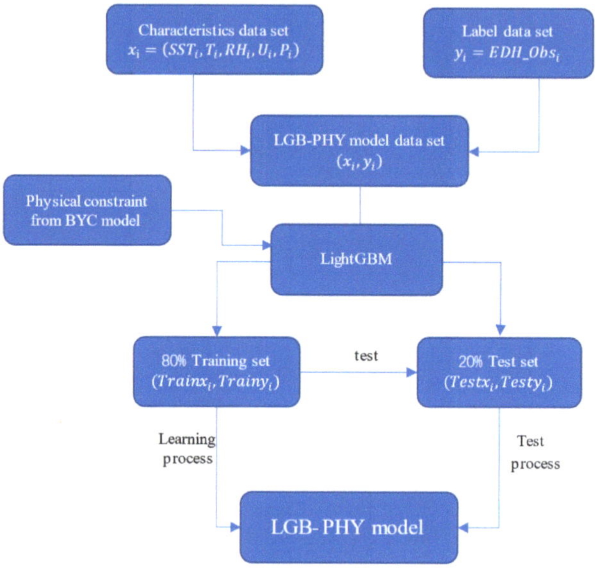

Figure 6. The model structure diagram.

4. Experiment and Result Analysis

4.1. Experimental Scheme

In this study, we obtain the data by deploying observation equipment on the ships, and different data sets correspond to different routes. The ship is equipped with six layers of hydrometeorological fundamental element measuring sensors on the ship deck from low to high and a set of infrared sea surface water temperature sensors. The sensors record the air temperature (°C), relative humidity (%), wind speed (m/s), air pressure (hPa), and SST(°C). For EDH measurement, preprocess the sensor measurement data of the six-layer to refractive index profile, and EDH corresponds to the point with a slope of 0. The method of EDH measurement is improved from Babin's method [6]. The sensor parameters are shown in Table 2.

Table 2. The parameters of sensors.

Sensor Category	Range	Precision	Resolution
Air temperature sensor	−40~+70 °C	±0.3 °C	0.1 °C
Relative humidity sensor	0~100%	±2%	1%
Pressure sensor	300~1100 hPa	±0.5 hPa	0.1 hPa
Wind speed sensor	0.1~60 m/s	±3%	0.01 m/s
Sea surface temperature sensor	−55 °C~+80 °C	±0.2 °C	0.1 °C

The observation voyage routes have been shown in Figure 7 and the data consists of the following parts:

Part 1: Voyage exploration experiment in the South China Sea (SCS) from 16 November 2020 to 21 November 2020. The corresponding data set is referred to as 5 DaysSCS.

Part 2: Short-distance voyage exploration experiment in the South China Sea. The corresponding data set is called Back, which represents the short-range return data of the South China Sea.

Part 3: Voyage exploration experiment in the Equator. The corresponding data set is referred to as ER.

Part 4: Sailing exploration experiment in the Sea of Sri Lanka. The corresponding data set is called SSL.

Part 5: Navigation exploration experiment in the North Bengal Bay. The corresponding data set is referred to as NBB.

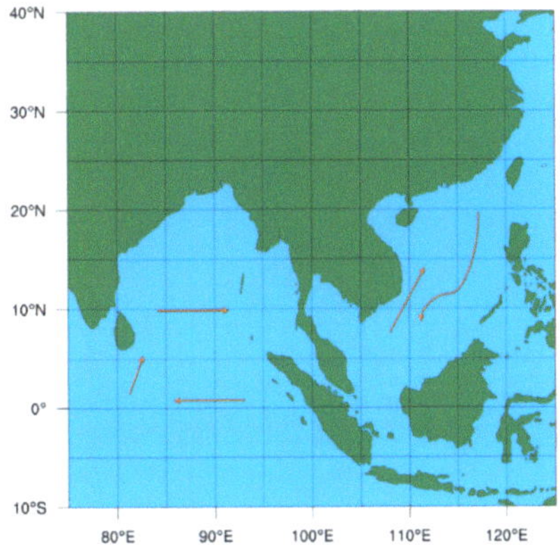

Figure 7. The voyage routes.

The characteristic values of the above meteorological and hydrological observation data samples are composed of sea surface temperature, air temperature, relative humidity, wind speed at 10 m height on the deck and air pressure. The corresponding EDH values are also collected as the data labels. Eliminating the missing and abnormal data, a total of about 83,000 groups of samples were made available. For the 5 DaySCS data set that includes a large number of samples, the data set was divided into days as a whole. Therefore, the data of the first four days can be used to predict the data of the fifth day. The experiment is divided into two parts. In the first part, 80% of the available data is used as the training set and 20% as the test set to evaluate the comprehensive effect of the LightGBM model. In the second part, the data of four different regions, including Back, ER, SSL and NBB, are used for cross-training. Here, all the data of one region is considered as the training set and the remaining three regions are considered as the test set. The goal is to test the generalization ability of the model area.

The root means square error (RMSE) and R-square (SCC) are also used as performance evaluation indexes. RMSE reflects the accuracy of the model in predicting the height of the evaporation duct. The lower the RMSE value, the higher the prediction accuracy of the model. The SCC represents the goodness of fit between the predicted result and the detected true value. The SCC value is in the range of (0, 1), and a value closer to 1 represents a better goodness of fit, i.e., the higher the degree of explanation of independent variables to dependent variables.

The calculation formulas of *RMSE* and *SCC* are as the following

$$RMSE = \sqrt{\frac{1}{n}\sum_{i=1}^{n}(f(x_i) - y_i)^2} \tag{27}$$

$$SCC = \frac{\left(n\sum_{i=1}^{n} f(x_i)y_i - \sum_{i=1}^{n} f(x_i)y_i \sum_{i=1}^{n} y_i\right)^2}{\left(n\sum_{i=1}^{n} f(x_i)^2 - \left(\sum_{i=1}^{n} f(x_i)\right)^2\right)\left(n\sum_{i=1}^{n} y_i{}^2 - \left(\sum_{i=1}^{n} y_i\right)^2\right)} \tag{28}$$

4.2. Data Augmentation

The 5 DaysSCS data set is a voyage observation data set and many labels are missing in this data set. To make up for this deficiency, we apply data enhancement based on the P-J model, which is the most widely used theoretical model of the evaporation duct. The P-J model has simple principles, a wide application range and provides stable calculated data. The percentage of data coming from data augmentation is 15%. During the observation, because of the observation equipment error and extreme natural conditions [39], sometimes the data label is missing, and the missing data accounts for about 15% of the total data volume. Here, for the data with missing labels, the value of the evaporation duct height model is calculated by using the P-J evaporation duct theoretical model. We then add Gaussian noise [40] to the obtained missing label values calculated by the P-J model to construct enhanced data. This provides samples that closely follow the natural law and meets the research requirements. Additionally, 15% data augmentation makes the data set more diverse on the premise of not reducing the influence of real observation data, the probability of overfitting in the training process is alleviated to some extent.

4.3. Evaluation of Model Prediction Effect

To test the prediction effect of the LGB-PHY model, the XGB model driven by pure data and the theoretical BYC model are used as references in the experiment. An 80% of the observed data of 5 DaysSCS obtained from the 5-day South China Sea cruise is also used as the training set, and the remaining 20% is used as the test set for verification. Figure 8 shows the evaporation duct height prediction in the test set which combines the LightGBM model and XGB model. The red line is the real observation value, and the blue line shows the predicted value of evaporation duct height using the LightGBM model with physical information.

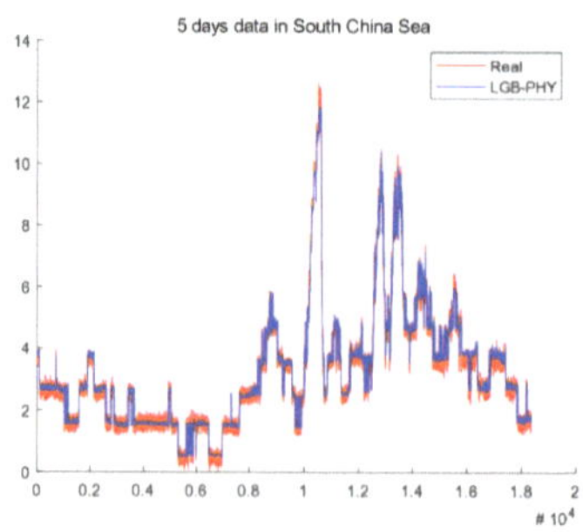

Figure 8. The fitting effect of LGB-PHY.

It is seen from the above that the predicted value of the evaporation duct by the LightGBM model with physical information closely follows the actual observation value. Figure 8 shows the effectiveness of the new model LightGBM model for the existing data sets.

Figure 9 further shows the difference distribution between the predicted value and the real value of the LGB-PHY and XGB models. The blue line shows the difference distribution between the predicted value and the real value of LGB-PHY, and the red line shows the difference distribution between the predicted value and the real value of the XGB model. It is seen that the LGB-PHY model provides better fitting performance of real values. Table 3 shows the RMSE and SCC values of the LGB-PHY and XGB models.

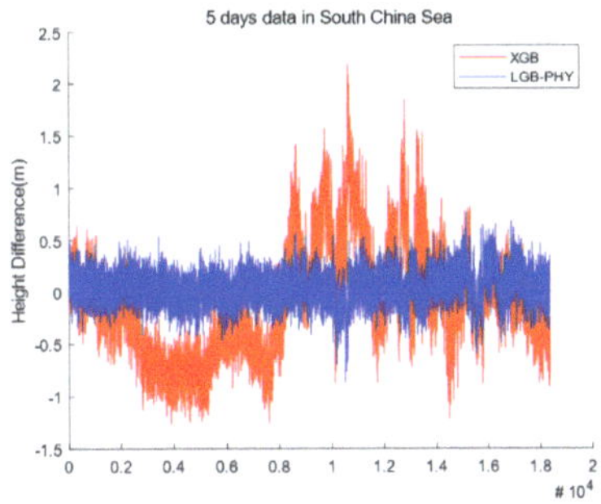

Figure 9. The difference distribution between the predicted and real values of the LGB-PHY and XGB models.

Table 3. The RMSE and SCC values of LGB-PHY and XGB models.

	RMSE	SCC
LGB-PHY	**0.166**	**0.993**
XGB	0.512	0.932

As is seen in Table 3, LGB-PHY model overperforms the XGB both in terms of RMSE and SCC.

4.4. Model Area Generalization Ability Test

To investigate the performance of the generalization ability, a sub-region data of each observed data set is used as the training set, and the rest of the sub-region data are used as the test set. The XGB model and LGB-PHY model are then compared in different combinations of experimental areas. This experiment shows that the LGB-PHY model overperforms the XGB model in terms of regional generalization.

Back, NBB, SSL, and ER data sets in four different regions, and all of the sub-data sets, are used as the training set each time. The trained model is then tested in the remaining four regions, and the prediction results are shown in Figure 10, where the blue line indicates the output result predicted by the LGB-PHY model combining physical information, and the red line indicates the output result predicted by XGB model. The green line also indicates the observation value of the evaporation duct height in this area.

The plots in Figure 10 show the comparison between the predicted value and the real value of the model using Back data as the training set and the data of the other three areas as the test set. The plots below also present the difference between the output value of the two models and the true value.

The plots above in Figure 11 demonstrate the comparison between the predicted value and the real value of the model using ER data set as the training set and the data of the other four areas as the test set. The plots below in Figure 11 also indicate the difference between the output value of the two models and the true value.

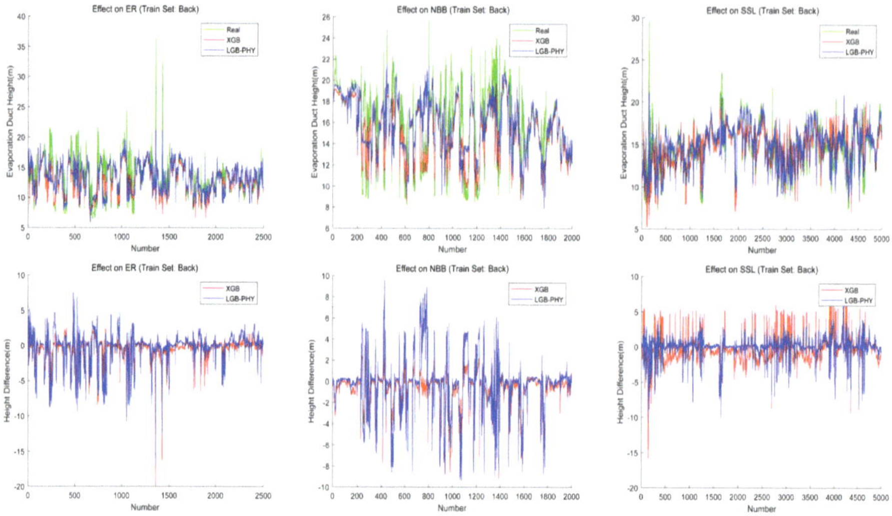

Figure 10. Comparison between the predicted value and the real value of the model using Back data as the training set and the data of the other three areas as the test set.

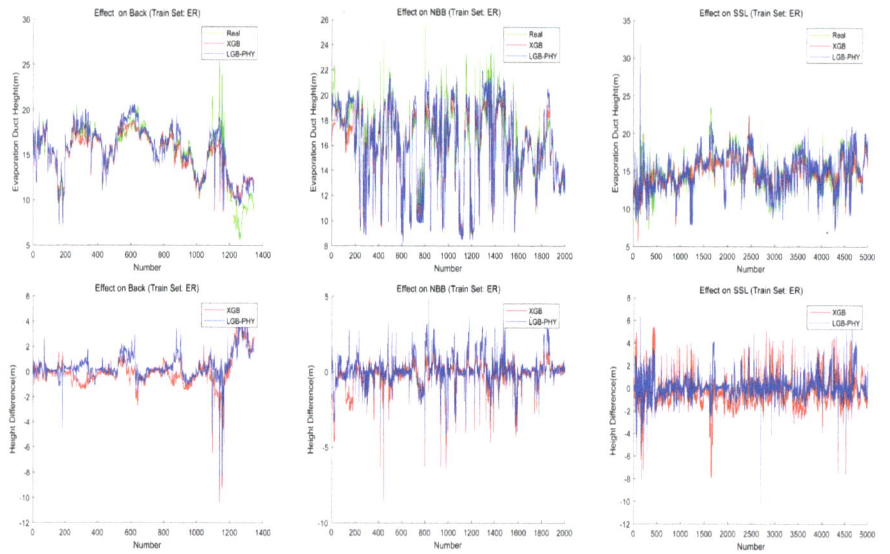

Figure 11. The comparison between the predicted and real values of the model using ER data as the training set and the data of the other three areas as the test set.

The plots above in Figure 12 indicate the comparison between the predicted value and the real value of the model using NBB data as the training set and the data of the other four regions as the test set. The plots below in Figure 12 show the difference between the output value of the two models and the true value.

In Figure 13, we also present the comparison between the predicted value and the real value of the model using SSL data as the training set and the data of the other four areas as the test set. The plots below in Figure 13 demonstrate shows the difference between the output value of the two models and the true value.

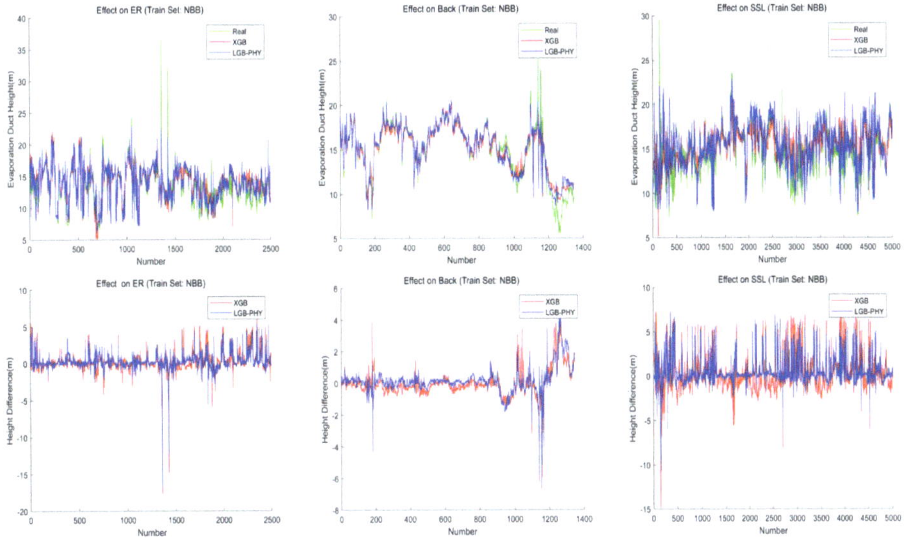

Figure 12. The comparison between the predicted and real values of the model using NBB data as the training set and the data of the other three areas as the test set.

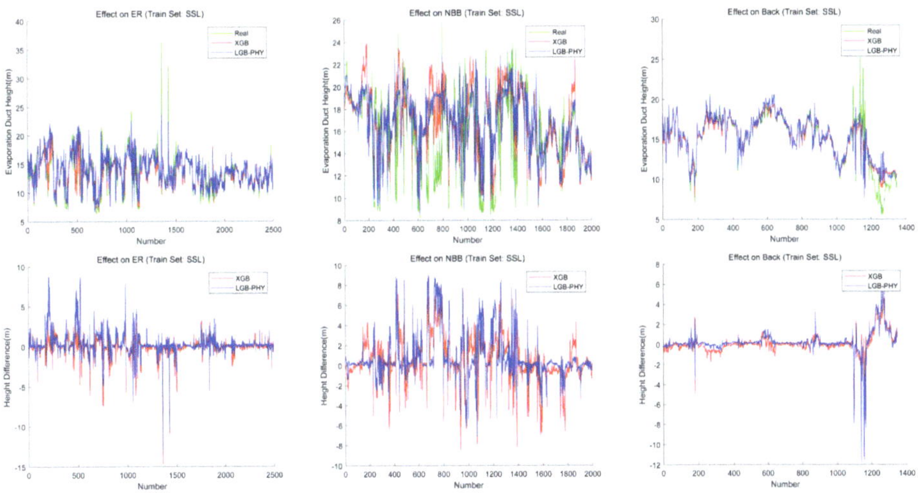

Figure 13. Comparison between the predicted value and the real value of the model using SSL data as the training set and the data of the other three areas as the test set.

The model performance differs among regions because the different data sets from different regions have different data characteristics. The machine learning models are pure data-driven evaporation duct models. The LightGBM algorithm and the physical constraints give the model a better ability to extract features of data and fit the true observation value of the evaporation duct height.

The data sets of different regions are from the observative experiment in different regions. Influenced by different meteorological and hydrological conditions in different regions, the different data sets from different areas have different data characteristics, and the different data characteristics affect the feature extraction of the model and the fitting effect of the true value.

Tables 4–7 show the root mean square error (RMSE) and R-square (SCC) values of different models under different experimental conditions.

Table 4. The RMSE values of LGB-PHY model in model area generalization ability test.

	Train Area	SSL	NBB	Back	ER
LGB-PHY (RMSE)	SSL	-	**2.41**	1.30	1.40
	NBB	**1.49**	-	**0.89**	**1.18**
	Back	**1.47**	2.75	-	2.39
	ER	**1.03**	**1.04**	**1.17**	-

Table 5. The RMSE values of XGB model in model area generalization ability test.

	Train Area	SSL	NBB	Back	ER
XGB (RMSE)	SSL	-	2.46	**1.04**	**1.20**
	NBB	1.95	-	0.92	1.48
	Back	1.84	**1.91**	-	**1.90**
	ER	1.44	1.07	1.35	-

Table 6. The SCC values of LGB-PHY model in model area generalization ability test.

	Train Area	SSL	NBB	Back	ER
LGB-PHY (SCC)	SSL	-	**0.47**	0.82	0.77
	NBB	**0.65**	-	**0.92**	**0.84**
	Back	**0.66**	0.31	-	0.34
	ER	**0.83**	**0.90**	**0.86**	-

Table 7. The SCC values of the XGB model in model area generalization ability test.

	Train Area	SSL	NBB	Back	ER
XGB (SCC)	SSL	-	0.44	**0.88**	**0.83**
	NBB	0.39	-	0.91	0.75
	Back	0.46	**0.66**	-	**0.58**
	ER	0.67	0.89	0.81	-

Based on the above results, the following observations are made:

(1) In most cases, the RMSE of the predicted value of the LGB-PHY model is smaller than that of the XGB model, and the SCC of the predicted value of the LGB-PHY model is larger than that of the XGB model. Hence, the LGB-PHY model has better generalization performance in most cases.

(2) In cases where the SSL data set is used as the training set, the RMSE of the LGB-PHY model is slightly larger than that of the XGB model where Back and ER are used as test sets, and SCC is slightly smaller than that of the XGB model. In cases where the Back data set is used as the training set, the RMSE of the LGB-PHY model is larger than that of the XGB model when NBB and ER are used as the test sets, and SCC is also smaller than that of the XGB model. According to the adaptability analysis of the BYC theoretical model, it can then be argued that the reason for the errors in some sea areas is that the LGB-PHY model, which combines the physical information of the BYC model, also inherits some of the properties of the BYC theoretical model. Note that there are empirical parameters from the actual observation in the BYC model (i.e., the data of the extracted empirical parameters come from the meteorological and hydrological data observed by the U.S. Navy in the middle and high latitudes) which are significantly different from the air and sea environments in the low latitudes around the equator. It is difficult for the BYC model to achieve good results in the waters near the equator, which leads to the error in some areas of the LGB-PHY model as it combines the physical information of the BYC model.

5. Conclusions

Influenced by the physical parameters that are not universal, the traditional evaporation duct theoretical models have limited accuracy and poor generalization ability. In this study, we address these issues by proposing the physically constrained LightGBM evaporation duct height prediction model (LGB-PHY). By comparing this model with the existing pure data-driven XGB evaporation duct height prediction model, the following conclusions are drawn:

(1) In the South China Sea, the LGB-PHY model has higher fitting accuracy than the XGB model. The experiments performed that the LGB-PHY model has higher performance in terms of RMSE and SCC than that of the XGBM model, where the RMSE index decreases by 68% and the SCC index increases by 6.5%.

(2) In the cross-comparison experiment of regional generalization, the LGB-PHY model shows better generalization ability than that of the XGB model in most cases. Nevertheless, for the case with the Back data set as the training set and the NBB data set being used as the test set, the LGB-PHY model demonstrates lower. It is attributed to the fact that some empirical parameters in the BYC model are derived from actual observation. However, the observation site is situated where the empirical parameters are mostly located in the sea area of the middle and high latitudes, which are quite different from the atmosphere and sea environment of the low latitudes around the equator. Affected by the poor universality of the physical experience parameters, the BYC model has difficulty achieving good results in the waters near the equator and at low latitudes. This results in lower accuracy of the LGB-PHY model, which combines the physical information of the BYC model. In general, our experiments confirm that in the middle and high latitudes, where the BYC model has strong adaptability, the LGB-PHY model has a stronger regional generalization performance.

Compared with the previous pure data-driven model based on machine learning, the proposed LGB-PHY model, which combines physical conditions, improves the accuracy and generalization ability. This model also inherits the characteristics of the theoretical model, hence being physically explainable. The proposed method constitutes a novel technique integrating physical constraints into machine learning methods.

Author Contributions: Conceptualization, J.L. and X.C.; methodology, X.C.; software, X.C.; validation, X.C. and W.W.; formal analysis, X.C. and J.L.; investigation, X.C.; resources, X.C. and J.L.; data curation, J.L. and J.Z.; writing—original draft preparation, X.C.; writing—review and editing, J.L. and J.Z.; visualization, X.C.; supervision, J.L. and J.Z.; project administration, X.C.; funding acquisition, J.L. and X.Z. All authors have read and agreed to the published version of the manuscript.

Funding: This research was funded by National Natural Science Foundation of China (grant number 41775027).

Data Availability Statement: Not applicable.

Acknowledgments: We thank all the editors and reviewers for their comments that greatly improved the presentation of this paper. We thank all the partners in laboratory for spiritual help.

Conflicts of Interest: The authors declare that they have no conflict of interest.

References

1. Kang, S.F.; Zhang, Y.S. *Atmospheric Duct in Troposphere Environment*; Wang, Z., Ed.; Science Press: Beijing, China, 2014.
2. Paulus, R.A. Practical Application of an Evaporation Duct Model. *Radio Sci.* **1985**, *20*, 887–896. [CrossRef]
3. Jeske, H. State and Limits of Prediction Methods of Radar Wave Propagation Conditions Over Sea. *Modern Radio Wave Propagation and Air-Sea interaction.* **1973**, *5*, 130–148.
4. Xie, L.X.; Tian, B. Study on Evaporation Duct RSHMU Model in Tropical Waters. *Ship Electron. Eng.* **2015**, *35*, 88–90, 100.
5. Luc, M.G.; Sylvie, G.; Eric, B. A Simple Method to Determine Evaporation Duct Height in the Sea Surface Boundary Layer. *Radio Sci.* **1992**, *27*, 635–644.
6. Steven, M.B.; George, S.Y.; James, A.C. A New Model of the Oceanic Evaporation Duct. *J. Appl. Meteorol.* **1997**, *36*, 193–204.

7. Frederickson, P.A. Estimating the Refractive Index Structure Parameter (C_2) over the Ocean Using Bulk Methods. *J. Appl. Meteorol.* **2000**, *39*, 10.
8. Liu, C.G.; Huang, J.Y. Modeling Evaporation Duct over Sea with Pseudo-Refractivity and Similarity Theory. *Acta Electron. Sin.* **2001**, *29*, 970–972.
9. Dai, F.S.; Li, Q. *Atmospheric Duct and Its Military Applications*; People's Liberation Army Publishing House: Beijing, China, 2002.
10. Ding, J.L.; Fei, J.F. Development and validation of an evaporation duct model. part i: Model establishment and sensitivity experiments. *J. Meteorol. Res.* **2015**, *29*, 467–481. [CrossRef]
11. Song, W.; Tian, B. Research on evaporation duct predicting model. *J. Huazhong Univ. Sci. Technol.* **2013**, *41*, 52–56.
12. Gerstoft, P.; Rogers, L.T. Inversion for refractivity parameters from radar sea clutter. *Radio Sci.* **2003**, *38*, 1–22. [CrossRef]
13. Yardim, C.; Gerstoft, P. Sensitivity analysis and performance estimation of refractivity from clutter techniques. *Radio Sci.* **2009**, *44*, 1008. [CrossRef]
14. Tian, S.S.; Cai, H. Erected antenna height of radars for searching over sea under evaporation duct. *J. Huazhong Univ. Sci. Technol.* **2010**, *38*, 16–19.
15. Zhang, J.P.; Wu, Z.S. Evaporation duct inversion based on radar sea clutters from different antenna heights. *Chin. J. Radio Sci.* **2011**, *26*, 422–430.
16. Zhao, X.; Huang, S. Atmospheric Duct Estimation Using Radar Sea Clutter Returns by the Adjoint Method with Regularization Technique. *J. Atmos. Ocean. Technol.* **2014**, *31*, 1250–1262. [CrossRef]
17. Wang, H.G.; Wu, Z.S. Retrieving evaporation duct heights from power of ground-based GPS occultation Signal. *Prog. Electromagn. Res.* **2013**, *30*, 189–194. [CrossRef]
18. Anderson, J.R.; Michalski, R.S.; Mitchell, T.M. *Machine learning. An Artificial Intelligence Approach*; Morgan Kaufmann Publishers Inc.: San Francisco, CA, USA, 1983; Volume 2.
19. Isaakids, S.; Dimou, I.; Xenos, T. An artificial neural network predictor for tropospheric surface duct phenomena. *Nonlinear Processes Geophys.* **2007**, *14*, 569–573. [CrossRef]
20. Douvenot, R.; Fabbro, V.; Bourlier, C. Radar coverage prediction over ocean: Duct mapping using least squares support vector machines. In Proceedings of the 1st European Conference on Antennas and Propagation, Nice, France, 1 June 2006.
21. Yang, C. A comparison of the machine learning algorithm for evaporation duct Estimation. *Radioengineering* **2013**, *22*, 657–661.
22. Zhu, X.Y.; Li, J.C. An Evaporation Duct Height Prediction Method Based on Deep Learning. *IEEE Geosci. Remote Sens. Lett.* **2018**, *15*, 1307–1311. [CrossRef]
23. Zhu, X.Y.; Zhu, M. An Optimization Research of Evaporation Duct Prediction Models Based on a Deep Learning Method. In Proceedings of the 2018 2nd IEEE Advanced Information Management, Communicates, Electronic and Automation Control Conference, Xi'an, China, 25–27 May 2018.
24. Allan, P. Approximation theory of the MLP model in neural networks. *Acta Numer.* **1999**, *8*, 143–195.
25. Zhao, W.P.; Li, J.C. Research on Evaporation Duct Height Prediction Based on Back Propagation Neural Network. *IET Microw. Antennas Propag. Inst. Eng. Technol.* **2020**, *14*, 1547–1554. [CrossRef]
26. Zhao, W.P.; Li, J.C. PDD_GBR: Research on Evaporation Duct Height Prediction Based on Gradient Boosting Regression Algorithm. *Radio Sci.* **2019**, *54*, 949–962. [CrossRef]
27. Zhao, W.P.; Li, J. XGB Model: Research on Evaporation Duct Height Prediction Based on XGBoost Algorithm. *Radioengineering* **2020**, *29*, 81–93. [CrossRef]
28. Anghel, A.; Papandreou, N. Benchmarking and Optimization of Gradient Boosted Decision Tree Algorithms. *Statistics* **2018**, *2*, 1467–1531.
29. Jie, H.; Jia, J.W. Evaporation Duct Height Nowcasting in China's Yellow Sea Based on Deep Learning. *Remote Sens.* **2021**, *13*, 1577.
30. Steven, M.B. LKB-based evaporation duct model comparison with buoy data. *J. Appl. Meteorol.* **2002**, *41*, 434–446.
31. Liu, W.T.; Blanc, T.V. *The Liu, Katsaros, and Businger (1979) Bulk Atmospheric Flux Computational Iteration Program in FOR-Tran and Basic*; Naval Research Lab: Washington, DC, USA, 1984.
32. Cook, J.A. A sensitivity study of weather data inaccuracies on evaporation duct height algorithms. *Radio Sci.* **2016**, *26*, 731–746. [CrossRef]
33. Breiman, L. Stacked Regressions. *Mach. Learn.* **1996**, *24*, 49–64. [CrossRef]
34. Mitchell, T. Machine Learning and Data Mining. *Commun. ACM* **1999**, *42*, 30–36. [CrossRef]
35. Jerome, H.F. Greedy function approximation: A gradient boosting machine. *Ann. Stat.* **2001**, *29*, 1187–1232.
36. Chen, T.; Carlos, C. XGBoost: A Scalable Tree Boosting System. In Proceedings of the KDD '16: The 22nd ACM SIGKDD International Conference on Knowledge Discovery and Data Mining, San Francisco, CA, USA, 13–17 August 2016.
37. Shi, H. *Best-First Decision Tree Learning*; University of Waikato: Hamilton, New Zealand, 2007.
38. Meng, Q. LightGBM: A Highly Efficient Gradient Boosting Decision Tree. In Proceedings of the Advances in Neural Information Processing Systems 30 (NIPS 2017), Long Beach, CA, USA, 4–9 December 2017; Volume 30.
39. Paulson, A.C. The Mathematical Representation of Wind Speed and Temperature Profiles in the Unstable Atmospheric Surface Layer. *J. Appl. Meteorol.* **1970**, *9*, 857–861. [CrossRef]
40. Hilarie, S.; Christopher, J. Gaussian Process Regression for Estimating EM Ducting Within the Marine Atmospheric Boundary Layer. *Radio Sci.* **2020**, *55*, 1–14.

future internet

Article

EBBA: An Enhanced Binary Bat Algorithm Integrated with Chaos Theory and Lévy Flight for Feature Selection

Jinghui Feng [1,2], Haopeng Kuang [1] and Lihua Zhang [1,*]

1 Academy for Engineering & Technology, Fudan University, Shanghai 200433, China; jhfeng20@fudan.edu.cn (J.F.); hpkuang19@fudan.edu.cn (H.K.)
2 Academy for Electromechanical, Changchun Polytechnic, Changchun 130033, China
* Correspondence: lihuazhang@fudan.edu.cn

Abstract: Feature selection can efficiently improve classification accuracy and reduce the dimension of datasets. However, feature selection is a challenging and complex task that requires a high-performance optimization algorithm. In this paper, we propose an enhanced binary bat algorithm (EBBA) which is originated from the conventional binary bat algorithm (BBA) as the learning algorithm in a wrapper-based feature selection model. First, we model the feature selection problem and then transfer it as a fitness function. Then, we propose an EBBA for solving the feature selection problem. In EBBA, we introduce the Lévy flight-based global search method, population diversity boosting method and chaos-based loudness method to improve the BA and make it more applicable to feature selection problems. Finally, the simulations are conducted to evaluate the proposed EBBA and the simulation results demonstrate that the proposed EBBA outmatches other comparison benchmarks. Moreover, we also illustrate the effectiveness of the proposed improved factors by tests.

Keywords: feature selection; bat algorithm; optimization; chaos theory; Lévy flight

Citation: Feng, J.; Kuang, H.; Zhang, L. EBBA: An Enhanced Binary Bat Algorithm Integrated with Chaos Theory and Lévy Flight for Feature Selection. *Future Internet* **2022**, *14*, 178. https://doi.org/10.3390/fi14060178

Academic Editor: Paolo Bellavista

Received: 10 May 2022
Accepted: 3 June 2022
Published: 9 June 2022

1. Introduction

Communication networks, computers, and artificial intelligence technologies provide a vast array of tools and techniques to improve efficiency. With the development of these technologies and tools, huge amounts of data are generated, stored, and utilized [1]. For example, a large number of IoT devices monitor, sense, and generate continuous data from the edge [2]. In addition, operators retain large amounts of historical user data in various transaction and computing platforms. Then again, the generation of rich media data such as short videos and motion pictures also makes the amount of data in the network grow exponentially [3]. Machine learning can effectively use these data and learn rules and patterns from them to help people make predictions and decisions. Machine learning algorithms have been successfully applied to various fields of life, such as medicine, materials science, and physics. Specifically, machine learning algorithms can extract features from various types of data and use the features to train models for classification, regression, and clustering operations [4].

Despite the advantages of the mentioned machine learning algorithms in terms of effectiveness, wide application, and malleability, there are still some challenges and urgent issues that machine learning algorithms need to address. First, the training of machine learning algorithms is a time-consuming, computationally intensive, and energy-intensive process, which can lead to limited applications of machine learning algorithms [5]. Second, the training datasets of machine learning algorithms are derived from the features they extract, which are mostly extracted using automated tools and human experience, and have many repetitive, meaningless, or even misleading features [6]. These features can slow down the training process of machine learning algorithms even more and reduce the effectiveness of classification, clustering, and regression of machine learning algorithms.

Therefore, the elimination of these useless and redundant features is important to improve the performance of machine learning algorithms and reduce their training consumption.

Feature selection is an effective means to solve the above problem. Feature selection can eliminate useless and redundant features in the dataset, thus reducing the number of features and improving the classification accuracy of machine learning algorithms. For a dataset with N features, there are 2^N feature selection schemes available, producing a combinatorial explosion. Therefore, selecting a subset of features with high classification accuracy and a low number of features can be regarded as an optimization problem. On the other hand, the feature selection problem is also proved to be an NP-hard problem. It is important to select and propose a suitable algorithm to solve the feature selection problem.

In general, feature selection methods can be classified into three categories, namely filter-based methods, wrapper-based methods, and embedded methods. Specifically, filter-based methods use a statistical measure that gives a score to the relevance of each feature in the dataset, by which the importance of the feature can be quantified. Subsequently, the decision-maker can set a threshold to remove features with scores below the threshold, thus achieving a reduction in the number of features. However, such methods do not consider the complementarity and mutual exclusivity among features, and therefore the classification accuracy obtained by the subset of features selected by such methods is low [7]. Embedded-based methods are a special kind of wrapper-based method, so they are not discussed in this paper. The wrapper-based methods introduce classifiers and learning algorithms. The learning algorithm continuously generates new feature subsets, while the classifier evaluates the generated feature subsets and selects the optimal one in continuous iterations [8]. In this type of method, the classification accuracy of feature selection is high but consumes more time due to the introduction of classifiers. On the other hand, this type of method has a great relationship with the performance of the learning algorithm.

Akin to some previous works [7–9], we aim to adopt the swarm intelligence algorithm as the learning algorithm in the wrapper-based feature selection method. Specifically, a swarm intelligence algorithm studies the complex behaviors of a swarm consisting of several simple agents. By iteratively updating the swarm, the agents will have a more powerful performance than before, so that the algorithm can provide a sub-optimal solution. Swarm intelligence has the benefits of high convergence and powerful solving ability. Moreover, swarm intelligence can also handle NP-hard problems such as feature selection. Thus, it can be seen as an effective method to overcome the challenges of feature selection. For instance, some well-known swarm intelligence, i.e., genetic algorithm (GA) [10], particle swarm optimization (PSO) [11], dragonfly algorithm (DA) [12], ant-lion optimizer (ALO) [13], and grey wolf optimizer (GWO) [14] has been applied in feature selection.

Bat algorithm (BA) and binary BA (BBA) are promising forms of swarm intelligence and have been demonstrated to be better than other algorithms in some applications due to their effectiveness and high performance. However, as suggested in no-free lunch (NFL) theory, there are no algorithms that can suitably solve all optimization problems. In addition, BA also has some shortcomings in solving feature selection problems. Thus, we aim to enhance the performance of BA for solving feature selection. The contributions of this work are summarized as follows:

- We show that the feature selection is a multi-objective optimization problem, and we present the decision variables and the optimization goals of the feature selection problem.
- We propose an enhanced binary BA (EBBA) for solving the feature selection problem. In EBBA, we propose Lévy flight-based global search method, which enables the algorithm to jump out of the local optimum. Moreover, we propose a population diversity boosting method so that the exploration capability of the algorithm can be further enhanced. In addition, we use a recently proposed chaotic mapping to assign values to the key parameter of the algorithm, thus enhancing the exploitation capability of the algorithm.
- Simulations are conducted based on open datasets of UC Irvine machine learning repository to verify the solving ability of the proposed EBBA. First, we introduce

some benchmark algorithms for comparisons. Then, we show the effectiveness of the proposed improved factors.

The rest of this work is arranged as follows. Section 2 reviews some key related works about swarm intelligence algorithms and feature selection. Section 3 gives the model of feature selection. Section 4 proposes the EBBA and details the improved factors. Section 5 provides simulation results and Section 6 concludes this work.

2. Related Works

In this work, we aim to use one of the swarm intelligence algorithms, i.e., BA, to solve the feature selection problem, and thus some key related works are briefly introduced in this section.

2.1. Swarm Intelligence Algorithms

Swarm intelligence algorithms refer to evolutionary theory and swarm behavior. In the past few years, a large number of researchers have proposed various types of swarm intelligence algorithms to solve optimization problems in different domains.

First, some representative classical swarm intelligence algorithms are presented as follows. PSO is another representative swarm intelligence algorithm, which is inspired by the behavior of bird/fish populations. Moreover, artificial bee colony (ABC) [15], ant colony optimization (ACO) [16], etc., are also well-known swarm intelligence algorithms. Second, swarm intelligence algorithms also contain various types of bio-inspired algorithms. For example, Meng et al. [17] proposed a chicken swarm optimization (CSO) to solve optimization problems by simulating the rank order and the behavior of chickens (including roosters, hens, and chicks) in a flock. Yang et al. [18] proposed a BA and validated the performance of BA using eight nonlinear engineering optimization problems. Third, certain swarm intelligence algorithms were proposed inspired by various natural phenomena of the universe. Jiang et al. [19] proposed a new metaheuristic method, artificial raindrop algorithm (ARA), from natural rainfall phenomena and used it for the identification of unknown parameters of chaotic systems. Kaveh et al. [20] proposed a ray optimization (RO) based on Snell's law of light refraction and the phenomenon of light refraction.

In summary, researchers have proposed a large number of effective swarm intelligence algorithms and applied them to various optimization problems. However, these algorithms are not necessarily applicable to all engineering fields. Accordingly, proposing an enhanced swarm intelligence algorithm version according to the characteristics of an optimization problem is a major challenge.

2.2. Ways of Feature Selection

There are several existing methods and ways have been proposed for the purpose of feature selection. First, some filter methods are widely used due to their simplicity and relatively high performance. For instance, some methods based on correlation criteria and mutual information are detailed in reference [21]. In this case, several effective filter-based algorithms including correlation-based feature selection (CFS) [22], fast correlation-based filter (FCBF) [23], wavelet power spectrum (Spectrum) [24], Information Gain (IG) [25], ReliefF [26], etc. Second, wrapper-based approaches are key methods in feature selection. This type of method can be categorized by the type of learning algorithms. For instance, exhaustive, random search and metaheuristic search methods. Due to their effectiveness, the metaheuristic search methods including swarm intelligence algorithms can be seen as the most popular methods [27]. Finally, there are several embedded methods. The main approach is to incorporate feature selection as part of the training process, e.g., [21,28].

2.3. Swarm Intelligence-Based Feature Selection

There are many swarm intelligence algorithms have been adopted or proposed as the learning algorithm in wrapper-based feature selection methods, and we review some key algorithms as follows.

Li et al. [29] proposed an improved binary GWO (IBGWO) algorithm for solving feature selection problems, in which an enhanced opposition-based learning (E-OBL) initialization and a local search strategy were proposed for improving the performance of the algorithm. Kale et al. [30] presented four different improved versions of the sine cosine algorithm (SCA), where the updating mechanism of SCA is the improvements and innovations. Ouadfel et al. [31] proposed a hybrid feature selection approach based on the ReliefF filter method and equilibrium optimizer (EO), which is composed of two phases and tested in some open datasets. Abdel-Basset et al. [14] proposed three variants of BGWO in addition to the standard variant, applying different transfer functions to tackle the feature selection problem. In [32], two different wrapper feature selection approaches were proposed based on farmland fertility algorithm (FFA), which denoted as BFFAS and BFFAG, and these methods are effective in solving feature selection problems. On the other hand, BA and some variants have been adopted for solving feature selection problems. Varma et al. [33] proposed a bat optimization algorithm for wrapper-based feature selection and conducted simulations based on the CICInvesAndMal2019 benchmark dataset. Naik et al. [34] proposed a feature selection method to identify the relevant subset of features for the machine-learning task using the wrapper approach via BA. Rodrigues et al. [35] presented a wrapper feature selection approach based on bat algorithm (BA) and optimum-path forest (OPF). In [36], the authors proposed an improved BPSO algorithm as an essential tool of pre-processing for solving classification problem, in which a new updating mechanism for calculating Pbest and Gbest were proposed. Moreover, the authors in [37] proposed a binary DA (BDA) and use it to solve the feature selection problems. Likewise, Nakamura et al. [38] proposed a binary version of the bat algorithm, i.e., BBA, and evaluate its performance in solving the feature selection problems. In addition, in [39], a new hybrid feature selection method was proposed by using the sine cosine algorithm (SCA) and genetic algorithm (GA), and the algorithm is used for solving feature selection problems. Furthermore, Nagpal et al. [40] proposed a feature selection method via binary gravitational search algorithms (BGSA) in medical datasets, in which they can reduce the number of features by an average of 66% and enhance the accuracy of prediction.

The aforementioned methods can solve feature selection problems in various applications. However, according to NFL theory, different swarm intelligence algorithms may have different performances in various applications. Therefore, the existing methods are insufficient to solve all feature selection problems, which motivates us to propose an EBBA to handle more feature selection problems in this work.

3. Feature Selection Model

As shown in [8,29,33], the feature selection problem can be seen as a binary optimization model, and in this section, we introduce it in details. Specifically, the main purpose of feature selection is to reduce the data dimension by retaining the most valuable features through feature selection methods. Thus, there are two possibilities for each feature, i.e., to be selected and to be discarded. Therefore, the feature selection problem can be regarded as an optimization problem with a binary solution space.

It is can be seen from Figure 1, the solution space of the considered feature selection problem is a binary. Each feature is represented by a binary number, and if that binary number is 1, it means that the feature is selected, and conversely, if that binary number is 0, it means that the feature is discarded. Thus, the feature selection of a dataset can be represented by a binary array as follow:

$$\text{(Decision variables)} \quad X = [x_1, x_2, x_3, \dots, x_{N_{dim}}], \tag{1}$$

where N_{dim} is the number of features, in other words, the dimension number of the dataset. Under this model, there are two main objectives of the feature selection, i.e., to reduce the classification error rate of the obtained feature subsets, and to reduce the feature number of feature subsets. Thus, the feature selection problem is a multi-objective problem in which the first objective can be expressed as follows:

$$(\text{Objective 1}) \quad f_1 = 1 - f_{acc}, \tag{2}$$

where f_{acc} is the classification accuracy of the obtained feature subsets. Note that we introduce the KNN as a classifier to evaluate the feature subsets and the reasons are analyzed in following section. Moreover, the second objective of this work is to reduce the feature number of feature subsets, which can be expressed as follows:

$$(\text{Objective 2}) \quad f_2 = \frac{N'_{dim}}{N_{dim}}, \tag{3}$$

where N'_{dim} is the feature number of the selected feature subsets. To simultaneously the aforementioned objectives, we introduce the fitness function as follows:

$$(\text{Fitness function}) \quad f_{fit} = a \times f_1 + b \times f_2, \tag{4}$$

where $a \in [0,1]$ and $b = 1 - a$ are constants that denote the weights of the two objectives f_1 and f_2, respectively. Specifically, we can increase a to obtain a higher classification accuracy or increase b to obtain a smaller dimensional feature subset.

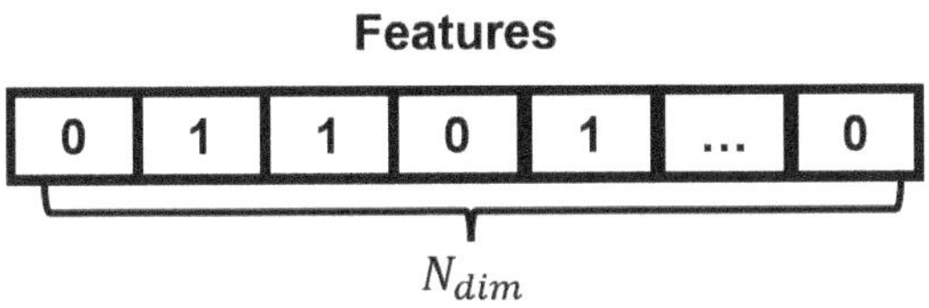

Figure 1. Feature selection model with binary solution space.

4. Proposed Algorithm

Based on the aforementioned feature selection model, we can optimize the decision variables shown in Equation (1) to obtain a better fitness function shown in Equation (4). Accordingly, we propose an EBBA in this section for solving the feature selection problem.

4.1. Conventional BA

BA is a swarm intelligence algorithm for global optimization, which is inspired by the echolocation behavior of bats. Specifically, bats look for prey by flying at a random velocity V_i at a random point X_i with a fixed frequency f_{min}, changing wavelength l, and loudness A_0. Depending on the proximity of their target, these bats can autonomously modify the wavelength (in other words, frequency) of their generated pulses as well as the rate of pulse emission r in the range of $[0,1]$. The corresponding mathematical model of BA can be detailed as follows. In the tth iteration, the frequency f_i of the the ith bat is expressed as follows.

$$f_i = f_{min} + (f_{max} - f_{min}) \times \beta, \tag{5}$$

where f_{max} and f_{min} are upper and lower bounds on the frequencies of all bats, respectively, and β is a random number between $[0,1]$.

Moreover, the velocity of the ith bat v_i can be modeled as follows:

$$V_i^t = V_i^{t-1} + (X_i^{t-1} - X^*) f_i, \tag{6}$$

where X^* is the bat with the highest fitness function value of the swarm.

In addition, the update method of the ith bat is shown as follows:

$$X_i^t = X_i^{t-1} + V_i^t, \tag{7}$$

where X_i^t is the position of the ith bat in the tth iteration.

Furthermore, BA also enhances search ability through local random walks. Specifically, BA asks the best bat in the swarm to conduct a local search with a certain probability, which can be expressed as follows:

$$X^N = X^* + \epsilon \times A^t, \tag{8}$$

where X^N is the newly generated bat after the random walk, A^t is the loudness of all bats in the t iteration, and ϵ is a random variable that ranges $[-1, 1]$.

Additionally, the loudness A_i and the rate r_i of pulse emission are also updated as the iterations proceed, which is shown as follows:

$$A_i^{t+1} = \alpha A_i^t, \quad r_i^{t+1} = r_i^0[1 - \exp(-\gamma t)], \tag{9}$$

where α is a parameter that ranges from 0 to 1, and $\gamma > 0$ is a parameter.

By using these mathematical models, the main steps of BA can be summarized as follows.

Step 1: Randomly generate population (bat swarm) $P = [X_1, X_2, \ldots, X_{N_{pop}}]$, where N_{pop} is the population size. Moreover, the velocity, pulse emissivity, and loudness of the bats are randomly generated. Then, the fitness values of all bats are calculated.

Step 2: Update the positions and velocities of bats by using Equations (5)–(9).

Step 3: A random number N_{rand} between 0 and 1 is firstly generated, and then if $N_{rand} > r_i$, a random walk will be performed by using Equation (8) to generate a new individual X^N around the current best individual X^*.

Step 4: Generate a random number N_{rand} of $[0, 1]$ again. If $N_{rand} < A_i$ and $f_{fit}(X^N) < f_{fit}(X_i)$, replace X_i with X^N and then update the loudness and pulse firing rate. If $f_{fit}(X^N) < f_{fit}(X^*)$, X^N is used to replace X^*.

Step 5: Repeat steps 2–4 until the terminal condition is reached.

Step 6: Return X^* as the final solution to the problem.

4.2. BBA

To make the BA can handle the binary solution space of the feature selection, Mirjalili et al. [41] introduce a binary operator. Specifically, the authors introduced a v-shape transfer function to map the continuous parameters into binary solution space, which can be expressed as follows:

$$V\left(v_{i,j}^t\right) = \left|\frac{2}{\pi}\arctan\left(\frac{\pi}{2}v_{i,j}^t\right)\right|, \tag{10}$$

$$x_{i,j}^{t+1} = \begin{cases} \left(x_{i,j}^t\right)^{-1} & \text{if rand } < V\left(v_{i,j}^{t+1}\right) \\ x_{i,j}^t & \text{if rand } \geq V\left(v_{i,j}^{t+1}\right) \end{cases}, \tag{11}$$

where $x_{i,j}^t$ and $v_{i,j}^t$ indicate the position and velocity of ith individual at tth iteration in jth dimension, and $(x_{i,j}^t)^{-1}$ is the complement of $x_{i,j}^t$. As such, the BBA can handle and update the binary decision variable reasonably.

4.3. EBBA

Conventional BA may confront some key challenges in solving the feature selection problems. First, when dealing with the big solution space of feature selection problem, BA may lack exploration ability, which may make the algorithm fall in local optima. Second, the bats of the BA are guided by the best bat of the swarm, i.e., X^*, which means that the population diversity is lower for the large scale datasets. Third, the exploration and exploration abilities of BA should be further balanced. Finally, BA is proposed for continuous problems whereas the feature selection problems are with a binary solution space. Thus, these reasons motivate us to enhance BA for better feature selection performance. The main steps of the proposed EBBA are detailed in Algorithm 1, and the correspondingly improved factors are as follows:

Algorithm 1 EBBA

Define the related parameters: population size N_{pop}, bat dimension N_{dim}, maximum iteration t_{max}, and fitness function, etc.;
Initialize the bat population, pulse frequency, pulse rates and loudness;
for $t = 1$ *to* t_{max} **do**
    Generate new bats by adjusting frequency, and updating velocities and locations by using Equations (5), (6) and (11);
    if $N_{rand} < r_i$ **then**
        if $N_{rand} < r_i'$ **then**
            Select the second best bat;
        end
        else
            Select the best bat;
        end
        Generate a new bat around the selected bat by using Equation (8) or (12);
    end
    if $N_{rand} < A_i$ *and* $f_{fit}(X_i) < f_{fit}(X^* or X^{**})$ **then**
        Accept the new bats and update r_i and A_i by using Equations (9) and (15);
    end
    Rank the bats and find the current best X^*;
    Generate $X^{new\prime}$ by using Equation (12);
    if $f_{fit}(X^*) > f_{fit}(X^{new})$ **then**
        $X^* = X^{new}$;
    end
end
Return X^*; //X^* **is the final feature selection result of the EBBA**

4.3.1. Lévy Flight-Based Global Search Method

Feature selection problems also are large-scale optimization problems since the dimension of some datasets is relatively large. In this case, the exploration ability of the optimization algorithm should be sufficient. However, the update of other bats in the swarm are determined by the best bat, and this mechanism will undoubtedly decrease the exploration ability of the algorithm. Thus, we introduce the Lévy flight to propose a global search method to improve the exploration ability of the algorithm. Specifically, a Lévy flight is a random walk in which the step-lengths have a Lévy distribution, a probability distribution that is heavy-tailed. By using the short-distance or long-distance searching alternately, the search scope can be extended.

First, mathematically, in each iteration, we generate a new bat according to the best bat X^* and Lévy flight, which can be expressed as follows:

$$X^{new} = X^* + \alpha \oplus \text{Lévy}~(\lambda), \tag{12}$$

where α is a parameter and its value is often assigned according to applications. Moreover, Lévy flight is taken from the Lévy distribution, which can be expressed as follows:

$$\text{Lévy}(\lambda) \sim u = t^{-\lambda}(1 < \lambda < 3), \tag{13}$$

Second, the newly generated bat X^{new} is evaluated to obtain its fitness value, and then we compare the X^{new} with the best bat X^* based on their fitness function values. If the X^{new} outmatches X^*, then $X^* = X^{new}$. By using this method, the best bat of the swarm is easy to jump out of local optima, thereby enhancing the exploration ability of EBBA.

4.3.2. Population Diversity Boosting Method

In BA, all the bats are guided by the best bat of the swarm, which may decrease the population diversity of the algorithm, thereby affecting the solving performance. In this case, we aim to propose a method for boosting population diversity. In bat swarm, the second best bat also is meaningful and with a strong value for guiding other bats. Thus, we use the second best bat X^{**} to generate a part of new bats as follows:

$$X^{N2} = X^{**} + \epsilon \times A^t, \tag{14}$$

where X^{N2} is the newly generated bat and whether the method used is determined by a parameter r' which can be expressed as follows:

$$r_i^{t\prime} = \frac{r_i^t}{2}, \tag{15}$$

By using this method, the bat swarm is simultaneously guided by the best bat and the second-best bat, so that enhancing the population diversity of the algorithm.

4.3.3. Chaos-Based Loudness Method

The loudness in BA can determine the weights of exploitation and exploration abilities of the BA. However, the loudness update method of conventional BA is linear, which may be unsuitable for the feature selection. Thus, we introduce a novel fractional one dimensional chaotic map to update loudness [42], which can be expressed as follows:

$$A_i^t = C^t, \tag{16}$$

where C^t is the tth dimension of the fractional one dimensional chaotic map, which can be expressed as follows:

$$C_{t+1} = f(C_t) = \begin{cases} \frac{1}{C_t^2+\alpha} - \beta C_t & \text{if } C_t \in \left[0, \frac{1}{\alpha}\right] \\ \frac{-1}{C_t^2+\alpha} - \beta C_t & \text{if } C_t \in \left[\frac{-1}{\alpha}, 0\right) \end{cases}, \tag{17}$$

where α and β are two real parameters, and they are assigned as 0.001 and 0.9 in this work, respectively. By using the high chaotic behavior of the method, the exploitation and exploration abilities of EBBA can be balanced.

4.3.4. Complexity Analysis of EBBA

The complexity of EBBA is analyzed in this part. In the proposed feature selection model, the most time-consuming step is the calculation of fitness function value since we introduce classifier, which is several orders of magnitude complex than other steps. In this case, other steps can be omitted. Accordingly, the complexity of EBBA is $\mathcal{O}(t_{max} \cdot N_{pop})$ when the maximum number of iteration and population size are denoted as t_{max} and N_{pop}, respectively.

5. Simulations

In this section, we conduct the simulations to evaluate the proposed EBBA. First, the datasets and setups are presented. Second, we compare the EBBA with some benchmark algorithms. Third, we illustrate the effectiveness of the improved factors.

5.1. Datasets and Setups

In this work, we introduce 10 typical UC Irvine Machine Learning Repository datasets. The main information of these datasets are shown in Table 1.

Table 1. Datasets.

	Dataset	Number of Features	Number of Instance
1	Breastcancer	10	699
2	BreastEW	30	569
3	Congress	16	435
4	Exactly	13	1000
5	Exactly2	13	1000
6	HeartEW	13	270
7	SonarEW	60	208
8	SpectEW	22	267
9	tic-tac-toe	9	958
10	Vote	16	300

Moreover, the used CPU is 11th Gen Intel(R) Core(TM) i7-11700 @ 2.50 GHz and the RAM is 16 GB. We use python to implement the simulation codes and adopt KNN (k = 5). Note that we use KNN classifier since it is simple, easy and highly accurate, it is also insensitive to outliers and no data entry settings. Moreover, using a simple and relatively cheap classification algorithm in a wrapper approach can obtain a good feature subset that is also suitable for complex classification algorithms. In contrast, if an advanced classification algorithm is used for wrapper-based feature selection, the obtained feature subset will be failed for simple classification algorithms. The reason is that when using advanced classification algorithms, the learning algorithm of the wrapper approach (e.g., the proposed EBBA) will capture the characteristics of the classification algorithm instead of the relationship of different features. In addition, a and b in the fitness function are set to 0.99 and 0.01, respectively. Furthermore, In this paper, binary PSO (BPSO) [36], BGWO [43], BDA [37], and BBA [41] are introduced as the benchmarks, and the key parameters of these algorithms are shown in Table 2. Note that the population size and iteration number of EBBA and other benchmarks are set as 24 and 100, respectively. Additionally, to avoid the experiment's random bias, each algorithm is performed 30 times independently in these selected datasets, as specified by the central limit theorem. What's more, 80% of the instances are utilized for training, while the remaining 20% are used for testing [7,44,45]. Note that the classifier only provides feedback to the EBBA, which means that the overfitting affected by the division of the dataset will not have too much impact on the feature selection method based on the wrapper and swarm intelligence.

Table 2. Key parameters of benchmark algorithms.

	Algorithm	Key Parameters
1	BPSO	$c1 = 2$, $c2 = 2$
2	BGWO	$\alpha = [2, 0]$
3	BDA	$w = [0.9, 0.4]$, $s = [0.2, 0]$, $a = [0.2, 0]$, $c = [0.2, 0]$, $f = [0.2, 0]$, $e = [0, 0.1]$
4	BBA	$A = 0.25$, $Q_{max} = 2$, $Q_{min} = 0$
5	EBBA	$Q_{max} = 2$, $Q_{min} = 0$

5.2. Simulation Results

Table 3 shows the optimization results of the accuracy rate, number of selected features, fitness function values, and CPU times achieved by various algorithms. Note that the best values among all comparison algorithms are highlighted in bold font. As can be seen, the proposed EBBA achieves the best accuracy rate on 7 datasets and achieves the best of selected feature number on 2 datasets. More intuitively, the proposed EBBA achieves the best fitness function values on 9 datasets, which means that the proposed EBBA is with the best performance among all benchmark algorithms. The reason may be that we enhance the EBBA by improving its exploration ability and balancing its exploration and exploitation abilities, which make the EBBA more suitable for solving feature selection problems.

Table 3. Optimization results achieved by various algorithms. (The best values are highlighted in bold).

		Accuracy	Feature #	Fitness Value	CPU Time
Breastcancer	BBA	**0.9786**	6.0000	**0.0272**	69.5141
	BDA	0.9767	**5.8000**	0.0289	68.7234
	BGWO	0.9767	6.5000	0.0296	75.3848
	BPSO	**0.9786**	6.0000	**0.0272**	69.2956
	EBBA	**0.9786**	6.0000	**0.0272**	**60.9926**
BreastEW	BBA	0.9613	6.6000	0.0405	73.3552
	BDA	0.9589	**5.1667**	0.0424	67.3963
	BGWO	0.9532	14.6667	0.0512	79.7631
	BPSO	0.9612	8.5000	0.0413	75.4869
	EBBA	**0.9614**	5.4333	**0.0400**	**63.4576**
Congress	BBA	**0.9793**	6.9000	0.0248	58.3419
	BDA	0.9743	**5.1667**	0.0287	55.9260
	BGWO	0.9750	8.8000	0.0303	61.5544
	BPSO	0.9785	6.7333	0.0255	59.3514
	EBBA	**0.9793**	6.4000	**0.0245**	**49.2745**
Exactly	BBA	**1.0000**	**6.0000**	**0.0046**	91.2861
	BDA	0.9197	6.6667	0.0846	100.3068
	BGWO	0.9040	8.0333	0.1012	116.4705
	BPSO	0.9999	6.0333	0.0048	103.6469
	EBBA	**1.0000**	**6.0000**	**0.0046**	**88.3184**
Exactly2	BBA	0.7854	2.2667	0.2142	86.8956
	BDA	0.7847	1.7667	0.2145	90.0558
	BGWO	0.7693	7.5333	0.2342	117.3080
	BPSO	**0.7890**	**1.0667**	**0.2097**	101.6048
	EBBA	0.7868	1.9667	0.2126	**81.4770**
HeartEW	BBA	0.8511	5.0333	0.1513	45.4301
	BDA	0.8391	**4.9333**	0.1631	44.0686
	BGWO	0.8357	6.6667	0.1678	45.2137
	BPSO	**0.8527**	5.0000	0.1497	44.8358
	EBBA	0.8520	5.0667	**0.1504**	**39.0150**
SonarEW	BBA	0.9100	27.3000	0.0937	49.9454
	BDA	0.9010	**20.5000**	0.1015	48.1458
	BGWO	0.8989	40.6667	0.1069	55.5145
	BPSO	0.9046	27.7333	0.0991	51.0521
	EBBA	**0.9117**	27.0667	**0.0919**	**42.8189**
SpectEW	BBA	0.7379	10.5000	0.2643	46.5456
	BDA	0.7185	**8.9667**	0.2827	44.8928
	BGWO	0.7219	14.3667	0.2819	47.5410
	BPSO	0.7335	10.5000	0.2687	46.3761
	EBBA	**0.7407**	10.7000	**0.2615**	**38.4262**
tic-tac-toe	BBA	0.8493	8.8667	0.1590	97.9119
	BDA	0.8099	7.1667	0.1962	97.8065
	BGWO	0.8465	8.8000	0.1618	91.4816
	BPSO	**0.8521**	**9.0000**	**0.1564**	88.2232
	EBBA	**0.8521**	**9.0000**	**0.1564**	**84.4753**
Vote	BBA	**0.9519**	5.2000	**0.0509**	46.4947
	BDA	0.9457	5.7667	0.0574	45.8363
	BGWO	0.9440	8.5333	0.0608	47.9505
	BPSO	0.9493	**4.9000**	0.0532	46.5221
	EBBA	0.9513	5.4000	0.0516	**38.9119**

In addition, Figure 2 shows the convergence rates obtained by different benchmark algorithms during the solving processes. As can be seen, the proposed EBBA achieves the best curves on most datasets, which performs the best convergence ability among all the comparison algorithms.

On the other hand, we also evaluate the effectiveness of the proposed improved factors. Specifically, we combine the proposed Lévy flight-based global search method, population diversity boosting method, and chaos-based loudness method with conventional BBA, namely, EBBA-IF1, EBBA-IF2, and EBBA-IF3, respectively. Table 4 shows the optimization results of the accuracy rate, number of selected features, fitness function values, and CPU times achieved by various algorithms. Moreover, Figure 3 shows the convergence rates obtained by different EBBA versions during the solving processes. As can be seen, the

EBBA, EBBA-IF1, EBBA-IF2, and EBBA-IF3 outperform conventional BBA, which means that the proposed improved factors are non-trivial and effective.

Table 4. Optimization results achieved by various EBBA versions. (The best values are highlighted in bold).

		Accuracy	Feature #	Fitness Value	CPU Time
Breastcancer	BBA	**0.9786**	**6.0000**	**0.0272**	69.5141
	BBA-IF1	**0.9786**	**6.0000**	**0.0272**	64.4441
	BBA-IF2	**0.9786**	**6.0000**	**0.0272**	67.0577
	BBA-IF3	**0.9786**	**6.0000**	**0.0272**	**60.1375**
	EBBA	**0.9786**	**6.0000**	**0.0272**	60.9926
BreastEW	BBA	0.9613	6.6000	0.0405	73.3552
	BBA-IF1	0.9613	**5.2333**	**0.0400**	65.2602
	BBA-IF2	**0.9614**	5.9000	0.0402	68.0782
	BBA-IF3	**0.9611**	5.4000	0.0403	**63.0563**
	EBBA	**0.9614**	5.4333	**0.0400**	63.4576
Congress	BBA	**0.9793**	6.9000	0.0248	58.3419
	BBA-IF1	0.9790	6.6667	0.0249	51.4119
	BBA-IF2	0.9783	**6.2333**	0.0253	54.6810
	BBA-IF3	0.9791	6.6667	0.0249	50.4441
	EBBA	**0.9793**	6.4000	**0.0245**	**49.2745**
Exactly	BBA	**1.0000**	**6.0000**	**0.0046**	91.2861
	BBA-IF1	**1.0000**	**6.0000**	**0.0046**	88.5090
	BBA-IF2	**1.0000**	**6.0000**	**0.0046**	92.2044
	BBA-IF3	**1.0000**	**6.0000**	**0.0046**	88.9417
	EBBA	**1.0000**	**6.0000**	**0.0046**	**88.3184**
Exactly2	BBA	0.7854	2.2667	0.2142	86.8956
	BBA-IF1	0.7859	2.0333	0.2136	**81.4199**
	BBA-IF2	0.7848	2.5333	0.2150	87.3399
	BBA-IF3	0.7848	2.4333	0.2149	84.0370
	EBBA	**0.7868**	**1.9667**	**0.2126**	81.4770
HeartEW	BBA	0.8511	5.0333	0.1513	45.4301
	BBA-IF1	0.8516	4.9000	0.1507	39.6087
	BBA-IF2	**0.8521**	5.1000	**0.1503**	40.7423
	BBA-IF3	0.8511	**4.8333**	0.1511	**38.9901**
	EBBA	0.8520	5.0667	0.1504	39.0150
SonarEW	BBA	0.9100	27.3000	0.0937	49.9454
	BBA-IF1	0.9097	27.0333	0.0939	42.1371
	BBA-IF2	0.9100	27.3000	0.0937	44.0751
	BBA-IF3	0.9105	**26.7000**	0.0931	**41.8531**
	EBBA	**0.9117**	27.0667	**0.0919**	42.8189
SpectEW	BBA	0.7379	10.5000	0.2643	46.5456
	BBA-IF1	0.7393	10.2333	0.2628	38.7934
	BBA-IF2	0.7393	**10.2000**	0.2628	40.4760
	BBA-IF3	0.7386	10.7000	0.2636	39.2009
	EBBA	**0.7407**	10.7000	**0.2615**	**38.4262**
tic-tac-toe	BBA	0.8493	8.8667	0.1590	97.9119
	BBA-IF1	0.8493	8.8667	0.1590	86.5623
	BBA-IF2	**0.8521**	**9.0000**	**0.1564**	89.9898
	BBA-IF3	**0.8521**	**9.0000**	**0.1564**	86.5650
	EBBA	**0.8521**	**9.0000**	**0.1564**	**84.4753**
Vote	BBA	**0.9519**	5.2000	**0.0509**	46.4947
	BBA-IF1	0.9518	5.3667	0.0511	39.7080
	BBA-IF2	0.9516	**5.1000**	0.0511	41.6823
	BBA-IF3	0.9513	5.4333	0.0516	40.1654
	EBBA	0.9513	5.4000	0.0516	**38.9119**

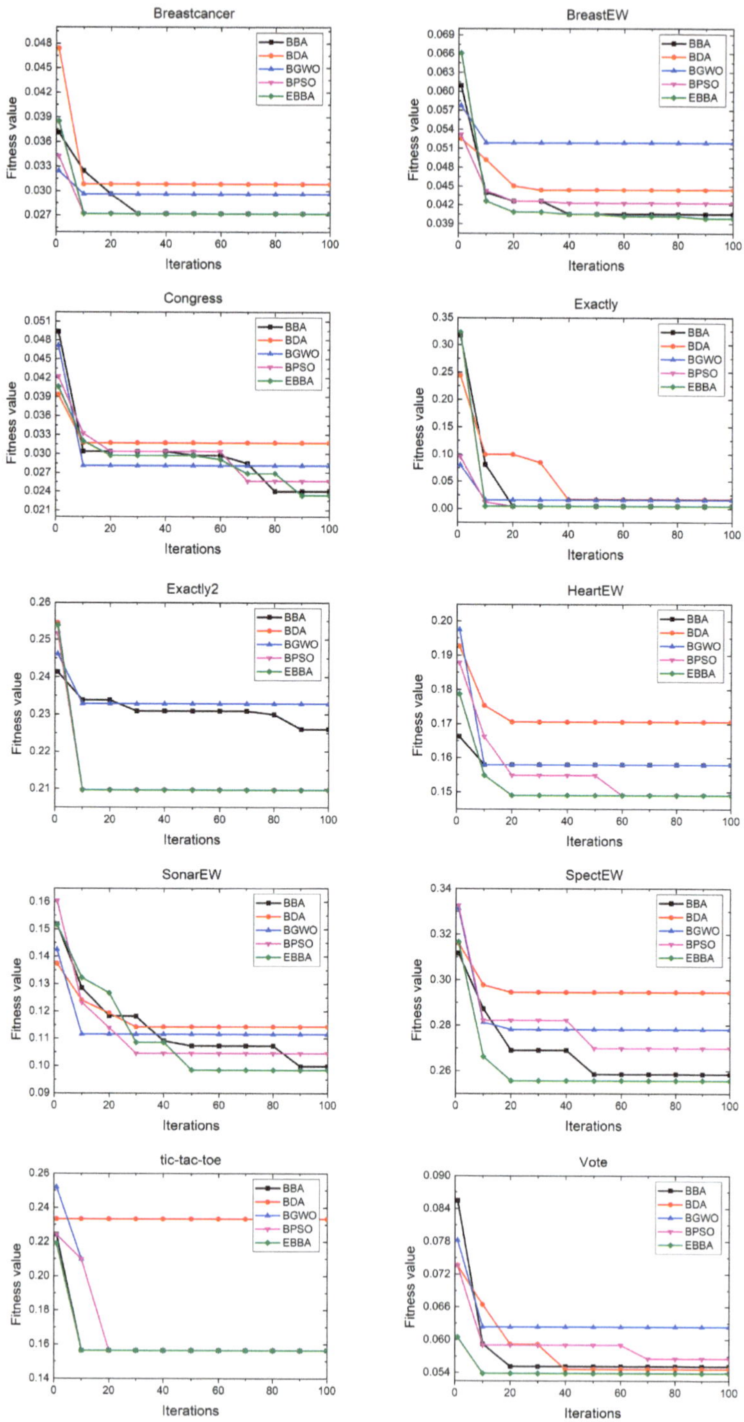

Figure 2. Convergence rates achieved by various algorithms.

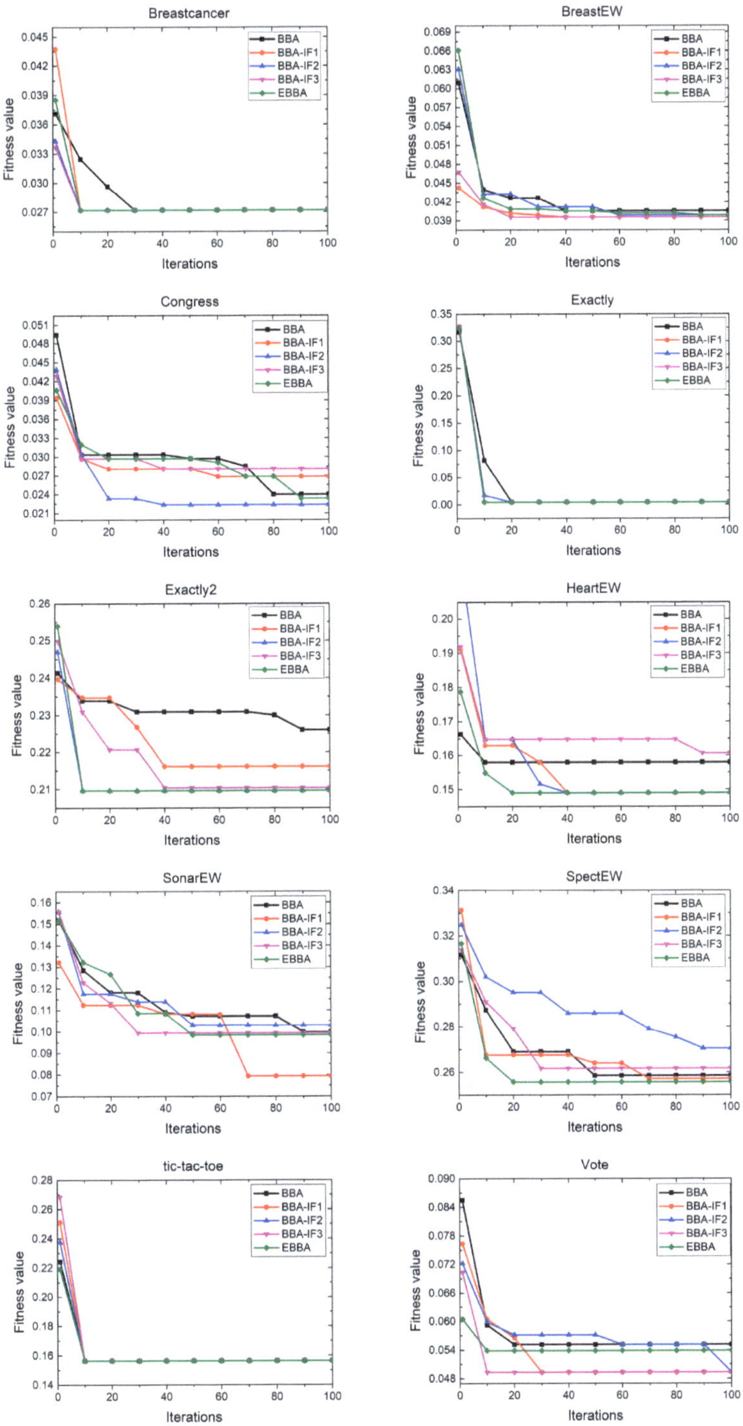

Figure 3. Convergence rates obtained by different EBBA versions.

5.3. Performance Evaluation under Different Classifiers

In this section, we consider two other classification algorithms which are decision tree and random forest. Specifically, the decision tree is easy to understand and visualized, requires only little data preparation, whereas it may be easy to be overfitting. Likewise, random forest is also extremely accurate and not prone to overfitting, and can run effectively on large datasets with good noise immunity. Moreover, due to the complexity of the classification algorithms, validating the newly introduced two classifiers using all datasets is a huge and time-consuming task. Thus, we only consider using the dataset SpectEW as the experimental dataset since it has the middle dimension number of all datasets, which is representative. In addition, other settings are similar to that of the KNN-based method.

In this case, Tables 5 and 6 provide the simulation results obtained by different algorithms in terms of accuracy, the number of features and fitness function value under decision tree and random forest, respectively. As can be seen, the proposed EBBA also outperforms other comparison algorithms under other classification algorithms, which shows that the improved factors are effective even if the classification algorithm is changed. Thus, the proposed method has good performance in both maybe overfitting and non-overfitting cases.

Table 5. Optimization results obtained by different algorithms under decision tree. (The best values are highlighted in bold).

Algorithms	Accuracy	The Number of Features	Fitness Value
BBA	0.7312	7.9333	0.2683
BDA	0.7211	**7.7333**	0.2783
BGWO	0.7115	12.4000	0.2886
BPSO	0.7281	9.1667	0.2726
EBBA	**0.7325**	8.5333	**0.2682**

Table 6. Optimization results obtained by different algorithms under random forest. (The best values are highlighted in bold).

Algorithms	Accuracy	The Number of Features	Fitness Value
BBA	0.7247	8.9333	0.2766
BDA	0.7058	**5.9333**	0.2940
BGWO	0.7035	12.4333	0.2992
BPSO	0.7235	9.0667	0.2779
EBBA	**0.7252**	9.0667	**0.2762**

6. Conclusions

In this paper, the feature selection problems which can enhance the classification and reduce data dimension are studied. First, we model the feature selection problem and then transfer it as a fitness function. Then, we propose an EBBA for solving the feature selection problem. In EBBA, we introduce Lévy flight-based global search method, population diversity boosting method and chaos-based loudness method to improve the BA and make it more applicable to feature selection problems. Finally, the simulations are conducted to evaluate the proposed EBBA and the simulation results demonstrate that the proposed EBBA outmatches other comparison benchmarks. Moreover, the non-trivial of the proposed improved factors is illustrated. In the future, we intend to use more realistic datasets to evaluate the proposed EBBA.

Author Contributions: Data curation, H.K.; Supervision, L.Z.; Writing—original draft, J.F. All authors have read and agreed to the published version of the manuscript.

Funding: This research was supported by National Natural Science Foundation of China under Grant 82090052, and Shanghai Municipal Science and Technology Major Project 2021SHZDZX0103.

Data Availability Statement: Not applicable.

Conflicts of Interest: The authors declare no conflict of interest.

References

1. Xia, Q.; Zhou, L.; Ren, W.; Wang, Y. Proactive and intelligent evaluation of big data queries in edge clouds with materialized views. *Comput. Netw.* **2022**, *203*, 108664. [CrossRef]
2. Berger, C.; Eichhammer, P.; Reiser, H.P.; Domaschka, J.; Hauck, F.J.; Habiger, G. A Survey on Resilience in the IoT: Taxonomy, Classification, and Discussion of Resilience Mechanisms. *ACM Comput. Surv.* **2022**, *54*, 147. [CrossRef]
3. Xu, L.; Tolmochava, T.; Zhou, X. Search History Visualization for Collaborative Web Searching. *Big Data Res.* **2021**, *23*, 100180. [CrossRef]
4. Notash, A.Y.; Bayat, P.; Haghighat, S.; Notash, A.Y. Evolutionary ensemble feature selection learning for image-based assessment of lymphedema arm volume. *Concurr. Comput. Pract. Exp.* **2022**, *34*, e6334. [CrossRef]
5. Abdulla, M.; Khasawneh, M.T. Integration of aggressive bound tightening and Mixed Integer Programming for Cost-sensitive feature selection in medical diagnosis. *Expert Syst. Appl.* **2022**, *187*, 115902. [CrossRef]
6. Alsahaf, A.; Petkov, N.; Shenoy, V.; Azzopardi, G. A framework for feature selection through boosting. *Expert Syst. Appl.* **2022**, *187*, 115895. [CrossRef]
7. Li, J.; Kang, H.; Sun, G.; Feng, T.; Li, W.; Zhang, W.; Ji, B. IBDA: Improved Binary Dragonfly Algorithm With Evolutionary Population Dynamics and Adaptive Crossover for Feature Selection. *IEEE Access* **2020**, *8*, 108032–108051. [CrossRef]
8. Ji, B.; Lu, X.; Sun, G.; Zhang, W.; Li, J.; Xiao, Y. Bio-Inspired Feature Selection: An Improved Binary Particle Swarm Optimization Approach. *IEEE Access* **2020**, *8*, 85989–86002. [CrossRef]
9. Agrawal, P.; Ganesh, T.; Oliva, D.; Mohamed, A.W. S-shaped and V-shaped gaining-sharing knowledge-based algorithm for feature selection. *Appl. Intell.* **2022**, *52*, 81–112. [CrossRef]
10. Lappas, P.Z.; Yannacopoulos, A.N. A machine learning approach combining expert knowledge with genetic algorithms in feature selection for credit risk assessment. *Appl. Soft Comput.* **2021**, *107*, 107391. [CrossRef]
11. Li, A.; Xue, B.; Zhang, M. Improved binary particle swarm optimization for feature selection with new initialization and search space reduction strategies. *Appl. Soft Comput.* **2021**, *106*, 107302. [CrossRef]
12. Too, J.; Mirjalili, S. A Hyper Learning Binary Dragonfly Algorithm for Feature Selection: A COVID-19 Case Study. *Knowl. Based Syst.* **2021**, *212*, 106553. [CrossRef]
13. Wang, M.; Wu, C.; Wang, L.; Xiang, D.; Huang, X. A feature selection approach for hyperspectral image based on modified ant lion optimizer. *Knowl. Based Syst.* **2019**, *168*, 39–48. [CrossRef]
14. Abdel-Basset, M.; Sallam, K.M.; Mohamed, R.; Elgendi, I.; Munasinghe, K.S.; Elkomy, O.M. An Improved Binary Grey-Wolf Optimizer With Simulated Annealing for Feature Selection. *IEEE Access* **2021**, *9*, 139792–139822. [CrossRef]
15. Bacanin, N.; Bezdan, T.; Venkatachalam, K.; Zivkovic, M.; Strumberger, I.; Abouhawwash, M.; Ahmed, A.B. Artificial Neural Networks Hidden Unit and Weight Connection Optimization by Quasi-Refection-Based Learning Artificial Bee Colony Algorithm. *IEEE Access* **2021**, *9*, 169135–169155. [CrossRef]
16. Zhao, H.; Zhang, C.; Zheng, X.; Zhang, C.; Zhang, B. A decomposition-based many-objective ant colony optimization algorithm with adaptive solution construction and selection approaches. *Swarm Evol. Comput.* **2022**, *68*, 100977. [CrossRef]
17. Meng, X.; Liu, Y.; Gao, X.Z.; Zhang, H. A New Bio-inspired Algorithm: Chicken Swarm Optimization. In *Advances in Swarm Intelligence*; Tan, Y., Shi, Y., Coello, C.A.C., Eds.; Lecture Notes in Computer Science; Springer: Berlin/Heidelberg, Germany, 2014; Volume 8794, pp. 86–94. [CrossRef]
18. Yang, X.S.; Gandomi, A.H. Bat algorithm: A novel approach for global engineering optimization. *Eng. Comput.* **2012**, *29*, 464–483. [CrossRef]
19. Jiang, Q.; Wang, L.; Hei, X. Parameter identification of chaotic systems using artificial raindrop algorithm. *J. Comput. Sci.* **2015**, *8*, 20–31. [CrossRef]
20. Kaveh, A.; Khayatazad, M. A new meta-heuristic method: Ray Optimization. *Comput. Struct.* **2012**, *112–113*, 283–294. [CrossRef]
21. Chandrashekar, G.; Sahin, F. A survey on feature selection methods. *Comput. Electr. Eng.* **2014**, *40*, 16–28. [CrossRef]
22. Doshi, M. Correlation based feature selection (CFS) technique to predict student Perfromance. *Int. J. Comput. Netw. Commun.* **2014**, *6*, 197. [CrossRef]
23. Senliol, B.; Gulgezen, G.; Yu, L.; Cataltepe, Z. Fast Correlation Based Filter (FCBF) with a different search strategy. In Proceedings of the 2008 23rd International Symposium on Computer and Information Sciences, Istanbul, Turkey, 27–29 October 2008; pp. 1–4.
24. Subramani, P.; Sahu, R.; Verma, S. Feature selection using Haar wavelet power spectrum. *BMC Bioinform.* **2006**, *7*, 432. [CrossRef] [PubMed]
25. Azhagusundari, B.; Thanamani, A.S. Feature selection based on information gain. *Int. J. Innov. Technol. Explor. Eng. (IJITEE)* **2013**, *2*, 18–21.
26. Spolaôr, N.; Cherman, E.A.; Monard, M.C.; Lee, H.D. ReliefF for Multi-label Feature Selection. In Proceedings of the Brazilian Conference on Intelligent Systems, BRACIS 2013, Fortaleza, Brazil, 19–24 October 2013; pp. 6–11. [CrossRef]
27. Rostami, M.; Berahmand, K.; Nasiri, E.; Forouzandeh, S. Review of swarm intelligence-based feature selection methods. *Eng. Appl. Artif. Intell.* **2021**, *100*, 104210. [CrossRef]
28. Dhal, P.; Azad, C. A comprehensive survey on feature selection in the various fields of machine learning. *Appl. Intell.* **2022**, *52*, 4543–4581. [CrossRef]

29. Li, W.; Kang, H.; Feng, T.; Li, J.; Yue, Z.; Sun, G. Swarm Intelligence-Based Feature Selection: An Improved Binary Grey Wolf Optimization Method. In *Knowledge Science, Engineering and Management*; Qiu, H., Zhang, C., Fei, Z., Qiu, M., Kung, S., Eds.; Lecture Notes in Computer Science; Springer: Berlin/Heidelberg, Germany, 2021; Volume 12817, pp. 98–110. [CrossRef]
30. Kale, G.A.; Yüzgeç, U. Advanced strategies on update mechanism of Sine Cosine Optimization Algorithm for feature selection in classification problems. *Eng. Appl. Artif. Intell.* **2022**, *107*, 104506. [CrossRef]
31. Ouadfel, S.; Elaziz, M.A. Efficient high-dimension feature selection based on enhanced equilibrium optimizer. *Expert Syst. Appl.* **2022**, *187*, 115882. [CrossRef]
32. Hosseinalipour, A.; Gharehchopogh, F.S.; Masdari, M.; Khademi, A. A novel binary farmland fertility algorithm for feature selection in analysis of the text psychology. *Appl. Intell.* **2021**, *51*, 4824–4859. [CrossRef]
33. Varma, P.R.K.; Mallidi, S.K.R.; Jhansi, S.J.; Dinne, P.L. Bat optimization algorithm for wrapper-based feature selection and performance improvement of android malware detection. *IET Netw.* **2021**, *10*, 131–140. [CrossRef]
34. Naik, A.K.; Kuppili, V.; Edla, D.R. Efficient feature selection using one-pass generalized classifier neural network and binary bat algorithm with a novel fitness function. *Soft Comput.* **2020**, *24*, 4575–4587. [CrossRef]
35. Rodrigues, D.; Pereira, L.A.M.; Nakamura, R.Y.M.; Costa, K.A.P.; Yang, X.; de Souza, A.N.; Papa, J.P. A wrapper approach for feature selection based on Bat Algorithm and Optimum-Path Forest. *Expert Syst. Appl.* **2014**, *41*, 2250–2258. [CrossRef]
36. Huda, R.K.; Banka, H. A group evaluation based binary PSO algorithm for feature selection in high dimensional data. *Evol. Intell.* **2021**, *14*, 1949–1963. [CrossRef]
37. Mafarja, M.M.; Eleyan, D.; Jaber, I.; Hammouri, A.; Mirjalili, S. Binary dragonfly algorithm for feature selection. In Proceedings of the 2017 International Conference on New Trends in Computing Sciences (ICTCS), Amman, Jordan, 11–13 October 2017; pp. 12–17.
38. Nakamura, R.Y.M.; Pereira, L.A.M.; Rodrigues, D.; Costa, K.A.P.; Papa, J.P.; Yang, X.S. Binary bat algorithm for feature selection. In *Swarm Intelligence and Bio-Inspired Computation*; Elsevier: Amsterdam, The Netherlands, 2013; pp. 225–237.
39. Abualigah, L.M.; Dulaimi, A.J. A novel feature selection method for data mining tasks using hybrid Sine Cosine Algorithm and Genetic Algorithm. *Clust. Comput.* **2021**, *24*, 2161–2176. [CrossRef]
40. Nagpal, S.; Arora, S.; Dey, S.; Shreya. Feature selection using gravitational search algorithm for biomedical data. *Procedia Comput. Sci.* **2017**, *115*, 258–265. [CrossRef]
41. Mirjalili, S.; Mirjalili, S.M.; Yang, X.S. Binary bat algorithm. *Neural Comput. Appl.* **2014**, *25*, 663–681. [CrossRef]
42. Talhaoui, M.Z.; Wang, X. A new fractional one dimensional chaotic map and its application in high-speed image encryption. *Inf. Sci.* **2021**, *550*, 13–26. [CrossRef]
43. Emary, E.; Zawbaa, H.M.; Hassanien, A.E. Binary grey wolf optimization approaches for feature selection. *Neurocomputing* **2016**, *172*, 371–381. [CrossRef]
44. Mafarja, M.M.; Aljarah, I.; Heidari, A.A.; Hammouri, A.I.; Faris, H.; Al-Zoubi, A.M.; Mirjalili, S. Evolutionary Population Dynamics and Grasshopper Optimization approaches for feature selection problems. *Knowl. Based Syst.* **2018**, *145*, 25–45. [CrossRef]
45. Tubishat, M.; Ja'afar, S.; Alswaitti, M.; Mirjalili, S.; Idris, N.; Ismail, M.A.; Omar, M.S. Dynamic Salp swarm algorithm for feature selection. *Expert Syst. Appl.* **2021**, *164*, 113873. [CrossRef]

remote sensing

MDPI

Article

Semi-Supervised Cloud Detection in Satellite Images by Considering the Domain Shift Problem

Jianhua Guo [1], Qingsong Xu [1], Yue Zeng [1], Zhiheng Liu [2] and Xiaoxiang Zhu [1,3,*]

1 Department of Aerospace and Geodesy, Data Science in Earth Observation, Technical University of Munich (TUM), 80333 Munich, Germany; jianhua.guo@tum.de (J.G.); qingsong.xu@tum.de (Q.X.); yue.zeng@tum.de (Y.Z.)

2 School of Aerospace Science and Technology, Xidian University, Xi'an 710126, China; liuzhiheng@xidian.edu.cn

3 Remote Sensing Technology Institute, German Aerospace Center (DLR), 82234 Weßling, Germany

* Correspondence: xiaoxiang.zhu@dlr.de

Abstract: In terms of semi-supervised cloud detection work, efforts are being made to learn a promising cloud detection model via a limited number of pixel-wise labeled images and a large number of unlabeled ones. However, remote sensing images obtained from the same satellite sensor often show a data distribution drift problem due to the different cloud shapes and land-cover types on the Earth's surface. Therefore, there are domain distribution gaps between labeled and unlabeled satellite images. To solve this problem, we take the domain shift problem into account for the semi-supervised learning (SSL) network. Feature-level and output-level domain adaptations are applied to reduce the domain distribution gaps between labeled and unlabeled images, thus improving predicted results accuracy of the SSL network. Experimental results on Landsat-8 OLI and GF-1 WFV multispectral images demonstrate that the proposed semi-supervised cloud detection network (SSCDnet) is able to achieve promising cloud detection performance when using a limited number of labeled samples and outperforms several state-of-the-art SSL methods.

Keywords: remote sensing imagery; cloud detection; semi-supervised learning; distribution drift; domain shift problem; domain adaptation

Citation: Guo, J.; Xu, Q.; Zeng, Y.; Liu, Z.; Zhu, X. Semi-Supervised Cloud Detection in Satellite Images by Considering the Domain Shift Problem. *Remote Sens.* **2022**, *14*, 2641. https://doi.org/10.3390/rs14112641

Academic Editors: Miltiadis D. Lytras and Andreea Claudia Serban

Received: 25 April 2022
Accepted: 30 May 2022
Published: 31 May 2022

1. Introduction

With the development of the Earth observation technology, an increasing number of optical satellites are launched for Earth observation missions. Remote sensing images acquired from the optical satellites can serve environment protection [1], global climate change [2], hydrology [3], agriculture [4], urban development [5], and military reconnaissance [6]. However, since 60% earth's surface is covered by clouds, the acquired optical remote sensing (RS) images are often contaminated by clouds [7]. In the field of meteorological, cloud information of RS images is useful in weather forecast [8], while, for earth surface observation missions, cloud coverage degrades the quality of satellite imagery. Therefore, it is important to improve RS images quality through cloud detection.

Over the past few decades, cloud detection from RS imagery has attracted much attention. Many advanced cloud detection technologies have been proposed. In this paper, we broadly categorize these methods into rule-based methods and machine learning-based methods. The rule-based methods are mostly developed from spectral/spatial domain [9–12]. These methods distinguish clouds from clear sky pixels by exploiting reflectance variations in visible, shortwave-infrared, and thermal bands. Rule-based methods have obvious flaws, i.e., they strongly depend on particular sensor models and have poor generalization performance. For example, Fmask algorithms [9–11] are developed for Sentinel-2 and Landsat 4/5/7/8 satellite images, while the multifeature combined (MFC) method [12] is developed for GF-1 wide field view (WFV) satellite images only. In addition, machine learning-based cloud detection methods have also attracted much attention due to their powerful data adaptability. The most representative machine learning-based cloud detection methods are maximum likelihood [13,14], support vector machine (SVM) [15,16], and neural network [17,18]. However, these methods heavily rely on hand-crafted features, such as color, texture, and morphological features, to distinguish clouds from clear sky pixels.

Recent years, with the development of deep learning, deep convolutional neural network (DCNN) methods have been rapidly developed and widely used for cloud detection from RS images. For example, U-Net and SegNet variants cloud detection frameworks [19–23], and multi-scale/level feature fusion cloud detection frameworks [24–28]. In addition, advanced convolutional neural network (CNN) models, such as CDnetV2 [7] and ACDnet [29], are developed for cloud detection from RS imagery with cloud–snow coexistence. To achieve real-time and onboard processing, lightweight neural networks, such as [30–32], are proposed for pixelwise cloud detection from RS images. However, most of the previous CNN-based cloud detection methods are based on supervised learning frameworks. Although these CNN-based cloud detection methods have achieved impressive performance, they heavily rely on a large number of training data with strong pixel-wise annotations. Some recent cloud detection works, such as unsupervised domain adaptation (UDA) [33,34] and domain translation strategy [35], have begun to explore how to avoid using pixel-wise annotations for cloud detection network training. However, these pixel-wise annotations free methods are not real label-free ones because they rely on other labeled datasets.

Obtaining data label is an expensive and time-consuming task, especially pixel-wise annotation. As illustrated in cityscapes dataset annotation work [36], it usually takes 1.5 h to label a pixel-wise annotation from a high-resolution urban scene image with pixel size of 1024×2048. For a remote sensing image with pixel size of $8\,\text{k} \times 8\,\text{k}$, it may take more hours to label a whole scene RS image according to such experience. Although it is easier to label the cloud pixel-wise samples individually, it may still take three to four hours for a tough case that contains a large number of tiny and thin clouds, which increases the heavy cost of manual labeling undoubtedly. In contrast, unlabeled RS images can be far more easily acquired than labeled ones [37]. Therefore, it desperately needs to exploit how to utilize a large number of unlabeled data to enhance the performance of cloud detection model.

In this paper, we proposed to use a semi-supervised learning (SSL) method [38,39] to train a cloud detection network. Because the SSL method is able to reduce the heavy cost of manual dataset labeling. In a semi-supervised segmentation framework, such as DAN [37] and s4GAN [40], the segmentation network (cloud detection network) is able to simultaneously take advantage of a large amount of unlabeled samples and a limited number of labeled examples for network's parameter learning. The core of SSL method is a self-training [41] strategy, which is able to leverage pseudo-label generated from a large amount of unlabeled samples to supervise the segmentation network training [42]. This also means that accurate pseudo-label labeling is the key of self-training. Therefore, most advanced SSL methods, such as [40,43], focus on improving pseudo-labels of unlabeled samples to improve the performance of the SSL network.

SSL networks developed for tradition natural image segmentation, such as [38–40,43], may not achieve a promising performance for satellite images cloud detection due to RS images are different from traditional camera natural images. In addition, the data drift problem may appear between a limited number of labeled examples and a large number of unlabeled samples due to different cloud shapes and land-cover types on different satellite images. The SSL network trained with labeled samples is difficult to generalize to unlabeled samples due to the data drift problem. During training, the SSL network may produce prediction results with lower certainty when the network input with unlabeled data [40]. The prediction results of unlabeled samples has shown lower certainty and cannot generate accurate pseudo-labels, which makes self-training unfavorable for providing supervision signals of network training.

To solve this problem, we take the domain shift problem into account for the SSL framework. In this paper, inspired by unsupervised domain adaptation (UDA) method [44], we apply domain adaptation at the feature-level [45] and output-level [46,47] in the proposed semi-supervised cloud detection network (SSCDnet) to solve the data drift problem. In this paper, we use two available cloud cover validation datasets, i.e., Landsat-8 OLI (Operational Land Imager) [48] cloud cover validation dataset (https://landsat.usgs.gov/landsat-8-cloud-cover-assessment-validation-data, accessed on 24 April 2022) and GF-1 WFV [12] cloud and cloud shadow cover validation dataset (http://sendimage.whu.edu.cn/en/mfc-validation-data/, accessed on 24 April 2022), to comprehensively evaluate the proposed SSCDnet. These two datasets have been widely used for evaluating the performance of supervised CNN-based cloud detection methods [19–23,27].

In summary, the main contributions of this work are summarized as follows:

(i) We propose a semi-supervised cloud detection framework, named SSCDnet, which learns knowledge from a limited number of pixel-wise labeled examples and a large number of unlabeled samples for cloud detection.
(ii) We take the domain shift problem into account between labeled and unlabeled images and propose the feature-level and output-level domain adaptation method to reduce domain distribution gaps.
(iii) We propose a double threshold pseudo-labeling method to obtain trustworthy pseudo label, which helps to avoid the effects of noise labels for self-training as much as possible and to further enhance the performance of SSCDnet.

This paper is organized as follows: in Section 2, we present the proposed SSCDnet in detail. Experimental datasets and networks training details are presented in Section 4. The experimental results and discussions are presented in Sections 4 and 5, respectively, followed by conclusions in Section 6.

2. The Proposed Method

In this section, we provide a detailed introduction of the proposed SSCDnet, including the traditional semi-supervised segmentation framework, the proposed overall workflow of SSCDnet, feature/out-level domain adaptation, trustworthy pseudo label labelling, cloud detection network, and discriminator network structures.

2.1. Traditional Semi-Supervised Segmentation Framework

In a traditional semi-supervised segmentation framework [37–40], as shown in Figure 1, the segmentation network G simultaneously takes advantage of a large number of unlabeled samples and a limited number of labeled examples for a network's parameter training. In this framework, there are two datasets, i.e., labeled dataset $M^l = \{x^l, y^l\}$ and unlabeled dataset $M^u = \{x^u\}$, where x^l and x^u are the input data of the segmentation network G and y^l is a pixel-wise label of x^l. p^l and p^u represent predicted results of x^l and x^u, respectively. $\hat{p}^u$ represents a pseudo label of p^u.

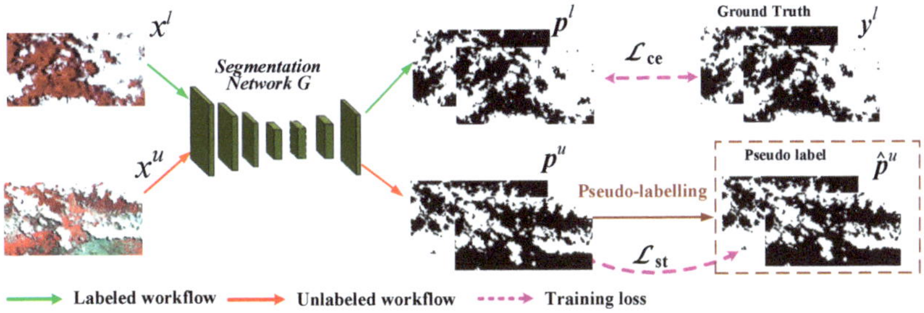

Figure 1. The traditional semi-supervised segmentation framework.

During training, given an input labeled image x^l, the segmentation network G is supervised by a standard cross-entropy loss $\mathcal{L}_{ce}$. When using the unlabeled data x^u, the segmentation network is further supervised by a self-training loss $\mathcal{L}_{st}$. That is, we use pseudo label $\hat{p}^u$ that generated from predicted result p^u as the "ground truth" for self-training to enhance the semantic segmentation network G. Therefore, total training objective $\mathcal{L}_G$ of the segmentation network G is defined as follows:

$$\mathcal{L}_G = \mathcal{L}_{ce} + \lambda_{st}\mathcal{L}_{st}, \tag{1}$$

where λ_{st} is weight used for minimizing the objective $\mathcal{L}_G$.

As shown in Figure 2, remote sensing images obtained from different places show large domain distribution gaps between each other due to different cloud shapes and land-cover types on Earth's surface. Therefore, there is data drift between labeled samples and unlabeled ones in training dataset of SSL. In SSL framework, segmentation network G trained with labeled samples is hard to generalize to unlabeled ones due to data drift problems. It is difficult to produce a highly certain prediction result p^u for unlabeled sample x^u. Predicted results with low certainty leads to low quality pseudo-labels, thus affecting the supervision signal provided from the self-training loss $\mathcal{L}_{st}$, and further affecting the performance of segmentation network G. Therefore, improving certainty of predicted results of unlabeled samples is the key to a semi-supervised learning framework.

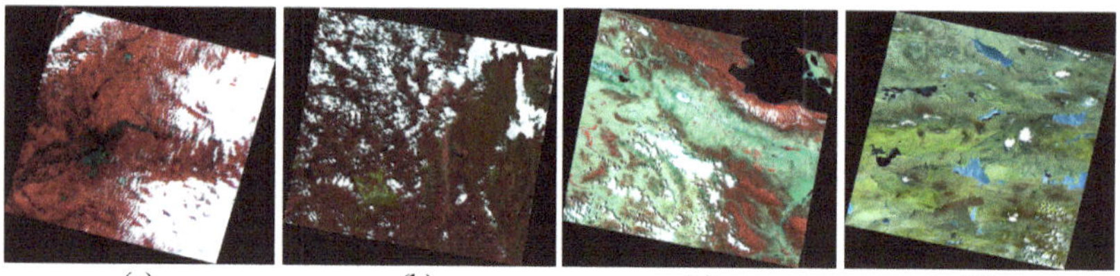

(a) (b) (c) (d)

Figure 2. Remote sensing images obtained from the same satellite sensor (GF-1 satellite) at different places. Where (**a**) GF1_WFV2_E127.2_N45.9_20140704_L2A00003099760, (**b**) GF1_WFV1_E102.0_N28.0_20140302_L2A00002804760, (**c**) GF1_WFV1_E99.4_N36.3_20140716_L2A00002789230, and (**d**) GF1_WFV3_E89.3_N35.6_20140702_L2A00008451540.

2.2. Proposed Semi-Supervised Cloud Detection Framework

In this paper, we take the domain shift problem into account for the semi-supervised learning (SSL) framework to improve generalization of the segmentation network to generate trustworthy pseudo-label for self training. We improve a standard SSL network with the unsupervised domain adaptation (UDA) strategy and propose an improved semi-supervised cloud detection network as shown in Figure 3. UDA methods are able to help semantic segmentation networks to learn domain-invariant features at different representation layers, such as input-level (pixel-level) [49], feature-level [45], or output-level [46,47].

Different from the traditional SSL framework, we apply feature-level and output-level domain adaptations at the intermediate layers and the end layer of network, respectively, to reduce domain distribution gaps and improve the generalization performance of SSCDnet. The highly generalized network is able to generate highly certain predicted results for unlabeled samples. To further improve the quality of pseudo-label, instead of directly using the predicted results of unlabeled samples for network training, we proposed a trustworthy pseudo label labelling method to obtain a high-quality pseudo label. To be specific, similar to [40,43], we take advantage of the feedback information from output-level domain adaptation to obtain high-quality candidate labels. Then, we use a threshold strategy to obtain trustworthy regions from the high-quality candidate labels. Finally, trustworthy regions are considered as the ground-truth labels for network training through the self-training loss.

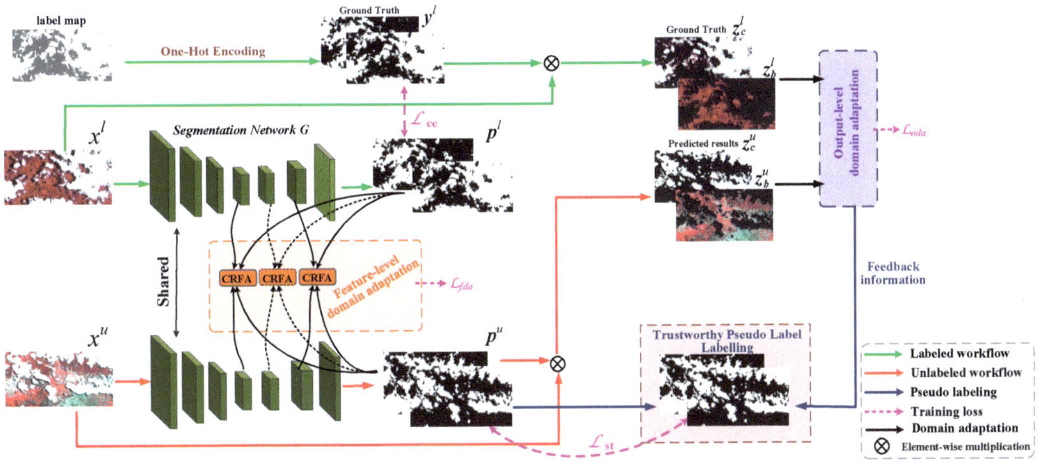

Figure 3. The detailed structure of the proposed SSCDnet.

During training, given an input labeled image x^l and its corresponding predicted result is p^l, the cloud detection network is supervised by a standard cross-entropy loss $\mathcal{L}_{ce}$ [27]. When inputting unlabeled data x^u, segmentation network G predicts its corresponding result p^u. Then, we generate the pseudo label $\hat{p}^u$ from predicted result p^u by using proposed trustworthy pseudo label labelling method. Pseudo label $\hat{p}^u$ as the "ground truth" for self-training to enhance the semantic segmentation network G. Therefore, the semantic segmentation network is further supervised by self-training loss $\mathcal{L}_{st}$ and feature-level domain adaptation loss $\mathcal{L}_{fda}$ as well as output-level domain adaptation loss $\mathcal{L}_{oda}$ when inputting unlabeled data. The total loss of the semi-supervised cloud detection network, named SSCDnet, is defined as follows:

$$\mathcal{L}_G = \mathcal{L}_{ce} + \lambda_{st}\mathcal{L}_{st} + \lambda_{oda}\mathcal{L}_{oda} + \lambda_{fda}\mathcal{L}_{fda}, \tag{2}$$

where λ_{st}, λ_{oda}, and λ_{fda} are three regulation parameters used for minimizing the objective $\mathcal{L}_G$.

During training, we minimize $\mathcal{L}_G$ for updating parameters of segmentation network G. The detailed information of the proposed feature/output-level domain adaptation strategies and trustworthy pseudo label labelling method will be introduced in following subsections.

2.3. Reducing Domain Distribution Gaps

2.3.1. Feature-Level Domain Adaptation

To reduce domain distribution gaps at the feature-level, we propose a class-relevant feature alignment (CRFA) strategy. As shown in Figure 4, we use predicted score maps of each class (i.e., cloud and background classes) as the attention maps to obtain class-relevant features. Then, we design a standard binary classification network as the discriminator to help segmentation network $\boldsymbol{G}$ to generate domain-invariant feature representations, thus helping to reduce domain distribution gaps between labeled samples and unlabeled ones at feature level. Therefore, the proposed CRFA domain adaptation consists of two stages: (i) class-relevant feature selection and (ii) class-relevant feature alignment.

To be specific, let $\boldsymbol{H}_l^k \in \mathbb{R}^{W\times H\times C}$ and $\boldsymbol{H}_u^k \in \mathbb{R}^{W\times H\times C}$ denote labeled and unlabeled samples' features extracted from the k-th intermediate hidden layer of network $\boldsymbol{G}$, respectively. Let $\tilde{p}_l^i(:,:,1)$ and $\tilde{p}_l^i(:,:,2)$ denote the spatial attention maps of cloud and background areas of labeled samples, respectively. Then, cloud-relevant feature $C_l^k \in \mathbb{R}^{W\times H\times C}$ and background-relevant feature $B_l^k \in \mathbb{R}^{W\times H\times C}$ of the labeled samples' feature $H_l^k \in \mathbb{R}^{W\times H\times C}$ are defined as follows:

$$C_l^k = H_l^k \otimes \mathcal{F}\left(\tilde{p}_l^i(:,:,1)\right), \tag{3}$$

and

$$B_l^k = H_l^k \otimes \mathcal{F}\left(\tilde{p}_l^i(:,:,2)\right), \tag{4}$$

where $\mathcal{F}(\cdot)$ represents sampling operator, which includes down-sampling or up-sampling operators. Similarly, we are able to obtain cloud-relevant feature $C_u^k \in \mathbb{R}^{W\times H\times C}$ and background-relevant feature $B_u^k \in \mathbb{R}^{W\times H\times C}$ of unlabeled samples' feature $H_u^k \in \mathbb{R}^{W\times H\times C}$.

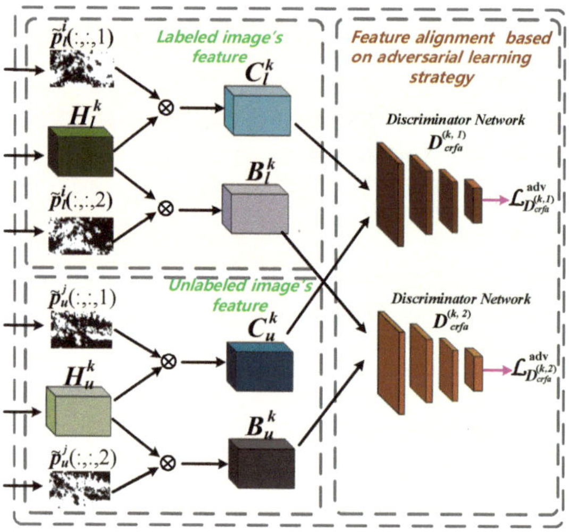

Figure 4. Structure of the proposed class-relevant feature alignment module.

After obtaining these class-relevant features, we input these features into discriminators as shown in Figure 4. For discriminator training, both discriminators $D_{\text{crfa}}^{(k,1)}$ and $D_{\text{crfa}}^{(k,2)}$ use cross-entropy domain classification loss [44] as the objective function. They are defined as follows:

$$\mathcal{L}_{D_{\text{crfa}}^{(k,1)}} = -\mathbb{E}[\log(D_{\text{crfa}}^{(k,1)}(C_l^k))] - \mathbb{E}[\log(1 - D_{\text{crfa}}^{(k,1)}(C_u^k))] \tag{5}$$

and

$$\mathcal{L}_{D_{\text{crfa}}^{(k,2)}} = -\mathbb{E}[\log(D_{\text{crfa}}^{(k,2)}(B_l^k))] - \mathbb{E}[\log(1 - D_{\text{crfa}}^{(k,2)}(B_u^k))]. \tag{6}$$

For segmentation network training, the adversarial objectives provided by discriminators ($D_{\text{crfa}}^{(k,1)}$ and $D_{\text{crfa}}^{(k,2)}$) are defined as follows:

$$\mathcal{L}_{D_{\text{crfa}}^{(k,1)}}^{adv} = -\mathbb{E}[\log D_{\text{crfa}}^{(k,1)}(C_u^k)] \tag{7}$$

and

$$\mathcal{L}_{D_{\text{crfa}}^{(k,2)}}^{adv} = -\mathbb{E}[\log D_{\text{crfa}}^{(k,2)}(B_u^k)]. \tag{8}$$

Therefore, the class-relevant feature alignment loss $\mathcal{L}_{crfa}^k$ of the k-th intermediate hidden feature is defined as follows:

$$\mathcal{L}_{crfa}^k = \mathcal{L}_{D_{\text{crfa}}^{(k,1)}}^{adv} + \mathcal{L}_{D_{\text{crfa}}^{(k,2)}}^{adv}. \tag{9}$$

Since the proposed feature-level domain adaptation is performed at multiple intermediate layers, the feature-level domain adaptation loss $\mathcal{L}_{fda}$ is provided from all intermediate layers' CRFA losses $\{\mathcal{L}^k_{crfa}\}^K_{k=1}$, i.e.,

$$\mathcal{L}_{fda} = \sum_{k=1}^{K} \mathcal{L}^k_{crfa}. \tag{10}$$

where $\mathcal{L}^k_{fda}$ is the feature domain adaptation loss of the k-th intermediate hidden feature.

2.3.2. Output-Level Domain Adaptation

Instead of directly aligning the predicted results, we align cloud and background objects images extracted from labeled and unlabeled samples, respectively, via a proposed class-relevant outputs alignment (CROA) method. Similar to the CRFA method, the CROA method is built on the standard adversarial learning framework. CROA consists of two discriminators, which are used to align cloud and background images, respectively. As shown in Figure 5, we use two standard binary classification networks as discriminators to reduce the domain distributions gaps between labeled and unlabeled datasets at the output level and improve the certainty of predicted results z^u for unlabeled sample x^u.

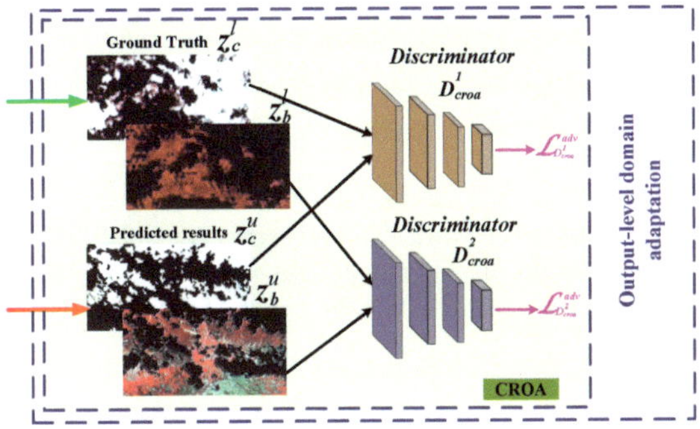

Figure 5. Structure of the proposed CROA.

As shown in Figure 5, we use the cloud and background objects extracted from original images as the input data of different discriminators. To be specific, let $z^u_c = p^u(:,:,1) \otimes x^u$ and $z^u_b = p^u(:,:,2) \otimes x^u$ denote extracted cloud and background objects of unlabeled sample. Similarly, let $z^l_c = p^l(:,:,1) \otimes x^l$ and $z^l_b = p^l(:,:,2) \otimes x^l$ denote extracted cloud and background objects of the labeled sample, where $\otimes$ represents element-wise multiplication. For discriminator training, both discriminators $\boldsymbol{D}^1_{\text{croa}}$ and $\boldsymbol{D}^2_{\text{croa}}$ use cross-entropy domain classification loss [44] as the objective function. They are defined as follows:

$$\mathcal{L}_{\boldsymbol{D}^1_{\text{croa}}} = -\mathbb{E}[\log(\boldsymbol{D}^1_{\text{croa}}(z^l_c)] - \mathbb{E}[\log(1 - \boldsymbol{D}^1_{\text{croa}}(z^u_c))], \tag{11}$$

and

$$\mathcal{L}_{\boldsymbol{D}^2_{\text{croa}}} = -\mathbb{E}[\log(\boldsymbol{D}^1_{\text{croa}}(z^l_b)] - \mathbb{E}[\log(1 - \boldsymbol{D}^1_{\text{croa}}(z^u_b))]. \tag{12}$$

For segmentation network training, adversarial objectives provided by these discriminators are defined as follows:

$$\mathcal{L}^{adv}_{\boldsymbol{D}^1_{croa}} = -\mathbb{E}[\log \boldsymbol{D}^1_{croa}(x^u_c)], \tag{13}$$

and

$$\mathcal{L}^{adv}_{\boldsymbol{D}^2_{croa}} = -\mathbb{E}[\log \boldsymbol{D}^2_{croa}(x^u_b)]. \tag{14}$$

Therefore, the output-level domain adaptation objective L_{oda} is defined as follows:

$$\mathcal{L}_{oda} = \mathcal{L}^{adv}_{D^1_{croa}} + \mathcal{L}^{adv}_{D^2_{croa}}. \quad (15)$$

As shown in Figure 6, we present the visualized experiment results on a GF-1 WFV image with or without applied domain adaptation. Results show that applying domain adaptation is able to improve the segmentation network to produce high certainly predicted results for unlabeled samples as shown in Figure 6c. High certainly predicted results ensure that we can obtain trustworthy pseudo label for self-training, thus improving the network's cloud detection performance. More detailed information can be found in Section 4.1 (Ablation Study).

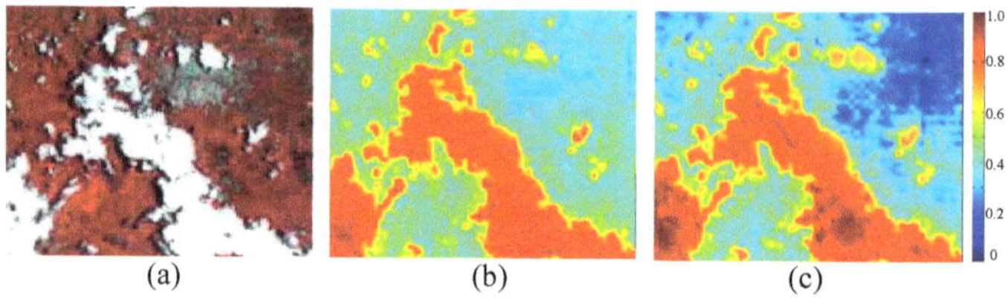

Figure 6. Experiment results on an unlabeled remote sensing image. (**a**) input GF-1 WFV image; (**b**) un-applied domain adaptation result, and (**c**) applied domain adaptation result.

2.4. *Trustworthy Pseudo Label Labelling*

Pseudo label labelling is the main work for self-training in an SSL framework [38–40,43,50]. During self-training, pseudo labels predicted from the segmentation model serve as the "ground truth" to provide additional supervisory signals for network training, which makes the segmentation network able to leverage the unlabeled data. In this paper, we propose a double threshold pseudo-labelling method to efficiently obtain trustworthy regions from pseudo label. We use trustworthy regions as the ground truth for network training.

2.4.1. Candidate Labels Selection

As shown in Figure 7, we sample an unlabeled image x^u into the segmentation network G, obtaining its corresponding predicted probability maps $p^u = \boldsymbol{G}(x^u)$ as well as extracted cloud and background objects z^u_c and z^u_b, respectively. We first use feedback information from the output-level domain adaptation to obtain candidate labels. Specifically, we select high-quality pseudo labels online based on two discriminator scores of output-level domain adaptation, i.e.,

$$\boldsymbol{D}_{croa}(z^u_c) > \tau_1, \boldsymbol{D}_{croa}(z^u_b) > \tau_1, \quad (16)$$

where τ_1 is the threshold. Equation (16) is feedback information from the output-level domain adaptation. We treat these predicted labels satisfied Equation (16) as candidate labels. If not satisfied, we directly set self-training loss $\mathcal{L}_{st} = 0$.

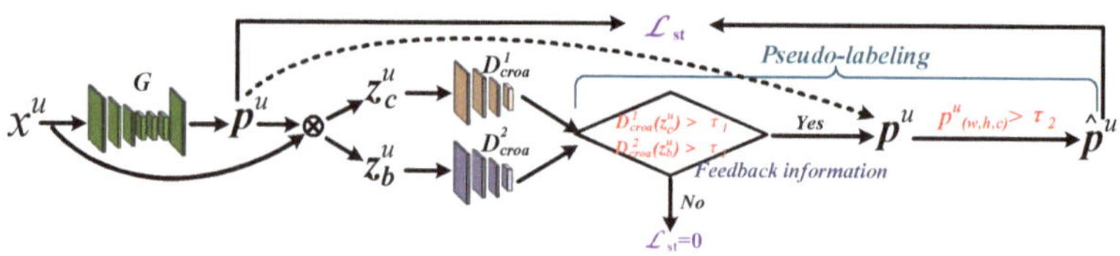

Figure 7. The double threshold pseudo-labelling method for self-training.

2.4.2. Trustworthy Regions Selection from Candidate Labels

After we obtained the candidate labels, we set a confidence threshold τ_2 to discover trustworthy regions from the selected candidate labels, i.e.,

$$\boldsymbol{D}^1_{croa}(z^u_c) > \tau_1, \boldsymbol{D}^2_{croa}(z^u_b) > \tau_1, and \quad p^u_{(w,h,c)} > \tau_2. \quad (17)$$

Equation (17) indicates that, if predicted probability of these regions is greater than τ_2, we use these trustworthy regions as the ground-truth for self-training.

2.4.3. Self-Training Loss $\mathcal{L}_{st}$

According to the above mentioned method, the self-training loss $\mathcal{L}_{st}$ is defined as:

$$\mathcal{L}_{st} = \begin{cases} -\sum \hat{p}^u_{(w,h,c)} \log p^u_{(w,h,c)}, & \mathbf{D}^1_{croa}(z^u_c) > \tau_1, \mathbf{D}^2_{croa}(z^u_b) > \tau_1, \\ & \textit{and} \quad p^u_{(w,h,c)} > \tau_2 \\ 0 & \text{otherwise.} \end{cases} \tag{18}$$

where $\hat{p}^u$ is the pseudo label generated from the prediction map p^u by using a one-hot encoding scheme. $p^u_{(w,h,c)}$ represents the predicted probability at the location (h,w) of the C-channel.

2.5. Network Architecture

2.5.1. Cloud Detection Network

Similar to most SSL approaches [40,43], we use DeepLabv2 [51] as our main cloud detection framework and resort to ResNet-101 [52] as the backbone to extract semantic segmentation information. DeepLabv2 uses atrous spatial pyramid pooling (ASPP) module [51], which incorporates multiple parallel dilated convolutional layers [53] with different sampling rates, to capture multi-scale features for robustness clouds detection. In this paper, we set the resolution of predicted probability map as $1/8 \times 1/8$ size of input image for fair comparison with previous semi-supervised works based on DeepLabv2, such as [37,40]. Then, we directly up-sample the predicted probability maps to the same size as the input images to obtain final predicted results.

2.5.2. Discriminator Network

In this paper, we design a standard binary classification network as the discriminator for both CRFA and CROA modules. Figure 8 illustrates the detailed structure of the designed discriminator. This discriminator contains four convolutional layers, a global average pooling layer, and a fully-connected layer. To be specific, there are four convolutional layers with 4×4 kernels. Four convolutional layers have $\{256, 192, 128, 64\}$ and $\{64, 128, 256, 512\}$ channels for CRFA and CROA modules, respectively. Each convolutional layer shares the same stride (stride = 2) and simultaneously followed by a Leaky-ReLU activation (slope = 0.2) and a dropout layer (dropout-rate = 0.5). After these convolution operations, a global average pooling layer and a fully-connected layer are designed to obtain confidence score for each input image.

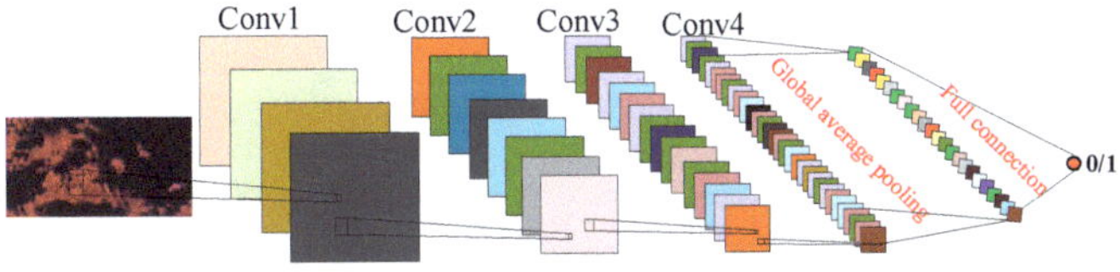

Figure 8. Structure of the proposed discriminator network.

3. Dataset and Experimental Settings

3.1. Experimental Dataset

In this paper, we use two available cloud cover validation datasets, i.e., Landsat-8 OLI cloud cover validation dataset [48] and GF-1 WFV cloud and cloud shadow cover validation dataset [12], to comprehensively evaluate the proposed SSCDnet. Table 1 shows the detailed information of Landsat-8 OLI and GF1-WFV multispectral images. Similar to [27], we follow the idea that the number of subimages with and without cloud in the training data should be balanced. Otherwise, the detection results would bias towards the majority. Therefore, we exclude some cloud-free and full cloud covered scenes. In addition, current cloud detection accuracy measurement is not robust when the cloud percentage is quite low [10]. A low cloud cover percentage in a scene may cause an apparent reduction in the cloud's producer accuracy and user accuracy. Therefore, original large size images with cloud percentage less than 5% are usually removed in training and testing. Finally, we select 40 and 20 scenes images from both Landsat-8 and Gaofen-1 WFV dataset for network training and evaluation, respectively.

Table 1. The detailed information of Landsat-8 OLI and GF1-WFV multispectral images.

Sensor	Spectral Band	Wavelength Range (μm)	Spatial Resolution (m)
Landsat-8 OLI	Band 1 (coastal)	0.433–0.453	30
	Band 2 (blue)	0.450–0.515	30
	Band 3 (green)	0.525–0.600	30
	Band 4 (red)	0.630–0.680	30
	Band 5 (NIR)	0.845–0.885	30
	Band 6 (SWIR1)	1.560–1.660	30
	Band 7 (SWIR2)	2.100–2.300	30
	Band 8 (Pan)	0.500–0.680	15
	Band 9 (Cirrus)	1.360–1.390	30
GF-1 WFV	Band 1 (blue)	0.450–0.520	16
	Band 2 (green)	0.520–0.590	16
	Band 3 (red)	0.630–0.690	16
	Band 4 (NIR)	0.770–0.890	16

As illustrated in [54], CNNs are strongly biased towards recognizing textures for object classification. Therefore, we can select a limited number of channels data of multispectral data to evaluate the proposed SSCDnet. In this paper, we select channels 3 (green), 4 (red), and 5 (near-infrared) of Landsat-8 OLI data and channels 2 (green), 3 (red), and 4 (near-infrared) of Gaofen-1 WFV data for segmentation network training and testing. During training, all training data are cropped into subimages of pixel size 321 × 321. There are about 30 k and 75 k annotated sub-images in Landsat-8 OLI and GF-1 WFV training dataset, respectively. We select a portion of the data set for network supervised training and the remaining de-annotated portion for network self-training (unsupervised training). During testing, we divide the whole RS image into a series of sub-images with image size of 1200 × 1200 for network evaluation. The final detection result is obtained by merging results of sub-images.

3.2. Network Training Details and Parameters Setting

SSCDnet is trained under PyTorch framework (https://pytorch.org/, accessed on 24 April 2022). The operating system is Ubuntu 14.04 equipped with NVIDIA GTX 1080 Ti GPU. We optimize the segmentation (generator) and discriminator networks by using SGD algorithm [55] and Adam optimizer [56], respectively. During training, these two datasets share the following parameters settings. That is, learning rates for generator and discriminator network are 2.5×10^{-4} and 1×10^{-4}, respectively. Training decay policy is "poly" [57]. The number of mini-batch size, momentum, and weight decay are 4, 0.9, and 5×10^{-4}, respectively. SSCDnet is trained for 20 and 15 epochs on Landsat-8 OLI and GF-1 WFV dataset, respectively.

In addition, in order to improve the performance, we use the model pre-trained on the ImageNet dataset [58] to fine-tune the parameters of the backbone network (ResNet-101). We conduct feature-level domain adaptation tasks at the end of *Conv4_x* and *Conv5_x* residual blocks, i.e., $K = 2$. We empirically set weight-parameters $\lambda_{st} = 1.0$, $\lambda_{fda} = 0.001$, and $\lambda_{oda} = 0.1$ for GF-1 dataset, set $\lambda_{st} = 1.0$, $\lambda_{fda} = 0.001$, and $\lambda_{oda} = 0.01$ for Landsat-8

OLI dataset. In addition, we empirically set threshold-parameters $\tau_1 = 0.6$ and $\tau_2 = 0.55$ to accurately generate trustworthy pseudo labels for both GF-1 and Landsat-8 OLI datasets.

4. Experimental Results

4.1. Ablation Study

To investigate the effectiveness of SSCDnet, we conduct a series of ablation studies on it. Five widely used quantitative metrics of RS images, i.e., mean intersection over union (MIoU), kappa coefficient (Kappa), overall accuracy (OA), producer accuracy (PA), and user accuracy (UA), are used to comprehensively measure the cloud detection results. All experiment results are obtained from the GF-1 WFV dataset.

4.1.1. Ablation Study on Loss Function

In this paper, the proposed SSCDnet is supervised by four loss functions, i.e., a standard cross-entropy loss L_{ce}, self-training loss L_{st}, output-level domain adaptation loss L_{oda}, and feature-level domain adaptation loss L_{fda}. In Table 2, we list ablation results under different proportion of labeled samples ($\frac{1}{200}$, $\frac{1}{100}$, $\frac{1}{40}$, $\frac{1}{20}$, and full labeled samples) to demonstrate the effects of each component loss.

Table 2. Cloud extraction accuracy (%) of different ablation networks on GF-1 WFV data.

Proportions	$\mathcal{L}_{ce}$ $\mathcal{L}_{st}$ $\mathcal{L}_{oda}$ $\mathcal{L}_{fda}$	√ × × ×	√ √ × ×	√ √ √ ×	√ √ √ √
$\frac{1}{200}$	OA	90.22	93.56	95.30	**95.51**
	MIoU	71.77	83.22	86.95	**87.54**
	Kappa	68.21	78.15	82.07	**82.69**
	PA	67.03	73.67	80.02	**81.28**
	UA	95.51	**96.40**	94.57	93.55
$\frac{1}{100}$	OA	91.34	93.89	95.73	**95.85**
	MIoU	74.10	84.11	88.05	**88.90**
	Kappa	71.51	78.90	83.34	**83.92**
	PA	69.71	74.32	81.52	**82.34**
	UA	94.93	**95.99**	94.28	93.36
$\frac{1}{40}$	OA	92.23	94.52	96.02	**96.24**
	MIoU	76.02	86.03	89.01	**89.30**
	Kappa	73.58	80.85	84.77	**85.01**
	PA	75.08	80.18	84.40	**85.49**
	UA	**93.43**	93.30	91.59	92.24
$\frac{1}{20}$	OA	93.19	95.11	96.55	**96.90**
	MIoU	79.88	87.16	90.46	**90.93**
	Kappa	75.75	83.77	86.25	**86.97**
	PA	80.69	82.22	84.87	**85.89**
	UA	92.30	92.10	**92.72**	92.24
Full	OA	96.91	97.11	97.17	**97.19**
	MIoU	90.15	91.50	91.67	**91.71**
	Kappa	88.64	89.49	89.35	**89.37**
	PA	87.28	88.67	88.84	**88.90**
	UA	92.05	**92.33**	92.51	92.32

where $\frac{1}{200}$, $\frac{1}{100}$, $\frac{1}{40}$, and $\frac{1}{20}$ are the fractions of the total training images in the dataset that are used as labeled data, and the rest of the data was used without labels. Bold indicates maximum value in this paper.

Results in Table 2 show that the best performance is achieved by a combination of all optimal loss terms (we use $\{\mathcal{L}_{ce}, \mathcal{L}_{st}, \mathcal{L}_{oda}, \mathcal{L}_{fda}\}$ to represent this combination.), while the baseline framework supervised only by the standard cross-entropy loss $\mathcal{L}_{ce}$ shows the worst performance. To be specific, a combination of $\mathcal{L}_{ce}$ and $\mathcal{L}_{st}$ (i.e., $\{\mathcal{L}_{ce}, \mathcal{L}_{st}\}$) and combination of $\mathcal{L}_{ce}$, $\mathcal{L}_{st}$, and $\mathcal{L}_{oda}$ (i.e., $\{\mathcal{L}_{ce}, \mathcal{L}_{st}, \mathcal{L}_{oda}\}$) obtain significant improvement of +9.44% and +13.86% in Kappa, respectively, compared with only $\mathcal{L}_{ce}$ loss when labeled sample proportion is $\frac{1}{200}$. In addition, $\{\mathcal{L}_{ce}, \mathcal{L}_{st}\}$ and $\{\mathcal{L}_{ce}, \mathcal{L}_{st}, \mathcal{L}_{oda}\}$ also outperform $\mathcal{L}_{ce}$

under other proportions of labeled sample settings, which shows that output-level domain adaptation and self-training strategies are able to improve the performance of segmentation network.

Based on $\{\mathcal{L}_{ce}, \mathcal{L}_{st}, \mathcal{L}_{oda}\}$, we further introduce feature domain adaptation loss $\mathcal{L}_{fda}$ to reduce feature-level domain distribution gaps between labeled and unlabeled datasets. Results in Table 2 show that $\{\mathcal{L}_{ce}, \mathcal{L}_{st}, \mathcal{L}_{oda}, \mathcal{L}_{fda}\}$ performs better than $\{\mathcal{L}_{ce}, \mathcal{L}_{st}, \mathcal{L}_{oda}\}$. In addition, it can be seen that the performance of $\{\mathcal{L}_{ce}, \mathcal{L}_{st}, \mathcal{L}_{oda}, \mathcal{L}_{fda}\}$ consistently outperforms those of other combinations under different proportions of labeled sample ($\{\frac{1}{200}, \frac{1}{100}, \frac{1}{40}, \frac{1}{20}\}$), which shows that the proposed $\{\mathcal{L}_{ce}, \mathcal{L}_{st}, \mathcal{L}_{oda}, \mathcal{L}_{fda}\}$ is a promising strategy for semi-supervised segmentation network.

In summary, results in Table 2 demonstrate that standard cross-entropy loss $\mathcal{L}_{ce}$, self-training loss $\mathcal{L}_{st}$, output-level domain adaptation loss $\mathcal{L}_{oda}$, and feature-level domain adaptation loss $\mathcal{L}_{fda}$ are beneficial for semi-supervised cloud detection work. Because applying domain adaptation strategy is able to reduce distribution gaps between labeled and unlabeled datasets and improve SSCDnet to generate trustworthy pseudo-labels for self-training, thus providing positive supervised signals for segmentation network learning. Combination of all these loss terms is able to achieve a state-of-the-art cloud detection performance when using a limited number of labeled samples.

4.1.2. Ablation Study on Feature-Level Domain Adaptation

In this paper, we use ResNet-101 [52] as the backbone network to extract feature maps and conduct domain adaptation study on the end features of *Conv4_x*, *Conv5_x*, *Conv3_x*, and *Conv2_x* residual blocks. In Table 3, we list the experiment results of ablation study on different intermediate layers. Experiment results show that applying feature domain adaptation on the end of *Conv4_x* and *Conv5_x* residual blocks, i.e., FDA_45, achieves the best performance. In contrast, applying feature domain adaptation on other layers' features always shows worse performance than on *Conv4_x* and *Conv5_x* layers. Hence, we conduct the domain adaptation tasks at the end of *Conv4_x* and *Conv5_x* residual blocks (two intermediate layers $K = 2$ in feature-level domain adaptation). Furthermore, we find that applying domain adaptation tasks at above mentioned intermediate layers is able to achieve promising performance on the Landsat-8 OLI dataset.

Table 3. Ablation study of domain adaptation on different intermediate layers (proportion = $\frac{1}{200}$).

Methods	OA	MIoU	Kappa	PA	UA
FDA_5	95.40	87.11	82.53	80.28	94.68
FDA_45	**95.51**	**87.54**	**82.69**	**81.28**	93.55
FDA_345	95.13	86.17	81.64	79.19	**94.89**
FDA_2345	95.41	87.06	82.93	81.73	93.54

4.1.3. Ablation Study on Self-Training's Double Threshold

In this paper, we proposed a double threshold method to obtain trustworthy pseudo labels for network self-training. In Table 4, we investigate effectiveness of the proposed pseudo-labeling method.

Table 4. Ablation study on double threshold (proportion = $\frac{1}{200}$).

Methods	OA	MIoU	Kappa	PA	UA
$\mathcal{L}_{st(N/A,N/A)}$	94.12	85.02	81.43	79.38	**94.74**
$\mathcal{L}_{st(\tau_1,N/A)}$	95.02	86.33	81.95	80.11	93.30
$\mathcal{L}_{st(\tau_1,\tau_2)}$	**95.30**	**86.95**	**82.07**	**80.02**	94.57

In Table 4, we list the experiment results under different pseudo-labeling strategies, i.e., pseudo-labeling directly transforms from probability maps $\mathcal{L}_{st(N/A,N/A)}$, pseudo-labeling based on a discriminator score of output-level domain adaptation $\mathcal{L}_{st(\tau_1,N/A)}$, and pseudo-labeling based on both discriminator score of output-level domain adaptation and trustworthy regions selection $\mathcal{L}_{st(\tau_1,\tau_2)}$, where τ_1 and τ_2 are used to obtain candidate labels and trustworthy regions, respectively. Results in Table 4 show that performance of the loss term $\mathcal{L}_{st(\tau_1,\tau_2)}$ is better than those of $\mathcal{L}_{st(\tau_1,N/A)}$ and $\mathcal{L}_{st(N/A,N/A)}$, which demonstrates that trustworthy regions selection based on high confidence candidate label is a promising strategy for pseudo label selecting.

4.1.4. Hyper-Parameter Analysis

The proposed training objective $\mathcal{L}_G$ has three balance weights, i.e., λ_{st}, λ_{oda}, and λ_{fda}. In this paper, since we use all the trustworthy regions' pseudo labels for self-training, we directly set $\lambda_{st} = 1.0$. Then, λ_{oda} and λ_{fda} are two important hyper-parameters that affect cloud detection results. In Table 5, we list a series of validation experimental results to investigate the impact of these hyper-parameters. As illustrated in Table 5, we first obtain the promising λ_{oda} by setting $\lambda_{fda} = N/A$ (N/A means Not Applicable). Then, we obtain the promising λ_{fda} based on obtained promising λ_{oda}. Experimental results in Table 5 show that setting $\lambda_{oda} = 0.10$ and $\lambda_{fda} = 0.001$ is able to achieve promising performance. Similarly, we can obtain the promising hyper-parameter setting based on above mentioned methods on Landsat-8 OLI data cloud detection task.

Table 5. The study on hyper-parameter λ_{st}, λ_{oda}, and λ_{fda} (proportion = $\frac{1}{200}$).

λ_{st}	λ_{oda}	λ_{fda}	OA	MIoU	Kappa	PA	UA
1.0	0.05	N/A	93.05	83.66	80.80	77.90	94.51
1.0	0.10	N/A	**93.48**	**84.74**	**81.51**	**79.02**	**93.61**
1.0	0.15	N/A	93.12	84.10	81.18	78.60	94.19
1.0	0.10	0.10	92.26	81.12	72.61	68.53	94.79
1.0	0.10	0.01	94.96	85.86	81.43	80.06	93.48
1.0	0.10	0.001	**95.51**	**87.54**	**82.69**	**81.28**	93.55
1.0	0.10	0.005	95.40	87.11	82.68	80.35	**94.86**

4.2. *Comparisons with State-of-the-Art Methods*

4.2.1. Comparison Methods

For comprehensive evaluation, we compare deep adversarial network DAN [37], which focuses on training with both un-labeled and labeled images simultaneously to improve segmentation performance. Moreover, we compare two semi-supervised CNN-based semantic segmentation methods, i.e., Hung et al. [43] and s4GAN [40]. These two SSL methods are based on an adversarial network and achieve promising performance on PASCAL VOC 2012 [59] and Cityscapes datasets [36]. For fair comparison with the proposed SSCDnet, we use DeepLabv2 [51] and ResNet-101 [52] as a segmentation network and backbone of above-mentioned competing methods, respectively. In addition, baseline network DeepLabv2 [51] is also used as the competing method. During training, we retrain these CNN-based methods under their optimal parameter settings on Landast-8 OLI and GF-1 WFV datasets.

4.2.2. Results on GF-1 WFV Data

Table 6 shows the quantitative results in terms of average OA, MIoU, Kappa, PA, and UA on the GF-1 WFV testing dataset. We show these comparison results on four different proportions of labeled samples ($\frac{1}{200}$, $\frac{1}{100}$, $\frac{1}{40}$, and $\frac{1}{20}$). Meanwhile, we also give comparison results on fully labeled samples. Experiment results show that our proposed SSCDnet consistently outperforms these competing methods at different proportions of labeled samples. Notably, SSCDnet achieves 83.08% Kappa and 86.96% MIoU using only 0.5% (1/200) training data with pixel-wise annotation. Results of SSCDnet significantly outperform those of competing methods. SSCDnet also achieves the best performance on fully labeled data and shows a larger gain in terms of OA, MIoU, Kappa, PA, and UA than competing methods.

Table 6. Cloud extraction accuracy (%) of different comparison networks on GF-1 WFV data. All results are the averaging results on all testing images.

Proportion	Methods	OA	MIoU	Kappa	PA	UA
$\frac{1}{200}$	DeepLabV2 [51]	90.22	71.77	68.21	67.03	**95.51**
	DAN [37]	92.59	79.93	74.65	72.08	94.59
	Hung et al. [43]	93.01	81.20	75.11	76.59	94.02
	s4GAN [40]	93.44	81.79	75.62	77.31	93.38
	SSCDnet	**95.51**	**87.54**	**82.69**	**81.28**	93.55
$\frac{1}{100}$	DeepLabV2 [51]	91.34	74.10	71.51	69.71	**94.93**
	DAN [37]	93.13	82.41	77.56	75.84	93.66
	Hung et al. [43]	94.20	83.88	78.56	76.11	93.02
	s4GAN [40]	94.60	84.91	79.14	76.47	92.63
	SSCDnet	**95.85**	**88.90**	**83.92**	**82.34**	93.36
$\frac{1}{40}$	DeepLabV2 [51]	92.23	76.02	73.58	75.08	**93.43**
	DAN [37]	93.89	83.63	80.15	81.29	92.53
	Hung et al. [43]	94.41	85.05	81.59	81.80	92.76
	s4GAN [40]	95.30	86.80	82.47	82.20	93.00
	SSCDnet	**96.24**	**89.30**	**85.01**	**85.49**	92.24
$\frac{1}{20}$	DeepLabV2 [51]	93.19	79.88	75.75	80.69	92.30
	DAN [37]	94.32	85.05	81.52	82.21	92.93
	Hung et al. [43]	94.94	86.12	82.53	82.44	93.09
	s4GAN [40]	95.61	87.18	83.63	82.96	**93.37**
	SSCDnet	**96.90**	**90.93**	**86.97**	**85.89**	92.24
Full	DeepLabV2 [51]	96.11	90.15	87.64	87.28	92.15
	DAN [37]	97.09	90.37	88.72	88.19	92.43
	Hung et al. [43]	97.12	90.57	88.79	88.50	92.71
	s4GAN [40]	97.16	90.88	88.83	88.57	92.60
	SSCDnet	**97.19**	**91.71**	**89.37**	**88.90**	**92.32**

where $\frac{1}{200}$, $\frac{1}{100}$, $\frac{1}{40}$, and $\frac{1}{20}$ are the fractions of the total training images in the dataset that are used as labeled data, and the rest of the data are used without labels.

It can be seen that the baseline network, DeepLabV2 [51], trained with only labeled data, shows the worst results. DAN [37] shows better results than DeepLabV2 [51] due to it being able to effectively utilize unlabeled data for training. These semi-supervised methods, Hung et al. [43] and s4GAN [40] show better results than DeepLabV2 [51] and DAN [37] due to being able to learn knowledge from a limited number of labeled examples and a large number of additional unlabeled samples. Although these two methods show competitive cloud detection performance, there is still a gap in performance between them and the proposed SSCDnet. In addition, it also can be seen that all of the methods show promising cloud detection performance on GF-1 WFV data when using fully labeled data for network training. Even the worst baseline method DeepLabV2 [51] achieves 90.15% mIoU and 88.64% Kappa.

In Figure 9, we show qualitative results when labeled samples proportion is $\frac{1}{200}$. These images contain typical land-cover types, i.e., Figure 9a includes mountains, wetlands, farmland, and grass/crops, Figure 9b includes village, forest and ice/snow, Figure 9c includes water. Experiment results show that detection results of SSCDnet show more consistency with the ground-truth than those of other competing methods. Results of competing methods show a lot of misclassification pixels in thin cloud areas, while SSCDnet shows less misclassification pixels, which indicates that SSCDnet is able to achieve more promising performance on these imageries when using a limited number of labeled images.

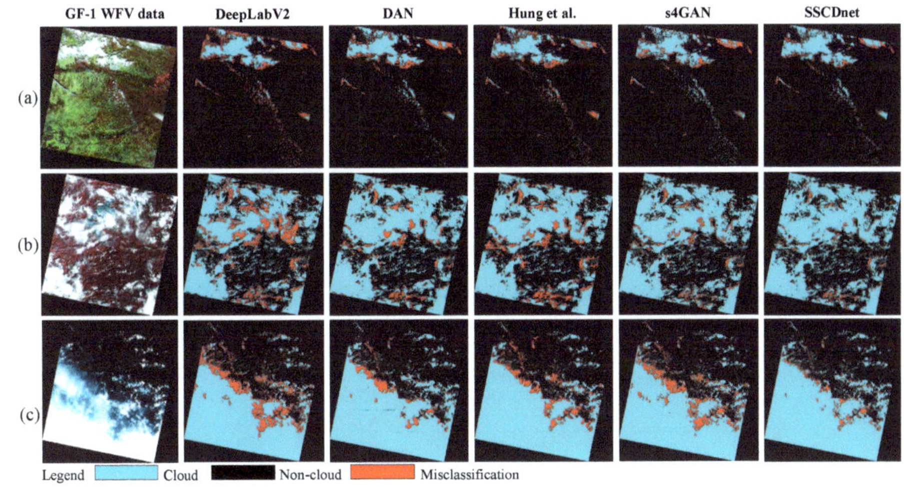

Figure 9. Comparison of cloud detection results of different methods on GF-1 WFV dataset with labeled sample proportion of 1/200. Image ID of (**a**), (**b**), and (**c**) are GF1_WFV2_W102.1_N37.6_20140517_L2A0000244678, GF1_WFV3_E114.1_N2.1_20151011_L2A0001094727, and GF1_WFV3_E87.8_N2.1_20140316_L2A0000184430, respectively.

4.2.3. Results on Landsat-8 OLI Data

In addition to the experiment results on the GF-1 data, we also conduct experiments on Landsat8 OLI data. In Table 7, we list quantitative results on the Landsat8 OLI testing dataset. Compared to other methods, the proposed SSCDnet also performs best on Landsat8 OLI data. For the low labeled sample's proportion, such as $\frac{1}{200}$, SSCDnet is still able to achieve satisfactory results (90.77% MIoU and 88.72% Kappa). These results show consistency with the ground-truth. In contrast, performance of these competing methods is less than that of SSCDnet. When increasing the proportion of labeled samples, competing methods can improve their performance, but they are still inferior to SSCDnet. From Table 7, we find that the performance of SSCDnet at a labeled sample proportion of $\frac{1}{20}$ approaches that of its full supervision and outperforms fully supervised DeepLabV2 [51] and DAN [37]. For fully labeled data, with the help of adversarial training and self-training strategies, SSCDnet still shows the best performance compared with other competing methods.

In Figure 10, we show qualitative results on three whole scene landsat-8 OLI images. These images contain typical land-cover types, i.e., Figure 10a includes mountains, forest, ice/snow, water, and wetlands areas. Figure 10b includes water, floating ice, urban, mountains, and forest areas. Figure 10c includes barren and desert areas. Experiment results are obtained when labeled sample proportion is $\frac{1}{200}$. Experiment results show that SSCDnet trained with a limited number of labeled samples can yield very competitive performance on Landsat-8 OLI data. It can be seen that results of these competing methods show a large number of misclassified pixels (red areas), especially in Figure 10c. There are a large number of misclassified pixels displayed in the thin cloud area. In contrast, SSCDnet works well on these images. Results of SSCDnet shows better consistency with the ground-truth and fewer misclassified pixels than competing methods. Overall, experiment results in Table 7 and Figure 10 show that the proposed SSCDnet is able to achieve promising cloud detection performance on Landsat-8 OLI data.

Table 7. Cloud extraction accuracy (%) of different comparison networks on Landsat-8 OLI data. All results are the averaged results on all the testing images.

Proportions	Method	OA	MIoU	Kappa	PA	UA
$\frac{1}{200}$	DeepLabV2 [51]	91.24	84.16	81.77	82.12	**94.53**
	DAN [37]	92.68	86.97	82.83	90.47	88.78
	Hung et al. [43]	93.14	87.02	84.49	91.55	90.13
	s4GAN [40]	93.88	88.76	85.91	93.38	91.05
	SSCDnet	**95.48**	**91.28**	**88.87**	**95.60**	90.62
$\frac{1}{100}$	DeepLabV2 [51]	92.54	88.76	85.06	88.30	89.41
	DAN [37]	93.99	89.32	85.75	91.11	89.45
	Hung et al. [43]	94.58	90.71	87.89	93.80	90.33
	s4GAN [40]	95.40	91.30	89.02	**94.17**	91.08
	SSCDnet	**95.63**	**91.75**	**89.38**	93.81	**92.01**
$\frac{1}{40}$	DeepLabV2 [51]	93.15	89.06	85.67	90.16	89.90
	DAN [37]	94.37	89.26	86.78	92.38	89.77
	Hung et al. [43]	94.79	90.93	88.11	**93.22**	91.05
	s4GAN [40]	95.54	91.33	89.19	93.15	92.00
	SSCDnet	**95.82**	**92.03**	**89.61**	93.18	**92.31**
$\frac{1}{20}$	DeepLabV2 [51]	94.12	89.33	86.45	91.85	90.36
	DAN [37]	95.29	89.61	86.99	**92.67**	90.14
	Hung et al. [43]	95.83	91.06	88.20	92.11	91.82
	s4GAN [40]	96.16	91.96	89.69	92.04	92.84
	SSCDnet	**96.67**	**92.19**	**90.13**	92.50	**93.26**
Full	DeepLabV2 [51]	96.56	91.03	89.25	91.10	91.26
	DAN [37]	96.60	92.09	89.53	91.62	93.11
	Hung et al. [43]	96.57	92.11	89.77	92.28	92.03
	s4GAN [40]	96.63	92.06	89.82	92.16	**92.41**
	SSCDnet	**97.19**	**92.65**	**90.34**	**92.75**	93.36

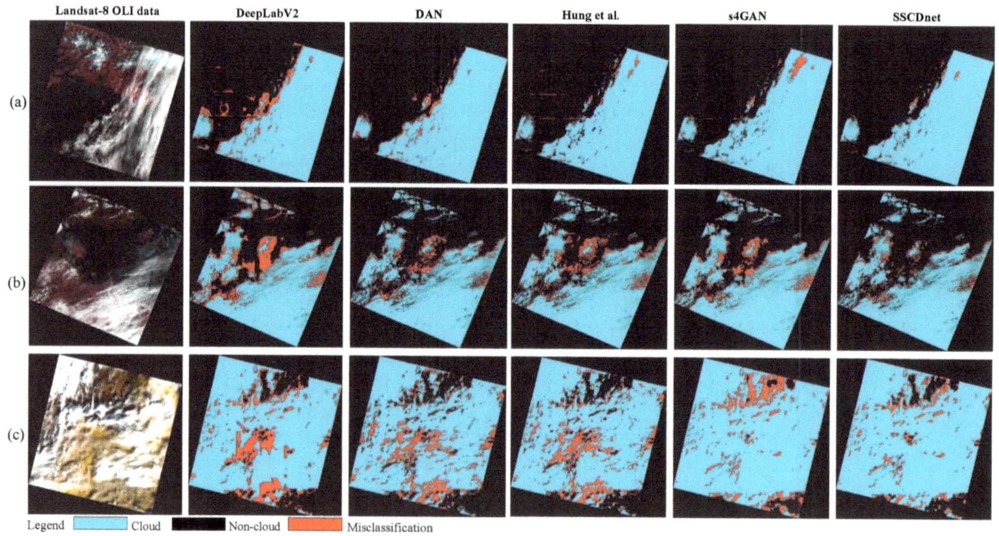

Figure 10. Comparison of cloud extraction results of different methods on Landsat-8 OLI dataset with a labeled sample proportion of 1/200. The image ID of (**a**), (**b**), and (**c**) are LC80650182013237LGN00, LC80430122014214LGN00, and LC81990402014267LGN00, respectively.

5. Discussion and Analysis

5.1. Robustness Analysis

To evaluate the robustness of the proposed method, we conducted a series of experiments as follows: (1) experiment results on the same area under different seasons and (2) experiment results on different land cover types.

5.1.1. Results on the Same Area under Different Seasons

In Figure 11, we present the cloud detection results on the same area under different seasons, i.e., Spring, Summer, Autumn, and Winter. Satellite images obtained from this area include different land cover types, such as mountain, village, urban, water, ocean, plant, and farmland areas. In Figure 11, overall accuracy (OA) of Spring, Summer, Autumn, and Winter images are 98.21%, 97.70%, 95.79%, and 96.03%, respectively. Experiment results show that the proposed semi-supervised cloud detection method SSCDnet is able to achieve a promising performance under different radiance (i.e., same area and different seasons). However, there is still a large number of misclassified pixels in the experiment results. These misclassified pixels are located mostly near the cloud object boundaries and thin cloud areas, which is also a difficult problem for most CNN-based cloud detection works, such as [7,19,27,33–35].

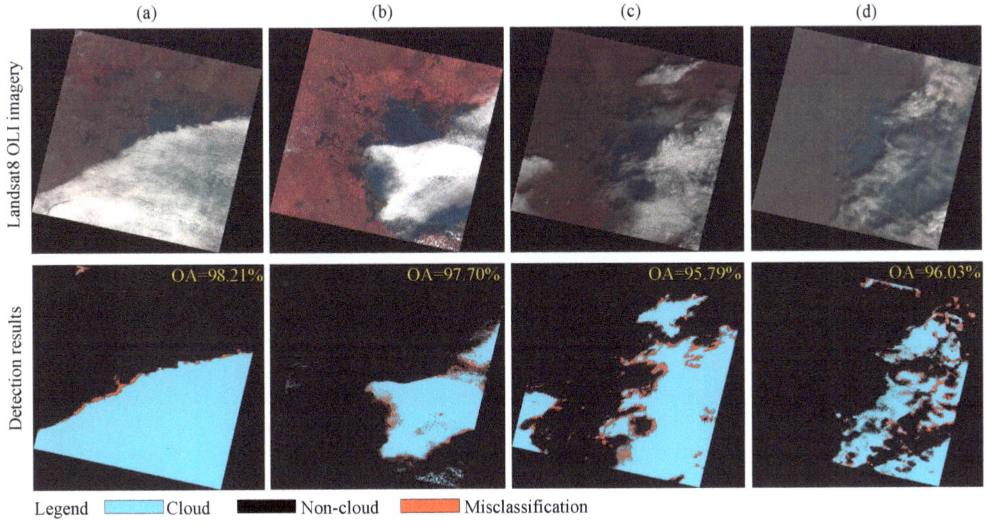

Figure 11. Cloud extraction results of SSCDnet on Landsat-8 OLI data under different seasons (labeled proportion is $\frac{1}{20}$). Where, (**a**) is the Spring season (ID: LC08_L1TP_122033_20160411_20170326_01_T1), (**b**) is the Summer season (ID: LC08_L1TP_122033_20180706_20180717_01_T1), (**c**) is the Autumn season(ID: LC08_L1TP_122033_20181111_20181127_01_T1), and (**d**) is the Winter season (ID: LC08_L1TP_122033_20150103_20170415_01_T1).

5.1.2. Results on Different Land Cover Types

In Figure 12, we present the experiment results on twelve sub images with different land cover types. Results show that these corresponding cloud detection results obtained by SSCDnet show consistency with the ground truth when our CNN model is trained with a limited number of labeled dataset (labeled proportion is $\frac{1}{10}$), except for some misclassified pixels located near cloud object boundaries and thin cloud areas.

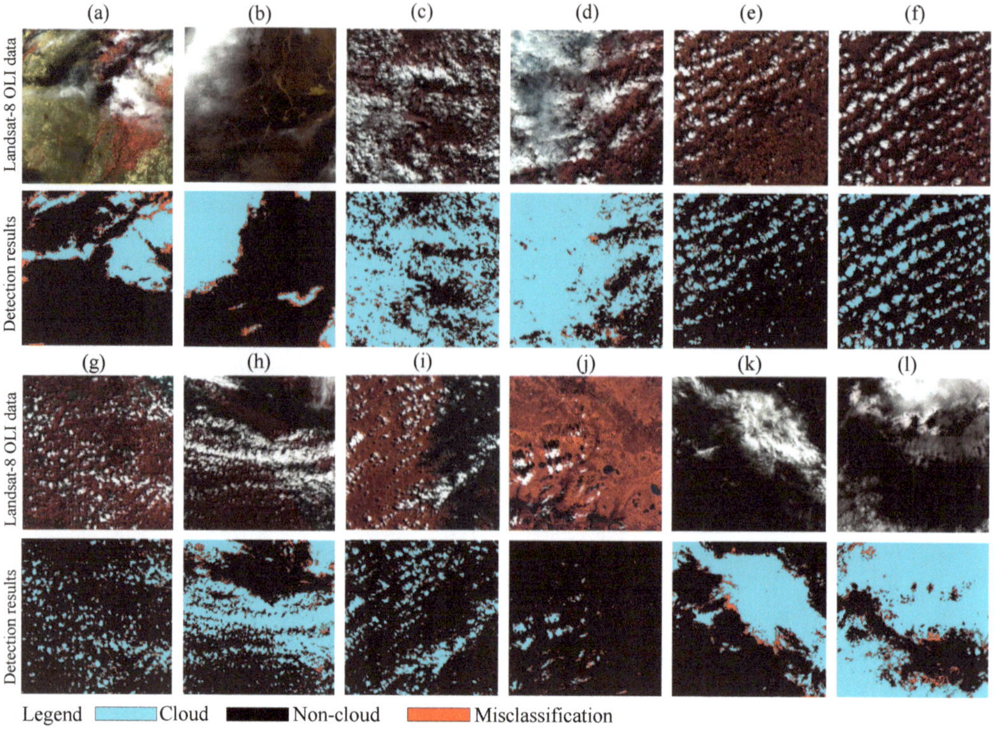

Figure 12. Cloud extraction results (1200 × 1200) of SSCDnet on Landsat-8 OLI data under different land cover types. Where, (**a**,**b**) are barren/desert areas. (**c**,**d**) are mountain/plant areas. (**e**,**f**) are farmland/villages areas. (**g**,**h**) are urban/river areas. (**i**,**j**) are mountain/lake areas. (**k**,**l**) are ocean areas.

To be specific, for tough cases, such as barren/desert areas (Figure 12a,b) and urban areas (Figure 12g,h), SSCDnet is able to achieve a promising performance on these land cover types. In addition, SSCDnet also shows promising performance on water areas, such as lake (Figure 12i,j), river (Figure 12h), and ocean (Figure 12k,l) areas. Except for snow, buildings, and some white objects, few ground objects affect cloud detection, and SSCDnet can easily obtain promising results in some general cases, such as mountain/plant (Figure 12e,f) and farmland/village (Figure 12g,h) areas. In general, results in Figure 12 demonstrate that SSCDnet has a robust cloud detection performance on different land cover types.

5.2. Computational Complexity Analysis

To analyze the computational complexity of SSCDnet, we evaluate computational complexity of these networks with six evaluation criterions, which are floating point operations (FLOPs), number of trainable parameters, training time, training GPU memory usage, testing time, and testing GPU memory usage. In Table 8, we list results of different competing methods. FLOPs are calculated from input data with image size of 321 × 321. Training times of all methods are obtained from 5000 iterations. Training GPU memory size is obtained by setting batch size of 4 and image size of 321 × 321. Testing time is obtained by testing 20 scene Landsat-8 OLI satellite images with image size of 8 k × 8 k. Training GPU memory size is obtained by setting batch size of 1 and image size of 1200 × 1200.

During training, both the segmentation network and discriminator network are trained simultaneously. Discriminators of different SSL methods have different raw input data channels, which results in these models having different model parameters, different computational complexity and training times, and different GPU memory requirements. In Table 8, we can see that FLOPs, number of parameters, and training GUP memory usage of SSCDnet are higher than those of competing methods. This is because SSCDnet performs intermediate feature map domain adaptation alignments, while competing methods have no such operation. Feature map alignment requires more computations and GPU memories. In addition, the discriminator network for feature alignment further increases the number of training parameters. Luckily, the longest training time is not the SSCDnet but Hung et al. [43].

During testing, we only need to use the segmentation network to detect clouds instead of the discriminator network. Since all the methods use the same baseline segmentation network, i.e., DeepLabV2 [51], all these methods share the same testing times and GPU memory usage. In Table 8, we can see that it takes about 400 s to detect 20 scenes Landsat 8-OLI satellite images with image size of 8 k × 8 k. In other words, it takes 20 s to detect an image. 2849 MB GPU memory is required to process an image with size of 1200 × 1200.

Table 8. Computational complexity analysis of different semi-supervised methods and the baseline method (DeepLabV2 [51]).

Methods	GFLOPs	Parameters	Training Time	Training GPU	Testing Time	Testing GPU
DeepLabV2 [51]	74.03	43.94 M	1118.9906 s	10,079 MB	400.7494 s	2849 MB
DAN [37]	76.67	46.70 M	2129.5402 s	10,297 MB	400.7494 s	2849 MB
Hung et al. [43]	76.61	46.70 M	**10,178.0476** s	10,297 MB	400.7494 s	2849 MB
s4GAN [40]	76.68	46.70 M	2224.8972 s	10,341 MB	400.7494 s	2849 MB
SSCDnet	**410.56**	**61.91** M	2771.0219 s	**10,435** MB	400.7494 s	2849 MB

The code for computational complexity is available from https://github.com/sovrasov/flops-counter.pytorch (accessed on 24 April 2022). 1 GFLOPs = 1×10^9 FLOPs. 1 M = 1×10^6. 1 MB = 1×10^6 bytes.

5.3. Limitations

Although SSCDnet achieves a promising cloud detection performance on both GF-1 WFV and Landsat-8 OLI data, there is still a large number of misclassified pixels in tough cases when using a limited number of labeled samples. Since urban and floating ice areas show the similar color or texture with the clouds, and thin cloud objects show few differences with underground objects, it is difficult for SSCDnet to handle these areas when using a limited number of labeled samples as shown in Figure 13, where Figure 13a,b are the results of GF-1 WFV data (GF1_WFV2_W70.8_N19.2_20140801_L2A0000292230), while Figure 13c,d are the results of Landsat-8 OLI dat (LC81180382014244LGN00).

Experiment results in Figure 13a,b show that there are many misclassified pixels at urban and floating ice areas when labeled sample proportion is $\frac{1}{200}$ and $\frac{1}{100}$. When labeled sample proportion is more than $\frac{1}{40}$, the qualitative results remain basically stable and show consistency with the ground-truth. Results in Figure 13c,d indicate that capturing sharp and detailed object boundaries in thin cloud areas is still very difficult even trained with fully labeled samples. Adding a sufficient of thin cloud samples for network training may obtain good detection performance. In future work, we will focus on this point to further improve this work.

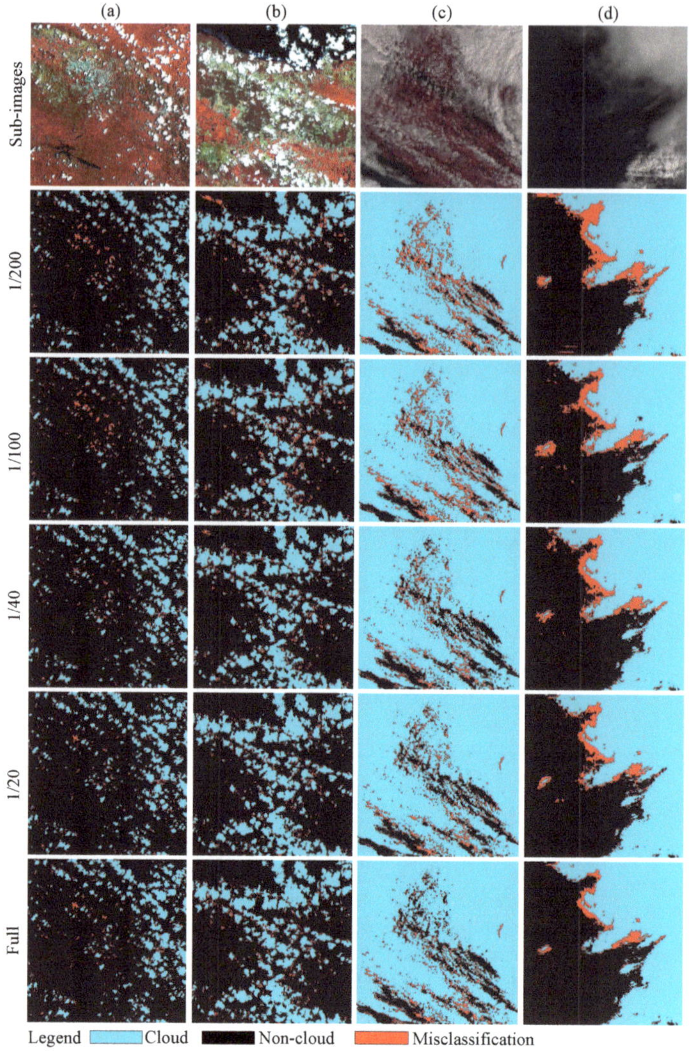

Figure 13. Cloud extraction results (1200 × 1200) of SSCDnet on GF-1 WFV and Landsat-8 OLI data under different labeled sample proportions. (**a**,**b**) are the results of GF-1 WFV data (GF1_WFV2_W70.8_N19.2_20140801_L2A0000292230), while (**c**,**d**) are the results of Landsat-8 OLI data (LC81180382014244LGN00).

6. Conclusions

Semi-supervised learning is an effective training strategy, which is able to train a segmentation network by using a limited number of pixel-wise labeled samples and a large number of unlabeled ones. In this paper, we present a semi-supervised cloud detection network, named SSCDnet. Since there are domain distribution gaps between the labeled and unlabeled datasets, we take the domain shift problem into account for the semi-supervised learning framework and propose feature-/output-level domain adaptation strategy to reduce domain distribution gaps, thus improving SSCDnet to generate trustworthy pseudo label for unlabeled data. A high certain pseudo label provides positive supervised signals for segmentation network learning through self-training. Experimental results on GF-1 WFV and Landsat-8 OLI datasets demonstrate that SSCDnet is able to achieve promising performance by using a limited number of labeled samples. It shows great promise for practical application on new satellite RS imagery in the presence of less labeled data available.

Although SSCDnet shows good performance, there is still much room for improvement, such as hyper-parameters setting of loss function and threshold setting of pseudo-labeling. Different cloud detection datasets have different domain distributions. We need to update these parameters to achieve a promising performance on different datasets. In addition, different ground objects have different characteristics, and the performance of SSCDnet on other objects detection also needs to be further evaluated. In our future work, we will further evaluate this method on other cloud detection datasets and other object detection tasks. In addition, SSCDnet performs poorly on cloud boundaries and thin cloud regions, which requires our future efforts to improve it. In our future work, we will also explore how to utilize some auxiliary information, such as land use and land cover (LULC) map, water index map, and vegetation index map, to improve the cloud detection performance.

In general, a semi-supervised learning training strategy provides us with an effective way for cloud detection from RS images in the presence of less labeled data available. In addition, this strategy may provide us a promising way for other object detection tasks such as water, vegetation, and building detection.

In order to promote understanding of the paper's technology, we released the code of SSCDnet. It is available at: https://github.com/nkszjx/SSCDnet (accessed on 24 April 2022).

Author Contributions: Conceptualization, J.G. and Z.L.; methodology, J.G.; validation, J.G. and Y.Z.; formal analysis, Z.L.; investigation, Q.X.; data curation, J.G.; writing—original draft preparation, J.G.; writing—review and editing, J.G. and Q.X.; supervision, X.Z.; project administration, X.Z. All authors have read and agreed to the published version of the manuscript.

Funding: This work was supported in part by the Sino-German (CSC-DAAD) Postdoc Scholarship Program under Grant 202006255045, in part by the German Federal Ministry of Education and Research (BMBF) in the framework of the international future AI Laboratory "AI4EO-Artificial Intelligence for Earth Observation: Reasoning, Uncertainties, Ethics and Beyond" under Grant 01DD20001, in part by the German Federal Ministry of Economics and Technology in the framework of the "National Center of excellence ML4Earth" under Grant 50EE2201C, in part by the European Research Council (ERC) through the European Union's Horizon 2020 Research and Innovation Programme under Grant ERC-2016-StG-714087, in part by the Helmholtz Association through the Framework of Helmholtz AI under Grant ZT-I-PF-5-01-Local Unit "Munich Unit at Aeronautics, Space and Transport (MASTr)", and in part by the Helmholtz Excellent Professorship "Data Science in Earth Observation Big Data Fusion for Urban Research" under Grant W2-W3-100.

Data Availability Statement: Not applicable.

Acknowledgments: We thank Huanfeng Shen's team from Wuhan University, China, for their providing a Gaofen-1 WFV cloud cover validation dataset. We are also thankful for Earth Explorer of the United States Geological Survey for providing the Landsat-8 OLI cloud cover validation dataset.

Conflicts of Interest: The authors declare no conflict of interest.

References

1. Zhao, S.; Wang, Q.; Li, Y.; Liu, S.; Wang, Z.; Zhu, L.; Wang, Z. An overview of satellite remote sensing technology used in China's environmental protection. *Earth Sci. Inform.* **2017**, *10*, 137–148. [CrossRef]
2. Yang, J.; Gong, P.; Fu, R.; Zhang, M.; Chen, J.; Liang, S.; Xu, B.; Shi, J.; Dickinson, R. The role of satellite remote sensing in climate change studies. *Nat. Clim. Chang.* **2013**, *3*, 875–883. [CrossRef]
3. Schmugge, T.J.; Kustas, W.P.; Ritchie, J.C.; Jackson, T.J.; Rango, A. Remote sensing in hydrology. *Adv. Water Resour.* **2002**, *25*, 1367–1385. [CrossRef]
4. Shanmugapriya, P.; Rathika, S.; Ramesh, T.; Janaki, P. Applications of remote sensing in agriculture—A Review. *Int. J. Curr. Microbiol. Appl. Sci.* **2019**, *8*, 2270–2283. [CrossRef]
5. Kabisch, N.; Selsam, P.; Kirsten, T.; Lausch, A.; Bumberger, J. A multi-sensor and multi-temporal remote sensing approach to detect land cover change dynamics in heterogeneous urban landscapes. *Ecol. Indic.* **2019**, *99*, 273–282. [CrossRef]

6. Bachagha, N.; Wang, X.; Luo, L.; Li, L.; Khatteli, H.; Lasaponara, R. Remote sensing and GIS techniques for reconstructing the military fort system on the Roman boundary (Tunisian section) and identifying archaeological sites. *Remote Sens. Environ.* **2020**, *236*, 111418. [CrossRef]
7. Guo, J.; Yang, J.; Yue, H.; Tan, H.; Hou, C.; Li, K. CDnetV2: CNN-Based Cloud Detection for Remote Sensing Imagery with Cloud-Snow Coexistence. *IEEE Trans. Geosci. Remote Sens.* **2020**, *59*, 700–713. [CrossRef]
8. Sliwa, B.; Falkenberg, R.; Liebig, T.; Piatkowski, N.; Wietfeld, C. Boosting vehicle-to-cloud communication by machine learning-enabled context prediction. *IEEE Trans. Intell. Transp. Syst.* **2019**, *21*, 3497–3512. [CrossRef]
9. Zhu, Z.; Wang, S.; Woodcock, C.E. Improvement and expansion of the Fmask algorithm: Cloud, cloud shadow, and snow detection for Landsats 4–7. 8, and Sentinel 2 images. *Remote Sens. Environ.* **2015**, *159*, 269–277. [CrossRef]
10. Qiu, S.; He, B.; Zhu, Z.; Liao, Z.; Quan, X. Improving Fmask cloud and cloud shadow detection in mountainous area for Landsats-8 images. *Remote Sens. Environ.* **2017**, *199*, 107–119. [CrossRef]
11. Qiu, S.; Zhu, Z.; He, B. Fmask 4.0: Improved cloud and cloud shadow detection in Landsats 4–8 and Sentinel-2 imagery. *Remote Sens. Environ.* **2019**, *231*, 111205. [CrossRef]
12. Li, Z.; Shen, H.; Li, H.; Xia, G.; Gamba, P.; Zhang, L. Multi-feature combined cloud and cloud shadow detection in GaoFen-1 wide field of view imagery. *Remote Sens. Environ.* **2017**, *191*, 342–358. [CrossRef]
13. Alonso-Montesinos, J.; Martínez-Durbán, M.; del Sagrado, J.; del Águila, I.; Batlles, F. The application of Bayesian network classifiers to cloud classification in satellite images. *Renew. Energy* **2016**, *97*, 155–161. [CrossRef]
14. Xu, L.; Wong, A.; Clausi, D.A. A novel Bayesian spatial–temporal random field model applied to cloud detection from remotely sensed imagery. *IEEE Trans. Geosci. Remote Sens.* **2017**, *55*, 4913–4924. [CrossRef]
15. Latry, C.; Panem, C.; Dejean, P. Cloud detection with SVM technique. In Proceedings of the 2007 IEEE International Geoscience and Remote Sensing Symposium, Barcelona, Spain, 23–27 July 2007; pp. 448–451.
16. Ishida, H.; Oishi, Y.; Morita, K.; Moriwaki, K.; Nakajima, T.Y. Development of a support vector machine based cloud detection method for MODIS with the adjustability to various conditions. *Remote Sens. Environ.* **2018**, *205*, 390–407. [CrossRef]
17. Hughes, M.J.; Hayes, D.J. Automated detection of cloud and cloud shadow in single-date Landsat imagery using neural networks and spatial post-processing. *Remote Sens.* **2014**, *6*, 4907–4926. [CrossRef]
18. Jang, J.d.; Viau, A.A.; Anctil, F.; Bartholomé, E. Neural network application for cloud detection in SPOT VEGETATION images. *Int. J. Remote Sens.* **2006**, *27*, 719–736. [CrossRef]
19. Jeppesen, J.H.; Jacobsen, R.H.; Inceoglu, F.; Toftegaard, T.S. A cloud detection algorithm for satellite imagery based on deep learning. *Remote Sens. Environ.* **2019**, *229*, 247–259. [CrossRef]
20. Wieland, M.; Li, Y.; Martinis, S. Multi-sensor cloud and cloud shadow segmentation with a convolutional neural network. *Remote Sens. Environ.* **2019**, *230*, 111203. [CrossRef]
21. Chai, D.; Newsam, S.; Zhang, H.K.; Qiu, Y.; Huang, J. Cloud and cloud shadow detection in Landsat imagery based on deep convolutional neural networks. *Remote Sens. Environ.* **2019**, *225*, 307–316. [CrossRef]
22. Drönner, J.; Korfhage, N.; Egli, S.; Mühling, M.; Thies, B.; Bendix, J.; Freisleben, B.; Seeger, B. Fast cloud segmentation using convolutional neural networks. *Remote Sens.* **2018**, *10*, 1782. [CrossRef]
23. Lopez, J.; Santos, S.; Atzberger, C.; Torres, D. Convolutional Neural Networks for Semantic Segmentation of Multispectral Remote Sensing Images. In Proceedings of the 2018 IEEE 10th Latin-American Conference on Communications (LATINCOM), Guadalajara, Mexico, 14–16 November 2018; pp. 1–5.
24. Zhan, Y.; Wang, J.; Shi, J.; Cheng, G.; Yao, L.; Sun, W. Distinguishing cloud and snow in satellite images via deep convolutional network. *IEEE Geosci. Remote Sens. Lett.* **2017**, *14*, 1785–1789. [CrossRef]
25. Morales, G.; Huamán, S.G.; Telles, J. Cloud Detection in High-Resolution Multispectral Satellite Imagery Using Deep Learning. In Proceedings of the International Conference on Artificial Neural Networks, Rhodes, Greece, 4–7 October 2018; pp. 280–288.
26. Li, Z.; Shen, H.; Cheng, Q.; Liu, Y.; You, S.; He, Z. Deep learning based cloud detection for medium and high resolution remote sensing images of different sensors. *ISPRS J. Photogramm. Remote Sens.* **2019**, *150*, 197–212. [CrossRef]
27. Yang, J.; Guo, J.; Yue, H.; Liu, Z.; Hu, H.; Li, K. CDnet: CNN-Based Cloud Detection for Remote Sensing Imagery. *IEEE Trans. Geosci. Remote Sens.* **2019**, *57*, 6195–6211. [CrossRef]
28. Shao, Z.; Pan, Y.; Diao, C.; Cai, J. Cloud Detection in Remote Sensing Images Based on Multiscale Features-Convolutional Neural Network. *IEEE Trans. Geosci. Remote Sens.* **2019**, *57*, 4062–4076. [CrossRef]
29. Chen, Y.; Weng, Q.; Tang, L.; Liu, Q.; Fan, R. An Automatic Cloud Detection Neural Network for High-Resolution Remote Sensing Imagery With Cloud–Snow Coexistence. *IEEE Geosci. Remote Sens. Lett.* **2022**, *19*, 6004205. [CrossRef]
30. Zhang, Z.; Iwasaki, A.; Xu, G.; Song, J. Cloud detection on small satellites based on lightweight U-net and image compression. *J. Appl. Remote Sens.* **2019**, *13*, 026502. [CrossRef]
31. Zhang, J.; Li, X.; Li, L.; Sun, P.; Su, X.; Hu, T.; Chen, F. Lightweight U-Net for cloud detection of visible and thermal infrared remote sensing images. *Opt. Quantum Electron.* **2020**, *52*, 397. [CrossRef]
32. Li, J.; Wu, Z.; Hu, Z.; Jian, C.; Luo, S.; Mou, L.; Zhu, X.X.; Molinier, M. A Lightweight Deep Learning-Based Cloud Detection Method for Sentinel-2A Imagery Fusing Multiscale Spectral and Spatial Features. *IEEE Trans. Geosci. Remote Sens.* **2021**, *60*, 1–19. [CrossRef]
33. Guo, J.; Yang, J.; Yue, H.; Li, K. Unsupervised Domain Adaptation for Cloud Detection Based on Grouped Features Alignment and Entropy Minimization. *IEEE Trans. Geosci. Remote Sens.* **2021**, 1–13. [CrossRef]

34. Guo, J.; Yang, J.; Yue, H.; Liu, X.; Li, K. Unsupervised Domain-Invariant Feature Learning for Cloud Detection of Remote Sensing Images. *IEEE Trans. Geosci. Remote Sens.* **2022**, *60*, 5405715. [CrossRef]
35. Guo, J.; Yang, J.; Yue, H.; Chen, Y.; Hou, C.; Li, K. Cloud Detection From Remote Sensing Imagery Based on Domain Translation Network. *IEEE Geosci. Remote Sens. Lett.* **2022**, *19*, 5000805. [CrossRef]
36. Cordts, M.; Omran, M.; Ramos, S.; Rehfeld, T. The Cityscapes Dataset for Semantic Urban Scene Understanding. *arXiv* **2016**, arXiv:1604.01685.
37. Zhang, Y.; Yang, L.; Chen, J.; Fredericksen, M.; Hughes, D.P.; Chen, D.Z. Deep Adversarial Networks for Biomedical Image Segmentation Utilizing Unannotated Images. In Proceedings of the Medical Image Computing and Computer Assisted Intervention (MICCAI 2017), Quebec City, QC, Canada, 11–13 September 2017; pp. 408–416.
38. Hong, S.; Noh, H.; Han, B. Decoupled Deep Neural Network for Semi-supervised Semantic Segmentation. *arXiv* **2015**, arXiv:1506.04924.
39. Mostafa S, I.; Arash, V.; Mani, R.; William G, M. Semi-Supervised Semantic Image Segmentation with Self-correcting Networks. *arXiv* **2020**, arXiv:1811.07073.
40. Mittal, S.; Maxim, T.; Thomas, B. Semi-Supervised Semantic Segmentation with High- and Low-level Consistency. *IEEE Trans. Pattern Anal. Mach. Intell.* **2021**, *43*, 1369–1379. [CrossRef] [PubMed]
41. Massih-Reza, A.; Vasilii, F.; Loic, P.; Emilie, D.; Yury, M. Self-Training: A Survey. *arXiv* **2022**, arXiv:2202.12040.
42. Rajat, R.; Alexis, B.; Honglak, L.; Benjamin, P.; Andrew, Y.N. Self-taught learning: Transfer learning from unlabeled data. In Proceedings of the Twenty-fourth International Conference on Machine Learning, Corvalis, OR, USA, 20–24 June 2007.
43. Hung, W.; Tsai, Y.; Liou, Y.; Lin, Y.Y.; Yang, M.H. Adversarial Learning for Semi-Supervised Semantic Segmentation. *arXiv* **2018**, arXiv:1802.07934.
44. Goodfellow, I.; Pouget-Abadie, J.; Mirza, M.; Xu, B.; Warde-Farley, D.; Ozair, S.; Courville, A.; Bengio, Y. Generative adversarial networks. In Proceedings of the Advances in Neural Information Processing Systems, Montreal, QC, Canada, 8–13 December 2014; pp. 2672–2680.
45. Ren, Z.; Lee, Y.J. Cross-Domain Self-Supervised Multi-task Feature Learning Using Synthetic Imagery. In Proceedings of the 2018 IEEE/CVF Conference on Computer Vision and Pattern Recognition, Salt Lake City, UT, USA, 18–22 June 2018; pp. 762–771.
46. Tuan-Hung, V.; Himalaya, J.; Maxime, B.; Matthieu, C.; Patrick, P. ADVENT: Adversarial Entropy Minimization for Domain Adaptation in Semantic Segmentation. *arXiv* **2019**, arXiv:1811.12833.
47. Tsai, Y.; Sohn, K.; Schulter, S.; Chandraker, M. Domain Adaptation for Structured Output via Discriminative Patch Representations. In Proceedings of the 2019 IEEE/CVF International Conference on Computer Vision, Seoul, Korea, 27 October–2 November 2019; pp. 1456–1465.
48. Foga, S.; Scaramuzza, P.L.; Guo, S.; Zhu, Z.; Dilley, R.D., Jr.; Beckmann, T.; Schmidt, G.L.; Dwyer, J.L.; Hughes, M.J.; Laue, B. Cloud detection algorithm comparison and validation for operational Landsat data products. *Remote Sens. Environ.* **2017**, *194*, 379–390. [CrossRef]
49. Chen, Y.; Lin, Y.; Yang, M.; Huang, J. CrDoCo: Pixel-Level Domain Transfer With Cross-Domain Consistency. In Proceedings of the 2019 IEEE/CVF Conference on Computer Vision and Pattern Recognition, Long Beach, CA, USA, 16–20 June 2019; pp. 1791–1800.
50. Souly, N.; Spampinato, C.; Shah, M. Semi Supervised Semantic Segmentation Using Generative Adversarial Network. In Proceedings of the 2017 IEEE International Conference on Computer Vision (ICCV), Venice, Italy, 22–29 October 2017; pp. 5689–5697.
51. Chen, L.; Papandreou, G.; Kokkinos, I.; Murphy, K.; Yuille, A.L. DeepLab: Semantic Image Segmentation with Deep Convolutional Nets, Atrous Convolution, and Fully Connected CRFs. *IEEE Trans. Pattern Anal. Mach. Intell.* **2018**, *40*, 834–848. [CrossRef] [PubMed]
52. He, K.; Zhang, X.; Ren, S.; Sun, J. Deep Residual Learning for Image Recognition. In Proceedings of the The IEEE Conference on Computer Vision and Pattern Recognition (CVPR), Las Vegas, NV, USA, 27–30 June 2016.
53. Yu, F.; Koltun, V. Multi-Scale Context Aggregation by Dilated Convolutions. *arXiv* **2016**, arXiv:1511.07122.
54. Geirhos, R.; Rubisch, P.; Michaelis, C.; Bethge, M.; Wichmann, F.A.; Brendel, W. ImageNet-trained CNNs are biased towards texture; increasing shape bias improves accuracy and robustness. *arXiv* **2018**, arXiv:1811.12231.
55. Bengio, Y. Practical recommendations for gradient-based training of deep architectures. In *Neural Networks: Tricks of the Trade*; Springer: Berlin/Heidelberg, Germany, 2012; pp. 437–478.
56. Kingma, D.; Ba, J. Adam: A Method for Stochastic Optimization. *arXiv* **2014**, arXiv:1412.6980v9.
57. Chen, L.C.; Papandreou, G.; Schroff, F.; Adam, H. Rethinking atrous convolution for semantic image segmentation. *arXiv* **2017**, arXiv:1706.05587.
58. Russakovsky, O.; Deng, J.; Su, H.; Krause, J.; Satheesh, S.; Ma, S.; Huang, Z.; Karpathy, A.; Khosla, A.; Bernstein, M.; et al. Imagenet large scale visual recognition challenge. *Int. J. Comput. Vis.* **2015**, *115*, 211–252. [CrossRef]
59. Everingham, M.; Van Gool, L.; Williams, C.K.I.; Winn, J.; Zisserman, A. The PASCAL Visual Object Classes Challenge 2012 (VOC2012) Results. Available online: http://host.robots.ox.ac.uk/pascal/VOC/voc2012/ (accessed on 24 April 2022).

sustainability MDPI

Article

Application of Combined Models Based on Empirical Mode Decomposition, Deep Learning, and Autoregressive Integrated Moving Average Model for Short-Term Heating Load Predictions

Yong Zhou [1,*], Lingyu Wang [2] and Junhao Qian [2]

[1] School of Management, Xi'an University of Architecture and Technology, No.13 Yanta Road, Xi'an 710055, China
[2] School of Building Services Science and Engineering, Xi'an University of Architecture and Technology, No.13 Yanta Road, Xi'an 710055, China; wanglingyu0827@163.com (L.W.); qianjunhao0402@gmail.com (J.Q.)
* Correspondence: zhouyong@xauat.edu.cn

Abstract: Short-term building energy consumption prediction is of great significance for the optimized operation of building energy management systems and energy conservation. Due to the high-dimensional nonlinear characteristics of building heat loads, traditional single machine-learning models cannot extract the features well. Therefore, in this paper, a combined model based on complete ensemble empirical mode decomposition with adaptive noise (CEEMDAN), four deep learning (DL), and the autoregressive integrated moving average (ARIMA) models is proposed. The DL models include a convolution neural network, long- and short-term memory (LSTM), bi-directional LSTM (bi-LSTM), and the gated recurrent unit. The CEEMDAN decomposed the heating load into different components to extract the different features, while the DL and ARIMA models were used for the prediction of heating load features with high and low complexity, respectively. The single-DL models and the CEEMDAN-DL combinations were also implemented for comparison purposes. The results show that the combined models achieved much higher accuracy compared to the single-DL models and the CEEMDAN-DL combinations. Compared to the single-DL models, the average coefficient of determination (R^2), root mean square error (RMSE), and coefficient of variation of the RMSE (CV-RMSE) were improved by 2.91%, 47.93%, and 47.92%, respectively. Furthermore, CEEMDAN-bi-LSTM-ARIMA performed the best of all the combined models, achieving values of R2 = 0.983, RMSE = 70.25 kWh, and CV-RMSE = 1.47%. This study provides a new guide for developing combined models for building energy consumption prediction.

Keywords: building energy consumption prediction; empirical mode decomposition; deep learning models; combined models

Citation: Zhou, Y.; Wang, L.; Qian, J. Application of Combined Models Based on Empirical Mode Decomposition, Deep Learning, and Autoregressive Integrated Moving Average Model for Short-Term Heating Load Predictions. *Sustainability* **2022**, *14*, 7349. https://doi.org/10.3390/su14127349

Academic Editors: Miltiadis D. Lytras and Andreea Claudia Serban

Received: 22 April 2022
Accepted: 10 June 2022
Published: 15 June 2022

1. Introduction

Building energy consumption accounts for more than one-third of the global total energy consumption and is expected to continue increasing with the acceleration of urbanization in developing countries [1]. Consequently, energy saving and sustainable buildings have become the top priority of all countries in the world. To solve this problem, researchers have put forward a series of building energy-saving measures, such as renewable energy utilization and high-performance building envelopes. Among them, efficient intelligent control is one of the most effective measures to reduce energy consumption during the building's lifecycle [2]. However, building energy consumption exhibits substantial randomness, so its accurate and reliable prediction is of great significance to predictive building systems' control and optimal scheduling [3–5].

Depending on the time-scale range, building energy consumption prediction (BECP) methods can be divided into short-term, medium-term, and long-term [6]. Short-term building energy forecasting is mainly used for the daily operation and management of building energy systems [7]; it can help users adjust the equipment in real time according to the predicted values so as to achieve the best match between energy supply and demand as well as formulate the operational and regulation strategy of building energy systems [8]. Medium-term BECP is mainly applied to the scheduling of energy systems [9], while long-term prediction is mainly used for capacity design [10].

BECP methods can be mainly divided into physical models and data-driven models [11]. The physical models are based on energy consumption analysis and make predictions according to the heat-transfer process and outdoor meteorological conditions [12]. A notable example of this approach is the EnergyPlus simulation tool [13]. Physical models require detailed physical building information, including parameters related to exterior walls, doors and windows, ground, and other details, as well as the heat sources in the building. This information is often difficult to obtain, which limits the wide adoption of physical models [14]. With the continuous development of the Internet of Things, building energy consumption monitoring technologies are gradually maturing, and a large amount of historical data have become available, which has allowed the rapid development of data-driven energy consumption prediction models [15].

Machine-learning models have become one of the most commonly used data-driven methods for BECP due to their high accuracy and excellent ability to deal with high-dimensional nonlinear problems [16]. Such methods include artificial neural networks (NN) [17], support vector machines (SVMs) [18], least square SVMs [19], decision trees [20], random forests (RFs) [21], extreme gradient boosting (XGBoost) [22], etc. For example, Ling et al. [23] adopted a back-propagation neural network (BPNN) and an SVM to select the optimal input parameters for BECP. Chen and Chou [24] reviewed the machine-learning models commonly used for this purpose. Yang et al. [25] compared an RF with a BPNN and an SVM for BECP and concluded that the RF exhibits obvious advantages in terms of performance. Liu et al. [26] developed the 24-h-ahead BECP model based on long- and short-term memory (LSTM) and demonstrated that their proposed models achieved high accuracy and reliable predictions. Lei et al. [27] adopted a rough dataset and deep learning (DL) models for short- and medium-term BECP using one-year data from a laboratory building of a university in Dalian. The comparison with a BPNN, an Elman-NN, and a fuzzy NN showed that the DL models using the rough dataset produced the highest accuracy. Zhou et al. [28] compared 15 machine-learning models for short-term building heat load prediction and concluded that the tree-based models that adopted decision tree as the base model easily countered over-fitting problems, while the SVM and Gaussian process regression models produced the highest accuracy. Li et al. [29] developed a genetic algorithm (GA) and an NN BECP model for the building design and summarized that the application of GA can improve the accuracy and efficiency of BECP.

As the accuracy of single models is to some degree limited, researchers have proposed different model combinations to improve the prediction accuracy of machine-learning models [30]. For example, Li et al. [31] compared five combined models, namely an LSTM-convolutional neural network (LSTM-CNN), a CNN-LSTM, LSTM-Attention, a CNN-Attention-LSTM, and an LSTM-Attention-CNN, using data from 60 buildings. The results showed that LSTM-Attention-LSTM performed the best, as the RMSE was decreased by 5.6% compared to the LSTM alone. Li et al. [32] developed different ensemble models, including a teaching learning-based optimization algorithm-BPNN (TLBO-BPNN), a TLBO-SVM, BPNN-Adaboost, an extreme learning machine, and an RF. The results showed that the ensemble models performed much better compared to the single models in all cases. Zhang et al. [33] proposed two combined models for short-term BECP. The first was based on LSTM and BPNN and was proposed to solve the time-lag problems, while the second was based on a dimensionless sensitivity index and a weighted Manhattan distance to allow the machine-learning models to be explained more easily using domain knowledge. The results showed that the combined models achieved much higher accuracy.

Overall, combined models offer an efficient way to improve the accuracy of BECP. Generally, the combined techniques used for BECP include ensemble-based models [32], cluster-based models [34], decomposition-based models, etc. The decomposition-based models are based on decomposition algorithm and machine-learning models and can process multi-resolution data by extracting the embedded information from non-stationary BECP data. It is also an efficient method for BECP. However, there is little research on decomposition-based models.

Due to the high-dimensional nonlinear characteristics of building load, traditional single models face difficulties in capturing the features of the original data accurately [35]. In recent years, researchers have proposed to decompose the time series using decomposition algorithms in an attempt to capture the different features of the original data. Such methods include the wavelet transform, empirical mode decomposition, ensemble empirical model decomposition, and so on [36]. Following this trend, ensemble empirical model decomposition with adaptive noise (CEEMDAN) has become one of the most widely used decomposition algorithms, as it does not require a predetermined mother wavelet basis and can be used to solve problems such as modal aliasing. Through the decomposition algorithm, different features of the building load data can be extracted effectively, and then, the most appropriate model for each component can be found, which leads to the improved prediction accuracy of single models [37].

Based on the above, in this study, a decomposition-based model is developed based on CEEMDAN, DL models (including CNN, LSTM, bi-directional LSTM, and gated recurrent unit (GRU)), and autoregressive integrated moving average (ARIMA) models for BECP. The one-step prediction model (single model) and two-step prediction models (CEEMDAN-DLs) are adopted for comparison. First, CEEMDAN is used to decompose the original data into n intrinsic mode functions (IMFs) and a residual (RES). Then, the sample entropy of each component is calculated, based on which the complexity of different components is determined. As for deep learning models, the deep structure can fit the highly nonlinear relationships; however, there are issues with overfitting and high computational complexity. In comparison, numerous studies have shown that the ARIMA model has been proven to be excellent in handling linear problems [28]. Furthermore, its structure is much simpler, and its computational complexity is much lower than that of DL models. Therefore, ARIMA is used to predict the low-complexity components, and DL models are used to predict the high-complexity components. Finally, the CEEMDAN-DL-ARIMA model is proposed as an integration of the prediction results of the two algorithms, and it is validated using real building energy consumption data. Consequently, in this paper, a novel approach to high-dimensional nonlinear building load forecasting is proposed.

2. Overview of the Combined Model

Figure 1 shows the development framework of the combined model. First, the original heating load data are decomposed into n IMFs and a RES through CEEMDAN, and the sample entropy of each component is calculated. For components with high complexity, a DL model is used for prediction, while the ARIMA model is used for components with low complexity. Finally, the prediction results of the different components are weighted linearly to obtain the final prediction results.

2.1. Empirical Model Decomposition

The decomposition algorithm can be used to extract the embedded information of non-stationary time series and obtain multi-resolution data. Generally, the time series is decomposed into several low-frequency components and one high-frequency component. Wavelet analysis is one of the most widely used decomposition algorithms. However, the wavelet basis and decomposition need to be selected manually [38], which limits the method's applicability. Empirical mode decomposition (EMD) can be used to select the wavelet basis and decomposition layers automatically [39] and has been widely adopted. However, EMD is prone to modal aliasing when there are jump changes in the time series. Therefore, Wu et al. [40] proposed the ensemble empirical model decomposition (EEMD) model, which can eliminate modal aliasing through the addition of white noise to the time series. However, redundant white noise may exist after the reconstruction [38]. Consequently, Torres et al. [41] proposed the complete ensemble empirical model decomposition with adaptive noise (CEEMDAN). By adding adaptive noise, CEEMDAN not only avoids the problem of modal aliasing but also eliminates the redundant white noise in the reconstructed time series.

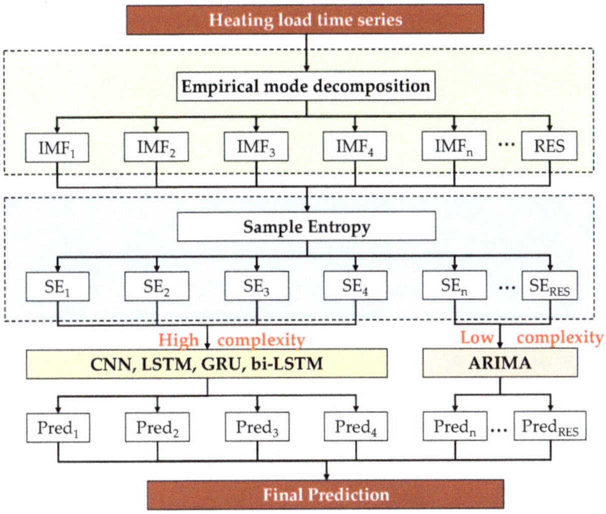

Figure 1. Combined model structure for building heating load prediction.

The calculation steps of CEEMDAN are as follows:

(1) White noise ω_i is added into the original signal x; then, the IMFs are obtained via decomposition of the signal using EMD. Thus, the first model is calculated using the following equation:

$$IMF_1 = \frac{1}{N}\sum\nolimits_{i=1}^{N} E_i[x + \varepsilon\omega_i] \tag{1}$$

where $E_i(\cdot)$ is the i-th IMF decomposed using EMD; ω_i is the white noise; and ε is the signal-to-noise ratio.

(2) Calculate the first residual (r_1) at the first stage:

$$r_1 = x - IMF_1 \tag{2}$$

(3) The signal ($r_1 + \varepsilon E_1[\omega_i]$) is further decomposed using EMD until IMF_1 is obtained. Then, IMF_2 can be calculated as follows:

$$IMF_2 = \frac{1}{N}\sum\nolimits_{i=1}^{N} E_i[r_1 + \varepsilon E_1[\omega_i]] \tag{3}$$

(4) The k-th residual is calculated.

$$r_k = r_{k-1} - IMF_k \tag{4}$$

(5) The realization signal ($r_k + \varepsilon E_k[\omega_i]$) is further decomposed using EMD, and IMF_{k+1} is calculated:

$$IMF_{k+1} = \frac{1}{N}\sum_{i=1}^{N} E_i[r_k + \varepsilon E_k[\omega_i]] \tag{5}$$

(6) Repeat the above steps until the residual satisfies the iteration termination condition, i.e., when there is no further decomposition possible. Then, the final residual is:

$$R = x - \sum_{k=1}^{K} IMF_k \tag{6}$$

The original signal can be reconstructed as follows:

$$x = \sum_{k=1}^{K} IMF_k + R \tag{7}$$

2.2. Deep Learning Models

2.2.1. Convolutional Neural Networks

CNNs consist of three main mapping layers types, namely convolutional, pooling, and fully-connected layers [42]. Convolutional layers are used to extract the features of input data; pooling layers are used to reduce the input data's dimensions, while fully-connected layers are used to produce the output. Overall, CNNs are a reliable technique for the efficient extraction of hidden features and the creation of filters according to these features. Weight sharing and local connections are two main characteristics of CNNs [43]. The topological structure of a CNN is shown in Figure 2.

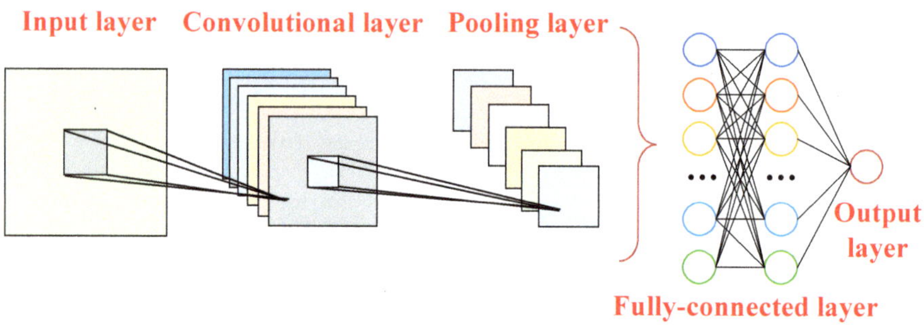

Figure 2. General topological structure of a CNN model.

The convolutional layer is defined as [44]:

$$h_{ij}^{k} = f\left(\left(W^{k} * x\right)_{ij} + b_k\right) \tag{8}$$

where f is the activation function; W^k is the weight vector; x is the input data; $*$ is the convolution operator; b_k is a bias matrix. For the activation function, the rectified linear unit (ReLU) is adopted and calculated as follows:

$$f(x) = \max(0, x) \tag{9}$$

2.2.2. Long–Short-Term Memory Networks

Long- and short-term memory (LSTM) has been proposed as a method to solve the time dependence problems of recursive neural networks. It is an excellent tool with strong memory for long time-series prediction. LSTMs consist of four gates, including the input, forget, control, and output gate [31]. The topological structure of the LSTM is shown in Figure 3.

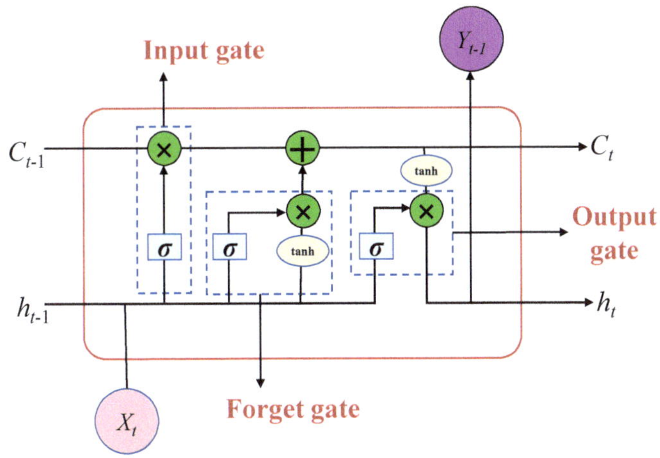

Figure 3. General topological structure of LSTM model.

The calculation steps of LSTM are as follows:

(1) Depending on the input x_t and the state of the last hidden layers h_{t-1}, the LSTM can determine the information to be thrown away by forget gate f_t:

$$f_t = \sigma\left(\omega_f[h_{t-1}, x_t] + b_f\right) \tag{10}$$

where σ is the activation function, ω_f is the weight matrix, and b_f is the bias vector.

(2) The last output and the current input value are transferred to the input gate. Then, the output i_t and candidate state $\widetilde{C}_t$ are generated.

$$i_t = \sigma(W_i \cdot [h_{t-1}, x_t] + b_i) \tag{11}$$

$$\widetilde{C}_t = \tanh(W_C \cdot [h_{t-1}, x_t] + b_C) \tag{12}$$

(3) The current state of cell C_t is updated through the integration of the input of the forget gate with the last state of the cell.

$$C_t = f_t \cdot C_{t-1} + i_t \cdot \widetilde{C}_t \tag{13}$$

(4) Finally, the output gate o_t is calculated using h_{t-1} and x_t. Then, the final output h_t is obtained from the output gate.

$$o_t = \sigma(W_o \cdot [h_{t-1}, x_t] + b_o) \tag{14}$$

$$h_t = \sigma_t \cdot \tanh(C_t) \tag{15}$$

2.2.3. Gated Recurrent Unit

The structure of the GRU is similar to that of LSTM, but there is no cell that stores the state. The GRU instead has a reset gate and an update gate, which are used to determine the information being stored or thrown away [45]. The reset gate allows the dismissal of irrelevant information from the hidden state, while the update gate controls how much information from the previous hidden state (h_{t-1}) can be stored in the current hidden state (h_t) [46]. The general topological structure of the GRU is shown in Figure 4.

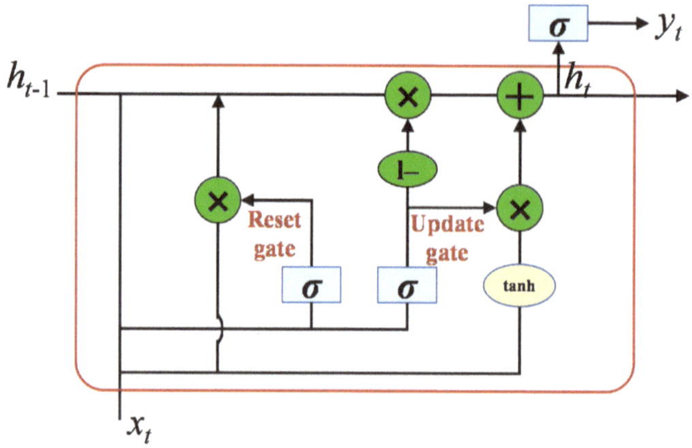

Figure 4. General topological structure of GRU model.

The reset gate r_t is defined as follows:

$$r_t = \sigma(W_r x_t + U_r h_{t-1} + b_r) \tag{16}$$

The update gate z_t is defined as follows:

$$z_t = \sigma(W_z x_t + U_z h_{t-1} + b_z) \tag{17}$$

where x_t is the new input; W_r, W_z, U_r, and U_z are the weight matrices; b_r, and b_z are the bias vectors. The output h_t is calculated as follows:

$$h_t = z_t \cdot h_{t-1} + (1 - z_t) \cdot \widetilde{h_t} \tag{18}$$

where $\widetilde{h_t}$ is the candidate state, calculated as follows:

$$\widetilde{h_t} = \tanh(W_h x_t + U_h (r_t \cdot h_{t-1}) + b_r) \tag{19}$$

2.2.4. Bidirectional Long–Short-Term Memory Network (Bi-LSTM)

Bi-LSTM transfers the input data at each time to the network for calculation. Each hidden layer contains two LSTM units. The first is the forward LSTM, which receives the output data corresponding to the previous instance and forwards it to the next instance after calculation to ensure that past information is taken into account. The other is the backward LSTM, which receives the output data of the next instance, calculates, and sends it back to the previous instance to capture future information in current calculations [47]. Bi-LSTM's topological structure is shown in Figure 5.

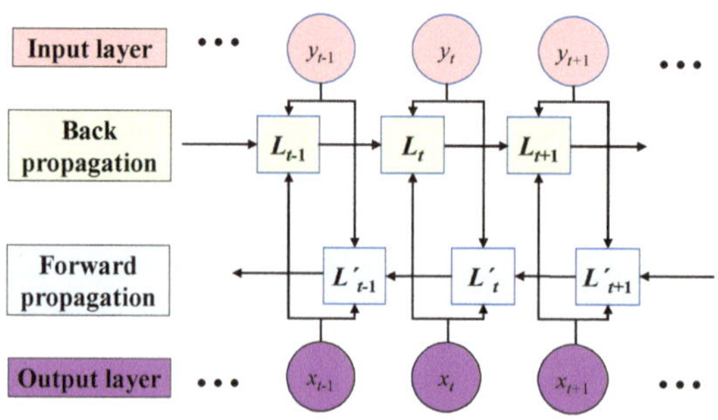

Figure 5. General topological structure of Bi-LSTM.

Defining the last hidden vector as $\overrightarrow{h}$ and the next hidden vector as $\overleftarrow{h}$, the output of bi-LSTM is calculated as follows:

$$h_t = \alpha \overrightarrow{h} + \beta \overleftarrow{h} \tag{20}$$

where α and β are the constants and $\alpha + \beta = 1$.

2.3. ARIMA Model

The autoregressive integrated moving average (ARIMA) model converts a non-stationary time series into a stationary one through the d-order difference between an autoregressive model (AR) and a moving average model (MA) [48]. Generally, it is represented as ARIMA (p, q, d), where p is the order of the AR model; q is the order of the MA model; and d is the difference order.

The AR model captures the relationship between the values at the current time and the previous p instances. It can be expressed as:

$$x_t = \mu + \sum_{i=1}^{p} \Upsilon_i x_{t-i} + \varepsilon_t \tag{21}$$

where x_t is the current value; μ is a constant; Υ_i is the correlation coefficients; and ε_t is the error.

The MA model is the weighted sum of white noises:

$$x_t = \mu + \sum_{i=1}^{q} \theta_i \varepsilon_{t-i} + \varepsilon_t \tag{22}$$

where θ is a coefficient. The combined AR (p) and MA (q) models can be obtained from the following equation:

$$x_t = \mu + \sum_{i=1}^{p} \Upsilon_i x_{t-i} + \varepsilon_t + \sum_{i=1}^{q} \theta_i \varepsilon_{t-i} \tag{23}$$

The difference order can be determined using Bayesian information criterion (BIC):

$$V = 2M + d \log N \tag{24}$$

where M is the optimal log likelihood value, and N is the number of samples.

3. Materials and Methods

3.1. Data Description

The research data were obtained from the monitoring data of the heating pipe network of Xi'an University of Architecture and Technology, China. The total heating area was 392,219.63 m^2 and included student dormitories, restaurants, teaching buildings, office buildings, and hospitals. The research data include hourly heating load data covering from 2 January to 2 March 2016. Figure 6 shows the heating load and the corresponding IMFs and RES decomposed using the CEEMDAN algorithm. The training and testing dataset are decomposed separately. First, the training dataset was decomposed, and then, the testing dataset was decomposed based on the training dataset. The total number of heating load samples was 1480, and 991 samples were used as training data, while the remaining 489 samples were taken as test data. The heating loads of the training dataset varied from 3743.9–5707.8 kWh with an average value of 7518.6 kWh, while the range of the test dataset was 3313.8–4764.0 kWh with an average value of 5612.9 kWh.

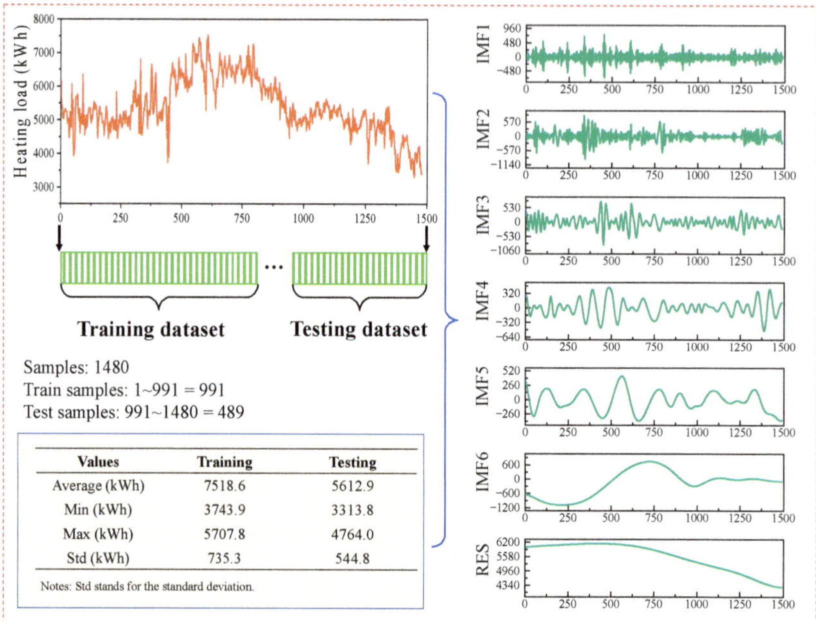

Figure 6. Heating load and the corresponding intrinsic mode functions and residuals decomposed using CEEMDAN.

3.2. Correlation Analysis

Methods based on correlation coefficients are the most widely used for feature selection. Among these methods, Pearson's correlation coefficient is one of the most popular options. For example, Zhang et al. [34] adopted Pearson's correlation coefficient to determine the key impact parameters of day-ahead and intra-day heating loads. However, generally, Pearson's correlation coefficient can only be used to describe linear correlations [9]. For the building heating load, due to the thermal inertial of buildings, the hourly-ahead heating load presents high nonlinear correlation with historic heating loads. As a result, Pearson's correlation coefficient's applicability is limited in building heating load prediction. In contrast, Spearman's correlation coefficient can describe the nonlinear correlation between two parameters. It is a nonparametric index to measure the dependence of two variables, and it has the advantages of universality, low complexity, and high efficiency. Therefore, in this study, Spearman's correlation coefficient was adopted. For a sample size N, the N original samples are converted into rank data, and the correlation coefficient ρ can be calculated through the following equation:

$$\rho = \frac{\sum_i (x_i - \overline{x})(y_i - \overline{y})}{\sqrt{\sum_i (x_i - \overline{x})^2 \sum_i (y_i - \overline{y})^2}} \tag{25}$$

where X_i and Y_i represent the independent and dependent variables, respectively. x_i and y_i represent the average values of independent and dependent variables. Figure 7 illustrates the Spearman correlation coefficients between the hour-ahead heating load with the 12-h historic heating loads. It is evident that the historic heating loads have a high correlation with the hour-ahead heating load, as the corresponding Spearman correlation coefficient values are higher than 0.70. For the 2-h historic heating load in particular, the corresponding values are above 0.90. As values further back in time are considered, the correlation coefficient decreases gradually. Considering the model complexity and accuracy comprehensively, the historical heating loads with correlation coefficients above 0.80 were selected as input parameters. That is, the 7-h historical heating loads were selected as model input parameters.

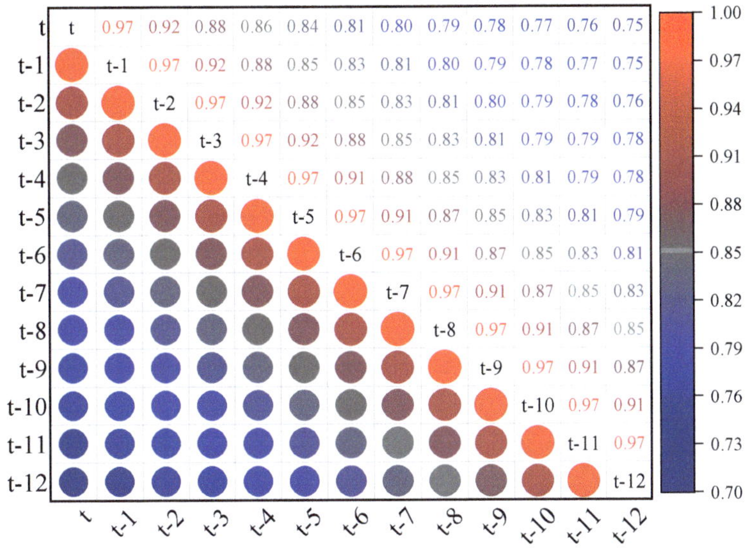

Figure 7. Correlation analysis between current heating load and historical heating loads.

3.3. Model Evaluation

Four evaluation indicators, namely the coefficient of determination (R^2), the root mean square error (RMSE), the coefficient of RMSE variation (CV-RMSE), and the mean bias error (MBE), were adopted in this study. The respective calculation equations are as follows [49]:

$$R^2 = \frac{\left[\sum_{i=1}^{n}(Y_{i,c} - \overline{Y}_{i,c})(Y_{i,m} - \overline{Y}_{i,m})\right]^2}{\sum_{i=1}^{n}(Y_{i,c} - \overline{Y}_{i,c})^2 \sum_{i=1}^{n}(Y_{i,m} - \overline{Y}_{i,m})^2} \quad (26)$$

$$\text{RMSE} = \sqrt{\frac{\sum_{i=1}^{n}(Y_{i,m} - Y_{i,c})^2}{n}} \quad (27)$$

$$\text{CV} - \text{RMSE} = \frac{\text{RMSE}}{\overline{Y_{i,m}}} \quad (28)$$

$$\text{MBE} = \frac{1}{n}\sum_{i=1}^{n}(Y_{i,m} - Y_{i,c}) \quad (29)$$

where $Y_{i,m}$ is the measured value, $Y_{i,c}$ is the predicted value, and n is the number of the observations. R^2 represents the ratio of the sum of regression squares to the total sum of squares, and values closer to 1 correspond to better performance. The RMSE describes the short-term performance of the model by comparing the deviation between the predicted and true values. CV-RMSE characterizes the performance of the model depending on its value as follows: when it is less than 10%, the accuracy of the model is considered to be excellent; between 10–30%, the model's accuracy is considered good; when it is greater than 30%, the accuracy of the model is considered poor [50]. MBE describes the long-term performance of the model.

3.4. Model Development

3.4.1. Sample Entropy Analysis

Sample entropy is a dimensionless parameter that measures the complexity of a time series through the measurement of the probability of new patterns occurring in the signal. Greater pattern-generation probabilities correspond to more complex sequences [51]. In this study, sample entropy was adopted to measure the complexity of the components. Figure 8 shows the values of sample entropy of the components decomposed using CEEMDAN.

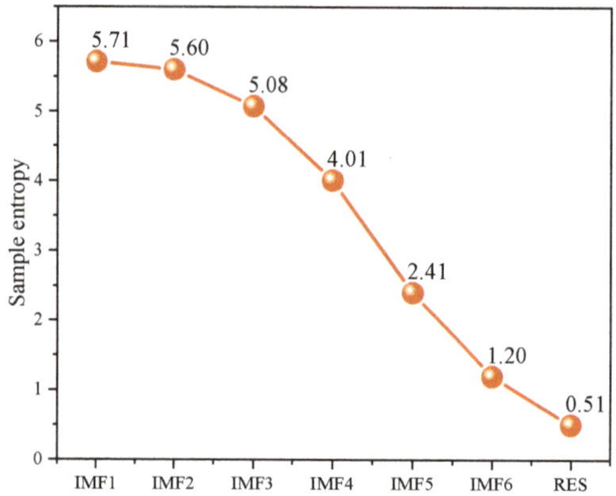

Figure 8. Sample entropy of the components decomposed using CEEMDAN.

Comparing the change trends of the sample entropies of the components, if the differences between the sample entropies of the IMFs and that of RES are large, or the change range of the IMFs' sample entropy suddenly becomes larger, then the last sequence between the two IMFs is used as the first sequence predicted by the ARIMA model, and the RES is used as the last sequence predicted by ARIMA model [52]. As seen in Figure 7, IMF5–RES are suitable for analysis using the ARIMA model, and IMF1−IMF4 are suitable for analysis using the DL models.

3.4.2. Model Development and Parameter Setting

In this study, four DL models, namely the CNN, LSTM, bi-LSTM, and GRU models, were adopted to develop the combined models. The process comprises three steps:

(1) Data preprocessing: The CEEMDAN algorithm is applied to decompose the heating load into six IMFs and one RES. Then, the sample entropies are calculated, and the corresponding prediction models of each component are determined.

(2) Model development: The four DL models are integrated with the ARIMA model to develop the combined models. That is, IMF1−IMF4 are used to train the DL models, and IMF5−RES are used to train the ARIMA model.

(3) Model prediction: The prediction of the four DL models and the ARIMA model are summed. Thus, the predictions obtained are from four combined models, namely CEEMDAN-CNN-ARIMA, CEEMDAN-LSTM-ARIMA, CEEMDAN-GRU-ARIMA, and CEEMDAN-bi-LSTM-ARIMA.

The adaptive moment estimation (Adam) method is adopted to optimize the weights of the CNN, LSTM, bi-LSTM, and GRU networks. The reader can refer to [45] for an introduction to Adam. The hyper-parameters optimized for the four DL models are shown in Table 1. In this study, the Tree of Parzen Estimators (TPE) [53] method was used to optimize the hyper parameters, where the number of maximum estimations was set to be 100.

Table 1. Hyper-parameters of DL models for building heating load prediction.

Models	Parameters	Maximum Values	Minimum Values
CNN	Filters	1	30
	Kernels	1	3
	Dropout rate	0.1	0.5
	Pooling size	1	2
	Layers	1	5
	Learning rate	0.01	0.1
	Epochs	10	100
	Batch size	12	72
LSTM	LSTM layers	1	5
	Dense layers	1	3
	LSTM units	10	100
	Dropout rate	0.1	0.5
	Dense units	10	100
	Batch size	12	72
	Epochs	10	100
	Learning rate	0.01	0.1
GRU	GRU layers	1	5
	Dense layer	1	3
	GRU units	10	100
	Dropout rate	0.1	0.5
	Dense units	10	100
	Batch size	12	72
	Epochs	10	100
	Learning rate	0.01	0.1
Bi-LSTM	Bi-LSTM layers	1	5
	Dense layers	1	3
	Dropout rate	0.1	0.5
	Dense units	10	100
	Batch size	12	72
	Epochs	10	100
	Learning rate	0.01	0.1

4. Results and Discussion

To verify the performance of the combined models proposed in this study, eight models, namely CNN, LSTM, GRU, bi-LSTM, CEEMDAN-CNN, CEEMDAN-LSTM, CEEMDAN-GRU, and CEEMDAN-bi-LSTM, were implemented and compared.

The CNN, LSTM, GRU, and bi-LSTM are one-step prediction models and were adopted to predict the heating load without any data preprocessing. CEEMDAN-CNN, CEEMDAN-LSTM, CEEMDAN-GRU, and CEEMDAN-bi-LSTM are the two-step prediction models based on the CEEMDAN algorithm, where the components of the heating load, including the IMFs and RES were decomposed using CEEMDAN algorithm and then predicted by the DL models. CEEMDAN-CNN-ARIMA, CEEMDAN-LSTM-ARIMA, CEEMDAN-GRU-ARIMA, and CEEMDAN-bi-LSTM-ARIMA are the three-step prediction models, which combine the DL models, the CEEMDAN algorithm, and the ARIMA model, where the ARIMA model is adopted to predict the components of low frequencies instead of the DL models.

4.1. Model Performance Comparison

Table 2 shows the evaluation indicators of the one-, two-, and three-step prediction models. Overall, all models achieved high accuracy. R^2, RMSE, CV-RMSE, and MBE varied from 0.917–0.988, 70.25–272.72 kWh, 1.47–5.72%, and −99.16–210.74 kWh with average values of 0.971, 151.73 kWh, 3.18%, and 78.70 kWh, respectively.

Table 2. Evaluation indicators of different models for short-term building heating load prediction.

Models	Model ID	R^2	RMSE (kWh)	CV-RMSE (%)	MBE (kWh)
CNN	C1	0.917	200.98	4.22	−99.16
CEEMDAN-CNN	C2	0.967	196.15	4.12	158.74
CEEMDAN-CNN-ARIMA	C3	0.967	104.63	2.20	34.12
LSTM	L1	0.942	272.72	5.72	210.74
CEEMDAN-LSTM	L2	0.981	94.04	1.97	44.13
CEEMDAN-LSTM-ARIMA	L3	0.984	74.53	1.56	16.32
Bi-LSTM	B1	0.975	177.06	3.71	120.60
CEEMDAN- Bi-LSTM	B2	0.983	92.27	1.94	−22.17
CEEMDAN- Bi-LSTM-ARIMA	B3	0.983	**70.25**	**1.47**	**14.25**
GRU	G1	0.979	192.40	4.03	174.81
CEEMDAN-GRU	G2	**0.988**	182.69	3.83	142.34
CEEMDAN-GRU-ARIMA	G3	**0.988**	163.00	3.42	149.66

Figure 9 shows the prediction results of the one-step prediction models. For the CNN model, it can be seen in the diagram of Figure 8 that before the 350th heating load sample, the predicted heating load showed good agreement with the measured values. However, after the 350th sample, the agreement between them gradually deteriorated, which was the cause of the lowest R^2 values. The main reason is that in the early stage of testing, the heating load change trend is similar to that in the training phases. Thus, the CNN model could fit the training set very well through its multi-layer structure. That is, when the change trend in the test phase is similar to that of the training phase, the CNN models will produce excellent predictions. However, as time progressed, the temperature gradually warmed up, the heating load gradually decreased, and the change trend of the heating load began to change. Because the CNN model does not have any memory, it is difficult to capture the change trend of long-lasting series. As a result, the coincidence of CNN prediction and measured values declined after that point.

Although the change trend of the LSTM model's heating load prediction was similar to that of the measured values at all stages, its deviation degree was high. Therefore, the R^2 value of LSTM was higher than that of CNN, while the RMSE and CV-RMSE were higher than those of CNN. Moreover, it can be seen from Figure 8 that the coincidence degree of the LSTM continued to improve as time progressed. This is because the LSTM has a memory function and continues to learn through the prediction, so its performance improves continuously. Similarly, the performance of bi-LSTM and GRU also improved continuously, and the agreement between heating load predictions and measured values was very high in all testing phases.

Comparing the GRU and the bi-LSTM, it can be seen that the GRU had higher R^2 value, while the bi-LSTM yielded lower RMSE, CV-RMSE, and MBE. Building heating load is related to climate conditions, historical heating loads, building characteristics, and other factors, and its randomness is strong. The bi-LSTM generates the current prediction value considering the past and future information of time series, so its prediction is relatively smooth, resulting in poor consistency of the change trend between the predicted and the measured values compared with the GRU. However, because the bi-LSTM integrates these additional aspects of the time series, its prediction deviation is smaller than that of the GRU, and the prediction error is smaller. Thus, the RMSE, CV-RMSE, and MBE of bi-LSTM were lower than the corresponding values of the GRU.

Figure 10 shows a comparison of predicted and measured heating load of the two-step prediction models, in which it can be easily seen that the performance was greatly improved compared with one-step prediction models. The average R^2, RMSE, CV-RMSE, and MBE were 0.980, 143.72 kWh, 3.02%, and 88.88 kWh, respectively, improved by 2.78%, 31.03%, 31.01%, and 5.08% compared to the corresponding one-step prediction models. Similarly, CEEMDAN-GRU produced the highest R^2 at a value of 0.988, while CEEMDAN-bi-LSTM yielded the lowest RMSE and CV-RMSE with values of 92.27 kWh and 1.94%, respectively.

Figure 11 shows a comparison of the predicted and measured heating load of the three-step prediction models. In a similar fashion to the previous comparison, compared with two-step prediction models, the performance was greatly improved, with average R^2 = 0.981, RMSE = 103.10 kWh, CV-RMSE = 2.16%, and MBE = 55.58 kWh. These values were improved by 0.08%, 28.26%, 28.28%, and 39.71%, respectively. CEEMDAN-CNN-ARIMA had worst performance; the R^2 value of CEEMDAN-GRU-ARIMA was the highest one; CEEMDAN-bi-LSTM-ARIMA produced the best performance with the lowest RMSE, CV-RMSE, and MBE.

4.2. Improvements of Combined Models

Figures 12–14 show the performance improvement percentages of the models investigated in this paper. It is evident that with regard to prediction accuracy, increasing the number of steps generally improved the prediction results. Furthermore, one-step models with worse performance showed higher performance improvement when used in combined models. For example, compared with one-step models, CNN models yielded the lowest R^2, while GRU models have highest R^2. The performance improvement percentages of the two- and three-step models were 5.45% and 5.45%, respectively. In contrast, the corresponding values for GRU were 0.92% and 0.92%, respectively.

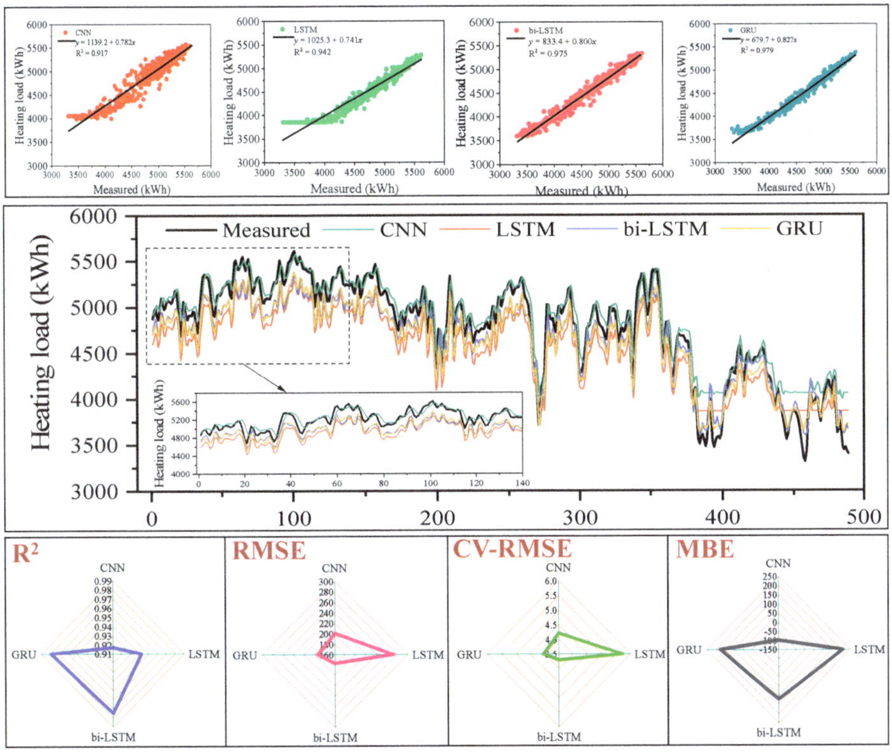

Figure 9. Comparison of prediction results of one-step prediction models.

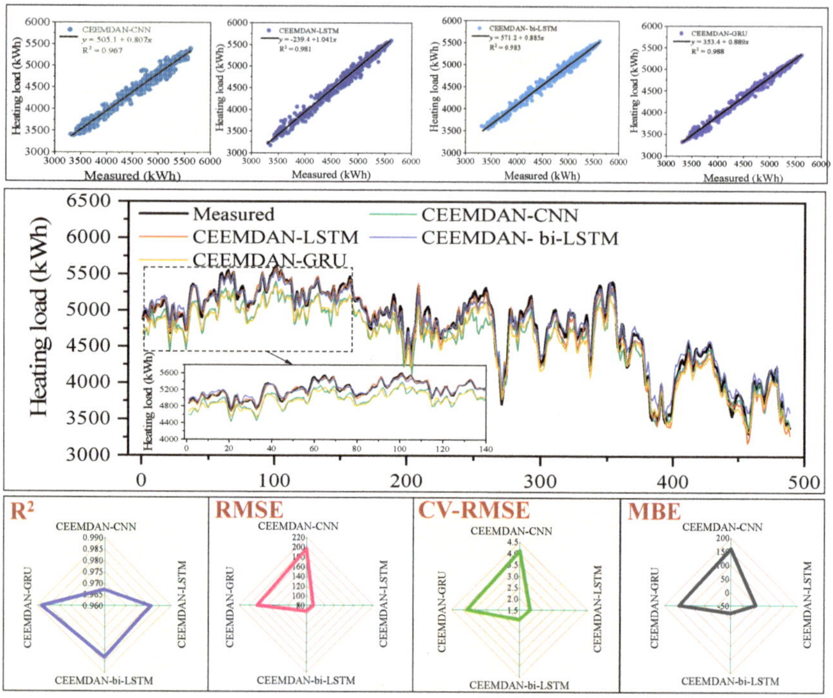

Figure 10. Comparison of prediction results of two-step prediction models.

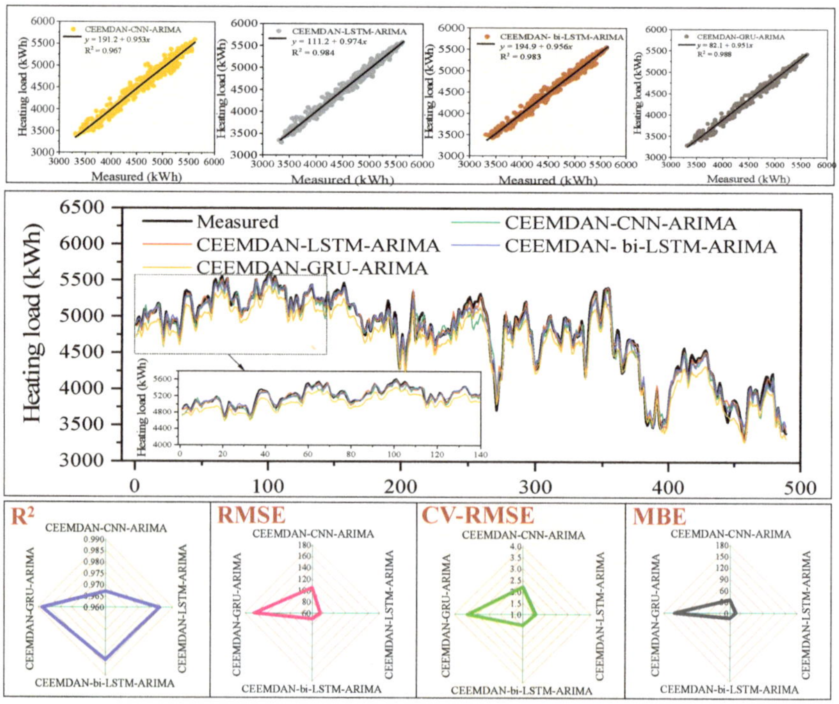

Figure 11. Comparison of prediction results of three-step prediction models.

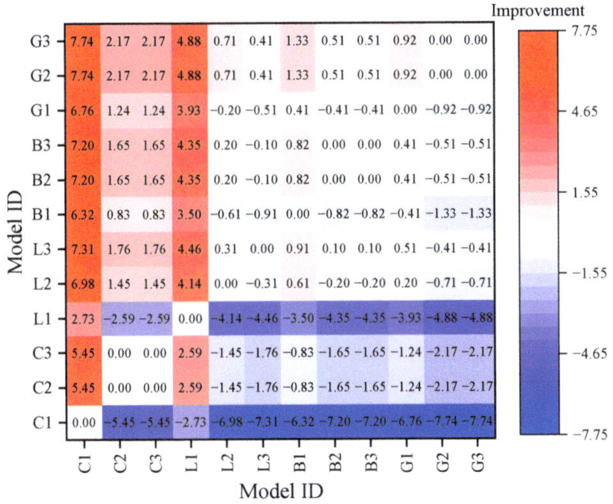

Figure 12. Improvement percentage of R^2 between different models investigated in this study. Notes: Improvement of R^2 = (R^2 in columns − R^2 in rows)/R^2 in columns.

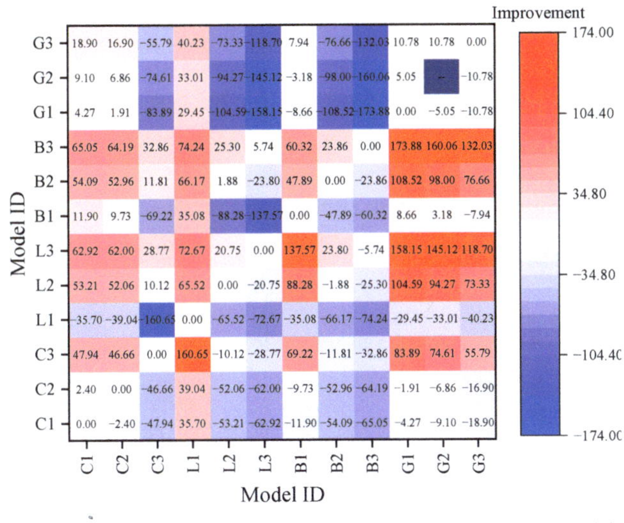

Figure 13. Improvement percentage of RMSE between different models investigated in this study (%). Notes: Improvement of RMSE = (RMSE in columns − RMSE in rows)/RMSE in columns.

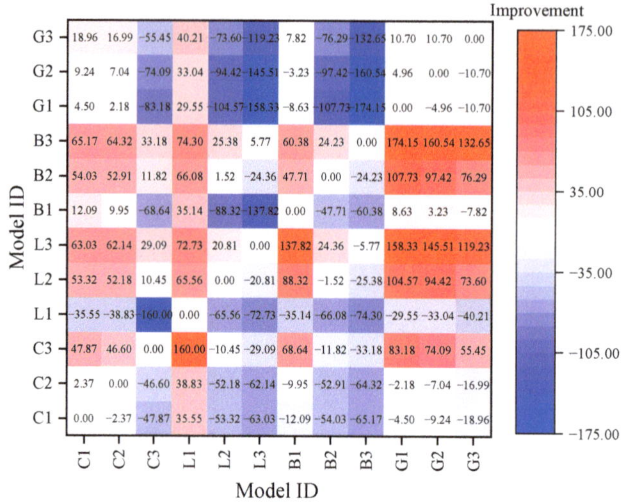

Figure 14. Improvement percentage of RMSE between different models investigated in this study (%). Notes: Improvement of CV-RMSE = (CV-RMSE in columns − CV-RMSE in rows)/CV-RMSE in columns.

Compared to one-step models, the performance improvement percentages of R^2, RMSE, and CV-RMSE of the two-step models were 2.83%, 30.22%, and 30.15%, respectively, and 2.91%, 47.93%, and 47.92% for the three-step models.

5. Conclusions

Accurate prediction of building energy consumption is of great significance to building energy management systems. In this paper, a model combining DL models (including CNN, LSTM, bi-LSTM, GRU) and the ARIMA model is proposed based on CEEMDAN. The CEEMDAN model was used to decompose the original heating load data, while the DL models were used to predict complex high-frequency subsequences; the ARIMA model was used to predict relatively stable low-frequency subsequences. Through the validation of models using actual heating load data, the following conclusions are obtained:

(1) At the initial stage of testing, the change trend of the heating load is similar to that observed during training phases, and the CNN model could follow the heating load due to its deep structure. Therefore, at the initial stage of testing, the CNN had the best performance. As time progressed, the heating load gradually decreased, the change trend became less familiar to the model, and the CNN prediction performance decreased. In comparison, the LSTM, bi-LSTM, and GRU have memory functions resulting in their prediction performance improving over time. The Bi-LSTM had the best comprehensive performance because it can integrate information from the previous and the following samples of the time series.

(2) Combined models performed better than single models. Compared with the single model, the average performance improvement percentages of R^2, RMSE, and CV-RMSE of the two-step models combining CEEMDAN and DL models were 2.83%, 30.22%, and 30.15%, respectively. The corresponding values of the three-step models, which combined CEEMDAN, DL, and ARIMA, were 2.91%, 47.93%, and 47.92%, respectively.

(3) Among all models, the CEEMDAN-Bi-LSTM-ARIMA model had the best performance. ARIMA model can predict the low-frequency subsequences decomposed by CEEMDAN effectively and reduce the prediction error of the bi-LSTM on these low-

frequency subsequences. This resulted in an improvement in the overall prediction accuracy of the hybrid model.

In conclusion, building energy consumption has high-dimensional nonlinearity and randomness. The CEEEMD-DL-ARIMA model can not only solve the problem of BECP but also solve other time series prediction problems, such as the solar radiation prediction problem. However, the different components of the model are still summed using a simple linear weighting. As the time series changes dynamically, the weight of each component also changes dynamically as time progresses. Therefore, in future research, a variable weight combination method should be considered to improve the prediction accuracy of building energy consumption.

Author Contributions: Methodology, Y.Z.; software, Y.Z.; validation, Y.Z.; investigation, L.W.; resources, J.Q.; data curation, L.W. and J.Q.; writing—original draft preparation, Y.Z.; writing—review and editing, Y.Z. All authors have read and agreed to the published version of the manuscript.

Funding: This research was funded by Basic research program of Natural Science in Shaanxi Province (No. 2021JQ-511), Key projects of Shaanxi Collaborative Innovation Center (No. 21JY024), Independent Research and Development project of State Key Laboratory of Green Building in Western China (No. LSZZ202106).

Data Availability Statement: The data were obtained from the monitoring system of the heating pipe network in Xi'an University of Architecture and Technology, China. Data are available on request to the authors.

Conflicts of Interest: The authors declare no conflict of interest.

Nomenclature

Symbols	
R^2	Coefficient of determination
Abbreviations	
Adam	Adaptive moment estimation
ANN	Artificial neural network
ARIMA	Autoregressive integrated moving average
Bi-LSTM	Bi-directional long- and short-term memory
BPNN	Back-propagation neural network
CEEMDAN	Complete ensemble empirical model decomposition with adaptive noise
CNN	Convolutional neural network
CV-RMSE	Coefficient of variation of the root mean square error
DL	Deep learning
EEMD	Ensemble empirical model decomposition
EMD	Empirical mode decomposition
GRU	Gated recurrent unit
IMF	Intrinsic mode function
LSSVM	Least square support vector machine
LSTM	Long- and short-term memory
MBE	Mean bias error
RF	Random forest
RMSE	Root mean square error
RES	Residuals
SVM	Support vector machine
TLBO	Teaching learning-based optimization algorithm
TPE	Tree of Parzen Estimators
XGBoost	Extreme gradient boosting

References

1. Lu, C.; Li, S.; Lu, Z. Building energy prediction using artificial neural networks: A literature survey. *Energy Build.* **2022**, *262*, 111718. [CrossRef]
2. Olu-Ajayi, R.; Alaka, H.; Sulaimon, I.; Sunmola, F.; Ajayi, S. Building energy consumption prediction for residential buildings using deep learning and other machine learning techniques. *J. Build. Eng.* **2022**, *45*, 103406. [CrossRef]
3. Li, A.; Xiao, F.; Zhang, C.; Fan, C. Attention-based interpretable neural network for building cooling load prediction. *Appl. Energy* **2021**, *299*, 117238. [CrossRef]
4. Razmjooy, N.; Ramezani, M. Uncertain Method for Optimal Control Problems with Uncertainties Using Chebyshev Inclusion Functions. *Asian J. Control* **2019**, *21*, 824–831. [CrossRef]
5. Razmjooy, N.; Ramezani, M. Interval structure of Runge-Kutta methods for solving optimal control problems with uncertainties. *J. Comput. Methods Differ. Equ.* **2019**, *7*, 235–251.
6. Jang, Y.; Byon, E.; Jahani, E.; Cetin, K. On the long-term density prediction of peak electricity load with demand side management in buildings. *Energy Build.* **2020**, *228*, 110450. [CrossRef]
7. Fang, X.; Gong, G.; Li, G.; Chun, L.; Li, W.; Peng, P. A hybrid deep transfer learning strategy for short term cross-building energy prediction. *Energy* **2021**, *215*, 119208. [CrossRef]
8. Dong, Z.; Liu, J.; Liu, B.; Li, K.; Li, X. Hourly energy consumption prediction of an office building based on ensemble learning and energy consumption pattern classification. *Energy Build.* **2021**, *241*, 110929. [CrossRef]
9. Liu, Y.; Hu, X.; Luo, X.; Zhou, Y.; Wang, D.; Farah, S. Identifying the most significant input parameters for predicting district heating load using an association rule algorithm. *J. Clean. Prod.* **2020**, *275*, 122984. [CrossRef]
10. Baglivo, C.; Congedo, P.M.; Murrone, G.; Lezzi, D. Long-term predictive energy analysis of a high-performance building in a mediterranean climate under climate change. *Energy* **2022**, *238*, 121641. [CrossRef]
11. Gassar, A.A.A.; Cha, S.H. Energy prediction techniques for large-scale buildings towards a sustainable built environment: A review. *Energy Build.* **2020**, *224*, 110238. [CrossRef]
12. Zhang, Q.; Tian, Z.; Ma, Z.; Li, G.; Lu, Y.; Niu, J. Development of the heating load prediction model for the residential building of district heating based on model calibration. *Energy* **2020**, *205*, 117949. [CrossRef]
13. Shabunko, V.; Lim, C.M.; Mathew, S. EnergyPlus models for the benchmarking of residential buildings in Brunei Darussalam. *Energy Build.* **2018**, *169*, 507–516. [CrossRef]
14. Wei, Y.; Zhang, X.; Shi, Y.; Xia, L.; Pan, S.; Wu, J.; Han, M.; Zhao, X. A review of data-driven approaches for prediction and classification of building energy consumption. *Renew. Sustain. Energy Rev.* **2018**, *82*, 1027–1047. [CrossRef]
15. Sun, Y.; Haghighat, F.; Fung, B.C.M. A review of the-state-of-the-art in data-driven approaches for building energy prediction. *Energy Build.* **2020**, *221*, 110022. [CrossRef]
16. Zhang, L.; Wen, J.; Li, Y.; Chen, J.; Ye, Y.; Fu, Y.; Livingood, W. A review of machine learning in building load prediction. *Appl. Energy* **2021**, *285*, 116452. [CrossRef]
17. Elbeltagi, E.; Wefki, H. Predicting energy consumption for residential buildings using ANN through parametric modeling. *Energy Rep.* **2021**, *7*, 2534–2545. [CrossRef]
18. Ahmad, A.S.; Hassan, M.Y.; Abdullah, M.P.; Rahman, H.A.; Hussin, F.; Abdullah, H.; Saidur, R. A review on applications of ANN and SVM for building electrical energy consumption forecasting. *Renew. Sustain. Energy Rev.* **2014**, *33*, 102–109. [CrossRef]
19. Chen, B.; Liu, Q.; Chen, H.; Wang, L.; Deng, T.; Zhang, L.; Wu, X. Multiobjective optimization of building energy consumption based on BIM-DB and LSSVM-NSGA-II. *J. Clean. Prod.* **2021**, *294*, 126153. [CrossRef]
20. Yu, Z.; Haghighat, F.; Fung, B.C.M.; Yoshino, H. A decision tree method for building energy demand modeling. *Energy Build.* **2010**, *42*, 1637–1646. [CrossRef]
21. Xiao, Z.; Gang, W.; Yuan, J.; Zhang, Y.; Fan, C. Cooling load disaggregation using a NILM method based on random forest for smart buildings. *Sustain. Cities Soc.* **2021**, *74*, 103202. [CrossRef]
22. Lu, H.; Cheng, F.; Ma, X.; Hu, G. Short-term prediction of building energy consumption employing an improved extreme gradient boosting model: A case study of an intake tower. *Energy* **2020**, *203*, 117756. [CrossRef]
23. Ling, J.; Dai, N.; Xing, J.; Tong, H. An improved input variable selection method of the data-driven model for building heating load prediction. *J. Build. Eng.* **2021**, *44*, 103255. [CrossRef]
24. Chen, Y.; Guo, M.; Chen, Z.; Chen, Z.; Ji, Y. Physical energy and data-driven models in building energy prediction: A review. *Energy Rep.* **2022**, *8*, 2656–2671. [CrossRef]
25. Liu, Y.; Chen, H.; Zhang, L.; Feng, Z. Enhancing building energy efficiency using a random forest model: A hybrid prediction approach. *Energy Rep.* **2021**, *7*, 5003–5012. [CrossRef]
26. Liu, J.; Zhang, Q.; Dong, Z.; Li, X.; Li, G.; Xie, Y.; Li, K. Quantitative evaluation of the building energy performance based on short-term energy predictions. *Energy* **2021**, *223*, 120065. [CrossRef]
27. Lei, L.; Chen, W.; Wu, B.; Chen, C.; Liu, W. A building energy consumption prediction model based on rough set theory and deep learning algorithms. *Energy Build.* **2021**, *240*, 110886. [CrossRef]
28. Zhou, Y.; Liu, Y.; Wang, D.; Liu, X. Comparison of machine-learning models for predicting short-term building heating load using operational parameters. *Energy Build.* **2021**, *253*, 111505. [CrossRef]
29. Li, X.; Liu, S.; Zhao, L.; Meng, X.; Fang, Y. An integrated building energy performance evaluation method: From parametric modeling to GA-NN based energy consumption prediction modeling. *J. Build. Eng.* **2022**, *45*, 103571. [CrossRef]

30. Wang, R.; Lu, S.; Feng, W. A novel improved model for building energy consumption prediction based on model integration. *Appl. Energy* **2020**, *262*, 114561. [CrossRef]
31. Li, G.; Zhao, X.; Fan, C.; Fang, X.; Li, F.; Wu, Y. Assessment of long short-term memory and its modifications for enhanced short-term building energy predictions. *J. Build. Eng.* **2021**, *43*, 103182. [CrossRef]
32. Li, K.; Tian, J.; Xue, W.; Tan, G. Short-term electricity consumption prediction for buildings using data-driven swarm intelligence based ensemble model. *Energy Build.* **2021**, *231*, 110558. [CrossRef]
33. Zhang, C.; Li, J.; Zhao, Y.; Li, T.; Chen, Q.; Zhang, X. A hybrid deep learning-based method for short-term building energy load prediction combined with an interpretation process. *Energy Build.* **2020**, *225*, 110301. [CrossRef]
34. Zhang, Q.; Tian, Z.; Ding, Y.; Lu, Y.; Niu, J. Development and evaluation of cooling load prediction models for a factory workshop. *J. Clean. Prod.* **2019**, *230*, 622–633. [CrossRef]
35. Gao, B.; Huang, X.; Shi, J.; Tai, Y.; Zhang, J. Hourly forecasting of solar irradiance based on CEEMDAN and multi-strategy CNN-LSTM neural networks. *Renew. Energy* **2020**, *162*, 1665–1683. [CrossRef]
36. Gao, Y.; Hang, Y.; Yang, M. A cooling load prediction method using improved CEEMDAN and Markov Chains correction. *J. Build. Eng.* **2021**, *42*, 103041. [CrossRef]
37. Karijadi, I.; Chou, S.-Y. A hybrid RF-LSTM based on CEEMDAN for improving the accuracy of building energy consumption prediction. *Energy Build.* **2022**, *259*, 111908. [CrossRef]
38. Zhou, Y.; Liu, Y.; Wang, D.; Liu, X.; Wang, Y. A review on global solar radiation prediction with machine learning models in a comprehensive perspective. *Energy Convers. Manag.* **2021**, *235*, 113960. [CrossRef]
39. Huang, N.E.; Shen, Z.; Long, S.R.; Wu, M.C.; Shih, H.H.; Zheng, Q.; Yen, N.-C.; Tung, C.C.; Liu, H.H. The empirical mode decomposition and the Hilbert spectrum for nonlinear and non-stationary time series analysis. *Proc. R. Soc. Lond. Ser. A Math. Phys. Eng. Sci.* **1998**, *454*, 903–995. [CrossRef]
40. Wu, Z.; Huang, N.E. Ensemble empirical mode decomposition: A noise-assisted data analysis method. *Adv. Adapt. Data Anal.* **2009**, *1*, 1–41. [CrossRef]
41. Torres, M.E.; Colominas, M.A.; Schlotthauer, G.; Flandrin, P. A complete ensemble empirical mode decomposition with adaptive noise. In Proceedings of the 2011 IEEE International Conference on Acoustics, Speech and Signal Processing (ICASSP), Prague, Czech Republic, 22–27 May 2011; pp. 4144–4147.
42. Song, J.; Zhang, L.; Xue, G.; Ma, Y.; Gao, S.; Jiang, Q. Predicting hourly heating load in a district heating system based on a hybrid CNN-LSTM model. *Energy Build.* **2021**, *243*, 110998. [CrossRef]
43. Kumari, P.; Toshniwal, D. Long short term memory–convolutional neural network based deep hybrid approach for solar irradiance forecasting. *Appl. Energy* **2021**, *295*, 117061. [CrossRef]
44. Li, X.; Chen, Q. Development of a novel method to detect clothing level and facial skin temperature for controlling HVAC systems. *Energy Build.* **2021**, *239*, 110859. [CrossRef]
45. Liu, M.-D.; Ding, L.; Bai, Y.-L. Application of hybrid model based on empirical mode decomposition, novel recurrent neural networks and the ARIMA to wind speed prediction. *Energy Convers. Manag.* **2021**, *233*, 113917. [CrossRef]
46. Yu, C.; Li, Y.; Bao, Y.; Tang, H.; Zhai, G. A novel framework for wind speed prediction based on recurrent neural networks and support vector machine. *Energy Convers. Manag.* **2018**, *178*, 137–145. [CrossRef]
47. Yan, X.; Guan, T.; Fan, K.; Sun, Q. Novel double layer BiLSTM minor soft fault detection for sensors in air-conditioning system with KPCA reducing dimensions. *J. Build. Eng.* **2021**, *44*, 102950. [CrossRef]
48. Khashei, M.; Bijari, M. A novel hybridization of artificial neural networks and ARIMA models for time series forecasting. *Appl. Soft Comput.* **2011**, *11*, 2664–2675. [CrossRef]
49. Liu, Y.; Zhou, Y.; Wang, D.; Wang, Y.; Li, Y.; Zhu, Y. Classification of solar radiation zones and general models for estimating the daily global solar radiation on horizontal surfaces in China. *Energy Convers. Manag.* **2017**, *154*, 168–179. [CrossRef]
50. Zhou, Y.; Liu, Y.; Wang, D.; De, G.; Li, Y.; Liu, X.; Wang, Y. A novel combined multi-task learning and Gaussian process regression model for the prediction of multi-timescale and multi-component of solar radiation. *J. Clean. Prod.* **2021**, *284*, 124710. [CrossRef]
51. Jia, S.; Ma, B.; Guo, W.; Li, Z.S. A sample entropy based prognostics method for lithium-ion batteries using relevance vector machine. *J. Manuf. Syst.* **2021**, *61*, 773–781. [CrossRef]
52. Ma, Z.; Chen, H.; Wang, J.; Yang, X.; Yan, R.; Jia, J.; Xu, W. Application of hybrid model based on double decomposition, error correction and deep learning in short-term wind speed prediction. *Energy Convers. Manag.* **2020**, *205*, 112345. [CrossRef]
53. Bergstra, J.; Yamins, D.; Cox, D.D. Making a Science of Model Search: Hyperparameter Optimization in Hundreds of Dimensions for Vision Architectures. In Proceedings of the International Conference on Machine Learning, Atlanta, GA, USA, 16–21 June 2013; pp. 115–123.

information

MDPI

Article

GenericConv: A Generic Model for Image Scene Classification Using Few-Shot Learning

Mohamed Soudy [1,*], Yasmine M. Afify [2] and Nagwa Badr [2]

[1] Bioinformatics Program, Faculty of Computer and Information Sciences, Ain Shams University, Cairo 11566, Egypt
[2] Information Systems Department, Faculty of Computer and Information Sciences, Ain Shams University, Cairo 11566, Egypt; yasmine.afify@cis.asu.edu.eg (Y.M.A.); nagwabadr@cis.asu.edu.eg (N.B.)
* Correspondence: m.soudy@cis.asu.edu.eg

Abstract: Scene classification is one of the most complex tasks in computer-vision. The accuracy of scene classification is dependent on other subtasks such as object detection and object classification. Accurate results may be accomplished by employing object detection in scene classification since prior information about objects in the image will lead to an easier interpretation of the image content. Machine and transfer learning are widely employed in scene classification achieving optimal performance. Despite the promising performance of existing models in scene classification, there are still major issues. First, the training phase for the models necessitates a large amount of data, which is a difficult and time-consuming task. Furthermore, most models are reliant on data previously seen in the training set, resulting in ineffective models that can only identify samples that are similar to the training set. As a result, few-shot learning has been introduced. Although few attempts have been reported applying few-shot learning to scene classification, they resulted in perfect accuracy. Motivated by these findings, in this paper we implement a novel few-shot learning model—GenericConv—for scene classification that has been evaluated using benchmarked datasets: MiniSun, MiniPlaces, and MIT-Indoor 67 datasets. The experimental results show that the proposed model GenericConv outperforms the other benchmark models on the three datasets, achieving accuracies of 52.16 ± 0.015, 35.86 ± 0.014, and 37.26 ± 0.014 for five-shots on MiniSun, MiniPlaces, and MIT-Indoor 67 datasets, respectively.

Keywords: scene classification; few-shot learning; computer vision; MIT-Indoor 67

Citation: Soudy, M.; Afify, Y.M.; Badr, N. GenericConv: A Generic Model for Image Scene Classification Using Few-Shot Learning. *Information* **2022**, *13*, 315. https://doi.org/10.3390/info13070315

Academic Editors: Miltiadis D. Lytras and Andreea Claudia Serban

Received: 5 April 2022
Accepted: 13 May 2022
Published: 28 June 2022

1. Introduction

Scene classification (SC) is a complex task that relies on other sub-tasks, including object detection (OD), object classification (OC), and texture classification. By employing object detection in the scene classification, accurate results could be achieved as prior knowledge about objects that exist in the scene will lead to an easier interpretation of the image content. In contrast, semantic areas and knowledge about objects present in the image may infer the scene type more precisely [1,2].

Machine learning is widely used in scene classification in both tasks: object detection and object classification. Even though machine learning and deep learning achieved optimal performance in simple tasks, such as object detection, which led to their usage in more complex tasks, such as image scene classification, there is still a wide area of improvement that could be performed. The models' training phases need a significant quantity of data, which is a challenging and time-consuming task. Furthermore, most models rely on data from the training set, which results in useless models that can only detect samples that are comparable to the training set. These limitations lead to the use of few-shot learning in computer-vision tasks. Given the optimal performance of few-shot learning in object detection, few attempts have been made in scene classification, including a few datasets for model evaluation. It is a fact that research in this area is still ongoing and rising by the day, but it still faces a number of obstacles.

In this work, we propose a few-shot learning model that tackles the scene classification challenge. By being generalized on three popular scene datasets, the model will overcome the constraints of previously described models in scene categorization research regarding the generalization of models and the classification accuracy.

Our proposed pipeline addresses the generalization of the scene classification task by implementing a novel model that achieved unprecedented performance compared to the previously reported models on three benchmarked datasets. Furthermore, the usability of a new dataset rather than the used datasets to confirm the generalizability and validity of our proposed model.

The rest of the paper is organized as follows. The following section gives a brief literature review that highlights the limitations of scene classification research work. Then, benchmark approaches, datasets, and evaluation metrics are presented. The proposed model is then discussed. Experimental results obtained are then described. Finally, the conclusion and direction for future work are presented.

2. Related Work

Scene classification is considered one of the most complex tasks in computer vision research as it involves two other subtasks: object detection and object classification. The interconnection between the aforementioned tasks makes it harder to achieve accurate results in scene classification. Researchers apply the logistic solution by solving the two subtasks with optimal performance, which will lead to accurate results in scene classification. Scene classification approaches are divided into two categories: machine/deep learning algorithms and few-shot learning algorithms. Machine learning and deep learning algorithms seek to optimize model accuracy on specific datasets, whereas few-shot learning techniques seek generalizability. The categories of scene classification approaches are presented in the following sub-sections.

2.1. Machine Learning Approaches

Many attempts have been made at object detection using machine learning and deep learning, as the goal of object detection is to create computational models and approaches that give one of the most fundamental bits of data required by computer vision applications. Furthermore, due to the adequateness of datasets and technologies in recent decades, machine learning, and deep learning have been employed in computer vision to extract knowledge and information from complex images [3]. Many attempts were made in the era of scene categorization using diverse methodologies, as shown in Figure 1.

The first subtask is object detection, in which algorithms can be classified into three main categories: traditional detectors, Conventional Neural Network (CNN)-based two-stage detectors, and CNN-based one-stage detectors [4]. Initially, classical detectors were inspired by P. Viola and M. Jones' work 20 years ago, which designed a sliding window to iterate over picture pixels and recognize faces in any window.

Furthermore, to increase accuracy, a Histogram of Oriented Gradient (HOG) is utilized, which is computed on a dense grid of regularly spaced cells with overlapping local contrast normalization (on "blocks") [5–7]. The deep learning approach relies on the architecture of Conventional Neural Networks (CNNs). It is an extension of the Neural Networks (NNs) but with modifications in the connections between layers to ignore less important features and include the vital features that will be used in the classification task. CNN is the most used algorithm in deep learning that extracts the vital features from the image using the Conventional layer that applies a filter to an image to produce a feature map that describes the detection of features in the input. Then, it passes the output features to the pooling layer that down-samples the feature map by sliding a two-dimensional filter over each channel of the feature map and summing the features inside the filter's region. The last layer is the fully connected layer which is the feed-forward layer that compiles the final output as a probability assigned to each class of the input data representing the weights.

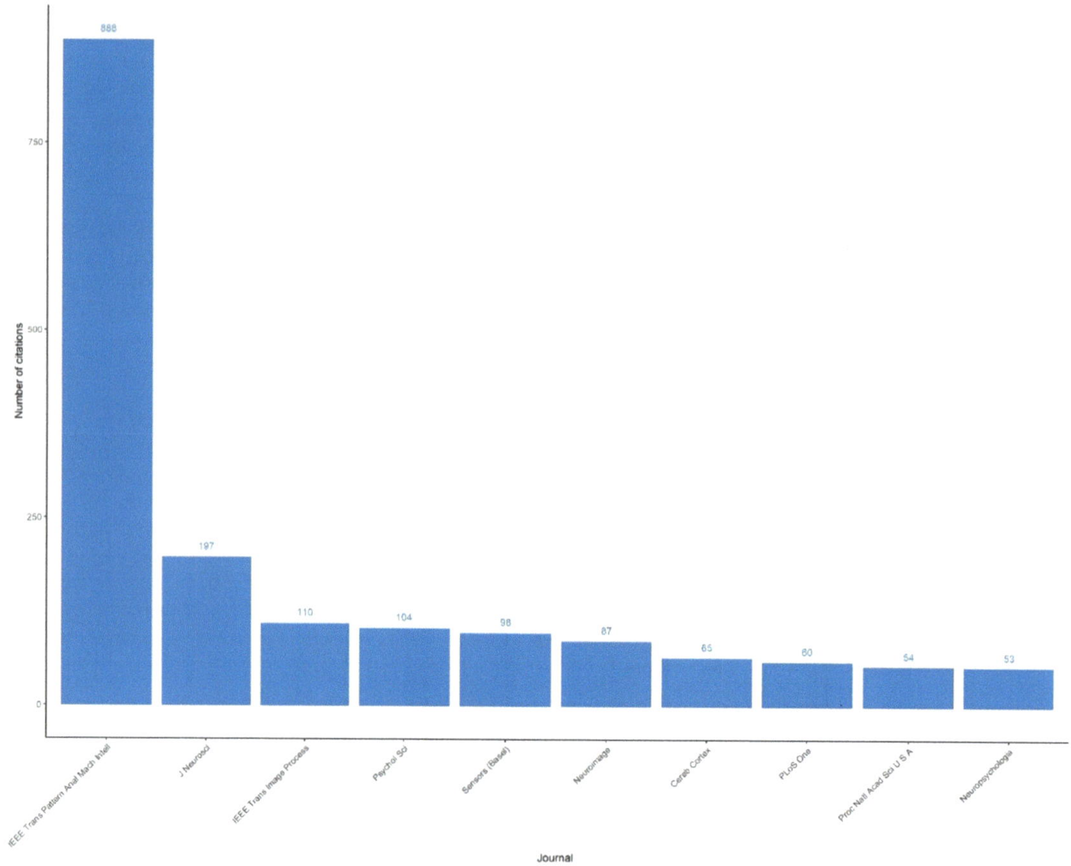

Figure 1. The distribution of publications and their citations over journals was collected using "Machine learning in scene classification" as a search query on the PubMed database.

Moreover, as data and technology increased, CNN-based two-stage detectors were adopted due to their capacity to extract critical information from complicated pictures. Eventually, CNN-based one-stage detectors were utilized to achieve optimal accuracy in object recognition algorithms such as Single Shot multi-box Detector (SSD) and You Only Look Once (YOLO), which were extensively employed owing to their accurate time and accuracy performance [8,9].

The second subtask is object classification, in which researchers seek to educate the computer to mimic human brain functions by learning from past knowledge or experience. The learning process begins with feeding the model training data including labels for each image, and the model attempts to learn the pattern that maps the input images to their labels. Then, we use unseen test images to evaluate the learned model.

For image classification, various models were employed initiated by Alex Krizhevsky that was the primary designer of AlexNet. AlexNet rose to prominence after competing in the ImageNet Large Scale Visual Recognition Challenge [10,11]. It had a top-5 mistake rate of 15.3%. This was 10.8% less than the runner-up. The original paper's main finding was that the depth of the model was definitely necessary for its excellent performance. This was highly computationally expensive, but it was made possible by GPUs, or Graphical Processing Units, during training. AlexNet has five convolutional layers, three max-pooling layers, two normalization layers, two fully connected layers, and one softmax layer. Convolutional filters and a nonlinear activation function ReLU are used in each convolutional layer. To maximize pooling, the pooling layers are utilized. The input size is mentioned at most of the places as 224 × 224 × 3 but due to some padding which happens it works out to be 227 × 227 × 3.

In 2013, ZFNet came out on top in the ImageNet Large Scale Visual Recognition Challenge (ILSVRC) with a substantial improvement over AlexNet. This work is a gold nugget that serves as a foundation for numerous concepts, including deep feature visualization, feature invariance, feature evolution, and feature significance. ZFNet is a modified version of AlexNet that provides more accuracy.

The techniques differed significantly in that ZF Net utilized 7 × 7 filters whereas AlexNet used 11 × 11 filters. The reasoning behind this is that they were losing a lot of pixel information by using larger filters, which can be kept by using lower filter sizes in the early conv layers. As they go deeper, the number of filters increases. This network, like others, uses ReLUs for activation and was trained using batch stochastic gradient descent [12]. In 2013, VGG was first proposed, and it finished second in the ImageNet competition in 2014. In comparison to AlexNet and ZFNet, it is commonly utilized as a basic design. VGG Net employed 3 × 3 filters, whereas AlexNet used 11 × 11 filters and ZFNet used 7 × 7 filters. The authors explain that having small repetitive fields of small filters as 3 × 3 filters offers an effective receptive field of 5 × 5, and 3–3 × 3 filters give a receptive field of 7 × 7 filters, but utilizing this we can train the network with a much less number of hyper-parameters [12].

In 2014, numerous excellent models were created, including VGG, however, GoogleNet emerged as the winner of the ImageNet competition. GoogleNet developed an inception module, which comprises skipping network connections to produce a tiny module, which is replicated throughout the network. GoogleNet employs nine inception modules and removes all fully connected layers by average pooling to reduce the size of the network from 7 × 7 × 1024 to 1 × 1 × 1024. This saves a large number of parameters [12].

The Microsoft ResNet protocol has 152 layers. The authors demonstrated empirically that when more layers are added, the error rate should decrease, in contrast to "simple nets," by adding a few layers that lead to larger training and test mistakes. On an 8 GPU computer, it took two to three weeks to train. One apparent reason why residual blocks increase classification is the direct step from one layer to the next, and intuitively employing all of these skip steps builds a gradient highway where the gradients computed may directly alter the weights in the first layer, making updates have a greater effect [13]. ResNet-50, another version of ResNet is a convolutional neural network that is 50 layers deep and developed by some researchers from the Microsoft research group. The architecture is designed to overcome the theory of vanishing gradient phenomena in models, such as AlexNet and VGG. The input layers of this network are made up of many residual blocks, and the operating idea is to optimize a residual function. Furthermore, ResNet won first place in ILSVRC and COCO 2015 competitions [12,13].

The pre-trained version of the network is trained on more than a million images from the ImageNet dataset. The pre-trained network can classify images into 1000 object categories, such as keyboard, mouse, pencil, and many animals. As a result, the network has learned rich feature representations for a wide range of images. The network has an image input size of 224-by-224.

Other versions of ResNet architecture were proposed including ResNet-50 (Places architecture) which is the same architecture as ResNet-50 but is trained on the places dataset, ResNet-101 which is a convolutional neural network that is 101 layers deep, and Squeeze-and-Excitation (SE) ResNext 101 is a convolutional neural network that is 101 layers deep. The pre-trained version of the network is trained on more than a million images from the ImageNet dataset. Recently, various approaches and variants of the ResNet model were implemented and show a better performance than the original architecture [14–16]. The architectures are shown in Table 1 below.

Table 1. Summary of machine learning models for image processing and scene classification.

Year	Model	Number of Layers	Developed by	Top5-Error Rate	# of Parameters
1998	LeNet	8	Yann LeCun et al.	NA	60 thousand
2012	AlexNet	7	Krizhevsky, Geoffrey Hinton, llya Sutskever	15.3%	60 million
2013	ZFNetO	NA	Matthew Zeiler and Rob Fergus	14.8%	NA
2014	GoogLeNet	19	Google	6.67%	4 million
2014	VGG Net	16	Simonyan, Zisserman	7.3%	138 million
2015	ResNet	152	Kaiming He	3.6%	60.4 million
2017	Inception-ResNet	NA	Szegedy, Christian et al.	3.08%	55.9 million
2018	ResNet	50	Jiang, Yun et al.	NA	25.6 million
2020	ReXNet_2.0	NA	Zhou, Daquan et al.	NA	19 million
2021	SENet	101	Srinivas, Aravind et al.	NA	49 million
2021	ResNet	152	Wightman, Ross et al.	NA	60 million

2.2. Few-Shot Learning Approaches

Few-shot learning arises to solve the existing limitations of machine learning approaches. Despite the optimal performance of the existing machine and deep learning models in image understanding and scene classification, there are still major issues. First, the training phase for the models necessitates a large amount of data, which is a difficult and time-consuming task. Furthermore, most models are reliant on data previously seen in the training set, resulting in ineffective models that can only identify samples that are similar to the training set.

Few-shot learning can be categorized into three main approaches: data, similarity, and learning. The data approach exploits prior knowledge about the structure and the variability of the data, which enables the construction of viable models through a few samples. This category involves the Pen-Stroke models which are the smart interface that automatically extracts and refines pen strokes from images of hand-drawn sketches, and neural statistician, a variational autoencoder extension that can learn a method for computing unsupervised representations, or statistics, of datasets. For each dataset, the network is trained to create statistics that embody a generative model. As a result, the network is capable of learning from fresh datasets efficiently for both unsupervised and supervised tasks. Researchers demonstrate that networks can learn statistics for grouping datasets, transferring generative models to new datasets, picking representative dataset samples, and categorizing previously unknown classes. The model is referred to as a neural statistician, which refers to a neural network that can learn to generate summary statistics of datasets without being supervised [17,18]. Although the similarity approaches tend to learn patterns in training data that, even when not visible, tend to distinguish various classes traditional machine learning models cannot distinguish between classes that are not present in training datasets, whereas few-dimensional ML models can. Few-shot learning strategies allow machine learning models to distinguish between classes that are not represented in the training data. This category can be classified into two major classes, models that discriminate between two unseen classes, and models that discriminate against multiple unseen classes. The models that discriminate between two unseen classes are Siamese networks, and triplet networks [19,20]. In Siamese networks, an input image of a person is used to determine the encodings of that image, after which the same network is utilized without any weights or biases updates to predict the encodings of an image of a different person. Then, researchers compare the two encodings to see if there's any similarity between the two images. These two encodings serve as a representation of the images' latent features. Similar features/encodings can be found in images of the same person. Researchers may use this to compare and determine whether the two images depict the same person or not. On the other hand, triplet networks rely on the concept of similarity and dissimilarity in the Triplet Loss architecture aids in learning distributed embedding. It is a type of neural network design in which many parallel networks with shared weights are trained. Input data are transmitted through one network during prediction time to produce distributed embedding representations of input data. The goal is to build three identical networks with the same neural net design and weights that they can share. All of the networks, I repeat, should have the same underlying weight vectors. The deep network's last layer contains a D-number of neurons for learning D-dimensional vector representation. Weight vectors are adjusted during backpropagation utilizing shared architecture, and anchor, positive, and negative images are sent over their respective networks. Any network is utilized to compute the vector representation of input data during prediction time.

For the category of the models that can disseminate among more than two classes, Matching networks in which each image from the support and the query set is fed to a CNN that outputs embedding for them, then each query image is classified using the softmax of the cosine distance from its embedding to the support-set embedding, and the Cross-Entropy Loss on the resulting classification is backpropagated through the CNN. Matching networks learn to compute image embedding in this way. MN can classify images using this method even if it has no prior knowledge of the classes. Matching networks compute attributes of the images that are relevant to discriminate across classes because the classes are distinct in each episode. In a normal classification, on the other hand, the algorithm learns the traits that are unique to each class [21].

The other approach in this category is prototypical networks, matching networks and prototypical networks (PN) are comparable. However, there are minor changes that aid the algorithm's performance. PN actually outperforms MN in terms of results. The PN procedure is nearly the same, except that the query image embedding is not compared to every picture embedding in the support set. Prototypical networks, on the other hand, present a different strategy. You must create class prototypes in PN. They are basically class embedding created by averaging the embedding of this class's images. Only these class prototypes are compared to the query image embedding. In addition, PN uses Euclidean distance rather than cosine distance. It is regarded as a critical component of the algorithm's advancements [22].

Finally, in the learning category in which meta-learning is introduced; meta-learning deciphers these limitations as it does not require a large number of training samples and it generalizes the model to be learned and evaluated in novel classes as never seen before [23].

Meta-learning is based on the premise that if a child has seen one or two pictures of a cat, he will be able to classify new pictures proficiently, reflecting the theory of learning by experience. Meta-learning also incorporates the concept of "learning to learn". The branch of meta-learning known as few-shot Learning (FSL) is observing a dramatic increase in research. Additionally, known as low-shot learning (LSL), it is a form of machine learning problem in which the training dataset contains only a small amount of data. The model is trained using well-defined episodes representing various classification tasks. The training set is split into two subsets (train; test) in each iteration to update the gradient and obtain the best weights for the learning process. Few-shot learning aims to generate a semi-generalized model that is able to classify novel classes using a low number of the training set and overcome the data collection and the time-consuming training process.

Based on the literature review, the few-shot learning algorithms development history is illustrated in Figure 2.

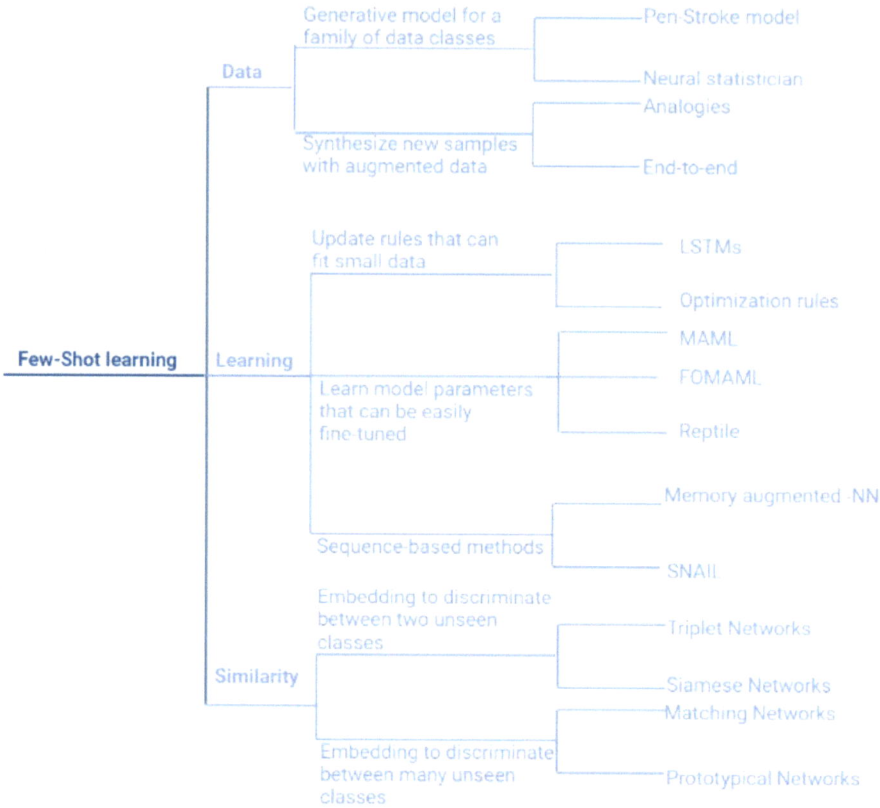

Figure 2. The development history of few-shot learning and its classes.

The best results in object recognition and classification reflect the accuracy of scene classification. Although machine learning is extensively used, it is still limited to data-driven models that cannot be generalized. The researchers' goal is to create a model that can do a task independently of the dataset utilized. A popular approach to this problem is few-shot learning. There have been a few attempts to use few-shot learning for scene categorization tasks. Although the best results were obtained, they lacked generalizability since the researchers employed a produced dataset instead of testing their model on a benchmark dataset [24,25]. Therefore, in this work, we directed our research to scene classification comparing benchmark models versus the proposed model over benchmarked datasets.

3. Materials and Methods

In this work, we provide some insights into the generalizability of few-shot learning models for the scene classification task. We assessed our models using several metrics, including accuracy, as expressed by Formula (1).

$$Accuracy = \frac{True\ positives + True\ Negatives}{True\ positives + False\ positives + True\ Negatives + False\ Negatives} \quad (1)$$

Additionally, the formula is implemented throughout 1000 test iterations, with the ultimate accuracy measured using Formula (2) [23].

$$\text{Accuracy} = avg(\Sigma_{n=1:1000}(\text{accuracy})) \pm 1.96 \times \text{std}(\Sigma_{n=1:1000}(\text{accuracy}))/\text{sqrt}(1000) \quad (2)$$

3.1. Datasets

The models were tested and evaluated using three benchmarked datasets: MiniSun, MiniPlaces, and MIT-Indoor 67 [25–27].

3.1.1. MiniSun Dataset

The MiniSun dataset contains 100 classes randomly chosen from Sun397 with 100 images of size 84 × 84 pixels per class. It is split into 64 base classes, 16 validation classes, and 20 novel classes [25].

3.1.2. MiniPlaces Dataset

The MiniPlaces dataset contains 100 classes randomly chosen from Places with 600 images of size 84 × 84 pixels per class. It is split into 64 base classes, 16 validation classes, and 20 novel classes [25].

3.1.3. MIT-Indoor 67 Dataset

The MIT-Indoor 67 dataset contains 67 indoor categories and a total of 15,620 images. The number of images varies across categories, but there are at least 100 images per category. All images are in jpg format. The images provided here are for research purposes only [27].

3.2. Benchmarked Models

3.2.1. Conv4

The Conv4 model's architecture consists of four conventional layers, four batch normalization, four activation layers, and flatten and softmax layers [28].

3.2.2. Conv6

The Conv6 model's architecture consists of six conventional layers, six batch normalization, six activation layers, and flatten and softmax layers [28].

3.2.3. Conv8

The Conv8 model's architecture consists of eight conventional layers, eight batch normalization, eight activation layers, and flatten and softmax layers [29].

3.2.4. ResNet12

The ResNet-12's architecture is made up of four depth three blocks with 3 × 3 kernels and shortcut connections. At the end of each block, a 2 × 2 max-pool is applied. The depth of the convolutional layer begins with 64 filters and is doubled after each max-pool [14,15].

3.2.5. MobileBlock1

The MobileBlock1 model's architecture is made up of a conventional layer, a batch normalization layer, and a Relu activation layer, which is then flattened with another Relu layer and lastly the final Softmax layer [25].

3.2.6. MobileConv

The MobileConv model's architecture consists of two conventional layers, two batch normalization, two Relu activation layers, and flatten and softmax layers [25].

3.2.7. Proposed Model Pipeline

The proposed data processing model *GenericConv* contains four critical sequential processes (data pre-processing; feature extraction; model training and model evaluation).

The first step is to read the data from the path directory and then perform feature-wise normalization to each image using Equations (3) and (4) [30].

$$Image = \frac{Image - mean(Image)}{adjusted\_stddev(Image)} \tag{3}$$

$$adjusted\_stddev\ (Image) = max(stddev(Image), \frac{1}{sqrt(Image.NumElements())}) \quad (4)$$

The second step is to apply feature extraction from the images using conventional neural networks that select and learn the crucial parameters from the input images. Furthermore, the model training process is applied by extracting the features from each image recursively and learning the pattern that matches the image to its label. Finally, the last step entails model evaluation by testing the model with unseen images and evaluating the results based on reference results, as shown in Figure 3.

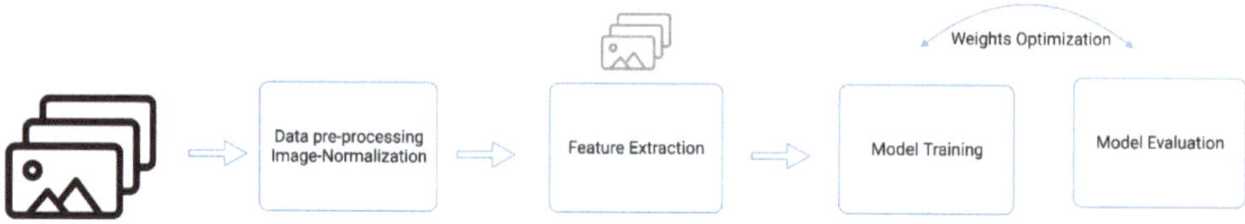

Figure 3. The pipeline of the proposed model *GenericConv*.

3.2.8. Proposed Model

The proposed model *GenericConv* architecture consists of three conventional layers, three max-pooling layers, a dropout layer followed by an average-pooling layer and a flatten layer then a dense layer with a relu activation, and finally a dense layer with softmax activation, as shown in Figure 4.

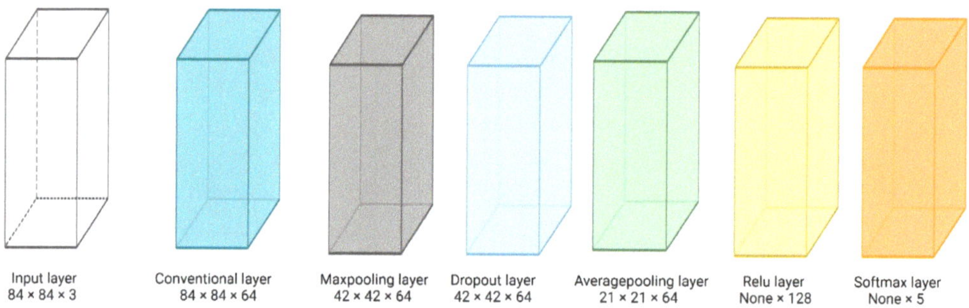

Figure 4. The architecture of the proposed model *GenericConv*.

The architecture of the proposed model is inspired by the best-performed architecture that was previously reported in scene classification MobileBlock1, and MobileConv. The architectures reported performing well on the benchmarked datasets. The architecture employs the CNN layers to extract the crucial features with the lowest parameters and depth compared to the aforementioned architectures, which will lead to gaining the most accurate results in the least training time taking memory management into consideration.

The architecture is designed by implementing a conventional layer that is a linear process that, like a regular neural network, involves multiplying a set of weights with the input. The multiplication is done between an array of input data and a two-dimensional array of weights, called a filter or a kernel, because the approach was created for two-dimensional input. The filter is smaller than the input data, and the dot product is the sort of multiplication used between a filter-sized patch of the input and the filter. A dot product is the element-wise multiplication of the input and filter's filter-sized patches, which are then summed to produce a single value. A popular method for ordering layers within a convolutional neural network that may be repeated one or more times in a given model is to add a pooling layer after the convolutional layer. A summary version of the features discovered in the input is the outcome of applying a pooling layer and constructing downsampled or pooled feature maps. They are beneficial because slight changes in the location of the feature in the input that the convolutional layer detects result in a pooled feature map with the feature in the same place. Although the training data are too small even one or five shots that may lead to model overfitting, a drop-out layer is added to prevent the model from overfitting. The drop-out is working by randomly setting the outgoing edges of hidden units (neurons that make up hidden layers) to 0 at each update of the training phase. This resulted novel combination of layers is benchmarked against other models to be tested and evaluated.

3.2.9. Proposed Model Hyperparameters

The proposed model hyperparameters are explained in Table 2.

Table 2. The proposed model's hyperparameters.

Learning Rate	Meta Step Size	Inner Batch Size	Evaluation Batch Size	Meta Iterations	Inner Iterations	Evaluation Iterations	Shots	Classes
0.003	0.25	25	25	2000	4	5	1/5	5

The hyperparameters are chosen using random hyperparameter optimization based on the comparison performed to other models (which use the same hyperparameters) to omit any variability in the experiment.

4. Results

Robust experiments were used to analyze and show the performance of the proposed *GenericConv* model in comparison to benchmarked models. Benchmarked datasets were utilized to demonstrate the generalizability of our model at various sizes. The accuracies were tracked across three datasets (MiniSun, MiniPlaces, and MIT Indoor-67) to monitor the model's performance as the model depth and the number of parameters increased.

4.1. Mini-Sun

The proposed model *GenericConv* outperformed the best-reported accuracy on MiniSun datasets as the MobileConv architecture achieved 47.5 $\pm$ 0.0158 as the best accuracy for five-shot five-ways classification, while MobileBlock1 achieved 30.86 $\pm$ 0.013 as the best accuracy for one-shot five-ways accuracy, our model achieved 52.16 $\pm$ 0.015 for five-shots five-ways and 32.72 $\pm$ 0.014 for one-shot five-ways accuracy with a significant increase in accuracy as 0.098, and 0.060 increase for five-shots and one-shot, respectively, as shown in Table 3.

Table 3. Five-ways accuracies on MiniSun.

Backbone Model	Parameters Fine-Tuning	Optimizer	5 Shots	1 Shot
MobileNetV2	Reptile	SGD	20.16 $\pm$ 0.011	
Conv4			39.14 $\pm$ 0.015	26.03 $\pm$ 0.013
Conv6			33.42 $\pm$ 0.0155	24.58 $\pm$ 0.012
Conv8			29.32 $\pm$ 0.012	21.48 $\pm$ 0.011
ResNet-12			20.16 $\pm$ 0.015	
MobileBlock1			40.12 $\pm$ 0.015	30.86 $\pm$ 0.013
MobileConv			47.5 $\pm$ 0.0158	30.72 $\pm$ 0.013
Proposed GenericConv			**52.16 $\pm$ 0.015**	**32.72 $\pm$ 0.014**

4.2. Mini-Places

The proposed model *GenericConv* outperformed the best-reported accuracy on Mini-Places datasets as the MobileConv architecture achieved 34.64 $\pm$ 0.014 as the best accuracy for five-shot five-ways classification, our model achieved 35.86 ± 0.014 for five-shots five-ways and 23.80 ± 0.012 for one-shot five-ways accuracy with a significant increase in accuracy as 0.035 for five-shots five-ways, while Conv4 is still the best accuracy in one-shot five-ways on the MiniPlaces dataset as mentioned in Table 4.

Table 4. Five-ways accuracies on MiniPlaces.

Backbone Model	Parameters Fine-Tuning	Optimizer	5 Shots	1 Shot
Conv4	Reptile	SGD	27.9 ± 0.014	**29.62 ± 0.013**
Conv6			19.84 ± 0.007	21.42 ± 0.009
Conv8			25.2 ± 0.011	21.14± 0.004
ResNet-12			20.16 ± 0.011	
MobileBlock1			20.1 ± 0.001	
MobileConv			34.64 ± 0.014	26.36 ± 0.013
Proposed GenericConv			**35.86 ± 0.014**	23.80 ± 0.012

4.3. MIT Indoor-67

The MIT-Indoor is used to ensure the benchmarking and generalization of our model compared to benchmarked models. Conv4 architecture achieved 28.7 ± 0.013 accuracies for five-shot five-ways classification, and 22.0 ± 0.012 for one-shot five-ways which decreased by 0.42, and 0.09 for five-shots, and one-shot, respectively, when we utilized Conv6 and deepened the model. To trace and confirm this behavior, we deepened the model one more fold by employing Conv8 architecture that achieved 22.18 ± 0.005 for five-shot five-ways, and 20.1 ± 0.003 for one-shot five-ways classification, a final confirmation step was performed by deepening the model more by employing ResNet-12 architecture which overfitted on the dataset. Meanwhile, the proposed model *GenericConv* outperformed all the aforementioned models by achieving 37.26 ± 0.014 accuracies on five-shot five-way classification, and 24.82 ± 0.013 accuracies on one-shot five-way classification with a variance of 0.92, and 0.0088 for five-shot and one-shot classification compared to the best-reported accuracy as shown in Table 5.

Table 5. Five-ways accuracies on MIT-Indoor 67.

Backbone Model	Parameters Fine-Tuning	Optimizer	5 Shots	1 Shot
Conv4	Reptile	SGD	28.7 ± 0.013	22.0 ± 0.012
Conv6			20.16 ± 0.003	20.0 ± 0.0014
Conv8			22.18 ± 0.005	20.10 ± 0.003
ResNet-12			20 ± 0.00011	20 ± 0.00011
MobileBlock1			34.1 ± 0.014	24.6 ± 0.012
MobileConv			33.18 ± 0.014	23.82 ± 0.012
Proposed GenericConv			**37.26 ± 0.014**	**24.82 ± 0.013**

5. Conclusions

Scene classification is considered one of the most complex tasks in computer vision research as it involves the interconnection of object detection and object classification tasks. Few attempts were made by researchers to apply few-shot learning for scene classification tasks. The finest findings were obtained, but they lacked generalizability. Our proposed pipeline addresses the generalization of the scene classification task by implementing a novel model that achieved unprecedented performance compared to the previously reported models on three benchmarked datasets.

The proposed model GenericConv achieved 52.16 ± 0.015 for five-shots five-ways and 32.72 ± 0.014 for one-shot five-ways accuracy with a significant increase in accuracy as 0.098, and 0.060 increase for five-shots and one-shot, respectively, than the reported results on the MiniSun dataset, while our model achieved 35.86 ± 0.014 for five-shots five-ways with a significant increase in accuracy as 0.035 for five-shots five-ways than the reported results on the MiniPlaces dataset. Furthermore, our proposed model outperformed all the aforementioned models by achieving 37.26 ± 0.014 accuracies on five-shot five-way classification, and 24.82 ± 0.013 accuracies on one-shot five-way classification with a variance of 0.92, and 0.0088 for five-shot and one-shot classification compared to the best-reported accuracy on the Indoor-67 dataset. Furthermore, we aim to develop a Graphical User Interface (GUI) that is able to perform scene classification regardless of user programming experience.

Author Contributions: Data curation, M.S.; Formal analysis, Y.M.A.; Methodology, M.S.; Project administration, N.B.; Software, M.S.; Supervision, Y.M.A. and N.B.; Writing—original draft, M.S.; Writing—review & editing, Y.M.A. and N.B. All authors have read and agreed to the published version of the manuscript.

Funding: This research received no external funding.

Institutional Review Board Statement: Not applicable.

Informed Consent Statement: Not applicable.

Data Availability Statement: To contribute to the advance of SR research, our work is made available at: Data-preprocessing and models: https://github.com/MohmedSoudy/A-generic-approach-for-image-scene-classification-using-few-shot-learning (access on 5 April 2022).

Acknowledgments: I would like to thank Islam Ibrahim for his insights and motivating assistance in producing this work.

Conflicts of Interest: The authors declare no conflict of interest.

References

1. Sonka, M.; Hlavac, V.; Boyle, R. *Image Processing, Analysis, and Machine Vision*; Cengage Learning: Boston, MA, USA, 2014.
2. Singh, V.; Girish, D.; Ralescu, A. Image Understanding-a Brief Review of Scene Classification and Recognition. *MAICS* **2017**, 85–91.
3. Yao, J.; Fidler, S.; Urtasun, R. Describing the scene as a whole: Joint object detection, scene classification, and semantic segmentation. In Proceedings of the 2012 IEEE Conference on Computer Vision and Pattern Recognition, Washington, DC, USA, 16–21 June 2012; pp. 702–709.
4. Zou, Z.; Shi, Z.; Guo, Y.; Ye, J. Object detection in 20 years: A survey. *arXiv* **2019**, arXiv:1905.05055.
5. Viola, P.; Jones, M. Rapid object detection using a boosted cascade of simple features. In Proceedings of the 2001 IEEE Computer Society Conference on Computer Vision and Pattern Recognition, Kauai, HI, USA, 8–14 December 2001; Volume 1, p. I.
6. Viola, P.; Michael, J. Fast and Robust Classification Using Asymmetric Adaboost and a Detector Cascade. Advances in Neural Information Processing Systems 14. 2001. Available online: https://www.researchgate.net/publication/2539888_Fast_and_Robust_Classification_using_Asymmetric_AdaBoost_and_a_Detector_Cascade (accessed on 12 May 2022).
7. Dalal, N.; Triggs, B. Histograms of Oriented Gradients for Human Detection. In Proceedings of the Computer Vision and Pattern Recognition, San Diego, CA, USA, 20–26 June 2005; pp. 886–893.
8. Liu, W.; Anguelov, D.; Erhan, D.; Szegedy, C.; Reed, S.; Fu, C.Y.; Berg, A.C. Ssd: Single shot multibox detector. In *European Conference on Computer Vision*; Springer: Cham, Switzerland, 2016; pp. 21–37.
9. Huang, R.; Pedoeem, J.; Chen, C. YOLO-LITE: A Real-Time Object Detection Algorithm Optimized for Non-GPU Computers. In Proceedings of the 2018 IEEE International Conference on Big Data, Seattle, WA, USA, 10–13 December 2018; pp. 2503–2510.
10. Szegedy, C.; Vanhoucke, V.; Ioffe, S.; Shlens, J.; Wojna, Z. Rethinking the inception architecture for computer vision. In Proceedings of the IEEE Conference on Computer Vision and Pattern Recognition, Las Vegas, NV, USA, 27–30 June 2016; pp. 2818–2826.
11. Russakovsky, O.; Deng, J.; Su, H.; Krause, J.; Satheesh, S.; Ma, S.; Huang, Z.; Karpathy, A.; Khosla, A.; Bernstein, M.; et al. Imagenet large scale visual recognition challenge. *Int. J. Comput. Vis.* **2015**, *115*, 211–252. [CrossRef]
12. He, K.; Zhang, X.; Ren, S.; Sun, J. Deep residual learning for image recognition. In Proceedings of the IEEE Conference on Computer Vision and Pattern Recognition, Las Vegas, NV, USA, 27–30 June 2016; pp. 770–778.
13. Simonyan, K.; Zisserman, A. Very deep convolutional networks for large-scale image recognition. *arXiv* **2014**, arXiv:1409.1556.
14. Zhou, B.; Lapedriza, A.; Khosla, A.; Oliva, A.; Torralba, A. Places: A 10 million image database for scene recognition. *IEEE Trans. Pattern Anal. Mach. Intell.* **2017**, *40*, 1452–1464. [CrossRef] [PubMed]
15. Chollet, F. Xception: Deep learning with depthwise separable convolutions. In Proceedings of the IEEE Conference on Computer Vision and Pattern Recognition, Honolulu, HI, USA, 21–26 July 2017; pp. 1251–1258.
16. Huang, G.; Liu, Z.; Van Der Maaten, L.; Weinberger, K.Q. Densely connected convolutional networks. In Proceedings of the IEEE Conference on Computer Vision and Pattern Recognition, Honolulu, HI, USA, 21–26 July 2017; pp. 4700–4708.

17. Srinivas, A.; Lin, T.Y.; Parmar, N.; Shlens, J.; Abbeel, P.; Vaswani, A. Bottleneck transformers for visual recognition. In Proceedings of the IEEE/CVF Conference on Computer Vision and Pattern Recognition, Nashville, TN, USA, 20–25 June 2021; pp. 16519–16529.
18. Wightman, R.; Touvron, H.; Jégou, H. Resnet strikes back: An improved training procedure in timm. *arXiv* **2021**, arXiv:2110.00476.
19. Koch, G.; Zemel, R.; Salakhutdinov, R. Siamese neural networks for one-shot image recognition. In *ICML Deep Learning Workshop*; ICML: Lille, France, 2015; Volume 2.
20. Hoffer, E.; Ailon, N. Deep metric learning using triplet network. In *International Workshop on Similarity-Based Pattern Recognition*; Springer: Cham, Switzerland, 2015; pp. 84–92.
21. Vinyals, O.; Blundell, C.; Lillicrap, T.; Wierstra, D. Matching networks for one shot learning. In *Advances in Neural Information Processing Systems 29 (NIPS 2016)*; Curran Associates: Red Hook, NY, USA, 2016; Available online: https://proceedings.neurips.cc/paper/2016/hash/90e1357833654983612fb05e3ec9148c-Abstract.html (accessed on 12 May 2022).
22. Snell, J.; Swersky, K.; Zemel, R. Prototypical networks for few-shot learning. *Adv. Neural Inf. Process. Syst.* **2017**, *30*.
23. Zhu, J.; Jang-Jaccard, J.; Singh, A.; Welch, I.; Ai-Sahaf, H.; Camtepe, S. A few-shot meta-learning based siamese neural network using entropy features for ransomware classification. *Comput. Secur.* **2022**, *117*, 102691. [CrossRef]
24. Sobti, P.; Nayyar, A.; Nagrath, P. EnsemV3X: A novel ensembled deep learning architecture for multi-label scene classification. *PeerJ Comput. Sci.* **2021**, *7*, e557. [CrossRef] [PubMed]
25. Soudy, M.; Yasmine, A.; Nagwa, B. Insights into few shot learning approaches for image scene classification. *PeerJ Comput. Sci.* **2021**, *7*, e666. [CrossRef] [PubMed]
26. Tripathi, A.S.; Danelljan, M.; Van Gool, L.; Timofte, R. Few-Shot Classification by Few-Iteration Meta-Learning. *arXiv* **2020**, arXiv:2010.00511.
27. Quattoni, A.; Antonio, T. Recognizing indoor scenes. In Proceedings of the 2009 IEEE Conference on Computer Vision and Pattern Recognition, Miami, FL, USA, 22–24 June 2009; pp. 413–420.
28. Hong, J.; Fang, P.; Li, W.; Zhang, T.; Simon, C.; Harandi, M.; Petersson, L. Reinforced attention for few-shot learning and beyond. In Proceedings of the IEEE/CVF Conference on Computer Vision and Pattern Recognition, Nashville, TN, USA, 20–25 June 2021; pp. 913–923.
29. Li, X.; Wu, J.; Sun, Z.; Ma, Z.; Cao, J.; Xue, J.H. BSNet: Bi-Similarity Network for Few-shot Fine-grained Image Classification. *IEEE Trans. Image Process.* **2020**, *30*, 1318–1331. [CrossRef] [PubMed]
30. Purkait, N. *Hands-On Neural Networks with Keras: Design and Create Neural Networks Using Deep Learning and Artificial Intelligence Principles*; Packt Publishing Ltd: Birmingham, UK, 2019.

remote sensing

MDPI

Article

MIMO: A Unified Spatio-Temporal Model for Multi-Scale Sea Surface Temperature Prediction

Siyun Hou [1,2], Wengen Li [1,*], Tianying Liu [1,2], Shuigeng Zhou [3,4], Jihong Guan [1], Rufu Qin [1,5] and Zhenfeng Wang [2]

1 Department of Computer Science and Technology, Tongji University, Shanghai 200082, China; housiyun@tongji.edu.cn (S.H.); liutianying@tongji.edu.cn (T.L.); jhguan@tongji.edu.cn (J.G.); qinrufu@tongji.edu.cn (R.Q.)
2 Project Management Office of China National Scientific Seafloor Observatory, Tongji University, Shanghai 200082, China; wangzhenfeng@cnsso.edu.cn
3 Shanghai Key Lab of Intelligent Information Processing, Shanghai 200438, China; sgzhou@fudan.edu.cn
4 School of Computer Science, Fudan University, Shanghai 200438, China
5 State Key Laboratory of Marine Geology, Tongji University, Shanghai 200082, China
* Correspondence: lwengen@tongji.edu.cn

Abstract: Sea surface temperature (SST) is a crucial factor that affects global climate and marine activities. Predicting SST at different temporal scales benefits various applications, from short-term SST prediction for weather forecasting to long-term SST prediction for analyzing El Niño–Southern Oscillation (ENSO). However, existing approaches for SST prediction train separate models for different temporal scales, which is inefficient and cannot take advantage of the correlations among the temperatures of different scales to improve the prediction performance. In this work, we propose a unified spatio-temporal model termed the Multi-In and Multi-Out (MIMO) model to predict SST at different scales. MIMO is an encoder–decoder model, where the encoder learns spatio-temporal features from the SST data of multiple scales, and fuses the learned features with a Cross Scale Fusion (CSF) operation. The decoder utilizes the learned features from the encoder to adaptively predict the SST of different scales. To our best knowledge, this is the first work to predict SST at different temporal scales simultaneously with a single model. According to the experimental evaluation on the Optimum Interpolation SST (OISST) dataset, MIMO achieves the state-of-the-art prediction performance.

Keywords: sea surface temperature (SST); multi-scale SST prediction; spatio-temporal model; data fusion

Citation: Hou, S.; Li, W.; Liu, T.; Zhou, S.; Guan, J.; Qin, R.; Wan, Z. MIMO: A Unified Spatio-Temporal Model for Multi-Scale Sea Surface Temperature Prediction. *Remote Sens.* **2022**, *14*, 2371. https://doi.org/10.3390/rs14102371

Academic Editor: Yukiharu Hisaki

Received: 29 March 2022
Accepted: 12 May 2022
Published: 14 May 2022

1. Introduction

Sea surface temperature (SST) refers to the temperature of the water from 1 millimeter to 20 meters below the sea surface. The ocean covers about three-quarters of the Earth's surface and greatly influences global climate [1] and human activities [2]. As one key factor of ocean environment, SST affects global climate when assimilating and releasing heat. For instance, global precipitation is influenced by ocean evaporation, which is highly dependent on SST [3]. The widely known El Niño–Southern Oscillation (ENSO) phenomenon is an irregular periodic variation of SST in the tropical eastern Pacific Ocean, and occurs every 3∼5 years [4]. Therefore, accurately predicting SST could benefit various applications, e.g., weather forecasting, global warming prevention and extreme climate tracking.

Most existing methods for SST prediction are either numerical models or data-driven models. Numerical models [5], e.g., General Circulation Model (GCM), Integrated Forecast System (IFS) and Global Forecast Systems (GFS), predict SST by using differential equations to describe the relations between SST and other oceanic factors (e.g., sea surface height and air temperature) according to the laws of physics. Numerical models are usually of high complexity due to the large number of parameters, thus resulting in high computation cost. Data-driven SST prediction models predict SST by discovering the hidden regularity and patterns in historical SST data, and can be further divided into three sub-categories, i.e., statistical models [6], shallow neural network models [7–9] and deep learning models [10–12]. In contrast to numerical models, data-driven models require less domain knowledge and usually can achieve better prediction performance.

The existing methods mentioned above all focus on single-scale SST prediction, e.g., predicting the SST of the next seven days, and predicting the monthly average SST in the next one year. However, single-scale SST prediction has some disadvantages. First, we have to train different models for predicting different scales of SST, which is inefficient and costs many computation resources. Second, single-scale SST prediction ignores the correlations among SST of different scales. In practice, SST has different scales of temporal regularity. For example, the short-term SST usually depends on the long-term trends and periodicity of SST. In this case, the correlations of different scales of SST could be used for improving SST prediction.

To overcome the disadvantages of existing SST prediction methods, we propose multi-scale SST prediction to predict daily, weekly and monthly SST simultaneously. To this end, we face some technical challenges. First, multi-scale SST prediction needs to learn the features that can capture the regularity of different temporal scales of SST. Second, it is nontrivial to exploit the correlations among SST data of different scales. Third, the prediction model should be able to adapt to different scales of SST prediction.

To address these challenges in multi-scale SST prediction, we propose a new spatio-temporal model, i.e., the Multi-In and Multi-Out (MIMO) model. Concretely, the MIMO model first learns the spatio-temporal features for different scales of SST with independent learning blocks. Then, the learned features are fused to obtain unified feature representation. Finally, the MIMO model adaptively predicts different scales of SST with different prediction components. This is the first work that achieves unified prediction for SST of different temporal scales. The proposed model can learn the spatio-temporal features of SST well and fully take advantage of the correlations among different scales of SST to enhance the prediction.

The main contributions of this work are highlighted as follows.

- We raise the multi-scale SST prediction problem and highlight its technical challenges.
- We propose the MIMO model that can predict SST at multiple scales simultaneously with a single model. MIMO learns different scales of temporal regularity in SST and can be adapted to the requirements of SST prediction at different scales.
- We conduct extensive experiments on real SST datasets to evaluate the proposed method, and the experimental results shows that MIMO outperforms existing SST prediction methods, including CNN, LeNet and ConvLSTM.

The remainder of this work is organized as follows. Section 2 reviews the related work on SST prediction and analyzes the disadvantages of existing methods. Section 3 formally defines the problem of multi-scale SST prediction, and Section 4 gives the technical details of the proposed MIMO model. Section 5 reports the experimental results. Finally, Section 6 concludes the work.

2. Background

As discussed previously, existing SST prediction methods can be classified into numerical models and data-driven models, where data-driven models can be further divided into statistical models, shallow neural network models and deep learning models. Figure 1 shows the taxonomy of existing SST prediction methods and their representative models.

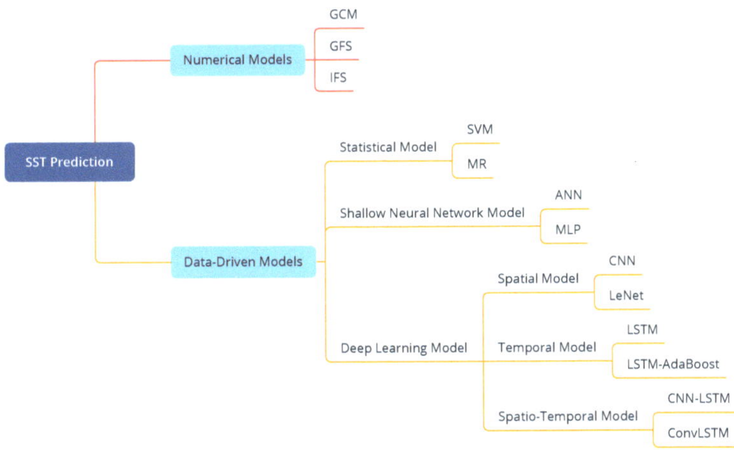

Figure 1. The taxonomy of existing SST prediction models.

2.1. Numerical Models

Numerical models utilize mathematical equations to describe the underlying regularity in SST and its correlations with other oceanic and atmospheric variables [13–15]. This requires a better understanding of the dynamics of SST to design the prediction models. Representative numerical SST prediction models include General Circulation Model (GCM), Integrated Forecast System (IFS) and Global Forecast Systems (GFS).

GCM simulates the changes of the global climate by calculating the hourly evolution of the atmosphere based on the conservation laws for atmospheric mass, momentum, total energy and water vapor [16]. For example, Krishnamurti et al. used 13 coupled atmosphere-ocean models to predict SST [17].

The IFS model and GFS model are developed by the European Centre for Medium-Range Weather Forecasts (ECMWF) and the Command National Centers for Environmental Prediction (NCEP), respectively [18]. The IFS model can predict the SST of the following 10–15 days while the GFS model can predict the SST of the following 16 days. Both models contain multiple GCMs and can also predict other variables of the ocean and atmosphere.

Numerical models are usually of high computational complexity due to the large number of mathematical equations they have. In addition, numerical models are widely used for analyzing the global trends of SST but cannot predict the SST of high spatial resolutions well.

2.2. Data-Driven Models

Different from numerical models, data-driven models directly learn knowledge from historical SST data with machine learning techniques to conduct the prediction. Therefore, data-driven models depend more on SST data than the domain knowledge in ocean climate and environment.

2.2.1. Statistical Models

Statistical models for SST prediction learn the regularity and hidden patterns of SST from historical data [19,20]. Representative statistical models include support vector machine (SVM) and multi-linear regression (MR) model. SVM constructs a set of hyperplanes in a high dimensional space and uses these hyperplanes for classification and regression. For example, Lins et al. developed an SVM model that uses historical daily SST as input to predict the SST across the Northeastern Brazilian Coast [21] and the tropical Atlantic [6]. The MR model combines the SST anomaly index with the tropical Pacific SST anomaly to predict SST. It takes 12 months of historical SST data as the input and assumes that the SST follows some statistical assumptions. Then, the MR model uses a linearly weighted sum method to predict the SST of the following month [22]. Statistical models need feature engineering to extract features from SST data, which may result in the loss of some important information for SST prediction.

2.2.2. Shallow Neural Network Models

Shallow neural network (SNN) models are of high flexibility in fitting raw SST data [23–25]. Compared with the statistical models, SNN models can capture the nonlinear correlations in SST data, thus achieving better prediction performance. For instance, Patil et al. proposed an Artificial Neural Network (ANN) to predict the SST of the following 5 days [26]. Aparna et al. proposed an SNN that comprises three layers, i.e., input layer, linear layer and output layer, to predict the SST of the next day at specific locations [18]. Wei et al. separated SST time series data into monthly mean SST and monthly anomaly SST, and constructed two multilayer perceptron (MLP) models to generate the predicted results of SST [7].

Most shallow neural network models predict SST at some specific locations and ignore the spatial correlations and temporal regularity in SST. In addition, due to the limited learning ability, SNN models cannot fully exploit the large volume of historical SST data to train prediction models.

2.2.3. Deep Learning Models

With the substantial increase of SST data, statistical models and SNN models are unable to learn comprehensive knowledge from big SST data to further improve the prediction performance. Therefore, deep learning (DL) models have been introduced for SST prediction [27]. DL models can be classified into three sub-categories, i.e., deep spatial models, deep temporal models and deep spatio-temporal (ST) models.

Deep spatial models focus on learning the spatial correlations in SST data to achieve prediction. Ham et al. proposed a CNN-based model to predict the *Niño* 3.4 index of the next 12 months based on the historical SST data of 12 months [28]. Zheng et al. proposed a model with stacked multiple CNN layers, Max Pooling layers and upsampling layers to predict the SST of the next day using the historical SST data of 14 days [11].

Deep temporal models focus on learning the temporal regularity in SST data to achieve prediction. Zhang et al. proposed an LSTM model to predict the SST in the next three days in the East China Sea with the historical SST data of 15 days [10]. Xiao et al. combined Ada-Boost and LSTM to predict the SST in the next 10 days in the East China Sea with the historical SST data of 40 days [29]. Yao et al. proposed an encoder–decoder model based on LSTM to predict the SST in the next 10 days using the historical data of 10 days [30].

Deep spatio-temporal models learn both spatial correlations and temporal regularity in SST to achieve prediction. Xiao et al. proposed a multi-layer convolutional LSTM model to predict the SST in the next 10 days with the historical SST data of 50 days in the East China Sea [12]. Weyn et al. combined the ConvLSTM model and the CNN model to predict the SST of the next three days with the historical SST data of 12 days [31]. Zhang et al. proposed a CNN-LSTM-based model to predict the SST in the next eight months using the historical SST data of 28 months [32]. In addition, deep graph models have also been used for SST prediction in recent years. Zhang et al. proposed a graph model MGCN that uses historical SST data of six days to predict the SST of the next three days. They used a temporal convolution to learn the temporal features of SST which are then put into the graph model to learn the spatial features [33].

In general, deep learning models can achieve better prediction performance than statistical models and shallow neural network models. However, all existing data-driven models only consider single-scale SST prediction, i.e., predicting daily or monthly SST separately, and cannot achieve multi-scale SST prediction.

3. Problem Definition

For SST prediction, we usually divide the target region of interest R into small grid regions of the same size along the latitude and longitude, and then predict the SST for each grid region. For example, as illustrated in Figure 2, the region *Niño* 3.4 can be divided into 40×200 grid regions of size $0.25° \times 0.25°$.

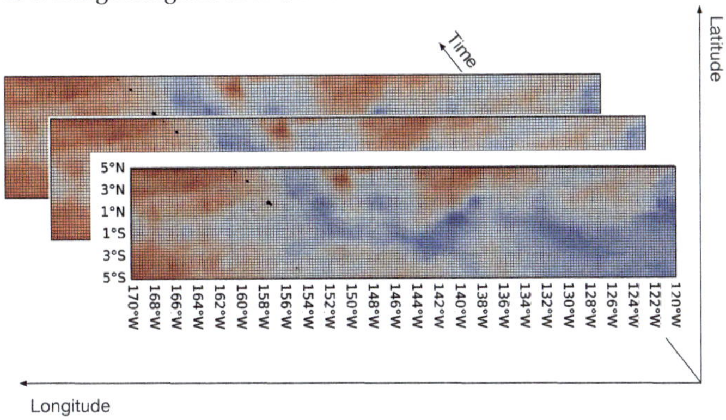

Figure 2. The daily SST sequence for the region of *Niño* 3.4, which is located within [5°N∼5°S, 120°W∼170°W] in the Pacific Ocean and divided into 40×200 grid regions of size $0.25° \times 0.25°$.

Assuming that the region of interest R is divided into $c \times d$ grid regions, the corresponding SST records at time slot t are denoted by $X_t \in \mathbb{R}^{c \times d}$. Data-driven SST prediction methods learn the underlying patterns of SST from historical SST data to predict the SST in the future. Specifically, the problem of single-scale SST prediction is defined as below.

Definition 1 (Single-Scale SST Prediction). *Given a sequence of historical SST records* $\mathbf{X}_{t-a+1\to t} = (\mathbf{X}_{t-a+1}, \mathbf{X}_{t-a+2}, \ldots, \mathbf{X}_t)$ *of length a, single-scale SST prediction aims to predict the SST sequence* $\mathbf{X}_{t+1\to t+b} = (\mathbf{X}_{t+1}, \mathbf{X}_{t+2}, \ldots, \mathbf{X}_{t+b})$ *of length b in the future, i.e.,*

$$\mathbf{X}_{t+1\to t+b} = \arg\max_{\mathbf{X}_{t+1\to t+b}} p(\mathbf{X}_{t+1\to t+b} | \mathbf{X}_{t-a+1\to t}) \tag{1}$$

In general, the time slot in single-scale SST prediction could be daily, weekly and monthly. For example, we can have the following SST prediction schemes:

- Using the historical daily SST of 10 days to predict the daily SST of the following seven days, where $a = 10$ and $b = 7$;
- Using the historical monthly SST of 36 months to predict the monthly SST of the following 12 months, where $a = 36$ and $b = 12$.

In existing studies on SST prediction, different scales of SST prediction are treated separately, i.e., training different prediction models for different scales of SST prediction. In this work, we aim to unify the prediction for multiple scales of SST and propose the problem of multi-scale SST prediction as defined below.

Definition 2 (Multi-scale SST prediction). *Given daily, weekly and monthly historical SST records* $\mathbf{X}^{daily}_{t-a+1\to t}$, $\mathbf{X}^{weekly}_{t-a+1\to t}$ *and* $\mathbf{X}^{monthly}_{t-a+1\to t}$, *respectively, multi-scale SST prediction aims to predict their corresponding records in the next b time slots, i.e.,*

$$\left\{\mathbf{X}^{daily}_{t+1\to t+b}, \mathbf{X}^{weekly}_{t+1\to t+b}, \mathbf{X}^{monthly}_{t+1\to t+b}\right\} = \arg\max_{\left\{\mathbf{X}^{daily}_{t+1\to t+b}, \mathbf{X}^{weekly}_{t+1\to t+b}, \mathbf{X}^{monthly}_{t+1\to t+b}\right\}} \\ p(\left\{\mathbf{X}^{daily}_{t+1\to t+b}, \mathbf{X}^{weekly}_{t+1\to t+b}, \mathbf{X}^{monthly}_{t+1\to t+b}\right\} | \left\{\mathbf{X}^{daily}_{t-a+1\to t}, \mathbf{X}^{weekly}_{t-a+1\to t}, \mathbf{X}^{monthly}_{t-a+1\to t}\right\}) \tag{2}$$

Figures 3 and 4 illustrate the difference between single-scale SST prediction and multi-scale SST prediction. Single-scale SST prediction uses the same scale of historical SST data to predict the corresponding future SST records. Therefore, in single-scale SST prediction, the input has three dimensions, i.e., latitude, longitude and time slots. In contrast, multi-scale SST prediction considers multiple scales of historical SST data and achieves the prediction for all the scales of SST together, thus covering short-term, mid-term and long-term SST prediction. Therefore, in multi-scale SST prediction, the input has four dimensions, i.e., latitude, longitude, time slots and scales.

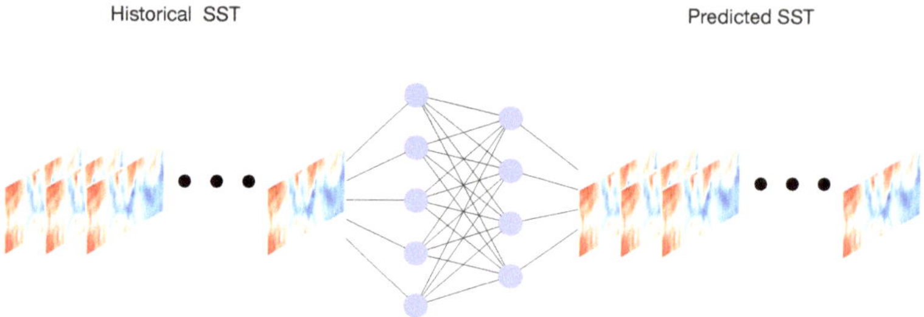

Figure 3. Single-scale SST prediction that uses the historical SST of single scale to predict the future single-scale SST.

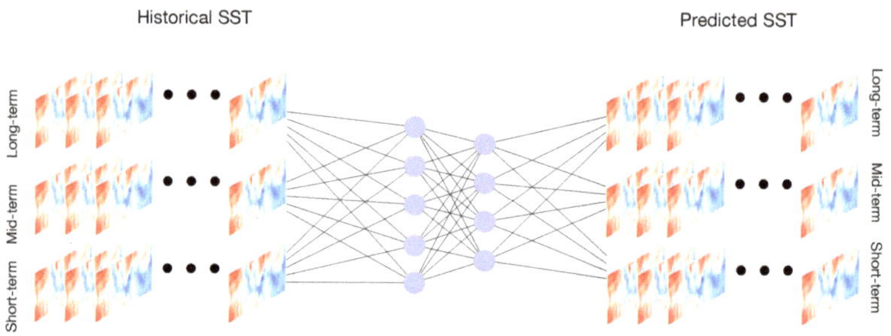

Figure 4. Multi-scale SST prediction that uses different scales, e.g., daily, weekly and monthly, of historical SST data to predict their future records simultaneously.

Considering that the prediction periods of most existing SST prediction methods are less than 12 time slots, we use the daily, weekly and monthly SST records of 36 time slots to predict the daily, weekly and monthly SST records in the next 12 time slots, i.e., $a = 36$ and $b = 12$. With such a setting, we can cover most of the prediction schemes in existing studies.

4. Methodology

Figure 5 presents the overall architecture of the MIMO model which consists of an input layer, an encoder and a decoder. The input layer contains multi-scale SST data and some external factors, where external factors, including short wave radiation (SWR) and long wave radiation (LWR), are regarded as important influence factors for SST and are thus introduced to enrich the information for SST prediction. Considering that the spatial resolutions of SST data and external factors are usually different due to the difference in the ways of data collection, we use Bicubic Convolutional Interpolation (BCI) to align the spatial resolutions of external factors to that of SST data. The technical details of BCI are discussed in Appendix A.

In the encoder, MIMO uses five independent Zoom In Spatio-Temporal (ZIST) blocks to learn the hierarchical spatio-temporal features from multi-scale SST data and external factors and uses Cross Scale Fusion (CSF) to fuse the learned features. In the decoder, MIMO designs three independent components for monthly, weekly and daily SST prediction, respectively. Each component comprises three Zoom Out Spatio-Temporal (ZOST) blocks and one Full Connection (FC) layer, where ZOST blocks decode the fused features from the encoder stage and the FC layer achieves the final prediction.

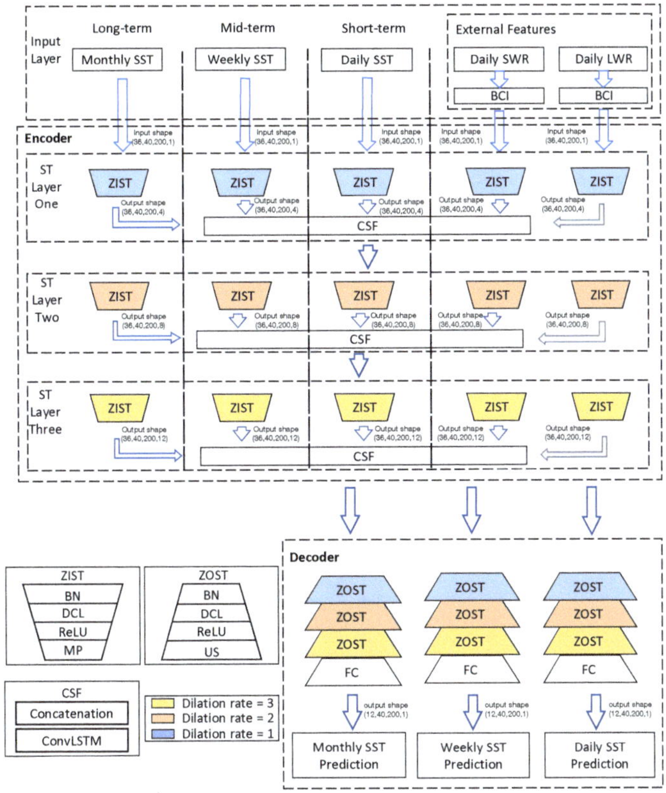

Figure 5. The architecture of MIMO model which consists of an input layer, an encoder and a decoder. The input layer contains multi-scale SST data and some external factors. The encoder contains three spatio-temporal layers and each layer contains five independent Zoom In Spatio-Temporal (ZIST) blocks, where each ZIST block consists of Batch Normalization (BN), Dilated ConvLSTM (DCL), Rectified Linear Unit (ReLU) and Max Pooling (MP). The decoder contains three components for monthly, weekly and daily SST prediction, respectively, and each component contains three Zoom Out Spatio-Temporal (ZOST) blocks and one Full Connection (FC) layer, where each ZOST block consists of BN, DCL, ReLU and Up Sampling (US).

4.1. Encoder

The encoder of the MIMO model has three spatio-temporal (ST) layers, and each ST layer consists of five Zoom In Spatio-Temporal (ZIST) blocks and one Cross Scale Fusion (CSF) sub-layer.

4.1.1. Zoom in Spatio-Temporal Block

Each ZIST block comprises Batch Normalization (BN), Dilated ConvLSTM (DCL), Rectified Linear Unit (ReLU) and Max Pooling (MP), where BN normalizes the data to avoid overfitting, DCL learns hierarchical features from the data using dilated convolution, ReLU filters the negative values to speed up the training process and MP reduces the number of model parameters and alleviates the position sensitivity.

Specifically, the DCL uses the dilated convolutional operation to replace the Hadamard product in LSTM to capture more spatial features. Figure 6 illustrates the 2D dilated convolutional operations that use intermittent connections between grids in traditional convolutional kernels. The dilation rate r is used to quantify the distance of intermittent connections between grids. Given a $k \times k$ convolutional kernel, the corresponding dilated kernel size is $k_d = k + (k-1)(r-1)$. For example, in Figure 6, $r = 1$ corresponds to a normal convolution kernel, $r = 2$ corresponds to a dilated convolution kernel of size $k_d = 5$, and $r = 3$ corresponds to a dilated convolution kernel of size $k_d = 7$.

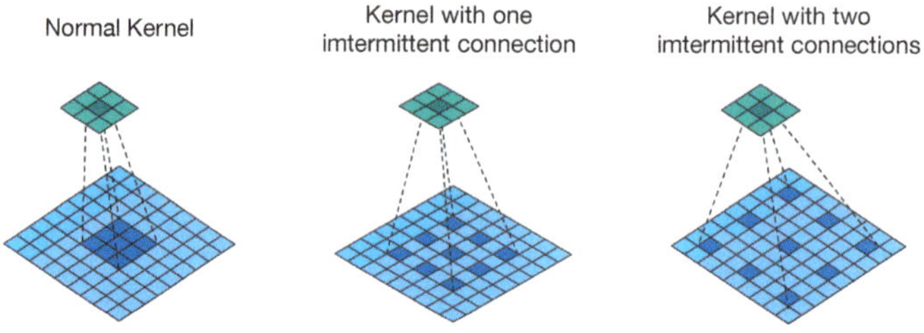

Figure 6. The normal convolutional kernel and dilation convolutional kernels.

DCL will face the "gridding" issue [34] when multiple ST layers are stacked. To address this issue, MIMO sets different dilation rates r, i.e., 1, 2 and 3, for three ST layers, respectively.

In addition, the features learned by DCL can be affected by the changes of position, which makes the prediction model sensitive to specific positions and shapes. Meanwhile, with multiple stacked DCLs, we may lose the features for the marginal areas of feature maps. Both issues will damage the generalization of the MIMO model. We introduce Max Pooling (MP) to address these two issues. MP can reduce the model size and indirectly increase the receptive field of the kernel size to better learn spatial features in the data. In addition, MP brings the advantage of feature invariance to the prediction model, which avoids the learning errors due to location changes.

4.1.2. Cross Scale Fusion

MIMO learns the spatio-temporal features of different scales of SST and external features separately, and then fuses them together. Concretely, in each ST layer, the features learned from five ZISTs are fused with the CSF sub-layer. Each CSF sub-layer is a ConvLSTM layer with kernel size of 1×1. As illustrated in Figure 7, CSF can reduce the dimensions of the outputs from five ZISTs to achieve the fusion.

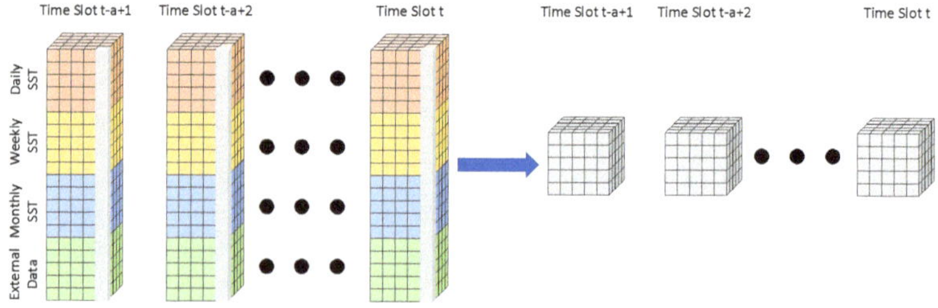

Figure 7. The structure of the Cross Scale Fusion (CSF) sub-layer, where the learned spatio-temporal features of different scales of SST and external features.

4.2. Decoder

The decoder of the MIMO model comprises three independent components, each consisting of three Zoom Out Spatio-Temporal (ZOST) blocks and one Full Connection (FC) layer. As illustrated in Figure 5, each ZOST comprises BN, DCL, ReLU and Up Sampling (US) and can decode the fused features from the encoder to the target scale of SST.

We use the nearest neighbor upsampling method to decode the features from the encoder. The nearest upsampling method directly enlarges the fused features in proportion to the input data. As illustrated in Figure 8, the 2 × 2 feature map is converted to a 4 × 4 feature map using upsampling with 2 × 2 filters.

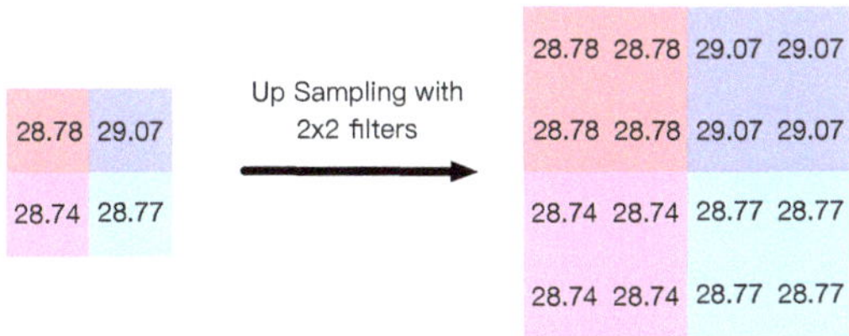

Figure 8. The upsampling that recovers the shape of the input data.

4.3. Loss Function

The objective of the MIMO model is to minimize the total error of the SST predictions of multiple scales, i.e.,

$$Loss = \sum_{i=1}^{3} \alpha_i L_i(\mathbf{X}_i^{obs}, \hat{\mathbf{X}}_i^{pre}) \tag{3}$$

where L_i is the loss function for the i-th scale of SST, α_i is the weight parameter for loss function L_i, $\mathbf{X}_i^{obs}$ is the ground truth and $\hat{\mathbf{X}}_i^{pre}$ is the predicted result of the MIMO model.

Since different scales of SST have different effects on the MIMO model, the loss functions corresponding to different scales of SST also have different effects on the total loss function. If we calculate the loss functions for SST predictions of different scales separately, the convergence speed of each loss function will be inconsistent. Hence, the MIMO model introduces weight parameters, i.e., α_i, to balance the convergence speed between the loss functions for different scales of SST prediction. According to the experimental analysis, we set the weights for the loss functions of daily, weekly and monthly scales of SST prediction to 0.6, 0.2 and 0.2, respectively.

The loss function L_i for the i-th scale of SST is the mean squared error (MSE), calculated as below.

$$L_i = \frac{1}{n} \sum (X^{obs} - \hat{X}^{pre})^2 \tag{4}$$

where n is the number of samples. During the training, the model is optimized using Nadam [35].

5. Model Evaluation

We used the SST data from the El Niño region to compare the accuracy of our model to SVM, CNN, LeNet, LSTM and ConvLSTM. The time range of the SST dataset is from 1 January 1982 to 31 December 2019.

5.1. Datasets

We use the Optimum Interpolation SST (OI-SST) data from the National Oceanic and Atmospheric Administration (NOAA) and select the region of *Niño* 3.4 as the target region for prediction. The NOAA 0.25° OI-SST is a long-term climate data record that incorporates the observations from different platforms, e.g., satellites, ships, buoys and Argo floats, into a regular global grid format. The *Niño* 3.4 region covers the area of [5°00′N∼5°00′S, 120°00′W∼170°00′W] and is divided into 40 × 200 grids of size 0.25° × 0.25°. In addition, the data for the external factors SWR and LWR are from NCEP/NCAR Reanalysis 1.

The whole dataset is divided into three subsets for training, validation and testing, respectively. The ratio of the three subsets is 72:1. Concretely, the training dataset contains 8956 samples, the validation dataset has 2239 samples and the testing dataset has 1244 samples. In the experiments, we aim to predict the daily, weekly and monthly SST records in the next 12 time slots. The number of iterations for model training is set to 1000 and the training process could end early if the loss function no longer changes for 10 consecutive rounds. The experiments run on a 64-core Intel Xeon processor with 256 GB RAM and 3 NVIDIA RTX 2080Ti GPUs. All SST prediction models are implemented based on TensorFlow 1.15.0.

5.2. Evaluation Metrics

We use four evaluation metrics to measure the performance of SST prediction models. The equations of the four evaluation metrics are listed as follows.

$$\text{MSE} = \frac{1}{n}\sum_{i}(x_i - \hat{x}_i)^2, \tag{5}$$

$$\text{RMSE} = \sqrt{\frac{1}{n}\sum_{i}(x_i - \hat{x}_i)^2}, \tag{6}$$

$$\text{MAE} = \frac{1}{n}\sum_{i}|x_i - \hat{x}_i|, \tag{7}$$

$$\text{MAPE} = \frac{100\%}{n}\sum_{i}|\frac{x_i - \hat{x}_i}{x_i}|. \tag{8}$$

where $\hat{x}$ and x are the predicted value and the observed value, respectively, and n is the total number of predicted samples.

5.3. Results

5.3.1. Weight Evaluation

To decide the values for the weights α_i $(i = 1, 2, 3)$ in the loss function, we evaluate the performance of MIMO models with different settings as below, where L_1, L_2 and L_3 correspond to the loss functions of daily, weekly and monthly predictions, respectively.

- **MIMO-122**: $L_1 = 1$, $L_2 = 2$ and $L_3 = 3$.
- **MIMO-911**: $L_1 = 0.9$, $L_2 = 0.1$ and $L_3 = 0.1$.
- **MIMO**: $L_1 = 0.6$, $L_2 = 0.2$ and $L_3 = 0.2$.
- **MIMO-433**: $L_1 = 0.4$, $L_2 = 0.3$ and $L_3 = 0.3$.
- **MIMO-244**: $L_1 = 0.2$, $L_2 = 0.4$ and $L_3 = 0.4$.

Table 1 presents the results of these models, where the best results are highlighted by boldface. As suggested by the table, MIMO-911 and MIMO achieve comparable performance and outperform all the other prediction models. Considering that MIMO has the best performance in predicting weekly and monthly SST and also achieves a high accuracy in predicting daily SST, we thus set $L_1 = 0.6$, $L_2 = 0.2$ and $L_3 = 0.2$ in the following experiments.

Table 1. The results of MIMO with different settings for the weights of loss function.

Model	Daily				Weekly				Monthly				Total MSE
	MSE	RMSE	MAE	MAPE	MSE	RMSE	MAE	MAPE	MSE	RMSE	MAE	MAPE	
MIMO-122	2.08	1.44	1.18	4.31%	1.88	1.37	1.11	4.07%	1.37	1.17	0.96	3.55%	1.99
MIMO-911	**0.37**	**0.61**	**0.47**	**1.76%**	0.61	0.78	0.60	2.22%	0.81	0.90	0.72	2.63%	**0.44**
MIMO	0.40	0.63	0.49	1.81%	**0.50**	**0.71**	**0.55**	**2.04%**	**0.58**	**0.76**	**0.60**	**2.24%**	0.45
MIMO-433	0.46	0.68	0.53	1.95%	0.58	0.76	0.60	2.20%	0.66	0.82	0.64	2.37%	0.56
MIMO-244	2.14	1.46	1.19	4.34%	2.01	1.42	1.14	4.19%	1.41	1.19	0.97	3.59%	1.80

5.3.2. Model Comparison

To verify the effectiveness of the MIMO model, we compare it with five representative forecasting models: CNN, ConvLSTM, LeNet, LSTM and SVM. The experimental results are shown in Table 2, where the best results are highlighted by boldface. The MIMO model has the best total MSE. LeNet and SVM also achieve good prediction performance and largely outperform CNN, ConvLSTM and LSTM. Concretely, the MIMO model has the best performance in predicting weekly and monthly SST. MIMO also achieves comparable performance with LeNet in predicting daily SST on MAE and MAPE. The MIMO model is designed for multi-scale SST prediction and needs to balance the prediction performance of different scales. Therefore, the MIMO model can achieve accurate predictions for all three scales of SST.

Table 2. The results of different SST prediction models.

Model	Daily				Weekly				Monthly				Total MSE
	MSE	RMSE	MAE	MAPE	MSE	RMSE	MAE	MAPE	MSE	RMSE	MAE	MAPE	
MIMO	**0.40**	**0.63**	0.49	1.81%	**0.50**	**0.71**	**0.55**	**2.04%**	**0.58**	**0.76**	**0.60**	**2.24%**	**0.45**
ConvLSTM	2.78	1.67	1.36	5.01%	2.86	1.69	1.40	5.11%	2.53	1.59	1.33	4.80%	2.75
LeNet	0.41	0.64	**0.47**	**1.77%**	0.99	0.99	0.76	2.86%	1.52	1.23	0.95	3.59%	0.58
CNN	1.20	1.10	0.84	3.09%	1.73	1.32	1.05	3.88%	2.82	1.68	1.35	5.09%	1.42
LSTM	2.04	1.41	1.15	4.24%	1.88	1.35	1.10	4.01%	1.39	1.16	0.95	3.53%	1.88
SVM	0.43	0.64	0.48	**1.77%**	0.82	0.90	0.71	2.62%	1.13	0.96	0.85	3.18%	0.61
LeNet with EXT	0.56	0.75	0.56	2.09%	1.62	1.27	1.00	3.69%	1.54	1.24	0.99	3.75%	0.97
LSTM with EXT	1.86	1.34	1.10	4.04%	1.80	1.32	1.08	3.98%	1.36	1.15	0.94	3.49%	1.75
SVM with EXT	1.33	1.14	0.91	3.35%	1.21	1.09	0.86	3.13%	1.04	1.01	0.82	3.00%	1.25

In addition, we add more experiments to evaluate the influence of external factors, i.e., short wave radiation (SWR) and long wave radiation (LWR). In accordance with Table 2, after including the external factors, the performance of LSTM has a small increase while the performance of LeNet and SVM decreases. This is because some models cannot handle external factors well and the contribution of these external factors is also limited. In addition, our MIMO model still outperforms these models with external factors, which indicates that the MIMO model can fully take advantage of the correlations among SSTs of different scales to improve the prediction.

5.3.3. Visualization

To better demonstrate the effectiveness of the MIMO model, we visualize the mean absolute errors of MIMO, ConvLSTM, LeNet and CNN for daily, weekly and monthly SST prediction in Figures 9–11, respectively.

As shown in Figure 9, MIMO and LeNet have similar MAE in daily SST prediction. The MAE of these two models is less than 0.5 for most grid regions. In contrast, the MAE of ConvLSTM is quite large. As for CNN, the predicted results for the 4th, 8th and 11th days are also bad.

Figure 10 illustrates the MAE for weekly SST prediction. MIMO has the best prediction performance among four models and its MAE errors are less than 0.5 for most grid regions. With the increase of time, the MAE of LeNet becomes large and reaches 1.25 for some grid regions. The MAEs of ConvLSTM and CNN are still very large but CNN performs better than ConvLSTM.

Figure 11 illustrates the MAE for monthly SST prediction. MIMO also achieves the best prediction performance. For LeNet, the MAE becomes larger than that of weekly SST prediction and still outperforms ConvLSTM and CNN. The MAE errors of ConvLSTM and CNN even reach 2.0 for some grid regions.

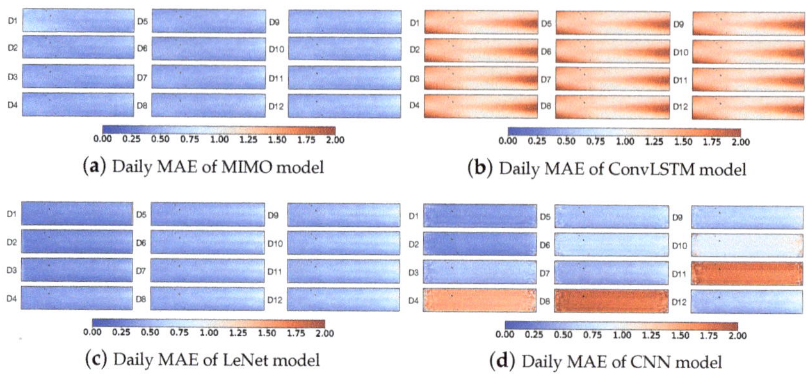

(**a**) Daily MAE of MIMO model (**b**) Daily MAE of ConvLSTM model

(**c**) Daily MAE of LeNet model (**d**) Daily MAE of CNN model

Figure 9. The results of daily SST prediction in *Niño* 3.4 region.

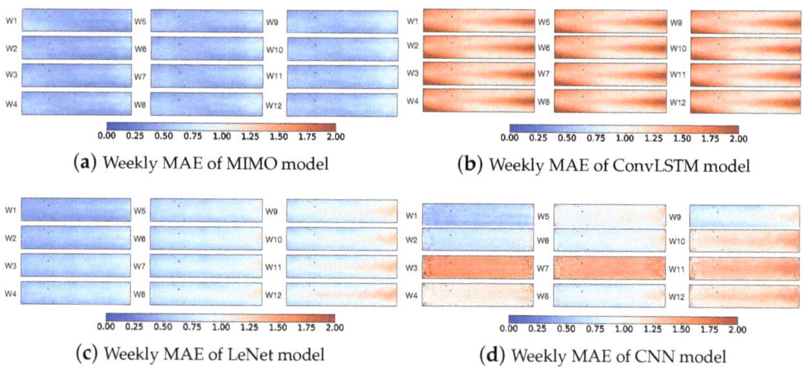

(**a**) Weekly MAE of MIMO model (**b**) Weekly MAE of ConvLSTM model

(**c**) Weekly MAE of LeNet model (**d**) Weekly MAE of CNN model

Figure 10. The results of weekly SST prediction in *Niño* 3.4 region.

In sum, in accordance with Figures 9–11, the MIMO model can achieve quite good prediction performance in all daily, weekly and monthly SST prediction. LeNet also achieves good prediction performance on daily SST prediction but has larger prediction errors than MIMO in weekly and monthly SST prediction. Both MIMO and LeNet largely outperform ConvLSTM and CNN.

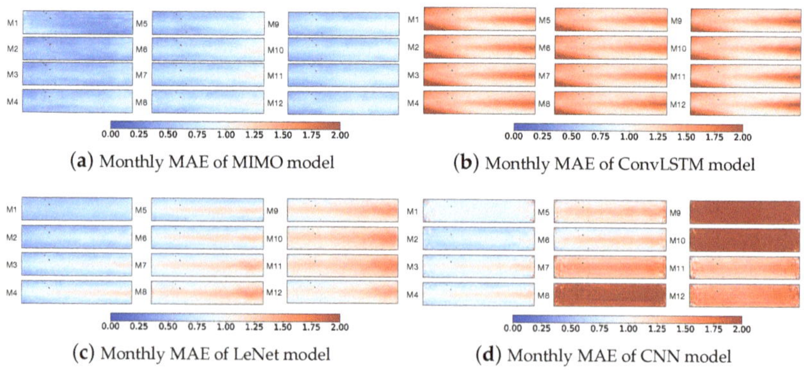

(**a**) Monthly MAE of MIMO model (**b**) Monthly MAE of ConvLSTM model

(**c**) Monthly MAE of LeNet model (**d**) Monthly MAE of CNN model

Figure 11. The results of monthly SST prediction in *Niño* 3.4 region.

5.4. Discussion

According to the experimental evaluation, the proposed MIMO model achieves good performance in predicting SSTs of different temporal scales and outperforms multiple classical forecasting methods. The underlying principles can be explained from two aspects.

On the one hand, the dynamics of SST has short-term temporal dependency, middle-term trend and long-term periodicity. For short-term temporal dependency, the SST in the next few days is usually similar to the that of past few days. For the middle-term trend, the SST may keep increasing within a few weeks. For long-term periodicity, the SST of the same month in two consecutive years is often similar. Therefore, integrating SST predictions of different temporal scales lets us have a better understanding of the dynamics of SST from an overall review and could further enhance the prediction performance.

On the other hand, the SSTs of different temporal scales are often correlated. For example, the increase of short-term SST causes the ocean to accumulate energy, which will result in the changes of long-term SST. Meanwhile, when the energy is accumulated to a certain extent, the ocean begins to release energy. Therefore, long-term SST changes will also affect short-term SST. The proposed MIMO model can fully take advantage of such correlations to improve SST prediction.

6. Conclusions

This work proposes the multi-scale SST prediction problem and develops a new model named Multi-In and Multi-Out (MIMO) to address this problem. MIMO can fuse multi-scale SST data and external data to learn comprehensive features and decode the learned features to adaptively predict daily, weekly and monthly SST. Experimental evaluation shows that MIMO can achieve much better prediction performance than existing SST prediction methods in predicting weekly and monthly SST, and comparable prediction performance with the state-of-the-art prediction method in predicting daily SST. The superiority of MIMO lies in that it can do multi-scale SST prediction in a unified model, thus improving the prediction accuracy by capturing the correlations among different scales of SST.

Author Contributions: Conceptualization: S.H., W.L.; Methodology: S.H., W.L., T.L.; validattion: S.H.; formal analyses: S.H., W.L.; data curation: S.H., T.L.; writing—original draft preparation: S.H.; writing—review and editing: W.L., T.L., S.Z.; visualization: S.H.; supervision: J.G., R.Q., Z.W. All authors have read and agreed to the published version of the manuscript.

Funding: This research was funded by National Natural Science Foundation of China under grants No. U1936205, 62172300, Shanghai Pujiang Program under grants No. 20PJ1414300, Open Reseach Program of Shanghai Key Lab of Intelligent Information Processing under grants No. IIPL201909, and China National Scientific Seafloor Observatory.

Data Availability Statement: The OISST data used in this work are available from https://www.ncei.noaa.gov/products/optimum-interpolation-sst, accessed on 27 February 2022. The NCEP/NCAR Reanalysis 1 data in this work are available at https://psl.noaa.gov/data/gridded/data.ncep.reanalysis.html, accessed on 27 February 2022.

Conflicts of Interest: The authors declare no conflict of interest.

Appendix A. Data Alignment

Due to the different data acquisition equipment and the different resolutions of various data, it is not easy to use the data. In this work, the resolution of SST data in the *Niño* 3.4 region is $0.25^\circ \times 0.25^\circ$ and the resolution of the external factors SWR and LWR are $1^\circ \times 1^\circ$, which means that the *Niño* 3.4 area can be represented by a 40×200 matrix based on SST resolution, and a 10×50 matrix based on external factor resolution.

BCI is introduced in MIMO to align external factors from low-resolution data to high-resolution data to overcome the issue, and it is an efficient 2D data interpolation method that has been validated in several fields. The target data resolution is increased using cubic polynomial interpolation along two axes on a two-dimensional plane. Compared with other interpolation methods, the interpolated results obtained by BCI are smoother.

Take the external data SWR as an example that needs to increase the resolution to the same size of the SST data, i.e., from 1° × 1° to 0.25° × 0.25°. Assume the input data is SW_t at time slot t, and the output data through BCI is SW'_t at time slot t. In this work, SW'_t is 4 times larger than SW_t, i.e., scale size $s = 4$. Take a grid $SW_t'^{(i,j)}$ in SW'_t as an example, where i, j represent the row and column where the grid is located in SW'_t. To calculate the output data $SW_t'^{(i,j)}$, the $SW_t'^{(i,j)}$ needs to be proportionally scaled into the input data matrix $SW_t'^{(\frac{i}{s},\frac{j}{s})}$, i.e., $(\frac{i}{s},\frac{j}{s})$ is the position of the scaled output data in the SW_t data.

As illustrated in Figure A1, a grid in SW'_t is scaled to the position $p = (\frac{i}{s},\frac{j}{s})$ in SW_t. Taking p as the original grid and assuming that the distance between adjacent grids is 1, the BCI obtains the value of 16 adjacent grids with a maximum distance of 2 from grid p to obtain the final value of p. This can be represented by Equation (A1).

$$g(x,y) = W(d_x)AW(d_y)^T \tag{A1}$$

where A is the matrix of the nearest 16 grid value represented as follows:

$$A = \begin{pmatrix} a_{-1,-1} & a_{-1,0} & a_{-1,1} & a_{-1,2} \\ a_{0,-1} & a_{0,0} & a_{0,1} & a_{0,2} \\ a_{1,-1} & a_{1,0} & a_{1,1} & a_{1,2} \\ a_{2,-1} & a_{2,0} & a_{2,1} & a_{2,2} \end{pmatrix}$$

d_x and d_y represent the distance between the p grid and nearest 4 grids on the x and y axes, respectively. the distance $d_{x_i} = [-1 - \frac{i}{s}, 0 - \frac{i}{s}, 1 - \frac{i}{s}, 2 - \frac{i}{s}]$, and $d_{y_j} = [-1 - \frac{j}{s}, 0 - \frac{j}{s}, 1 - \frac{j}{s}, 2 - \frac{j}{s}]$. $W(\cdot)$ is the Bicubic Convolutional Kernel (BCK), i.e.,

$$W(d) = \begin{cases} (a+2)|d|^3 - (a+3)|d|^2 + 1 & |d| \leq 1 \\ a(|d|^3 - 5a|d|^2 + 8a|d| - 4d & 1 < |d| \leq 2 \\ 0 & |d| > 2 \end{cases} \tag{A2}$$

where a is a hyperparameter. According to the experience of other related studies [36], the value of a is set to -0.5.

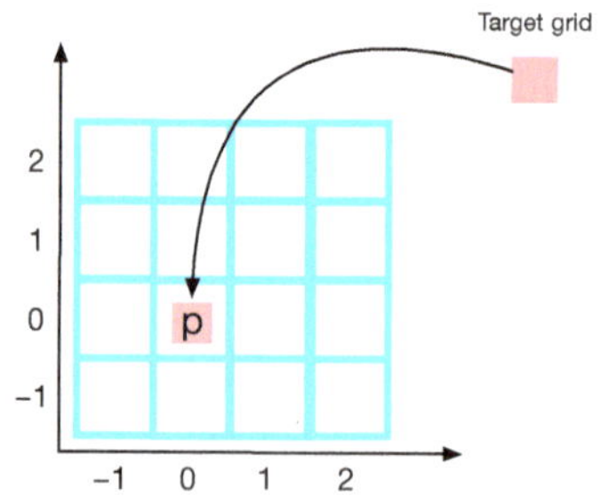

Figure A1. The bicubic mapping in BCI.

References

1. Pisano, A.; Marullo, S.; Artale, V.; Falcini, F.; Yang, C.; Leonelli, F.E.; Santoleri, R.; Buongiorno Nardelli, B. New Evidence of Mediterranean Climate Change and Variability from Sea Surface Temperature Observations. *Remote Sens.* **2020**, *12*, 132. [CrossRef]
2. Patricola, C.M.; Wehner, M.F. Anthropogenic influences on major tropical cyclone events. *Nature* **2018**, *563*, 339–346. [CrossRef] [PubMed]
3. Aemisegger, F.; Sjolte, J. A Climatology of Strong Large-Scale Ocean Evaporation Events. Part II: Relevance for the Deuterium Excess Signature of the Evaporation Flux. *J. Clim.* **2018**, *31*, 7313–7336. [CrossRef]

4. Song, Y.; Zhenning, L.; Jin-Yi, Y.; Xiaoming, H.; Wenjie, D.; Shan, H. El Niño–Southern Oscillation and its impact in the changing climate. *Natl. Sci. Rev.* **2018**, *5*, 840-857. [CrossRef]
5. Stockdale, T.N.; Balmaseda, M.A.; Vidard, A. Tropical Atlantic SST Prediction with Coupled Ocean–Atmosphere GCMs. *J. Clim.* **2006**, *19*, 6047–6061. [CrossRef]
6. Lins, I.D.; Araujo, M.; Moura, M.d.C.; Silva, M.A.; Droguett, E.L. Prediction of sea surface temperature in the tropical Atlantic by support vector machines. *Comput. Stat. Data Anal.* **2013**, *61*, 187–198. [CrossRef]
7. Wei, L.; Guan, L.; Qu, L. Prediction of Sea Surface Temperature in the South China Sea by Artificial Neural Networks. *IEEE Geosci. Remote Sens. Lett.* **2020**, *17*, 558–562. [CrossRef]
8. Garcia-Gorriz, E.; Garcia-Sanchez, J. Prediction of Sea Surface Temperatures in the Western Mediterranean Sea by Neural Networks Using Satellite Observations. *Geophys. Res. Lett.* **2007**, *34*, L11603. Available online: https://onlinelibrary.wiley.com/doi/pdf/10.1029/2007GL029888 (accessed on 17 February 2022). [CrossRef]
9. Patil, K.; Deo, M.C. Basin-Scale Prediction of Sea Surface Temperature with Artificial Neural Networks. *J. Atmos. Ocean. Technol.* **2018**, *35*, 1441–1455. [CrossRef]
10. Zhang, Q.; Wang, H.; Dong, J.; Zhong, G.; Sun, X. Prediction of Sea Surface Temperature Using Long Short-Term Memory. *IEEE Geosci. Remote Sens. Lett.* **2017**, *14*, 1745–1749. [CrossRef]
11. Zheng, G.; Li, X.; Zhang, R.H.; Liu, B. Purely satellite data–driven deep learning forecast of complicated tropical instability waves. *Sci. Adv.* **2020**, *6*, eaba1482. [CrossRef] [PubMed]
12. Xiao, C.; Chen, N.; Hu, C.; Wang, K.; Xu, Z.; Cai, Y.; Xu, L.; Chen, Z.; Gong, J. A spatiotemporal deep learning model for sea surface temperature field prediction using time-series satellite data. *Environ. Model. Softw.* **2019**, *120*, 104502. [CrossRef]
13. Carton, J.A.; Cao, X.; Giese, B.S.; Silva, A.M.D. Decadal and Interannual SST Variability in the Tropical Atlantic Ocean. *J. Phys. Oceanogr.* **1996**, *26*, 1165–1175. [CrossRef]
14. Arakawa, O.; Kitoh, A. Comparison of Local Precipitation–SST Relationship between the Observation and a Reanalysis Dataset. *Geophys. Res. Lett.* **2004**, *31*, L12206. Available online: https://onlinelibrary.wiley.com/doi/pdf/10.1029/2004GL020283 (accessed on 21 February 2022). [CrossRef]
15. Zafarparandeh, I.; Lazoglu, I. 4-Application of the finite element method in spinal implant design and manufacture. In *The Design and Manufacture of Medical Devices*; Davim, J.P., Ed.; Woodhead Publishing Reviews: Mechanical Engineering Series; Woodhead Publishing: Sawston, UK, 2012; pp. 153–183. [CrossRef]
16. Grotch, S.L.; MacCracken, M.C. The Use of General Circulation Models to Predict Regional Climatic Change. *J. Clim.* **1991**, *4*, 286–303. [CrossRef]
17. Krishnamurti, T.N.; Chakraborty, A.; Krishnamurti, R.; Dewar, W.K.; Clayson, C.A. Seasonal Prediction of Sea Surface Temperature Anomalies Using a Suite of 13 Coupled Atmosphere–Ocean Models. *J. Clim.* **2006**, *19*, 6069–6088. [CrossRef]
18. Aparna, S.G.; D'Souza, S.; Arjun, N.B. Prediction of daily sea surface temperature using artificial neural networks. *Int. J. Remote Sens.* **2018**, *39*, 4214–4231. [CrossRef]
19. Khemchandani, R.; Goyal, K.; Chandra, S. TWSVR: Regression via Twin Support Vector Machine. *Neural Netw.* **2016**, *74*, 14–21. [CrossRef]
20. Kumar, C.; Podestá, G.; Kilpatrick, K.; Minnett, P. A machine learning approach to estimating the error in satellite sea surface temperature retrievals. *Remote Sens. Environ.* **2021**, *255*, 112227. [CrossRef]
21. Lins, I.; Moura, M.; Silva, M.; Droguett, E.; Veleda, D.; Araujo, M.; Jacinto, C. Sea surface temperature prediction via support vector machines combined with particle swarm optimization. In Proceedings of the 10th International Probabilistic Safety Assessment & Management Conference, Seattle,WA, USA, 7–11 June 2010.
22. Dommenget, D.; Jansen, M. Predictions of Indian Ocean SST Indices with a Simple Statistical Model: A Null Hypothesis. *J. Clim.* **2009**, *22*, 4930–4938. [CrossRef]
23. Wu, A.; Hsieh, W.W.; Tang, B. Neural network forecasts of the tropical Pacific sea surface temperatures. *Neural Netw.* **2006**, *19*, 145–154. [CrossRef] [PubMed]
24. Chaudhari, S.; Balasubramanian, R.; Gangopadhyay, A. Upwelling Detection in AVHRR Sea Surface Temperature (SST) Images using Neural-Network Framework. In Proceedings of the IGARSS 2008—2008 IEEE International Geoscience and Remote Sensing Symposium, Boston, MA, USA, 8–11 July 2008; Volume 4, pp. IV–926–IV–929; ISSN 2153-7003. [CrossRef]
25. Wu, Z.; Jiang, C.; Conde, M.; Deng, B.; Chen, J. Hybrid improved empirical mode decomposition and BP neural network model for the prediction of sea surface temperature. *Ocean. Sci.* **2019**, *15*, 349–360. [CrossRef]
26. Patil, K.; Deo, M.C. Prediction of daily sea surface temperature using efficient neural networks. *Ocean. Dyn.* **2017**, *67*, 357–368. [CrossRef]
27. LeCun, Y.; Bengio, Y.; Hinton, G. Deep learning. *Nature* **2015**, *521*, 436–444. [CrossRef]
28. Ham, Y.G.; Kim, J.H.; Luo, J.J. Deep learning for multi-year ENSO forecasts. *Nature* **2019**, *573*, 568–572. [CrossRef]
29. Xiao, C.; Chen, N.; Hu, C.; Wang, K.; Gong, J.; Chen, Z. Short and mid-term sea surface temperature prediction using time-series satellite data and LSTM-AdaBoost combination approach. *Remote Sens. Environ.* **2019**, *233*, 111358. [CrossRef]
30. Yao, G.; Liu, Z.; Guo, X.; Wei, C.; Li, X.; Chen, Z. Prediction of Weather Radar Images via a Deep LSTM for Nowcasting. In Proceedings of the 2020 International Joint Conference on Neural Networks (IJCNN), Glasgow, UK, 19–24 July 2020; pp. 1–8; ISSN 2161-4407. [CrossRef]

31. Weyn, J.A.; Durran, D.R.; Caruana, R. Can Machines Learn to Predict Weather? Using Deep Learning to Predict Gridded 500-hPa Geopotential Height From Historical Weather Data. *J. Adv. Model. Earth Syst.* **2019**, *11*, 2680–2693. Available online: https://agupubs.onlinelibrary.wiley.com/doi/pdf/10.1029/2019MS001705 (accessed on 3 March 2022). [CrossRef]
32. Zhang, K.; Geng, X.; Yan, X.H. Prediction of 3-D Ocean Temperature by Multilayer Convolutional LSTM. *IEEE Geosci. Remote Sens. Lett.* **2020**, *17*, 1303–1307. [CrossRef]
33. Zhang, X.; Li, Y.; Frery, A.C.; Ren, P. Sea Surface Temperature Prediction With Memory Graph Convolutional Networks. *IEEE Geosci. Remote Sens. Lett.* **2021**, *19*, 1–5. [CrossRef]
34. Yu, F.; Koltun, V. Multi-Scale Context Aggregation by Dilated Convolutions. In Proceedings of the 4th International Conference on Learning Representations, ICLR 2016, San Juan, Puerto Rico, 2–4 May 2016.
35. Timothy, D. Incorporating nesterov momentum into adam. *Nat. Hazards* **2016**, *3*, 437–453.
36. Keys, R. Cubic convolution interpolation for digital image processing. *IEEE Trans. Acoust. Speech Signal Process.* **1981**, *29*, 1153–1160. [CrossRef]

remote sensing

Technical Note

Rescaling-Assisted Super-Resolution for Medium-Low Resolution Remote Sensing Ship Detection

Huanxin Zou [1,*], Shitian He [1], Xu Cao [1], Li Sun [1], Juan Wei [1], Shuo Liu [1] and Jian Liu [2]

1 College of Electronic Science and Technology, National University of Defense Technology, Changsha 410073, China; heshitian19@nudt.edu.cn (S.H.); cx2020@nudt.edu.cn (X.C.); sunli20@nudt.edu.cn (L.S.); weijuan20@nudt.edu.cn (J.W.); liushuo21@nudt.edu.cn (S.L.)

2 Unit 32032 of PLA, Beijing 100094, China; 201208191109@mail.sdu.edu.cn

* Correspondence: zouhuanxin@nudt.edu.cn; Tel.: +86-731-8700-3289

Abstract: Medium-low resolution (M-LR) remote sensing ship detection is a challenging problem due to the small target sizes and insufficient appearance information. Although image super resolution (SR) has become a popular solution in recent years, the ability of image SR is limited since much information is lost in input images. Inspired by the powerful information embedding ability of the encoder in image rescaling, in this paper, we introduce image rescaling to guide the training of image SR. Specifically, we add an adaption module before the SR network, and use the pre-trained rescaling network to guide the optimization of the adaption module. In this way, more information is embedded in the adapted M-LR images, and the subsequent SR module can utilize more information to achieve better performance. Extensive experimental results demonstrate the effectiveness of our method on image SR. More importantly, our method can be used as a pre-processing approach to improve the detection performance.

Keywords: image super-resolution; image rescaling; ship detection; medium-low resolution remote sensing images

Citation: Zou, H.; He, S.; Cao, X.; Sun, L.; Wei, J.; Liu, S.; Liu, J. Rescaling-Assisted Super-Resolution for Medium-Low Resolution Remote Sensing Ship Detection. *Remote Sens.* **2022**, *14*, 2566. https://doi.org/10.3390/rs14112566

Academic Editors: Gemine Vivone and Józef Lisowski

Received: 27 March 2022
Accepted: 23 May 2022
Published: 27 May 2022

1. Introduction

Medium-low resolution (M-LR) remote sensing images (with the ground sample distance (GSD) larger than 10 m/pixel) play an important role in marine monitoring [1], rescuing [2] and management [3]. Meanwhile, ship detection is a significant component of remote sensing image interpretation. However, the lower resolution means the ship targets have smaller sizes and less appearance information, and this introduces challenges for ship detection task in M-LR images. With the ability of recovering image detail information, image super-resolution (SR) has become a popular solution for small object detection in recent years [4–8]. However, the ability of image SR is limited since it is an ill-posed problem and much information is lost in input images. In this paper, we propose a rescaling assisted image SR method (RASR) to better recover the lost information, and thus improve the performance of image SR and ship detection.

Image rescaling tries to learn an invertible down-sampling encoder, and the original images can be recovered greatly by using the corresponding up-sampling decoder. Different from image SR, image rescaling considers image down-sampling and up-sampling as a dual problem, and jointly optimizes them to minimize the lost information during the down-sampling process. Compared to only a decoder (i.e., an SR model), image rescaling can preserve more information during the down-sampling process, and thus achieves a better restoration performance. Inspired by the powerful information embedding ability of the rescaling encoder, we introduce image rescaling to assist the image SR process. Specifically, we add an adaption module before the SR network, and use the pre-trained rescaling network to guide the optimization of the adaption module. In this way, more information is embedded in the adapted M-LR images and transferred to the SR module.

Consequently, our rescaling assisted image SR method (named RASR) can achieve better SR performance. Furtherly, we take RASR as a pre-processing approach, and develop RASR-Det to improve the detection performance on M-LR images.

In summary, the contributions of our work are as follows:

- Inspired by the powerful information embedding ability of the encoder in image rescaling, we propose a rescaling assisted image SR method to better restore the lost information in M-LR images.
- We conduct extensive ablation studies to investigate the effectiveness of RASR. Moreover, we compare our RASR with several SR methods on three public datasets. Comparative results demonstrate the competitive performance of our method.
- Taking RASR as a pre-processing approach, we develop RASR-Det to improve the ship detection performance on M-LR images. Experimental results demonstrate the effectiveness of our RASR-Det in handling the M-LR remote sensing ship detection problem.

2. Related Works

2.1. Image Super Resolution

Image super-resolution (SR) aims at reconstructing a high-resolution (HR) image from one or multiple low-resolution observations. Recently, deep learning has been successfully applied to image SR and has achieved continuously improving performance. Readers can refer to recent surveys [9,10] for a comprehensive overview of image SR. Here, we review several major works in the literature. Dong et al. [11] proposed the first CNN-based single image SR method to reconstruct HR images by using a three-layer CNN. Kim et al. [12] proposed a deeper network named VDSR to improve the reconstruction accuracy. Zhang et al. [13] combined residual connection [14] with dense connection [15], and proposed residual dense network (i.e., RDN) to fully exploit hierarchical feature representations for image SR. Li et al. [16] proposed a multiscale residual network to fully exploit the hierarchical feature representation for image super-resolution. Wang et al. [17] explored the sparsity prior in image SR and used sparse convolutions to achieve accurate and efficient image SR. Subsequently, Wang et al. [18] proposed a degradation-aware network and achieved image SR with arbitrary blur kernels and noise levels. Although these SR methods have shown their effectiveness, the ability of image SR is limited since much information is lost in input degradation images.

2.2. Image Rescaling

Image Rescaling is usually used to fit the displays of various resolutions and save the storage. Kim et al. [19] proposed an auto-encoder-based framework that enables joint learning of the down-scaling network and the up-scaling network to maximize the restoration performance. Sun et al. [20] proposed a learned image downscaling method based on content adaptive resampler with consideration of the up-scaling process. Xiao et al. [21] proposed an invertible rescaling net that can produce visually-pleasing low-resolution images and meanwhile capture the distribution of the lost information using a latent variable following a specified distribution in the down-scaling process. In this paper, we introduce an image rescaling method to assist image SR in recovering more detail information.

2.3. SR-Based Detectors

In recent years, image SR has become a popular solution in remote sensing small object detection tasks. Dong et al. [4] proposed a second-order multi-scale SR network to make full use of the multi-scale information, and demonstrated its effectiveness on object classification. Rabbi et al. [5] proposed an edge-enhanced SR generative adversarial network (EESRGAN), and combined it with Faster-RCNN [22] and SSD [23] in an end-to-end manner to improve the detection accuracy. Courtrai et al. [6] tailored a GAN-based SR network with a detection network to improve the performance of both image SR and object detection. Wang et al. [7] designed a feature-driven loss function to make the SR network more suitable for the detection task. Noh et al. [8] enhanced the features of small regions of interest using the feature-level SR technique, and thus improved the detection performance. Zou et al. [24] fed both the super-solved images and the intermediate features to detection network to improve the detection performance. He et al. [25] introduced feature distillation to fully exploit the information in ground-truth HR images to handle M-LR ship detection. Similarly to some SR-based detectors [4,5], we use our proposed RASR as a pre-processing approach, and develop RASR-Det to further improve the performance of the M-LR ship detection task.

3. Methodology

In this section, we introduce our proposed RASR method in detail. As shown in Figure 1, our method consists of two parts, including image rescaling and image SR. Image rescaling uses an encoder and a decoder to perform down-sampling and up-sampling operations, respectively. We name the down-sampled images as pseudo LR images, since they carry more information that is beneficial to image SR than the original M-LR images. Image SR consists of an image adaption module and an SR module, and the output of the two modules are adapted images and SR images, respectively. In our method, image rescaling is pre-trained first to generate informative pseudo LR images, and thus, to guide the training of image SR. The pre-training of image rescaling is under the supervision of a rescaling loss between rescaled images and HR images. Then, image rescaling guides the adaption module through a specific adaption loss between the pseudo LR images and the adapted images, and thus the adaption module can learn to transform the original LR images to a more informative version (i.e., adapted images). In this way, the subsequent SR module can utilize more information and achieves better performance. In the following subsections, we introduce the structure of each module in detail.

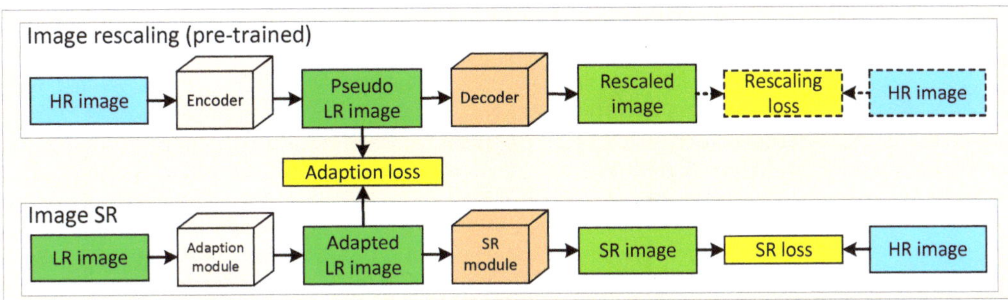

Figure 1. An overview of our proposed RSAR method.

3.1. Image Rescaling

Image rescaling takes an HR image $I_{HR} \in \mathbb{R}^{\alpha H \times \alpha W \times 3}$ as its input and produces an intermediate pseudo LR image $I_{LR}^{pseudo} \in \mathbb{R}^{H \times W \times 3}$ and a rescaled image $I_{rescaled} \in \mathbb{R}^{\alpha H \times \alpha W \times 3}$, where H and W represent the height and width of the input M-LR image (i.e., $I_{LR} \in \mathbb{R}^{H \times W \times 3}$), and α denotes the upscaling factor. We followed [26] to develop our encoder module. Specifically, as shown in Figure 2b, the encoder module consists of four convolution layers. All the convolution layers in our encoder have a stride of 1 except for the second one (which has a stride α). The kernel sizes of the first two and the last two convolutions are 5×5 and 3×3, respectively. The decoder module has the same architecture as the SR module (as shown in Figure 2a), which will be introduced in the following subsections.

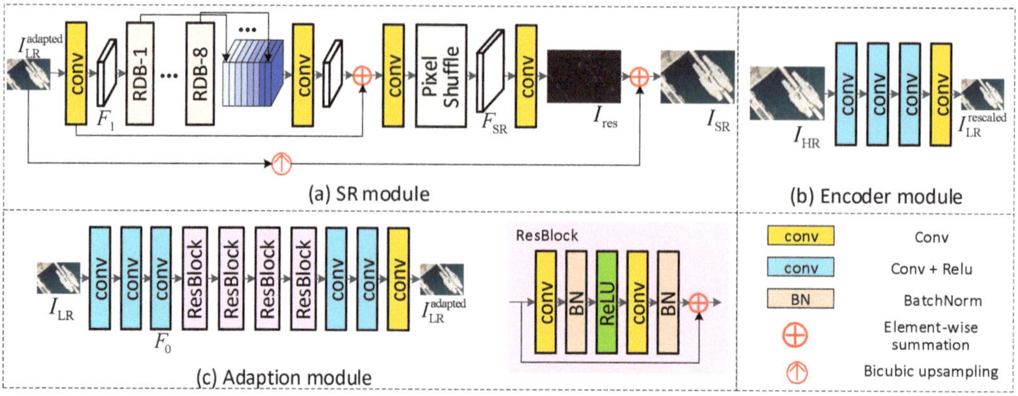

Figure 2. The structure of each module in our proposed method.

3.2. *Rescaling Assisted Image SR*

Since image rescaling is an invertible dual problem, the intermediate pseudo LR images have more detail information and can be used to guide the training of image SR. Specifically, we add an adaption module before the SR module to transfer an original M-LR image $I_{LR} \in \mathbb{R}^{H\times W\times 3}$ to an informative one, and then the SR module takes the adapted M-LR image $I_{LR}^{adapted} \in \mathbb{R}^{H\times W\times 3}$ instead as input to perform image SR. The details of the adaption and SR modules are as follows.

Adaption Module. As shown in Figure 2c, the adaption module consists of six convolutions and four residual blocks. Specifically, the input image I_{LR} is first fed to three 3×3 convolutions to generate initial feature $F_0 \in \mathbb{R}^{H\times W\times 64}$. Then, F_0 is fed to residual blocks for deep feature extraction, and the extracted features are then fed to three 3×3 convolutions to produce the adapted LR image (i.e., $I_{LR}^{adapted}$). The structure of the residual block is shown in Figure 2c. Note that we use PReLU [27] as the non-linear layer.

SR Module. As shown in Figure 2a, the SR module takes the adapted image $I_{LR}^{adapted}$ as input, and generates the super-resolved image I_{SR}. Specifically, the input image is first fed to a 3×3 convolution and eight residual dense blocks (RDBs) to generate initial feature F_1 and eight deep features. Then, the extracted deep features are concatenated and fed to a 3×3 convolution for global feature fusion. After that, the fused feature is added with the initial feature F_1 and fed to a sub-pixel layer [28] to generate residual feature F_{SR}. Finally, the residual feature F_{SR} is fed to a 3×3 convolution to produce residual image I_{Res}, and the final output image I_{SR} is the pixel sum of I_{res} and the up-sampling version of $I_{LR}^{adapted}$.

3.3. *Loss Function*

The training process of our RASR consists of two stages: the pre-training of image rescaling and the training of RASR.

In the first stage, we pre-train the image rescaling part with a rescaling loss, which can be defined as Equation (1), where $L_{rescaling}$ represents the L1 distance (i.e., $\|\cdot\|_1$) between output rescaled image $I_{rescaled}$ and HR image I_{HR}.

$$L_{rescaling} = \| I_{rescaled} - I_{HR} \|_1 . \quad (1)$$

In the second stage, we use the generated pseudo LR images from image rescaling to guide the training of RASR method. The loss function of our RASR is defined as:

$$L_{RASR} = L_{SR} + \lambda L_{adaption}, \quad (2)$$

where L_{SR} and $L_{adaption}$ represent the SR loss and adaption loss, respectively. λ is a hyper-parameter to balance the SR loss and adaption loss, and is empirically set to 0.1 in our implementation. In our method, L_{SR} is the L1 distance between the ground-truth image L_{HR} and the super-resolved image I_{SR}; $L_{adaption}$ is the L1 distance between the pseudo LR image I_{LR}^{pseudo} and the adapted LR image $I_{LR}^{adapted}$.

4. Experiments

In this section, we first introduce the datasets and implementation details, then conduct ablation studies to validate the effectiveness of RASR for bothimage SR and ship detection. Moreover, we compare RASR with several SR methods to investigate the superiority of RASR. Finally, we use RASR as a pre-processing approach and choose Faster-RCNN as the detector to develop RASR-Faster-RCNN, and compare it with several state-of-the-art detection methods.

4.1. Datasets and Implementation Details

4.1.1. Datasets

We used the HRSC2016 [29], DOTA [30] and NWPU VHR-10 [31] datasets in our experiments.

- **HRSC2016**: HRSC2016 is a public ship dataset in remote sensing images, and contains 436 images for training, 181 images for validation and 438 images for test. In our experiment, we use both the train and validation dataset for training and use the test dataset for test. The GSD of the HRSC2016 dataset is between 0.4 and 2 m/pixel, and the image sizes are between 300 × 300 pixels and 1500 × 900 pixels. As most image sizes are about 1000 × 600, we resized original images to 800 × 512 as the ground-truth HR images.
- **DOTA**: DOTA is a large-scale dataset used for multi-class object detection in remote sensing images. The sizes of images are between 800 × 800 and 4000 × 4000 pixels. Objects in the dataset have various proportions, directions and shapes. In our experiment, the large-scale images are cropped to patches of 512 × 512 pixels for training and validation. For the ship detection task, we discarded patches without ship targets and retained 4163 patches for training and 1411 patches for test.
- **NWPU VHR-10**: NWPU VHR-10 dataset is a multi-class object detection dataset with GSD smaller than 2 m/pixel. We performed the same operations as in DOTA dataset to generate training and validation samples. The image sizes are from 533 × 358 to 1728 × 1028 pixels. In our experiment, we cropped the original images into patches of 512 × 512 pixels and discarded patches without ship targets. Our customized NWPU VHR-10 dataset contains 249 images for training and 52 images for test.

In our experiments, the aforementioned images (patches) were used as the ground-truth HR images, and their bicubicly down-sampled versions were used as the M-LR images. Note that since the aligned HR images of real LR images are unavailable, here we use bicubical down-sampling technology to produce aligned HR and LR image pairs for the training of image SR. Although bicubical down-sampling is a rather simpler degradation model than the real scenery, many researches have proved that a model can achieve promising SR performance on real LR images when it can achieve promising SR performance on bicubicly down-sampled images. Moreover, our method can also be used to super-resolve real LR images if the degradation model can be well formulated.

4.1.2. Implementation Details

We implement our RASR and RASR-Det using PyTorch 1.11.0 with an Nvidia RTX 2080Ti GPU.

The training of RASR. We trained the RASR following a two stage pipeline. In the first stage, we pre-trained the image rescaling network. During training, we randomly cropped the ground-truth HR images into patches of size 256 × 256 pixels, and performed random flipping for data augmentation. The image rescaling was trained with an L1 loss between the rescaled images and original HR images. The batch size was set to 16 and the learning rate was set to 1×10^{-4}. The rescaling model was optimized using the Adam method [32], and the training was stopped after 1.2×10^5 iterations. In the second stage, we loaded the pre-trained rescaling model, and trained our RASR network with the guide of SR loss and adaption loss. The training settings are the same as the first stage, but the total training iterations was 2×10^5.

The training of Detector. All the detectors in our method are implemented by using the MMDetection [33] framework. When training the detector, the learning rate was initially set to 1×10^{-4} and decreased by a factor of 0.1 when the epoch was 14 and 20. The total number of training epochs was 24, and the batch size was set to 4. We chose SGD [34] as the optimizer. We performed the following data augmentations for all detectors: random flipping, random rotation, random color transformation, random brightness and contrast transformation. Other hyper-parameter settings, such as anchor scales and anchor ratios, are set as the default settings in MMDetection. Furthermore, in our experiments, we can use the training dataset for training, and use the test dataset for performance evaluation.

4.1.3. Evaluation Metrics

Image quality evaluation. We use the Peak Signal-to-Noise Ratio (PSNR) and Structure Similarity Index Measure (SSIM) to measure the quality of super-resolved images. Given a ground-truth image I_{HR} (signed as $\mathbf{x}$) and its corresponding super-resolved image I_{SR} (signed as $\mathbf{y}$), the PSNR value is:

$$\mathrm{PSNR}(\mathbf{x},\mathbf{y}) = 10log_{10}\frac{255^2}{\mathrm{MSE}(\mathbf{x},\mathbf{y})}, \tag{3}$$

where $MSE(\mathbf{x},\mathbf{y})$ indicates the mean squared error between I_{HR} and I_{SR}. The SSIM value is:

$$\mathrm{SSIM}(\mathbf{x},\mathbf{y}) = \frac{(2\mu_{\mathbf{x}}\mu_{\mathbf{y}} + c_1)(2\sigma_{\mathbf{xy}} + c_2)}{\left(\mu_{\mathbf{x}}^2 + \mu_{\mathbf{y}}^2 + c_1\right)\left(\sigma_{\mathbf{x}}^2 + \sigma_{\mathbf{y}}^2 + c_2\right)} \tag{4}$$

where $\mu_{\mathbf{x}}$ and $\sigma_{\mathbf{x}}$ are the mean and variance of image $_{\mathbf{x}}$, respectively; $\mu_{\mathbf{y}}$ and $\sigma_{\mathbf{y}}$ are the mean and variance of image $_{\mathbf{y}}$, respectively. $\sigma_{\mathbf{xy}}$ is the covariance between $\mathbf{x}$ and $\mathbf{y}$. c_1 and c_2 are two constants. PSNR and SSIM are calculated in RGB channel space, and we use the average value on three channels as the final results. The SSIM value is between 0 and 1, and higher PSNR and SSIM scores indicate better image SR performance.

In our experiments, we use the average PSNR and SSIM values of whole images on specific datasets to represent the quality of super-resolved images.

Detection performance evaluation. For detection performance evaluation, we followed [33] to use the average precision (AP) as the quantitative metric with the Intersection over Union (IoU) being set to 0.5 (i.e., $AP50$). We further analyze AP on ship targets of different sizes: AP_s for small targets, AP_m for medium targets and AP_l for large targets. We use COCO division standard to divide the datasets. Small and large targets means objects with areas smaller than 32×32 pixels and larger than 96×96 pixels, respectively. Medium targets means the other objects (areas between 32×32 and 96×96 pixels.)

4.2. Ablation Study

In this section, we conduct extensive ablation studies on the HRSC2016 [29], DOTA [30] and NWPU VHR-10 [31] datasets to validate the effectiveness of our proposed RASR for both image SR and ship detection tasks.

4.2.1. Effectiveness of RASR for Image SR

To validate the effectiveness of our proposed RASR for SR problem, we compare three SR variants: bicubic up-sampling, our SR module and the proposed RASR method. The quantitative evaluation results of image quality achieved by different variants are shown in Table 1. Here, we use the Bicubic down-sampling approach and Bicubic + Gaussian blur degradation approach to generate LR images.

Table 1. Quantitative evaluation results of image quality achieved by different variants of our proposed RASR. We use SSIM and PSNR as the evaluation indices.

Down-Sampling Metrics	Datasets	Bicubic		SR		RASR	
		SSIM	PSNR	SSIM	PSNR	SSIM	PSNR
Bicubic	HRSC2016 [29]	0.659	23.55	0.687	24.31	0.698	24.44
	DOTA [30]	0.672	25.89	0.698	26.12	0.718	26.31
	NWPU VHR-10 [31]	0.856	30.18	0.857	30.48	0.870	30.72
Bicubic + Gaussian Blur	HRSC2016 [29]	0.627	22.65	0.664	23.42	0.697	24.25
	DOTA [30]	0.642	24.21	0.654	25.32	0.694	25.02
	NWPU VHR-10 [31]	0.841	28.71	0.838	29.57	0.856	30.38

We first validate our method on Bicubic down-sampling degradation. On the HRSC2016 datasets, the PSNR value achieved by RASR is 0.89 and 0.13 dB higher than Bicubic up-sampling and SR module, respectively. Similarly, the SSIM value of our RASR is 0.699, and is also higher than the results of two variants. That is because, with the assistance of image rescaling, the SR module in our RASR can acquire more information from the adapted LR images, and thus improve the SR performance. We also verify the effectiveness of our method on the DOTA and NWPU VHR-10 datasets. Specifically, on the DOTA dataset, the PSNR and SSIM of RASR is 26.31 dB and 0.718, and are higher than the bicubic up-sampling (25.89 dB and 0.672) and SR module (26.12 dB and 0.698). On the NWPU VHR-10 dataset, the PSNR and SSIM of RASR is 30.72 dB and 0.870, and are higher than the bicubic up-sampling (30.18 dB and 0.861) and the SR module (30.48 dB and 0.857). The above results demonstrate the effectiveness of our RASR method.

Then, we extend our ablation experiments on more complex scenarios. Specifically, besides bicubicly down-sampling operation, we add Gaussian blur to original HR images when generating LR images. The sizes of Gaussian kernel are set to 21 $\times$ 21, and sigma is set to 8. In this way, LR images are more difficult to be super-resolved. The results are shown in Table 1. It can be observed that the PSNR and SSIM values of different variants on Bicubic + Gaussian blur down-sampling degradation are lower than the bicubic down-sampling degradation. That is because, Bicubic + Gaussian blur degradation lead to more severe information lost in the LR images. Furthermore, on Bicubic + Gaussian blur degradation, PSNR and SSIM values of different variants (i.e., Bicubic, SR module and our RASR) are gradually increasing, which demonstrates the effectiveness and generality of our method.

Figure 3 shows the visualization results of the original HR images and the restored images by different variants, where I_{bic}, I_{SR}, I_{RASR} and I_{HR} mean HR image, bicubic up-sampling image, output of SR module and output of the proposed RASR. Taking bicubic up-sampling image as baseline, we show the residual image between bicubic up-sampling image and the output of SR module (i.e., $|I_{SR} - I_{bic}|$). It can be observed that the major differences between the SR image and bicubicly up-sampled image are located at the edges. Moreover, we show the residual image between the output of RASR and SR module (i.e., $|I_{RASR} - I_{SR}|$). The differences are still at the edges. That is, the SR module can super-resolve detail information in edges and our RASR can further restore these edge detail information with the assistance of image rescaling.

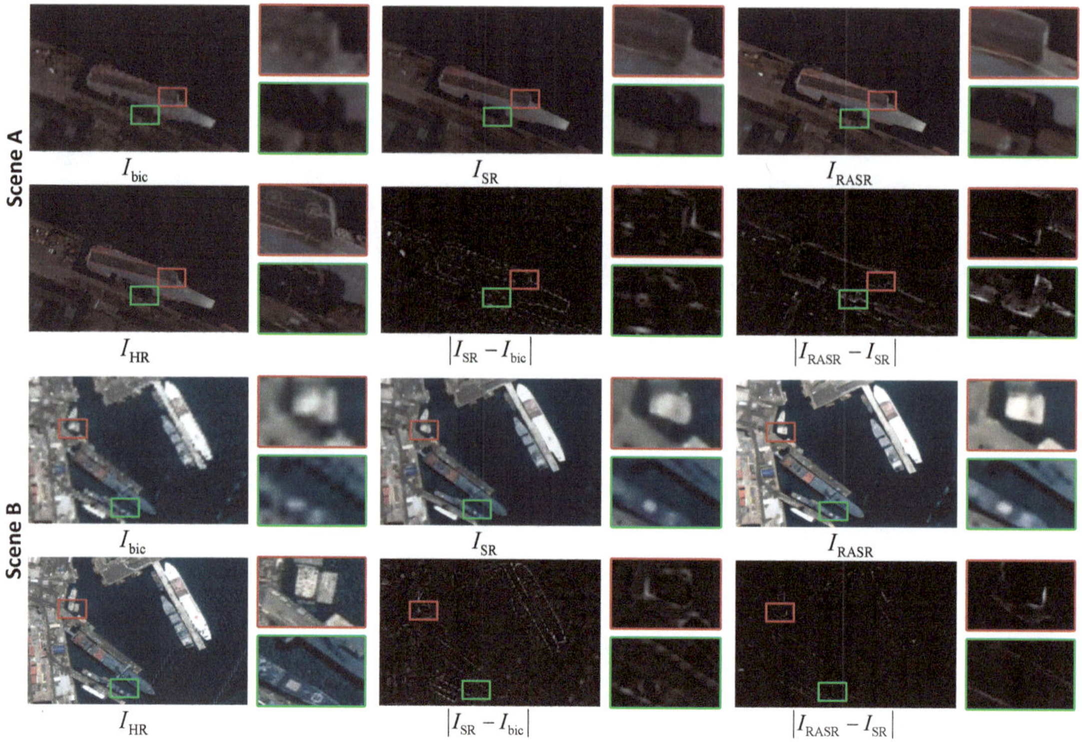

Figure 3. Visualizations of the original HR image (I_{HR}) and the super-resolved images using different SR methods: bicubicly up-sampling (I_{bic}), SR module (I_{SR}) and RASR method (I_{RASR}). Furthermore, the absolute difference of I_{RASR}, I_{SR} and I_{bic} is visualized.

To qualitatively illustrate the assistance of image rescaling to image SR, we provide the amplitude-frequency curves of original HR images, outputs of RASR and its variants. We use Fast Fourier Transform to acquire the frequency images. Specifically, we first transfer an RGB image to gray version, then we compute the 2-dimensional Discrete Fourier Transform of the gray image and shift zero-frequency terms to the center of the frequency domain, finally we use the decadic logarithm of frequency amplitude to generate the frequency image. Since the frequency map is symmetrical around the center, we can group it according to the distance between its elements and its center uniformly. We use the pixel distance between each group and the image center as the abscissa value, and use the summation of all amplitude values in each group as the ordinate values to generate the amplitude-frequency graph. Note that we use the decadic logarithm of amplitude to narrow the range of frequency values, and thus show the difference between curves more clearly. As shown in Figure 4, the frequency can be divided into low, medium and high frequency bands. Information in the low-frequency band is rarely lost, and different SR methods contribute to restoring the information in the medium-frequency band. From Figure 4 we can find that the outputs of image rescaling is closer to HR images than the outputs of SR module. That is because pseudo M-LR images in image rescaling has more beneficial information than original M-LR images. With the assistance of image rescaling, the amplitude-frequency curve of RASR is beyond the curve of the SR module and is closer to that of image rescaling. Although RASR can restore partial information in the medium-frequency band, information in the high-frequency band is hard to be restored due to the irreversible information lost in image degradation. Moreover, there are still spaces to investigate more innovative SR methods since there are still distances between the curve of HR images and RASR outputs in the medium-frequency band.

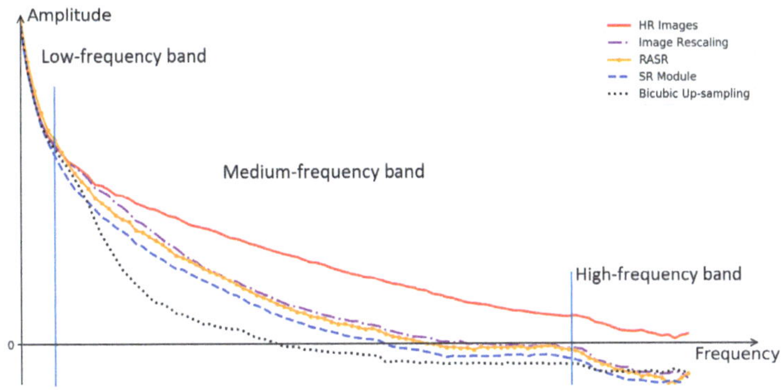

Figure 4. Amplitude–frequency curves of original HR images and the output images of RASR and its variants.

4.2.2. Effectiveness of RASR for Ship Detection

To demonstrate the effectiveness of our proposed RASR method for ship detection, we use the outputs of different RASR variants as the inputs of detector, and choose Faster-RCNN (with ResNet50 as backbone) and FCOS (with ResNet101 as backbone) to detect ship targets. The detection performance is shown in Table 2. Consistent with Section 4.2.1, we validate on Bicubic and Bicubic + Gaussian blur down-sampling metrics, separately.

Table 2. Quantitative evaluation results of detection performance achieved by different variants of our proposed RASR. We use Faster-RCNN (with ResNet50 as backbone) and FCOS (with ResNet101 as backbone) as the evaluation detectors, and use $AP50$ as the evaluation indices. Note that, the improvement of RASR compared to SR module is shown in the table.

Down-Sampling Metrics	Detector	Inputs	Datasets		
			HRSC2016 [29]	**DOTA** [30]	**NWPU VHR-10** [31]
Bicubic	Faster-RCNN [22]	Bicubic	0.788	0.633	0.861
		SR	0.838	0.655	0.875
		RASR	0.859 (0.021↑)	0.746 (0.091↑)	0.894 (0.019↑)
		HR images	0.894	0.847	0.921
	FCOS [34]	Bicubic	0.478	0.513	0.822
		SR	0.626	0.597	0.852
		RASR	0.735 (0.109↑)	0.695 (0.098↑)	0.898 (0.046↑)
		HR images	0.839	0.792	0.892
Bicubic + Gaussian Blur	Faster-RCNN [22]	Bicubic	0.785	0.675	0.856
		SR	0.876	0.772	0.877
		RASR	0.884 (0.008↑)	0.780 (0.008↑)	0.907 (0.030↑)
		HR images	0.894	0.847	0.921
	FCOS [34]	Bicubic	0.149	0.521	0.807
		SR	0.772	0.715	0.852
		RASR	0.780 (0.008↑)	0.716 (0.001↑)	0.898 (0.046↑)
		HR images	0.839	0.792	0.892

We first validate the effectiveness of our method on Bicubic down-sampling degradation. When taking Faster-RCNN as the evaluation detector, on the HRSC2016 dataset, directly using bicubically up-sampled images achieves the lowest detection accuracy. When the SR module is introduced, the $AP50$ value is significantly improved from 0.788 to 0.838. This demonstrates that image SR can provide more useful appearance information than the bicubic up-sampling operation. Furthermore, using the outputs of RASR to detect ship targets can further improve the detection performance, and achieves 0.859 in terms of $AP50$. That is because, the adaption module can transform the original LR images to informative ones with the guide of image rescaling, and more information is embedded in the outputs of RASR and transferred to the detection network. Note that using the outputs of RASR cannot achieve good performance as HR images (i.e., 0.894) due to the information lost. Similarly, when taking FCOS as the evaluation detector, on HRSC2016 dataset, directly using bicubic up-sampling images achieves 0.478 in terms of $AP50$. When the SR module is introduced, the $AP50$ value is improved to 0.626. Using the outputs of RASR to detect ship targets can further improve the detection performance to 0.735.

We extend our ablation studies on the DOTA and NWPU VHR-10 datasets. With the bicubicly up-sampled images as input, Faster-RCNN achieves 0.633 and 0.861 in terms of $AP50$ on DOTA and NWPU VHR-10 datasets, respectively. When the outputs of the SR module are fed to the detector, the detection accuracy improves to 0.655 and 0.875 on DOTA and NWPU VHR-10 datasets, respectively. It can be observed that the detection performance can be further improved when the outputs of RASR are used, but is still inferior to the results on HR images. The detection results of FCOS are consistent with Faster-RCNN on the DOTA and NWPU VHR-10 datasets. FCOS achieves 0.513 and 0.822 when taking bicubicly up-sampled images as inputs on DOTA and NWPU VHR-10 datasets, respectively. When the SR module is introduced, the $AP50$ values are improved to 0.597 and 0.852 on DOTA and NWPU VHR-10 datasets, respectively. Using the outputs of RASR to detect ship targets can further improve the detection performance to 0.695 and 0.898 on two datasets, respectively. These consistent results on two detectors and three datasets validate the effectiveness of our RASR method on the ship detection task.

Furthermore, we validate the effectiveness of our method on Bicubic + Gaussian blur down-sampling degradation. It can be observed form Table 2 that two evaluation detectors with RASR results as inputs both achieve better detection performance than with Bicubic and SR module results as inputs. Take Faster-RCNN as an evaluation detector example, our SR module achieves 0.876 in terms of $AP50$, and is 0.091 better than the Bicubic method, and our RASR achieves 0.008 improvements over the SR module. For DOTA and NWPU VHR-10 datasets, Faster-RCNN on RASR results also achieves 0.008 and 0.030 improvements as compared to the results of the SR module, respectively. This demonstrates the effectiveness and generality of our method on more complex scenarios.

We further show the detection performance on targets with different sizes, as shown in Table 3. Note that here we use Faster-RCNN with ResNet50 as the backbone to validate the influence of our method for targets with different scales. Furthermore, we use bicubic as the down-sampling approach to generate the LR images. It can be observed that the performance improvement on small targets is the largest. Taking the DOTA dataset as an example, compared with using the outputs of SR module, using the outputs of RASR can achieve a 0.086 improvement on small ships, 0.065 improvement on medium targets and 0.053 improvement on large targets, respectively. For the HRSC2016 dataset, Faster-RCNN achieves 0.052, 0.021 and 0.027 improvements in terms of AP_s, AP_m and AP_l when using RASR as pre-processing compared to using the SR module. Moreover, the improvement on small targets is still higher than medium and large targets, respectively. That is because there is more information lost for smaller targets than larger ones. Thus, the recovered detail information using our RASR is more beneficial for small target detection.

Table 3. Quantitative evaluation results of detection performance achieved by different variants of our proposed RASR. We use Faster-RCNN (with ResNet50 as backbone) as the evaluation detector, and use $AP50$, AP_s, AP_m and AP_l as the evaluation indices. Note that, since there is no large ship on NWPU VHR-10 dataset, the AP_l on this dataset is −1.

Dataset	Inputs	Evaluation Indices			
		$AP50$	AP_s	AP_m	AP_l
HRSC2016 [29]	Bicubic	0.788	0.016	0.817	0.865
	SR	0.838	0.050	0.863	0.891
	RASR	0.859 (0.021↑)	0.102 (0.052↑)	0.884 (0.021↑)	0.918 (0.027↑)
	HR Images	0.894	0.399	0.925	0.917
DOTA [30]	Bicubic	0.633	0.631	0.739	0.258
	SR	0.655	0.647	0.755	0.285
	RASR	0.746 (0.091↑)	0.733 (0.086↑)	0.820 (0.065↑)	0.338 (0.053↑)
	HR Images	0.847	0.842	0.892	0.460
NWPU VHR-10 [31]	Bicubic	0.861	0.036	0.896	−1
	SR	0.875	0.011	0.910	−1
	RASR	0.894 (0.019↑)	0.162 (0.151↑)	0.913 (0.003↑)	−1
	HR Images	0.921	0.230	0.946	−1

4.3. Comparison to the State-of-the-Art Methods

We compare our RASR with several state-of-the-art SR models (SelfExSR [35], SR-CNN [11], VDSR [12], EDSR [36] and RCAN [37]) on three public remote sensing datasets to investigate the superiority of our method on SR task. Then, we use RASR as a pre-processing approach and choose Faster-RCNN as detector to develop RASR-Faster-RCNN. We compare RASR-Faster-RCNN with four state-of-the-art detection methods: GFL [38], Reppoints [39], HTC [40] and DetectoRS [41]. Note that we use the bicubicly up-sampled images as the inputs of the compared detectors to ensure the identical input sizes of different detectors.

4.3.1. Comparison with SR Methods

We compare our RASR to several competitive SR methods in terms of the number of parameters, SSIM and PSNR on three datasets. Among these competitive SR methods, SelfExSR is based on non-machine learning, and others are based on CNN. Comparative results are shown in Table 4. It can be observed that SelfExSR consumes the most inference time, since it need more iterations to achieve comparable results. Furthermore, our RASR outperforms SelfExSR, SRCNN and VDSR on all three datasets. Compared with EDSR, our RASR has fewer parematers and faster inference speed, and is superior to EDSR except for the PSNR score on the DOTA dataset. Furthermore, with only one sixth of parameters, our RASR achieves more competitive SR performance than RCAN. The inference speed of RASR is also faster than RCAN. Although SRCNN and VDSR take fewer time to super-solve images, the SR quality is not as good as our RASR method. These experimental results demonstrate the superiority of the proposed RASR.

Figure 5 shows the qualitative results achieved by different SR methods. The selected scenes are from the three datasets, respectively. Scene A is from the HRSC2016 dataset, and it can be observed that bicubic up-sampling, SelfExSR, SRCNN and VDSR can not recover the shape of bow and cannot distinguish bow and port areas. Although EDSR and RCAN can recover the bow, the contrast between bow and background is lower than RASR. That is because, RASR can better restore the detail information at the edge areas due to the assistance of image rescaling. Furthermore, the super-resolved results of SelfExSR have more visual artifacts, that is because the limited SR ability of SelfExSR on lager scale factor (the scale factor is 8 in our experiments). Scene B is from the NWPU VHR-10 dataset, it can be found that RASR can better recover the shape information (as shown in the red bounding box in scene B) and color information (as shown in the green bounding box in scene B) of ship targets. In fact, there is more detail information embedded and transferred to the SR module in our RASR method. Scene C is from the DOTA dataset, and the ships in this scene are tiny and clustered. In this scene, EDSR, RCAN and RASR can generate SR images with distinguished ship instances, but only RASR can further preserve the shape and color information. Furthermore, the hexagonal star architecture in scene C brings a challenge for image SR, since much corner information is lost in the degraded images. SRCNN, VDSR, RDSR, RCAN restore the hexagonal star to quadrilateral or oval by mistake. In contrast, SelfExSR and our RASR can recover the six corners of the architecture. However, due to the strong visual artifacts, the suer-resolved ability of SelfExSR is inferior to RASR. Note that, our RASR cannot recover the complete hexagonal star due to the information lost. It is worth noting that the super-resolved results of SelfExSR have more visual artifacts, that is because the limited SR ability of SelfExSR on lager scale factor (the scale factor is 8 in our experiments, and the max factor is 4 in SelfExSR). These result images on different datasets demonstrate the effectiveness and superiority of our method.

Table 4. Comparisons of the number of parameters (#Param.), SSIM score and PSNR score for different SR methods on HRSC2016, DOTA and NWPU VHR-10 datasets. The best results are in bold faces, and the second results are ubderlined. Note that, the Time is the average processing time per image of each method on the DOTA validation dataset.

Method	#Param.	Time	HRSC2016		DOTA		NWPU VHR-10	
			SSIM	PSNR	SSIM	PSNR	SSIM	PSNR
SelfExSR [35]	None	37.878 s	0.670	24.09	0.625	24.64	0.856	30.48
RCAN [37]	15.66 M	69.746 ms	<u>0.689</u>	**24.30**	<u>0.704</u>	**26.63**	0.861	30.29
SRCNN [11]	0.12 M	**15.944 ms**	0.672	23.82	0.682	26.06	0.861	30.59
EDSR [36]	39.17 M	42.114 ms	0.683	24.01	0.701	<u>26.51</u>	<u>0.863</u>	30.62
VDSR [12]	0.67 M	<u>29.978 ms</u>	0.677	23.92	0.686	23.11	<u>0.863</u>	<u>30.71</u>
RASR	2.50 M	38.626 ms	**0.698**	<u>24.10</u>	**0.718**	26.31	**0.870**	**30.72**

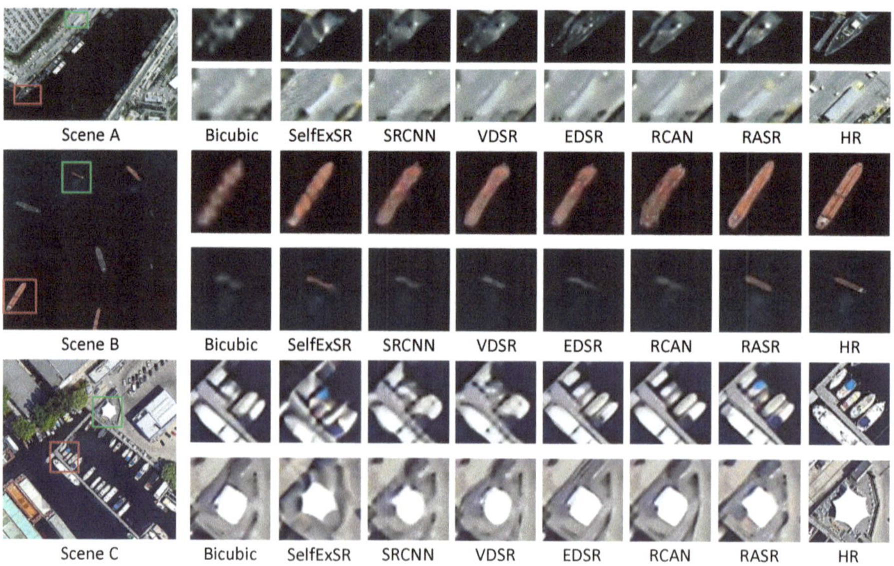

Figure 5. Visual results achieved by different SR methods.

4.3.2. Comparison with Detection Methods

Comparative results achieved by RASR-Faster-RCNN and other methods are shown in Table 5. Compared to original Faster-RCNN, our RASR-Faster-RCNN achieves 0.053, 0.086 and 0.031 improvements in terms of $AP50$ on the HRSC2016, DOTA and NWPU VHR-10 datasets, respectively. The significant improvements are at a low cost: about 2.50 MB parameters for image pre-processing approach. It can be observed that our RASR-Faster-RCNN achieves the best detection performance on three datasets. Note that HTC and DetectoRS are developed based on Faster-RCNN and have more parameters, and their performance is better than the original Faster-RCNN. However, compared to RASR-Faster-RCNN, HTC and DetectoRS are less competitive due to the missing details in the input bicubicly up-sampled images. In contrast, our RASR-Faster-RCNN uses RASR as the pre-processing and takes the super-resolved images as the input of Faster-RCNN, and thus achieves better detection performance.

Table 5. Quantitative results (i.e., mAP50) achieved by different methods (based on ResNet101) on the HRSC2016, DOTA and NWPU VHR-10 datasets. Our RASR-Faster-RCNN achieves state-of-the-art detection performance. The best results are in bold faces.

Method	Datasets		
	HRSC2016 [29]	**DOTA** [30]	**NWPU VHR-10** [31]
GFL [38]	0.826	0.575	0.856
Reppoints [39]	0.792	0.592	0.875
HTC [40]	0.634	0.668	0.863
DetectoRS [41]	0.860	0.677	0.808
Faster-RCNN [22]	0.820	0.651	0.853
RASR-Faster-RCNN	**0.873**	**0.737**	**0.884**

Figure 6 shows the detection results achieved by different methods on three scenes. Scene A is from HRSC2016, and the ship targets have various scales. It can be observed that GFL, Reppoints, HTC and DetectoRS cannot successfully detect the smallest ship. Although Faster-RCNN can detect the smallest ship, there are extra false alarms detected by mistake. Our RASR-Faster-RCNN can successfully detect the tiny ship, since the pre-processing RASR can restore more detail information, and thus improve the detection accuracy. Scene B is from the NWPU VHR-10 dataset, in which the shadow and reflection make some ships (indicated by red arrow) have lower contrast, and introduce challenges to the detection task. All the comparative detectors cannot detect the ship due to the lost information in the bicubicly up-sampled images. However, our RASR-Faster-RCNN can successfully detect these targets with the assistance of RASR. However, when handling some extremely challenging scenes, our method will produce some false alarms due to the similar detail structure. For example, in scene C (from the DOTA dataset), the big trace produces many areas that have similar detail structures with tiny ships, and thus leads to many false alarms. GFL, Reppoints, HTC and our RASR all generate some false alarms. Only detectoRS can successfully detect ships in this scene accurately. In the future, we will improve the performance of our method on these challenging scenes.

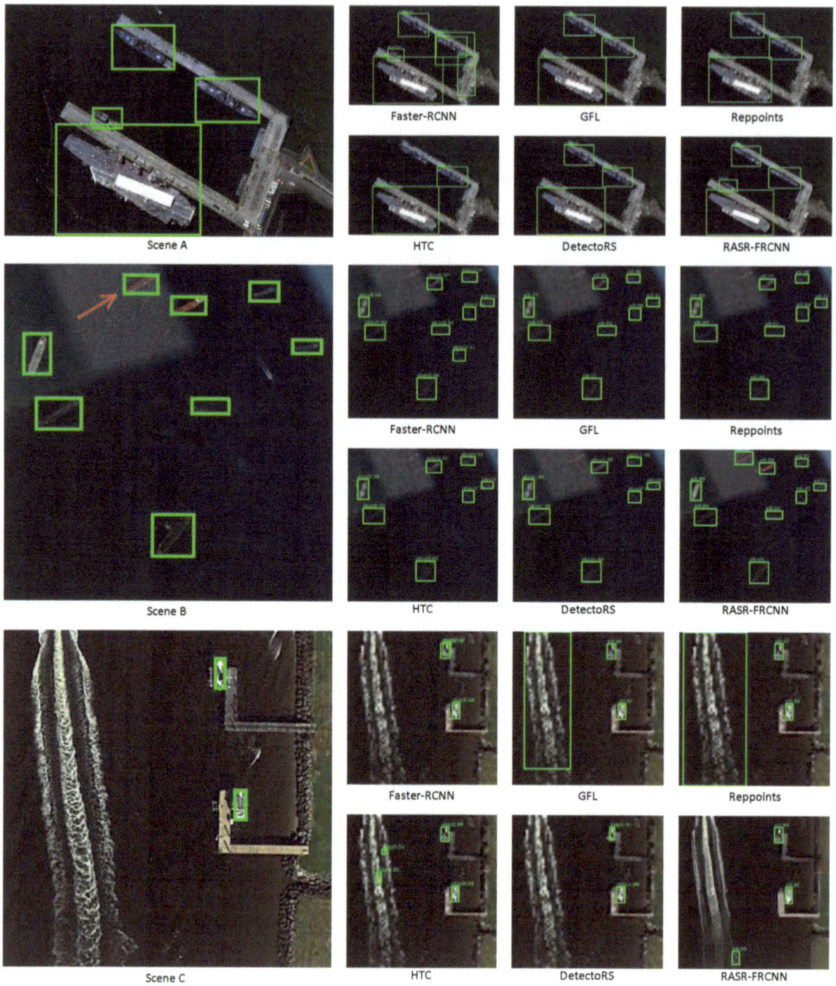

Figure 6. Qualitative results achieved by different methods on three example scenes. We use green bounding boxes to mark the ground-truth labels and detection results.

5. Conclusions

In this paper, we propose an image rescaling assisted SR method RASR to super-resolve remote sensing images more effectively. In our method, more detail information can be embedded and transferred to the SR process, and thus enhance the SR performance. Extensive ablation studies and visualizations have demonstrated the effectiveness of our RASR method for both image SR and ship detection. Moreover, we take RASR as a pre-processing approach to the detection network, and develop RASR-Det to further improve the detection performance on medium-low resolution remote sensing ship detection. Note that, although our method could recover some texture information, it may hallucinate fake textures and brings some false alarms. In the future, we would continue to work on this challenge.

Author Contributions: H.Z. determined the research direction and modified the article expression; S.H. conceived the innovative ideas, and designed the RASR framework, conducted experiments and completed the first version of manuscript. X.C. helped to modify the conception, and revised the manuscript and provided suggestions in expression. L.S., J.W., S.L. and J.L. checked out the article's writing and helped to build datasets. All authors have read and agreed to the published version of the manuscript.

Funding: This work was supported by the National Natural Science Foundation of China under Grant 62071474.

Data Availability Statement: Not applicable.

Conflicts of Interest: The authors declare no conflict of interest.

Abbreviations

The following abbreviations are used in this manuscript:

M-LR	Medium-low resolution
GSD	Ground sample distance
SR	Super-resolution
RASR	Rescaling assisted super-resolution
RASR-Det	Detection framework with RASR as a pre-processing approach
HR	High resolution
CNN	Convolution neutral network
VDSR	Very deep super-resolution network
RDN	Residual dense network
EESRGAN	Edge-enhanced super-resolution generative adversarial network
Faster-RCNN	Method in paper "Faster-RCNN: Towards real-time object detection with region proposal networks"
FCOS	Fully Convolutional One-Stage Object Detection
ResNet50	Deep residual network with 50 convolution and fully connection layers
ResNet101	Deep residual network with 101 convolution and fully connection layers
GAN	Generative adversarial network
RDB	Residual dense block
HRSC2016	High resolution ship collections 2016
DOTA	A Large-scale Dataset for Object Detection in Aerial Images
NWPU VHR-10	A challenging 10-class geospatial object detection dataset
SGD	Stochastic gradient descent
MMDetection	Open MMLab Detection Toolbox and Benchmark
PSNR	Peak signal-to-noise ratio
SSIM	Structural similarity
MSE	Mean squared error
RGB	Red-green-blue
AP	Average precision
IoU	Intersection over union
$AP50$	Average precision with intersection over union being 0.50
AP_s	Average precision for small targets
AP_m	Average precision for medium targets
AP_l	Average precision for large targets
COCO	Microsoft COCO: Common Objects in Context
dB	Decibel
SRCNN	Super-Resolution Convolutional Neural Network
EDSR	Enhanced deep residual super-resolution network
RCAN	Residual channel attention network
RASR-Faster-RCNN	Detection method integrating RASR and Faster-RCNN
GFL	Generalized focal loss
Reppoints	Point Set Representation for object detection
HTC	Hybrid task cascade for instance segmentation
DetectoRS	Detecting objects with recursive feature pyramid and switchable atrous convolution
MB	Million bytes

References

1. Zhang, X.; Wang, G.; Zhu, P.; Zhang, T.; Li, C.; Jiao, L. GRS-Det: An anchor-free rotation ship detector based on Gaussian-mask in remote sensing images. *IEEE Trans. Geosci. Remote. Sens.* **2020**, *59*, 3518–3531. [CrossRef]
2. Yu, Y.; Yang, X.; Li, J.; Gao, X. A cascade rotated anchor-aided detector for ship detection in remote sensing images. *IEEE Trans. Geosci. Remote. Sens.* **2020**, *60*, 1–14. [CrossRef]
3. Li, L.; Zhou, Z.; Wang, B.; Miao, L.; Zong, H. A novel CNN-based method for accurate ship detection in HR optical remote sensing images via rotated bounding box. *IEEE Trans. Geosci. Remote. Sens.* **2020**, *59*, 686–699. [CrossRef]
4. Dong, X.; Wang, L.; Sun, X.; Jia, X.; Gao, L.; Zhang, B. Remote Sensing Image Super-Resolution Using Second-Order Multi-Scale Networks. *IEEE Trans. Geosci. Remote. Sens.* **2020**, *59*, 3473–3485. [CrossRef]
5. Rabbi, J.; Ray, N.; Schubert, M.; Chowdhury, S.; Chao, D. Small-Object Detection in Remote Sensing Images with End-to-End Edge-Enhanced GAN and Object Detector Network. *Remote Sens.* **2020**, *12*, 1432. [CrossRef]
6. Courtrai, L.; Pham, M.; Lefèvre, S. Small Object Detection in Remote Sensing Images Based on Super-Resolution with Auxiliary Generative Adversarial Networks. *Remote Sens.* **2020**, *12*, 3152. [CrossRef]
7. Wang, B.; Lu, T.; Zhang, Y. Feature-Driven Super-Resolution for Object Detection. In Proceedings of the IEEE CRC, Wuhan, China, 16–18 October 2020; pp. 211–215.
8. Noh, J.; Bae, W.; Lee, W.; Seo, J.; Kim, G. Better to follow, follow to be better: Towards precise supervision of feature super-resolution for small object detection. In Proceedings of the IEEE ICCV, Seoul, Korea, 27 October–2 November 2019; pp. 9725–9734.
9. Wang, Z.; Chen, J.; Hoi, S.C. Deep learning for image super-resolution: A survey. *IEEE Trans. Pattern Anal. Mach. Intell.* **2020**, *43*, 3365–3387. [CrossRef] [PubMed]
10. Li, J.; Pei, Z.; Zeng, T. From Beginner to Master: A Survey for Deep Learning-based Single-Image Super-Resolution. *arXiv* **2021**, arXiv:2109.14335.
11. Dong, C.; Loy, C.C.; He, K.; Tang, X. Learning a deep convolutional network for image super-resolution. In Proceedings of the ECCV, Zurich, Switzerland, 6–12 September 2014; Springer: Berlin/Heidelberg, Germany, 2014; pp. 184–199.
12. Kim, J.; Kwon Lee, J.; Mu Lee, K. Accurate image super-resolution using very deep convolutional networks. In Proceedings of the IEEE CVPR, Las Vegas, NV, USA, 27–30 June 2016; pp. 1646–1654.
13. Zhang, Y.; Tian, Y.; Kong, Y.; Zhong, B.; Fu, Y. Residual dense network for image super-resolution. *arXiv* **2018**, arXiv:1802.08797v2.
14. He, K.; Zhang, X.; Ren, S.; Sun, J. Deep residual learning for image recognition. In Proceedings of the IEEE CVPR, Las Vegas, NV, USA, 27–30 June 2016; pp. 770–778.
15. Huang, G.; Liu, Z.; Van Der Maaten, L.; Weinberger, K.Q. Densely connected convolutional networks. In Proceedings of the IEEE CVPR, Honolulu, HI, USA, 21–26 July 2017; pp. 4700–4708.
16. Li, J.; Fang, F.; Mei, K.; Zhang, G. Multi-scale residual network for image super-resolution. In Proceedings of the ECCV, Munich, Germany, 8–14 September 2018; pp. 517–532.
17. Wang, L.; Dong, X.; Wang, Y.; Ying, X.; Lin, Z.; An, W.; Guo, Y. Exploring Sparsity in Image Super-Resolution for Efficient Inference. In Proceedings of the IEEE CVPR, Nashville, TN, USA, 20–25 June 2021; pp. 4917–4926.
18. Wang, L.; Wang, Y.; Dong, X.; Xu, Q.; Yang, J.; An, W.; Guo, Y. Unsupervised Degradation Representation Learning for Blind Super-Resolution. In Proceedings of the IEEE CVPR, Nashville, TN, USA, 20–25 June 2021; pp. 10581–10590.
19. Kim, H.; Choi, M.; Lim, B.; Lee, K.M. Task-aware image downscaling. In Proceedings of the ECCV, Munich, Germany, 8–14 September 2018; pp. 399–414.
20. Sun, W.; Chen, Z. Learned image downscaling for upscaling using content adaptive resampler. *IEEE Trans. Image Process.* **2020**, *29*, 4027–4040. [CrossRef] [PubMed]
21. Xiao, M.; Zheng, S.; Liu, C.; Wang, Y.; He, D.; Ke, G.; Bian, J.; Lin, Z.; Liu, T.Y. Invertible image rescaling. In Proceedings of the ECCV; Glasgow, UK, 23–28 August 2020; Springer: Berlin/Heidelberg, Germany, 2020; pp. 126–144.
22. Ren, S.; He, K.; Girshick, R.; Sun, J. Faster R-CNN: Towards real-time object detection with region proposal networks. *IEEE Trans. Pattern Anal. Mach. Intell.* **2016**, *39*, 1137–1149. [CrossRef] [PubMed]
23. Liu, W.; Anguelov, D.; Erhan, D.; Szegedy, C.; Reed, S.; Fu, C.Y.; Berg, A.C. SSD: Single shot multibox detector. In Proceedings of the ECCV; Amsterdam, The Netherlands, 11–14 October 2016; Springer: Berlin/Heidelberg, Germany 2016; pp. 21–37.
24. Zou, H.; He, S.; Wang, Y.; Li, R.; Cheng, F.; Cao, X. Ship detection based on medium-low resolution remote sensing data and super-resolved feature representation. *Remote Sens. Lett.* **2022**, *13*, 323–333. [CrossRef]
25. He, S.; Zou, H.; Wang, Y.; Li, R.; Cheng, F.; Cao, X.; Li, M. Enhancing Mid–Low-Resolution Ship Detection With High-Resolution Feature Distillation. *IEEE Geosci. Remote. Sens. Lett.* **2021**, *19*, 1–5. [CrossRef]
26. Lee, W.; Lee, J.; Kim, D.; Ham, B. Learning with privileged information for efficient image super-resolution. In Proceedings of the ECCV, Virtual, 15 August 2020; Springer: Berlin/Heidelberg, Germany 2020; pp. 465–482.
27. He, K.; Zhang, X.; Ren, S.; Sun, J. Delving deep into rectifiers: Surpassing human-level performance on imagenet classification. In Proceedings of the IEEE ICCV, Santiago, Chile, 7–13 December 2015; pp. 1026–1034.
28. Shi, W.; Caballero, J.; Huszár, F.; Totz, J.; Aitken, A.P.; Bishop, R.; Rueckert, D.; Wang, Z. Real-time single image and video super-resolution using an efficient sub-pixel convolutional neural network. In Proceedings of the IEEE CVPR, Las Vegas, NV, USA, 27–30 June 2016; pp. 1874–1883.

29. Liu, Z.; Yuan, L.; Weng, L.; Yang, Y. A High Resolution Optical Satellite Image Dataset for Ship Recognition and Some New Baselines. In Proceedings of the International Conference on Pattern Recognition Applications and Methods, Porto, Portugal, 24–26 February 2017; Volume 2, pp. 324–331.
30. Xia, G.S.; Bai, X.; Ding, J.; Zhu, Z.; Belongie, S.; Luo, J.; Datcu, M.; Pelillo, M.; Zhang, L. DOTA: A large-scale dataset for object detection in aerial images. In Proceedings of the IEEE CVPR, Salt Lake City, UT, USA, 18–22 June 2018; pp. 3974–3983.
31. Cheng, G.; Han, J.; Zhou, P.; Guo, L. Multi-class geospatial object detection and geographic image classification based on collection of part detectors. *J. Photogramm. Remote. Sens.* **2014**, *98*, 119–132. [CrossRef]
32. Kingma, D.P.; Ba, J. Adam: A method for stochastic optimization. *arXiv* **2014**, arXiv:1412.6980.
33. Everingham, M.; Winn, J. The pascal visual object classes challenge 2012 (VOC2012) development kit. *Pattern Anal. Stat. Model. Comput. Learn. Tech. Rep* **2011**, *8*, 5.
34. Tian, Z.; Shen, C.; Chen, H.; He, T. Fcos: Fully convolutional one-stage object detection. In Proceedings of the IEEE ICCV, Seoul, Korea, 27 October–2 November 2019; pp. 9627–9636.
35. Huang, J.B.; Singh, A.; Ahuja, N. Single Image Super-resolution from Transformed Self-Exemplars. In Proceedings of the Proceedings of the IEEE Conference on Computer Vision and Pattern Recognition, Boston, MA, USA, 7–12 June 2015; pp. 5197–5206.
36. Lim, B.; Son, S.; Kim, H.; Nah, S.; Mu Lee, K. Enhanced deep residual networks for single image super-resolution. In Proceedings of the IEEE CVPRW, Honolulu, HI, USA, 21–26 July 2017; pp. 136–144.
37. Zhang, Y.; Li, K.; Li, K.; Wang, L.; Zhong, B.; Fu, Y. Image super-resolution using very deep residual channel attention networks. In Proceedings of the ECCV, Munich, Germany, 8–14 September 2018; pp. 286–301.
38. Li, X.; Wang, W.; Wu, L.; Chen, S.; Hu, X.; Li, J.; Tang, J.; Yang, J. Generalized Focal Loss: Learning Qualified and Distributed Bounding Boxes for Dense Object Detection. *arXiv* **2020**, arXiv:2006.04388.
39. Yang, Z.; Liu, S.; Hu, H.; Wang, L.; Lin, S. Reppoints: Point set representation for object detection. In Proceedings of the IEEE ICCV, Seoul, Korea, 27 October–2 November 2019; pp. 9657–9666.
40. Chen, K.; Pang, J.; Wang, J.; Xiong, Y.; Li, X.; Sun, S.; Feng, W.; Liu, Z.; Shi, J.; Ouyang, W.; et al. Hybrid task cascade for instance segmentation. In Proceedings of the IEEE CVPR, Long Beach, CA, USA, 15–20 June 2019; pp. 4974–4983.
41. Qiao, S.; Chen, L.C.; Yuille, A. Detectors: Detecting objects with recursive feature pyramid and switchable atrous convolution. In Proceedings of the IEEE CVPR, Nashville, TN, USA, 20–25 June 2021; pp. 10213–10224.

future internet

MDPI

Article

Unsupervised Anomaly Detection and Segmentation on Dirty Datasets

Jiahao Guo, Xiaohuo Yu and Lu Wang *

School of Computer Engineering and Science, Shanghai University, Shanghai 200444, China; g747173965@shu.edu.cn (J.G.); xiaohuoyu@shu.edu.cn (X.Y.)
* Correspondence: luwang@shu.edu.cn

Abstract: Industrial quality control is an important task. Most of the existing vision-based unsupervised industrial anomaly detection and segmentation methods require that the training set only consists of normal samples, which is difficult to ensure in practice. This paper proposes an unsupervised framework to solve the industrial anomaly detection and segmentation problem when the training set contains anomaly samples. Our framework uses a model pretrained on ImageNet as a feature extractor to extract patch-level features. After that, we propose a trimming method to estimate a robust Gaussian distribution based on the patch features at each position. Then, with an iterative filtering process, we can iteratively filter out the anomaly samples in the training set and re-estimate the Gaussian distribution at each position. In the prediction phase, the Mahalanobis distance between a patch feature vector and the center of the Gaussian distribution at the corresponding position is used as the anomaly score of this patch. The subsequent anomaly region segmentation is performed based on the patch anomaly score. We tested the proposed method on three datasets containing the anomaly samples and obtained state-of-the-art performance.

Keywords: anomaly detection and segmentation; unsupervised method; noise resistance

Citation: Guo, J.; Yu, X.; Wang, L. Unsupervised Anomaly Detection and Segmentation on Dirty Datasets. *Future Internet* **2022**, *14*, 86. https://doi.org/10.3390/fi14030086

Academic Editor: Paolo Bellavista

Received: 28 February 2022
Accepted: 10 March 2022
Published: 13 March 2022

1. Introduction

In the manufacturing production process, quality control is one of the most important aspects. Finding and locating surface defects in industrial products using machine vision is much more efficient than manual quality inspection [1]. Thus, vision-based detection and segmentation of surface defects in industrial products have become a key research topic [2,3]. As a rule, products without surface defects are defined as normal samples, and products with surface defects are defined as anomaly samples. The anomaly detection task is to determine whether a sample contains defects or not. The anomaly segmentation task is to locate the defective region.

Anomaly detection and segmentation have the following challenges. Firstly, anomalies present in various ways, such as irregular, inconsistent, faulty, unnatural, atypical, etc. [4]. Secondly, the camera rarely captures anomaly products in practice leading to data imbalance, i.e., there are many normal samples and a few anomaly samples. Thirdly, anomaly samples are rare entities, hence it is inefficient to manually label them [5].

Due to the abovementioned points, unsupervised methods are used for anomaly detection, and these methods assume that all samples in the training set are normal samples [4]. This assumption is hard to satisfy in practice. The performance of most of the existing unsupervised anomaly detection methods degrades severely in the case of "dirty" datasets, i.e., the training set contains anomaly samples [2,4,5].

We propose a framework to address anomaly detection and segmentation in the case of dirty datasets. We consider samples in the low-density region of the probability model as anomaly samples. Firstly, inspired by the trimming method in robust statistics [6], we fit a Gaussian distribution using patch-level features for each position where the patch is located and exclude the samples that contain patches in low-density regions, as shown in Figure 1. Secondly, we obtain the final selected normal samples through the iteration module. Lastly, we fit patch-level Gaussian distributions based on the final selected normal samples. In the testing phase, the Mahalanobis distance between each patch of the test sample and the center of the corresponding Gaussian distribution is used as the anomaly score to perform anomaly detection and segmentation. This paper calls this the framework PaIRE (Patch Iterative Robust Estimation) method. The main contributions of our work are as follows:

- Unsupervised anomaly detection methods degrade in performance due to noise. We propose a framework PaIRE to address unsupervised anomaly detection when the training set contains noise;
- A slope sliding window method is designed to reduce the effect of noise for improving the noise resistance of the proposed model;
- An iterative module is proposed to make the model estimate the distribution of normal samples more accurately from the dirty dataset;

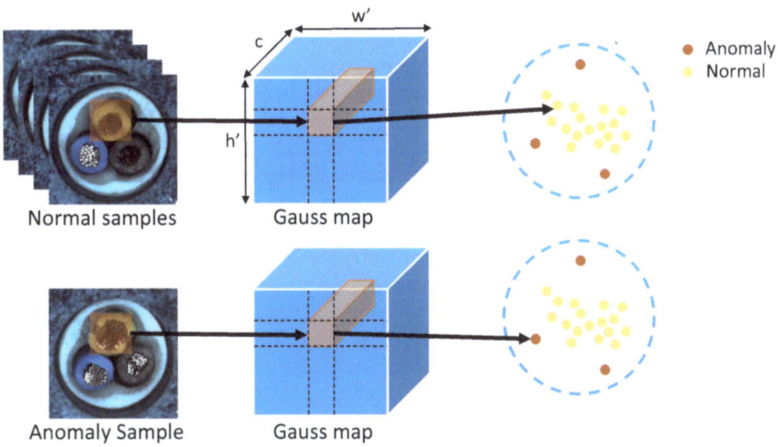

Figure 1. The feature vector corresponding to the image patch can be obtained from the feature map extracted from the pre-trained network. In the feature space, the normal patches are in the high-density region and the anomaly patches are in the low-density region.

2. Related Works

Unsupervised industrial anomaly detection and segmentation methods can be classified into classification methods, probabilistic methods, and reconstruction methods. In addition to this, a few unsupervised anomaly detection methods are based on dirty datasets. The above methods will be reviewed below.

Most of the unsupervised classification anomaly-detection methods are based on normal samples. OC-SVM [7] treats the coordinate origin as anomaly samples and trains a classifier with the coordinate origin and the normal samples. SVDD [8,9] finds the minimum volume of hypersphere that can enclose most of the samples and considers an anomaly sample outside the hypersphere. Recently, self-supervised classification has been used for anomaly detection. Geometric transformations [10] (rotation, flip, translation, etc.) train the classifier by predicting which transformation the image has performed. Anomaly scores are obtained from the softmax activation of the classifier. RotNet [11] uses rotation for data augmentation. After that, a one-class classifier is obtained by contrast learning [12] for anomaly detection. NSA [13] uses seamless cloning to create anomaly samples and trains a binary classifier to distinguish artificial anomaly samples from normal samples. CutPaste [14] performs anomaly detection by training a classifier with normal samples and anomaly samples constructed by special data augmentation. Anomaly detection using unsupervised classification methods is sensitive to noise in the training set. When the training set contains noise, both one-class classification methods [7,9] and self-supervised classification methods [10,11,13,14] produce incorrect labels, which seriously affect the detection performance of the model.

In the probabilistic model, anomalies are defined as low probability samples. Gaussian-AD [15] fits a Gaussian distribution at each feature hierarchy of the pretrained network, and the sum of the Mahalanobis distances between the test samples and the center of the corresponding Gaussian distribution at each hierarchy is calculated as the anomaly score. PaDim [16] uses a pre-trained network to extract features, fits a Gaussian distribution at patch-level, and uses Mahalanobis distances as anomaly scores. FYD [17] uses an auxiliary alignment network to align the objects in the images and uses a Patch-level Gaussian classifier for anomaly detection. SPADE [18] also uses a pretraining network to extract features. It uses a memory module to store the features of all training samples, and the Euclidean distance sum of the test sample and the K-NN samples in the memory module as the anomaly score. STPM [19] uses knowledge distillation for anomaly detection: the student network learns the pretrained teacher network based on normal samples. It utilizes the difference between student and teacher extracted features of the anomaly samples for anomaly detection. For parametric probabilistic models, noisy samples affect the estimation of parameters. For nonparametric models, noisy samples are stored in the memory module and mistaken for normal samples, affecting the model performance.

Reconstruction is one of the mainstream anomaly-detection methods based on normal samples. The reconstruction model is trained only on normal samples, which leads to low reconstruction error for normal samples and high reconstruction error for anomaly samples. Generative adversarial networks (GAN) [20–25], autoencoders (AE) [26–30], and variants are the most commonly used reconstruction models. InTra [31] uses a transformer to inpaint the artificial missing image patches. The patches containing anomalies will have high reconstruction errors. The reconstruction error of the test sample is used as the anomaly score. The drawback of the reconstruction method is that when the training set contains anomaly samples, it leads to low reconstruction error for the anomaly samples.

There are few studies on industrial anomaly detection and segmentation in the case of dirty datasets. TrustMAE [32] filters anomaly samples in the autoencoder latent space based on the distance of the sample from the center of the batch samples. Samples with a distance greater than a threshold from the centroid are considered anomaly samples. The disadvantage of TrustMAE is that at the early stage of model training, there are no labels to guide the encoder to extract distinguishing features, which can lead to the possibility that samples close to the center of the batch sample also be anomaly samples. Li et al. [33] use the feature vector of the latent space of autoencoder as a proxy to perform clustering and select clusters with low variance as normal samples. This process is cycled in the training process of the autoencoder. These two methods have a drawback: early in the autoencoder training process, the extracted feature vectors may not satisfy their assumptions. DFR [34] extracts the ImageNet pretrained network features, determines the optimal latent space dimension of AE by PCA, and then uses AE to reconstruct the extracted features. DFR is robust to noise because of the noise-reduction effect of PCA.

3. Proposed Method

As shown in Figure 2, our proposed PaIRE (Patch Iterative Robust Estimation) method consists of the following parts: feature extractor, trimming patch-level gaussian distribution (Trimming Module), iterative filtering(Iteration Module), anomaly detection and segmentation (Inference) based on anomaly scores. Each part will be described in the following.

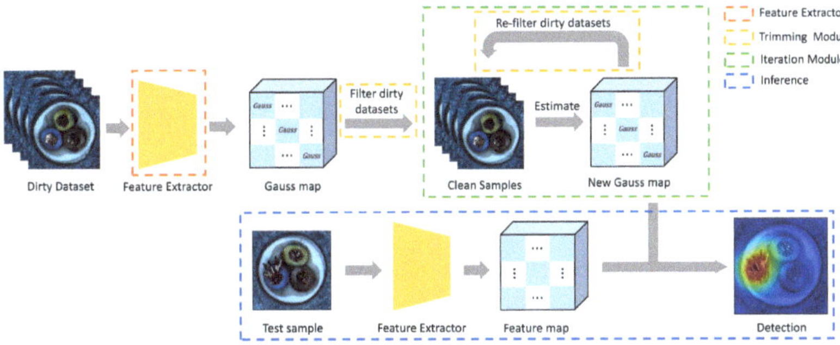

Figure 2. The structure of our proposed method PaIRE.

3.1. Patch Feature Extraction

For industrial anomaly-detection tasks, a good feature extractor should distinguish the difference between normal and anomaly samples. When the dataset contains anomaly samples without any labels, it is hard to train a good feature extractor. Some recent work uses networks that are pretrained on ImageNet for anomaly detection [15,16,18,34], and these works illustrate the ability of pre-trained networks to extract discriminative features for anomaly detection. We use a network pretrained on ImageNet to generate the patch-embedding vector.

Different blocks produce feature maps containing different hierarchical semantic information in the ResNet-like architecture. Since anomalies may occur at different scales, and to avoid features being too biased towards ImageNet classification, we choose the feature map of blocks 1,2,3 outputs of WideResnet-50 to extract the patched-level features as shown in Figure 3. We use $\phi_q, q \in \{1,2,3\}$ to denote the output of block q.

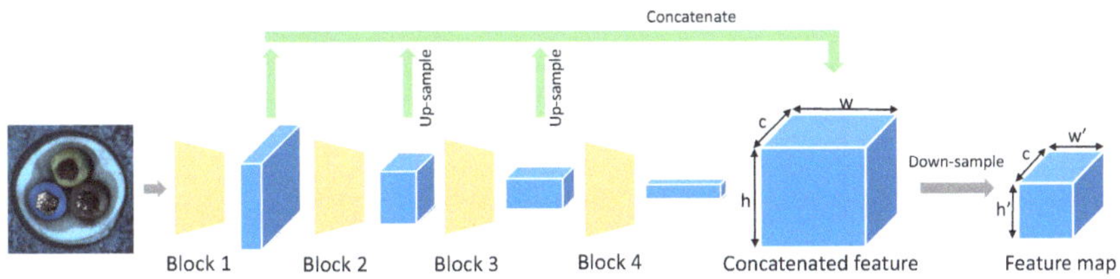

Figure 3. Extracting feature map from WideResnet-50 pretrained on ImageNet.

To fuse the information from different hierarchies and align feature maps spatially, we use bilinear interpolation to resize ϕ_2 and ϕ_3 to the same resolution on h and w as ϕ_1, as shown in Equation (1). Here, $\phi_q \in \mathbb{R}^{c\times h\times w}$is a three-dimensional tensor of channel c, height h, width w.

$$\hat{\phi}_q = resize(\phi_q), q \in \{2,3\} \tag{1}$$

Then we concatenate ϕ_1, ϕ_2, ϕ_3 on the channel, as shown in Equation (2).

$$\hat{\phi}_{\{1:3\}} = cat(\phi_1, \hat{\phi}_2, \hat{\phi}_3) \tag{2}$$

We fuse the feature vectors with the neighborhoods to reduce the resolution while retaining information. avg is the average pooling with kernel size k and striding s, as shown in Equation (3).

$$\hat{\phi} = avg(\hat{\phi}_{\{1:3\}}, k, s) \tag{3}$$

Each c-dimensional feature slice at positions $i \in \{1, ..., h'\}$ and $j \in \{1, ..., w'\}$ is the patch feature vector of the corresponding image-patch at the spatial position. c, h', w' are the channel, height, width of $\hat{\phi}$.

3.2. Trimming Patch-Level Gaussian Distribution

For all samples in the training set, we obtain the patch feature vectors $x_{i,j}$ at patch position $(i,j) \in [1,h'] \times [1,w']$, and the set of feature vectors $X_{i,j} = \{x_{i,j}^n | n \in [1,N]\}$, N is the number of samples in the training set. We can estimate a Gaussian distribution with parameters $\mu_{i,j}$ and $\Sigma_{i,j}$ based on $X_{i,j}$, where $\mu_{i,j}$ is the sample mean in $X_{i,j}$ and $\Sigma_{i,j}$ is the covariance matrix shown in Equations (4) and (5). In order to ensure that the covariance matrix is invertible for the subsequent Mahalanobis distance calculation, we added the term λI. I is an identity matrix, $\lambda = 1 \times 10^{-6}$.

$$\mu_{i,j} = \frac{1}{N}\sum_{n=1}^{N} x_{i,j}^n \tag{4}$$

$$\Sigma_{i,j} = \frac{1}{N-1}\sum_{n=1}^{N}(x_{i,j}^n - \mu_{i,j})(x_{i,j}^n - \mu_{i,j})^T + \lambda I \tag{5}$$

After estimating a Gaussian distribution at each patch position(i,j), we obtain a Gaussian map. As shown in Figure 1, in the patch-level distribution, the patches containing anomalies are in the low-density region. In robust statistics, trimming methods [6] remove part of the maximum–minimum samples to obtain a statistical model with better robustness. Inspired by [15], we use the Mahalanobis distance in Equation (6) to measure the distance of a patch feature $x_{i,j}$ to the center of the Gaussian distribution as the anomaly score. A high anomaly score indicates that the patch is in a low-density region, and a low anomaly score indicates that the patch is in a high-density region. Inspired by the trimming method in robust statistics, we exclude the samples that contain patches in low-density regions.

$$M(x_{i,j}) = \sqrt{(x_{i,j} - \mu_{i,j})^T \Sigma_{i,j}^{-1} (x_{i,j} - \mu_{i,j})} \tag{6}$$

$$M_{i,j}^{norm} = \frac{M_{i,j} - M_{i,j}^{min}}{M_{i,j}^{max} - M_{i,j}^{min}} \tag{7}$$

As shown in Figure 4, sorting the obtained Mahalanobis distance from smallest to largest, a line segment with a sudden increase in slope can be found. We can calculate a distance threshold from this line segment. All patches with Mahalanobis distance greater than this threshold are considered anomaly patches.

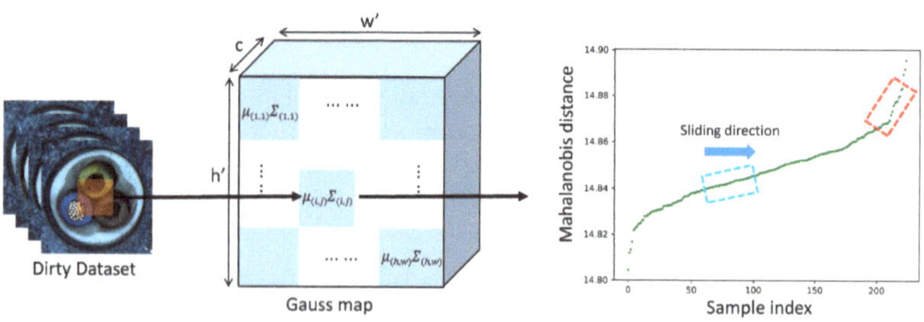

Figure 4. At position (i,j), a Gaussian distribution is fitted with the patch feature vector of all samples. The dashed box represents the sliding window and the red dashed box represents the target window.

It is important to choose an appropriate threshold of the Mahalanobis distance to perform trimming operations. If the threshold is too low, too many normal samples will be excluded, and anomaly samples will not be effectively excluded if the threshold is too high. To make the distances on a similar scale, we use Min–Max scaling in Equation (7) for each group of Mahalanobis distances.

This paper uses the sliding-window method to implement an adaptive threshold. A window slides from left to right, and the slope of the line connecting the two endpoints of the window is considered the slope of the line segment inside the window. As the window slides to the right, we save the slope of each line segment in the window and calculate the mean value of the saved slope $Slope_{mean}$. We define the window with a slope larger than $v \times Slope_{mean}$ as the target window. The value of v will be described in Section 3.3.2.

In order to find the threshold of the Mahalanobis distance more accurately, the target window occurs U times consecutively, and these target windows are saved as sequence SEQ. We use the mean Mahalanobis distance of all samples within the first window of SEQ as the threshold.

$$L = \alpha N \tag{8}$$

$$S = \frac{L}{\beta} \tag{9}$$

$$U = \frac{L}{S} + 1 \tag{10}$$

$$T_{i,j} = \frac{1}{L} \sum_{l=1}^{L} W_{i,j}^{l} \tag{11}$$

As shown in Equations (8)–(11), L is the length of the sliding window, N is the number of samples in the dataset, S is the sliding window step. U is the number of consecutive occurrences of the target window, which can be used to ensure that the sliding window can pass the jump points in the scatter plot. α and β are hyperparameters, $T_{i,j}$ is the Mahalanobis distance threshold at patch position(i,j). $W_{i,j}$ is the first window in the target window sequence SEQ at patch position (i,j).

$$Label^n = \begin{cases} 0, & \text{if } M(x_{i,j}^n) < T_{i,j}, i \in [1,h'], j \in [1,w'] \\ 1, & \text{else} \end{cases}, n \in [1,...,N] \tag{12}$$

$$D_{clean} = \{x^n | Label^n = 0, n \in [1,N]\} \tag{13}$$

After determining distance thresholds, we can select normal samples from the dirty dataset. In Equation (12), $Label^n$ is the label of sample n in the dataset, 0 indicates normal sample, 1 indicates anomaly sample. Considering that anomaly regions have different scales, fixed-scale patch regions do not apply to all anomaly region scales. If an anomaly region is slightly larger than the patch scale, it is difficult to ensure that adjacent patches containing subtle anomalies are filtered out. It is insurance that we exclude this sample once a sample contains an anomaly patch. When the Mahalanobis distance of all patches of the sample x^n is less than the corresponding threshold, we consider sample x^n as a normal sample. Otherwise, we consider sample x^n as an anomaly sample. According to this criterion, we can select some samples from the dirty dataset, and these selected samples are considered normal samples which constitute the clean set D_{clean} shown in Equation (13).

3.3. *Iterative Filtering*

3.3.1. Iterative Processing

As shown in Figure 5, the Gaussian map is estimated based on the dirty dataset. After filtering, we can obtain a clean sample set D_{clean}. The new Gaussian map is estimated based on D_{clean} and used to refilter the dirty dataset. The above iteration is executed until the set of clean samples D_{clean} no longer changes or the number of D_{clean} is more than half of the total number of the dirty dataset.

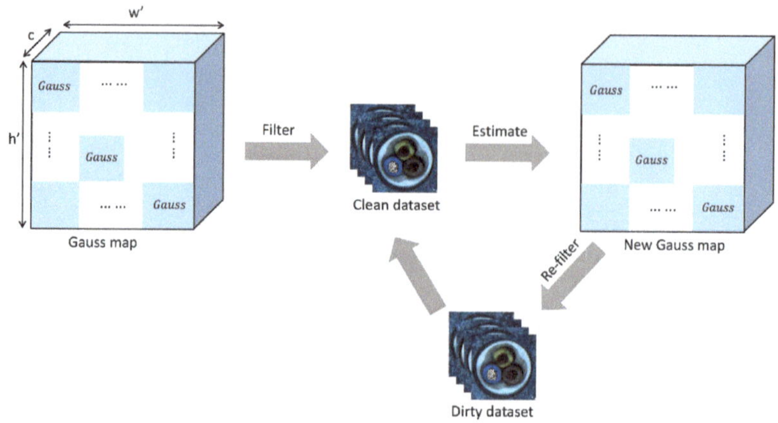

Figure 5. After estimating Gaussian distribution for each position on the dirty dataset, the iterative process is started.

The above iterative filtering process shown in Figure 2 can select normal sample set D_{clean} from the dirty dataset. The new Gaussian map estimated based on D_{clean} is used for anomaly segmentation.

3.3.2. Different Strategy for Sliding Window

In the first and other iterations, we used slightly different strategies to determine the slope threshold of the target window.

As mentioned before, we define the window with a slope greater than a multiplier of the historical mean of the slope as the target window. The value of this multiplier directly affects the determination of the target window. In order to determine this multiplier reasonably, we need to know the trend of the slope in the scatter plot. As shown in Figure 6, the scatter plot at the first iteration is not the same as the scatter plot at a certain iteration. For the first iteration, the starting point is the point with the index at $N \times 1/3$. For a certain iteration, the starting point is the point with a small offset from the jump point. The points between the start point and the endpoint are divided into three segments. We define Equation (14) to measure the trend in the slope of the scatter plot. In Section 3.2, the parameter v concerning the slope threshold $v \times Slope_{mean}$ is defined as Equation (15).

$$R = \frac{Slope_{line3}}{(Slope_{line1} + Slope_{line2}) \times \frac{1}{2}} \tag{14}$$

$$v = \frac{R - \tau}{\eta} + \tau \tag{15}$$

In Equation (14), we use the ratio of the slope of line3 to the mean of the slopes of line1 and line2 to judge the slope trend. If the slope ratio R is close to 1, the line between the starting point and endpoint can be considered a straight line, and we can directly choose the endpoint as the Mahalanobis distance threshold. If the slope ratio R is much greater than 1, it means that the slope of the line between the starting point and the endpoint grows large. It indicates the existence of some patches far away from the Gaussian distribution. In Equation (15), τ is the hyperparameter. If the slope ratio R is less than τ, the slope ratio R can be considered close to 1. Otherwise, a value between τ and R is taken as the slope threshold v to search for the target window. The hyperparameter η is greater than 1, ensuring that v is between τ and R. We tested the effect of different values of η on the performance in Section 5.2.

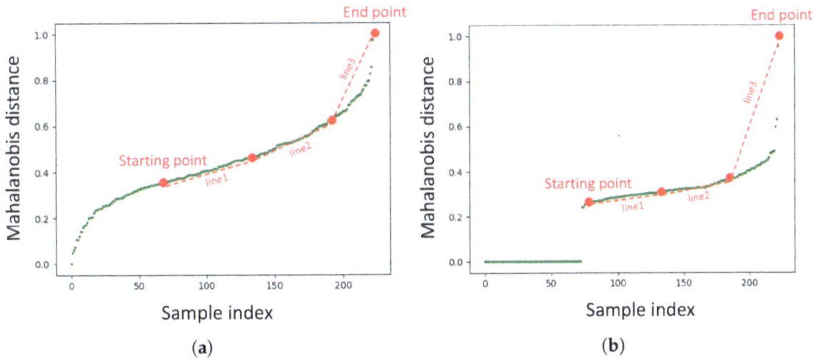

Figure 6. (**a**) The sorted Mahalanobis distance plot for a certain patch in the first iteration. (**b**) The sorted Mahalanobis distance plot of a certain patch at a certain iteration.

3.4. Anomaly Detection and Segmentation

For a test sample x_{test}, the Mahalanobis distance between its patch feature vector $x_{i,j}$ at position (i,j) and the center of the Gaussian distribution at the position (i,j) is used as the anomaly score of the image patch corresponding to the position (i,j). A two-dimensional matrix anomaly map is obtained.

We select the maximum value of the anomaly map as the image-level anomaly score. The anomaly score of each pixel is obtained by restoring the anomaly map to the size of the original image using bilinear interpolation, as shown in Figure 7.

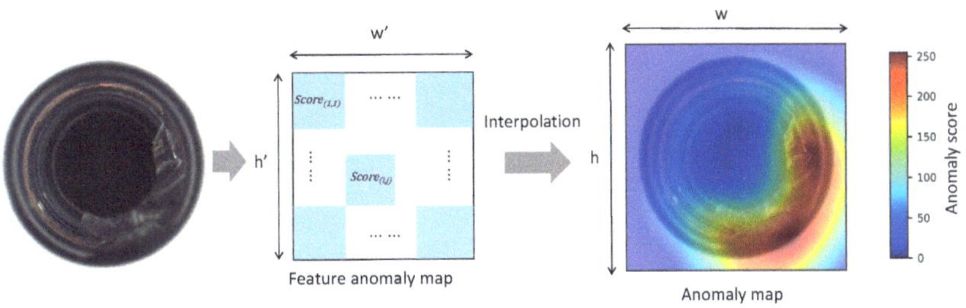

Figure 7. Using Mahalanobis distance as the anomaly score yields the feature anomaly map. The anomaly map is obtained by upsampling the feature anomaly map.

3.5. Flowchart of PaIRE

In Figure 8, all training samples are samples in the dirty dataset. The patch-level Gaussian distributions are estimated based on all samples. For the first iteration, we use the sliding window for the first iteration to compute the Mahalanobis distance threshold. We can obtain clean samples from the dirty dataset based on the Mahalanobis distance threshold. We re-estimate the patch-level Gaussian distribution based on clean samples, and we use the sliding window for other iterations to compute the Mahalanobis distance threshold. After that, we can select clean samples from the dirty dataset. The above iterative process stops when the number of selected clean samples is unchanged, or the number of clean samples is more than half of the number of samples in the dirty dataset. The patch-level Gaussian distributions estimated based on the final clean samples are used for anomaly detection.

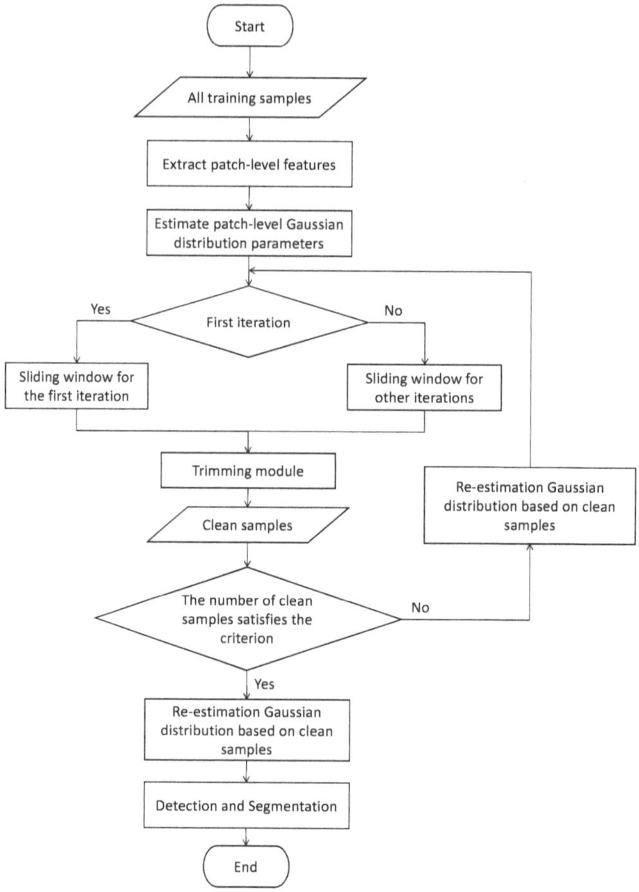

Figure 8. Flowchart of PaIRE.

4. Experiment

This part is divided into three sections: construction of the dirty datasets, comparison experiments, and discussion. Each section will be described below.

4.1. Construction of the Dirty Datasets

In order to construct dirty datasets, we modified the three datasets. These datasets are MVTec, BTAD, and KolektorSDD2. The way to modify each dataset into a dirty dataset will be described below.

- MVTec [35]: An industrial anomaly detection dataset containing 10 object categories and 5 texture categories. Each category has a test set with multiple real anomaly samples and a training set with only normal samples.
  To simulate the dirty dataset, we randomly selected a certain number of anomaly samples from the test set to be added to the training set and removed an equal number of normal samples from the training set. The test set remained unchanged. The number of randomly selected anomaly samples occupy 5%, 10%, 15%, 20%, 25%, and 30% ratio of the training set. We use the data augmentation by rotation and flip if there are insufficient anomaly samples in the test set. The details of the composition of the dirty MVTec dataset are shown in Table 1.

- BTAD [36]: BeanTech Anomaly Detection Dataset. The dataset contains a total of 2830 real-world images of three industrial products, which showcased body and surface defects with the size of 600 × 600. The training set only contains normal samples. The same approach simulates the dirty dataset as MVTec was used. If the real anomaly samples are insufficient, the same rotation and flip data augmentation are used.
- KolektorSDD2 [37]: A surface-defect detection dataset with over 3000 images containing several types of defects obtained while addressing a real-world industrial problem. In the original KSDD2 dataset, the training set contains 10% of the anomaly samples. On the KSDD2 dataset, we use the same strategy as before to simulate the dirty dataset with noise ratios of 15%, 20%, 25%, and 30%.

Table 1. Components of the dirty MVTec dataset. The upper row is the number of normal samples and the lower row is the number of anomaly samples. The bolded font means that the number of the anomaly samples in the test set were insufficient and we used data augmentation for them.

Category	5%	10%	15%	20%	25%	30%
bottle	198	188	178	167	157	146
	11	21	31	42	52	63
cable	213	201	190	179	168	157
	11	23	34	45	56	67
capsule	208	197	186	175	164	153
	11	122	33	44	55	66
carpet	266	252	238	224	210	196
	14	28	42	56	70	84
grid	250	238	224	211	198	185
	14	26	40	53	**66**	**79**
hazelnut	371	352	332	313	293	274
	20	39	59	**78**	**98**	**117**
leather	233	221	208	196	184	172
	12	24	37	49	61	73
metal nut	209	198	187	176	165	154
	11	22	33	44	55	66
pill	254	240	227	214	200	187
	13	27	40	53	67	80
screw	304	288	272	256	240	224
	16	32	48	64	80	96
tile	219	207	196	184	173	161
	11	23	34	46	57	69
toothbrush	57	54	51	48	45	42
	3	6	9	12	15	18
transistor	202	192	181	170	160	149
	11	21	32	**43**	**53**	**64**
wood	235	222	210	198	185	173
	12	25	37	49	**62**	**74**
zipper	228	216	204	192	180	168
	12	24	36	48	60	72

4.2. Comparison Experiments

4.2.1. Experiments on Dirty MVTec Dataset

We compared our method with three others: CutPaste, PaDim, DFR. The image was resized to 256 × 256 for training and inference. In formula (3), we set $k = 10$, $s = 4$. In Equations (8) and (9), we set $\alpha = 0.05$, $\beta = 3$. In Equation (15), we set $\tau = 1.15$ and $\eta = 2.3$.

We tested the Image-level detection performance of the compared methods at noise ratios of 5%, 10%, 15%, 20%, 25%, and 30%, respectively. As shown in Figure 9, the y-axis is the mean image-level AUROC value for all categories, and the x-axis is the different noise ratios. Our proposed PaIRE method outperforms all comparison methods when the dataset contains noise. PaDim and CutPaste are sensitive to noise and suffered from severe performance degradation. Detailed comparison data with different noise ratios are shown in Table 2. At each noise ratio, the bold format indicates the best performance.

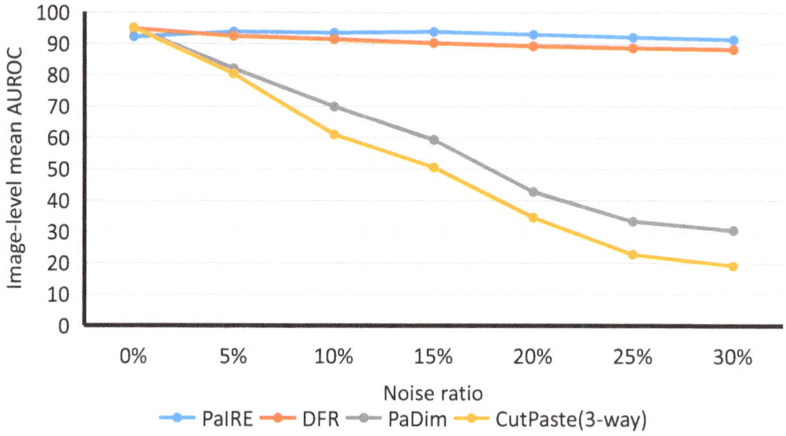

Figure 9. Comparison on MVTec dataset.

Table 2. Comparison with SOTA methods at different noise ratios.

Noise Ratio	PaIRE	PaDim	DFR	CutPaste
0%	92.19	95.06	94.90	**95.2**
5%	**93.86**	82.11	92.51	80.48
10%	**93.50**	69.95	91.49	61.06
15%	**93.79**	59.27	90.25	50.58
20%	**92.94**	42.75	89.30	34.71
25%	**92.05**	33.31	88.70	22.94
30%	**91.36**	30.5	88.26	19.26

The anomaly segmentation performance of PaIRE at different noise ratios is shown in Figure 10. As shown in Table 2 and Figure 10a, it can be seen that both image-level detection and pixel-level detection have only a slight performance degradation when the noise ratio is larger than 15%. As shown in Figure 10b, PaIRE can precisely locate the anomaly area in the image.

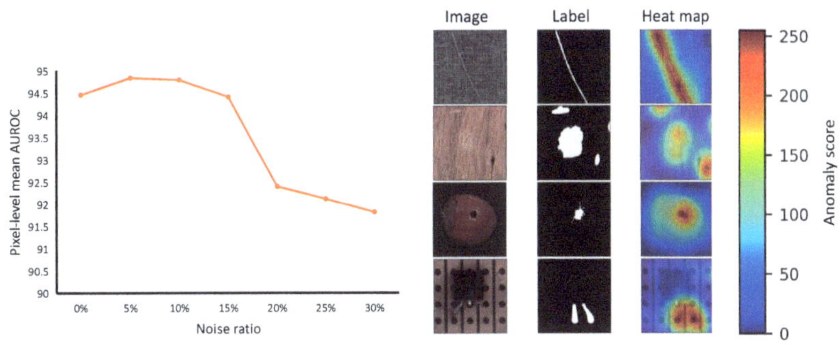

(a) (b)

Figure 10. (**a**) Pixel-level mean AUROC at different noise ratios. (**b**) Segmentation result at a noise ratio of 15%.

4.2.2. Experiments on Dirty BTAD Dataset

For the comparison on the BTAD dataset, all hyperparameters are the same as for the MVTec dataset. The image is resized to 256 × 256 for training and inference. The comparison results are shown in Figure 11.

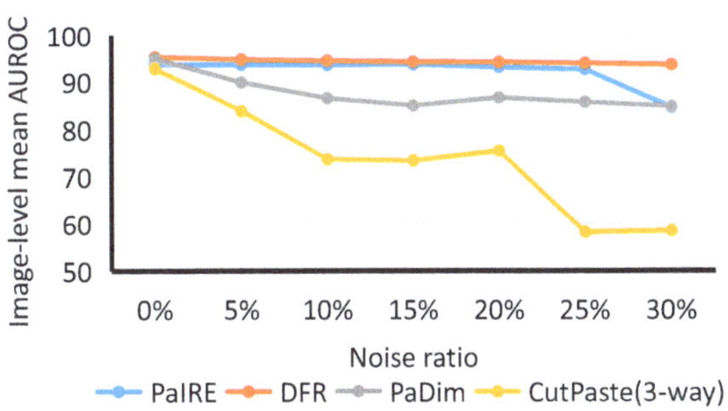

Figure 11. Comparison on BTAD dataset.

As shown in Figure 11, PaIRE achieved state-of-the-art results except at a 30% noise ratio. Both PaIRE and DFR show good noise resistance on the BTAD dataset. PaDim and CutPaste are heavily affected by noise, the same as for the performance on the MVTec dataset.

4.2.3. Experiments on Dirty KSDD2 Dataset

For the comparison on the KSDD2 dataset, we used the same hyperparameters as before, except for the image size. The image is resized to 224 × 224 for training and inference.

The original KSDD2 dataset contains 10% anomaly samples, so we started our comparison from a 10% noise ratio. As shown in Figure 12, PaIRE also achieved a state-of-the-art performance on the KSDD2 dataset. The hyperparameters of PaIRE based on MVTec tuning achieved a state-of-the-art performance on both BTAD and KSDD2 datasets, indicating the generality of the hyperparameters.

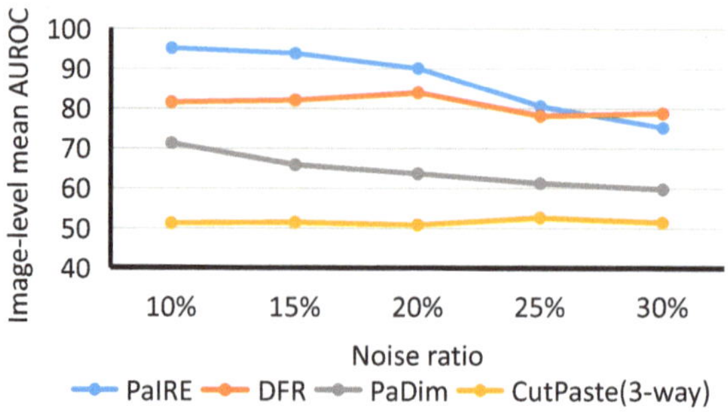

Figure 12. Comparison on KSDD2 dataset.

4.3. Discussion

We will discuss the results of the comparison experiments on MVTec, BTAD, and KSDD2 datasets.

PaDim is a parametric probabilistic model, and it needs to estimate the Gaussian distribution of normal samples. However, PaDim does not consider the noisy samples in the dataset when estimating the parameters of the Gaussian distribution, and the estimated Gaussian distribution deviates severely from the distribution of normal samples, resulting in a degradation of the model performance.

The self-supervised method CutPaste is sensitive to noisy samples because it considers all samples in the dataset as normal samples, creates artificial anomaly samples by applying special data augmentation to the samples in the dataset, and trains a classifier to discriminate normal samples from artificial anomaly samples. When the dataset contains noise, the noisy samples will generate false labels to mislead the classifier resulting in a degradation of the model performance.

The reconstruction method DFR uses a pretrained network to extract the feature map and reconstruct it using an autoencoder. DFR uses PCA to determine the size of the latent space of the autoencoder. Benefiting from the noise reduction effect of PCA, DFR has better noise resistance than PaDim and CutPaste.

Our proposed PaIRE belongs to the probabilistic model, which can estimate the Gaussian distribution of normal samples from the dirty dataset by filtering the noisy samples and increasing the number of clean samples. PaIRE achieves advanced results in all three datasets.

5. Ablation Study

5.1. Trimming Method and Iterative Filtering

Our proposed framework PaIRE has two modules for anomaly samples in the training set: trimming method in robust statistics and iterative filtering. We add the trimming module and the iterative filtering module one by one to explore the contribution of these modules. We gradually tested on MVTec to explore the impact of the module on the average AUROC of all categories at all noisy ratios.

As shown in Figure 13, When no module is added, the anomaly samples severely affect the model performance. After adding the trimming module, the model performance is improved tremendously. This shows that the trimming module can effectively exclude noisy samples so that the estimated Gaussian distribution is closer to the distribution of normal samples. After adding the iteration module, more clean samples are selected so that the estimated distribution is further closer to the distribution of normal samples.

As shown in Figure 14, The x-axis indicates the different noise ratios, and the y-axis indicates the mean of the number of clean samples of all categories. After adding the iteration module, more clean samples were selected from the dirty dataset, but the model performance increased only slightly, as shown in Figure 13. This is because most of the normal samples are similar, and more similar samples do not increase the model performance significantly.

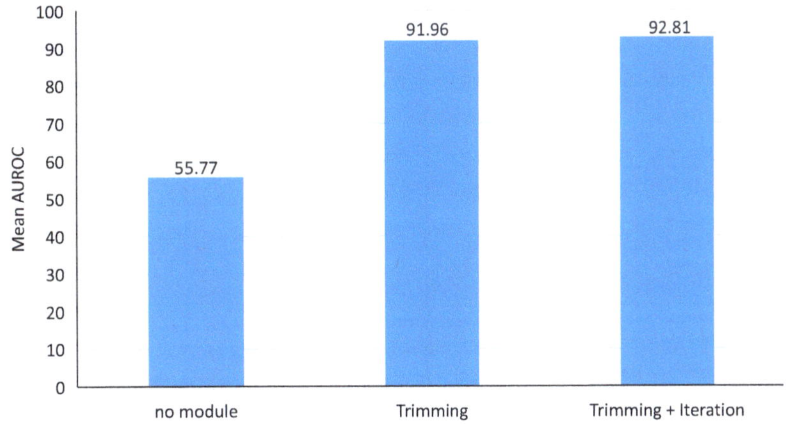

Figure 13. Trimming and iteration module performance testing.

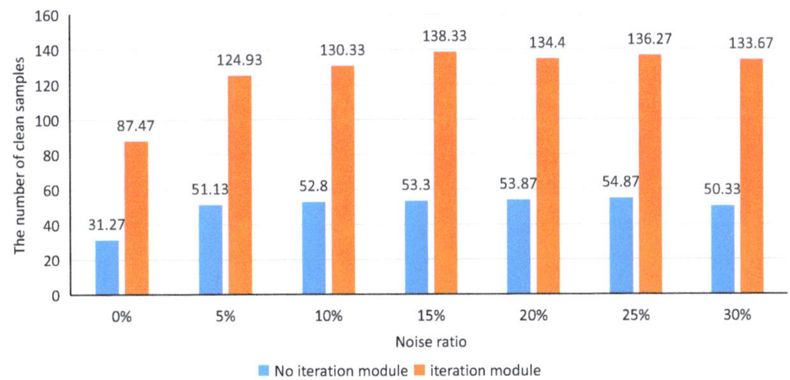

Figure 14. Comparison of the number of clean samples with and without the iteration module.

5.2. Effect of Mahalanobis Distance Threshold on Performance

We performed experiments to investigate the effect of hyperparameter η in Equation (15), which relates to the slope threshold $v * Slope_{mean}$. The $v * Slope_{mean}$ is used to determine the target window. With $k = 10$, $s = 4$ in Equation (3), we tested the performance of image-level detection for different η. The experiments are based on the MVTec dataset.

As shown in Figure 15, The x-axis represents the different hyperparameter η, and the y-axis represents the mean value of image-level AUROC at 5%, 15%, and 25% noise ratio. In Equation (15), when η is small, the Mahalanobis distance threshold is large. It leads to anomaly samples not being effectively excluded, and the estimated Gaussian distribution deviates from the distribution of normal samples. When η is large, the Mahalanobis distance threshold is small, leading to a small number of selected clean samples, and the estimated Gaussian distribution cannot accurately match the distribution of normal samples. The ablation experiment shows that the average performance is highest when $\eta = 2.3$.

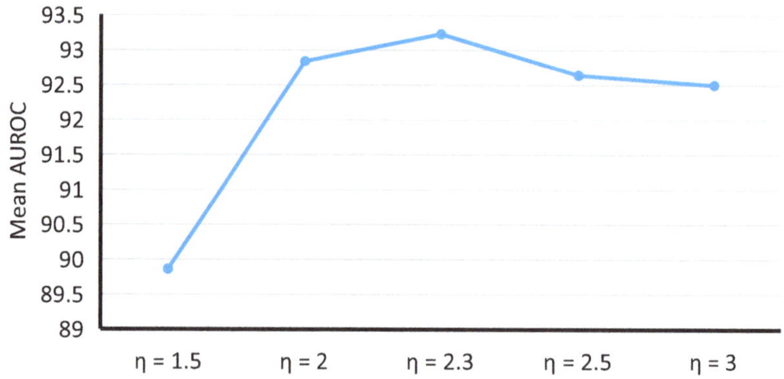

Figure 15. Effect of hyperparameter η on performance.

6. Conclusions

We propose an unsupervised anomaly detection and segmentation framework, PaIRE, in the case of training datasets containing noise samples. PaIRE reduces the effect of noisy samples on the estimation of Gaussian distribution parameters by filtering them from the dirty dataset. Compared to the state-of-the-art methods, PaIRE has better noise resistance and anomaly-detection performance.

The methods based on image patches have a limitation. These methods assume that patches at the same position should be independently identically distributed. The texture category and aligned category satisfy this assumption, but the nonalignment category may not satisfy it. Aligning targets in image space or feature space to improve anomaly-detection performance will be our next work.

Author Contributions: Supervision, L.W.; methodology, J.G.; data preparation, X.Y.; writing—original draft, J.G.; writing—review, L.W. All authors have read and agreed to the published version of the manuscript.

Funding: This research received no external funding.

Data Availability Statement: All datasets used in this paper are publicly available, MVTec at https://www.mvtec.com/company/research/datasets/mvtec-ad/ (accessed on 10 May 2021), BTAD at http://avires.dimi.uniud.it/papers/btad/btad.zip (accessed on 3 November 2021), KSDD2 at https://www.vicos.si/resources/kolektorsdd2/ (accessed on 9 November 2021).

Acknowledgments: No external contributions or funding.

Conflicts of Interest: The authors declare no conflict of interest.

References

1. Kumar, A. Computer-vision-based fabric defect detection: A survey. *IEEE Trans. Ind. Electron.* **2008**, *55*, 348–363. [CrossRef]
2. Pang, G.; Shen, C.; Cao, L.; Hengel, A.V.D. Deep learning for anomaly detection: A review. *ACM Comput. Surv.* **2021**, *54*, 1–38. [CrossRef]
3. Czimmermann, T.; Ciuti, G.; Milazzo, M.; Chiurazzi, M.; Roccella, S.; Oddo, C.M.; Dario, P. Visual-based defect detection and classification approaches for industrial applications—A survey. *Sensors* **2020**, *20*, 1459. [CrossRef] [PubMed]
4. Ruff, L.; Kauffmann, J.R.; Vandermeulen, R.A.; Montavon, G.; Samek, W.; Kloft, M.; Dietterich, T.G.; Müller, K.-R. A unifying review of deep and shallow anomaly detection. *Proc. IEEE* **2021**, *109*, 756–795. [CrossRef]
5. Chalapathy, R.; Chawla, S. Deep learning for anomaly detection: A survey. *arXiv* **2019**, arXiv:1901.03407.
6. Wilcox, R.R.; Rousselet, G.A. A guide to robust statistical methods in neuroscience. *Current Protoc. Neurosci.* **2018**, *82*, 8–42. [CrossRef]
7. Schölkopf, B.; Platt, J.C.; Shawe-Taylor, J.; Smola, A.J.; Williamson, R.C. Estimating the support of a high-dimensional distribution. *Neural Comput.* **2001**, *3*, 1443–1471. [CrossRef]
8. Yi, J.; Yoon, S. Patch svdd: Patch-level svdd for anomaly detection and segmentation. In Proceedings of the Asian Conference on Computer Vision, Kyoto, Japan, 30 November–4 December 2020.

9. Tax, D.M.; Duin, R.P. Support vector data description. *Mach. Learn.* **2004**, *54*, 45–66. [CrossRef]
10. Golan, I.; El-Yaniv, R. Deep anomaly detection using geometric transformations. *arXiv* **2018**, arXiv:1805.10917.
11. Sohn, K.; Li, C.-L.; Yoon, J.; Jin, M.; Pfister, T. Learning and evaluating representations for deep one-class classification. *arXiv* **2020**, arXiv:2011.02578.
12. Chen, T.; Kornblith, S.; Norouzi, M.; Hinton, G. A simple framework for contrastive learning of visual representations. In *Proceedings of the International Conference on Machine Learning, Virtual, 13–18 July 2020*; PMLR: Cambridge, MA, USA, 2020; pp. 1597–1607.
13. Schlüter, H.M.; Tan, J.; Hou, B.; Kainz, B. Self-supervised out-of-distribution detection and localization with natural synthetic anomalies (nsa). *arXiv* **2021**, arXiv:2109.15222.
14. Li, C.-L.; Sohn, K.; Yoon, J.; Pfister, T. Cutpaste: Self-supervised learning for anomaly detection and localization. In Proceedings of the IEEE/CVF Conference on Computer Vision and Pattern Recognition, Nashville, TN, USA, 19–25 June 2021; pp. 9664–9674.
15. Rippel, O.; Mertens, P.; Merhof, D. Modeling the distribution of normal data in pre-trained deep features for anomaly detection. In Proceedings of the 2020 25th International Conference on Pattern Recognition (ICPR), Milan, Italy, 10–15 January 2021; pp. 6726–6733.
16. Defard, T.; Setkov, A.; Loesch, A.; Audigier, R. Padim: A patch distribution modeling framework for anomaly detection and localization. In Proceedings of the ICPR 2020-25th International Conference on Pattern Recognition Workshops and Challenges, Milan, Italy, 10–11 January 2021.
17. Zheng, Y.; Wang, X.; Deng, R.; Bao, T.; Zhao, R.; Wu, L. Focus your distribution: Coarse-to-fine non-contrastive learning for anomaly detection and localization. *arXiv* **2021**, arXiv:2110.04538.
18. Cohen, N.; Hoshen, Y. Sub-image anomaly detection with deep pyramid correspondences. *arXiv* **2020**, arXiv:2005.02357.
19. Wang, G.; Han, S.; Ding, E.; Huang, D. Student-teacher feature pyramid matching for anomaly detection. *arXiv* **2021**, arXiv:2103.04257.
20. Schlegl, T.; Seeböck, P.; Waldstein, S.M.; Schmidt-Erfurth, U.; Langs, G. Unsupervised anomaly detection with generative adversarial networks to guide marker discovery. In *International Conference on Information Processing in Medical Imaging, Boone, NC, USA, 25–30 June 2017*; Springer: Berlin/Heidelberg, Germany, 2017; pp. 146–157.
21. Schlegl, T.; Seeböck, P.; Waldstein, S.M.; Langs, G.; Schmidt-Erfurth, U. f-anogan: Fast unsupervised anomaly detection with generative adversarial networks. *Med. Image Anal.* **2019**, *54*, 30–44. [CrossRef] [PubMed]
22. Deecke, L.; Vandermeulen, R.; Ruff, L.; Mandt, S.; Kloft, M. Image anomaly detection with generative adversarial networks. In *Poceedings of the Joint European Conference on Machine Learning and Knowledge Discovery in Databases, Riva del Garda, Italy, 19–23 September 2018*; Springer: Berlin/Heidelberg, Germany, 2018; pp. 3–17.
23. Sabokrou, M.; Khalooei, M.; Fathy, M.; Adeli, E. Adversarially learned one-class classifier for novelty detection. In Proceedings of the IEEE Conference on Computer Vision and Pattern Recognition, Salt Lake City, UT, USA, 18–23 June 2018; pp. 3379–3388.
24. Akcay, S.; Atapour-Abarghouei, A.; Breckon, T.P. Ganomaly: Semi-supervised anomaly detection via adversarial training. In *Proceedings of the Asian conference on computer vision, Perth, Australia, 2–6 December 2018*; Springer: Berlin/Heidelberg, Germany, 2018; pp. 622–637.
25. Venkataramanan, S.; Peng, K.-C.; Singh, R.V.; Mahalanobis, A. Attention guided anomaly localization in images. In *Proceedings of the European Conference on Computer Vision, Glasgow, UK, 23–28 August 2020*; Springer: Berlin/Heidelberg, Germany, 2020; pp. 485–503.
26. Bergmann, P.; Löwe, S.; Fauser, M.; Sattlegger, D.; Steger, C. Improving unsupervised defect segmentation by applying structural similarity to autoencoders. *arXiv* **2018**, arXiv:1807.02011.
27. Gong, D.; Liu, L.; Le, V.; Saha, B.; Mansour, M.R.; Venkatesh, S.; Hengel, A.v.d. Memorizing normality to detect anomaly: Memory-augmented deep autoencoder for unsupervised anomaly detection. In Proceedings of the IEEE/CVF International Conference on Computer Vision, Seoul, Korea, 27–28 October 2019; pp. 1705–1714.
28. Li, Z.; Li, N.; Jiang, K.; Ma, Z.; Wei, X.; Hong, X.; Gong, Y. Superpixel masking and inpainting for self-supervised anomaly detection. *BMVC*, **2020**. Available online: https://www.bmvc2020-conference.com/assets/papers/0275.pdf (accessed on 27 February 2022).
29. An, J.; Cho, S. Variational autoencoder based anomaly detection using reconstruction probability. *Spec. Lect. IE* **2015**, *2*, 1–18.
30. You, S.; Tezcan, K.C.; Chen, X.; Konukoglu, E. Unsupervised lesion detection via image restoration with a normative prior. In Proceedings of the International Conference on Medical Imaging with Deep Learning, London, UK, 8–10 July 2019; pp. 540–556.
31. Pirnay, J.; Chai, K. Inpainting transformer for anomaly detection. *arXiv* **2021**, arXiv:2104.13897.
32. Tan, D.S.; Chen, Y.-C.; Chen, T.P.-C.; Chen, W.-C. Trustmae: A noise-resilient defect classification framework using memory-augmented auto-encoders with trust regions. In Proceedings of the IEEE/CVF Winter Conference on Applications of Computer Vision, Waikoloa, HI, USA, 4–8 January 2021; pp. 276–285.
33. Li, T.; Wang, Z.; Liu, S.; Lin, W.-Y. Deep unsupervised anomaly detection. In Proceedings of the IEEE/CVF Winter Conference on Applications of Computer Vision, Waikoloa, HI, USA, 4–8 January 2021; pp. 3636–3645.
34. Yang, J.; Shi, Y.; Qi, Z. Dfr: Deep feature reconstruction for unsupervised anomaly segmentation. *arXiv* **2020**, arXiv:2012.07122.
35. Bergmann, P.; Fauser, M.; Sattlegger, D.; Steger, C. Mvtec ad–a comprehensive real-world dataset for unsupervised anomaly detection. In Proceedings of the IEEE/CVF Conference on Computer Vision and Pattern Recognition, Long Beach, CA, USA, 15–20 June 2019; pp. 9592–9600.

36. Božič, J.; Tabernik, D.; Skočaj, D. Mixed supervision for surface-defect detection: from weakly to fully supervised learning. *Comput. Ind.* **2021**, *129*, 103459. [CrossRef]
37. Mishra, P.; Verk, R.; Fornasier, D.; Piciarelli, C.; Foresti, G.L. Vt-adl: A vision transformer network for image anomaly detection and localization. *arXiv* **2021**, arXiv:2104.10036.

information

Article

A LiDAR–Camera Fusion 3D Object Detection Algorithm

Leyuan Liu [1], Jian He [1,2,*], Keyan Ren [1,2,*], Zhonghua Xiao [3] and Yibin Hou [1,2]

1 Faculty of Information Technology, Beijing University of Technology, Beijing 100124, China; chinaleyuan@emails.bjut.edu.cn (L.L.); ybhou@bjut.edu.cn (Y.H.)
2 Beijing Engineering Research Center for IOT Software and Systems, Beijing University of Technology, Beijing 100124, China
3 Suzhou Exinova Robot Technology Co., Ltd., Suzhou 215163, China; xiaozhonghua@exinova.cn
* Correspondence: jianhee@bjut.edu.cn (J.H.); keyanren@bjut.edu.cn (K.R.)

Abstract: 3D object detection with LiDAR and camera fusion has always been a challenge for autonomous driving. This work proposes a deep neural network (namely FuDNN) for LiDAR–camera fusion 3D object detection. Firstly, a 2D backbone is designed to extract features from camera images. Secondly, an attention-based fusion sub-network is designed to fuse the features extracted by the 2D backbone and the features extracted from 3D LiDAR point clouds by PointNet++. Besides, the FuDNN, which uses the RPN and the refinement work of PointRCNN to obtain 3D box predictions, was tested on the public KITTI dataset. Experiments on the KITTI validation set show that the proposed FuDNN achieves AP values of 92.48, 82.90, and 80.51 at easy, moderate, and hard difficulty levels for car detection. The proposed FuDNN improves the performance of LiDAR–camera fusion 3D object detection in the car category of the public KITTI dataset.

Keywords: 3D object detection; LiDAR–camera fusion; LiDAR point cloud; KITTI benchmark

Citation: Liu, L.; He, J.; Ren, K.; Xiao, Z.; Hou, Y. A LiDAR–Camera Fusion 3D Object Detection Algorithm. *Information* **2022**, *13*, 169. https://doi.org/10.3390/info13040169

Academic Editor: Stefano Berretti

Received: 21 February 2022
Accepted: 24 March 2022
Published: 26 March 2022

1. Introduction

Object detection is one of the issues that has received much attention in autonomous driving. In the process of autonomous driving, cars need to detect and track objects in real-time, such as cars, bicycles, pedestrians, etc. [1]. According to the types of perception sensors, object detection can be divided into camera-based object detection, LiDAR-based 3D object detection, and LiDAR–camera fusion object detection [2].

Over recent years, camera-based 2D object detection had achieved unprecedented progress. Starting from the RCNN (Region with CNN Feature) [3], camera-based 2D object detection algorithms began to develop rapidly. A series of studies [4,5] started to use two-stage approaches, using a Region Proposal Network (RPN) to propose candidate proposals and then refining the candidate proposals for classification. Redmon et al. [6] originally proposed the one-stage object detection architecture, which provided fast and simple architecture, but the effect was not as good as the two-stage approaches. Lin et al. [7] employed a focal loss function that enables the one-stage approach to outperform the two-stage approaches in both accuracy and efficiency. Although camera-based 2D object detection algorithms had achieved excellent results, they were easily affected by factors such as lighting and weather in autonomous driving scenarios [8]. CaDDN [9] generated depth distributions for each pixel of the image and combined them with image features for camera-based 3D object detection. GUPNet [10] alleviated the error amplification problem of inference by computing deep uncertainty in monocular 3D object detection. As the depth obtained from the image was not as accurate as LiDAR, the detection performance improvement was limited.

In recent years, with the continuous improvement of LiDAR hardware, LiDAR-based 3D object detection research has increased. LiDAR-based 3D object detection mainly includes three categories: Voxel-based methods, point-based methods, and graph-based methods [11], as shown in Table 1. The voxel-based methods first divided the point clouds into voxels and then input the voxelized point clouds into the backbone network for feature extraction. For instance, VoxelNet [12] extracted features from equidistant 3D voxelized point clouds and fed them to RPN to predict 3D bounding boxes. However, VoxelNet was very slow due to the 3D convolutions. SECOND [13] improved the efficiency of VoxelNet with a sparse convolutional network (SparsConv) and improved the detection performance through data augmentation. Shi et al. [14] combined the point method with SECOND and proposed PV-RCNN, which achieved high detection results. After PV-RCNN, the voxel-based methods could not be improved on a large scale. The point-based methods directly extracted features from point clouds. Since PointNet [15] and PointNet++ [16] could effectively extract point cloud features, they were used as a backbone by many methods. PointRCNN [17] used PointNet++ as the backbone to extract the features of point clouds and generate candidate boxes. The candidate boxes were then refined to generate the final detection results. STD [18], which had a high recall and low computational time, adopted a spherical anchor mechanism to generate proposals and implemented a parallel intersection-over-union (IoU) branch to improve positioning accuracy and detection performance. The graph-based methods used a graph neural network (GNN) to extract features of point clouds. For example, Point-GNN [19] efficiently encoded point clouds in a fixed-radius nearest-neighbor graph and used a GNN to obtain the 3D box predictions. Since the LiDAR point cloud of distant objects is very sparse, it encounters difficulties in detecting distant or small objects [20]. Moreover, the LiDAR point cloud provides little color and texture information, which limits further performance improvements [21].

Recently, some studies have explored LiDAR–camera fusion object detection methods, as shown in Table 1. Some multi-view methods transformed point clouds into a bird's-eye view (BEV), front view (FV), or range view (RV), and then fused the features of these views with image features [22,23]. For instance, MV3D [24] generated 3D object proposals from BEVs, which were projected to BEVs, FVs, and image views. The features of different views were then fused to predict 3D object-bounding boxes. AVOD [25] extended MV3D to extract equal-sized features from BEVs and image views and then fused these features using an average pooling operation. The disadvantages of multi-view methods are computationally intensive and time-consuming. Some frustum-based methods utilized existing 2D detectors to generate frustum proposals in the image and then learned the corresponding frustum point cloud features for 3D box predictions. For instance, Frustum PointNet [26] generated 2D object region proposals in images via a CNN, and then extruded each 2D region to a 3D frustum. Finally, PointNet was used to predict a 3D bounding box of each object from the points in the frustum. The performances of frustum-based methods are easily limited by 2D image detectors. Later methods tended to design independent backbone networks to extract features from RGB images and raw LiDAR point clouds, and then the two kinds of features were fused for 3D box predictions. For instance, 3D-CVF [27] used an image backbone to extract image features and converted them to BEV domain features. Next, a gated feature fusion network was utilized to combine image features with point cloud features. At last, a fusion network based on the 3D region of interest (ROI) was used to refine proposals and predict the output. EPNet [28] fused image features with point cloud features five times, which improved the prediction performance. PI-RCNN [29] proposed a multi-task LiDAR–camera fusion object detection method that combined image segmentation and 3D detection. These methods using two independent feature extractors have made progress, but the performance needs to be further improved [30].

Table 1. Classification of different methods using LiDAR.

Modality	Method Category	Method
LiDAR-based	Voxel-based methods Point-based methods Graph-based methods	VoxelNet [12], SECOND [13], PV-RCNN [14], PointNet [15], PointNet++ [16], PointRCNN [17], STD [18] Point-GNN [19]
LiDAR–camera fusion	Multi-view methods Frustum-based methods Independent backbone methods	MV3D [24], AVOD [25] Frustum PointNet [26] 3D-CVF [27], EPNet [28], PI-RCNN [29]

Focusing on the problem that the accuracy of LiDAR–camera fusion object detection methods is not high enough, this work proposes a new deep neural network (namely FuDNN) for LiDAR–camera fusion 3D object detection. First, a 2D backbone is proposed to learn 2D features from camera images. Second, an attention-based fusion sub-network is designed to fuse the 2D features and the 3D features extracted from 3D LiDAR point clouds. Finally, the FuDNN is verified in the experiments using the KITTI [31] 3D object detection benchmark dataset. The major contributions of this work are two-fold. First, the proposed 2D backbone has a more compact structure than the commonly used 2D backbones but has better performance. Second, the proposed attention-based fusion sub-network only needs to fuse point cloud features and image features once to achieve better results. The follow-up content is organized as follows: Section 2 introduces the architecture of FuDNN and the overall loss function. Section 3 elaborates the KITTI dataset, the FuDNN training, the performance metrics, and the analysis of the results. Section 4 provides a summary of the entire work and the follow-up research directions.

2. FuDNN for 3D Object Detection

2.1. FuDNN Architecture

A deep learning network based on PointRCNN, named FuDNN, is designed to realize 3D object detection. The architecture of FuDNN is shown in Figure 1, including a 2D backbone, a 3D backbone, an attention-based fusion sub-network, an RPN, and a 3D box refinement network. The 2D backbone is designed to learn 2D features from camera images automatically, and the attention-based fusion sub-network is proposed to fuse the 2D features with the 3D features extracted by PointNet++. The RPN and 3D box refinement network of PointRCNN are used to generate 3D proposals and refine the 3D box locations, respectively.

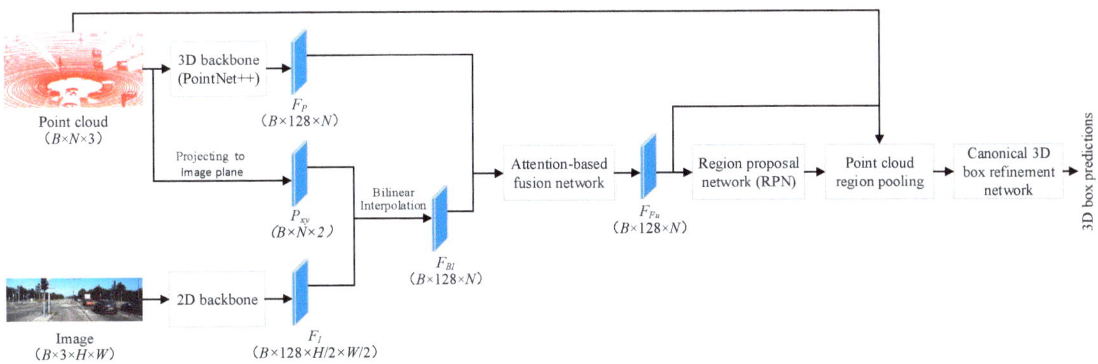

Figure 1. Network architecture of the FuDNN. *B*, *N*, *H*, and *W* represent the batch size of FuDNN, the number of LiDAR points, the height, and width of image, respectively.

The input of FuDNN is point clouds and RGB images. The shape of input images is a matrix of $B \times 3 \times H \times W$, where B, H, and W represent the batch size of FuDNN, the height, and the width of the image, respectively. The shape of input point clouds is a matrix of $B \times 3 \times N$, where N is the number of LiDAR points.

The architecture of the 2D backbone is shown in Figure 2. The first layer (Conv1) of the 2D backbone is a 2D convolutional layer with 128 convolution kernels, each 7×7, using a stride of 1. The output of Conv1 is a matrix of $B \times 128 \times H \times W$. The research of Ioffe and Szegedy [32] showed that the batch normalization method could speed up network convergence and reduce training difficulty, so a batch normalization layer (BN1) was used after Conv1. A rectified linear unit layer (ReLU1) is added after the BN1 layer to avoid gradient disappearance. The output shape of BN1 and ReLU1 is the same as that of Conv1. To reduce the amount of calculation, a max-pooling layer (S1) with a kernel size of 2×2 is adopted after ReLU1, and its output shape is $B \times 128 \times H/2 \times W/2$. Next, the similar structures repeat two times: The 2D convolutional layer (Conv2) with 256 convolution kernels, each 5×5, is followed by the batch normalization layer (BN2). Then there is the ReLU2. At last, the 2D convolutional layer (Conv3) with 128 convolution kernels, each 3×3, is followed by the ReLU3. The 2D backbone outputs the image feature matrix F_I of the shape $B \times 128 \times H/2 \times W/2$.

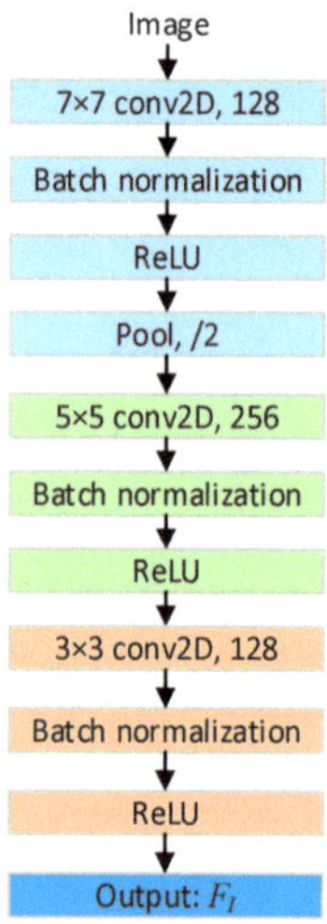

Figure 2. Network architecture of the 2D backbone.

Since point cloud features and image features lack correspondence, it is necessary to establish the correspondence between image features and point cloud features before projecting the point clouds to the image plane. The point clouds (x, y, z) are mapped to the image plane (u, v) via a transformation matrix $\mathbf{M}$, as shown in Equations (1) and (2):

$$(u,\ v,\ 1)^T = \mathbf{M}\cdot(x,\ y,\ z,\ 1)^T \quad (1)$$

$$\mathbf{M} = \mathbf{P}_{rect}\cdot\begin{pmatrix} \mathbf{R}_{LiD}^{cam} & \mathbf{t}_{LiD}^{cam} \\ 0 & 1 \end{pmatrix} \quad (2)$$

where $\mathbf{P}_{rect}$ is the projection matrix, $\mathbf{R}_{LiD}^{cam}$ is the 3×3 rotation matrix, and $\mathbf{t}_{LiD}^{cam}$ is the 1×3 translation vector from the LiDAR to the camera, respectively. Projecting the point clouds to the image plane obtains a $B \times N \times 2$ matrix P_{xy}.

The point cloud feature F_P and the image feature F_I need to have the same dimension to be fused, so the bilinear interpolation method is used to sample F_I to obtain the same dimension as F_P. First, P_{xy} is expanded from $B \times N \times 2$ to obtain a $B \times 1 \times N \times 2$ matrix. Second, with the $B \times 1 \times N \times 2$ matrix as the guide map, F_I is sampled by bilinear interpolation to obtain the feature map of $B \times 128 \times 1 \times N$, which is then removed by the dimension of 1 to obtain the final feature map F_{BI} of $B \times 128 \times N$.

Figure 3 shows the structure of the attention-based fusion sub-network. Since pictures are easily affected by light and occlusion and introduce interference information, an attention mechanism is introduced to weight image features according to their importance. Before applying the attention mechanism, F_{BI} is fed to the fully connected layer to obtain the vector F_{FC} with the shape $BN \times 128$. Then, the attention mechanism is applied to F_{FC}, as shown in Equations (3)–(5). Among them, w and b are parameters learned during training to obtain the hidden representation h from the input F_{FC}. The shape of h is $BN \times 128$. The multiplication result of F_{FC} and w is a matrix product. The vector u_w, which is randomly initialized and learned during training, is introduced to capture context. The similarity is used as a measure of importance and is obtained by the matrix product of h and u_w. The similarity is processed by the sigmoid function to obtain the normalized attention weight vector a with the shape $BN \times 1$. Then a is transformed into $B \times 1 \times N$. The output of the attention mechanism is F_{Att}, which is obtained by the dot product of a and F_{FC}. Finally, F_{Att} is concatenated with F_P to obtain the fused feature F_{Fu}.

$$h = \tanh(wF_{FC} + b) \tag{3}$$

$$a = \text{sigmoid}\left(h^T u_w\right) \tag{4}$$

$$F_{Att} = aF_{FC} \tag{5}$$

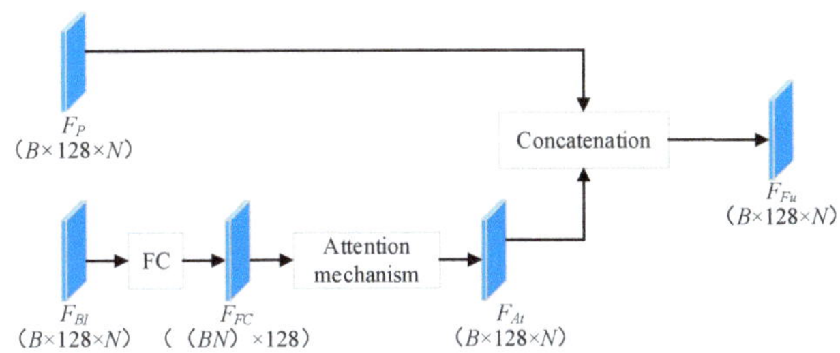

Figure 3. Architecture of the fusion network.

The RPN of PointRCNN is utilized to generate 3D bounding box proposals. Given the images and point cloud fused features F_{Fu}, a segmentation head is used to estimate the foreground mask, and a box regression head is used to generate 3D proposals. The 3D proposals are generated from segmented foreground points, which avoids using a large number of pre-defined 3D boxes in 3D space. The 3D points and corresponding point features are pooled according to the location of each 3D proposal to eliminate proposals without interior points. The canonical 3D box refinement network of PointRCNN is employed to refine the box locations and finally obtain 3D box predictions.

2.2. Overall Loss Function

The overall loss function is mainly based on PointRCNN, and the consistency enforcing te loss of EPNet is introduced. The proposed FuDNN is optimized with a loss function $\mathcal{L}_{total}$, which is the sum of the RPN loss $\mathcal{L}_{rpn}$ and the refinement network loss $\mathcal{L}_{refine}$. The loss function $\mathcal{L}_{total}$ can be formulated as:

$$\mathcal{L}_{total} = \mathcal{L}_{rpn} + \mathcal{L}_{refine} \tag{6}$$

The RPN loss $\mathcal{L}_{rpn}$ consists of the classification loss $\mathcal{L}_{cls}$, regression loss $\mathcal{L}_{reg}$, and consistency enforcing loss $\mathcal{L}_{ce}$, as shown in Equation (7). The focal loss [33] is used as $\mathcal{L}_{cls}$ to balance positive and negative samples for classification, as shown in Equation (8). The parameters of focal loss keep the default values of $c = 0.25$ and $r = 2$. The definition of $\mathcal{L}_{ce}$ is shown in Equation (9), where D denotes the predicted bounding box, G denotes the ground truth, and k denotes the classification confidence of D.

$$\mathcal{L}_{rpn} = \mathcal{L}_{cls} + \mathcal{L}_{reg} + \lambda \mathcal{L}_{ce} \tag{7}$$

$$\mathcal{L}_{cls} = -c(1 - p_t)^r \log p_t \tag{8}$$

$$\mathcal{L}_{ce} = -\log\left(k \times \frac{Area(D \cap G)}{Area(D \cup G)}\right) \tag{9}$$

The regression loss $\mathcal{L}_{reg}$ is used to constrain the bounding box $(x, y, z, h, w, l, \theta)$, where (x, y, z) is the center, (h, w, l) is the size, and θ is the orientation. PointRCNN divides each foreground point-surrounding area into bins along the X and Z axes, and $\mathcal{L}_{reg}$ is the bin-based regression loss. The calculation process of $\mathcal{L}_{reg}$ is as follows:

$$\mathcal{L}_{bin} = \sum_{u \epsilon \{x,z,\theta\}} \left(E\left(\hat{b}_u, b_u\right) + S(\hat{r}_u, r_u)\right) \tag{10}$$

$$\mathcal{L}_{res} = \sum_{v \epsilon \{y,h,w,l\}} S(r_v, \hat{r}_v) \tag{11}$$

$$\mathcal{L}_{reg} = \frac{1}{N_{pos}} \sum_{p \epsilon pos} (\mathcal{L}_{bin} + \mathcal{L}_{res}) \tag{12}$$

where E represents the cross-entropy loss, S represents the smooth L1 loss, $\hat{b}_u$ and b_u are the predicted bin assignments and the ground-truth bin assignments, $\hat{r}_u$ and r_u are the residuals and the ground-truth residuals, and N_{pos} is the number of foreground points. Each residual is a variable for further optimizing the position of the specified bin, and its calculation process is detailed in PointRCNN.

3. Experiments and Analysis

3.1. Dataset

The KITTI 3D object detection benchmark dataset is used as the experimental dataset. KITTI was collected by an equipment platform fixed on the top of the car, including two grayscale cameras, two color cameras, a Velodyne 64-line 3D LiDAR, four optical lenses, and one GPS/IMU system. KITTI was sampled and synchronized at 10 Hz, including 7481 training samples and 7581 testing samples. Each sample provides both the point cloud and the camera image. The labels of KITTI were divided into three subsets: Easy, moderate, and hard, according to the heights of their 2D bounding boxes, occlusion levels, and truncation levels. The easy difficulty level has a minimum 2D bounding box height of 40 Px, an occlusion level of full visibility, and a maximum truncation level of 15%. The minimum 2D bounding box height, occlusion level, and maximum truncation level of moderate difficulty level are 25 Px, partial occlusion, and 30%, respectively. The minimum 2D bounding box height for the hard difficulty level is 25 Px, the occlusion level is hard to see, and the maximum truncation level is 50%.

The image data of color camera 2 and the point cloud data of Velodyne LiDAR are used in this work. Since the test samples of KITTI are not labeled, the approach of Shi et al. [17] is adopted to further divide the training samples into 3712 samples for training and 3769 samples for validation. Only visible objects in the image are labeled, so according to the general practice of previous studies [19,24], only the LiDAR points projected into the image are used. The LiDAR points are mapped at the camera coordinates X (right) axis $[-40, 40]$ meters, Y (down) axis $[-1, 3]$ meters, and Z (forward) axis $[0, 70.4]$ meters. The range of point clouds is $[-\pi, \pi]$.

3.2. FuDNN Training

The experiments were carried out on a server with the Ubuntu 16.04.1 LTS system. The server GPU was GeForce RTX 3090 24 GB, the CPU was Intel E5-2620V4@2.10 GHz, and the RAM size was 256 GB. The main installation packages required by the running environment of the program are shown in Table 2.

Table 2. Main installation packages required by the running environment.

Package Name	Version
Python	3.7.6
CUDA	11.3
PyTorch	1.10.1
TensorboardX	2.4.1
Numpy	1.21.2
Pillow	8.4.0
Numba	0.54.1
Opencv-python	4.5.5.62
Torchvision	0.11.2

The FuDNN was trained in an end-to-end manner. During the data preprocessing stage, the images were scaled to a uniform size of 384 × 1280. Each point cloud was subtracted or supplemented to ensure that the number of points in each point cloud was 16,384. Several data augmentation techniques were employed on the point clouds to avoid over-fitting, which are widely used in [1,12,13], including flip (along the forwarding axis), random scaling (scale factor 0.95∼1.05 for all 3 axes), and rotation (−10∼10 degrees along the vertical axis). The data augmentation was only used for training. During network training, the network parameters were optimized by the Adam method [34]. The initial learning rate and weight decay of the network were set to 0.002 and 0.001, respectively. The model was trained for 150 epochs with a batch size of 4 and saved every 5 epochs. The threshold of IoU was set to 0.7 to distinguish between true positives and false positives.

3.3. Performance Metrics

According to the latest rules of KITTI, the 40-point Interpolated Average Precision metric was used as the performance metric, as shown in Equation (13):

$$AP|_{R_{40}} = \frac{1}{|R_{40}|} \sum_{r \in R_{40}} \rho_{interp}(r) \tag{13}$$

where $R_{40} = \{1/40, 2/40, 3/40, \ldots, 1\}$ and $|R_{40}| = 40$. The interpolation function ρ_{interp} is defined as:

$$\rho_{interp}(r) = \max_{r':r' \geq r} \rho(r') \tag{14}$$

where $\rho(r)$ is the precision at recall r. $\rho_{interp}(r)$ is the value at recall r on the smoothed precision–recall curve, not the actual precision values $\rho(r)$. Furthermore, $\rho_{interp}(r)$ is greater than or equal to $\rho(r)$.

3.4. Results and Discussion

Since the car category of KITTI provided enough samples, the experiments in this work were all conducted in the car category. KITTI divided the samples of the car category into three difficulty levels, easy, moderate, and hard, so the verification experiments of this work evaluated the three levels of difficulty, respectively. Figure 4 shows the relationship between the AP of FuDNN and the number of training epochs. When the epoch number increased from 5 to 45, the AP values increased rapidly. The AP values increased flatly after epoch 45. When the epoch number was 135, the AP values reached the maximum at the three difficulty levels of easy, moderate, and hard, which were 92.48, 82.90, and 80.51, respectively. When the epoch number was greater than 135, the AP values did not exceed that of epoch 135. Therefore, the model of epoch 135 was used as the best-trained model.

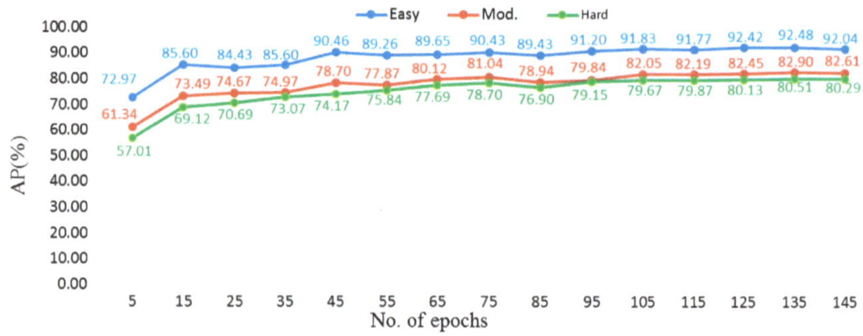

Figure 4. The AP values for different training epoch numbers.

Table 3 shows the AP comparison between the proposed FuDNN and the previous methods on the car class of KITTI validation set with the IoU threshold of 0.7. The AP values of FuDNN in the moderate and hard difficulty levels were higher than those of other methods. At the easy difficulty level, the AP value of FuDNN reached 92.48, which was 4.73 higher than the lowest PointPillars [1], but slightly lower than the best PointRCNN [17]. At the moderate difficulty level, the AP value of FuDNN showed a 4.51 improvement over the lowest PointPillars and a 0.31 improvement over the best EPNet [28]. At the hard difficulty level, the AP value of FuDNN outperformed the lowest PointPillars and the best EPNet by 5.33 and 0.37, respectively. Although the AP value of FuDNN was 0.06 lower than PointRCNN at the easy difficulty level, it was 0.74 and 2.63 higher than PointRCNN at moderate and hard difficulty levels, respectively. PointPillars achieved a speed of 62.0 fps, but its AP values were too low. SECOND was also fast, reaching 26.3 fps, but its AP values were lower than the proposed FuDNN. The speed of FuDNN was similar to PointRCNN, 3D-CVF, EPNet, and PI-RCNN. Therefore, the overall effect of FuDNN was better than that of PointRCNN. The above comparisons demonstrated the effectiveness of FuDNN. Since the car samples in the KITTI dataset all have different degrees of occlusion and truncation, the AP of object detection will be limited. Especially for the hard difficulty level, the occlusion level is hard to see, and the cutoff level is 50%, so it is very important to improve the AP value.

Table 3. AP comparison of different 3D object detection algorithms in the car class of KITTI validation set with IoU threshold 0.7.

Method	Modality	Easy	Moderate	Hard	Speed (fps)
PointPillars [1]	LiDAR-based	87.75	78.39	75.18	62.0
SECOND [13]	LiDAR-based	90.97	79.94	77.09	26.3
PointRCNN [17]	LiDAR-based	92.54	82.16	77.88	10.0
3D-CVF [27]	LiDAR–camera fusion	89.67	79.88	78.47	13.3
EPNet [28]	LiDAR–camera fusion	92.28	82.59	80.14	10.0
PI-RCNN [29]	LiDAR–camera fusion	88.27	78.53	77.75	11.1
FuDNN (Proposed)	LiDAR–camera fusion	92.48	82.90	80.51	10.5

Table 4 shows a set of ablation experiments to compare the effects of different 2D backbones in the car class of the KITTI validation set with an IoU threshold of 0.7. Models A, B, C, and D were obtained by replacing the 2D backbone of FuDNN with Resnet50 [35], Resnet101 [35], VGG16 [36], and DensNet121 [37], respectively. The AP values of model B and model D were higher than those of model A and model C at all three difficulty levels. The reason may be that model B and model D had more network layers to extract richer feature information. The proposed 2D backbone of FuDNN not only had the simplest structure but also had the highest AP values across all three difficulty levels. This set of experiments demonstrated the effectiveness of the proposed 2D backbone.

Table 4. Ablation experiments of different 2D backbones in the car class of KITTI validation set with IoU threshold 0.7.

Model	2D backbone	Easy	Moderate	Hard
A	Resnet50 [35]	91.75	81.22	79.620
B	Resnet101 [35]	92.24	81.87	80.03
C	VGG16 [36]	91.70	80.80	79.11
D	DensNet121 [37]	92.46	82.05	79.97
FuDNN	Proposed	92.48	82.90	80.51

Another set of ablation experiments was designed to analyze the effect of different fusion modes, as shown in Table 5. Model G removed the attention mechanism based on FuDNN, and model F changed the concatenation of FuDNN to point-wise addition. Model E removed the attention mechanism based on model F. The AP values of FuDNN across three difficulty levels were all higher than those of model G, and the AP values of model F across three difficulty levels were also higher than those of model E, which proved that the attention mechanism could improve the object detection effect of the model. The comparison between FuDNN and model F, as well as the comparison between model G and model E, showed that the effects of the two fusion modes of addition and concatenation were similar.

Table 5. Ablation experiments of different fusion methods in the car class of KITTI validation set with IoU threshold 0.7.

Model	Fusion Method	Easy	Moderate	Hard
E	Addition	91.99	82.26	80.17
F	Attention-based Addition	92.74	82.76	80.50
G	Concatenation	92.13	81.91	79.96
FuDNN	Attention-based Concatenation	92.48	82.90	80.51

4. Conclusions

This work proposed a LiDAR–camera fusion 3D object detection deep neural network FuDNN. Table 3 shows that the proposed FuDNN has the highest mean AP value across the three difficulty levels in the car category of the public KITTI dataset, and thus outperforms other LiDAR–camera fusion 3D object detection approaches in terms of the mean AP value. Although the speed of FuDNN is not as fast as PointPillars and SECOND, it is comparable to several other popular methods. Two further sets of ablation experiments verified the effectiveness of the proposed 2D backbone sub-network and the attention-based fusion sub-network, respectively. In the future, efforts will be made to further improve the detection AP and reduce network complexity.

Author Contributions: Conceptualization, L.L., J.H., K.R. and Z.X.; methodology, L.L., J.H., K.R. and Z.X.; software, L.L.; validation, L.L. and K.R.; formal analysis, J.H. and K.R.; investigation, L.L., K.R. and J.H.; resources, L.L. and K.R.; data curation, L.L. and J.H.; writing—original draft preparation, L.L.; writing—review and editing, L.L., Y.H., J.H. and K.R.; visualization, L.L.; supervision, Y.H., J.H. and K.R.; project administration, Y.H. and J.H.; funding acquisition, Y.H. All authors have read and agreed to the published version of the manuscript.

Funding: This research received no external funding.

Institutional Review Board Statement: Not applicable.

Informed Consent Statement: Not applicable.

Data Availability Statement: A publicly available dataset was analyzed in this study. This dataset can be found here: http://www.cvlibs.net/datasets/kitti/eval_object.php?obj_benchmark=3d, accessed on 16 February 2022.

Conflicts of Interest: The authors declare no conflict of interest.

References

1. Lang, A.H.; Vora, S.; Caesar, H.; Zhou, L.; Yang, J.; Beijbom, O. Pointpillars: Fast encoders for object detection from point clouds. In Proceedings of the IEEE/CVF Conference on Computer Vision and Pattern Recognition, Long Beach, CA, USA, 16–17 June 2019; pp. 12697–12705.
2. Wang, Y.; Mao, Q.; Zhu, H.; Zhang, Y.; Ji, J.; Zhang, Y. Multi-modal 3d object detection in autonomous driving: A survey. *arXiv* **2021**, arXiv:2106.12735.
3. Girshick, R.; Donahue, J.; Darrell, T.; Malik, J. Rich feature hierarchies for accurate object detection and semantic segmentation. In Proceedings of the IEEE Conference on Computer Vision and Pattern Recognition, Columbus, OH, USA, 23–28 June 2014; pp. 580–587.
4. He, K.; Gkioxari, G.; Dollár, P.; Girshick, R. Mask r-cnn. In Proceedings of the IEEE International Conference on Computer Vision, Venice, Italy, 22–29 October 2017; pp. 2961–2969.
5. Ren, S.; He, K.; Girshick, R.; Sun, J. Faster r-cnn: Towards real-time object detection with region proposal networks. *Adv. Neural Inf. Process. Syst.* **2015**, *28*, 91–99. [CrossRef] [PubMed]
6. Redmon, J.; Divvala, S.; Girshick, R.; Farhadi, A. You only look once: Unified, real-time object detection. In Proceedings of the IEEE Conference on Computer Vision and Pattern Recognition, Honolulu, HI, USA, 21–26 July 2016; pp. 779–788.
7. Liu, W.; Anguelov, D.; Erhan, D.; Szegedy, C.; Reed, S.; Fu, C.-Y.; Berg, A.C. Ssd: Single shot multibox detector. In Proceedings of the European Conference on Computer Vision, Amsterdam, The Netherlands, 11–14 October 2016; Springer: Berlin/Heidelberg, Germany, 2016; pp. 21–37.
8. An, P.; Liang, J.; Yu, K.; Fang, B.; Ma, J. Deep structural information fusion for 3D object detection on LiDAR–camera system. *Comput. Vis. Image Underst.* **2022**, *214*, 103295. [CrossRef]
9. Reading, C.; Harakeh, A.; Chae, J.; Waslander, S.L. Categorical depth distribution network for monocular 3d object detection. In Proceedings of the IEEE/CVF Conference on Computer Vision and Pattern Recognition, Nashville, TN, USA, 20–25 June 2021; pp. 8555–8564.
10. Lu, Y.; Ma, X.; Yang, L.; Zhang, T.; Liu, Y.; Chu, Q.; Yan, J.; Ouyang, W. Geometry uncertainty projection network for monocular 3d object detection. In Proceedings of the IEEE/CVF International Conference on Computer Vision, Montreal, BC, Canada, 11–17 October 2021; pp. 3111–3121.
11. Guo, Y.; Wang, H.; Hu, Q.; Liu, H.; Liu, L.; Bennamoun, M. Deep learning for 3d point clouds: A survey. *IEEE Trans. Pattern Anal. Mach. Intell.* **2020**, *43*, 4338–4364. [CrossRef] [PubMed]
12. Zhou, Y.; Tuzel, O. Voxelnet: End-to-end learning for point cloud based 3d object detection. In Proceedings of the IEEE Conference on Computer Vision and Pattern Recognition, Salt Lake City, UT, USA, 18–23 June 2018; pp. 4490–4499.
13. Yan, Y.; Mao, Y.; Li, B. Second: Sparsely embedded convolutional detection. *Sensors* **2018**, *18*, 3337. [CrossRef] [PubMed]
14. Shi, S.; Guo, C.; Jiang, L.; Wang, Z.; Shi, J.; Wang, X.; Li, H. Pv-rcnn: Point-voxel feature set abstraction for 3d object detection. In Proceedings of the IEEE/CVF Conference on Computer Vision and Pattern Recognition, Seattle, WA, USA, 13–19 June 2020; pp. 10529–10538.
15. Qi, C.R.; Su, H.; Mo, K.; Guibas, L.J. Pointnet: Deep learning on point sets for 3d classification and segmentation. In Proceedings of the IEEE Conference on Computer Vision and Pattern Recognition, Honolulu, HI, USA, 21–26 July 2017; pp. 652–660.
16. Qi, C.R.; Yi, L.; Su, H.; Guibas, L.J. Pointnet++: Deep hierarchical feature learning on point sets in a metric space. *Adv. Neural Inf. Process. Syst.* **2017**, *30*, 5105–5114.
17. Shi, S.; Wang, X.; Li, H. Pointrcnn: 3d object proposal generation and detection from point cloud. In Proceedings of the IEEE/CVF Conference on Computer Vision and Pattern Recognition, Long Beach, CA, USA, 15–20 June 2019; pp. 770–779.
18. Yang, Z.; Sun, Y.; Liu, S.; Shen, X.; Jia, J. Std: Sparse-to-dense 3d object detector for point cloud. In Proceedings of the IEEE/CVF International Conference on Computer Vision, Seoul, Korea, 27–28 October 2019; pp. 1951–1960.
19. Shi, W.; Rajkumar, R. Point-gnn: Graph neural network for 3d object detection in a point cloud. In Proceedings of the IEEE/CVF Conference on Computer Vision and Pattern Recognition, Seattle, WA, USA, 13–19 June 2020; pp. 1711–1719.
20. Mao, J.; Niu, M.; Bai, H.; Liang, X.; Xu, H.; Xu, C. Pyramid r-cnn: Towards better performance and adaptability for 3d object detection. In Proceedings of the IEEE/CVF International Conference on Computer Vision, Montreal, BC, Canada, 11–17 October 2021; pp. 2723–2732.
21. Wen, L.-H.; Jo, K.-H. Fast and accurate 3D object detection for lidar-camera-based autonomous vehicles using one shared voxel-based backbone. *IEEE Access* **2021**, *9*, 22080–22089. [CrossRef]
22. Lu, H.; Chen, X.; Zhang, G.; Zhou, Q.; Ma, Y.; Zhao, Y. SCANet: Spatial-channel attention network for 3D object detection. In Proceedings of the ICASSP 2019–2019 IEEE International Conference on Acoustics, Speech and Signal Processing (ICASSP), Brighton, UK, 12–17 May 2019; IEEE: New York, NY, USA, 2019; pp. 1992–1996.

23. Liang, M.; Yang, B.; Wang, S.; Urtasun, R. Deep continuous fusion for multi-sensor 3d object detection. In Proceedings of the European Conference on Computer Vision (ECCV), Munich, Germany, 8–14 September 2018; pp. 641–656.
24. Chen, X.; Ma, H.; Wan, J.; Li, B.; Xia, T. Multi-view 3d object detection network for autonomous driving. In Proceedings of the IEEE Conference on Computer Vision and Pattern Recognition, Honolulu, HI, USA, 21–26 July 2017; pp. 1907–1915.
25. Ku, J.; Mozifian, M.; Lee, J.; Harakeh, A.; Waslander, S.L. Joint 3d proposal generation and object detection from view aggregation. In Proceedings of the 2018 IEEE/RSJ International Conference on Intelligent Robots and Systems (IROS), Madrid, Spain, 1–5 October 2018; IEEE: New York, NY, USA, 2018; pp. 1–8.
26. Qi, C.R.; Liu, W.; Wu, C.; Su, H.; Guibas, L.J. Frustum pointnets for 3d object detection from rgb-d data. In Proceedings of the IEEE Conference on Computer Vision and Pattern Recognition, Salt Lake City, UT, USA, 18–22 June 2018; pp. 918–927.
27. Yoo, J.H.; Kim, Y.; Kim, J.; Choi, J.W. 3d-cvf: Generating joint camera and lidar features using cross-view spatial feature fusion for 3d object detection. In Proceedings of the European Conference on Computer Vision, Glasgow, UK, 23–28 August 2020; Springer: Berlin/Heidelberg, Germany, 2020; pp. 720–736.
28. Huang, T.; Liu, Z.; Chen, X.; Bai, X. Epnet: Enhancing point features with image semantics for 3d object detection. In Proceedings of the European Conference on Computer Vision, Glasgow, UK, 23–28 August 2020; Springer: Berlin/Heidelberg, Germany, 2020; pp. 35–52.
29. Xie, L.; Xiang, C.; Yu, Z.; Xu, G.; Yang, Z.; Cai, D.; He, X. PI-RCNN: An efficient multi-sensor 3D object detector with point-based attentive cont-conv fusion module. In Proceedings of the AAAI Conference on Artificial Intelligence, New York, NY, USA, 7–12 February 2020; pp. 12460–12467.
30. Wen, L.-H.; Jo, K.-H. Three-attention mechanisms for one-stage 3-d object detection based on LiDAR and camera. *IEEE Trans. Ind. Inform.* **2021**, *17*, 6655–6663. [CrossRef]
31. Geiger, A.; Lenz, P.; Urtasun, R. Are we ready for autonomous driving? The kitti vision benchmark suite. In Proceedings of the 2012 IEEE Conference on Computer Vision and Pattern Recognition, Providence, RI, USA, 16–21 June 2012; IEEE: New York, NY, USA, 2012; pp. 3354–3361.
32. Ioffe, S.; Szegedy, C. Batch normalization: Accelerating deep network training by reducing internal covariate shift. In Proceedings of the International Conference on Machine Learning, Lille, France, 6–11 July 2015; PMLR: New York, NY, USA, 2015; pp. 448–456.
33. Lin, T.-Y.; Goyal, P.; Girshick, R.; He, K.; Dollár, P. Focal loss for dense object detection. In Proceedings of the IEEE International Conference on Computer Vision, Venice, Italy, 22–29 October 2017; pp. 2980–2988.
34. Da, K. A method for stochastic optimization. *arXiv* **2014**, arXiv:1412.6980.
35. He, K.; Zhang, X.; Ren, S.; Sun, J. Deep residual learning for image recognition. In Proceedings of the IEEE Conference on Computer Vision and Pattern Recognition, Las Vegas, NV, USA, 27–30 June 2016; pp. 770–778.
36. Simonyan, K.; Zisserman, A. Very deep convolutional networks for large-scale image recognition. *arXiv* **2014**, arXiv:1409.1556.
37. Huang, G.; Liu, Z.; Van Der Maaten, L.; Weinberger, K.Q. Densely connected convolutional networks. In Proceedings of the IEEE Conference on Computer Vision and Pattern Recognition, Honolulu, HI, USA, 21–26 July 2017; pp. 4700–4708.

remote sensing

MDPI

Article

A Two-Stage Low-Altitude Remote Sensing Papaver Somniferum Image Detection System Based on YOLOv5s+DenseNet121

Qian Wang [1,2,3], Chunshan Wang [1,2,3,*], Huarui Wu [2], Chunjiang Zhao [2], Guifa Teng [1,3], Yajie Yu [4] and Huaji Zhu [2]

1 School of Information Science and Technology, Hebei Agricultural University, Baoding 071001, China; hebau_wangqian@163.com (Q.W.); tguifa@hebau.edu.cn (G.T.)
2 National Engineering Research Center for Information Technology in Agriculture, Beijing 100097, China; wuhr@nercita.org.cn (H.W.); zhaocj@nercita.org.cn (C.Z.); zhuhj@nercita.org.cn (H.Z.)
3 Hebei Key Laboratory of Agricultural Big Data, Baoding 071001, China
4 School of Software, Tsinghua University, Beijing 100084, China; yuyajie130@mail.tsinghua.edu.cn
* Correspondence: wangcs007@nercita.org.cn

Abstract: *Papaver somniferum* (opium poppy) is not only a source of raw material for the production of medical narcotic analgesics but also the major raw material for certain psychotropic drugs. Therefore, it is stipulated by law that the cultivation of *Papaver somniferum* must be authorized by the government under stringent supervision. In certain areas, unauthorized and illicit *Papaver somniferum* cultivation on private-owned lands occurs from time to time. These illegal *Papaver somniferum* cultivation sites are dispersedly-distributed and highly-concealed, therefore becoming a tough problem for government supervision. The low-altitude inspection of *Papaver somniferum* cultivation by unmanned aerial vehicles has the advantages of high efficiency and time saving, but the large amount of image data collected needs to be manually screened, which not only consumes a lot of manpower and material resources but also easily causes omissions. In response to the above problems, this paper proposed a two-stage (target detection and image classification) method for the detection of *Papaver somniferum* cultivation sites. In the first stage, the YOLOv5s algorithm was used to detect *Papaver somniferum* images for the purpose of identifying all the suspicious *Papaver somniferum* images from the original data. In the second stage, the DenseNet121 network was used to classify the detection results from the first stage, so as to exclude the targets other than *Papaver somniferum* and retain the images containing *Papaver somniferum* only. For the first stage, YOLOv5s achieved the best overall performance among mainstream target detection models, with a Precision of 97.7%, Recall of 94.9%, and mAP of 97.4%. For the second stage, DenseNet121 with pre-training achieved the best overall performance, with a classification accuracy of 97.33% and a Precision of 95.81%. The experimental comparison results between the one-stage method and the two-stage method suggest that the Recall of the two methods remained the same, but the two-stage method reduced the number of falsely detected images by 73.88%, which greatly reduces the workload for subsequent manual screening of remote sensing *Papaver somniferum* images. The achievement of this paper provides an effective technical means to solve the problem in the supervision of illicit *Papaver somniferum* cultivation.

Keywords: *Papaver somniferum* inspection; unmanned aerial vehicle (UAV); small target detection; YOLOv5s; two-stage detection and classification

Citation: Wang, Q.; Wang, C.; Wu, H.; Zhao, C.; Teng, G.; Yu, Y.; Zhu, H. A Two-Stage Low-Altitude Remote Sensing Papaver Somniferum Image Detection System Based on YOLOv5s+DenseNet121. *Remote Sens.* **2022**, *14*, 1834. https://doi.org/10.3390/rs14081834

Academic Editors: Miltiadis D. Lytras and Andreea Claudia Serban

Received: 9 February 2022
Accepted: 8 April 2022
Published: 11 April 2022

1. Introduction

Papaver somniferum (opium poppy) is an annual herb cultivated in many parts of Asia. On the one hand, the *Papaver somniferum* seed is an important food product, which contains amino acids such as glycine, alanine, tyrosine, phenylalanine, and aspartic acid, which are beneficial to health. It is widely used in making all kinds of baked foods and salads. On the other hand, *Papaver somniferum* fruit extracts are the source of a variety of sedatives, such as morphine, thebaine, codeine, papaverine, and noscapine [1]. In particular, *Papaver somniferum* fruit is the major raw material for the production of opium. Thus, it is stipulated by law in many countries and regions that the cultivation and sales of *Papaver somniferum* must be authorized by the government and are subject to official supervision throughout the entire industry chain.

Despite stringent legal constraints, there are still people who cultivate *Papaver somniferum* in private fields or yards without authorization, and then sell and eat *Papaver somniferum* illegally in certain countries and regions. Illicit *Papaver somniferum* cultivation causes potential harm to the social order. Once *Papaver somniferum* related products are out of regulation and flow into the market, they pose a serious threat to people's physical and mental health. Hence, it has become an urgent issue for local regulatory authorities to investigate illicit *Papaver somniferum* cultivation. In view of this, the rapid and accurate detection of illicit *Papaver somniferum* cultivation sites is an important means of combating such crimes. Nakazawa et al. [2] identified the differences in spectral characteristics between opium poppy and other crops by utilizing hyperspectral imaging and developed a method to detect illicit opium poppy fields based on hyperspectral data. However, hyperspectral instruments are generally expensive, so the rapid detection of opium poppy from remote-sensing images is not the first choice for researchers. At present, the use of unmanned aerial vehicle (UAV) equipped with high-definition cameras to inspect illicit *Papaver somniferum* cultivation is deemed an accurate and effective approach. Zhou et al. [3] and Wang et al. [4] realized the detection of low-altitude remote sensing opium poppy images by applying a target detection algorithm. Nevertheless, in large-scale high-resolution remote sensing images, the accurate detection of opium poppy is actually a very challenging task due to a variety of problems (e.g., blurred images due to natural shooting conditions; existence of easily-confused objects such as green onion, white flowers, wine bottle, plastic toy, green plastic film, etc.; complex backgrounds; etc.) Currently, the complete process of opium poppy screening and elimination is as follows: (1) after the UAV captures the images, anti-drug experts will conduct a naked eye search on a large number of collected images to label and record the suspicious targets in the images; (2) the relevant government departments will collect coordinate points for the suspicious targets and then visit the sites to uproot the *Papaver somniferum* plants. Although this method can accurately locate opium poppy cultivation sites, it has many obvious shortcomings. For example, the number of images collected by UAV can be incredibly huge, which is often counted on the order of terabytes. Screening of these images relying on the naked eye is extremely laborious and time-consuming, and the accuracy of the results is highly dependent on the proficiency and skills of the inspectors.

With the rapid development of artificial intelligence, target detection algorithms have been successfully applied to various fields such as medicine [5], agriculture [6], forestry [7], physics [8], and remote sensing communities [9].The target detection algorithm can be traced back to the cascade classifier framework proposed by in 2001 [10], which brought target detection into reality for the first time. In 2005, Dalal N. and Triggs B [11] achieved human detection using Histogram of Gradients (HOG) and Support Vector Machines (SVM). Early scholars mainly relied on expert experience to describe and extract the shape, color, texture, and other features of objects and then to implement target detection using artificial feature extraction algorithms. Until 2012, Krizhevsky [12] innovatively applied AlexNet to the field of image classification [13]. Since then, researchers have turned their attention from artificial feature extraction methods to deep learning methods. The core of deep learning lies in the fact that the feature extraction network can realize automatic target feature extraction, which greatly reduces the difficulty and improves the accuracy and efficiency of target detection.

Target detection methods based on deep learning can be generally divided into one-stage methods and two-stage methods. The two-stage method is implemented in the following steps: (1) select region proposals from the input images that may contain the target object; (2) classify and perform location regression on the region proposals to obtain the detection results. The one-stage method omits the step of generating region proposals and directly computes anchor boxes of different sizes and aspect ratios in the input images. The commonly used two-stage detectors include R-CNN [14], Fast R-CNN [15], Faster R-CNN [16], Mask R-CNN [17], R-FCN [18], and Cascade R-CNN [19]. The one-stage target detection algorithms mainly include the SSD [20–24] series and the YOLO series [25–28]. Compared with two-stage detectors, the inference speed of one-stage detectors is generally faster.

The target detection technology based on deep learning has increasingly wide applications in the field of remote sensing. Xu et al. [29] proposed an improved YOLOv3 algorithm by using DenseNet (Densely Connected Network) to enhance the network performance. On the basis of the original YOLOv3 network, the improved method increased the detection scale to 4 and thereby improved the mAP from 77.10% to 88.73% on the RSOD remote sensing dataset. Especially, for the detection of small targets such as aircrafts, the mAP was increased by 12.12%, suggesting that large-scale detection mechanism can help improve the accuracy of small target detection to a certain extent. By comparing the detection performance between one-stage and two-stage detectors on a single UAS-based image, Ezzy et al. [30] concluded that the self-trained YOLOv3 detector could achieve accurate small target detection of rodent caves. In view of the difficulty in small target detection, researchers have made consistent efforts to improve the network structure from different perspectives so as to increase the accuracy of small target detection [31]. Miura et al. [32] utilized an improved CNN network and the post-earthquake aerial images in Japan to detect collapsed buildings, non-collapsed buildings, and buildings covered with blue tarps. Shi et al. [33] proposed to use the improved classic YOLOv4 model to quickly detect collapsed buildings from aerial images after the Wenchuan and Beichuan earthquakes in China and used the K-means algorithm to cluster the optimal anchor boxes from the images. By improving the CSPDarkNet53 module and the loss function in the original network, the proposed method effectively increased the average precision of collapsed building detection. Zhu et al. [34] improved the YOLOv5 algorithm based on the attention mechanism and designed a pyramid-based method for boulder detection in planetary images. The proposed network was equipped with the Convolutional Block Attention Module (CBAM) and the Efficient Channel Attention Network (ECA-Net) to improve the overall detection performance. Yan et al. [35] proposed to use the improved YOLOv5s model to distinguish apples that can be grasped from those that cannot be grasped. Specifically, the authors upgraded the CSP module to CSP-2 module and inserted the SE module in the attention mechanism network into the modified backbone network. The improved model was more accurate, faster, and lighter than the original YOLOv5s and other detection models. Mehdi et al. [36] improved yolov3 to detect of hazardous and noxious substances of vital transportation by sea. It can be seen from the research status described above that, in the field of remote sensing image detection, the one-stage YOLO series is more in line with the real-time and accurate detection requirements at the current stage.

YOLOv5 is the latest version of the YOLO series of networks and is also the most advanced network of the one-stage method. YOLOv5 treats target detection as a regression problem so as to perform localization and classification concurrently to achieve fast and accurate detection. Wang et al. [4] proposed an improved YOLOv3 algorithm, which realized fast and accurate low-altitude remote sensing opium poppy image detection. Their algorithm effectively improved the accuracy of opium poppy detection and reduced the parameter size of the model, but they did not provide a good solution for distinguishing easily confused plants and objects and for the accurate detection of opium poppy cultivation. Hu et al. [37] relied on the two-stage idea to realize the identification of individual cows. First, the YOLO algorithm was used to locate the parts of each cow of interest, and then the segmented parts were sent to three convolutional neural networks for feature extraction. After the fusion of parts features, the SVM was used to achieve individual cow identification.

In order to address the problems of labor intensity and time-consuming of manual screening opium poppy inspection images, this paper proposes a two-stage automatic detection method of "detection + classification". YOLOv5 was chosen as the first-stage detector in our study considering its characteristics of high detection accuracy and real-time performance, for the purpose of improving the recall and detection speed. Then, the detection results from the first stage were input into the second stage network for further classification, so as to improve the overall accuracy of *Papaver somniferum* detection.

2. Materials and Methods

2.1. Data Acquisition

In mainland China, the best time to plant and harvest *Papaver somniferum* is from March to August. When an UAV is used to capture images, the flying height is usually set between 130 m and 145 m in order to avoid the interference of uncertain factors such as power lines, buildings, and birds. The dataset used in the present study is composed of data obtained in 2019, 2020, and 2021, and the UAV model is fixed-wing UAV with a wingspan of 2.4 m, which is equipped with a Sony a7RIV camera. Samples of the original images collected are shown in Figure 1, and the metadata of the images are shown in Table 1. Photoshop was used to crop the original images containing papaver somniferum. Since the areas occupied by the *Papaver somniferum* in the original images are different, in order to ensure the complete appearance of the *Papaver somniferum* plot, the width and height of the cropped image are set to 640 × 640 pixels and 800 × 800 pixels, respectively. Finally, a total of 643 images were labelled. The process of labeling data is that the professionals manually screen these images and select all the images of suspected papaver somniferum, and finally, the supervisors go to the scene according to the GPS information to verify whether the images of suspected opium poppy are real papaver somniferum. If it is a real *Papaver somniferum* target, it is marked as a ground truth sample.

Figure 1. Low-altitude remote sensing images in the village area captured by UAV.

Table 1. The metadata of the image.

Image File Name	Resolution	Ground Resolution/cm	File Type
DSC000001	7952 × 5304	2	JPG

2.2. Data Processing

The collection of *Papaver somniferum* images is very challenging, and the images are often affected by light, so it is difficult to accurately extract features from the images. Meanwhile, easily-confused plants, single-plants, and complex backgrounds may also negatively impact the detection precision. Therefore, appropriate image enhancement strategies need to be applied to improve the quality, color and contrast of the original images. In this paper, three enhancement methods were adopted, which are contrast enhancement, brightness enhancement, and random angle rotation with brightness enhancement. The contrast and color enhancement can simulate different weather and light conditions during image acquisition and can effectively enhance the images of opium poppy planted in the shaded area. The random angle rotation with brightness enhancement does not change the image itself but only alters the position and direction of the image so as to increase the richness of the target. When UAV collects images, it is necessary to set a certain side overlap between routes to ensure that the image acquisition can completely cover the whole survey area. Although the overlapping part contains the ground at the same position, the ground at the same position is not exactly the same in the two images. The ground at the same position often has a certain deflection angle in the two images. Therefore, random angle rotation enhancement is taken as image data augmentation. After enhancing the contrast by 100%, enhancing the color by 100%, and enhancing the rotation angle with brightness by 200%, the total number of enhanced images is 3215. The images were labelled using the LabelImg tool, and an xml file containing the location information was generated. Table 2 presents the original and enhanced images for different planting strategies.

Table 2. Examples of preprocessed *Papaver somniferum* images for different planting strategies.

1				
2				
3				
4				
	a	b	c	d

1–4 on the left represent different planting strategies (1, large-scale planting; 2, small-scale planting; 3, shaded planting; 4, mixed planting); a, b, c, d represent the original image, contrast enhancement, color enhancement, and random rotation with brightness enhancement, respectively. The yellow circle in the original image refers to the specific location of the *Papaver somniferum*.

In the classification task, the opium poppy category was from the ground truth cropping of the 643 labeled images in the original dataset. The cropped images were resized to 224 × 224 pixels. The enhancement methods included brightness enhancement, contrast enhancement, and flip enhancement. The name for the opium poppy category is "*Papaver somniferum*". The other two categories are images that are falsely detected or wrongly detected by the detector, namely, "plant" (e.g., weeds and other plants) and "ground" (e.g., roof, green field and land). Figure 2 shows the classification dataset, and Table 3 presents the specific number of data items in the classification dataset.

Figure 2. Classification dataset.

Table 3. The specific number of data items in the dataset.

Category	Number of Training Images	Number of Test Images	Total
Papaver somniferum	927	379	1306
plant	736	315	1051
ground	787	337	1124

3. Experiment

3.1. YOLOv5 Network Structure

The YOLOv5 network is mainly composed of three parts, the backbone network (Backbone), the neck network (Neck), and the output prediction part (Prediction). The Backbone functions as the main feature extraction module of the entire network, which utilizes the convolutional neural network to perform feature map convolution calculation so as to generate four feature maps of different sizes. The Backbone adopts the Focus structure. The main idea of the Focus structure is to crop the input image through the Slice operation. The Slice and Concat operations can reduce the size of the feature map by half to better extract the feature information. The Neck fuses the corresponding feature maps that are generated for the purpose of linking the context information to reduce information loss. In the fusion process, the feature pyramid structure of FPN+PAN is adopted. The FPN structure is a top-down feature map that fuses a high-level feature information with low-level features through up-sampling for prediction. It can enhance the semantic information of the pyramid. On the other hand, the PAN structure transfers the positioning feature information from the bottom to the top, so as to aggregate the parameters of different feature layers from different Backbone layers.

In the overall structure of YOLOv5, the CBL module serves as the basic module, which consists of the convolution module, the normalization module, and the Leaky relu activation function [38]. Two different Cross-Stage Partial (CSP) networks [39] are used in the Backbone and Neck network, respectively. The CSP network connects the front and rear layers of the network through cross-stage connection. After fusing shallow features with deep features, the semantic information of key areas in the image can be better transmitted, which reduces the size of the model and improves the inference speed while ensuring the detection precision. The network structure of YOLOv5 is shown in Figure 3.

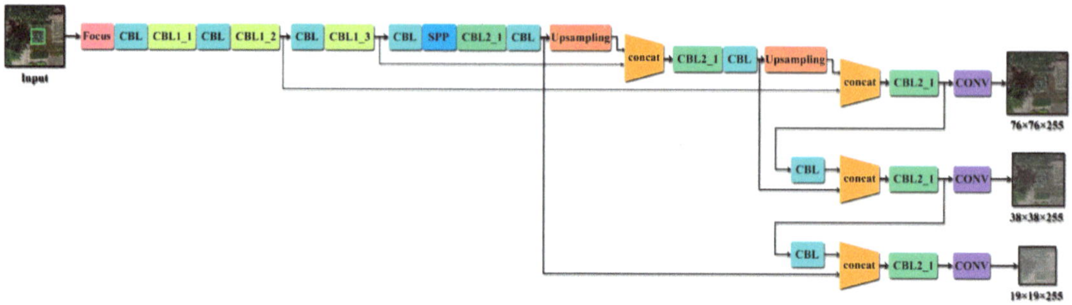

Figure 3. Network structure of YOLOv5.

Target detection networks with different structures have their respective advantages when dealing with different detection tasks, and most of these networks are composed of the convolutional layer, pooling layer, and fully connected layer. In our experiment, the YOLOv5 detection network delivered the best comprehensive performance and was therefore chosen for the detection task. Given the image I, the image label L and the real coordinates X_1, X_2, X_3, and X_4 as the network input, the output of the first stage detection $[Y_1, Y_2, Y_3, Y_4, C, P]_{img}$ can be obtained through the Backbone, the Neck, and the Prediction, as shown in Equation (1).

$$[Y_1, Y_2, Y_3, Y_4, C, P]_{img} = D(I, L, X_1, X_2, X_3, X_4) \quad (1)$$

where $D(\cdot)$ is the detect ion function of YOLOv5; C is the target category; P is the target confidence; Y_1, Y_2, Y_3, and Y_4 represent the coordinates of the output.

3.2. Principle of Classification Network

The input of image classification is the original images and the image labels provided by experts. Image classification networks are mostly composed of the convolutional layer, the pooling layer, the batch normalization layer, the activation function, and the fully-connected layer, and finally, the Softmax function is used to obtain the image label probability. Different classification network structures have different feature extraction capabilities and may show different effects in different recognition tasks. In order to extract more detailed features, this study regarded the second-stage task as a three-category (i.e., "*Papaver somniferum*", "plant", and "ground") image recognition task. The images were cropped and processed into the size of 224 × 224 pixels, and the DenseNet121 classification network [40] was used to extract features from the processed images. Given image I and label L as the input of DenseNet121, the output P_n can be obtained through feature extraction, as shown in Equation (2).

$$P_n = S(C(I), L) \quad (2)$$

where $C(\cdot)$ represents the extraction results of the feature extraction network DenseNet121; $S(\cdot)$ is the Softmax function; P represents the prediction probability of different categories; n represents the category.

3.3. Two-Stage Method

In the *Papaver somniferum* detection task, the detection results of a single detector cannot meet the actual requirements. When the model is applied in the real environment, it must have a good fit not only for the training dataset but also for the unknown dataset. For *Papaver somniferum* detection, the data collection across different times of a year is often limited by the weather conditions and collection equipment, so that the dispersion of the data collected from different batches raises high requirements on the generalization ability of the model. In order to improve the robustness and the generalization ability of the model, this study adopted a two-stage judgment method. In the first stage, YOLOv5s was used as the detector to detect the suspicious targets, which were resized into images of 224 × 224 pixels. Then, a trained classification model was used to assist with the judgment to generate the final judgment results. The two-stage network structure is shown in Figure 4.

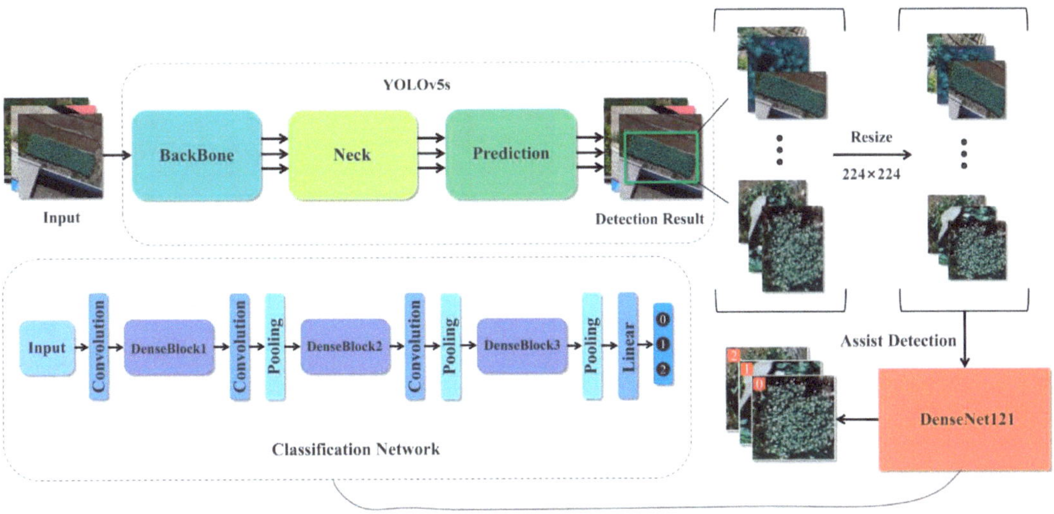

Figure 4. Two-stage network structure.

The input of YOLOv5s is images and the target location information; BackBone refers to the backbone network of YOLOv5s; Neck refers to the neck network of YOLOv5s; Prediction refers to the prediction stage of YOLOv5s. In the DenseNet121 classification network structure, Convolution refers to the convolutional operation; DenseBlock1, DenseBlock2 and DenseBlock3 are composed of BN, ReLu, and different numbers of 1 × 1 and 3 × 3 convolutional layers.

4. Experiment and Analysis

4.1. Experiment Dataset

The experiment in this paper consisted of two stages. In the detection task, there were a total of 643 original images, which were segmented into 3215 images after enhancement. According to the ratio of 8:2 between the training set and the test set, 2572 images were used for training, and 643 images were used for testing. The image labelling was manually processed using the LabelImg tool, and the labelling information was exported as an xml file, which contains the category information and the $x_{\min}$, $y_{\min}$, $x_{\max}$, and $y_{\max}$ information of the image. The dataset was saved in the COCO format for the training and testing purpose of the detection network model. In the classification task, the data were sorted and divided into three categories: "*Papaver somniferum*", "plant" and "ground"; the image size is 224 × 224 pixels.

4.2. Model Training

In this experiment, the hardware configuration for training and testing is as follows: Intel® Core (TM) i9-10920X CPU @ 3.00GHz, 24G memory, NVIDIA GeForce RTX 3090 graphics card, 64-bit Ubuntu 20.04.3 LTS, CUDA version 11.4, and Pytorch version 1.7.0. For the detection task, the learning rate was set to 0.01; the momentum was set to 0.937; the weight decay was set to 0.0005; the batch_size was set to 16; and the epoch was set to 300. For the classification task, the learning rate was set to 0.0001; the optimizer adopted Adam; the batch_size was set to 16; and the epoch was set to 50.

4.3. Evaluation Indicators

4.3.1. Evaluation Indicators for Target Detection

The YOLOv5 model will assign a confidence threshold for each image containing the target area. In this experiment, the confidence threshold was set to 0.25. Any prediction result with the confidence threshold higher than 0.25 would be judged as a correct prediction by the model. There will be a gap between the actually labelled box and the predicted box. The smaller the gap, the more accurate the model detection results are. The Intersection over Union (IoU) is a key parameter to measure the accuracy of detection, which is expressed as Equation (3). In this experiment, IoU was set to 0.45. In Equation (3), DR stands for Detection Result, representing the prediction result; GT stands for Ground Truth, representing the actual label.

$$IoU = \frac{DR \cap GT}{DR \cup GT} \times 100\% \tag{3}$$

The most commonly-used indicators for target detection include Precision, Recall, F1 Harmonic Mean, Average Precision (AP), and mean Average Precision (mAP). Precision is defined as the proportion of samples that are positively labelled out of all the samples predicted to be positive. Recall is defined as the proportion of samples that are successfully predicted as positive out of all the samples that are positively labelled. Precision and Recall are calculated by Equations (4) and (5), respectively. F1, defined by Equation (6), refers to the harmonic mean of Precision and Recall. AP, defined by Equation (7), refers to the area of the closed area enclosed by the (P-R) curve. mAP, defined by Equation (8), is the main evaluation indicator of target detection, which measures the overall performance of the network. In Equation (8), n represents the detection category. In this experiment, there was only one category, i.e., "*Papaver somniferum*", so n = 1.

$$Precision = \frac{TP}{TP + FN} \times 100\% \tag{4}$$

$$Recall = \frac{TP}{TP + FP} \times 100\% \tag{5}$$

$$F1 = \frac{2PR}{P + R} \tag{6}$$

$$AP = \int_0^1 P(R)dR \tag{7}$$

$$mAP = \frac{\sum_{i=1}^{n} AP_i}{n} \tag{8}$$

4.3.2. Evaluation Indicators for Image Classification

The performance of the second-stage classification model was evaluated by four evaluation indicators: recognition accuracy, recognition precision, model sensitivity and model specificity, which are expressed by Equations (9)–(12), respectively.

$$Accuracy = \frac{TP + TN}{TP + TN + FP + FN} \times 100\% \tag{9}$$

$$Precision = \frac{TP}{TP + FP} \times 100\% \tag{10}$$

$$Sensitivity = \frac{TP}{TP + FN} \times 100\% \tag{11}$$

$$Specificity = \frac{TN}{FP + TN} \times 100\% \quad (12)$$

where TP refers to the number of samples that belong to category C and are correctly classified by the classifier; FP refers to the number of samples that do not belong to category C but are misclassified by the classifier; TN refers to the number of samples that do not belong to category C and are correctly classified by the classifier; FN refers to the number of samples that belong to category C but are misclassified by the classifier.

4.4. Comparison and Analysis of Different Target Detection Models

The YOLOv5 algorithm consists of four different model sizes, i.e., YOLOv5s, YOLOv5m, YOLOv5l, and YOLOv5x, which are used to address different problems and different application scenarios. Table 4 presents and compares the testing results of different model structures on the same dataset, and the network structure delivering the best comprehensive performance was chosen as the final detector.

Table 4. Comparison between different YOLOv5 models.

Model	Precision/%	Recall/%	F1/%	mAP/%	Total Parameter
YOLOv5s	97.7	94.9	96.3	97.4	7,022,326
YOLOv5m	98.6	92.8	95.6	96.3	20,871,318
YOLOv5l	98.1	92.2	95.1	96.0	46,138,294
YOLOv5x	97.7	94.0	95.8	96.7	86,217,814

By testing the four models of YOLOv5 with different sizes and structures, it can be seen that YOLOv5m had the highest Precision, and its mAP value was 0.963. However, in the *Papaver somniferum* detection task, the higher the Recall, the better the model is. Therefore, out of the four YOLOv5 structures, YOLOv5s, which had the highest Recall, was chosen as the detector for the first stage. The mAP value of YOLOv5s was 0.974, and its overall performance was the best.

The original YOLOv5s structure was used for the training and testing of the first stage. Table 5 presents the comparison results between YOLOv5s and other existing networks in terms of the comprehensive performance of P, R, F1, mAP, and the parameter size.

Table 5. Comparison between different target detection algorithms.

Model	Precision/%	Recall/%	F1/%	mAP/%	Total Parameter
Faster rcnn	49.8	66.4	57.0	62.2	137,078,239
Mask-RCNN	70.7	87.7	78.3	82.6	63,733,406
RetinaNet	88.8	64.8	75.0	75.8	37,968,692
CenterNet	98.6	60.3	74.8	84.8	32,665,432
SSD	94.7	66.0	78.0	76.8	26,151,824
YOLOv3	97.7	91.9	95.0	93.3	61,949,149
YOLOv4	96.0	86.3	91.0	89.3	64,363,101
YOLOv5s	97.7	94.9	96.3	97.4	7,022,326

As can be seen from Table 5, YOLOv5s delivered the best results in terms of Recall, F1 score and mAP. The Precision, Recall and mAP of Faster RCNN were all at a low level. The Precision of other detection algorithms was generally higher than 88%. The Recall of RetinaNet, CenterNet, and SSD were lower than 70%, while that of YOLOv3 and YOLOv5s were higher than 90%. Based on the comparison between mainstream target detection algorithms, it was found that YOLOv5s delivered the best comprehensive performance with stronger robustness, so it is a more suitable option for solving practical problems. In order to further analyze the robustness of different models, different *Papaver somniferum* cultivation strategies were selected from the test results for comparative analysis, and the results are presented in Table 6.

Table 6. Detection results.

Model	Detection Result			
	Sample 1	**Sample 2**	**Sample 3**	**Sample 4**
Faster RCNN				
Mask-RCNN				
RetinaNet				
CenterNet				
SSD				
YOLOv3				
YOLOv4				
YOLOv5s				

From the comparison results between different detection algorithms, it can be found that YOLOv3, YOLOv4, and YOLOv5s successfully identified all the targets, while Faster

RCNN, RetinaNet, and SSD failed to detect single-plant *Papaver somniferum* in test picture 1. The main reason for the missed detection is attributed to the small size of the target, making it indistinguishable from the green field. In general, the detector had a weak learning ability for small targets in the feature extraction process. Mask R-CNN successfully detected the *Papaver somniferum* target, but there was a false detection in the test sample 2. The color of the false detection target is similar to that of a *Papaver somniferum*. The detection algorithms in the control group showed different abilities in detecting *Papaver somniferum* in different environments. Specifically, Faster RCNN and RetinaNet had relatively low confidence in the detection task, which conforms to the result of mAP as shown in Table 5.

Practice shows that the detection accuracy is positively correlated with the area of *Papaver somniferum* cultivation plot. On the existing data sets, the relationship between the detection accuracy and the area of *Papaver somniferum* cultivation plot is shown in Table 7.

Table 7. The detection accuracy and area of *Papaver somniferum* cultivation field.

Type of Plot	Detection Accuracy/%	Area of Field S/m^2
Large area plot	98.4	$S > 9$
Medium area plot	97.5	$1 \leq S \leq 9$
Small area plot	80.0	$S < 1$

It can be seen from Table 7 that the detection accuracy of *Papaver somniferum* plots planted in large and medium areas remains at a high level, but the detection accuracy of opium poppy plots planted in small areas is low. This is mainly because the characteristics of poppies in small plots are ambiguous, so detection is difficult. In addition, it is also related to the lower number of small area plot image samples in the dataset.

4.5. Comparison between Different Classification Models

In this section, VGG16, ResNet18, ResNet50, DenseNet121, DenseNet169, MobileNet, AlexNet, EfficientNet, Xception, and InceptionResNetv2 were chosen as the control group. The training accuracy curves and loss curves are shown in Figure 5a, and the comparison of test results is presented in Table 8. In order to confirm whether overfitting occurred due to too few data samples during the training process on the single-image branch, network retraining was performed for all the networks on the basis of pre-training of the ImageNet dataset, and the training accuracy curves and loss curves with the pre-training weight were obtained, as shown in Figure 5b. The comparison of test results is presented in Table 9.

As shown in Table 9, DenseNet121 derived the best results in terms of Accuracy, Precision, Sensitivity, and Specificity. It can be found that the comprehensive performance of the model using residual module and dense connection is at a high level. The characteristics of these two methods are the combination of shallow semantic features and deep semantic features in feature extraction, which enhances the feature extraction ability of the network. Both MobileNet and EfficientNet are lightweight networks, but the comprehensive performance gap is large. MobileNet network contains deep separable convolution. This convolution operation can greatly reduce the number of parameters while ensuring accuracy. Deep separable convolution is also used in Xception network structure, so its performance is better than EfficientNet. Its confusion matrix is shown in Figure 6. It can be seen that category 1 (plant) had the highest error rate. Our experiment showed that category 1 had the highest error rate in other control networks as well. From the original test set, it was found that the misclassified images were all green plants, which are very similar to the shape and color of papaver somniferum. According to Figure 5a,b, after retraining the model on the *Papaver somniferum* dataset on the basis of ImageNet pre-training, the accuracy and loss values can be largely optimized as compared to the initial state. The results in Tables 8 and 9 also show that the test accuracy had been greatly improved. On the one hand, it proves that a sufficiently large sample size allowed the classifier to fully learn the features of papaver somniferum. On the other hand, it also proves that overfitting did not occur when using the image classifier.

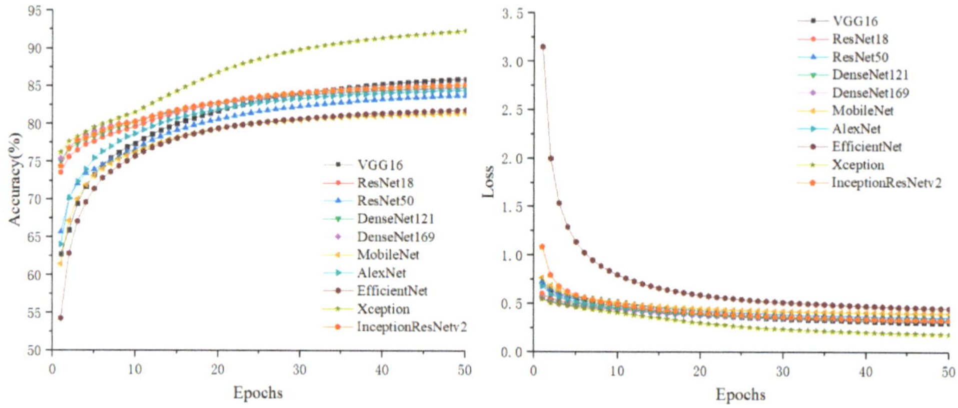

(**a**) The experiment process without pre-training.

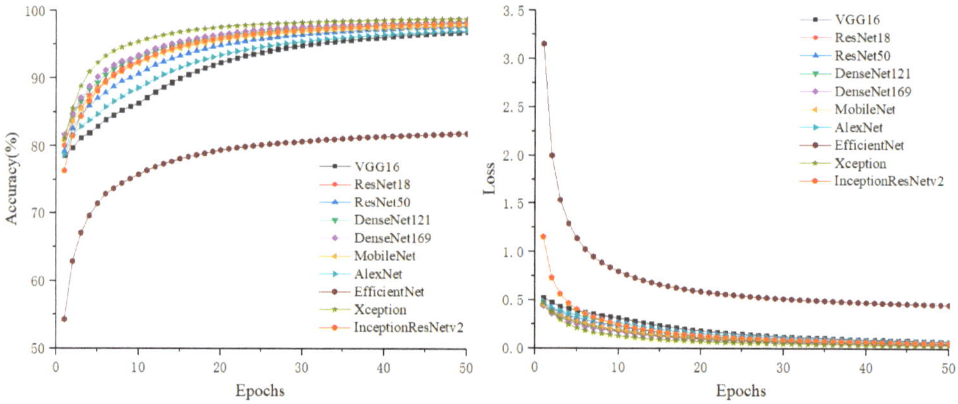

(**b**) The experiment process with pre-training.

Figure 5. Comparison of training accuracy and loss in the classification control group.

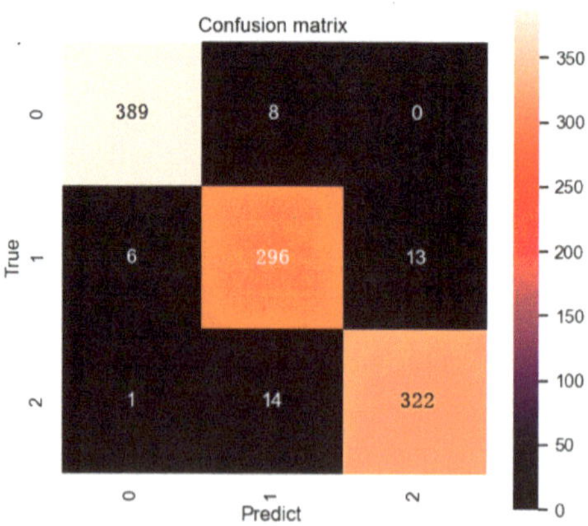

Figure 6. The recognition result confusion matrix of DenseNet121.

Table 8. Comparison of results on image classification without pre-training in the control group.

Models	Accuracy (%)	Precision (%)	Sensitivity (%)	Specificity (%)
VGG16	89.77	85.41	83.71	92.07
ResNet18	90.59	85.80	85.62	92.85
ResNet50	88.43	82.73	81.96	91.11
DenseNet121	88.43	82.72	82.17	91.20
DenseNet169	88.69	83.65	82.18	91.29
MobileNet	88.18	82.49	81.37	90.90
AlexNet	89.77	85.43	83.71	92.06
EfficientNet	89.26	84.07	83.36	91.76
Xception	90.59	86.05	85.39	92.79
InceptionResNetv2	89.51	84.72	83.50	91.89

Table 9. Comparison of results on image classification with pre-training in the control group.

Models	Accuracy (%)	Precision (%)	Sensitivity (%)	Specificity (%)
VGG16	93.52	90.18	90.18	95.17
ResNet18	96.57	94.68	94.75	97.46
ResNet50	96.00	93.89	93.75	96.99
DenseNet121	97.33	95.81	95.83	98.03
DenseNet169	97.01	95.45	95.22	97.75
MobileNet	96.44	94.46	94.46	97.36
AlexNet	92.88	89.29	89.08	94.60
EfficientNet	89.26	84.07	83.36	91.76
Xception	95.36	92.80	92.81	96.58
InceptionResNetv2	97.20	95.59	95.70	97.96

In Figure 6, "0" represents "*Papaver somniferum*", "1" represents "plant", and "2" represents "ground".

4.6. Comparison between One-Stage Method and Two-Stage Method

This section compares the detection performance between the one-stage method and two-stage method. A total of 40 original remote sensing images were selected, consisting of 18 images containing *Papaver somniferum* and 22 images not containing papaver somniferum. The images were segmented into the size of 640 × 640 pixels, and 4920 images were obtained after segmentation, including 4900 images without *Papaver somniferum* and 20 images with papaver somniferum. The segmented images were input into the trained model for testing, and the prediction results were compared with the actual results. Among the 4920 images, the one-stage and two-stage methods both detected 19 images containing *Papaver somniferum* (Recall, 95%). The one-stage method had 291 false detections, and the two-stage method had 76 false detections, suggesting a reduction of false detection by 73.88%. The number of test images and the detection results are presented in Table 10. Figure 7a,b show the detection results of the one-stage method and the two-stage method, respectively. In the figure, True means correct detection, False means wrong detection.

Table 10. Comparison of results between one-stage and two-stage methods.

<table>
<tr><th rowspan="3">Model</th><th colspan="2">Total Number of Original Remote Sensing Images</th><th colspan="2">Total Number of Image Patches after Segmentation</th><th colspan="2">Detection Result</th></tr>
<tr><th colspan="2">40</th><th colspan="2">4920</th><th rowspan="2">Correct Detection</th><th rowspan="2">Wrong Detection</th></tr>
<tr><th>Contain Papaver somniferum</th><th>No Papaver somniferum</th><th>Contain Papaver somniferum</th><th>No Papaver somniferum</th></tr>
<tr><td>YOLOv5s</td><td>18</td><td>22</td><td>20</td><td>4900</td><td>19</td><td>291</td></tr>
<tr><td>YOLOv5s+DenseNet121</td><td>18</td><td>22</td><td>20</td><td>4900</td><td>19</td><td>76</td></tr>
</table>

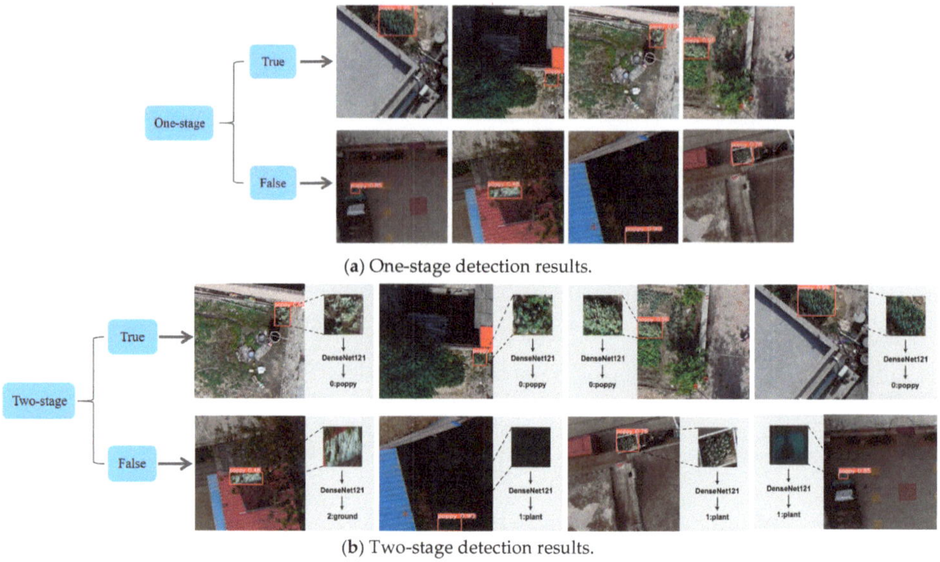

Figure 7. One-stage and Two-stage detection results.

It can be seen from Figure 7 that the falsely judged *Papaver somniferum* images of the one-stage detector were subject to the problem of high confidence. The method of improving the confidence in the detection stage cannot solve this problem effectively. By applying the two-stage method, which means using a classifier to perform secondary judgment after the detector, the non-*Papaver somniferum* category with high confidence can be excluded so as to reduce the workload of subsequent manual screening. However, there were still some green crop images that are difficult to distinguish.

5. Discussion

(1) It is technically feasible to use target detection to monitor illicit opium *Papaver somniferum* cultivation. Researchers explored opium poppy detection in satellite images and low-altitude images of drones. The images detected in the literature [41] come from satellites, so the method in the paper is only effective for detecting large opium poppy cultivation fields. Zhou et al. [3] proved through experiments that reducing the flying height of a UAV can improve the detection accuracy because reducing the flying height can improve the ground resolution of pixels in the image and make the opium poppy features in the image clearer. However, low-altitude flight is easily affected by flight safety problems caused by tall buildings. Wang et al. [4] improved the original YOLOv3 by adding an ASPP module, replacing ordinary convolution with ResNext grouping convolution, and adding a wider range of detection frames to improve the accuracy of the model. However, the false positive rate of this method is high in practice. The two-stage method of detection and recognition proposed in this paper can effectively reduce the false positive rate. A comparison between our work and that of other authors is shown in Table 11.

Table 11. The comparison between our work and that of other authors.

Methods	Stages	Height/m	Precision/%	Recall/%
Liu et al. [41]	one	506,000	95.00	85.00
Zhou et al. [3]	one	30/60/150	96.37	-
Wang et al. [4]	one	120	94.30	90.70
Our	two	130–145	97.70	94.90

(2) While collecting data and detecting *Papaver somniferum* from the images in the early stage, it was found that *Papaver somniferum* plants might be in different growth stages. In remote sensing images, the *Papaver somniferum* plants in the seedling stage, flowering stage and fruiting stage were of different characteristics. Specifically, the *Papaver somniferum* plants in the seedling stage were generally low in size, with the leaves in emerald color. The *Papaver somniferum* plants in the flowering stage had grown taller; the leaves became pale green with a serrated shape, and the flower color could be purple or white (the follower color of mutated *Papaver somniferum* might be more difficult to detect). In the fruiting stage, the leaves and fruits of *Papaver somniferum* were both in pale green color. The above factors make the "intra class difference" of poppy targets larger and improve the difficulty of detection. The key to solve this problem is to obtain as many *Papaver somniferum* samples as possible, so that the model can learn enough *Papaver somniferum* characteristics.

(3) The plot area of opium poppy cultivation is closely related to the detection accuracy. However, due to the size of the data set, the functional relationship between detection accuracy and plot area is still unclear. In the future, we should collect more detection accuracy of different plots and establish a quantitative description function between plot area and detection accuracy.

6. Conclusions

The use of UAV has become the main approach to combat illicit *Papaver somniferum* cultivation. However, at present, the acquired images mainly rely on manual screening through naked eyes. In view of this, we proposed a two-stage method for the detection of *Papaver somniferum* cultivation based on target detection and classification assistance. The results show that the two-stage method proposed in this paper has the same recall rate of 95% with the one-stage method, but the false detection rate is reduced by 73.88%. This achievement greatly reduces the workload for subsequent manual screening.

Author Contributions: Q.W.: writing—original draft preparation; C.W. and H.W.: writing—review and editing; C.Z.: methodology; G.T.: supervision, data curation; Y.Y.: visualization, software; H.Z.: investigation, validation. All authors have read and agreed to the published version of the manuscript.

Funding: This research was funded by National Key Research and Development Program of China, grant number 2019YFD1101105, and in part by the National Natural Science Foundation of China, grant number 61871041, and in part by the Hebei Province Key Research and Development Program, grant number 20327402D, 19227210D.

Acknowledgments: We are grateful to our colleagues at Hebei Key Laboratory of Agricultural Big Data and National Engineering Research Center for Information Technology in Agriculture for their help and input, without which this study would not have been possible.

Conflicts of Interest: The authors declare no conflict of interest.

References

1. Yadav, H.K.; Shukla, S.; Singh, S.P. Genetic variability and interrelationship among opium and its alkaloids in opium papaver somniferum (*Papaver somniferum* L.). *Euphytica* **2006**, *150*, 207–214. [CrossRef]
2. Nakazawa, A.; Kim, J.H.; Mitani, T.; Odagawa, S.; Takeda, T.; Kobayashi, C.; Kashimura, O. A study on detecting the papaver somniferum field using hyperspectral remote sensing techniques. In Proceedings of the 2012 IEEE International Geoscience and Remote Sensing Symposium, Munich, Germany, 22–27 July 2012; pp. 4829–4832.
3. Zhou, J.; Tian, Y.; Yuan, C.; Yin, K.; Yang, G.; Wen, M. Improved uav opium papaver somniferum detection using an updated yolov3 model. *Sensors* **2019**, *19*, 4851. [CrossRef] [PubMed]
4. Wang, C.; Wang, Q.; Wu, H.; Zhao, C.; Teng, G.; Li, J. Low-Altitude Remote Sensing Opium papaver somniferum Image Detection Based on Modified YOLOv3. *Remote Sens.* **2021**, *13*, 2130. [CrossRef]
5. Ozturk, T.; Talo, M.; Yildirim, E.A.; Baloglu, U.B.; Yildirim, O.; Acharya, U.R. Automated detection of COVID-19 cases using deep neural networks with X-ray images. *Comput. Biol. Med.* **2020**, *121*, 103792. [CrossRef] [PubMed]
6. Weiss, M.; Jacob, F.; Duveiller, G. Remote sensing for agricultural applications: A meta-review. *Remote Sens. Environ.* **2020**, *236*, 111402. [CrossRef]

7. Dasgupta, B.; Gupta, A.; Ray, A. Dark matter capture in celestial objects: Light mediators, self-interactions, and complementarity with direct detection. *J. Cosmol. Astropart. Phys.* **2020**, *2020*, 023. [CrossRef]
8. Lei, Y.; Yang, B.; Jiang, X.; Jia, F.; Li, N.; Nandi, A.K. Applications of machine learning to machine fault diagnosis: A review and roadmap. *Mech. Syst. Signal Process.* **2020**, *138*, 106587. [CrossRef]
9. Ma, L.; Liu, Y.; Zhang, X.; Ye, Y.; Yin, G.; Johnson, B.A. Deep learning in remote sensing applications: A meta-analysis and review. *ISPRS J. Photogramm. Remote Sens.* **2019**, *152*, 166–177. [CrossRef]
10. Viola, P.; Jones, M. Rapid object detection using a boosted cascade of simple features. In Proceedings of the 2001 IEEE Computer Society Conference on Computer Vision and Pattern Recognition. CVPR 2001, Kauai, HI, USA, 8–14 December 2001; Volume 1, p. I-I.
11. Dalal, N.; Triggs, B. Histograms of oriented gradients for human detection. In Proceedings of the 2005 IEEE Computer Society Conference on Computer Vision and Pattern Recognition (CVPR'05), San Diego, CA, USA, 20–25 June 2005; Volume 1, pp. 886–893.
12. Krizhevsky, A.; Sutskever, I.; Hinton, G.E. Imagenet classification with deep convolutional neural networks. *Adv. Neural Inf. Process. Syst.* **2012**, *25*, 1097–1105. [CrossRef]
13. Deng, J.; Dong, W.; Socher, R.; Li, L.-J.; Li, K.; Fei-Fei, L. Imagenet: A large-scale hierarchical image database. In Proceedings of the 2009 IEEE Conference on Computer Vision and Pattern Recognition, Miami, FL, USA, 20–25 June 2009; pp. 248–255.
14. Girshick, R.; Donahue, J.; Darrell, T.; Malik, J. Rich feature hierarchies for accurate object detection and semantic segmentation. In Proceedings of the IEEE Conference on Computer Vision and Pattern Recognition, Columbus, OH, USA, 23–28 June 2014; pp. 580–587.
15. Girshick, R. Fast r-cnn. In Proceedings of the IEEE International Conference on Computer Vision, Quebec City, QC, Canada, 27–30 September 2015; pp. 1440–1448.
16. Ren, S.; He, K.; Girshick, R.; Sun, J. Faster r-cnn: Towards real-time object detection with region proposal networks. *arXiv* **2015**, arXiv:1506.01497. [CrossRef]
17. He, K.; Gkioxari, G.; Dollár, P.; Girshick, R. Mask r-cnn. In Proceedings of the IEEE International Conference on Computer Vision, Venice, Italy, 22–29 October 2017; pp. 2961–2969.
18. Dai, J.; Li, Y.; He, K.; Sun, J. R-fcn: Object detection via region-based fully convolutional networks. In Proceedings of the 30th International Conference on Neural Information Processing Systems, Barcelona Spain, 5–10 December 2016; pp. 379–387.
19. Cai, Z.; Vasconcelos, N. Cascade r-cnn: Delving into high quality object detection. In Proceedings of the IEEE Conference on Computer Vision and Pattern Recognition, Salt Lake City, UT, USA, 18–23 June 2018; pp. 6154–6162.
20. Liu, W.; Anguelov, D.; Erhan, D.; Szegedy, C.; Reed, S.; Fu, C.Y.; Berg, A.C. Ssd: Single shot multibox detector. In *European Conference on Computer Vision*; Springer: Cham, Switzerland, 2016; pp. 21–37.
21. Fu, C.Y.; Liu, W.; Ranga, A.; Tyagi, A.; Berg, A.C. Dssd: Deconvolutional single shot detector. *arXiv* **2017**, arXiv:1701.06659.
22. Li, Z.; Zhou, F. FSSD: Feature fusion single shot multibox detector. *arXiv* **2017**, arXiv:1712.00960.
23. Zheng, L.; Fu, C.; Zhao, Y. Extend the shallow part of single shot multibox detector via convolutional neural network. Tenth International Conference on Digital Image Processing (ICDIP 2018). *Int. Soc. Opt. Photonics* **2018**, *10806*, 1080613.
24. Cui, L.; Ma, R.; Lv, P.; Jiang, X.; Gao, Z.; Zhou, B.; Xu, M. MDSSD: Multi-scale deconvolutional single shot detector for small objects. *arXiv* **2018**, arXiv:1805.07009. [CrossRef]
25. Redmon, J.; Divvala, S.; Girshick, R.; Farhadi, A. You only look once: Unified, real-time object detection. In Proceedings of the IEEE Conference on Computer Vision and Pattern Recognition, Las Vegas, NV, USA, 27–30 June 2016; pp. 779–788.
26. Redmon, J.; Farhadi, A. YOLO9000: Better, faster, stronger. In Proceedings of the IEEE Conference on Computer Vision and Pattern Recognition, Honolulu, HI, USA, 21–26 July 2017; pp. 7263–7271.
27. Redmon, J.; Farhadi, A. Yolov3: An incremental improvement. *arXiv* **2018**, arXiv:1804.02767.
28. Bochkovskiy, A.; Wang, C.Y.; Liao, H.Y.M. Yolov4: Optimal speed and accuracy of object detection. *arXiv* **2020**, arXiv:2004.10934.
29. Xu, D.; Wu, Y. Improved YOLO-V3 with DenseNet for multi-scale remote sensing target detection. *Sensors* **2020**, *20*, 4276. [CrossRef]
30. Ezzy, H.; Charter, M.; Bonfante, A.; Brook, A. How the Small Object Detection via Machine Learning and UAS-Based Remote-Sensing Imagery Can Support the Achievement of SDG2: A Case Study of Vole Burrows. *Remote Sens.* **2021**, *13*, 3191. [CrossRef]
31. Lim, J.S.; Astrid, M.; Yoon, H.J.; Lee, S.I. Small object detection using context and attention. In Proceedings of the 2021 International Conference on Artificial Intelligence in Information and Communication (ICAIIC), Jeju Island, Korea, 13–16 April 2021; IEEE: Piscataway, NJ, USA, 2021; pp. 181–186.
32. Miura, H.; Aridome, T.; Matsuoka, M. Deep learning-based identification of collapsed, non-collapsed and blue tarp-covered buildings from post-disaster aerial images. *Remote Sens.* **2020**, *12*, 1924. [CrossRef]
33. Shi, L.; Zhang, F.; Xia, J.; Xie, J.; Zhang, Z.; Du, Z.; Liu, R. Identifying Damaged Buildings in Aerial Images Using the Object Detection Method. *Remote Sens.* **2021**, *13*, 4213. [CrossRef]
34. Zhu, L.; Geng, X.; Li, Z.; Liu, C. Improving YOLOv5 with Attention Mechanism for Detecting Boulders from Planetary Images. *Remote Sens.* **2021**, *13*, 3776. [CrossRef]
35. Yan, B.; Fan, P.; Lei, X.; Liu, Z.; Yang, F. A Real-Time Apple Targets Detection Method for Picking Robot Based on Improved YOLOv5. *Remote Sens.* **2021**, *13*, 1619. [CrossRef]

36. Mehdi, S.R.; Raza, K.; Huang, H.; Naqvi, R.A.; Ali, A.; Song, H. Combining Deep Learning with Single-Spectrum UV Imaging for Rapid Detection of HNSs Spills. *Remote Sens.* **2022**, *14*, 576. [CrossRef]
37. Hu, H.; Dai, B.; Shen, W.; Wei, X.; Sun, J.; Li, R.; Zhang, Y. Cow identification based on fusion of deep parts features. *Biosyst. Eng.* **2020**, *192*, 245–256. [CrossRef]
38. Mastromichalakis, S. ALReLU: A different approach on Leaky ReLU activation function to improve Neural Networks Performance. *arXiv* **2020**, arXiv:2012.07564.
39. Kim, D.; Park, S.; Kang, D.; Paik, J. Improved Center and Scale Prediction-Based Pedestrian Detection Using Convolutional Block. In Proceedings of the 2019 IEEE 9th International Conference on Consumer Electronics (ICCE-Berlin), Berlin, Germany, 8–11 September 2019; pp. 418–419.
40. Huang, G.; Liu, Z.; Van Der Maaten, L.; Weinberger, K.Q. Densely connected convolutional networks. In Proceedings of the IEEE Conference on Computer Vision and Pattern Recognition, Honolulu, HI, USA, 21–26 July 2017; pp. 4700–4708.
41. Liu, X.; Tian, Y.; Yuan, C.; Zhang, F.; Yang, G. Opium papaver somniferum detection using deep learning. *Remote Sens.* **2018**, *10*, 1886. [CrossRef]

big data and cognitive computing

MDPI

Article

An Emergency Event Detection Ensemble Model Based on Big Data

Khalid Alfalqi * and Martine Bellaiche *

Computer and Software Engineering Department, École Polytechnique de Montréal, Montréal, QC H3T 1J4, Canada
* khalid.alfalqi@polymtl.ca (K.A.); martine.bellaiche@polymtl.ca (M.B.)

Abstract: Emergency events arise when a serious, unexpected, and often dangerous threat affects normal life. Hence, knowing what is occurring during and after emergency events is critical to mitigate the effect of the incident on humans' life, on the environment and our infrastructures, as well as the inherent financial consequences. Social network utilization in emergency event detection models can play an important role as information is shared and users' status is updated once an emergency event occurs. Besides, big data proved its significance as a tool to assist and alleviate emergency events by processing an enormous amount of data over a short time interval. This paper shows that it is necessary to have an appropriate emergency event detection ensemble model (EEDEM) to respond quickly once such unfortunate events occur. Furthermore, it integrates Snapchat maps to propose a novel method to pinpoint the exact location of an emergency event. Moreover, merging social networks and big data can accelerate the emergency event detection system: social network data, such as those from Twitter and Snapchat, allow us to manage, monitor, analyze and detect emergency events. The main objective of this paper is to propose a novel and efficient big data-based EEDEM to pinpoint the exact location of emergency events by employing the collected data from social networks, such as "Twitter" and "Snapchat", while integrating big data (BD) and machine learning (ML). Furthermore, this paper evaluates the performance of five ML base models and the proposed ensemble approach to detect emergency events. Results show that the proposed ensemble approach achieved a very high accuracy of 99.87% which outperform the other base models. Moreover, the proposed base models yields a high level of accuracy: 99.72%, 99.70% for LSTM and decision tree, respectively, with an acceptable training time.

Keywords: machine learning; big data; social networks; classification; emergency event detection; snapchat; twitter; hotspot map; ensemble

Citation: Alfalqi, K.; Bellaiche, M. An Emergency Event Detection Ensemble Model Based on Big Data. *Big Data Cogn. Comput.* **2022**, *6*, 42. https://doi.org/10.3390/bdcc6020042

Academic Editors: Miltiadis D. Lytras and Andreea Claudia Serban

Received: 7 February 2022
Accepted: 14 April 2022
Published: 16 April 2022

Publisher's Note: MDPI stays neutral with regard to jurisdictional claims in published maps and institutional affiliations.

1. Introduction

Emergency events arise when a serious, unexpected, and often dangerous threat affects normal life. Both natural catastrophes (such as earthquakes, cyclones and floods) and man-made casualties (like fires, explosions and moving vehicle accidents) are examples of emergency events. Hence, knowing what is happening during and after an emergency is critical to mitigate the consequences of the event on humans' lives, infrastructures, environment and finances [1]. As a result, obtaining information rapidly and precisely is crucial for immediate responses. One of the most trustworthy sources of information remains social media platforms. Nowadays, social media platforms have emerged as prominent sources of real-time information regarding emergency events. Individuals at the incident location, and those nearby, may immediately upload valuable information on such platforms—the usage of which has exploded due to the prevalence of cellphones and internet connectivity [2]. It is thus not surprising that social networks are online communication platforms that many people use worldwide [3]. There are different types of social network platforms. One of which is Twitter: a platform that enable people to communicate by sending short messages of no more than 280 characters. Another platform is Snapchat: an application that allows users to share their personal stories by sending, to their friends and relatives, images and videos which are removed 24 h after they are posted [4].

Emergency events create spikes in social networks' information rates which then requires fast data processing to handle such events. During emergency events, the information on social network platforms generates massive data that cannot be handled by traditional data processing methods, thus requiring the use of big data processing techniques [5]. Big data is a collection of more comprehensive methods that seek to study, save, and supervise massive data within an adequate time frame [6]. It provides decision-makers with unique capabilities to analyze and understand a set of circumstances to ensure they choose the most appropriate decisions when emergencies strike [7].

A robust emergency event detection system is needed to limit the disastrous impacts of such events. Social networks and big data play a critical role to expedite the discovery of emergency events. Leveraging data from social networks can help to manage, monitor, analyze and detect emergency events. However, when gathering real-time data from social networks, the main research gap concerns identifying the exact location of the emergency events.

Several research articles attempted to tackle this drawback by presenting promising suggestions. Alomari E et al. [8] utilized Twitter, big data, and social media platforms to develop an auto labeling system for traffic-related event detection. Their approach detected traffic events automatically from tweets using distributed machine learning and Apache Spark. However, this paper does not suggest the use of Snapchat maps to detect the exact location of traffic-related events. Instead, it presents multiple methods to extract the event location: from the tweet's message, from the hashtag if it mentions a specific place, from a predefined list of account names specialized in posting about traffic conditions, or by checking coordinates of a tweet. In addition, while this paper focuses only on traffic-related event detection, our approach provides a broader overview and can detect different types of emergency events.

Furthermore, the novelty of our approach is using the Snapchat hotspot maps as a reliable source to detect precise emergency event locations without prior knowledge of the location or the emergency event itself. Accordingly, our approach first identifies the emergency event locations through the Snapchat heat map. Then, after detecting the exact location of such events, we leverage Twitter stream API to collect the tweets in the identified locations in near real-time.

This paper proposes a new and efficient emergency event detection ensemble model (EEDEM) to address the aforementioned research gap. The main objective of this model is to find the exact location of an emergency event by utilizing the collected data from social networks (SN) with the integration of big data (BD) and machine learning (ML) [9].

The suggested approach aspires to acquire the best performance in terms of accuracy while reducing the processing time to fulfill the goals of this paper. More particularly, the contributions of this paper can be stated as follow:

- Propose a data collection model to collect the data from social networks by leveraging the Snapchat API and Twitter API.
- Develop a new and efficient emergency event detection ensemble model (EEDEM) by utilizing the collected data from social networks (SN) with the integration of big data (BD) and machine learning (ML).
- Identify the exact location of an emergency event by observing the Snapchat hotspot map to detect any potential emergency events and then utilizing Twitter API to collect the dataset of such events.
- Propose an ensemble voting model to evaluate the performance of our approach and improve the accuracy of the base models.

The rest of the paper is organized as follows: Section 2 "Related Works" explains the related works of three different research domains studied in this research. Section 3 "Background" presents a brief background of Big data layer architecture, the Apache spark streaming and the machine learning models. Section 4 "Methodology" describes the research methodology steps, techniques and models used for the experimentation. Section 5 "Evaluation approach", provides the details of the selected case study which highlights the reasons behind this selection and experimental environment setup, evaluation metrics, the performance analysis for our proposed approach and discusses study results. Finally, Section 6 "Conclusion and Discussion", concludes the research work and provides future research directions.

2. Related Work

This section illustrates a literature review pertaining to three research fields studied in this paper. First, we shed light on up-to-date studies that address machine learning classification methods in emergency events detection using social networks. Then, we elaborate on current big data methods in emergency events detection using social networks. After that, we address the recent usage of Snapchat in emergency events detection.

2.1. Machine Learning Classification Methods in Emergency Events Detection Using Social Networks

Septianto et al. [10] presented an innovative idea to collect traffic flow data from Twitter in Jakarta city utilizing a machine learning classifier called 'Nave Bayes Classifier (NBC). They built a software that can display Jakarta's traffic conditions in real time and then sort the data to be injected into Google Map. In addition, they utilised an illustrative forecasting models relying on up-to-date data to oversee Jakarta's roads during designated times using NBC to encourage drivers to take different paths instead of heading towards traffic jams.

Toujani et al. [11] proposed a unique approach that identifies event information following a natural catastrophe by employing social networks as the main reference. Then, they cluster individuals into levels founded on the phase of hazard. This clustering procedure is profitable for reporters to simplify the method of deriving information in emergencies. Furthermore, they utilized the fuzzy theory steps on these events to promote clustering excellency and eradicate opacity in the collected data.

Kumara et al. [12] suggested a procedure to bolster the after-disaster response methods by specifying the right location and the type of crisis founded on three stages. Nevertheless, this procedure has some limitations and requires further optimization.

The above approaches deliver ambitious perspectives on detecting emergency events by applying machine learning to the data from social networks. However, none of the above-mentioned papers have used big data technologies.

2.2. Big Data Techniques in Emergency Events Detection Using Social Networks

Similar works have used big data with data collected from social networks. Hagras et al. [13] employed the Latent Dircherilet Allocation (LDA) topic analysis method to categorize and assess tweets correlating with the Japan Tsunami. Yet, this method can be expanded by using additional datasets to boost the accuracy and expedite the processing in real-time.

Ragini et al. [14] introduced a strategy for crisis governance by incorporating emotion analysis and ML algorithms. They used a Support Vector Machine (SVM) to analyze the data obtained from Twitter via human sentiments, both optimistic and pessimistic, and then categorize them based on their necessities. Although the suggested strategy simplifies emergency crews' mission to recognize the disastrous case and take suitable measures directly, it has some problems in employing social network data for crisis mitigation-specifically the unclarity in obtaining crisis data through various sources and the lack of the suitable criterion. However, these problems can be expiated by gathering data from diverse venues to categorize the data efficiently and enhance precision.

Salas et al. [15] employed Apache Spark to gather tweets and build a model using the SVM classification approach to classify them. Then, they used Name Entity Recognition (NER) and Wikipedia to obtain the location information.

Lau, R.Y. et al. [16] suggested the Latent Dirichlet Allocation (LDA) to classify datasets retrieved from Twitter and Weibo. Then, they used SVM, KNN and NB to classify the data by utilizing Spark MLib. In addition, they created a classification ensemble technique to automatically detect specific crowded traffic events.

Bhuvaneswari et al. [17] presented an end-to-end framework to enhance emergency events detection rate by deploying topic modelling to the data collected from Twitter stream API. They built their real-time framework based on Apache Spark and the LDA technique. The event detection approach reached a high accuracy of %96 and the event could be detected after only 75-100 milliseconds of its occurrence.

Alomari E et al. [18] utilized Twitter and the big data approach to design a dictionary that facilitates the detection of traffic events in Saudi Arabia. In addition, they used sentiment analysis based on the lexicon approach for Arabic and Saudi dialect words.

The aforementioned studies present novel ideas that apply big data techniques to the data collected from social network platforms following emergency events. However, none of the previous papers consider using both Twitter along with Snapchat hotspot maps to pinpoint the exact location of emergency events. By leveraging the Snapchat hotspot map and Twitter, and combining it with big data techniques, we can precisely identify the locations of emergency events and thus accelerate the detection processing.

2.3. Using Snapchat in Emergency Events Detection

On the other hand, some recent work proposed using Snapchat maps to solve similar issues. For example, Al-ghamdi N et al. [19] suggested using Snapchat heat maps to analyze the crowd behavior at the Grand Holy Mosque in Mecca, Saudi Arabia. Furthermore, a similar approach was proposed by Alageeli N et al. [20] to analyze the sentiment of Riyadh's season visitors. They predefined the locations of three major events in Riyadh's season and then used Snapchat maps to extract crowd behavior patterns. However, these studies utilized the Snapchat map as a social sensor to track and analyze the visitors' activities. Moreover, Juhasz et al. [21] utilized Snapchat and different social network platforms to compare their activity patterns. Likewise, Lamba et al. [22] crawled the Snapchat map API to collect data in predefined cities to annotate them into dangerous driving or not dangerous driving.

Even though aforementioned papers utilized the Snapchat map in their approaches, they used it in predefined locations. Unlike our approach, we propose using the Snapchat heat maps as a reliable source of precise emergency event location detection without prior knowledge of the location.

3. Background

This section provides a brief overview of relevant background concepts inherent to this study, which include big data, Apache Spark and machine learning algorithms.

3.1. Big Data

Big Data architecture facilitates the data flow development to suit both batch processing and stream processing. This architecture comprises four layers that promote safe data flow. These layers are merely a method of categorizing components that conduct specific tasks. Figure 1 shows the overall layers of big data.

- The data resource layer: This layer is responsible for managing and collecting all potential data sources to strengthen the scheme [6].
- The data aggregation layer: Adding multiple heterogeneous data sources inevitably improves the efficiency and usefulness of the framework to allow taking the most appropriate decisions. However, it may also raise framework instability and add certain difficulties. The main responsibility for this layer consists of gathering all the data from different channels before injecting it into a multi-source database [23].

- The data analytic and processing layer: The main data processing layer includes a range of specific resources to acquire, store, retrieve, search and analyse data. A mix of various big data analytic platforms can be used to build a near real-time system.
- The application and support services layer: An integrated web-based computer framework will offer decision-makers (e.g. emergency services, public safety staff, police, fire departments) the necessary information. The main goal of this layer consists of enhancing the decision-making cycle with a continuous stream of the required information, as well as the latest trends for further perspective [23].

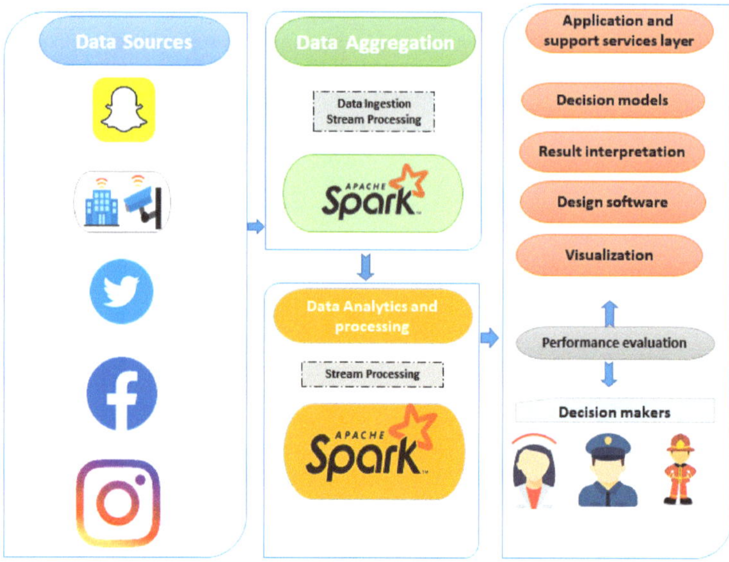

Figure 1. Overall of big data layers.

3.2. Machine Learning

Machine learning is described as the phases that involve learning given data in order to obtain adequate knowledge and then create the targeted outcome. Machine learning can be supervised, unsupervised or semi-supervised [24]. The following algorithms are used in our approach: long short-term memory (LSTM), decision tree (DT), support vector machine (SVM), K nearest neighbor (KNN) and Naïve Bayes (NB).

- **Long Short-Term Memory (LSTM)**: is a kind of neural network layer that is usually adopted in Recurrent Neural Networks (RNN) which overcomes RNN's vanishing gradient problem. A cell, an input gate, an output gate, and a forget gate comprise a standard LSTM unit. The cell retains data for unlimited time frames, and the three gates monitor data transmission to and from the cell [25].
- **The decision tree (DT)**: is a kind of supervised classifiers technique used for classification and regression. Due to its strength and effectiveness, it became a standard tool for machine learning and big data problems. It contains three types of nodes: decision nodes, chance nodes and end nodes [26].
- **Support vector machine (SVM)**: is used for both classification and regression problems. The goal of SVM is to create the most appropriate hyperplane and split the dataset into two classes. SVM is a robust machine learning technique which can be deployed in a broad range of subjects such as linear, nonlinear classification, and regression. It is among the most widely used machine learning techniques [27].
- **K Nearest Neighbor (KNN)**: is a machine learning approach that matches raw data to data that has already been categorized with a stipulated training data class. The

disparity between the raw and trained data is used to assess this comparison. The nearest neighbors are detected with the data set with the lowest distance estimates [28].
- **Naïve Bayes (NB)**: is beneficial to define data sets with a significant volume of knowledge since it performs quickly and is easily enforced. The NB algorithm is a Bayesian classification approach; hence, it is founded on the Bayes probability concept and generates probability tables for each variable individually [29].

4. Research Methodology

In this section, to tackle the research gap and fulfill our goals, the core aspects of this research must be elaborated by describing the required steps. First, Apache Spark Streaming is integrated to empower stream processing. Second, a data collection model is proposed to collect and label "Snapchat" and "Twitter" data to prepare the dataset. Third, data preprocessing steps are applied so that the machine learning algorithms can recognize the data. Fourth, feature extraction algorithms are utilized to help construct machine learning classifiers. Fifth, machine learning classifiers for emergency event detection models are proposed to facilitate the classification process. Finally, we deploy an ensemble based emergency events detection approach and conduct a performance evaluation to depict the detection accuracy of the base models and the proposed voting ensemble model.

4.1. Build Big Data Processing—Apache Spark Streaming

This research is developed on the Apache Spark platform, which empowers the stream processing of different data sources. Spark streaming is widely used to process near real-time data from numerous sources. Spark streaming is an enhancement of the main Spark API that provides extensible, elevated, fault-tolerant continuous data stream processing. Spark collects live, raw data feeds and separates them into packets, which are subsequently processed by the Spark engine to provide the latest version of results. Its fundamental concept is the Discretized Stream, which depicts a flow of data separated into small batches. DStreams are founded on the basis of Resilient Distributed Datasets (RDDs). This allows Spark streaming to collaborate with other Spark components including MLlib and Spark SQL [30]. Spark ML and Spark SQL are the primary libraries that have been used. Spark.ML, an extension package integrated into core Spark, provides a higher application programming interface stacked on top of data to develop and enhance machine learning [31]. Furthermore, the python built-in libraries were used to process the dataset.

4.2. Data Collection Model

The most popular social network platforms Twitter and Snapchat are used in the proposed emergency event detection model. The collected dataset can take the shape of pictures, videos or texts. Figure 2 shows the proposed data collection model for both Snapchat and Twitter platforms.

4.2.1. Snapchat Data Collection

To identify the precise location from the Snapchat platform, we need to develop a classification model using snaps from the Snapchat Map. This interactive map enables us to share our location with others. It is a distinctive characteristic whereby any content can be publically available, and immediate updates are generated while Snapchat is accessible. Furthermore, content posted on a Snapchat Map is geo-tagged and indicated in a precise area [22]. Thus, by monitoring the Snapchat Map, any hotspot area on the map could indicate an emergency event zone [32]. Accordingly, snaps in that area are scrapped to be labelled and marked as either emergency event-content or non-emergency event-content. Then, each hotspot location is ranked based on the highest number of snaps in that location, to prioritize the locations. For this purpose, we developed a wrapper for Snap Map's API to create Node.js JavaScript that collect snaps posted at the precise hotspot location.

The collected snaps can be image or video; therefore, the images and the videos will be classified using an image classification technique.

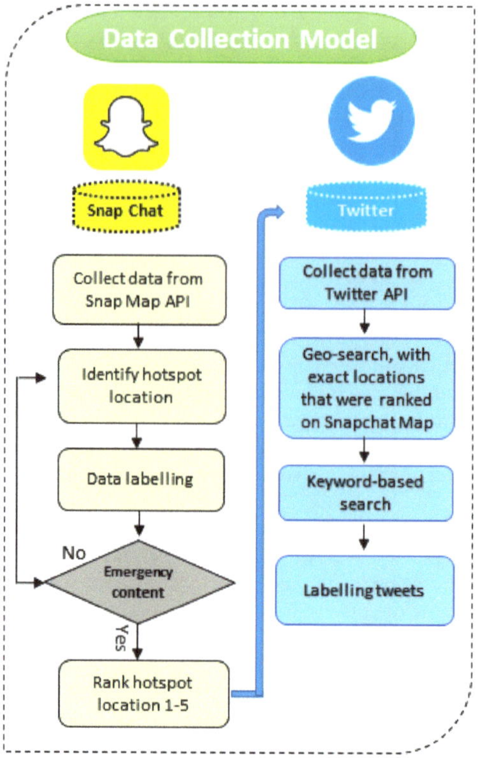

Figure 2. Data collection model.

4.2.2. Twitter Data Collection

After identifying and ranking hotspot zones on the Snapchat Map, the top five hotspots are used to collect Twitter data. The benefit of this step is to narrow down the tweets' geolocation search based on the high-ranked hotspot areas and reduce processing times [21]. In this research, tweets will be collected through the Twitter Search API; geo-tagged tweets are collected in the exact locations that were identified and ranked on Snapchat Map. To illustrate, the collection process is carried out with the following steps:

1. Geographical Search "Geo-Search": it searches within a bounding box of the most highly ranked locations from a Snapchat Map, to collect all tweets within the identified zone.
2. Keyword-Based Search: it searches, within the collected tweets, keywords and hashtags which are considered relevant to the emergency event. This step is intended to extract tweets that include at least one keyword or hashtag connected to the emergency situation to minimize non-relevant data [12].
3. Labelling Tweets: according to the type of emergency event (by matching keywords and hashtags data to the emergency event detected on the Snapchat), tweets are associated with one of two labels: related or not to the case study [33,34].

4.3. Data Preprocessing

The collected tweets are stored in MongoDB, one of the most prominent document-oriented databases [35]. Thus, we utilized MongoDB to store the tweets acquired through the Twitter API before it is exported to the Apache Spark Data Frame. The datasets is preprocessed at this stage to ensure machine learning algorithms can recognize the data in the next steps. Specific processing tasks are conducted on the acquired Twitter datasets. The main preprocessing steps used are: tokenization, normalization, lower case text conversion, as well as the removal of stop words, punctuation, white space and repeated characters. Algorithm 1 shows the preprocessing steps

Algorithm 1: Prepossessing

```
Data: TweetList; stopWordsList; punctuationList;
Result: Normalized tokens
spark ← SparkConnector()
tweets_DF['fullText'] ← spark.load(TweetList)
for tweet in tweets_DF do
    clean_Tweet ← remove_repeating_char(tweet)
    clean_Tweet ← casefoldingText(clean_Tweet)
    clean_Tweet ← remove_mentioins(clean_Tweet)
    clean_Tweet ← remove_links(clean_Tweet)
    clean_Tweet ← remove_RTs(clean_Tweet)
    clean_Tweet2 - [ ]
    for char in clean_Tweet do
        if char not in punctuation List AND not char.isdigit() AND not char is English
         Char() then
            clean_Tweet2.append(char)
        end
    end
    tokens_List ← tokenizingText(clean_Tweet2)
    normalizedTokens - [ ]
    for token in tokens_List do
        if token not in stop Words List then
            normalizedTokens.append(token)
        end
    end
    normalizedTokens ← tweets_DF ['Tokens']
end
```

4.4. Feature Extraction

Feature extraction is an essential component to develop machine learning algorithms. Its objective is to transform raw data into controllable parameters (a collection of features) while retaining data accuracy. Additionally, it empowers us to choose the essential features while developing a classifier model. Different feature extraction techniques such as TF-IDF and Bag of Words [36] were used in this research, are defined as follows:

1. **TF-IDF:** It is an acronym for Term Frequency-Inverse Document Frequency. It considers the number of occurrences of a phrase, in a document, to determine its significance [37].
2. **Bag of Words:** The full text is displayed as a list of words in the Bag of Words (BOW) feature. The frequency of each word is employed as a feature while training an algorithm. Substantial preprocessing is required to ensure that the bag of words feature provides acceptable accuracy. The preparation stage excludes extraneous words from the database, allowing the classifier to avoid any redundant features throughout the process of learning [36].

4.5. Applying Machine Learning Classifiers

According to the rules developed throughout the learning or training stages, classification in machine learning helps predict a group or class of an input dataset. To filter tweets into related or non-related emergency events, a classifier is built using ML classification algorithms [36]. We have a balanced dataset as the number of samples for the negative class (not related to the case study) is 49.6%, while the positive class (related to the case study) is 50.4%.

Also, we need to tune the parameters after data processing and feature extraction to reach the optimal performance modelling. Grid search is a method to find the optimal tuning parameter values. To achieve this, a collection of tuning-parameter possibility values must be defined and assessed [18]. Table 1 shows the hyper-parameters of the models.

Table 1. Hyper-parameters of the models.

Models	Hyper-Parameters
LSTM	dropout_rate = 0.2, embed_dim = 32, hidden_unit= 16 , optimizers = RMSprop, output = 3 neurons, activation = tanh + linear , epochs = 10, batch_size = 128
SVM	C = 1, gamma = 0.5 , kernel = rbf, probability is False, shrinking = True, tol = 0.001
KNN	n_neighbors = 8.
Naive Bayes	var_smoothing = 1×10^8
Decision Tree	criterion ='entropy', splitter ="best"
The Proposed Ensemble	Estimators = LSTM, SVM, KNN, Naive Bayes and Decision Tree, Voting = hard

Then, a Spark ML library was utilized to develop and train the models. In order to classify the dataset, five base-models were built with LSTM, SVM, Naive Bayes, decision tree and K Nearest Neighbor algorithms [38].

4.6. Proposed Voting Ensemble for Emergency Event Detection Model

In this step, we use the five trained models combined together to deploy the voting ensemble approach. We applied the 5-Fold cross-validation to split the dataset and train the models, with data split into five parts. Each round uses 5-1 parts to train the classifier and the residual part as a validation set. The 5-Fold cross-validation is depicted in Figure 3.

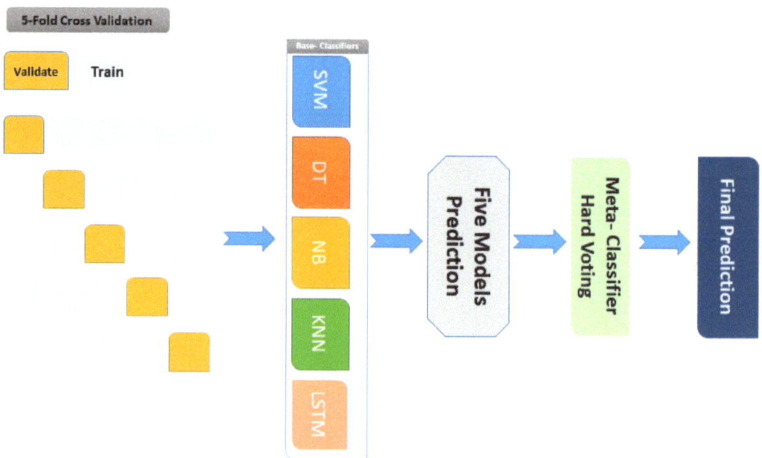

Figure 3. The 5-Fold cross validation.

The evaluation is based on a majority voting (Hard Voting) approach of the five base classifiers (LSTM, SVM, Naive Bayes, decision tree, and K Nearest Neighbor). As a result, the proposed hard voting ensemble model combines five base classifiers predictions to give the final prediction [39].

After that, we will conduct a performance evaluation to depict the detection accuracy of all models.

Figure 4 shows the proposed steps in Sections 4.3–4.6 of the emergency event detection ensemble model.

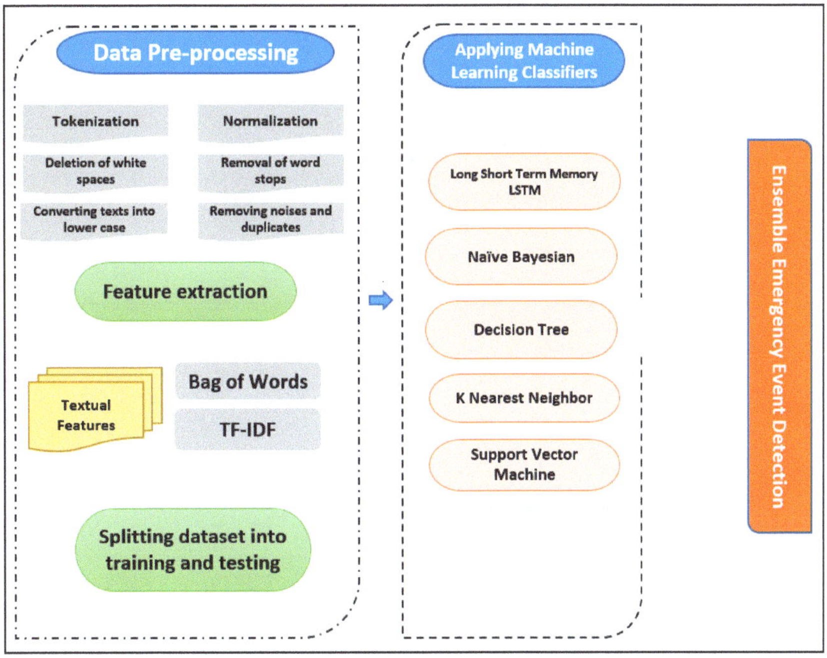

Figure 4. The proposed steps of the emergency event detection ensemble model.

5. Evaluation Approach

This section provides the details regarding the experimental environment setup, the evaluation metrics, and the performance analysis for the proposed approach.

5.1. Case Study

The Beirut Port Explosion

On 4 August 2020, a major explosion, one of the most enormous non-nuclear blasts in history, damaged a large section of Beirut. Roughly 2750 tons of ammonium nitrate (which equals to 1.1 kilotons of TNT) stockpiled in a facility near the port detonated, inflicting significant damage throughout the capital. The shock was felt in countries as far away as Turkey, Syria and Cyprus. The explosion killed at least 200 people, wounded over 6000 and an estimated 300,000 inhabitants became homeless after the $15 billion in property destruction [40]. Figure 5 shows the words cloud for the tweets related to the case study.

Using the developed Node.js Google function wrapper, the search for the hotspot location with a Snapchat map revealed more than five hotspots on 4 August 2020. However, based on the ranking criteria of the Snapchat data collection in Section 4.2.1, the hotspot location which has the most snaps at that time was the Beirut port explosion. Figure 6 shows the location of Beirut Port explosion on the Snapchat hotspot map. After Identifying the location of this emergency event, we collected snaps in this location and then classify them using Convolutional Neural Network(CNN) with the help of the transfer learning approach; by utilizing the pre-trained model as a feature extractor. The pre-train model is ResNet50, a variant of the ResNet model that has 48 convolution layers and 1 MaxPool, and 1 Average Pool layer. Then we have modified the output dense layer to suit our case study. Table 2 present the structure of this model.

Figure 5. Words cloud of tweets.

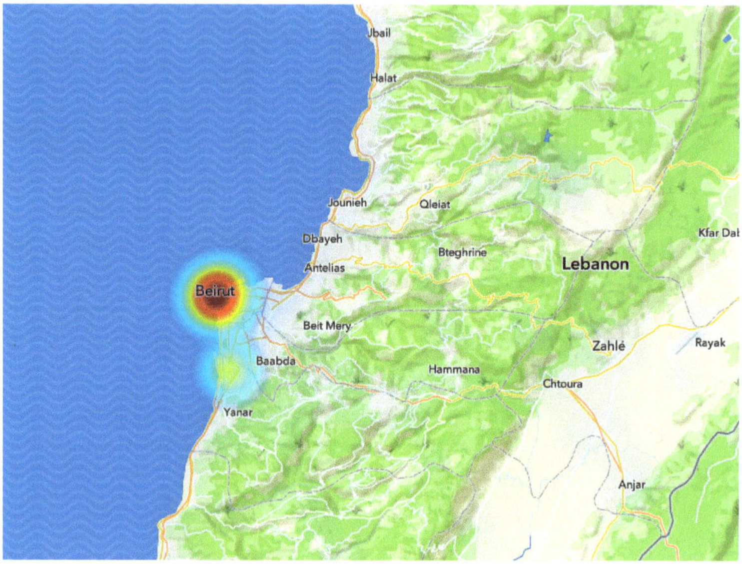

Figure 6. Explosion location of Beirut Port.

Table 2. Structure of the CNN model.

Models	Hyper-Parameters
CNN	optimizer='adam', loss='binary_crossentropy', metrics = ['accuracy'], activation = 'sigmoid' input_2 (InputLayer), Output Shape [(None, 224, 224, 3)] resnet50(Functional), Output Shape (None, 7, 7, 2048) global_average_pooling2d (GlobalAveragePooling2D) Output Shape, (None, 2048) dense (Dense), Output Shape (None, 6)

5.2. Experimental Setup

Our models were built with different Python packages, including Scikit-Learn, Keras and TensorFlow. Furthermore, experiments and results were compiled and the TensorFlow environment was implemented on Google Colabotary, with 16GB RAM and a 108GB disk. The experimental platform is a laptop with 11th Generation Intel® Core™ i7 processor

NVIDIA® GeForce® MX450 (2 GB GDDR6 dedicated) 16 GB RAM; 64-bit operating system, x64-based processor, Windows 11.

To evaluate the performance of our approach, the dataset was first collected. The approach used the most common social networks platforms, Snapchat and Twitter, to collect the data in order to identify the location of an emergency event without any prior knowledge. Therefore, after identifying the exact location of the emergency event in Beirut from the Snapchat map, tweets were collected from Twitter API on 4 August 2020. The geographical search was applied to collect all tweets within the chosen area.

After all tweets in the location of this very case study were collected, we automatically applied the selected keywords search to retain solely the tweets that include keywords related to the case study using the Twitter API. The following specific keywords were used: "Lebanon Explosion, Explosion in Beirut, Huge explosion Lebanon, Beirut port exploded, Beirut Blasts, Lebanon Blasts, Disaster in Beirut, Disaster in Lebanon and Beirut". The Twitter API collects tweets and formats them using JavaScript Object Notation (JSON). Every tweet has a distinct Identifier, the actual text, a tweet timestamp (indicating the time tweet was posted), and numerous additional attributes including "username" and "location".

Finally, a total of 50,244 tweets were collected. However, since the obtained dataset contained both English and non-English tweets, only the English tweets were kept for the experiment. Moreover, replies to tweets, retweets as well as quoted tweets were also removed. Consequently, after deleting the aforementioned tweets, the total number of tweets went down to 20,144.

Furthermore, the dataset was labelled as either Beirut-related or not related using python script. After we collected the dataset and then filtered it using the related keyword search in the previous step, we stored the filtered tweets that contain keywords related to our case study in a separate JSON file. Then we labelled them as related to our case study using our script, which will add a new column to the dataset and assign the label to all tweets in this file. Moreover, we created another JSON file and stored the tweets that were excluded from the first JSON file, which did not contain any keyword related to our case study. Then we labelled them as not-related to our case study using our script, which will add a new column to the dataset and assign the label to all tweets in this file. Subsequently, both files were merged into a single JSON file to prepare our dataset for further processing. Furthermore, we applied the 5-Fold cross-validation approach to split the dataset. The dataset was randomly partitioned into 5-sized subsamples using a K-fold cross-validation method. One k section was utilized for validation assessment, while the residual k-1 sections were used for classifier training. In the future, we plan to improve our study by the using the 10-fold cross-validation approach.

Moreover, the preprocessing steps were applied to extract the most important features from the data to reduce the computational overhead. Then, detection models were built using the trained dataset. Finally, the trained models were integrated into the Apache Spark streaming and used the test dataset to evaluate the performance of the proposed approach.

5.3. Performance Evaluation

The evaluation metrics play a substantial aspect in achieving the desired classification during the training stage. Furthermore, collecting appropriate measurement criteria is crucial for differentiate and acquire the ideal classifier [41]. After applying the machine learning classifiers mentioned above for emergency events detection, the accuracy of the proposed model was tested through diverse performance evaluation criteria such as [42]:

- **Precision (P)**: It is used to subtract the number of accurately predicted positive patterns from the total number of expected positive patterns in a positive category.
  Precision can be calculated using the following equation.

$$Precision = \frac{TP}{TP + FP} \quad (1)$$

- **Recall (R)**: It is used to determine the percentage of correctly classified positive patterns. Recall can be calculated using the following equation.

$$Recall = \frac{TP}{TP + FN} \tag{2}$$

- **F-Measure (FM)**: Also known as the F1-score, which is a metric to determine accuracy on a given data. It is used to examine binary classification algorithms that label data as "positive" or "negative". The F1-score can be calculated using the following equation.

$$F1 = 2 \times \frac{(precision \times recall)}{precision + recall} \tag{3}$$

- **Accuracy (ACC)**: It is used to calculate the proportion of accurate classification compared to the total number of instances examined. Accuracy can be calculated using the following equation.

$$Accuracy = \frac{(TP + TN)}{TP + FP + TN + FN} \tag{4}$$

  True Positive (TP): It refers to the model's correct classified number of instances [43].
  False Positive (FP) refers to the number of negative instances identified incorrectly as positive instances [43].
  alse Negative (FN): refers to the number of positive instances identified incorrectly as negative instances [43].
  True Negative (TN): The true negative values refer to the number of negative instances that were correctly classified by the model [43].
- **Throughput**: It represents the number of results delivered over a given period of time. In the framework of this study, this is measured in units of flow.

5.4. Results

The experiments to evaluate our approach were conducted locally, in the environment set up described in the previous subsection. The proposed Snapchat classification model achieved a high accuracy of 93.17%. Figure 7 shows the performance evaluation of the Snapchat classification model.

Also, Table 3 shows the results of the Twitter classification models. The experimental results proved that our proposed ensemble approach achieved a very high accuracy of 99.87%, outperforming the other base models. In addition, it can be noted from Table 3 that three base models (LSTM, decision tree and KNN) achieved high accuracy of 99.72%, 99.70% and 85.22%, respectively. Table 3 shows that the LSTM base model achieved higher results than the other base models except for recall, where decision tree outperforms it with 99.71%. LSTM achieved 99.72% for accuracy, F1-score, 99.72% for precision and 99.69% for recall. The difference between the results achieved by LSTM and decision tree is negligible. However, the LSTM performed better in training time (35 s). Even though KNN required less training time (24 s) than decision tree (36 s), decision tree outperformed it with 99.71% for precision and recall, 99.72% for f1-score and 99.70% for accuracy. Moreover, SVM required less training time (18 s.) among all other base models, but it achieved a very low accuracy of 60.67%. In addition, Table 3 shows that Naive Bayes achieved the lowest accuracy among all base models with the most training time needed. Figure 8 shows the performance evaluation for each model separately.

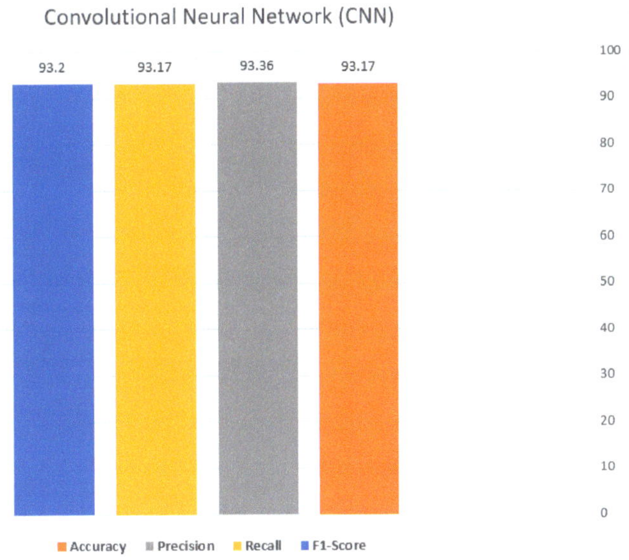

Figure 7. The performance evaluation of the Snapchat classification model.

Table 3. Results of Models Classification.

Model	Precision	Recall	F1-Score	Accuracy	Training Time (s)
LSTM	99.75	99.69	99.72	99.72	35
SVM	56.15	60.12	71.91	60.67	18
KNN	85.04	85.72	85.38	85.22	24
Naive Bayes	52.24	52.44	68.10	53.86	40
Decision Tree	99.71	99.71	99.72	99.70	36
The Proposed Ensemble	**99.87**	**99.87**	**99.87**	**99.87**	**60**

Figure 8. The performance evaluation of each model separately.

The models were integrated into Apache Spark streaming to assess the processing time using a test dataset. As shown in Figure 9, different time windows were used to stream the data from 1 to 100 s. Obviously, the Naive Bayes model processes the data much faster than all other base models in all window frame times—except for window 1. Indeed, in window 1, the LSTM model takes less time to process the data. However, as the window size increases, the processing time of the LSTM model decreases. Moreover, as we can observe from Figure 9, the proposed ensemble model outperforms the other models except in window 1. However, as the window size increases, the processing time of the proposed ensemble model increases.

By analyzing the impacts of the selected keywords which were applied to the collected tweets, Figure 10 shows that the keyword "Lebanon Explosion" yields a high score of 0.837. The whole list of the keywords are presented in Figure 10.

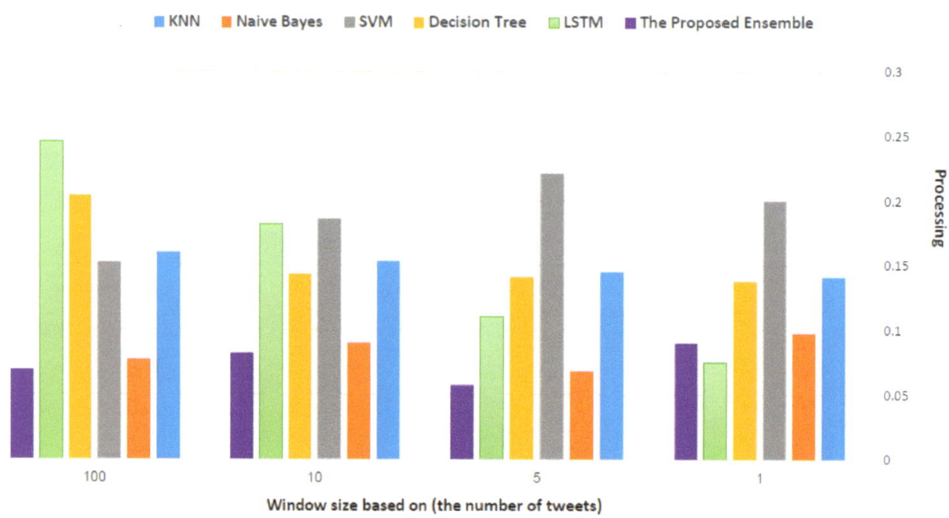

Figure 9. Processing time of models classification based on window size.

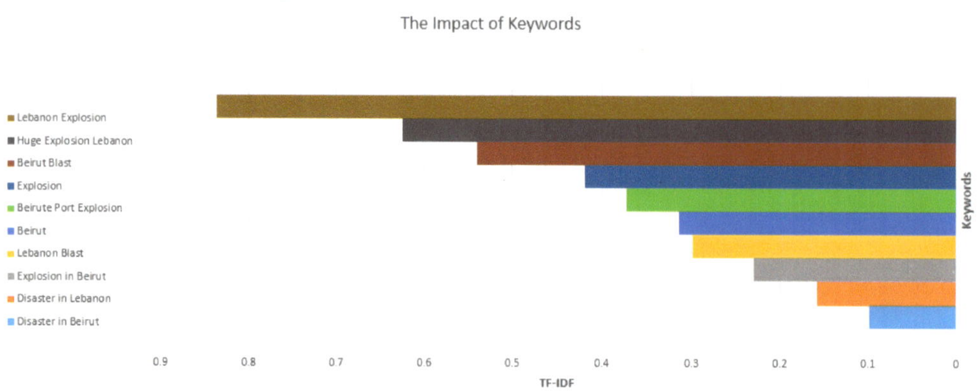

Figure 10. The impacts of the selected keywords.

6. Conclusions and Discussion

This paper indicated that it is necessary to have an accurate event detection model to react quickly and efficiently to emergencies. This paper's main contribution lies in the fact that it uses a Snapchat map to pinpoint the precise location of an emergency event. Monitoring the Snapchat map allows detecting the hotspot locations and analyzing them by labelling every snap in that particular hotspot as either associated or not to an emergency event. Then, a data collection model was proposed to collect the data from the Twitter API. Subsequently, the preprocessing steps were applied so that the machine learning algorithms could recognize the data, and feature extraction algorithms were utilized to help construct a machine learning classifier. Then, five machine learning base classifiers for emergency event detection models are proposed to construct our proposed ensemble approach. Finally, a performance evaluation of the proposed models was conducted to evaluate the model in terms of accuracy detection achievement.

Apache Spark streaming was integrated into our data to assess the model processing time using the aforementioned evaluation metrics. As a result, the proposed ensemble approach achieved accuracy levels of 99.87% which outperformed the other base models. Moreover, LSTM and decision tree yield 99.72%, 99.70% for accuracy, respectively, with an adequate training time. Therefore, the LSTM model is considered superior to the other base models. However, in our approach we applied only one deep learning model "LSTM" to the collected dataset from Twitter. Because the LSTM model is very effective in the sequential problem of NLP, also, as sequential data such as text and time series are the backbone of the Twitter platform, we utilized LSTM in this study. Moreover, even though the ensemble learning technique is beneficial for improving quality, accuracy, and decreasing predictions errors compared to the single learning machine model, the complexity of the learning process has significantly increased, which eventually led to the increase in the training time of the model.

Future work expected that the performance would improve if more feature extractions and more deep learning classifiers were added to the approach. Furthermore, our system shows good evidence of using the Snapchat hotspot map as a reliable source to detect the location of emergency events. However, to improve the scope and profundity of the work concerning what can be detected using the Snapchat platform, we need to enrich our approach. Therefore, we plan to enhance this work by employing computer vision methods to the collected dataset from Snapchat API. Accordingly, by applying such techniques, we can have a broader analysis of the dataset to detect emergency events and assess the severity of such events. Moreover, our approach focuses on the data collected from social network platforms without considering other data sources. Therefore, we will boost our work by exploring IoT sensors and satellite imagery sources.

Author Contributions: Conceptualization, K.A.; methodology, K.A.; software, K.A.; validation, K.A.; formal analysis, K.A.; investigation, K.A.; resources, K.A.; data curation, K.A.; writing original draft preparation, K.A.; writing review and editing, K.A.; visualization, K.A.; supervision, M.B.; project administration, K.A. and M.B.; funding acquisition, K.A. and M.B. All authors have read and agreed to the published version of the manuscript.

Funding: This research received no external funding.

Institutional Review Board Statement: Not applicable.

Informed Consent Statement: Not applicable.

Data Availability Statement: Not applicable.

Conflicts of Interest: The authors declare no conflict of interest.

References

1. Rong, Y.; Liu, Y.; Pei, Z. A novel multiple attribute decision-making approach for evaluation of emergency management schemes under picture fuzzy environment. *Int. J. Mach. Learn. Cybern.* **2021**, *13*, 633–661. [CrossRef]
2. Lee, J.; Wood, J.; Kim, J. Tracing the Trends in Sustainability and Social Media Research Using Topic Modeling. *Sustainability* **2021**, *13*, 1269. [CrossRef]
3. Usf.edu. Introduction to Social Media, University Communications and Marketing. 2020. Available online: https://www.usf.edu/ucm/marketing/intro-social-media.aspx (accessed on 1 January 2021).
4. Koch, J. *Teach Introduction to Education*; SAGE Publications: New York, NY, USA, 2018; p. 119.
5. Oussous, A.; Benjelloun, F.; Lahcen, A.A.; Belfkih, S. Big Data technologies: A survey. *J. King Saud Univ. Comput. Inf. Sci.* **2018**, *30*, 431–448. [CrossRef]
6. Bhadani, A.; Jothimani, D. Big data: Challenges, opportunities and realities. In *Effective Big Data Management and Opportunities for Implementation*; Singh, M.K., Kumar, D.G., Eds.; IGI Global: Pennsylvania, PA, USA, 2016; pp. 1–24.
7. Horita, F.E.; de Albuquerque, J.P.; Marchezini, V.; Mendiondo, E.M. Bridging the gap between decision-making and emerging big data sources: An application of a model-based framework to disaster management in Brazil. *Decis. Support Syst.* **2017**, *97*, 12–22. [CrossRef]
8. Alomari, E.; Katib, I.; Albeshri, A.; Yigitcanlar, T.; Mehmood, R. Iktishaf+: A Big Data Tool with Automatic Labeling for Road Traffic Social Sensing and Event Detection Using Distributed Machine Learning. *Sensors* **2021**, *21*, 2993. [CrossRef] [PubMed]
9. Alfalqi, K.; Bellaiche, M. IoT-Based Disaster Detection Model Using Social Networks and Machine Learning. In Proceedings of the 2021 4th International Conference on Artificial Intelligence and Big Data (ICAIBD), Chengdu, China, 28–31 May 2021; pp. 92–97. [CrossRef]
10. Septianto, G.R.; Mukti, F.F.; Nasrun, M.; Gozali, A.A. Jakarta congestion mapping and classification from twitter data extraction using tokenization and naïve bayes classifier. In Proceedings of the 2015 Asia Pacific Conference on Multimedia and Broadcasting, Bali, Indonesia, 23–25 April 2015. [CrossRef]
11. Toujani, R.; Akaichi, J. Event news detection and citizens community structure for disaster management in social networks. *Online Inf. Rev.* **2019**, *43*, 113–132. [CrossRef]
12. Banujan, K.; Kumara, T.G.S.B.; Paik, I. Twitter and Online News analytics for Enhancing Post-Natural Disaster Management Activities. In Proceedings of the 2018 9th International Conference on Awareness Science and Technology (iCAST), Fukuoka, Japan, 19–21 September 2018; pp. 302–307. [CrossRef]
13. Hagras, M.; Hassan, G.; Farag, N. Towards Natural Disasters Detection from Twitter Using Topic Modelling. In Proceedings of the 2017 European Conference on Electrical Engineering and Computer Science (EECS), Bern, Switzerland, 17–19 November 2017.
14. Ragini, J.R.; Anand, P.R.; Bhaskar, V. Big data analytics for disaster response and recovery through sentiment analysis. *Int. J. Inf. Manag.* **2018**, *42*, 13–24. [CrossRef]
15. Salas, A.; Georgakis, P.; Nwagboso, C.; Ammari, A.; Petalas, I. Traffic event detection framework using social media. In Proceedings of the 2017 IEEE International Conference on Smart Grid and Smart Cities (ICSGSC), Singapore, 23–26 July 2017; pp. 303–307.
16. Lau, R.Y. Toward a social sensor based framework for intelligent transportation. In Proceedings of the 2017 IEEE 18th International Symposium on A World of Wireless, Mobile and Multimedia Networks (WoWMoM), Hong Kong, China, 12–15 June 2017; pp. 1–6. [CrossRef]
17. Bhuvaneswari, A.; Jayanthi, R.; Meena, A.L. Improving Crisis Event Detection Rate in Online Social Networks Twitter Stream using Apache Spark. *J. Phys.: Conf. Ser.* **2021**, *1950*, 012077. [CrossRef]
18. Alomari, E.; Mehmood, R.; Katib, I. Sentiment Analysis of Arabic Tweets for Road Traffic Congestion and Event Detection. In *Smart Infrastructure and Applications*; Springer: Berlin/Heidelberg, Germany, 2019; pp. 37–54._2. [CrossRef]
19. Alghamdi, N.; Alrajebah, N.; Al-Megren, S. Crowd Behavior Analysis using Snap Map: A Preliminary Study on the Grand Holy Mosque in Mecca. In Proceedings of the 2019 on Computer Supported Cooperative Work and Social Computing (CSCW'19), Austin, TX, USA, 9–13 November 2019; pp. 137–141. [CrossRef]
20. Alghamdi, N.; Alageeli, N.; Abu Sharkh, D.; Alqahtani, M.; Al-Razgan, M. An Eye on Riyadh Tourist Season: Using Geo-tagged Snapchat Posts to Analyse Tourists Impression. In Proceedings of the 2020 2nd International Conference on Computer and Information Sciences (ICCIS), Sakaka, Saudi Arabia, 13–15 October 2020; pp. 1–6. [CrossRef]
21. Juhász, L.; Hochmair, H.H. Analyzing the spatial and temporal dynamics of Snapchat. In Proceedings of the AnaLysis, Integration, Vision, Engagement (VGI-ALIVE) Workshop, Lund, Sweden, 12 June 2018.
22. Lamba, H.; Srikanth, S.; Pailla, D. *Driving the Last Mile: Characterizing and Understanding Distracted Driving Posts on Social Networks*; Association for the Advancement of Artificial Intelligence: Menlo Park, CA, USA, 2019.
23. Shah, S.A.; Seker, D.Z.; Rathore, M.M.; Hameed, S.; Ben Yahia, S.; Draheim, D. Towards Disaster Resilient Smart Cities: Can Internet of Things and Big Data Analytics Be the Game Changers? *IEEE Access* **2019**, *7*, 91885–91903. [CrossRef]
24. Kwekha-Rashid, A.S.; Abduljabbar, H.N.; Alhayani, B. Coronavirus disease (COVID-19) cases analysis using machine-learning applications. *Appl. Nanosci.* **2021**. [CrossRef]
25. Yang, J.; Qu, J.; Mi, Q.; Li, Q. A CNN-LSTM Model for Tailings Dam Risk Prediction. *IEEE Access* **2020**, *8*, 206491–206502. [CrossRef]

26. Zhou, H.; Zhang, J.; Zhou, Y.; Guo, X.; Ma, Y. A feature selection algorithm of decision tree based on feature weight. *Expert Syst. Appl.* **2020**, *164*, 113842. [CrossRef]
27. Olowononi, F.O.; Rawat, D.B.; Liu, C. Resilient Machine Learning for Networked Cyber Physical Systems: A Survey for Machine Learning Security to Securing Machine Learning for CPS. *IEEE Commun. Surv. Tutorials* **2020**, *23*, 524–552. [CrossRef]
28. Bout, E.; Loscri, V.; Gallais, A. How Machine Learning Changes the Nature of Cyberattacks on IoT Networks: A Survey. *IEEE Commun. Surv. Tutorials* **2021**, *24*, 248–279. [CrossRef]
29. Gu, J.; Lu, S. An effective intrusion detection approach using SVM with naïve Bayes feature embedding. *Comput. Secur.* **2020**, *103*, 102158. [CrossRef]
30. Qolomany, B.; Al-Fuqaha, A.; Gupta, A.; Benhaddou, D.; Alwajidi, S.; Qadir, J.; Fong, A.C. Leveraging Machine Learning and Big Data for Smart Buildings: A Comprehensive Survey. *IEEE Access* **2019**, *7*, 90316–90356. [CrossRef]
31. Podhoranyi, M. A comprehensive social media data processing and analytics architecture by using big data platforms: A case study of twitter flood-risk messages. *Earth Sci. Inform.* **2021**, *14*, 913–929. [CrossRef]
32. Juhász, L.; Hochmair, H. Comparing the Spatial and Temporal Activity Patterns between Snapchat, Twitter and Flickr in Florida. *GIForum* **2019**, *1*, 134–147._01_s134. [CrossRef]
33. Hernandez-Suarez, A.; Sanchez-Perez, G.; Toscano-Medina, K.; Perez-Meana, H.; Portillo-Portillo, J.; Sanchez, V.; Villalba, L.J.G. Using Twitter Data to Monitor Natural Disaster Social Dynamics: A Recurrent Neural Network Approach with Word Embeddings and Kernel Density Estimation. *Sensors* **2019**, *19*, 1746. [CrossRef]
34. Said, N.; Ahmad, K.; Riegler, M.; Pogorelov, K.; Hassan, L.; Ahmad, N.; Conci, N. Natural disasters detection in social media and satellite imagery: A survey.*Multimedia Tools Appl.* **2019**, *78*, 31267–31302. [CrossRef]
35. Eyada, M.M.; Saber, W.; El Genidy, M.M.; Amer, F. Performance Evaluation of IoT Data Management Using MongoDB Versus MySQL Databases in Different Cloud Environments. *IEEE Access* **2020**, *8*, 110656–110668. [CrossRef]
36. Wijeratne, S.; Sheth, A.; Bhatt, S.; Balasuriya, L.; Al-Olimat, H.S.; Gaur, M.; Yazdavar, A.H.; Thirunarayan, K. *Feature Engineering for Twitter-Based Applications*; CRC Press: Boca Raton, FL, USA, 2018; pp. 359–393. [CrossRef]
37. De Pablo, Á.; Araque, O.; Iglesias, C.A. Transfer Learning with Social Media Content in the Ride-Hailing Domain by Using a Hybrid Machine Learning Architecture. *Electronics* **2022**, *11*, 189. [CrossRef]
38. Hasan, A.; Moin, S.; Karim, A.; Shamshirband, S. Machine Learning-Based Sentiment Analysis for Twitter Accounts. *Math. Comput. Appl.* **2018**, *23*, 11. [CrossRef]
39. Awan, F.M.; Saleem, Y.; Minerva, R.; Crespi, N. A Comparative Analysis of Machine/Deep Learning Models for Parking Space Availability Prediction. *Sensors* **2020**, *20*, 322. [CrossRef] [PubMed]
40. Ramzy, A.; Peltier, E. What We Know about the Beirut Explosions. *The New York Times*. Available online: https://www.nytimes.com/2020/08/05/world/middleeast/beirut-explosion-what-happened.html (accessed on 4 October 2021).
41. Miguel, J.; Caballé, S.; Xhafa, F.; Prieto, J. A massive data processing approach for effective trustworthiness in online learning groups. *Concurr. Comput. Pr. Exp.* **2014**, *27*, 1988–2003. [CrossRef]
42. Hossin, M.; Sulaiman, M.N. A Review on Evaluation Metrics for Data Classification Evaluations. *Int. J. Data Min. Knowl. Manag. Process* **2015**, *5*, 1–11. [CrossRef]
43. Grandini, M.; Bagli, E.; Visani, G. Metrics for Multi-Class Classification: An Overview. *arXiv* **2020**, arXiv:abs/2008.05756.

remote sensing

Article

Learning Spatio-Temporal Attention Based Siamese Network for Tracking UAVs in the Wild

Junjie Chen [1,†], Bo Huang [1,†], Jianan Li [1], Ying Wang [1], Moxuan Ren [1] and Tingfa Xu [1,2,*]

1 Image Engineering & Video Technology Lab, School of Optics and Photonics, Beijing Institute of Technology, Beijing 100081, China; 3120190516@bit.edu.cn (J.C.); 3120185333@bit.edu.cn (B.H.); lijianan@bit.edu.cn (J.L.); 3120215325@bit.edu.cn (Y.W.); renmoxuanx@bit.edu.cn (M.R.)
2 Big Data and Artificial Intelligence Laboratory, Beijing Institute of Technology Chongqing Innovation Center (BITCQIC), Chongqing 401135, China
* Correspondence: ciom_xtf1@bit.edu.cn
† These authors contributed equally to this work.

Abstract: The popularity of unmanned aerial vehicles (UAVs) has made anti-UAV technology increasingly urgent. Object tracking, especially in thermal infrared videos, offers a promising solution to counter UAV intrusion. However, troublesome issues such as fast motion and tiny size make tracking infrared drone targets difficult and challenging. This work proposes a simple and effective spatio-temporal attention based Siamese method called SiamSTA, which performs reliable local searching and wide-range re-detection alternatively for robustly tracking drones in the wild. Concretely, SiamSTA builds a two-stage re-detection network to predict the target state using the template of first frame and the prediction results of previous frames. To tackle the challenge of small-scale UAV targets for long-range acquisition, SiamSTA imposes spatial and temporal constraints on generating candidate proposals within local neighborhoods to eliminate interference from background distractors. Complementarily, in case of target lost from local regions due to fast movement, a third stage re-detection module is introduced, which exploits valuable motion cues through a correlation filter based on change detection to re-capture targets from a global view. Finally, a state-aware switching mechanism is adopted to adaptively integrate local searching and global re-detection and take their complementary strengths for robust tracking. Extensive experiments on three anti-UAV datasets nicely demonstrate SiamSTA's advantage over other competitors. Notably, SiamSTA is the foundation of the 1st-place winning entry in the 2nd Anti-UAV Challenge.

Keywords: anti-UAV; single object tracking; thermal infrared videos; Siamese network; correlation filter; re-detect; motion feature

Citation: Chen, J.; Huang, B.; Li, J.; Wang, Y.; Ren, M.; Xu, T. Learning Spatio-Temporal Attention Based Siamese Network for Tracking UAVs in the Wild. *Remote Sens.* **2022**, *14*, 1797. https://doi.org/10.3390/rs14081797

Academic Editor: Francesco Nex

Received: 24 January 2022
Accepted: 29 March 2022
Published: 8 April 2022

1. Introduction

In recent years, the rapid development of unmanned aerial vehicles (UAVs) has promoted a large number of applications, such as aerial photography [1–3], video surveillance [4], and biological monitoring [5]. On the contrary, the potential abuses of this technology could lead to significant negative impacts on society. Thus, anti-UAV techniques are of great importance and in urgent need of practical research, among which vision-based approaches are more widely adopted due to their higher efficiency, lower power consumption, and easier deployment.

Given a UAV target specified by a bounding box in the first frame, visual object tracking aims to determine the exact state of the target sequentially in a video, which servers as a fundamental step for anti-UAV task. As is obvious, Thermal Infrared (TIR) tracking technique is better suited to the low-light scenarios, thus catering to all-weather requirements. However, tracking UAVs in TIR video, compared to tracking objects in visible video, further introduces significant challenges, e.g., small object, fast motion, thermal clutter background. How to tackle these problems remain challenging and ill-solved.

Siamese-based algorithms play a dominant role in the field of visual object tracking. SiamFC [6] first treats the tracking task as a similarity matching problem between target template and the search region. Later on, numerous improvements have been done in terms of adding auxiliary branch [7], digging deeper feature [8], improving embedding strategy [9], augmenting online update mechanism [10–12], etc. While considering the tracking object is UAV, which contains a wide range of fast motion and out-of-view situations, unlike the improvements mentioned above, the global re-detection mechanism and trajectory modeling could play critical roles in accurate tracking.

SiamRCNN [13] introduces global re-detection mechanism into Siamese networks and achieves outstanding tracking performance, which is thus used as our baseline algorithm. But on the other hand, as the UAV targets themselves severely lack semantic features, continuous global detection mechanism could be more likely to induce tracking drift, especially when the UAV targets are drowned in thermal clutter background. At this time, it seems more appropriate to detect targets in the local neighborhoods. Obviously, local detection and global re-detection are the two opposite strategies that can cope with different challenging situations during tracking. For this reason, this work elaborately designs a framework adaptively switching these two strategies, achieving robust tracking through performing reliable designed local tracking and wide-range re-detection alternatively.

This paper proposes a simple yet effective spatio-temporal attention-based Siamese network, called SiamSTA, to track UAVs in TIR videos robustly. SiamSTA follows the typical Siamese framework that consists a template branch and a detection branch. The template branch extracts features for the template target specified in the first frame, while the detection branch takes the search image as input and selects target candidates from redundant Region Proposal Network (RPN) proposals. To tackle the key challenges, i.e., small scale and fast movement, commonly faced in anti-UAV tracking scenarios, SiamSTA integrates both a reliable local tracking and a wide-range global re-detection mechanism, and takes their complementary advantages in an alternative-performing fashion.

Specifically, to better perceive small targets that easily be distracted by background clusters, the local tracking strategy incorporates spatial and temporal constraints to limit the position and aspect ratio of generated candidate proposals in a local neighborhood, so as to suppress background distractors and locate the target accurately. Meanwhile, in case the target runs out of the local region due to rapid movement, a three-stage global re-detection mechanism is designed to redetect the target: (i) provides re-detections using the first-frame template, (ii) implements re-detections of high-confidence predictions from previous frames, and (iii) adopts correlation filter based on change detection, short for CDCF, to exploit beneficial motion features to better locate fast-moving target in a wide range. Finally, a switching policy is adopted to apply local tracking and global re-detection adaptively depending on varying target states to make optimal predictions, hence achieving robust tracking.

To verify the performance of SiamSTA, comprehensive experiments are conducted on three challenging UAV infrared tracking datasets, i.e., Anti-UAV2020 test-dev [14], Anti-UAV2021 test-dev [15], Anti-UAV [16]. Detailed experimental comparisons show that SiamSTA has advantages over its competing counterparts in addressing the key challenges faced by anti-UAV tracking, including but not limited to small scale and fast movements. In addition, SiamSTA serves as the foundation of the 1st-place winning entry in the 2nd Anti-UAV Challenge, further evidencing its robustness in real-world scenarios.

To sum up, this paper makes the following contributions:

- This paper proposes a novel Siamese based tracker that integrates local tracking and global re-detection mechanisms in a unified framework and perform them adaptively depending on varying target states.
- This paper designs a spatio-temporal attention based local tracking strategy to eliminate background clusters and better perceive small targets.
- A three-stage global re-detection strategy to recapture targets in a wide range is proposed.

- Our method establishes state-of-the-art performance in Anti-UAV2020 test-dev [14], Anti-UAV2021 test-dev [15] and Anti-UAV [16] datasets.

2. Related Works

This section first reviews the development of two mainstream tracking frameworks, i.e., correlation filters and Siamese networks. Then, some representative tracking algorithms in thermal infrared videos are introduced.

2.1. Correlation Filter

Correlation filter (CF) based tracking methods have been widely applied to visual tracking due to the high computational efficiency. MOSSE [17] tracker is the first to validate the feasibility of the correlation filter in tracking. After MOSSE, CF trackers obtain a wide attention. Refs. [18,19] introduce circulant matrix to produce enough samples for training, while maintaining a fast speed. However, the resulting periodic repetitions at boundary positions limit the discriminative ability of the tracker. To mitigate this issue, Danelljan et al. [20] adopt a spatial regularization term which allows the tracker focus on the target center, Galoogahi et al. [21] apply a binary matrix to crop real samples for training, both of which are promising in lifting the performance of trackers. Besides addressing boundary effect, some CF trackers introduce scale estimation [22–24] to improve the tracking performance. In addition, A powerful feature extraction such as HOG [18,19], CN [25] and deep feature [26] will enhance the feature representation ability and increase the tracking accuracy. However, traditional CF trackers mostly apply nearby search, which is difficult to capture the target again when the target is out of view for a while and then re-enters the field of view, which limits the application of CF trackers in anti-UAV mission.

2.2. Siamese Network

Recently, the Siamese network based trackers have gained a lot of attention for their great success in multiple video object tracking benchmarks and competitions. Bertinetto et al. [6] propose the initial SiamFC tracker, which formulates visual tracking as a cross-correlation problem and expects to learn a similarity evaluation map based fully-convolutional network in an end-to-end manner. Li et al. [27] significantly enhance the tracking performance of SiamFC by introducing a Region Proposal Network (RPN), which allows estimation of target locations, sizes, and ratios by enumerating multiple anchors. However, these trackers implement nearby search, which is difficult to recapture target after it lost. Recently, Voigtlaender et al. [13] unleash the full power of global searching by a two-stage Siamese re-detection architecture, which makes full use of both the first-frame template and previous-frame predictions for the optimal decision. Ref. [13] not only solves the problem of update, but also improves the probability of re-detection after target lost. However, global searching also introduces too much distractors which hurts the performance of tracking small target in complex background. To this end, this paper proposes a spatio-temporal attention based Siamese network to enhance the tracking robust of global re-detection.

2.3. TIR Tracking

Recently, more attentions [28–30] have been paid to TIR tracking for the rapidly development of infrared sensors in resolution and quality. Due to the poor semantic information in TIR images, how to extract effective features is crucial to distinguish targets from background. Refs. [31–33] compute motion features by thresholding the absolute difference between the current and the previous frame in pixel-wise as an extra feature channel, which is beneficial for identifying moving objects. Yu et al. [34] propose structural learning on dense samples around the object. Their tracker uses edge and HOG features which is suitable for UAV tracking. With the development of deep learning, Convolutional Neural Networks (CNN) have shown competitive performance compared to handcrafted feature. However, due to the limited semantic information in TIR images, traditional

RGB backbones performs poorly, and a number of works [35,36] start to design networks specialized for TIR images. However, the limited data in the infrared dataset and the large amount of data required for training the deep network are not friendly to anti-UAV tasks. Wu et al. [36] exploit the information in the initial frame to train a feature extraction network for correlation filter. However, in long-term tracking, the target and background in the initial frame and subsequent frames differ significantly, which is not general to long-term anti-UAV tasks. Based on our previous work [37], our method first extracts motion features using a Gaussian mixed model to select candidate regions where small targets are likely to exist, and then combines the Histograms of Oriented Gradients (HOG) features of the candidate regions to train a Correlation Filter based tracker to assist the Siamese network in making joint decisions, thus better perceiving small targets in TIR images. Finally, a novel distractor-aware regularization term is further proposed to learn the distractor information in the background, thus leading to better tracking robust against thermal clutter background.

3. Method

This section first briefly reviews the baseline tracker SiamRCNN [13]. Then the design of the proposed SiamSTA is described, which consists of spatio-temporal constraints, global motion estimation and change detection based CFs. Next, the online tracking and updating strategy integrating both local search and global detection is further presented.

3.1. Revisiting SiamRCNN

SiamRCNN [13] is a two-stage Siamese tracker with elaborate re-detection mechanism. Its network architecture is sequentially composed of three modules: (1) A backbone feature extraction module containing a template branch for extracting ground-truth features in the target area, and a test branch for preparing possible RPN proposals in the search area; (2) A re-detection head module which performs a two-stage re-detection to learns a similarity evaluation using the initial template and previous predictions; (3) An online dynamic programming module that implicitly tracks both the target and potential similar-looking distractors based on spatio-temporal cues. In the vital third module, SiamRCNN preserves plenty of discontinuous trajectories for making the most comprehensive decisions. Suppose one tracking trajectory consists of N non-overlapping sub-trajectories, $A = (a_1, a_2, \ldots, a_N)$, each sub-trajectory $a_i, \forall i \in \{1, 2, \ldots, N-1\}$ satisfies $end(a_i) < start(a_{i+1})$, where *start* and *end* denote the start and end times of a sub-trajectory, respectively. The overall measuring score of such trajectory is computed by,

$$score(A) = \sum_{i=1}^{N} sim\_eva(a_i) + \sum_{i=1}^{N-1} w_l loc\_eva(a_i, a_{i+1}), \tag{1}$$

where the similarity evaluation *sim_eva* and location consistency evaluation *loc_eva* are defined as following,

$$sim\_eva(a_i) = \sum_{t=start(a_i)}^{end(a_i)} [w_r sim(a_{i,t}, gt) + (1 - w_r) sim(a_{i,t}, a_{i,start})], \tag{2}$$

$$loc\_eva(a_i, a_{i+1}) = -|end\_box(a_i) - start\_box(a_{i+1})|_1, \tag{3}$$

here w_l and w_r are the complementary ratios. $a_{i,t}$ denotes the detection of sub-trajectory a_i at time t, and $a_{i,start}$ means the first detection of a_i. $sim(a_{i,t}, gt)$ and $sim(a_{i,t}, a_{i,start})$ return the re-detection confidence of $a_{i,t}$ using the first-frame ground truth reference and the initial detection of the current sub-trajectory, respectively. As presented in Equation (3), the location consistency evaluation between two adjacent sub-trajectory is computed using the negative L_1 norm of the difference between the last bounding box of a_i and the first bounding box of a_{i+1}.

SiamRCNN backs up a lot of trajectories to ensure the success rate of re-detection. However, due to the lack of semantic target features and the presence of complex thermal background in TIR video, these trajectories introduce a large number of similar-looking distractors, thus causing too much disruptions during tracking and eventually leading to tracking drift. To address such issue, finer exploitation of spatio-temporal prior knowledge is a feasible solution.

3.2. SiamSTA Framework

Inspired by SiamRCNN, SiamSTA is built based on a three-stage re-detection mechanism that first retains template information in the initial frame, then integrates predictive information from historical frames, and finally lifts discriminative capability to perceive tiny objects with a change detection based CF, as shown in Figure 1. To deal with background distractors, several practical guidelines using spatio-temporal attention are introduced to regulate candidate proposals. SiamSTA further incorporates a collaborative strategy that combines local search and global detection to facilitate online tracking.

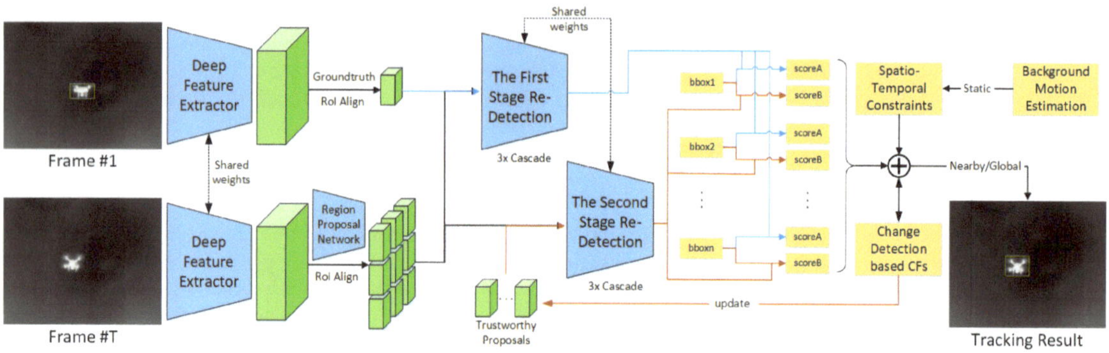

Figure 1. Overall architecture of SiamSTA. It consists of a Siamese backbone that extracts deep features from the template and the search image, followed by a three-stage re-detector that first re-detects the first-frame template, then re-locates historical predictions from previous frames, and finally fixes potential tracking failures using a change detection based CF. The symbol $\oplus$ indicates an ensemble classifier that conditionally switches between local track and global detection to make optimal decisions upon predictions from the three-stage re-detector.

3.2.1. Spatio-Temporal Constraints for Local Tracking

UAV targets in practical TIR tracking are typically very small and without salient texture or fixed shapes, making them extremely hard to be distinguished. To alleviate this, a novel spatio-temporal constraint is introduced. From the spatial perspective, considering the drastic position changes of the targets are unlikely to occur in two adjacent frames captured by a long-range static camera, reliable tracking results can be obtained by searching for targets in local neighborhoods rather than detecting it globally. From the temporal perspective, SiamSTA introduces a memory bank to store valuable historical states of the targets, i.e., target size and aspect ratio, learned from all previous frames to better distinguish potential distractors.

Concretely, SiamSTA records the historical minimum and maximum size and aspect ratio of the target appeared in all previous frames, denoted as (S_{min}, S_{max}) and (R_{min}, R_{max}), respectively, to indicate its range of potential scale variation. Initialize $S_{min} = S_{max} = S$, $R_{min} = R_{max} = R$ with the size S and aspect ratio R of the ground-truth target bounding box specified in the first frame. For an arbitrary frame c, we specify a local neighborhood around the previous target center as the search region where the target is most likely to appear. Only if a high-confidence proposal has been found within the specified search region, whose size S_c and aspect ratio R_c meets the constrain below $S_c \in [0.8 * S_{min}, 1.2 * S_{max}]$,

$R_c \in [0.8 * R_{min}, 1.2 * R_{max}]$, we regard the detection result to be reliable and the trajectory to be continuous. Then SiamSTA updates the stored target state as following:

$$S_{min} = min(S_{min}, S_c), S_{max} = max(S_{max}, S_c), \tag{4}$$

$$R_{min} = min(R_{min}, R_c), R_{max} = max(R_{max}, R_c). \tag{5}$$

The above process lasts until the end of a trajectory. Define the trusted trajectory as $C = (c_1, c_2, \ldots, c_L)$, and compute the evaluation score of a candidate proposal cc as,

$$score(cc) = w_r sim(cc, gt) + (1 - w_r)\frac{1}{L}\sum_{i=1}^{L} sim(cc, c_{i,start}) + w_l iou(cc, c_{L,end}), \tag{6}$$

where $iou(cc, c_{L,end})$ is the intersection over union (IoU) of $bbox(cc)$ and $bbox(c_{L,end})$. Thanks to the spatio-temporal constraints, the number of remaining candidate proposals can be very small, or even unique, which greatly alleviates the interference of distractors.

However, if the target is temporarily lost, the local search strategy may cause the tracker to fail completely. To mitigate the effect of target loss, especially severe occlusion or out-of-view, global re-detection techniques associated with a mutual compensation mechanism that conditionally switches between local tracking and global search is essential, as detailed below.

3.2.2. Global Motion Estimation

Targets in anti-UAV tracking are typically very small with little semantic information, which easily leads to early tracking failures. Fortunately, background scenes in such tracking scenario commonly remain fixed throughout an entire sequence, which provides feasibility to employ motion features to re-capture lost targets.

Motivated by this, this section establishes a global motion estimation model to reveal dynamic change of background scenes. To be specific, SiamSTA extracts the ShiTomasi [38] key points from background regions and track these points to estimate the motion of background scenes. Let $I(x, y)$ denotes the intensity value of pixel (x, y) on input image I. Key points should have a significant gradient change in gray values, such as corner points. Let $[u, v]$ be the local displacement, and the gradient change vector in the local neighborhood can be calculated as,

$$E(u, v) = \sum_{x,y} \tau(x, y)[I(x + u, y + v) - I(x, y)]^2, \tag{7}$$

where $\tau(x, y)$ is a Gaussian window function. Equation (7) can be further simplified as,

$$E(u, v) \cong [u, v]M\begin{bmatrix} u \\ v \end{bmatrix}, \tag{8}$$

where M is a 2×2 matrix:

$$M = \sum_{x,y} w(x, y)\begin{bmatrix} I_x^2 & I_x I_y \\ I_x I_y & I_y^2 \end{bmatrix}, \tag{9}$$

where I_x and I_y represent the derivatives of image I in the horizontal and vertical direction, respectively. We can obtain two eigenvalues λ_1, λ_2 of M, and the key point response function is defined as,

$$G = \lambda_1 \lambda_2 - k(\lambda_1 + \lambda_2)^2, \tag{10}$$

point (x, y) is consider as a key point if $G > 0$, more details can be found in [38].

The number of key points is set to 5 to 100. Then Lucas-Kanade (L-K) optical flow algorithm [39] is applied to track these key points, with forward-backward (F-B) error [39] employed to evaluate the matching accuracy of key points between two consecutive frames.

Key points with F-B error less than a preset threshold are regarded as successful tracking points. If the average spatial displacement of all successful tracking points is less than 0.5 pixel across 5 consecutive frames, we consider the background scene is static without camera jitters.

3.2.3. Change Detection Based CFs for Three-Stage Re-Detection

Based on the accurate motion estimation of background, change detection based correlation filter (CDCF) tracker is further proposed to take advantage of target's motion features. CDCF is coupled with SiamRCNN's two-stage re-detection to form a three-stage re-detection framework. The pipeline of CDCF module is shown in Figure 2. The red area on the left side indicates the Gaussian Mixture Model (GMM) construction process, which is used to describe the background and updated every frame. The blue area shows the computation process of correlation filter, and the green area depicts the combination of motion features and the correlation filter tracker to finally obtain a credible target state output.

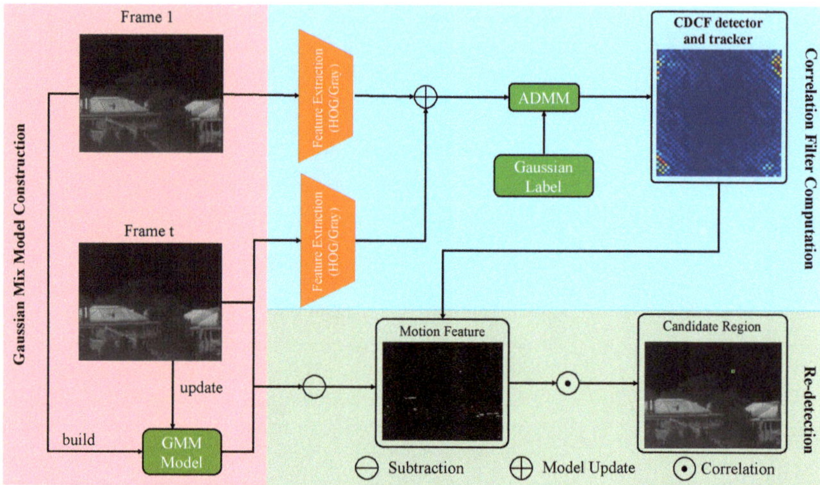

Figure 2. Pipeline of CDCF. We use the first frame of sequence to construct a Gaussian Mix Model (GMM), and update GMM model every frame. CDCF extracts hand-drafted feature to build a correlation filter, which can track target and redetect target when target lost.

Background Modeling. When the background is static, each pixel is normally distributed in the time domain, pixels within a certain threshold are judged as background and vice versa as moving targets. Based on this assumption, a Gaussian mixed model (GMM) is built to capture moving targets. Denote X_t as the intensity value of pixel (x, y) in frame t, and the GMM model is computed as,

$$P(X_t) = \sum_{i=1}^{K} \kappa_{i,t} * \eta(X_t, \mu_{i,t}, \Sigma_{i,t}), \tag{11}$$

where K is the number of Gaussian components (usually ranges from 3 to 5), $\kappa_{i,t}$ is the weight of component i in frame t, $\mu_{i,t}$ and $\Sigma_{i,t}$ are the mean and variance matrix of component i, respectively. Gaussian probability density function $\eta(X_t, \mu_{i,t}, \Sigma_{i,t})$ is defined as,

$$\eta(X_t, \mu_{i,t}, \Sigma_{i,t}) = \frac{1}{(2\pi)^{\frac{n}{2}} |\Sigma|^{\frac{1}{2}}} e^{-\frac{1}{2}(X_t - \mu_t)^T \Sigma^{-1} (X_t - \mu_t)}. \tag{12}$$

For a pixel value, X_t, it will be checked from the existing K Gaussian components, until a match is found. The match is defined as success if the pixel value X_t is within 2.5 times the standard deviations of a component. Then, GMM model is updated as,

$$\kappa_{i,t} = (1-\alpha)\kappa_{i,t-1} + \alpha Q_{i,t}, \tag{13}$$

where α is the learning rate, $Q_{i,t}$ equals 1 when the matching is successful and 0 otherwise. Keep the parameters μ, Σ for unmatched components unchanged, and update matched component as,

$$\mu_t = (1-\rho)\mu_{t-1} + \rho X_t, \tag{14}$$

$$\Sigma_t = (1-\rho)\Sigma_{t-1} + \rho(X_t - \mu_t)^{\mathrm{T}}(X_t - \mu_t), \tag{15}$$

where ρ is learning rate:

$$\rho = \alpha\eta(X_t|\mu_k, \sigma_k). \tag{16}$$

If X_t does not match any of the K components, we classify the pixel as motion target (As we can see in the right subfigures of Figure 3, the pink region is the visualization of motion feature, represents the presence probability of the moving target). Based on the motion feature, SiamSTA can perceive tiny moving target with little semantic information in dynamic background.

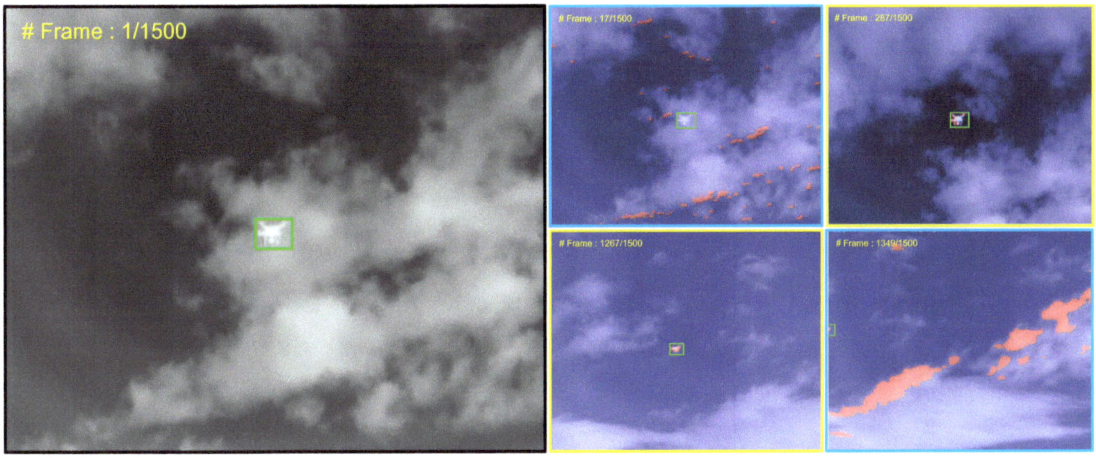

Figure 3. Motion features of CDCF. The areas with motion features are marked in pink, the rest in blue. When background is static e.g., frame 287, 1267, the motion feature is quite distinct, which is suitable for CDCF to perceive target, even the target is tiny and with little semantic information. When dynamic background occurs e.g., frame 17, 1349, the moving clouds in the scene will cause a heavy disturbance in discerning the real target.

Change Detection Correlation Filter. CDCF is trained using the initial frame and perform correlation operation to track the target in subsequent frames. Previous work [37] exploits motion feature as a re-detection method, which simply treats the motion feature as candidate region of target, and restart tracking on the new position based on the motion feature. However, motion feature is not accurate enough to describe target state. As shown in Figure 3 (frame 17), when there are moving distractors in the background and the target remains static, the motion feature will produce a strong response value in the background distractors, while the response at the target position is zero, which will easily lead to tracking drift.

To make better use of motion feature, this work proposes a novel correlation filter which is not only used as tracking, but also used as re-detection. To construct a robust correlation filter, a novelty objective function is proposed, which can perceive background distractors [40]:

$$E(f_k, \boldsymbol{H}) = \frac{1}{2}\left\|\sum_{d=1}^{D}(\boldsymbol{B}x_k^d * f_k^d) - \boldsymbol{H}\right\|_2^2 + \frac{1}{2}\sum_{d=1}^{D}\left\|f_k^d\right\|_2^2 + \frac{\alpha}{2}\|\boldsymbol{H} - \boldsymbol{y}\|_2^2, \tag{17}$$

where $x_k^d, f_k^d \in \mathbb{R}^T$ denote the d-th channel of the vectorized image and filter of frame k, respectively. D is the total channel number. $\boldsymbol{y} \in \mathbb{R}^T$ is the expected response (with Gaussian distribution). $*$ indicates convolution operator. $\boldsymbol{B} \in \mathbb{R}^{M \times N}$ is a cropping matrix to select central M elements in x_k^d, N is the length of x. Usually $N >> M$, $\boldsymbol{H}$ is expected response output. α is a adaptive parameter during tracking, which can be calculated from APCE [41]:

$$\alpha = APCE = \frac{1}{WH} \cdot \frac{|F_{max} - F_{min}|^2}{\sum_{i=1}^{N}\sum_{j=1}^{H}(F_{i,j} - F_{min})^2}. \tag{18}$$

APCE is a confidence evaluation method, when the target state is quite credible, the APCE value is high, otherwise, the value will decrease. We use it as the regularization coefficient, when background is clear, APCE is very high, we make the expected output closer to the Gaussian response y, and when there are cluttered disturbances in the background, we choose to tolerate them so that the expected output has a non-zero value response at the background.

Then Alternating Direction Method of Multipliers (ADMM) algorithm is applied to minimize Equation (17) to achieve a local optimal solution. The augmented Lagrangian form of the equation can be formulated as:

$$\begin{aligned} L_k(f_k, \hat{g}_k^d, \xi^T, \hat{\boldsymbol{H}}) &= E(f_k, \hat{g}_k^d, \hat{\boldsymbol{H}}) + \hat{\xi}^T(\hat{g}_k - \sqrt{N}(\boldsymbol{F}\boldsymbol{B}^T \otimes I_D)f_k) \\ &+ \frac{\rho}{2}\left\|\hat{g}_k - \sqrt{N}(\boldsymbol{F}\boldsymbol{B}^T \otimes I_D)f_k\right\|_2^2, \end{aligned} \tag{19}$$

where $\hat{\xi}^T = [\hat{\xi}_1^T, \hat{\xi}_2^T, \ldots, \hat{\xi}_D^T]$ is the $1 \times DN$ Lagrangian vector in Fourier domain, ρ is a penalty factor. Then, ADMM is applied by alternately solving the following sub-problems:

$$\begin{aligned} \hat{g}_{k+1} = argmin\frac{1}{2}\left\|\sum_{d=1}^{D}(\hat{x}_k^d \odot \hat{g}_k^d) - \hat{\boldsymbol{H}}\right\|_2^2 + \hat{\xi}^T(\hat{g}_k - \sqrt{N}(\boldsymbol{F}\boldsymbol{B}^T \otimes I_D)f_k) \\ + \frac{\rho}{2}\left\|\hat{g}_k - \sqrt{N}(\boldsymbol{F}\boldsymbol{B}^T \otimes I_D)f_k\right\|_2^2, \end{aligned} \tag{20}$$

$$f_{k+1} = argmin\frac{1}{2}\sum_{d=1}^{D}\left\|f_k^d\right\|_2^2 + \hat{\xi}^T(\hat{g}_k - \sqrt{N}(\boldsymbol{F}\boldsymbol{B}^T \otimes I_D)f_k) + \frac{\rho}{2}\left\|\hat{g}_k - \sqrt{N}(\boldsymbol{F}\boldsymbol{B}^T \otimes I_D)f_k\right\|_2^2, \tag{21}$$

$$\hat{\boldsymbol{H}} = \underset{\hat{\boldsymbol{H}}}{argmin}\frac{1}{2}\left\|\sum_{d=1}^{D}(\hat{x}_k^d \odot \hat{g}_k^d) - \hat{\boldsymbol{H}}\right\|_2^2 + \frac{\alpha}{2}\|\hat{\boldsymbol{H}} - \hat{y}\|_2^2, \tag{22}$$

which can be easily solved as follows:

$$\hat{g}_{k+1}(t) = \frac{1}{\rho N}(I - \frac{\hat{x}(t)\hat{x}(t)^T}{\rho N + \hat{x}(t)\hat{x}(t)^T})(\hat{\boldsymbol{H}}(t)\hat{x}(t) - N\xi^T + N\rho\hat{f}_k(t)). \tag{23}$$

$$f_{k+1} = \frac{\xi^T N + \rho N g_{k+1}}{1 + \rho N}. \tag{24}$$

$$\hat{\boldsymbol{H}} = \frac{\sum_{d=1}^{D}(\hat{g}_k^d * \hat{x}_k^d) + \alpha\hat{y}}{1 + \alpha}. \tag{25}$$

Beside used for tracking target, the correlation filter calculated from (23) is also served as a detector in CDCF.

Then the appearance model $\hat{x}_k^{model}$ is updated as follows:

$$\hat{x}_k^{model} = (1-\eta)\hat{x}_{k-1}^{model} + \eta\hat{x}_k, \tag{26}$$

where η is the learning rate of the appearance model.

Remark 1. *To the best of our knowledge, no correlation filters have been used for re-detection. This is mainly because the correlation filter is very sensitive to the position offset, and when there is a large position deviation between the correlation filter model and the target, the correlation filter will get a very low response value, and the target will be detected only when the position deviation is very small. Therefore, how to judge the location where the target reappears is crucial for whether the correlation filter can find the target again. Thanks to the robust motion feature, CDCF will obtain a suitable target candidate region when background is static. Then, CDCF will correct the target position and scale based on correlation operation.*

3.3. Online Tracking and Updating

As aforementioned, local tracking equipped with spatial-temporal constraints helps to locate small targets with limited semantic information. Global re-detection, instead, could be more reliable when faced with challenges like occlusion and out-of-view in long-term tracking. Hence, it is crucial to learn to switch adaptively between local tracking and global re-detection to leverage their complementary strengths.

Local Tracking: Suppose $c_i = [c_{i,start}, c_{i,start+1}, \ldots, c_{i,t-1}]$ is a continuous sub-trajectory from frame $start(c_i)$, and $c_{i,t-1}$ is a trustworthy tracking result in frame $t-1$. For frame t, previous trusted predictions in $[c_1, c_2, \ldots, c_i]$ are fed into the second stage of the re-detector. For static background, only proposals with an overlap greater than 0.01 with the bounding box in $c_{i,t-1}$ are considered as target candidates. If the re-detector finds a proposal with a confidence score over 0.5, local tracking is believed to be valid, and its corresponding result $c_{i,t}$ is added to c_i. Otherwise, local tracking is paused, indicating the end of this continuous trajectory.

Global Re-detection: Starting from the failed frame, global re-detection is performed. Like SiamRCNN, SiamSTA also track potential similar-looking distractors and record their trajectories A. Then the results of CDCF is introduced to guide the global re-detection. SiamSTA compares the bounding box size S_{CD} predicted by CDCF with the target size of previous frames. When the background is static and $S_{CD} \in [S_{min}, S_{max}]$, CDCF's results are judged to be credible, initialize a new sub-trajectory c_{i+1}, and restart local tracking. Otherwise, SiamSTA treats it as a reference result to facilitate selecting the suitable proposal as output for the current frame. When the predicted bounding box of CDCF has a large overlap with a high-confidence proposal, SiamSTA considers the target has been successfully recaptured and initialize a new sub-trajectory c_{i+1}, then restart local tracking. If there is no overlap between CDCF and high-confidence proposals, SiamSTA chooses the proposal with highest score as current frame output. For dynamic background, the continuous sub-trajectory is terminated directly, and the tracker relies on global re-detection to estimate the position and size of the target.

4. Experiments

4.1. Experimental Setup

4.1.1. Evaluation Metrics

This experiment uses three widely-used metrics to evaluate, including precision plot, success rate plot and average overlap accuracy. The first metric computes the percentages of frames in which the estimated target location is within a given distance threshold to the ground-truth. The second one measures the fractions of successful frames where the Intersection over Union (IoU) between the predicted bounding box and ground-truth is greater than a certain threshold varied from 0 to 1. The last one is the evaluation metric given in the Anti-UAV benchmark [16]. It calculates the mean IoU of all videos. In this experiment, an error threshold of 20 pixels are adopted in the precision plot, and the area under the curve (AUC) of the success plot is used to evaluate tracking performance.

4.1.2. Network Parameters

Our SiamSTA is built upon SiamRCNN network, and SiamSTA also borrows its trained weights. The max corners, min distance and block size for computing the background key points are set to 500, 7 and 7 respectively. For optical flow, SiamSTA utilizes a two-level pyramid with a 15 $\times$ 15 sliding window. The F-B error threshold for selecting the correct key points is set to 1.0. If the average moving distance of these selected key points for 5 consecutive frames is less than 0.5, SiamSTA considers the background to be static. A 5 $\times$ 5 median filter is used to remove the tiny foreground noises in the change detection. The weight w_r, for the first stage of re-detection is set to 0.1, hence the weight for the second re-detection stage is 0.9. The location score weight w_l is set to 5.5. In the global detection phase, the settings are consistent with SiamRCNN. As for CDCF, the learning rate η is set as 0.02.

Remark 2. *Due to the competition restrictions of the 2nd Anti-UAV challenge, we did not perform additional training on the model when testing the Anti-UAV2020 [14] and Anti-UAV2021 [15] test-dev dataset, and directly exploit the model trained on RGB datasets given by SiamRCNN. While the Anti-UAV [16] dataset contains a training subset, SiamSTA is retrained on the train subset of the Anti-UAV dataset, with a total of 160 sequences. During training, we also apply motion blur, grayscale, gamma, flip and scale augmentations. The other parameters about training is the same as the SiamRCNN [13].*

4.1.3. Details about UAV Platform

In the evaluation datasets, the picture sizes include 640 $\times$ 512, 640 $\times$ 480 and 1280 $\times$ 720 (in pixels). The detection range (distance from UAVs) varies from 0.1 km to 2.5 km. The UAV size in the image ranges from 5 $\times$ 5 to 60 $\times$ 90 (in pixels), including a variety of UAV platform such as DJI-Phantom4 (196 $\times$ 289.5 $\times$ 289.5 mm), DJI-Mavic-Air (168 $\times$ 184 $\times$ 64 mm), DJI-Spark (143 $\times$ 143 $\times$ 55 mm), DJI-Mavic-Pro (322 $\times$ 242 $\times$ 84 mm), DJI-Inspire (438 $\times$ 451 $\times$ 301 mm) and Parrot.

Remark 3. *The evaluation videos include a variety of UAV platforms with detection distances ranging from 0.1 to 2.5 km, which fully demonstrates the generality of our approach.*

4.2. Comparing with State-of-the-Arts Trackers

Comprehensive experiments are conducted to compare our SiamSTA with some of the currently best performing deep trackers, i.e., SiamRCNN [13], SiamRPN++ [8], Globaltrack [42], PrDiMP [11], DiMP [43], ATOM [10], KYS [44], SiamRPN [27] and other recent CF trackers including AutoTrack [45], ECO [46], ARCF [47], STRCF [48], KCF [18], CSRDCF [49]. For a fair comparison, these compared algorithms are reproduced on our platform with their default parameter settings maintained. The details of different comparison results on different datasets are listed below.

Anti-UAV2020 test-dev datasets [14]: The datasets contains 100 high quality IR videos and 100 RGB videos, spanning multiple occurrences of multi-scale UAVs with complex backgrounds such as clouds, urban buildings, etc. The results of the precision plots and success plots which compare the trackers mentioned above on Anti-UAV2020 test-dev datasets are shown in Figure 4. It is obviously that the proposed SiamSTA can perform better than the other trackers. SiamSTA outperforms the previous best tracker SiamRCNN [13] by 2.32% and 0.97% in terms of precision and success, respectively.

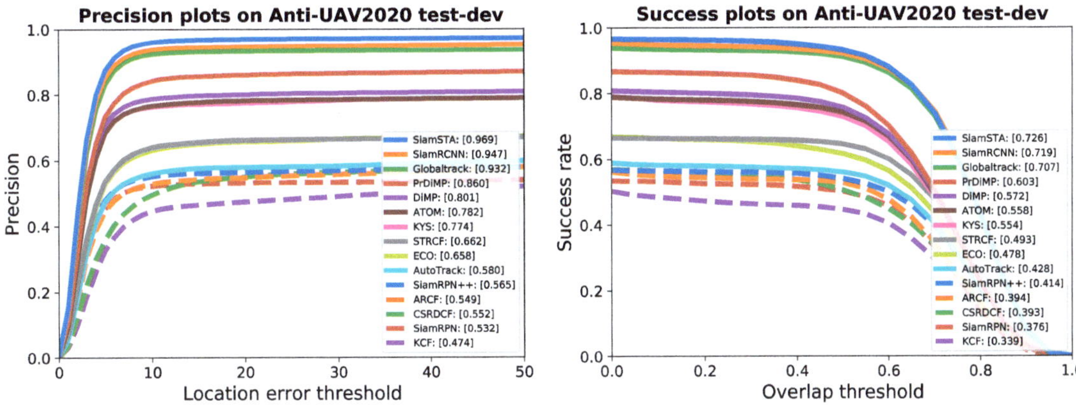

Figure 4. Precision and success plots of our SiamSTA and state-of-the-art trackers on the Anti-UAV2020 test-dev dataset. The mean precision and AUC scores are reported for each tracker. Best viewed in color and zoom.

Anti-UAV2021 test-dev datasets [15]: Based on Anti-UAV2020 test-dev, the 2021 version [15] abandons RGB videos and extends the IR data of the former to 140 videos. Furthermore, the dataset incorporates more complex scenarios such as sea, forest, mountain, and more challenging issues such as tiny objects, weak targets, which makes the tracker easily overwhelmed in the clustered backgrounds. Figure 5 reports the comparison results on Anti-UAV2021 test-dev. As one can see, SiamSTA yields the best precision score 0.888, which surpasses the second-best (SiamRCNN [13]) and third-best (GlobalTrack [42]) trackers by 6.09% and 9.09%, respectively. What is more, SiamSTA is also the best tracker in terms of success with a score of 0.655. Notably, the performance gains of our algorithm in the Anti-UAV2021 test-dev dataset are more impressive than that of Anti-UAV2020. This is mainly because the Anti-UAV2021 test-dev introduces many tiny and weak target videos, while our spatio-temporal attention and change detection are exactly designed to address such challenges, thus leading to a higher accuracy.

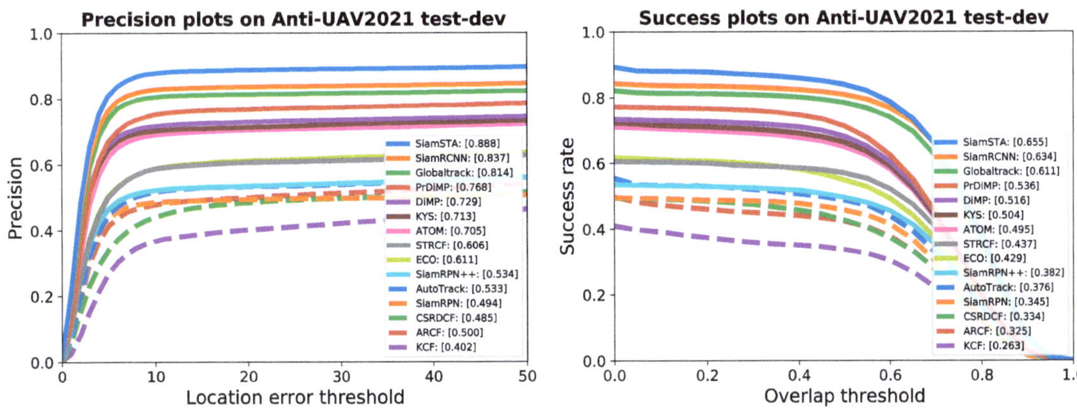

Figure 5. Precision and success plots of our SiamSTA and state-of-the-art trackers on the Anti-UAV2021 test-dev dataset. The mean precision and AUC scores are reported for each tracker. Best viewed in color and zoom.

Anti-UAV [16]: The dataset contains 318 high quality thermal infrared sequences, including train (160 sequences), test (91 sequences) and validation (67 sequences) sub-datasets. Besides the 14 trackers mentioned above, this experiment also introduce 15 state-of-the-art trackers for comparison, including, KeepTrack [50], Stark [51], HiFT [52], STMTrack [53], TransT [54], TransformerTrack [55], ROAM [56], SiamBAN [57], SiamCAR [58], SiamFC++ [59], TADT [60], DaSiamRPN [61], BACF [21], SiamFC [6], SRDCF [20] and DSST [23]. The results are shown in Table 1, SiamSTA performs best in both test and validation sets, suppressing the previous best tracker SiamRCNN by 4.76% and 1.81%, respectively.

Table 1. The overall average accuracy (%) of state-of-the-art trackers on Anti-UAV test and validation sets. The top three results are highlighted in red, green and blue, respectively.

Method	Source	Test	Validation
DSST [23]	BMVC14	33.11	39.86
KCF [18]	T-PAMI15	33.33	38.53
SRDCF [20]	ICCV15	41.00	46.99
SiamFC [6]	ECCVW16	37.51	45.14
BACF [21]	ICCV17	40.94	45.74
ECO [46]	CVPR17	43.68	51.95
STRCF [48]	CVPR18	44.89	50.63
SiamRPN [27]	CVPR18	41.64	43.87
DaSiamRPN [61]	ECCV18	39.61	44.64
ARCF [47]	ICCV19	40.55	43.82
ATOM [10]	CVPR19	49.98	59.82
TADT [60]	CVPR19	43.52	55.20
SiamRPN++ [8]	CVPR19	42.58	45.88
DiMP50 [43]	ICCV19	49.33	62.48
PrDiMP50 [11]	CVPR20	55.70	62.61
AutoTrack [45]	CVPR20	38.70	47.49
SiamFC++ [59]	AAAI20	44.92	50.44
KYS [44]	ECCV20	46.70	60.35
GlobalTrack [42]	AAAI20	64.31	73.84
SiamCAR [58]	CVPR20	46.59	54.79
SiamBAN [57]	CVPR20	39.53	42.42
Siam R-CNN [13]	CVPR20	65.16	74.76
ROAM [56]	CVPR20	45.15	56.15
TransformerTrack [55]	CVPR21	54.75	65.21
TransT [54]	CVPR21	52.14	60.86
STMTrack [53]	CVPR21	40.86	46.41
HiFT [52]	ICCV21	37.87	47.41
Stark [51]	ICCV21	59.08	69.03
KeepTrack [50]	ICCV21	61.05	67.95
SiamSTA	Ours	68.26	76.11

4.3. *Qualitative Evaluation*

Figure 6 shows qualitative comparisons between SiamSTA and other state-of-the-art trackers. SiamSTA shows clear superiority over other trackers in handing challenging tracking situations, including Motion Blur, Fast Motion, Thermal Crossover, Out-of-view and Occlusion. With the help of the proposed spatio-temporal attention mechanism and change detection correlation filter, SiamSTA performs excellent against target loss and thermal crossover, which are common in TIR tracking.

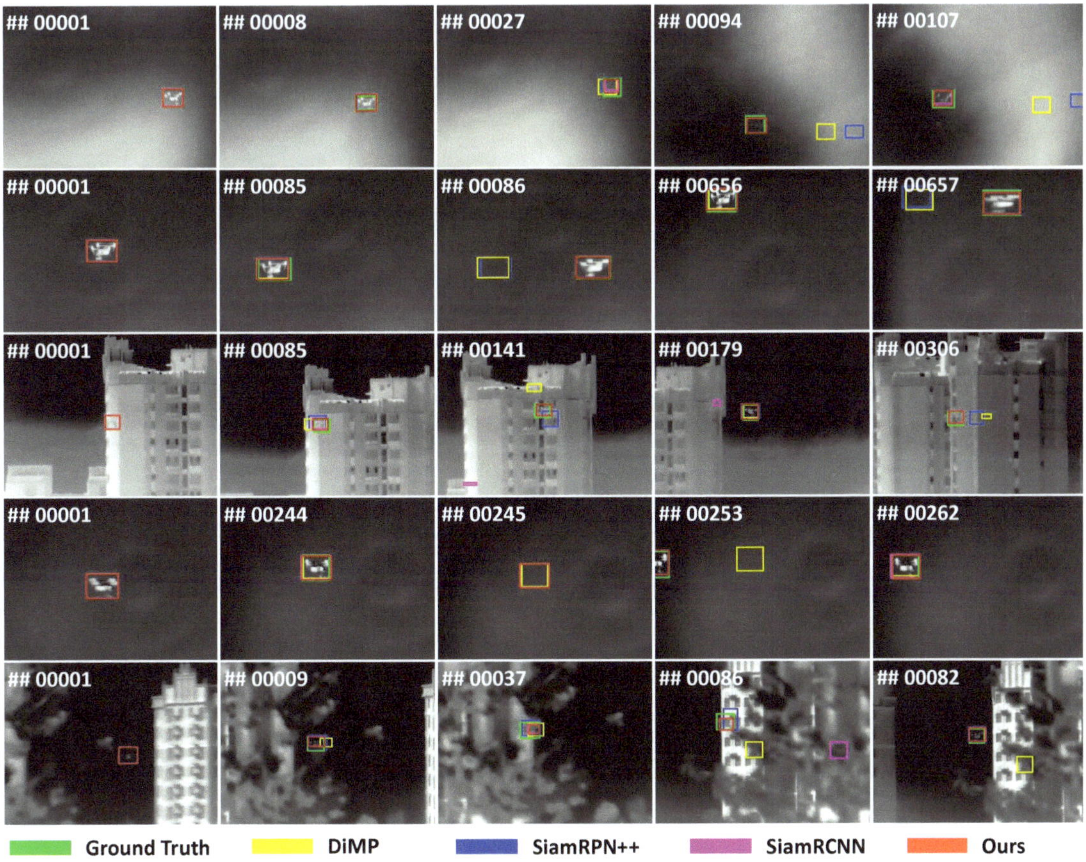

Figure 6. Qualitative comparison of SiamSTA with other state-of-the-art trackers in handling different challenging scenarios. From top to bottom are Motion Blur (Sequence 20190925_222534_1_7_1), Fast Motion (Sequence 20190925_210802_1_3_1), Thermal Crossover (Sequence 20190925_133630_1_9_1), Out-of-view (Sequence 20190925_210802_1_8_1) and Occlusion (Sequence 20190925_143900_1_5_1). The tracking video can be found in https://youtu.be/_l4hP1ZWG3w (accessed on 17 March 2022).

4.4. Attribute-Based Evaluation

For a comprehensive evaluation of SiamSTA, in addition to the 5 attributes (Thermal crossover, scale variation, out-of-view, fast motion and occlusion) defined in Anti-UAV [16], this experiment has added a new attribute, tiny scale (with a diagonal length less than 10 pixels). Figure 7 reports the comparison results (precision and success plots) of SiamSTA and other state-of-the-art trackers in different attributes. It can be seen that SiamSTA achieves the highest scores on all attributes, which fully validates the superior performance of SiamSTA in tackling the challenges of various attributes. Especially noteworthy is the fact that SiamSTA performs particularly well in Thermal Crossover and Tiny Scale attributes, surpassing the second-best tracker SiamRCNN by 7.76% and 7.45% in terms of precision, 7.24% and 9.51% in terms of success rate, respectively.

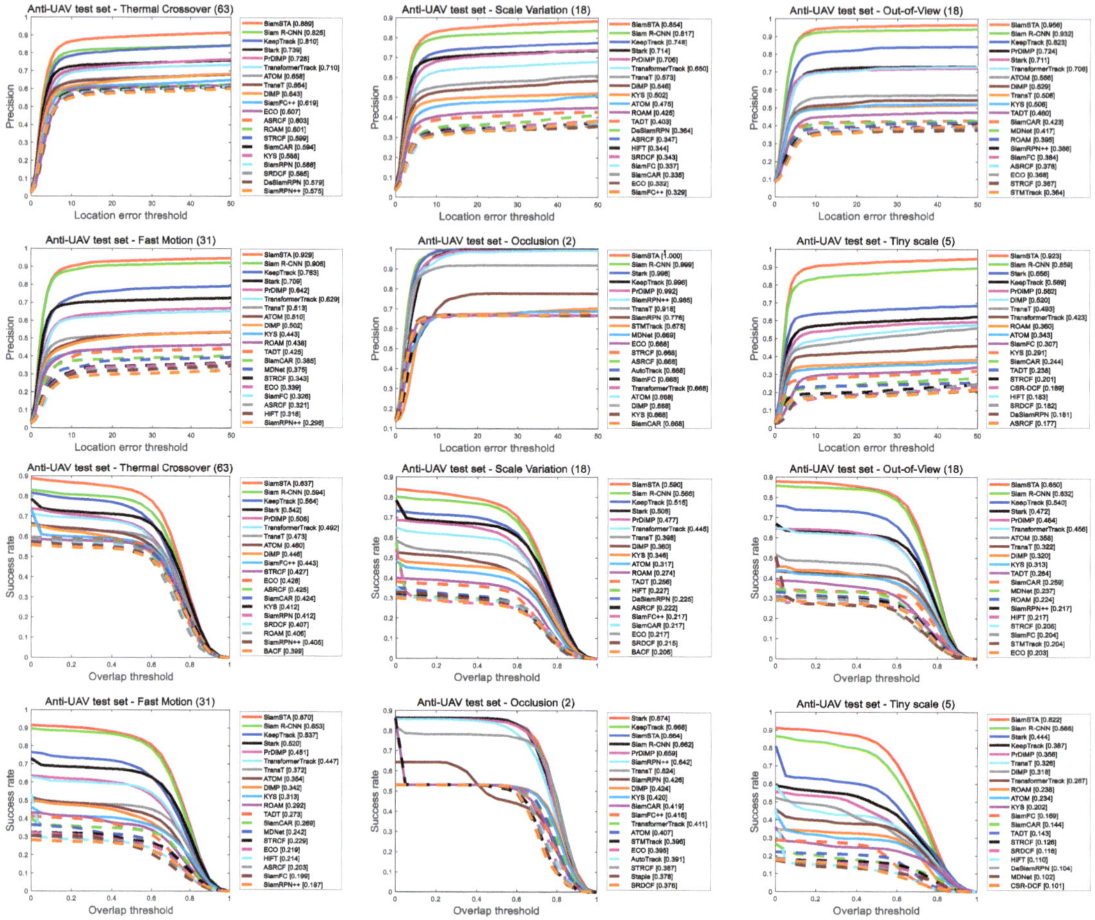

Figure 7. Precision and success plots of our SiamSTA and state-of-the-art trackers on the different attributes defined in Anti-UAV [16] test set. The mean precision and AUC scores are reported for each tracker. Best viewed in color and zoom.

4.5. Tracker Robustness Testing against Weather Challenges

Weather conditions have a critical impact on tracker performance, especially for anti-UAV tracking in real-world scenarios where complex weather conditions such as foggy days, cloudy days, and nighttime bring new challenges to tracking. For this purpose, we selected video sequences from the Anti-UAV [16] dataset under different weather conditions for visualization and analysis. As shown in the Figure 8, the left side is the scene captured by the visible camera, and the right side is the infrared imaging scene at the same moment and under the same shooting angle. It can be seen that in extreme weather conditions, the infrared imaging quality is superior to the visible imaging quality and thus allows a clearer view of the UAV target, so the infrared anti-UAV tracking is more adaptable to the environment. At the same time, SiamSTA can still stably track the target under complex weather conditions, providing a strong guarantee for anti-UAV tracking.

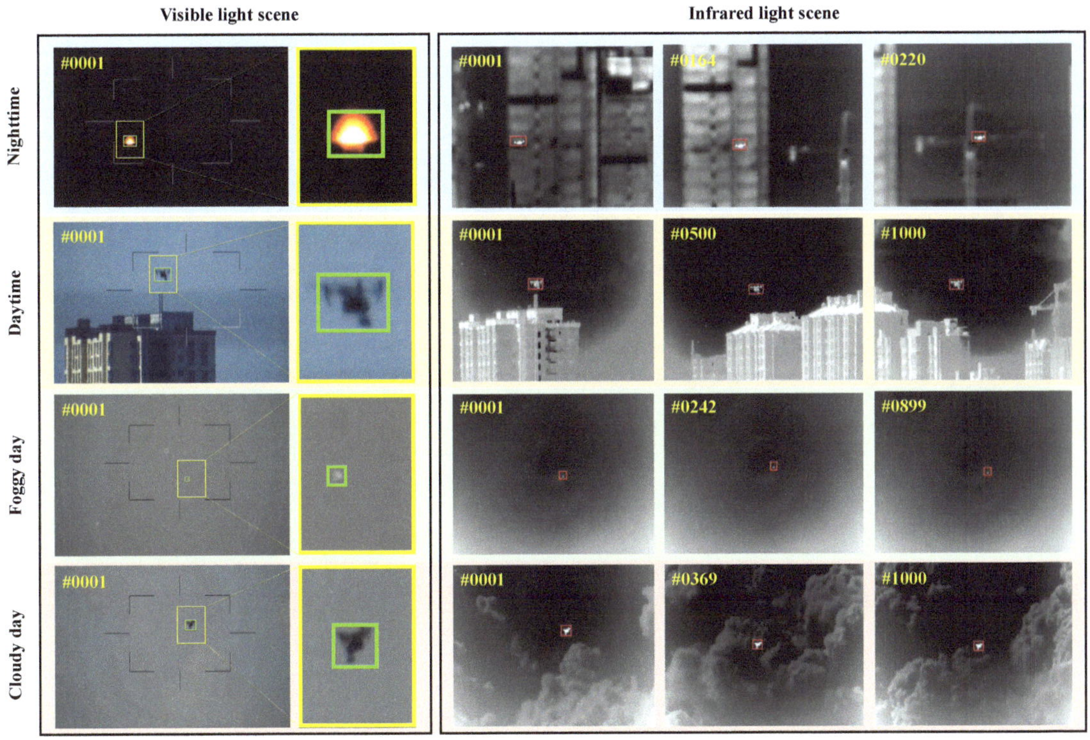

Figure 8. Visual analysis of SiamSTA for different weather challenges. From top to bottom are Sequence 20190926_195921_1_9, Sequence 20190925_101846_1_8, Sequence 20190925_133630_1_2, Sequence 20190925_131530_1_4. The red rectangle indicates the predicted results of SiamSTA.

4.6. Ablation Study

We perform an ablation study to demonstrate the impact of each component in the proposed SiamSTA method on Anti-UAV2021 test-dev [15]. Average tracking accuracy defined in Anti-UAV [16] is adpoted as the evaluation criteria. The baseline method is the original SiamRCNN [13] method.

Effects of Lost Estimation. SiamSTA treats the target state as lost when the confidence score falls below 0. As shown in Table 2, integrating lost estimation brings an improvement of 0.71% over the SiamRCNN baseline, validating that this simple operation is quite effective.

Table 2. Ablation studies on components of SiamSTA. Lost: lost estimation, STA: spatio-temporal attention, CD: change detection.

	Lost	STA	CD	Score (%)
Baseline				64.29
	✓			64.70
		✓		65.61
			✓	66.44
	✓	✓	✓	67.30
GLCF				37.01
		✓	✓	56.04

Effects of Spatio-Temporal Attention. To verify the effect of Spatio-Temporal attention, a variant is created by adding spatio-temporal attention (STA) to baseline. Result in Table 2 shows the effectiveness of Spatio-Temporal Attention (STA) that leads to 2.05% improvement in average tracking accuracy. This can be attributed to the precise switching between global re-detection and nearby tracking, which suppresses the disturbance of cluttered background and thus improves the robustness of tracking.

Effects of Change Detection. We further explore the effectiveness of Change detection (CD). Through purely adding CD to the baseline, the tracking result achieves a performance lift of 2.15% (from 64.29% to 66.44%), the best among all three components, which can be mainly credited to the precise perception ability of tiny and weak target.

To further demonstrate the universality of our approach, we incorporate CD and STA into GLCF tracker [40], which achieves a score of 56.04%, a 51.42% performance improvement over the original GLCF tracker. This indicates that motion feature used in CDCF is generic and applicable for various TIR trackers.

5. Conclusions

This paper proposes a novel algorithm called SiamSTA, which fully exploits the prior knowledge to inspire the current tracker to make optimal decisions. SiamSTA first employs a spatio-temporal attention mechanism to limit the candidate proposals focus on the validate regions and reduce the interference caused by background distractors. Then a CDCF re-detection submodule is introduced into SiamSTA to combat the challenges of target occlusion and out of view. Finally, SiamSTA achieves high-precision online tracking and high-confidence feedback updates by combining local search and global detection. Extensive experiments on three anti-UAV datasets have demonstrated the effectiveness of our SiamSTA, and we strongly believe that our work can promote the development of visual tracking in remote sensing and its application in anti-UAV missions.

Author Contributions: Conceptualization, J.C. and B.H.; methodology, J.C. and B.H.; software, J.C. and B.H.; validation, J.C., B.H. and J.L..; formal analysis, J.L. and T.X.; investigation, J.C., B.H., Y.W. and M.R.; resources, J.L. and T.X.; data curation, J.L. and T.X.; writing—original draft preparation, J.C. and B.H.; writing—review and editing, J.L., Y.W., M.R. and T.X.; visualization, J.C., B.H., Y.W. and M.R.; supervision, J.L. and T.X.; project administration, J.L. and T.X.; funding acquisition, J.L. and T.X. All authors have read and agreed to the published version of the manuscript.

Funding: This research was funded by Key Laboratory Foundation under Grant TCGZ2020C004 and Grant 202020429036.

Institutional Review Board Statement: Not applicable.

Informed Consent Statement: Not applicable.

Data Availability Statement: All data used in this paper are publicly available and can be accessed at https://anti-uav.github.io/dataset/ (accessed on 20 January 2022) for Anti-UAV2021 test dev, https://github.com/ucas-vg/Anti-UAV (accessed on 20 January 2022) for Anti-UAV.

Acknowledgments: This work was supported by the Key Laboratory Foundation under Grant TCGZ2020C004 and Grant 202020429036.

Conflicts of Interest: The authors declare no conflict of interest.

References

1. Mueller, M.; Smith, N.; Ghanem, B. A benchmark and simulator for uav tracking. In Proceedings of the Computer Vision—ECCV 2016—14th European Conference, Amsterdam, The Netherlands, 11–14 October 2016.
2. Fu, C.; Lin, F.; Li, Y.; Chen, G. Correlation filter-based visual tracking for uav with online multi-feature learning. *Remote Sens.* **2019**, *11*, 549. [CrossRef]
3. Xue, X.; Li, Y.; Dong, H.; Shen, Q. Robust correlation tracking for UAV videos via feature fusion and saliency proposals. *Remote Sens.* **2018**, *10*, 1644. [CrossRef]
4. Huang, B.; Xu, T.; Jiang, S.; Chen, Y.; Bai, Y. Robust visual tracking via constrained multi-kernel correlation filters. *IEEE Trans. Multimed.* **2020**, *22*, 2820–2832. [CrossRef]

5. Cliff, O.M.; Saunders, D.L.; Fitch, R. Robotic ecology: Tracking small dynamic animals with an autonomous aerial vehicle. *Sci. Robot.* **2018**, *3*, eaat8409. [CrossRef] [PubMed]
6. Bertinetto, L.; Valmadre, J.; Henriques, J.F.; Vedaldi, A.; Torr, P.H.S. Fully-Convolutional Siamese Networks for Object Tracking. In Proceedings of the Computer Vision—ECCV 2016 Workshops, Amsterdam, The Netherlands, 8–10 and 15–16 October 2016.
7. Wang, Q.; Zhang, L.; Bertinetto, L.; Hu, W.; Torr, P.H. Fast online object tracking and segmentation: A unifying approach. In Proceedings of the IEEE Conference on Computer Vision and Pattern Recognition, CVPR 2019, Long Beach, CA, USA, 16–20 June 2019.
8. Li, B.; Wu, W.; Wang, Q.; Zhang, F.; Xing, J.; Yan, J. SiamRPN++: Evolution of Siamese Visual Tracking With Very Deep Networks. In Proceedings of the IEEE Conference on Computer Vision and Pattern Recognition, CVPR 2019, Long Beach, CA, USA, 16–20 June 2019.
9. Guo, D.; Shao, Y.; Cui, Y.; Wang, Z.; Zhang, L.; Shen, C. Graph Attention Tracking. In Proceedings of the IEEE Conference on Computer Vision and Pattern Recognition, CVPR 2021, Virtual, 19–25 June 2021.
10. Danelljan, M.; Bhat, G.; Khan, F.S.; Felsberg, M. ATOM: Accurate Tracking by Overlap Maximization. In Proceedings of the IEEE Conference on Computer Vision and Pattern Recognition, CVPR 2019, Long Beach, CA, USA, 16–20 June 2019.
11. Danelljan, M.; Gool, L.V.; Timofte, R. Probabilistic Regression for Visual Tracking. In Proceedings of the IEEE/CVF Conference on Computer Vision and Pattern Recognition, CVPR 2020, Seattle, WA, USA, 13–19 June 2020.
12. Huang, B.; Xu, T.; Shen, Z.; Jiang, S.; Zhao, B.; Bian, Z. SiamATL: Online Update of Siamese Tracking Network via Attentional Transfer Learning. *IEEE Trans. Cybern.* **2021**, 1–14. [CrossRef] [PubMed]
13. Voigtlaender, P.; Luiten, J.; Torr, P.H.; Leibe, B. Siam R-CNN: Visual Tracking by Re-Detection. In Proceedings of the IEEE/CVF Conference on Computer Vision and Pattern Recognition, CVPR 2020, Seattle, WA, USA, 13–19 June 2020.
14. Anti-UAV Challenge Dataset. Available online: https://anti-uav.github.io/ (accessed on 1 May 2021).
15. Zhao, J.; Wang, G.; Li, J.; Jin, L.; Fan, N.; Wang, M.; Wang, X.; Yong, T.; Deng, Y.; Guo, Y.; et al. The 2nd Anti-UAV Workshop & Challenge: Methods and Results. *arXiv* **2021**, arXiv:2108.09909.
16. Jiang, N.; Wang, K.; Peng, X.; Yu, X.; Wang, Q.; Xing, J.; Li, G.; Guo, G.; Zhao, J.; Han, Z. Anti-UAV: A Large Multi-Modal Benchmark for UAV Tracking. *arXiv* **2021**, arXiv:2101.08466.
17. Bolme, D.; Beveridge, J.; Draper, B.; Lui, Y.M. Visual object tracking using adaptive correlation filters. In Proceedings of the Twenty-Third IEEE Conference on Computer Vision and Pattern Recognition, CVPR 2010, San Francisco, CA, USA, 13–18 June 2010.
18. Henriques, J.F.; Caseiro, R.; Martins, P.; Batista, J. High-Speed Tracking with Kernelized Correlation Filters. *IEEE Trans. Pattern Anal. Mach. Intell.* **2015**, *37*, 583–596. [CrossRef] [PubMed]
19. Henriques, J.F.; Caseiro, R.; Martins, P.; Batista, J.P. Exploiting the Circulant Structure of Tracking-by-Detection with Kernels. In Proceedings of the Computer Vision—ECCV 2012—12th European Conference on Computer Vision, Florence, Italy, 7–13 October 2012.
20. Danelljan, M.; Häger, G.; Khan, F.; Felsberg, M. Learning Spatially Regularized Correlation Filters for Visual Tracking. In Proceedings of the IEEE International Conference on Computer Vision, ICCV 2015, Santiago, Chile, 7–13 December 2015.
21. Galoogahi, H.K.; Fagg, A.; Lucey, S. Learning Background-Aware Correlation Filters for Visual Tracking. In Proceedings of the IEEE International Conference on Computer Vision, ICCV 2017, Venice, Italy, 22–29 October 2017.
22. Li, Y.; Zhu, J. A Scale Adaptive Kernel Correlation Filter Tracker with Feature Integration. In Proceedings of the Computer Vision—ECCV 2014 Workshops, Zurich, Switzerland, 6–7 and 12 September 2014.
23. Danelljan, M.; Häger, G.; Khan, F.; Felsberg, M. Accurate scale estimation for robust visual tracking. In Proceedings of the British Machine Vision Conference, BMVC 2014, Nottingham, UK, 1–5 September 2014.
24. Li, F.; Yao, Y.; Li, P.; Zhang, D.; Zuo, W.; Yang, M.H. Integrating Boundary and Center Correlation Filters for Visual Tracking with Aspect Ratio Variation. In Proceedings of the IEEE International Conference on Computer Vision Workshops, ICCV Workshops 2017, Venice, Italy, 22–29 October 2017.
25. Danelljan, M.; Khan, F.; Felsberg, M.; van de Weijer, J. Adaptive Color Attributes for Real-Time Visual Tracking. In Proceedings of the IEEE Conference on Computer Vision and Pattern Recognition, CVPR 2014, Columbus, OH, USA, 23–28 June 2014.
26. Ma, C.; Huang, J.B.; Yang, X.; Yang, M.H. Hierarchical Convolutional Features for Visual Tracking. In Proceedings of the IEEE International Conference on Computer Vision, ICCV 2015, Santiago, Chile, 7–13 December 2015.
27. Li, B.; Yan, J.; Wu, W.; Zhu, Z.; Hu, X. High Performance Visual Tracking With Siamese Region Proposal Network. In Proceedings of the IEEE Conference on Computer Vision and Pattern Recognition, CVPR 2018, Salt Lake City, UT, USA, 18–22 June 2018.
28. Wan, M.; Gu, G.; Qian, W.; Ren, K.; Chen, Q.; Zhang, H.; Maldague, X. Total variation regularization term-based low-rank and sparse matrix representation model for infrared moving target tracking. *Remote Sens.* **2018**, *10*, 510. [CrossRef]
29. Zingoni, A.; Diani, M.; Corsini, G. A flexible algorithm for detecting challenging moving objects in real-time within IR video sequences. *Remote Sens.* **2017**, *9*, 1128. [CrossRef]
30. Wan, M.; Gu, G.; Qian, W.; Ren, K.; Chen, Q.; Maldague, X. Infrared image enhancement using adaptive histogram partition and brightness correction. *Remote Sens.* **2018**, *10*, 682. [CrossRef]
31. Zhang, L.; Gonzalez-Garcia, A.; Van De Weijer, J.; Danelljan, M.; Khan, F.S. Synthetic data generation for end-to-end thermal infrared tracking. *IEEE Trans. Image Process.* **2018**, *28*, 1837–1850. [CrossRef] [PubMed]

32. Felsberg, M.; Berg, A.; Hager, G.; Ahlberg, J.; Kristan, M.; Matas, J.; Leonardis, A.; Cehovin, L.; Fernandez, G.; Vojir, T. The Thermal Infrared Visual Object Tracking VOT-TIR2015 Challenge Results. In Proceedings of the IEEE International Conference on Computer Vision Workshop, ICCV Workshops 2015, Santiago, Chile, 7–13 December 2015.
33. Cao, Y.; Wang, G.; Yan, D.; Zhao, Z. Two algorithms for the detection and tracking of moving vehicle targets in aerial infrared image sequences. *Remote Sens.* **2016**, *8*, 28. [CrossRef]
34. Yu, X.; Yu, Q. Online structural learning with dense samples and a weighting kernel. *Pattern Recognit. Lett.* **2017**, *105*, 59–66. [CrossRef]
35. Li, M.; Peng, L.; Yingpin, C.; Huang, S.; Qin, F.; Peng, Z. Mask Sparse Representation Based on Semantic Features for Thermal Infrared Target Tracking. *Remote Sens.* **2019**, *11*, 1967. [CrossRef]
36. Wu, S.; Zhang, K.; Li, S.; Yan, J. Learning to Track Aircraft in Infrared Imagery. *Remote Sens.* **2020**, *12*, 3995. [CrossRef]
37. Huang, B.; Chen, J.; Xu, T.; Wang, Y.; Jiang, S.; Wang, Y.; Wang, L.; Li, J. SiamSTA: Spatio-Temporal Attention based Siamese Tracker for Tracking UAVs. In Proceedings of the IEEE/CVF International Conference on Computer Vision Workshops, ICCVW 2021, Montreal, BC, Canada, 11–17 October 2021.
38. Shi, J.; Tomasi, G. Good features to track. In Proceedings of the Conference on Computer Vision and Pattern Recognition, CVPR 1994, Seattle, WA, USA, 21–23 June 1994.
39. Kalal, Z.; Mikolajczyk, K.; Matas, J. Tracking-learning-detection. *IEEE Trans. Pattern Anal. Mach. Intell.* **2011**, *34*, 1409–1422. [CrossRef] [PubMed]
40. Chen, J.; Xu, T.; Li, J.; Wang, L.; Wang, Y.; Li, X. Adaptive Gaussian-Like Response Correlation Filter for UAV Tracking. In Proceedings of the Image and Graphics—11th International Conference, ICIG 2021, Haikou, China, 6–8 August 2021.
41. Wang, M.; Liu, Y.; Huang, Z. Large Margin Object Tracking with Circulant Feature Maps. In Proceedings of the IEEE Conference on Computer Vision and Pattern Recognition, CVPR 2017, Honolulu, HI, USA, 21–26 July 2017.
42. Huang, L.; Zhao, X.; Huang, K. GlobalTrack: A Simple and Strong Baseline for Long-Term Tracking. In Proceedings of the Thirty-Fourth AAAI Conference on Artificial Intelligence, AAAI 2020, New York, NY, USA, 7–12 February 2020.
43. Bhat, G.; Danelljan, M.; Gool, L.V.; Timofte, R. Learning Discriminative Model Prediction for Tracking. In Proceedings of the IEEE/CVF International Conference on Computer Vision, ICCV 2019, Seoul, Korea, 27 October–2 November 2019.
44. Bhat, G.; Danelljan, M.; Van Gool, L.; Timofte, R. Know Your Surroundings: Exploiting Scene Information for Object Tracking. In Proceedings of the Computer Vision—ECCV 2020—16th European Conference, Glasgow, UK, 23–28 August 2020.
45. Li, Y.; Fu, C.; Ding, F.; Huang, Z.; Lu, G. AutoTrack: Towards High-Performance Visual Tracking for UAV with Automatic Spatio-Temporal Regularization. In Proceedings of the IEEE/CVF Conference on Computer Vision and Pattern Recognition, CVPR 2020, Seattle, WA, USA, 13–19 June 2020.
46. Danelljan, M.; Bhat, G.; Shahbaz Khan, F.; Felsberg, M. ECO: Efficient Convolution Operators for Tracking. In Proceedings of the IEEE Conference on Computer Vision and Pattern Recognition, CVPR 2017, Honolulu, HI, USA, 21–26 July 2017.
47. Huang, Z.; Fu, C.; Li, Y.; Lin, F.; Lu, P. Learning Aberrance Repressed Correlation Filters for Real-Time UAV Tracking. In Proceedings of the IEEE/CVF International Conference on Computer Vision, ICCV 2019, Seoul, Korea, 27 October–2 November 2019.
48. Li, F.; Tian, C.; Zuo, W.; Zhang, L.; Yang, M.H. Learning Spatial-Temporal Regularized Correlation Filters for Visual Tracking. In Proceedings of the IEEE Conference on Computer Vision and Pattern Recognition, CVPR 2018, Salt Lake City, UT, USA, 18–22 June 2018.
49. Lukežič, A.; Vojíř, T.; Čehovin Zajc, L.; Matas, J.; Kristan, M. Discriminative Correlation Filter with Channel and Spatial Reliability. In Proceedings of the IEEE Conference on Computer Vision and Pattern Recognition, CVPR 2017, Honolulu, HI, USA, 21–26 July 2017.
50. Mayer, C.; Danelljan, M.; Paudel, D.P.; Van Gool, L. Learning Target Candidate Association to Keep Track of What Not to Track. *arXiv* **2021**, arXiv:2103.16556.
51. Yan, B.; Peng, H.; Fu, J.; Wang, D.; Lu, H. Learning spatio-temporal transformer for visual tracking. *arXiv* **2021**, arXiv:2103.17154.
52. Cao, Z.; Fu, C.; Ye, J.; Li, B.; Li, Y. HiFT: Hierarchical Feature Transformer for Aerial Tracking. *arXiv* **2021**, arXiv:2108.00202.
53. Fu, Z.; Liu, Q.; Fu, Z.; Wang, Y. STMTrack: Template-Free Visual Tracking With Space-Time Memory Networks. In Proceedings of the IEEE Conference on Computer Vision and Pattern Recognition, CVPR 2021, Virtual, 19–25 June 2021.
54. Chen, X.; Yan, B.; Zhu, J.; Wang, D.; Yang, X.; Lu, H. Transformer tracking. In Proceedings of the IEEE Conference on Computer Vision and Pattern Recognition, CVPR 2021, Virtual, 19–25 June 2021.
55. Wang, N.; Zhou, W.; Wang, J.; Li, H. Transformer Meets Tracker: Exploiting Temporal Context for Robust Visual Tracking. In Proceedings of the IEEE Conference on Computer Vision and Pattern Recognition, CVPR 2021, Virtual, 19–25 June 2021.
56. Yang, T.; Xu, P.; Hu, R.; Chai, H.; Chan, A.B. ROAM: Recurrently optimizing tracking model. In Proceedings of the IEEE/CVF Conference on Computer Vision and Pattern Recognition, CVPR 2020, Seattle, WA, USA, 13–19 June 2020.
57. Chen, Z.; Zhong, B.; Li, G.; Zhang, S.; Ji, R. Siamese box adaptive network for visual tracking. In Proceedings of the IEEE/CVF Conference on Computer Vision and Pattern Recognition, CVPR 2020, Seattle, WA, USA, 13–19 June 2020.
58. Guo, D.; Wang, J.; Cui, Y.; Wang, Z.; Chen, S. SiamCAR: Siamese fully convolutional classification and regression for visual tracking. In Proceedings of the IEEE/CVF Conference on Computer Vision and Pattern Recognition, CVPR 2020, Seattle, WA, USA, 13–19 June 2020.

59. Xu, Y.; Wang, Z.; Li, Z.; Yuan, Y.; Yu, G. SiamFC++: Towards Robust and Accurate Visual Tracking with Target Estimation Guidelines. In Proceedings of the Thirty-Fourth AAAI Conference on Artificial Intelligence, AAAI 2020, New York, NY, USA, 7–12 February 2020.
60. Li, X.; Ma, C.; Wu, B.; He, Z.; Yang, M.H. Target-Aware Deep Tracking. In Proceedings of the IEEE Conference on Computer Vision and Pattern Recognition, CVPR 2019, Long Beach, CA, USA, 16–20 June 2019.
61. Zhu, Z.; Wang, Q.; Li, B.; Wu, W.; Yan, J.; Hu, W. Distractor-Aware Siamese Networks for Visual Object Tracking. In Proceedings of the Computer Vision—ECCV 201—15th European Conference, Munich, Germany, 8–14 September 2018.

remote sensing

Article

Graph-Based Embedding Smoothing Network for Few-Shot Scene Classification of Remote Sensing Images

Zhengwu Yuan [1], Wendong Huang [1,*], Chan Tang [1], Aixia Yang [2] and Xiaobo Luo [1]

1 Chongqing Engineering Research Center for Spatial Big Data Intelligent Technology, School of Computer Science and Technology, Chongqing University of Posts and Telecommunications, Chongqing 400065, China; yuanzw@cqupt.edu.cn (Z.Y.); s200231016@stu.cqupt.edu.cn (C.T.); luoxb@cqupt.edu.cn (X.L.)

2 State Key Laboratory of Remote Sensing Science, Aerospace Information Research Institute, Chinese Academy of Sciences, Beijing 100101, China; yangax@radi.ac.cn

* Correspondence: s190201036@stu.cqupt.edu.cn

Abstract: As a fundamental task in the field of remote sensing, scene classification is increasingly attracting attention. The most popular way to solve scene classification is to train a deep neural network with a large-scale remote sensing dataset. However, given a small amount of data, how to train a deep neural network with outstanding performance remains a challenge. Existing methods seek to take advantage of transfer knowledge or meta-knowledge to resolve the scene classification issue of remote sensing images with a handful of labeled samples while ignoring various class-irrelevant noises existing in scene features and the specificity of different tasks. For this reason, in this paper, an end-to-end graph neural network is presented to enhance the performance of scene classification in few-shot scenarios, referred to as the graph-based embedding smoothing network (GES-Net). Specifically, GES-Net adopts an unsupervised non-parametric regularizer, called embedding smoothing, to regularize embedding features. Embedding smoothing can capture high-order feature interactions in an unsupervised manner, which is adopted to remove undesired noises from embedding features and yields smoother embedding features. Moreover, instead of the traditional sample-level relation representation, GES-Net introduces a new task-level relation representation to construct the graph. The task-level relation representation can capture the relations between nodes from the perspective of the whole task rather than only between samples, which can highlight subtle differences between nodes and enhance the discrimination of the relations between nodes. Experimental results on three public remote sensing datasets, UC Merced, WHU-RS19, and NWPU-RESISC45, showed that the proposed GES-Net approach obtained state-of-the-art results in the settings of limited labeled samples.

Keywords: few-shot learning; scene classification; embedding smoothing; graph neural network; task-level relation representation

Citation: Yuan, Z.; Huang, W.; Tang, C.; Yang, A.; Luo, X. Graph-Based Embedding Smoothing Network for Few-Shot Scene Classification of Remote Sensing Images. *Remote Sens.* **2022**, *14*, 1161. https://doi.org/10.3390/rs14051161

Academic Editors: Miltiadis D. Lytras and Andreea Claudia Serban

Received: 13 January 2022
Accepted: 22 February 2022
Published: 26 February 2022

1. Introduction

Scene classification, as a vital part of remote sensing image processing and analysis, divides the scene images into the corresponding scene classes according to the differences of their content. It has been extensively applied in various scenarios, including land use, land cover [1–3], urban planning, geological hazard monitoring, and traffic management [4–6]. Recently, approaches adopting deep learning have made significant advances in the remote sensing scene classification task due to the availability of large amounts of label data and the powerful learning capability of deep neural networks [7–11].

However, approaches based on deep learning typically depend on a sea of labeled samples to obtain outstanding performance. In a wide range of realistic scenarios, collecting large-scale remote sensing images and labeling them are quite time-consuming and painstaking tasks. The approaches on the basis of deep learning are subject to overfitting when the labeled data are limited, which poses a sharp degeneration in the performance of the model. Moreover, the trained model cannot accurately identify the unseen scene category due to the deviation in the data distribution. The most common approach is to expand the dataset, that is to spend a mass of resources to collect samples and label them. In addition, the high computational cost also limits the application range of approaches adopting deep learning.

To imitate the way of human learning, an emerging research direction has appeared, namely few-shot learning, which aims to enable the model to rapidly learn to identify new categories from a small amount of labeled data [12,13]. Recently, some work adopted transfer learning or meta-learning ideas to solve scene classification tasks with a small amount of labeled data. Zhai et al. [14] proposed a meta-learning model for the scene classification task, referred to as LLSR, to quickly achieve scene classification with limited samples. There is also some work exploring the usage of graph representation to address the image classification problem with limited labeled data. To generalize matching-based methods, Garcia et al. [15] introduce a neural network model based on the graph, which aims to view learning as the transfer of messages from the training data to the test data.

The methods mentioned above mainly focus on applying transfer knowledge or meta-knowledge to address the task of few-shot scene classification, while overlooking the significance of learning distinctive feature representations. Differing from common natural images, remote sensing images have some unique attributes, such as some differences between images belonging to the same category and certain similarities between images belonging to diverse categories, as indicated in Figure 1. It shows a four-way two-shot classification scenario, which contains four different scenes, freeway, runway, lake, and wetland, and two images for each scene in the support set. For few-shot scene classification tasks, due to illumination, background, distance, angle, and other imaging factors, there are various class-irrelevant noises and there are few samples available for each class, which can easily cause confusion in scene classification, while increasing the difficulty of classification. Accordingly, increasing the discrimination and robustness of scene features is a vital line of thinking to improve scene classification performance with finite labeled samples.

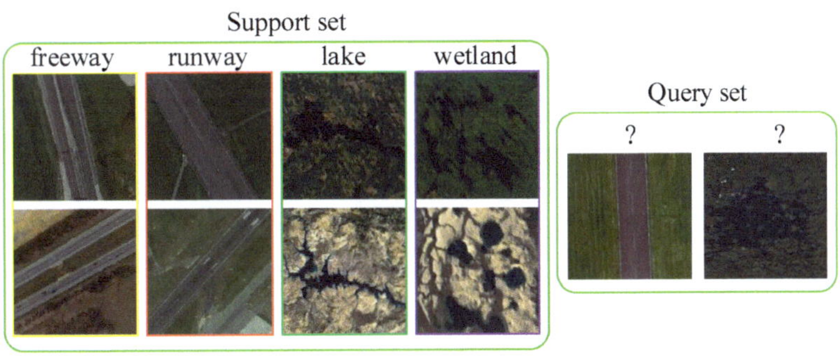

Figure 1. Illustration of adopting few-shot learning to learn scene information from only two labeled images.

To this end, a novel graph-based embedding smoothing network, called GES-Net, is presented in this paper for remote sensing scene classification, which not only has the ability to learn from few samples, but also addresses the aforementioned issues. First, a new regularization method, called embedding smoothing, is proposed in our framework, which can yield a cluster of interpolations from embedding features by means of the relations between them. Since embedding smoothing has no parameters that need to be trained, it can be combined with the embedding module to construct a regularized embedding space for the sample set at a small cost. The assumption of this improvement is on the basis of the actuality that interpolated embedding features are of smoother decision boundaries than the unsmoothed embedding features, and they are more robust to class-irrelevant noises. These characteristics have been shown to be vital for generalization performance [16–18]. Second, to learn to identify unseen samples from limited samples, humans typically compare the target sample with all seen samples rather than only one of them. Hence, instead of using common sample-level distances, such as the cosine distance or Euclidean distance, the task-level relation representation is adopted to construct the graph in GES-Net through the attention mechanism. It can associate the target sample with all the samples in the task and yield a more discriminative relation representation between different scene categories. According to the constructed graph, label matching can iteratively yield prediction labels for samples in the query set by transductive learning until the optimal solution is obtained. Then, the cross-entropy loss between the ground truth labels and the predicted labels of the query samples can be calculated. Finally, all learnable parameters can be updated through back-propagation in an end-to-end manner. In addition, the whole model is trained in an episode-by-episode manner from meta-learning to avoid over-fitting. The proposed approach was validated on three publicly available remote sensing datasets, and the experimental results illustrated that GES-Net largely surpassed existing approaches and obtained new state-of-the-art results in the case of limited labeled samples.

Overall, our main contributions in this paper can be summarized as follows:

- A novel graph neural network, referred to as GES-Net, is presented to enhance the performance of scene classification in few-shot settings. GES-Net adopts a new regularization technology to urge the model to learn discriminative and robust embedding features;
- The attention mechanism is further adopted to measure the relation representation at the task level. It can consider the relations between samples from the task level and improve the discrimination of the relation representation;
- The experimental results obtained on three publicly available remote sensing datasets showed that our proposed GES-Net method significantly outperformed state-of-the-art approaches in few-shot settings and obtained new state-of-the-art results in the case of limited labeled samples.

The rest of our paper is organized as follows. A concise review of related work is introduced in Section 2. In Section 3, our proposed approach is described in detail. The setup and results of the experiments are shown in Section 4. In the end, the conclusion of our paper is given in Section 5.

2. Related Work

In the following section, the related work from four aspects is specifically reviewed: metric learning, few-shot learning, regularization for generalization, and transductive learning.

2.1. Remote Sensing Scene Classification

Recently, approaches on the basis of deep learning have been extensively adopted in a broad variety of fields because of their vigorous representation learning capability [19–23], where a large quantity of research work relates to solving remote sensing scene classification problems [24–27]. For instance, Lu et al. [28] designed a network that uses multiple complementary source domains to address the problem that existing methods cannot capture the distribution of various source domains well, which carries out cross-domain scene classification. To improve scene classification, Lu et al. [29] introduced a convolutional neural network (CNN) that can exploit the semantic information of the label to aggregate the intermediate features. Similarly, Nogueira et al. [30] adopted the CNN to extract features and achieved scene classification through the linear SVM. Cheng et al. [31] designed the discriminative CNNs (D-CNNs) to resolve the issue of large intra-class differences and small inter-class differences in scene images. Although the above work boosts the performance of scene classification, most of these methods only consider employing CNNs to extract embedding features and neglect refining the embedding features to reduce class-irrelevant noises. Differing from the above work, our research attempted to enhance the distinguishability of embedding features and extend the decision boundaries through embedding smoothing.

2.2. Few-Shot Learning

At present, most popular few-shot learning algorithms adopt the meta-learning framework [32–36]. Recently, few-shot learning has gained tremendous attention and made significant advances in the area of remote sensing. For example, Liu et al. [37] and Gao et al. [38] introduced a series of few-shot learning approaches to resolve the issue of hyperspectral image classification in few-shot scenarios. Rostami et al. [39] proposed a novel model for synthetic aperture radar image classification with limited labeled data. Li et al. [40] introduced a meta-learning approach, which learns a measurement rule to address the problems relevant to scene classification with limited labeled samples. Li et al. [41] proposed an end-to-end few-shot learning network consisting of a feature extractor and a matcher for improving scene classification. Guo et al. [42] proposed a cross-domain classification benchmark in few-shot scenarios, which consists of datasets from various domains. Gong et al. [43] proposed a coarse-to-fine approach to deal with unseen classes in domain adaptation. In these methods, some are based on transfer learning and others are based on meta-learning. Transfer learning aims to derive prior knowledge that can be applied to a similar target task from source tasks with large-scale datasets. The goal of meta-learning is to make the machine possess the capacity to learn how to learn, that is to learn meta-knowledge on multiple tasks, which can make the machine quickly adapt to new tasks. Even though the above work proves that transfer learning can enhance classification performance when labeled data are scarce, the work in [44] showed that the improvement brought by transfer learning will decrease, as the gap between the target task distribution and the source task distribution increases, whereas meta-learning, as human learning, can employ a few samples to perform effective inference. This shows that meta-learning is more effective than transfer learning. Hence, our proposed method addresses the issues relevant to scene classification with limited labeled data by combining meta-learning. The embedding features learned by earlier methods are not sufficiently distinguishable for remote sensing scene images and only learn the sample-level relation representation, which cannot robustly measure the similarity between scene images. Therefore, a task-level relation representation is proposed to learn the task-specific relations between samples. Compared with previous approaches (e.g., RS-MetaNet [40] and DLA-MatchNet [41]), our proposed approach is not only robust to diverse class-irrelevant noises, but it can also consider the relations between samples from the task level.

2.3. Regularization for Generalization

Regularization is an essential method for enhancing the generalization performance of neural networks. Typically adopted methods include batch normalization [45], dropout [46], and spatial dropout [47], which attempt to enhance the robustness to various differences from inputs. Others are related to how to regularize the weights [48,49]. Another line of work is related to manifold regularization, which focuses on smoothing the decision boundaries and refining the category representations [18,50,51]. Existing research work shows that perturbing feature representation tends to achieve better generalization performance [45,46]. The smoothing method that we propose is the closest to feature interpolation. For example, Zhao et al. [52] proposed to employ the interpolation of nearest neighbors to conduct predictions to augment adversarial robustness. To obtain better generalization, manifold mixup is introduced to smooth feature representations of the model [16,17]. Different from the above methods, our proposed smoothing method, called embedding smoothing, smooths the manifold by interpolating embedding features end-to-end and is only applied at the end of the backbone, which can capture higher-level interactions between embedding features in an unsupervised manner and obtain better classification performance.

2.4. Transductive Learning

Transductive learning was first proposed by Vapnik [53], which performs label prediction on only test samples. In comparison, inductive learning aims to yield a prediction function in a pre-defined space. When there are only a few labeled samples, transductive learning can achieve better performance than inductive learning [54,55], which makes transductive learning quite suitable for solving few-shot classification problems. Liu et al. [56] introduced a few-shot method, called TPN, through transductive learning. Likewise, our proposed method also adopts transductive learning for few-shot scene classification. However, their method does not consider the effect of regularization terms and task-level relation representation on model performance.

3. Proposed Method

A novel graph neural network is presented for few-shot scene classification by transductive learning in this section, referred to as GES-Net. It is made up of four components: embedding learning, embedding smoothing, graph constructing, and label matching. As shown in Figure 2, GES-Net first extracts scene embedding features by a embedding learning module. Second, a new embedding smoothing method is presented to transform the embedding features into a cluster of interpolated features, which are referred to as smoothed embedding features. The smoothed embedding features are then adopted to calculate the task-level relation representation between nodes to carry out graph construction. Finally, given support samples, label matching is applied to label query samples on the constructed graph. On the one hand, GES-Net applies embedding smoothing to embedding features to enhance the smoothness of the embedding space by reducing various class-irrelevant noises, which was shown to improve generalization [16–18]. On the other hand, the proposed approach adopts the attention mechanism to measure task-level relation representation to perform graph constructing rather than common linear metrics.

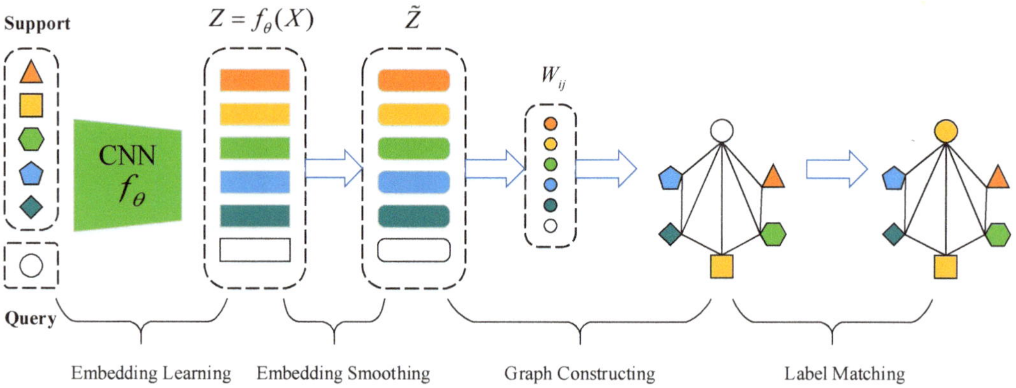

Figure 2. Graph-based embedding smoothing network framework for the 5-way 1-shot scenario with one query sample.

3.1. Few-Shot Setting Setup

In our work, three datasets are given: a training set (D_{train}), a test set (D_{test}), and a validation set (D_{val}). The training set D_{train} was made up of a significant quantity of labeled data $D_{train} = (x_i, y_i)_{i=1}^{N_{train}}$, where sample x_i corresponds to category $y_i \in Y_{train}$. The test set $D_{test} = (x_j, y_j)_{j=1}^{N_{test}}$, where x_j derives from unseen category $y_j \in Y_{test}$ in the training set, that is $Y_{train} \cap Y_{test} = \emptyset$, was adopted to estimate the generalization capability of the model. The validation set D_{val} comprised the classes that do not exist in D_{train} and D_{test} and was adopted to adjust the hyperparameters.

Moreover, episode training was adopted to mimic the few-shot setting, that is an episode is referred to as a task. Each episode (task) is made up of N categories that are sampled uniformly without replacement from all categories, a support set S (K samples per category), and a query set Q (a total of T samples for all categories), which is termed N-way K-shot learning.

3.2. Embedding Learning

A CNN $f_\theta(x_i;\theta)$ was employed to extract the embedding features z_i of the input x_i, where θ serves as the parameter of the network. By adopting the same backbone utilized in previous work [32,57,58], fair comparisons can be offered in the experiments, focusing on the effectiveness of the methods. The backbone contains four convolution modules, where each convolution module starts at a two-dimensional convolution layer containing a 3×3 convolution kernel and kernel size of 64. After each convolution layer, there is a batch normalization layer [45], a ReLU activation function [59], and a 2×2 max-pooling layer. The backbone $f_\theta(x;\theta)$ is applied to the support set S and the query set Q.

3.3. Embedding Smoothing

Embedding smoothing is designed to process a set of embedding features $z_i \in \mathbb{R}^m$ that is obtained by inputting samples x of an episode into the embedding learning module $f_\theta(x;\theta)$. A group of new embedding features $\widetilde{z_i} \in \mathbb{R}^m$ is then yielded by adopting the following steps. First, for the pairwise features (i,j), the proposed approach calculates the distance as $d_{ij} = \|z_i - z_j\|_2$ and the corresponding adjacent matrix as $B_{ij} = \exp\left(-d_{ij}/2\sigma\right)$, where σ is a scale parameter and $B_{ii} = 0$ for $\forall i$. $\sigma = Std(d_{ij})$ was selected, which empirically made the training stage stable.

Next, the Laplacian of the adjacency matrix is calculated as,

$$S = D^{-\frac{1}{2}} B D^{-\frac{1}{2}} \tag{1}$$

where:

$$D_{ii} = \Sigma_j B_{ij} \tag{2}$$

Subsequently, adopting the label propagation formulation described in [60], propagating matrix M is obtained as,

$$M = (I - \beta S)^{-1} \tag{3}$$

where $\beta \in \mathbb{R}$ denotes the scale parameter and I refers to the identity matrix. The smoothed embedding features are then yielded by the following operation,

$$\widetilde{z_i} = \sum M_{ij} z_j \tag{4}$$

Since $\widetilde{z_i}$ is obtained by the weighted combination of its neighbors, embedding smoothing can effectively reduce the impact of noises from class-irrelevant features.

3.4. Graph Constructing

The embedded low-dimensional subspace in the data is uncovered by manifold learning, where selecting a suitable adjacency graph is crucial. To construct an appropriate adjacency graph in the few-shot setting, a novel graph constructing module is proposed, which was built around the attention mechanism. Specifically, the attention mechanism was adopted to convert smoothed embedding features to relation representations with regard to task-specific features; that is, the relation representation not only denotes the distance between nodes, but also contains the task-specific relation. The proposed graph constructing module can effectively avoid directly comparing irrelevant local relations.

As shown in Figure 3, given smoothed embedding features $\widetilde{Z} \in \mathbb{R}^{(N\times K+T)\times m}$, the task-level relation representation W can be achieved by Equation (7).

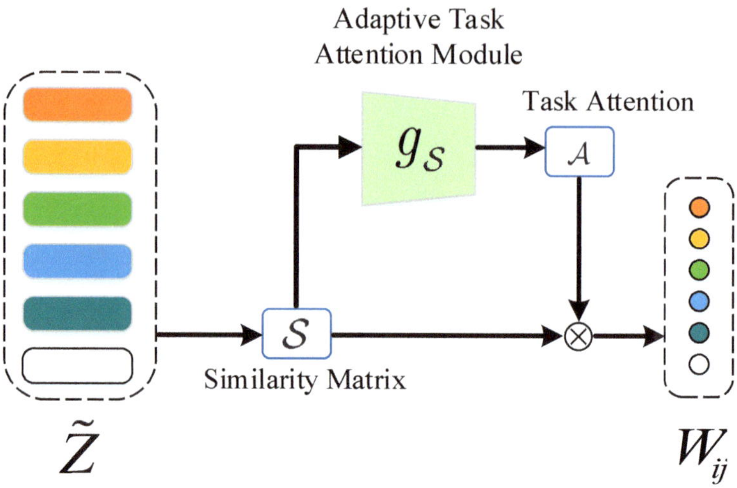

Figure 3. Schematic of the task-level relation module.

For node i, the corresponding attention value between the target embedding feature and features of all other samples in the task can be yielded by applying a common method in the attention mechanism. The corresponding attention value is acquired by the adaptive task attention module, which can be formulated as:

$$A(\widetilde{z}_i, \widetilde{z}_j) = \frac{exp(s_{ij})}{\sum_{k=1}^{N\times K+T} exp(s_{ik})} \tag{5}$$

where s_{ij} represents the similarity degree of node i to node j and $A \in \mathbb{R}^{(N\times K+T)\times(N\times K+T)}$ denotes the task-level similarity between nodes after comparing to all other nodes in the task. Therefore, the similarity degree between nodes is higher and A_{ij} is greater. The implementation of the similarity degree s_{ij} is obtained as follows:

$$s_{ij} = \frac{e(\widetilde{z}_i, \widetilde{z}_j{}^T)}{\sqrt{m}} \tag{6}$$

where the smoothed embedding feature $\widetilde{z}_j \in \mathbb{R}^{1\times m}$ of the target sample is reshaped to $\widetilde{z}_j{}^T \in \mathbb{R}^{m\times 1}$ adopting the matrix inversion operation and $e(\widetilde{z}_i, \widetilde{z}_j{}^T)$ is the pairwise distance operation (e.g., radial basis function). Then, $A_{i,j}$ is employed to incorporate task-level information, and the relation representation for the current task can be acquired, which can be represented as follows:

$$W_{ij} = s_{ij} \otimes A(\widetilde{z}_i, \widetilde{z}_j) \tag{7}$$

The relation representation W_{ij} between node i and node j can be weighted through Equation (7), which denotes a task-level relation of node i to j after comparing to all the other nodes. While the relation representations of task-irrelevant regions are restrained, the relation representations of task-relevant regions are strengthened. To construct the

k-nearest neighbor graph, the k-max values are reserved for each row of W. Then, the normalized graph Laplacian [61] is applied on W, that is,

$$L = O^{-\frac{1}{2}} W O^{-\frac{1}{2}} \tag{8}$$

where:

$$O_{ii} = \sum_j A_{ij} \tag{9}$$

To simulate few-shot scenarios, the episodic paradigm for meta-training was followed; that is, the graph was separately constructed for each episode in each task, as shown in Figure 2. In general, in the setting of the five-way one-shot scenario, $N = 5, K = 1, T = 75$, the shape of W is 80×80, which is quite efficient.

3.5. Label Matching

First, the prediction process of the labels of the query set Q is described. Assume $\mathcal{G}$ represents the set of matrices, where each matrix is composed of nonnegative values and has a shape of $(N \times K + T) \times N$. A label matrix $Y \in \mathcal{G}$ is defined with $Y_{ij} = 1$ if x_i belongs to the support set and marked as $y_i = j$, otherwise $Y_{ij} = 0$. Given the label matrix Y, label matching iteratively identifies the unseen labels of samples from $S \cup Q$ on the constructed graph adopting the label propagation formula, i.e.,

$$G_{t+1} = \gamma L G_t + (1 - \gamma) Y \tag{10}$$

where $G_t \in \mathcal{G}$ represents the inferred label matrix at the t-th round, L is the normalized graph weight, that is Equation (8), and $\gamma \in (0, 1)$ weighs the amount of information from neighbors and Y. When t is large enough, there is a closed-form solution for the modified sequence,

$$G^* = (I - \gamma L)^{-1} Y \tag{11}$$

where I denotes the identity matrix. Since this solution is directly applied to label prediction, the episodewise learning procedure becomes more effective.

Subsequently, the classification loss, between the predicted labels and the ground truth labels, is calculated. Specifically, to train all learnable parameters in an end-to-end manner, the cross-entropy loss was adopted in the experiments. Therein, ground truth labels derived from $S \cup Q$ and the predicted scores G^* were taken as corresponding inputs, where G^* was transformed into prediction scores by the softmax function:

$$P(\hat{y}_i = j \mid x_i) = \frac{exp(G^*_{ij})}{\sum_{j=1}^{N} exp(G^*_{ij})} \tag{12}$$

where $\hat{y}_i$ is the ultimate predicted label for the i-th sample from $S \cup Q$ and G^*_{ij} represents the j-th element of $\hat{y}_i$. Hence, the corresponding loss function can be written as:

$$L_{CE} = \sum_{i=1}^{N \times K + T} \sum_{c=1}^{N} -\mathbb{I}(y_i == c) ln(P(\hat{y}_i = c \mid x_i)) \tag{13}$$

where $\mathbb{I}(u)$ denotes the indicative function, that is $\mathbb{I}(u) = 0$ if u is false and $\mathbb{I}(u) = 1$ otherwise, and y_i denotes the ground truth label corresponding to the sample x_i. Note that, to imitate few-shot scenarios, all learnable parameters were iteratively updated by meta-learning in an end-to-end fashion.

4. Results and Discussion

In this section, our GES-Net approach is evaluated on three public remote sensing datasets for scene classification. The datasets used in the experiments are first described. The experimental settings and evaluation metrics are then illustrated in detail. Finally, the results of the experiments and analysis are presented.

4.1. Dataset Description

The validity of the proposed model was verified on three publicly available datasets, including UC Merced [62], WHU-RS19 [63], and NWPU-RESISC45 [64]. The dataset segmentation in [41] was followed, and the specific details are described below.

The UC Merced dataset contains 21 land use categories. There are 100 images in total per scene class, with a spatial resolution of 0.3 m. Each image consists of 256 × 256 pixels. In 2010, the dataset was released by the UC Merced Computer Vision Laboratory. In the dataset, 10 classes are split as the training set, 5 classes are deemed as the validation set, and 6 classes are selected as the test set.

The WHU-RS19 dataset is also a remote sensing scene dataset, which was released by Wuhan University. There are 19 diverse scene categories, and the image size of each category is at least 50 images. It includes a total of 1005 images with 600 × 600 pixels in size. For the dataset, 9 classes are split as the training set, 5 classes are divided as the validation set, and 5 classes are selected as the test set.

The NWPU-RESISC45 dataset is composed of 45 different classes in a total of 31,500 images. Each class consists of 700 images in the RGB color space. The pixel size per image is 256 × 256. In the dataset, the spatial resolution of most categories ranges from 0.2 m to 30 m for each pixel value. In the area of scene classification, NWPU-RESISC45 is the biggest in the matter of the total number of images and the quantity of the scene classes, which makes it abound with image variation, increasing classification difficulty. In our experiments, the dataset was split into 25, 10, and 10 categories as the training set, validation set, and test set, respectively.

4.2. Experimental Settings

For a fair comparison, four modules with the 3 × 3 convolution kernel and kernel size of 64 were selected for embedding learning. For each module, there was padding = 1 to retain abundant feature information, as well as the batch normalization layer [45], the ReLU nonlinearity activation function [59], as well as the 2 × 2 max pooling layer. The hyperparameter β of embedding smoothing, γ of label matching, and k of the k-nearest neighbor graph were set to 0.5, 0.2, and 20, respectively, as suggested by Zhou et al. [60]. The proposed model was implemented in the PyTorch [65] framework, and the GPU was Tesla V100. Before training, all images are resized to 84 × 84 pixels. During the training period, the initial learning rate was set to 1×10^{-3}, and it was halved every 100 epochs. ADAM [66] was selected as the optimizer for our model.

4.3. Evaluation Metrics

For the performance evaluation, the proposed GES-Net approach was compared with the state-of-the-art approaches by the overall accuracy (OA) and confusion matrix (CM). OA is explicitly defined as:

$$\mathrm{OA} = \frac{1}{U}\sum_{j=1}^{U}\frac{V_j}{T} \tag{14}$$

where U denotes the number of tasks, that is the number of episodes, and V_j is the number of samples that are rightly predicted in the j-th task. Specifically, the overall accuracy refers to the proportion of rightly classified query samples to the total quantity of query samples, that is the average of multiple episode results. In this work, all experimental results were averaged over 600 episodes by keeping the protocol in [57]. CM is an information table adopted to analyze the confusions and errors between various categories, where the item in the i-th row and j-th column refers to the proportion of test samples from the i-th category that are classified as the j-th category.

4.4. Time Complexity Analysis

In most scenarios, the overall accuracy is employed to evaluate the classification performance of deep neural networks. When practical issues need to be resolved, the time complexity required for the model should also be fully considered, which will significantly affect the cost of resolving the issues. For this purpose, floating-point operations (FLOPs) are typically adopted to estimate the time complexity of neural network models. For example, for the ResNet [67] series of networks, as the network depth continuously increases, the overall accuracy will also be improved, but it is noteworthy that the time and hardware costs will also rise significantly. Excessive parameters are liable to make model optimization more difficult, and it is hard to reach the anticipated classification effect.

When the classification effect of the model is relatively approximate, the model with a smaller time complexity is typically preferable. Therefore, the time complexity of various few-shot learning models on NWPU-RESISC45 in the setting of the five-way one-shot scenario is presented in Table 1. It can be seen that in the setting of the five-way one-shot scenario, the time complexity of our proposed approach was lower than that of RelationNet by 0.19×10^8, and the classification accuracy exceeded RelationNet by at least 4.4%. Compared to other approaches, the approach we propose also achieved better results in terms of time complexity and classification accuracy. This is because only the embedding learning module contains learnable parameters, and other modules, such as embedding smoothing, graph constructing, and label matching, are non-parametric in our proposed framework. Therefore, our proposed model can achieve significant improvement at a relatively small cost.

Table 1. Time complexity analysis of different approaches on NWPU-RESISC45 in the same few-shot setting (%).

Method	5-Way 1-Shot Accuracy	FLOPs
ProtoNet	40.33 ± 0.18	1.98×10^8
MatchingNet	37.61	1.99×10^8
MAML	48.40 ± 0.82	1.98×10^8
RelationNet	66.43 ± 0.73	2.16×10^8
GES-Net (ours)	70.83 ± 0.85	1.97×10^8

4.5. Embedding Space Analysis

As indicated in Figure 1, the proposed method was carefully designed to solve the scene classification problem of only a few labeled samples and learn an embedding space that enables maximization of the gap between features of diverse categories and minimization of the gap between features of the same category to enhance the discrimination between sample features. Consequently, when there are unseen scene categories with one or five samples, scene classification can be achieved through the above idea. To verify this viewpoint, the five-way one-shot scenario was selected as the testing environment, and the test set was input into GES-Net and TPN [56]. TPN is one of the most influential few-shot learning approaches based on graph representation, which adopts label propagation to calculate label information for query samples. To illustrate the experimental results more intuitively, three subsets with five classes and one sample per class were individually sampled from the three datasets. T-SNE [68] was then adopted to visualize the experimental results on three datasets, namely UC Merced, WHU-RS19, and NWPU-RESISC45, as indicated in Figures 4–6. It was noticed that the proposed GES-Net showed excellent results and was more discriminative than TPN in the embedding space, even though it had the phenomenon of separation between classes. Especially for the experiments on NWPU-RESISC45, the advantages of our GES-Net were more clearly shown;

that is, samples belonging to diverse categories were more scattered and samples from the same category were more aggregated in the embedding space, which made the decision boundary more prominent. By comparison, samples from different categories embedded by the TPN algorithm were randomly distributed across the entire embedding space, and the decision boundaries among the diverse categories were quite blurred.

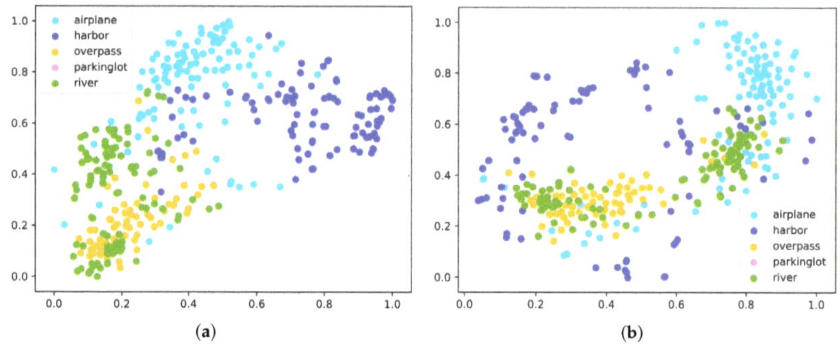

Figure 4. Embedding feature visualization of GES-Net (**a**) and TPN (**b**) on UC Merced in the 5-way 1-shot scenario.

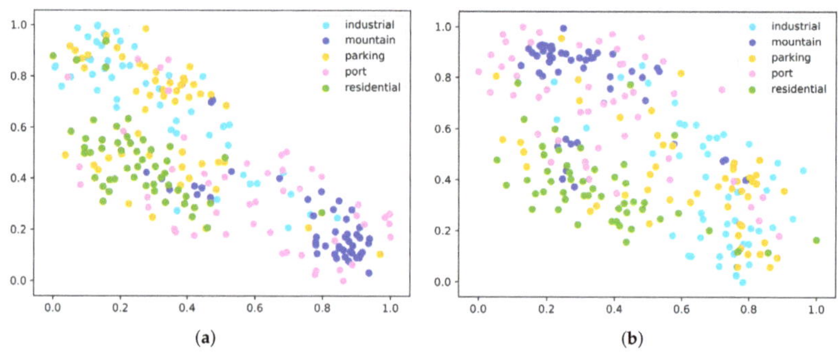

Figure 5. Embedding feature visualization of GES-Net (**a**) and TPN (**b**) on WHU-RS19 in the 5-way 1-shot scenario.

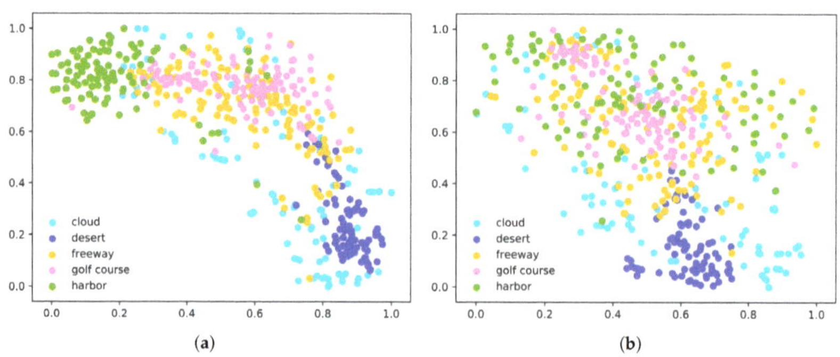

Figure 6. Embedding feature visualization of GES-Net (**a**) and TPN (**b**) on NWPU-RESISC45 in the 5-way 1-shot scenario.

In addition, to measure the similarity degree between samples in the embedding space, Equation (6) was adopted in our proposed method. To illustrate the advantages of the metric we adopted, two commonly used metrics, that is the Euclidean distance and cosine distance, were utilized for comparison. Comparative experiments on NWPU-RESISC45 in the setting of the five-way one-shot scenario were carried out, as shown in Figure 7. It was observed that our proposed approach had a faster convergence rate, while obtaining higher classification performance as the number of iterations increased. Hence, the distance metric that we adopted was significantly superior to the more commonly used Euclidean distance and cosine distance.

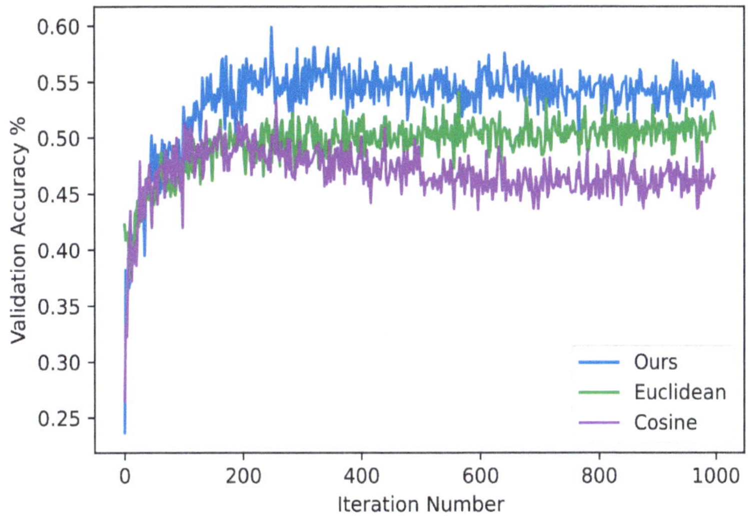

Figure 7. Validation accuracies with diverse distance metrics on the NWPU-RESISC45 dataset in the setting of the 5-way 1-shot scenario.

4.6. Ablation Study

In this section, embedding smoothing and task-level relation representation are explored. Table 2 tabulates the experimental results obtained by ablating the embedding smoothing and task-level relation representation on the NWPU-RESISC45 dataset, where the following scenarios were considered.

Table 2. Ablation study of embedding smoothing (ES) and task-level relation (TR). The comparison experiments were conducted on NWPU-RESISC45 (%).

Model	5-Way	
	1-Shot	5-Shot
Baseline	66.51 ± 0.87	78.50 ± 0.56
Baseline+ES	68.93 ± 0.91	81.73 ± 0.59
Baseline+TR	67.40 ± 0.85	80.93 ± 0.61
GES-Net (ours)	**70.83 ± 0.85**	**82.27 ± 0.55**

4.6.1. Baseline

Embedding smoothing was excluded from the proposed model, and the task-level relation representation was replaced with the sample-level relation representation, that is TPN, which was implemented through the radial basis function.

4.6.2. Baseline+ES

Embedding smoothing was included in the proposed model, while task-level relation representation was not adopted.

4.6.3. Baseline+TR

The task-level relation representation was adopted, while embedding smoothing was not included.

From the experimental results, it was observed that combining the two components, namely embedding smoothing and task-level relation representation, could make our model achieve a superior classification performance. In the setting of the five-way one-shot scenario, the overall accuracies of Baseline+ES and Baseline+TR were 68.93% and 67.40%, respectively, higher than the Baseline, while in the setting of the five-way one-shot scenario, Baseline+ES obtained 81.73% overall accuracy, which was 3.23% higher than the Baseline. Moreover, in two scenarios, Baseline+ES achieved a higher overall accuracy than Baseline+TS, which indicates that ES (i.e., embedding smoothing) plays a more significant role in our model.

To further illustrate the effect of the above two components on the performance of the proposed GES-Net, five-way one-shot classification experiments were conducted on the NWPU-RESISC45 dataset. Figure 8 presents the specific experimental results, including the training loss and validation accuracy of our approach and the Baseline, that is TPN. The quantity of iterations was changed from 0–1000 with a step size of one. As can be observed in Figure 8, in the training stage, the training loss generated by our approach was smaller than that of the Baseline, and the highest validation accuracy achieved by our approach was greater than that of the Baseline. Moreover, to intuitively illustrate the advantages of the proposed embedding smoothing (ES), two commonly used regularization methods were compared with it, including spatial dropout (SD) and manifold mixup (MM). T-SNE was employed to visualize the experimental results of various regularization approaches on NWPU-RESISC45 in the five-way one-shot scenario, as shown in Figure 9, where the five colors correspond to the five scene categories. It was observed that the manifold of Baseline+ES was more concentrated and compact than Baseline+SD and Baseline+MM, which may reduce the impact of class-irrelevant noises on embedding representations.

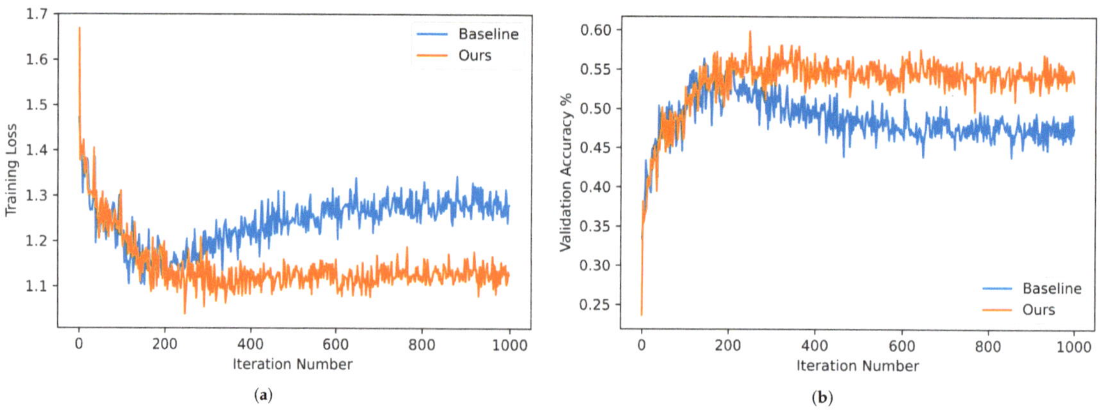

Figure 8. Training loss and validation accuracy on the NWPU-RESISC45 dataset in the setting of the 5-way 1-shot scenario. The training loss is shown on the left, and the validation accuracy is shown on the right. (**a**) Training loss; (**b**) validation accuracy.

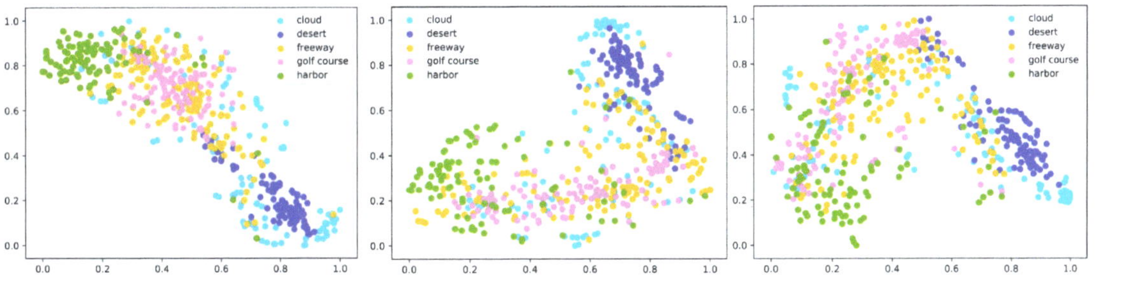

(a) Baseline + ES (b) Baseline + SD (c) Baseline + MM

Figure 9. Visualization of embedding features with different regularization approaches on the same dataset.

Additionally, the experimental results were also measured in the confusion matrix. Figure 10 shows the confusion matrices for the three public datasets in the settings of the five-way one-shot and five-way five-shot scenarios. By comparing the left and right confusion matrices, it was observed that as the quantity of labeled samples per class increased, the confusion errors decreased and the accuracy of each class increased. This shows that as the quantity of labeled samples per class increased, the performance of our model also increased accordingly.

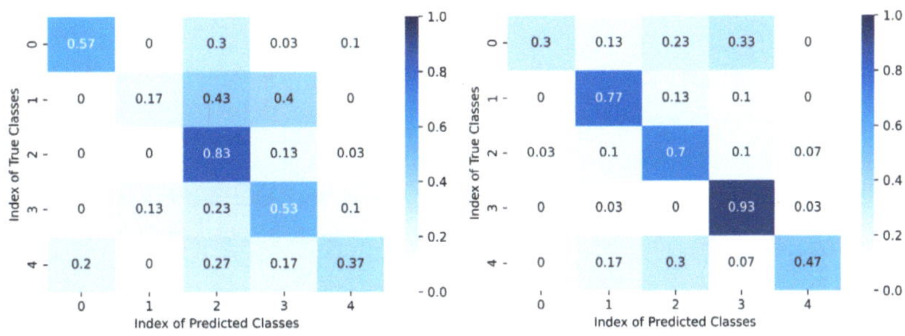

(a) (b) (c) (d)

Figure 10. *Cont.*

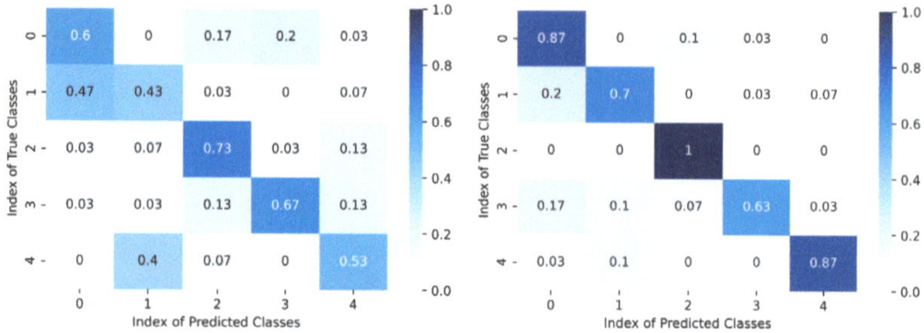

(e) (f)

Figure 10. Confusion matrices on UC Merced (**top**), WHU-RS19 (**middle**), and NWPU-RESISC45 (**bottom**). The experiment results with only one sample per class (one-shot) are shown on the left, and those with five samples per class (five-shot) are shown on the right. (**a**) One-shot on UC Merced; (**b**) five-shot on UC Merced; (**c**) one-shot on WHU-RS19; (**d**) five-shot on WHU-RS19; (**e**) one-shot on NWPU-RESISC45; (**f**) five-shot on NWPU-RESISC45.

4.7. Comparison with the State-of-the-Art Approaches

To verify the validity of the proposed scheme, our proposed approach was compared with several recent approaches, including TPN [56], ProtoNet [57], MatchingNet [32], MAML [33], Meta-SGD [69], LLSR [14], RelationNet [58], RS-MetaNet [40], and DLA-MatchNet [41]. The experimental results of our approach and the above-mentioned approaches are shown in Tables 3–5, which consists of five-way one-shot and five-way five-shot results on three public benchmark datasets: UC Merced, WHU-RS19, and NWPU-RESISC45. All experiment results, presented in Tables 3–5, were averaged on 600 randomly sampled episodes in the test set. This shows that our proposed approach exceeded other approaches by a significant margin and obtained state-of-the-art results on the three public benchmark datasets. Even though our proposed approach slightly surpassed RS-MetaNet in the five-way one-shot scenario on UC Merced, it exceeded RS-MetaNet by at least 5.58% in other scenarios. Overall, our proposed approach showed great superiority in multiple scenarios, e.g., 1-shot (1.65%) on UC Merced, 1-shot (7.57%) and 5-shot (1.67%) on WHU-RS19, and 1-shot (2.03%) on NWPU-RESISC45.

Table 3. Few-shot overall accuracies and standard deviations (%) on UC Merced.

Model	**5-Way**	
	1-Shot	**5-Shot**
TPN	53.36 ± 0.77	68.23 ± 0.52
ProtoNet	52.27 ± 0.20	69.86 ± 0.15
MatchingNet	34.70	52.71
MAML	48.86 ± 0.74	60.78 ± 0.62
Meta-SGD	50.52 ± 2.61	60.82 ± 2.00
LLSR	39.47	57.40
RelationNet	48.08 ± 1.67	61.88 ± 0.50
RS-MetaNet	57.23 ± 0.56	76.08 ± 0.28
DLA-MatchNet	53.76 ± 0.60	63.01 ± 0.51
GES-Net (ours)	**58.88 ± 0.81**	**81.66 ± 0.50**

Table 4. Few-shot overall accuracies and standard deviations (%) on WHU-RS19.

Model	5-Way	
	1-Shot	5-Shot
TPN	59.28 ± 0.72	71.20 ± 0.55
ProtoNet	58.01 ± 0.16	80.70 ± 0.11
MatchingNet	50.13	54.10
MAML	49.13 ± 0.65	62.49 ± 0.51
Meta-SGD	51.54 ± 2.31	61.74 ± 2.02
LLSR	57.10	70.65
RelationNet	60.92 ± 1.86	79.75 ± 1.19
DLA-MatchNet	68.27 ± 1.83	79.89 ± 0.33
GES-Net (ours)	**75.84 ± 0.78**	**82.37 ± 0.38**

Table 5. Few-shot overall accuracies and standard deviations (%) on NWPU-RESISC45.

Model	5-Way	
	1-Shot	5-Shot
TPN	66.51 ± 0.87	78.50 ± 0.56
ProtoNet	40.33 ± 0.18	63.82 ± 0.56
MatchingNet	37.61	47.10
MAML	48.40 ± 0.82	62.90 ± 0.69
Meta-SGD	60.63 ± 0.90	75.75 ± 0.65
LLSR	51.43	72.90
RelationNet	66.43 ± 0.73	78.35 ± 0.51
RS-MetaNet	52.78 ± 0.09	71.49 ± 0.81
DLA-MatchNet	68.80 ± 0.70	81.63 ± 0.46
GES-Net (ours)	**70.83 ± 0.85**	**82.27 ± 0.55**

5. Conclusions

A novel graph neural network framework, referred to as GES-Net, was presented for the few-shot scene classification of remote sensing images in this paper. GES-Net regularizes the embedding space through a non-parametric embedding smoothing strategy. Embedding smoothing can constrain the embedding features, which enables the embedding learning module to extract more discriminative and robust scene embedding features to deal with complex and realistic scenarios. Moreover, GES-Net adopts an attention mechanism to capture the task-level relation representation. Considering the embedding features of all samples in a task, our method can obtain the task-level relation representation between nodes to construct the graph. Comparative experiments on three remote sensing scene datasets remarkably demonstrated the validity of GES-Net and outperformed state-of-the-art approaches by a considerable margin. Additionally, several ablation studies were conducted to analyze the influence of embedding smoothing and the task-level relation representation. However, there is still a great deal of unlabeled remote sensing data that are not utilized in the real world. Therefore, in future work, a large amount of unlabeled data will be introduced into the training procedure of the model to further improve the performance of scene classification in the settings of finite labeled samples.

Author Contributions: Conceptualization, Z.Y., W.H., C.T. and X.L.; methodology, W.H., A.Y. and C.T.; software, Z.Y. and A.Y.; validation, Z.Y., W.H. and C.T.; formal analysis, C.T.; resources, Z.Y. and X.L.; investigation, X.L.; data curation, C.T. and X.L.; writing—original draft preparation, Z.Y. and W.H.; writing—review and editing, W.H. and A.Y.; supervision, Z.Y. and C.T.; visualization, Z.Y. and W.H.; funding acquisition, X.L.; project administration, W.H. and C.T. All authors have read and agreed to the published version of the manuscript.

Funding: This work was supported by the National Natural Science Foundation of China (Grant No. 41871226).

Institutional Review Board Statement: Not applicable.

Informed Consent Statement: Not applicable.

Data Availability Statement: The data provided in this work are available from the corresponding author.

Conflicts of Interest: The authors declare no conflict of interest.

References

1. Zhao, B.; Zhong, Y.; Xia, G.S.; Zhang, L. Dirichlet-derived multiple topic scene classification model for high spatial resolution remote sensing imagery. *IEEE Trans. Geosci. Remote Sens.* **2015**, *54*, 2108–2123. [CrossRef]
2. Yao, X.; Han, J.; Cheng, G.; Qian, X.; Guo, L. Semantic annotation of high-resolution satellite images via weakly supervised learning. *IEEE Trans. Geosci. Remote Sens.* **2016**, *54*, 3660–3671. [CrossRef]
3. Huang, X.; Wang, Y. Investigating the effects of 3D urban morphology on the surface urban heat island effect in urban functional zones by using high-resolution remote sensing data: A case study of Wuhan, Central China. *ISPRS J. Photogramm. Remote Sens.* **2019**, *152*, 119–131. [CrossRef]
4. Zhao, W.; Du, S. Learning multiscale and deep representations for classifying remotely sensed imagery. *ISPRS J. Photogramm. Remote Sens.* **2016**, *113*, 155–165. [CrossRef]
5. Huang, X.; Han, X.; Ma, S.; Lin, T.; Gong, J. Monitoring ecosystem service change in the City of Shenzhen by the use of high-resolution remotely sensed imagery and deep learning. *Land Degrad. Dev.* **2019**, *30*, 1490–1501. [CrossRef]
6. Zhu, Q.; Zhong, Y.; Zhang, L.; Li, D. Adaptive deep sparse semantic modeling framework for high spatial resolution image scene classification. *IEEE Trans. Geosci. Remote Sens.* **2018**, *56*, 6180–6195. [CrossRef]
7. Wu, Z.; Li, Y.; Plaza, A.; Li, J.; Xiao, F.; Wei, Z. Parallel and distributed dimensionality reduction of hyperspectral data on cloud computing architectures. *IEEE J. Sel. Top. Appl. Earth Obs. Remote Sens.* **2016**, *9*, 2270–2278. [CrossRef]
8. Chen, J.; Wang, C.; Ma, Z.; Chen, J.; He, D.; Ackland, S. Remote sensing scene classification based on convolutional neural networks pre-trained using attention-guided sparse filters. *Remote Sens.* **2018**, *10*, 290. [CrossRef]
9. Zhu, Q.; Zhong, Y.; Wu, S.; Zhang, L.; Li, D. Scene classification based on the sparse homogeneous–heterogeneous topic feature model. *IEEE Trans. Geosci. Remote Sens.* **2018**, *56*, 2689–2703. [CrossRef]
10. Liu, Y.; Zhong, Y.; Qin, Q. Scene classification based on multiscale convolutional neural network. *IEEE Trans. Geosci. Remote Sens.* **2018**, *56*, 7109–7121. [CrossRef]
11. Zhao, B.; Huang, B.; Zhong, Y. Transfer learning with fully pretrained deep convolution networks for land-use classification. *IEEE Geosci. Remote Sens. Lett.* **2017**, *14*, 1436–1440. [CrossRef]
12. Gidaris, S.; Bursuc, A.; Komodakis, N.; Pérez, P.; Cord, M. Boosting few-shot visual learning with self-supervision. In Proceedings of the 2019 IEEE/CVF International Conference on Computer Vision, ICCV 2019, Seoul, Korea, 27 October–2 November 2019; pp. 8059–8068.
13. Chu, W.H.; Li, Y.J.; Chang, J.C.; Wang, Y.C.F. Spot and learn: A maximum-entropy patch sampler for few-shot image classification. In Proceedings of the IEEE/CVF Conference on Computer Vision and Pattern Recognition, Long Beach, CA, USA, 16–20 June 2019; pp. 6251–6260.
14. Zhai, M.; Liu, H.; Sun, F. Lifelong learning for scene recognition in remote sensing images. *IEEE Geosci. Remote Sens. Lett.* **2019**, *16*, 1472–1476. [CrossRef]
15. Garcia, V.; Bruna, J. Few-shot learning with graph neural networks. In Proceedings of the 6th International Conference on Learning Representations, ICLR 2018, Vancouver, BC, Canada, 30 April–3 May 2018; p. 1–13.
16. Bartlett, P.; Shawe-Taylor, J. Generalization performance of support vector machines and other pattern classifiers. In *Advances in Kernel Methods: Support Vector Learning*; MIT Press: Cambridge, MA, USA, 1999; pp. 43–54.
17. Lee, W.S.; Bartlett, P.L.; Williamson, R.C. Lower bounds on the VC dimension of smoothly parameterized function classes. *Neural Comput.* **1995**, *7*, 1040–1053. [CrossRef]
18. Verma, V.; Lamb, A.; Beckham, C.; Najafi, A.; Mitliagkas, I.; Lopez-Paz, D.; Bengio, Y. Manifold mixup: Better representations by interpolating hidden states. In Proceedings of the 36th International Conference on Machine Learning (PMLR), Long Beach, CA, USA, 9–15 June 2019; pp. 6438–6447.
19. Wang, S.; Wang, X.; Zhang, L.; Zhong, Y. Auto-AD: Autonomous Hyperspectral Anomaly Detection Network Based on Fully Convolutional Autoencoder. *IEEE Trans. Geosci. Remote Sens.* **2021**, *60*, 1–4. [CrossRef]
20. Zhan, T.; Song, B.; Xu, Y.; Wan, M.; Wang, X.; Yang, G.; Wu, Z. SSCNN-S: A Spectral-Spatial Convolution Neural Network with Siamese Architecture for Change Detection. *Remote Sens.* **2021**, *13*, 895. [CrossRef]
21. Wang, Y.; Hou, J.; Hou, X.; Chau, L.P. A Self-Training Approach for Point-Supervised Object Detection and Counting in Crowds. *IEEE Trans. Image Process.* **2021**, *30*, 2876–2887. [CrossRef] [PubMed]
22. Zhu, S.; Du, B.; Zhang, L.; Li, X. Attention-Based Multiscale Residual Adaptation Network for Cross-Scene Classification. *IEEE Trans. Geosci. Remote Sens.* **2021**, *60*. [CrossRef]

23. Chen, J.; Qiu, X.; Ding, C.; Wu, Y. CVCMFF Net: Complex-Valued Convolutional and Multifeature Fusion Network for Building Semantic Segmentation of InSAR Images. *IEEE Trans. Geosci. Remote Sens.* **2021**, *60*. [CrossRef]
24. Xu, C.; Zhu, G.; Shu, J. A Lightweight and Robust Lie Group-Convolutional Neural Networks Joint Representation for Remote Sensing Scene Classification. *IEEE Trans. Geosci. Remote Sens.* **2021**, *60*. [CrossRef]
25. Wang, X.; Wang, S.; Ning, C.; Zhou, H. Enhanced Feature Pyramid Network With Deep Semantic Embedding for Remote Sensing Scene Classification. *IEEE Trans. Geosci. Remote Sens.* **2021**, *59*, 7918–7932. [CrossRef]
26. Penatti, O.A.; Nogueira, K.; Dos Santos, J.A. Do deep features generalize from everyday objects to remote sensing and aerial scenes domains? In Proceedings of the 2015 IEEE Conference on Computer Vision and Pattern Recognition Workshops (CVPRW), Boston, MA, USA, 7–12 June 2015; pp. 44–51.
27. Lu, X.; Zheng, X.; Yuan, Y. Remote sensing scene classification by unsupervised representation learning. *IEEE Trans. Geosci. Remote Sens.* **2017**, *55*, 5148–5157. [CrossRef]
28. Lu, X.; Gong, T.; Zheng, X. Multisource compensation network for remote sensing cross-domain scene classification. *IEEE Trans. Geosci. Remote Sens.* **2019**, *58*, 2504–2515. [CrossRef]
29. Lu, X.; Sun, H.; Zheng, X. A feature aggregation convolutional neural network for remote sensing scene classification. *IEEE Trans. Geosci. Remote Sens.* **2019**, *57*, 7894–7906. [CrossRef]
30. Nogueira, K.; Penatti, O.A.; Dos Santos, J.A. Towards better exploiting convolutional neural networks for remote sensing scene classification. *Pattern Recognit.* **2017**, *61*, 539–556. [CrossRef]
31. Cheng, G.; Yang, C.; Yao, X.; Guo, L.; Han, J. When deep learning meets metric learning: Remote sensing image scene classification via learning discriminative CNNs. *IEEE Trans. Geosci. Remote Sens.* **2018**, *56*, 2811–2821. [CrossRef]
32. Vinyals, O.; Blundell, C.; Lillicrap, T.; Kavukcuoglu, K.; Wierstra, D.; Matching Networks for One Shot Learning. *Proc. Neural Inf. Process. Syst.* **2016**, *29*, 3630–3638.
33. Finn, C.; Abbeel, P.; Levine, S. Model-agnostic meta-learning for fast adaptation of deep networks. In Proceedings of the International Conference on Machine Learning PMLR, Sydney, Australia, 6–11 August 2017; pp. 1126–1135.
34. Santoro, A.; Bartunov, S.; Botvinick, M.; Wierstra, D.; Lillicrap, T. Meta-learning with memory-augmented neural networks. In Proceedings of the International Conference on Machine Learning PMLR, New York, NY, USA, 19–24 June 2016; pp. 1842–1850.
35. Tokmakov, P.; Wang, Y.X.; Hebert, M. Learning compositional representations for few-shot recognition. In Proceedings of the IEEE/CVF International Conference on Computer Vision (ICCV), Seoul, Korea, 27 October–2 November 2019; pp. 6372–6381.
36. Li, H.; Dong, W.; Mei, X.; Ma, C.; Huang, F.; Hu, B.G. Lgm-net: Learning to generate matching networks for few-shot learning. In Proceedings of the International Conference on Machine Learning PMLR, Long Beach, CA, USA, 9–15 June 2019; pp. 3825–3834.
37. Liu, B.; Yu, X.; Yu, A.; Zhang, P.; Wan, G.; Wang, R. Deep few-shot learning for hyperspectral image classification. *IEEE Trans. Geosci. Remote Sens.* **2018**, *57*, 2290–2304. [CrossRef]
38. Gao, K.; Liu, B.; Yu, X.; Qin, J.; Zhang, P.; Tan, X. Deep relation network for hyperspectral image few-shot classification. *Remote Sens.* **2020**, *12*, 923. [CrossRef]
39. Rostami, M.; Kolouri, S.; Eaton, E.; Kim, K. Deep transfer learning for few-shot sar image classification. *Remote Sens.* **2019**, *11*, 1374. [CrossRef]
40. Li, H.; Cui, Z.; Zhu, Z.; Chen, L.; Zhu, J.; Huang, H.; Tao, C. RS-MetaNet: Deep Metametric Learning for Few-Shot Remote Sensing Scene Classification. *IEEE Trans. Geosci. Remote Sens.* **2020**, *59*, 6983–6994. [CrossRef]
41. Li, L.; Han, J.; Yao, X.; Cheng, G.; Guo, L. DLA-MatchNet for Few-Shot Remote Sensing Image Scene Classification. *IEEE Trans. Geosci. Remote Sens.* **2020**, *59*, 7844–7853. [CrossRef]
42. Guo, Y.; Codella, N.C.; Karlinsky, L.; Codella, J.V.; Smith, J.R.; Saenko, K.; Rosing, T.; Feris, R. A broader study of cross-domain few-shot learning. In *European Conference on Computer Vision*; Vedaldi, A., Bischof, H., Brox, T., Frahm, J.M., Eds.; Springer: Cham, Switzerland, 2020; pp. 124–141.
43. Gong, T.; Zheng, X.; Lu, X. Cross-Domain Scene Classification by Integrating Multiple Incomplete Sources. *IEEE Trans. Geosci. Remote Sens.* **2020**, *59*, 10035–10046. [CrossRef]
44. Yosinski, J.; Clune, J.; Bengio, Y.; Lipson, H. How transferable are features in deep neural networks? *Proc. Adv. Neural Inf. Process. Syst.* **2014**, *27*, 3320–3328.
45. Ioffe, S.; Szegedy, C. Batch normalization: Accelerating deep network training by reducing internal covariate shift. In Proceedings of the International Conference on Machine Learning PMLR, Lille, France, 6–11 July 2015; pp. 448–456.
46. Srivastava, N.; Hinton, G.; Krizhevsky, A.; Sutskever, I.; Salakhutdinov, R. Dropout: A simple way to prevent neural networks from overfitting. *J. Mach. Learn. Res.* **2014**, *15*, 1929–1958.
47. Tompson, J.; Goroshin, R.; Jain, A.; LeCun, Y.; Bregler, C. Efficient object localization using convolutional networks. In Proceedings of the IEEE Conference on Computer Vision and Pattern Recognition, Boston, MA, USA, 7–12 June 2015; pp. 648–656.
48. Rodríguez, P.; Gonzalez, J.; Cucurull, G.; Gonfaus, J.M.; Roca, X. Regularizing cnns with locally constrained decorrelations. *arXiv* **2016**, arXiv:1611.01967.
49. Salimans, T.; Kingma, D.P. Weight normalization: A simple reparameterization to accelerate training of deep neural networks. *Proc. Adv. Neural Inf. Process. Syst.* **2016**, *29*, 901–909.
50. Zhang, H.; Cisse, M.; Dauphin, Y.N.; Lopez-Paz, D. mixup: Beyond empirical risk minimization. *arXiv* **2017**, arXiv:1710.09412.
51. Belkin, M.; Niyogi, P.; Sindhwani, V. Manifold regularization: A geometric framework for learning from labeled and unlabeled examples. *J. Mach. Learn. Res.* **2006**, *7*, 2399–2434.

52. Cho, K.; Zhao, J. Retrieval-augmented convolutional neural networks against adversarial examples. In Proceedings of the IEEE/CVF Conference on Computer Vision and Pattern Recognition (CVPR), Long Beach, CA, USA, 16–20 June 2019; pp. 11563–11571.
53. Vapnik, V.N. An overview of statistical learning theory. *IEEE Trans. Neural Netw.* **1999**, *10*, 988–999. [CrossRef] [PubMed]
54. Iscen, A.; Tolias, G.; Avrithis, Y.; Chum, O. Label propagation for deep semi-supervised learning. In Proceedings of the IEEE/CVF Conference on Computer Vision and Pattern Recognition (CVPR), Long Beach, CA, USA, 16–20 June 2019; pp. 5070–5079.
55. Liu, B.; Wu, Z.; Hu, H.; Lin, S. Deep metric transfer for label propagation with limited annotated data. In Proceedings of the 2019 IEEE/CVF International Conference on Computer Vision Workshops, ICCV Workshops 2019, Seoul, Korea, 27–28 October 2019; pp. 1–10.
56. Liu, Y.; Lee, J.; Park, M.; Kim, S.; Yang, E.; Hwang, S.J.; Yang, Y. Learning to propagate labels: Transductive propagation network for few-shot learning. *arXiv* **2018**, arXiv:1805.10002.
57. Snell, J.; Swersky, K.; Zemel, R. Prototypical networks for few-shot learning. In Proceedings of the Advances in Neural Information Processing Systems, Long Beach, CA, USA, 4–9 December 2017; pp. 4080–4090.
58. Sung, F.; Yang, Y.; Zhang, L.; Xiang, T.; Torr, P.H.; Hospedales, T.M. Learning to compare: Relation network for few-shot learning. In Proceedings of the IEEE/CVF Conference on Computer Vision and Pattern Recognition, Salt Lake City, UT, USA, 18–22 June 2018; pp. 1199–1208.
59. Glorot, X.; Bordes, A.; Bengio, Y. Deep sparse rectifier neural networks. In Proceedings of the International Conference Artificial Intelligence and Statistics, JMLR Workshop and Conference Proceedings, Fort Lauderdale, FL, USA, 11–13 April 2011; pp. 315–323.
60. Zhou, D.; Bousquet, O.; Lal, T.N.; Weston, J.; Schölkopf, B. Learning with local and global consistency. In Proceedings of the Advances in Neural Information Processing Systems, Vancouver, BC, Canada, 13–18 December 2004; pp. 321–328.
61. Chung, F.R.; Graham, F.C. *Spectral Graph Theory*; Number 92; AMS, American Mathematical Society: Providence, RI, USA, 1997.
62. Yang, Y.; Newsam, S. Bag-of-visual-words and spatial extensions for land-use classification. In Proceedings of the 18th ACM SIGSPATIAL International Conference on Advances in Geographic Information Systems (ACM SIGSPATIAL GIS 2010), San Jose, CA, USA, 2–5 November 2010; pp. 270–279.
63. Sheng, G.; Yang, W.; Xu, T.; Sun, H. High-resolution satellite scene classification using a sparse coding based multiple feature combination. *Int. J. Remote Sens.* **2012**, *33*, 2395–2412. [CrossRef]
64. Cheng, G.; Han, J.; Lu, X. Remote sensing image scene classification: Benchmark and state of the art. *Proc. IEEE* **2017**, *105*, 1865–1883. [CrossRef]
65. Paszke, A.; Gross, S.; Chintala, S.; Chanan, G.; Yang, E.; DeVito, Z.; Lin, Z.; Desmaison, A.; Antiga, L.; Lerer, A. Automatic differentiation in pytorch. In Proceedings of the Workshop Advances Neural Information Processing Systems, Long Beach, CA, USA, 4–9 December 2017; pp. 1–4.
66. Kingma, D.P.; Ba, J. Adam: A Method for Stochastic Optimization. In Proceedings of the 3rd International Conference for Learning Representations, San Diego, CA, USA, 7–9 May 2015.
67. He, K.; Zhang, X.; Ren, S.; Sun, J. Deep residual learning for image recognition. In Proceedings of the IEEE/CVF Conference on Computer Vision and Pattern Recognition (CVPR), Honolulu, HI, USA, 21–26 July 2016; pp. 770–778.
68. Van der Maaten, L.; Hinton, G. Visualizing data using t-SNE. *J. Mach. Learn. Res.* **2008**, *9*, 2579–2605.
69. Li, Z.; Zhou, F.; Chen, F.; Li, H. Meta-sgd: Learning to learn quickly for few-shot learning. *arXiv* **2017**, arXiv:1707.09835.

sustainability

MDPI

Article

A Deep Learning Framework for Multimodal Course Recommendation Based on LSTM+Attention

Xinwei Ren [1,2], Wei Yang [1,*], Xianliang Jiang [2], Guang Jin [2] and Yan Yu [2]

1 College of Science & Technology, Ningbo University, Ningbo 315300, China
2 College of Electrical Engineering and Computer Sciences, Ningbo University, Ningbo 315211, China; akon_ren@163.com (X.R.); jiangxianliang@nbu.edu.cn (X.J.); jinguang@nbu.edu.cn (G.J.); 1911082064@nbu.edu.cn (Y.Y.)
* Correspondence: yangwei1@nbu.edu.cn

Abstract: With the impact of COVID-19 on education, online education is booming, enabling learners to access various courses. However, due to the overload of courses and redundant information, it is challenging for users to quickly locate courses they are interested in when faced with a massive number of courses. To solve this problem, we propose a deep course recommendation model with multimodal feature extraction based on the Long- and Short-Term Memory network (LSTM) and Attention mechanism. The model uses course video, audio, and title and introduction for multimodal fusion. To build a complete learner portrait, user demographic information, explicit and implicit feedback data were added. We conducted extensive and exhaustive experiments based on real datasets, and the results show that the AUC obtained a score of 79.89%, which is significantly higher than similar algorithms and can provide users with more accurate recommendation results in course recommendation scenarios.

Keywords: multimodal; deep learning; course recommendation; multimodal recommendation

Citation: Ren, X.; Yang, W.; Jiang, X.; Jin, G.; Yu, Y. A Deep Learning Framework for Multimodal Course Recommendation Based on LSTM+Attention. *Sustainability* **2022**, *14*, 2907. https://doi.org/10.3390/su14052907

Academic Editor: Miltiadis D. Lytras and Andreea Claudia Serban

Received: 13 January 2022
Accepted: 18 February 2022
Published: 2 March 2022

1. Introduction

With the rapid development of the Internet and online education, traditional education methods are constantly being transformed, and students are no longer satisfied with the original rigid curriculum and tend to choose courses they are interested in through online education platforms. Platforms can provide high-quality, targeted teaching courses, a complete curriculum, and supporting assignments, while having the advantages of free courses, rich content, and flexible choices. However, the exponential growth of platforms and courses leads to "information overload", and users are easily confused and have difficulty in choosing courses when facing many classes. Therefore, it is crucial to help users quickly choose the right course that interests them. When users use the online education platform to learn new courses, although the courses are classified, in the face of many courses, quickly locating exciting and personalized valuable courses is of great significance. To recommend courses to users, a recommendation model based on course content was proposed [1]. At present, to recommend appropriate courses, it is necessary to shorten the selection distance between users' learning ability and a large number of courses.

Currently, there are many problems with course recommendations. Only considering the user's nearest neighbor recommendation method and integrating basic personal information and social preferences for the recommendation, the user's implicit feedback data is ignored [2]. However, the multimodal graph convolution model is used to generate the specific modal representation of users and videos to capture better the implicit feedback data of users [3]. In addition, Gong et al. [4] used the knowledge concept of graph neural networks for course recommendation, considering the fusion of heterogeneous information to grasp the differences in the interests of different students. However, the implicit and explicit feedback information of learners should be fully regarded in the extraction of course features; moreover, course-related information can further enhance the recommendation efficiency. At the same time, because the traditional course recommendation system and algorithm mainly focus on the experiment and test of single-mode information and lack effective utilization and research of multi-mode information, the system cannot obtain the deep-seated preferences of users' construct and complete the user portrait. Therefore, this paper proposes to use LSTM and Attention deep learning to process multimodal data for efficient course recommendations.

When users use the platform to learn a new course, the first search for the course name and read the course description to understand the course details. Users also consider the total review scores and the number of reviews about the course on the platform. Additionally, demographic information [5,6] represents the user's characteristic data, so that course text and numerical information can be used as recommendation data. When viewing a course, users mark the course according to favorites, collections, comments, and so forth, and such behaviors can also be used as explicit feedback data; users' browsing behaviors just for the course can be used as implicit feedback data. In addition, image and acoustic information can be extracted from the course videos by using deep learning models. Therefore, digital, textual, video, and audio information of the course can be used when building the course recommendation model, and users' explicit and implicit feedback data can be used as inputs. Short video recommendations have developed rapidly in the industry and academia in recent years. Videos contain rich modal features, including video, audio, text, tags, and other multimodal data. Through these multimodal features, users' deep interests can be mined. Therefore, it is essential to propose a multimodal in-depth learning recommendation system, which can be applied to the online education platform and improve the accuracy of personalized course recommendations.

The main contributions of this paper are as follows: first, multimodal data were used as model input, mainly including course text, video, and audio information; meanwhile, user demographic information, explicit and implicit feedback are also considered. Second, this paper designs a new multimodal deep learning course recommendation model for extracting the modal features of learners and courses; the model improves and optimizes the traditional multimodal model to enhance its effectiveness in the field of course recommendation. The practical significance of this research is to improve recommendation accuracy and increase the users' click-through rate of courses by merging various modal information.

The rest of this paper is organized as follows. Section 2 is about related work, summarizing the current literature related to online course platforms and multimodal course recommendation research. Section 3, the algorithm framework, describes the extraction of course features and the overall framework of the algorithm. The experimental section in Section 4 compares the impact on course recommendation results by different modal course information and learners. Finally, the conclusion and future work section gives possible research directions in multimodal course recommendations.

2. Related Works

2.1. Traditional Course Recommendation

Recommendation systems have been widely used in music, movies, videos, and other fields in recent years. Traditional course recommendations mainly include content-based recommendations, collaborative filtering recommendations, and hybrid recommendations combining both advantages. Content-based recommendation mainly considers the demographic information of learners, such as gender, age, major, and the course text description, overall rating, and number of reviews. Khribi et al. [7] proposed an online learning platform in which the offline module is loaded to pre-process data to build a model with a learner–content relationship, and the online module dynamically identifies learner needs and goals and predicts the recommendation list. However, the cold start of the system is not considered [8]. For the cold start problem, Sengottuvelan et al. [9] proposed to use multiple attributes of learners to model their preferences, use data mining techniques to model learners' history records, and finally sort them according to the similarity threshold. The K-means clustering algorithm effectively reduces data sparsity and cold start problems and increases the diversity of ecological annotation lists. In addition, demographic information, learning emotion, motivation, and learners' style can also be studied as modal information (Fu et al. [10]). The advantage of content-based recommendation is understanding user preferences fully and not considering data sparsity. At present, the main advantages of content-based course recommendation methods are: do not consider the sparse data, and the recommended content depends on user preferences. However, its main disadvantage is that it requires the feature content to have a good structure and only considers the user's preferences, ignoring the situation of other users.

Collaborative filtering algorithm recommendations can be divided into user-based, and item-based depending on the object. Firstly, based on the user's collaborative filtering algorithm, the user behavior logs are collected for analysis, and the user interest model vector is constructed; secondly, a label is created for each resource in the system, and the user score of the resource is collected in the process of user use; finally, according to the user interest model and resource characteristics, combined with the personalized recommendation algorithm, the resources that meet the needs of users are recommended to the target users [11]. In order to effectively alleviate the problems of cold start and sparse data, a nearest-neighbor collaborative filtering algorithm based on weight optimization according to user history score is proposed [12]. In item-based collaborative filtering, the similarity matrix among courses is first calculated, and then the similar courses are further ranked and recommended based on the user's positive feedback history of similar courses. Pang et al. [13] proposed a multilayer Bucketing Model (MLBR) for MOOC course recommendation, which first transforms the learner vector into the same dimension and disperses it into buckets containing similar learners that have more common courses. Collaborative filtering can fully consider similar learner or course attributes without concerning about the content attributes of the courses themselves compared to content-based recommendation algorithms. In short, compared with the content-based recommendation algorithm, the advantage of collaborative filtering is that it can fully consider the attributes of similar learners or courses without considering the content attributes of the course itself. However, the actual situation is that users have little evaluation of the course, which leads to the problem of sparse data, and there is no user using data in the initial stage of the platform. Such problems will also be encountered after the new course is uploaded. At the same time, traditional collaborative filtering is effective in dealing with small data sets, but in the face of massive data sets, the accuracy of the recommendation system will decline.

However, all the above recommendation algorithms have corresponding disadvantages, such as cold start problems in content-based recommendations and sparse data and cold start for new users in collaborative filtering recommendations. Therefore, deployments often combine the advantages of different algorithms and models for hybrid recommendations. For example, Burke et al. [14] classified hybrid recommendation combination strategies into weighting, switching, partitioning, hierarchy, waterfall, feature blending, and feature enhancement, and introduced hybrid recommendation systems based on knowledge and collaborative filtering. Jannach et al. [15] classified the hybrid recommendation approach into holistic, parallel, and pipelined, where firstly, artificial neural networks are used to classify learners and users can get course recommendations based on learners' opinions; then when relevant interest groups are established, data mining techniques can be used to elicit the best learning paths. To sum up, the current traditional recommendation algorithms mostly use the learners' preferences and the unimodal information of demographic information in the representation of the curriculum model, ignoring the text information and video information of the curriculum itself. In hybrid recommendations, collaborative filtering and content-based recommendation model fusion are mostly used, but the audio mode of the course and the explicit feedback and implicit feedback of users are not considered.

2.2. Course Recommendation in the Field of Deep Learning

With the application of deep learning in the field of recommendation, it is gradually applied to the field of course recommendation. The deep course recommendation system can be divided into personalized learning recommendations based on convolutional neural networks and recurrent neural networks. Among them is the personalized learning recommendation based on the convolutional neural network: firstly, the learning behavior and learning history of learners are represented as feature vectors; then the correlation is estimated based on the difference between the estimated value and the actual value by using the attention mechanism; finally, the course is recommended to learners by training the recommendation model.

Personalized learning recommendation based on recurrent neural networks: Zhou et al. [16] clustered learners and used the LSTM model to predict learning paths and achievement, and finally recommended personalized and complete learning paths for learners. To extract the emotional factors expressed in the text, Ange et al. used the user-sensitive deep multimodal structure and extracted the rich potential data representation of users, which improved the effect of text classification [17]. Wang et al. [18] proposed to extract features by using learners' behavior and history, combining attention mechanisms and the difference between predicted and actual values of neural networks to improve recommendation performance. At the same time, to improve the attention mechanism, Zhu et al. introduced the double-layer attention mechanism into the summary of the parallel mental network recommendation model through data preprocessing and standardization, which effectively improved the ability of the model to mine the characteristics of users and courses [19]. Liu et al. [20] predicted the list of courses that students are good and bad at in the next semester by a hybrid model of deep learning and collaborative filtering. On the other hand, Ni et al. [21] first constructed a graph convolution network using the bipartite graph of user project interaction, and then integrated multi-task learning into the convolution neural network learning framework using multimodal auxiliary information, improving the classification effect multi-task.

At present, in the process of improving the model structure, the course recommendation in the field of in-depth learning continuously excavates the hidden meaning between the course and the user data, which increases the network depth, increases the calculation time of the model, and reduces the timeliness of recommendations while increasing the recommendation accuracy. Therefore, in mining user characteristics, we need to fully consider the computational performance and the complete modal characteristics of users.

2.3. Multimodal-Based Course Recommendation in Deep Learning Domain

With a large number of applications of deep learning in the fields of image recognition and sentiment analysis, researchers have gradually fused multi-domain data to achieve complementarity between heterogeneous information for more comprehensive information. For example, three kinds of information, such as image, video, and text, are fused in cross-modal embedding [22]. Tamura et al. [23] suggests a multimodal response synchronization measurement system with an eye tracker and electroencephalography signals, through which the eye tracker can obtain information from the learner's attention and the brain signal can provide clues to estimate the mental state in learning. Wang et al. [24] used an online learning system to collect multimodal behavioral data from three dimensions: psychological, physiological, and behavioral, thus providing a more comprehensive evaluation of the overall situation. Xu et al. [25] proposed a multimodal deep learning framework to extract multimodal course information, such as course video, text, and audio, as well as explicit and implicit feedback from users. Chango et al. [26], on the other hand, focused on improving the recommendation performance using different fusion algorithms for four types of multimodal information about students' theory classes, practical classes, online courses, and final grades. However, the multi-modal data of the course need to capture representative modal features in the acquisition process, and the implicit feedback data of users is difficult to obtain. The hidden behavior of users watching videos has essential research significance in the platform. Users' implicit feedback data can be effectively obtained through users' pause times, playback times, and playback time nodes.

Multimodal fusion can provide more information for model decision-making, which can improve the accuracy and precision of the overall decision result. The difficulty lies in handling heterogeneous information, selecting the fusion method, and adjusting the modal alignment.

3. Methods

We design and implement a course recommendation system based on a multimodal deep learning framework, which uses different modal information in the course for fusion recommendation. The user's operation while watching the course video involves explicit and implicit feedback, and therefore, it needs to be differentiated according to the user's behavior. This system mainly includes six key components: data collection, data processing, feature extraction, profiling, deep learning course recommendation and recommended course result presentation. The overall process of this multimodal course recommendation framework is shown in Figure 1.

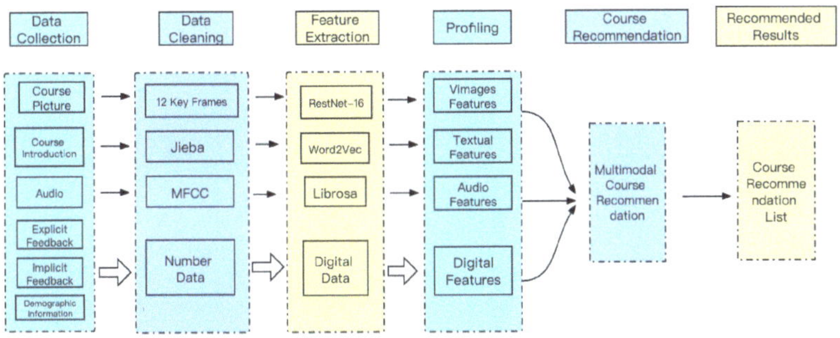

Figure 1. Multimodal course recommendation framework.

3.1. Data Collection Module

The online platform collects data from two aspects: learners and course data. The learner data mainly includes explicit and implicit feedback data. The explicit feedback data is reflected in the comments, collections, and thumbs of the course; the implicit feedback is reflected in the data of only browsing the course without marking. The course data mainly includes the course name and introduction, video data, and audio data extracted from the course video. Figure 2 depicts the main process of acquiring data.

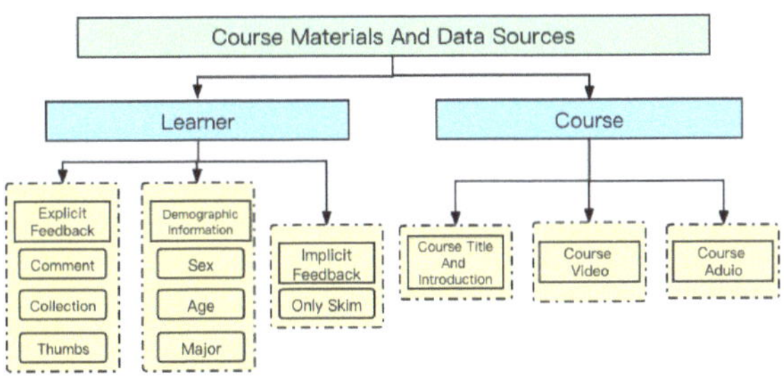

Figure 2. Data collection module.

3.2. Data Processing

The course information collected directly has interference information, so the data needs to be cleaned before extracting the course features. The video information of the course is described by video keyframes. However, due to the long duration of course videos (usually more than 10 min), acquiring video frames frame by frame will cause a sudden increase in data volume. Therefore, In this paper, video description is performed by capturing keyframes. We extracted the audio from the course video to obtain the course audio information and analyzed the obtained audio information. Additionally, the title, introduction, and comments of the course need to be divided into words to clean up the deactivated parts, such as "of" and "the", as well as the markers.

3.3. Feature Extraction Module

3.3.1. Video Feature Extraction

The course video contains visual modality information of the course. To obtain the course video keyframes, Mahasseni et al. [27] implemented the selection of keyframes without keyframe annotation through a per-frame score. Inspired by the above idea, in this paper, the video stream is considered as a series of images, and keyframes are extracted from the video at intervals by the FFmpeg tool. To obtain the high-level features of video frames, for extracting the content of each frame a ResNe-16 model is used to extract visual features, which can provide stable initialization in ensuring semantic recognition. The complete video is transformed into 12 keyframes by initializing the ImageNet dataset with pre-training weights, where each keyframe contains 4096-dimensional features.

3.3.2. Audio Feature Extraction

Due to diverse contents and some missing information in the video part, audio modal information is extracted to supplement the information. The audio modal information includes the tone, intonation, and pauses of the lecturer. For audio modality extraction, we use the Librosa module to process the audio content by executing a 10 s window and 80% overlapping spectrograms, and for obtaining 512-dimensional features of audio modality.

3.3.3. Text Feature Extraction

The text modal information mainly includes the course name and introduction. In this paper, the course name and introduction are processed using the Jieba word separation model, and the information that does not contain specific content is eliminated according to the deactivation word rule. The collection dataset is obtained on a MOOC platform with complex data dimensions. Therefore, a large Chinese corpus is used to capture the semantic information between different words [28]. Besides, we use the Word2Vec tool to convert the text corpus as input and convert the title and course profile content into a 300-dimensional feature vector.

3.3.4. Digital Feature Extraction

To obtain learners' comments, collections, and thumbs data of the course, we marked them by one-hot coding. At the same time, learners' browsing of the detailed interface of the course was marked by the platform script.

Based on the above four kinds of feature data extraction, the course multimodal information extracted in this paper is summarized as follows, which contains 4096-dimensional visual modal information, 512-dimensional audio modal information, 300-dimensional text modal information, and 4-dimensional digital features. The course multimodal information is shown in Table 1.

Table 1. Summary of course multimodal information.

Model Information	Dimension
Course Video	4096
Course Audio	512
Course Text	300
Digital Features	4

3.4. Profiling Module

After obtaining the multimodal features of the course through the feature extraction module, the different modal features need to be reasonably fused. Given that there is an interaction between the demographic information of learners and explicit and implicit feedback data, this paper integrates the above three data into digital modal features. The specified features are first extracted from the sequential units of each modality, and then these features are input into the parallel LSTM separately, and finally, the sequential structures in the visual, audio, text, and digital modal features are captured.

Since the output of the LSTM is a vector sequence model, the variable-length vectors need to be transformed into fixed-length vectors to facilitate the later model training. This paper uses an attention-based pooling operation to achieve dynamic adjustment of weights according to the importance of time steps based on the weighted sum sequences of all time step vector representations and the content relevance of the weights to each time step [29]. The output of the fully connected layer is used as the result of course recommendation. Figure 3 illustrates the schematic diagram of the multimodal course recommendation algorithm model.

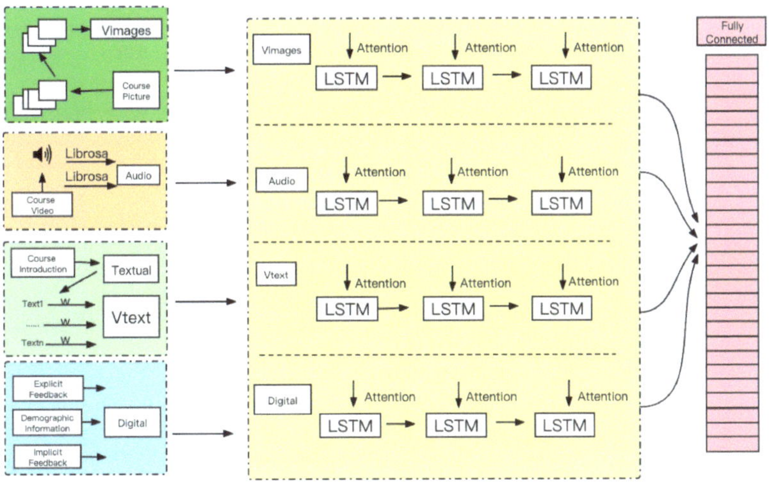

Figure 3. Multimodal course recommendation algorithm model.

3.5. Course Recommendation and Recommendation Results

By extracting and processing the features of different modalities, the output of the fully connected layer is obtained. In this paper, multiple modal information is processed by LSTM and Attention, and visual, audio, text, and digital modal feature vectors are connected vertically to finally obtain a 4912-dimensional course vector, which is used as the input of the recommendation model. The platform will discriminate whether the learner has completed the course, and if the output target is 1 then the course is completed, otherwise, the learning is not completed.

In the deep learning-based course recommendation method, Salampasis et al. [30] presented the use of Recurrent Neural Networks (RNN) to consider the similarity between the user's recently viewed items and precomputed items. Pradhan et al. [31] proposed a bidirectional Long- and Short-Term Memory model (LSTM) and an integrated framework of a recommender system based on an attention mechanism for recommending suitable academic papers. Li et al. [32] suggested a neural network model of a Bayes Personalized Ranking Network (BPRN), which learns paired course preferences based on each user's history of course registration. Current deep learning applications in the field of course recommendation focus on improving and enhancing the model, but usually consider only one kind of modal information without mining data features from the text, image, and video information of the course itself as the representation of course recommendation modeling.

Therefore, by obtaining the video, audio, text, and digital modal features of the course, this paper improves the accuracy of recommendation from the perspectives of course information and learners and increases the click rate of users. Keyframes are extracted from a series of video frames in the course video, and video modal features are extracted for each frame by a pre-trained ResNet-16 model; audio modal features are extracted in the course video by using Librosa; text modal features are extracted by Word2Vec in the course name and introduction text information; and finally, digital modal features are extracted by statistical learners' demographic information, explicit feedback, and implicit feedback data.

The course recommendation results will be based on learners' collections, favorites, and comments on the platform, and the multimodal course features will be integrated to improve the recommendation accuracy and learners' course click rate.

4. Experimental Settings

4.1. Datasets

The course data and learner data in this paper were obtained from a real online learning MOOC platform [33]. We mainly collect course data including video information, audio information, the course name, and introduction of the course. The duration of the course video was about 10 min. The learner data mainly included the learners' comments, collections, and thumbs on the course in the explicit feedback data; the behavioral data of the learners who only browsed the course content but did not operate were mainly recorded in the implicit feedback data. This dataset was collected from March 2020, with a total of 32,413 learners, 813 courses, 63,936 course-explicit feedback data, and 81,537 course-implicit feedback data.

In the experimental process, the data set was further divided into three disjoint sets in this paper, which were 80%, 10%, and 10% of the course data randomly selected as the training set, validation set, and test set, respectively. The validation set was used for model optimization. Finally, the final model performance was verified on the test set.

4.2. Evaluation Indicators

The Area Under the Curve (AUC) [34] represents the area under the ROC curve, and its practical meaning is to correctly predict the ratio of learner and course pairs to all learner and course pairs in the recommender system, that is, the probability that a positive sample ranks ahead of a negative sample based on its rating. Suppose the list of recommended courses contains a_0 positive samples and a_1 negative samples, where n positive samples are predicted to be larger than the negative samples; then, the AUC calculation process can be expressed as:

$$AUC = \frac{n}{a_0 \cdot a_1} \quad (1)$$

The Hit Radio (HR) [35] represents the hit rate, which is the proportion of learners who have K correct recommendations in the recommended course list. The denominator GT represents all test sets, and the numerator *NumberHits@K* represents the sum of the number of test sets in each learner's Top-N list. *HR@K* can be expressed as:

$$HR@K = \frac{NumberHits@K}{GT} \quad (2)$$

Recall that the Ref. [36] refers to how many of the positive examples in the recommended course were predicted correctly.

Normalized Discounted Cumulative Gain (*NDCG*) [37] means the normalized discounted cumulative gain, which is often used as an evaluation of the recommended ranking results. *NDCG* can be expressed as:

$$A_{NDCG_u@K} = \frac{B_{DCG_u@K}}{B_{IDCG_u}} \quad (3)$$

$$C_{NDCG@K} = \frac{B_{NDCG_u@K}}{|u|} \quad (4)$$

$A_{NDCG_u@K}$ represents the discounted cumulative of the user's true list; $B_{DCG_u@K}$ represents the score of the user's true list, and $C_{NDCG@K}$ represents the average of each user.

4.3. Experimental Parameter Settings

We processed the experimental data by Python and built the course recommendation algorithm framework by PyTorch. In the initialized embedding layer mixed with hidden layer parameters, a Gaussian distribution was used to set randomly (mean value of 0 and standard deviation of 0.1). Relu was used as the activation function in the feature extraction of visual, audio, and text modalities, and the Sigmoid activation function was used in the fully connected layer. The three modalities were extracted by LSTM, and their hidden units were set to 300, 200, and 50, respectively. The loss function was "binary cross-entropy" and the evaluation was "accuracy". At last, the batch size was set to 12, the number of iterations is 200, and the learning rate was 0.001.

4.4. Experimental Results

4.4.1. Experimental Results in Different Recommendation Models

This section evaluates the impact of the feature extraction module on the results of the LSTM+ATTENTION network model using the LSTM and the Attention mechanism networks, and the combination of both. The LSTM model was used to extract multi-modal information, which can mine the potential temporal details of the data but has disadvantages in parallel processing. When the Attention model is used to extract the modal information, it can adjust the weights dynamically according to the extracted information; however, it is unable to learn the temporal relationships in the sequence. The LSTM+ATTENTION model can combine the advantages of the two, mine the potential timing information of the data, and dynamically adjust the weight according to the extracted information. The above analysis is also demonstrated by the experimental results, which show that the LSTM+ATTENTION model can combine the advantageous parts of the LSTM and Attention mechanisms with improving the extraction of modal features. The LSTM+ATTENTION model proposed in this paper outperforms the baseline model in both the AUC and the four click-through evaluation metrics. The effects of different extraction methods on the model are shown in Table 2.

Table 2. Effects of different extraction methods on the model.

	LSTM	ATTENTION	LSTM+ATTENTION
AUC	75.93	78.07	79.89
HR@5	0.68	0.71	0.78
HR@10	1.53	1.54	1.62
HR@15	2.41	2.48	2.53
HR@20	3.03	3.14	3.21

Table 3 compares the experimental results of the baseline model comparison. From the table, it is shown that the model in this paper achieves better results in both recall rate and NDCG, which is significantly better than other methods. The main reason is that LSTM+ATTENTION can fully exploit the temporal features and important dimensional features in multimodality. Meanwhile, the improved performance of AUC further supports the effectiveness of the attention mechanism to effectively distinguish important features in each modality. As shown in Table 3, a comparison of the performance under different recommended methods is presented.

Table 3. Performance comparison of different methods.

Methods	k = 5		k = 10		k = 15		k = 20	
	Recall	**NDCG**	**Recall**	**NDCG**	**Recall**	**NDCG**	**Recall**	**NDCG**
LSTM	0.5145	0.4176	0.6723	0.4623	0.7321	0.4832	0.7541	0.5215
ATTENTIONN	0.5132	0.4259	0.6671	0.4672	0.7143	0.4875	0.7583	0.5145
LSTM+ATTENTION	0.5573	0.4531	0.6801	0.4881	0.7412	0.5102	0.7591	0.5257

4.4.2. Experimental Results in Different Course Modalities

To verify the effectiveness of the multimodal features proposed above, we compare and test the effects of different course feature combinations on the results. In this paper, the different course modal combinations are as follows: C1 (digital modality), C2 (digital modality and text modality), C3 (digital modality, text modality, and audio modality), C4 (digital modality, text modality, audio modality, and visual modality), and C5 (all modal information of the course). Figure 4 represents the comparison of information hit rates under different combinations of course modalities.

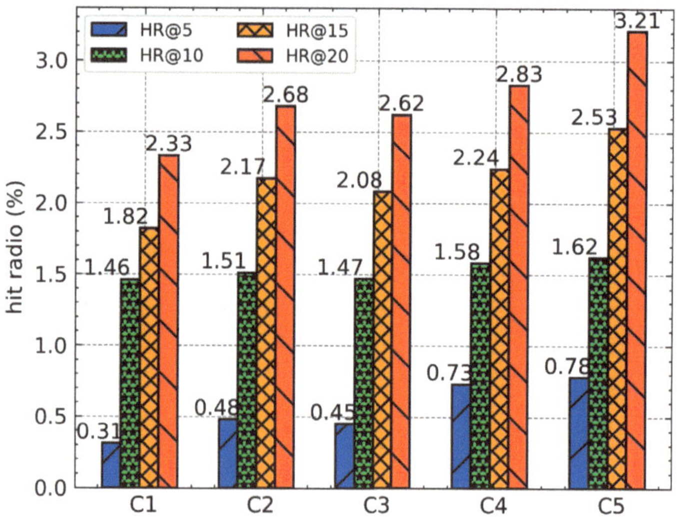

Figure 4. Information hit rate of different course modalities.

The experimental results show that the hit rate of C1 (numerical modality) is the lowest, which means that it represents the least information of the course. The highest hit rate of the user for the course is reached in C5 (all modal information of the course), which improves the hit rate by 0.47% compared with HR@5. In comparing the hit rate of C3 (digital modality, text modality, and audio modality) and C4 (digital modality, text modality, audio modality, and visual modality), it can be found that visual modality information of the course effectively improves the accuracy of the recommendation, which further indicates that the video modality contains more information of the course. Thus, it is shown that multimodal course information has a positive impact on recommendation accuracy.

4.4.3. Experimental Results with Different Learner Modal Information

In this section, we verify the impact of four information sets (L1, L2, L3, and L4) on the results, which represent demographic information (gender, age, major, etc.), learner-explicit feedback information, learner-implicit feedback information, and the sum information of all modal characteristics. These information sets mainly describe the users' operational behavior data and their attributes. Figure 5 depicts the hit rate of different learner modalities.

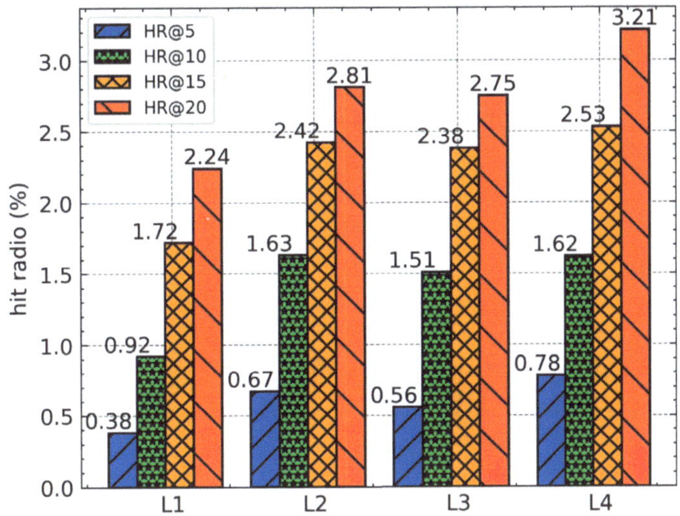

Figure 5. Hit radio of different learner modalities.

The relatively small amount of information represented by demographics leads to a low hit rate on courses. Adding learners' explicit feedback data (comments, collections, and thumbs) and implicit feedback data (browsing operations on the course) can effectively improve the user hit rate. L4 achieves better performance in the hit rate in all results, with hit rates of 2.24%, 2.81%, 2.75%, and 3.21% for four different K values, mainly because L4 incorporates all modalities of this paper, and can better obtain more modal information which is beneficial to improve the recommendation performance. The above experimental results show that the fusion of multiple modal user information effectively increases the course hit rate and can improve the course recommendation effect.

4.5. Discussion

When learners browse courses on the platform, they can click on the courses they are interested in through the course name and recommended courses, and they will comment, collect and thumb the courses they are interested in. Otherwise, there will be no operation behavior for the courses they are not interested in. The platform will record various operation behaviors of learners, and can collect explicit and implicit feedback data of learners to further explore users' viewing habits. When users watch the course video, the video modality, audio modality, course name, and introduction of the course can effectively build the user portrait. Therefore, by fusing the above modal information, we can effectively improve the performance of the course recommendation model, and LSTM and Attention can effectively explore the important contents in different modal information, as well as explore the temporal information between modalities to effectively improve the recommendation performance. The experimental results reveal that the framework of this paper can achieve better course recommendation results, in which the AUC score reaches 79.89% and the hit rate reaches 0.78%, 1.62%, 2.53%, and 3.21% at Top-N values from 5 to 20, respectively, indicating that the framework can effectively improve the accuracy of course recommendation.

In summary, the use of multimodal data in course recommendations can effectively improve the accuracy rate. The LSTM+ATTENTION model used in this paper is more effective in multimodal feature extraction compared with traditional neural networks, and the visual modality contains more effective information than the audio modality.

5. Conclusions and Future Work

With the development of the Internet and cloud computing, online education platforms have proliferated, and it has become especially important for learners to find suitable courses among the massive courses. Previous studies only consider unimodal features and build or improve deep learning models for a recommendation; however, less attention is being paid to multimodal features in courses. Moreover, most of the current course recommendations focus on digital modal features, and there is less research on video mode, audio mode, and text mode. When constructing user portraits, researchers pay more attention to the explicit feedback data of users and less capture the implicit feedback data of users in the operating platform. In addition, the effective combination of user demographic information, explicit feedback data, and implicit feedback data can improve the construction of the user model.

The research contributions of this paper are as follows: first, a deep learning framework for course recommendation based on multimodality is proposed, which can effectively improve the accuracy of course recommendation. Among them, the multimodal features involved in this paper mainly considered the video modality, audio modality, text modality, feedback, and implicit feedback of users and demographic information. Secondly, the LSTM+Attention model was constructed to build the course recommendation system modeling, and the temporal features and important dimensional features in the multimodal features were effectively mined. Finally, the validity of the model was verified by real data.

The practical significance of this paper lies in the following. The multimodal recommendation system based on deep learning realizes the course recommendation of an online education platform, effectively reducing the time for users to choose courses and realizing the Personalized Course recommendation service. In fully acquiring the modal features of users and the three modal features of courses, the in-depth learning framework of this paper effectively considers the integration of multiple modal features, so the recommended personalized course content can meet the needs of users. Because this paper uses the operation of users browsing the course to obtain the implicit feedback data, most of the implicit feedback data of users in daily life are difficult to capture, which needs to be improved in the next step of this paper.

In future work, based on obtaining learners' implicit feedback data, we will further delve into modeling learners' operations, such as the number of pauses, the number of plays, and the number of repeated views of videos while watching videos. Given that the degree of preference for the course is represented in the user's evaluation of the course, we will subsequently research information on the learner's evaluation of the course and conduct sentiment analysis on the learners' evaluation, modeling and verifying the relationship between learners' course preferences and ratings to improve the effectiveness of course recommendation.

Author Contributions: Data curation, X.R. and Y.Y.; Formal analysis, X.J.; Funding acquisition, W.Y; Project administration, G.J.; Software, X.R.; Visualization, Y.Y.; Writing—original draft, X.R.; Writing—review & editing, X.R., X.J. and W.Y. All authors have read and agreed to the published version of the manuscript.

Funding: This research was funded by the Talent Project of College of Science and Technology (RC202016), the Natural Science Foundation of Zhejiang Province grant number LY20F020008, and the Ningbo Natural Science Foundation grant number 202003N4085.

Institutional Review Board Statement: Not applicable.

Informed Consent Statement: Not applicable.

Data Availability Statement: Not applicable.

Conflicts of Interest: The authors declare no conflict of interest.

References

1. Huang, R.; Lu, R. Research on Content-based MOOC Recommender Model. In Proceedings of the 2018 5th International Conference on Systems and Informatics (ICSAI), Nanjing, China, 10–12 November 2018; pp. 676–681.
2. da S Dias, A.; Wives, L.K. Recommender system for learning objects based in the fusion of social signals, interests, and preferences of learner users in ubiquitous e-learning systems. *Pers. Ubiquitous Comput.* **2019**, *23*, 249–268. [CrossRef]
3. Wei, Y.; Wang, X.; Nie, L.; He, X.; Hong, R.; Chua, T.S. MMGCN: Multi-modal graph convolution network for personalized recommendation of micro-video. In Proceedings of the 27th ACM International Conference on Multimedia, Nice, France, 21–25 October 2019; pp. 1437–1445.
4. Gong, J.; Wang, S.; Wang, J.; Feng, W.; Peng, H.; Tang, J.; Yu, P.S. Attentional graph convolutional networks for knowledge concept recommendation in moocs in a heterogeneous view. In Proceedings of the 43rd International ACM SIGIR Conference on Research and Development in Information Retrieval, Virtual Event, China, 25–30 July 2020; pp. 79–88.
5. Aljohani, T.; Cristea, A.I. Predicting Learners' Demographics Characteristics: Deep Learning Ensemble Architecture for Learners' Characteristics Prediction in MOOCs. In Proceedings of the 2019 4th International Conference on Information and Education Innovations, Durham, UK, 10–12 July 2019; pp. 23–27.
6. Hsu, C.M.; Efendi, H.; Junedi. Perspectives of Online Learners: Demographic Characteristics on Synchronous Learning Environment in Taiwan. In *International Conference on Educational Sciences and Teacher Profession (ICETeP 2020)*; Atlantis Press: Dordrecht, The Netherlands, 2021; pp. 261–267.
7. Khribi, M.K.; Jemni, M.; Nasraoui, O. Automatic recommendations for e-learning personalization based on web usage mining techniques and information retrieval. In Proceedings of the 2008 Eighth IEEE International Conference on Advanced Learning Technologies, Santander, Spain, 1–5 July 2008; pp. 241–245.
8. Viniski, A.D.; Barddal, J.P.; de Souza Britto, A., Jr.; Enembreck, F.; de Campos, H.V.A. A case study of batch and incremental recommender systems in supermarket data under concept drifts and cold start. *Expert Syst. Appl.* **2021**, *176*, 114890. [CrossRef]
9. Sengottuvelan, P.; Gopalakrishnan, T.; Lokesh Kumar, R.; Kavya, M. A recommendation system for personal learning environments based on learner clicks. *Int. J. Appl. Eng. Res.* **2015**, *10*, 15316–15321.
10. Fu, D.; Liu, Q.; Zhang, S.; Wang, J. The undergraduate-oriented framework of MOOCs recommender system. In Proceedings of the 2015 International Symposium on Educational Technology (ISET), Wuhan, China, 27–29 July 2015; pp. 115–119.
11. Zhao, X.; Liu, B. Application of Personalized Recommendation Technology in MOOC System. In Proceedings of the 2020 International Conference on Intelligent Transportation, Big Data Smart City (ICITBS), Vientiane, Laos, 11–12 January 2020; pp. 720–723.
12. Salehi, M.; Kamalabadi, I.N.; Ghoushchi, M.B.G. An effective recommendation framework for personal learning environments using a learner preference tree and a GA. *IEEE Trans. Learn. Technol.* **2013**, *6*, 350–363. [CrossRef]
13. Pang, Y.; Jin, Y.; Zhang, Y.; Zhu, T. Collaborative filtering recommendation for MOOC application. *Comput. Appl. Eng. Educ.* **2017**, *25*, 120–128. [CrossRef]
14. Burke, R. Hybrid recommender systems: Survey and experiments. *User Model. User-Adapt. Interact.* **2002**, *12*, 331–370. [CrossRef]
15. Jannach, D.; Zanker, M.; Felfernig, A.; Friedrich, G. *Recommender Systems: An Introduction*; Cambridge University Press: Cambridge, UK, 2010.
16. Zhou, Y.; Huang, C.; Hu, Q.; Zhu, J.; Tang, Y. Personalized learning full-path recommendation model based on LSTM neural networks. *Inf. Sci.* **2018**, *444*, 135–152. [CrossRef]
17. Ange, T.; Roger, N.; Aude, D.; Claude, F. Semi-supervised multimodal deep learning model for polarity detection in arguments. In Proceedings of the 2018 International Joint Conference on Neural Networks (IJCNN), Rio de Janeiro, Brazil, 8–13 July 2018; pp. 1–8.
18. Wang, J.; Xie, H.; Au, O.T.S.; Zou, D.; Wang, F.L. Attention-based CNN for personalized course recommendations for MOOC learners. In Proceedings of the 2020 International Symposium on Educational Technology (ISET), Bangkok, Thailand, 21–23 July 2020; pp. 180–184.
19. Zhu, Q. Network Course Recommendation System Based on Double-Layer Attention Mechanism. *Sci. Program.* **2021**, *2021*, 7613511. [CrossRef]
20. Liu, J.; Yin, C.; Li, Y.; Sun, H.; Zhou, H. Deep Learning and Collaborative Filtering-Based Methods for Students' Performance Prediction and Course Recommendation. *Wirel. Commun. Mob. Comput.* **2021**, *2021*, 2157343. [CrossRef]
21. Ni, J.; Huang, Z.; Hu, Y.; Lin, C. A two-stage embedding model for recommendation with multimodal auxiliary information. *Inf. Sci.* **2022**, *582*, 22–37. [CrossRef]
22. Pan, Y.; Mei, T.; Yao, T.; Li, H.; Rui, Y. Jointly modeling embedding and translation to bridge video and language. In Proceedings of the IEEE Conference on Computer Vision and Pattern Recognition, Las Vegas, NV, USA, 27–30 June 2016; pp. 4594–4602.

23. Tamura, K.; Lu, M.; Konomi, S.; Hatano, K.; Inaba, M.; Oi, M.; Okamoto, T.; Okubo, F.; Shimada, A.; Wang, J.; et al. Integrating multimodal learning analytics and inclusive learning support systems for people of all ages. In Proceedings of the International Conference on Human-Computer Interaction, Orlando, FL, USA, 26–31 July 2019; pp. 469–481.
24. Wang, L.; He, Y. Online Learning Engagement Assessment Based on Multimodal Behavioral Data. In *Transactions on Edutainment XVI*; Springer: Berlin/Heidelberg, Germany, 2020; pp. 256–265.
25. Xu, W.; Zhou, Y. Course video recommendation with multimodal information in online learning platforms: A deep learning framework. *Br. J. Educ. Technol.* **2020**, *51*, 1734–1747. [CrossRef]
26. Chango, W.; Cerezo, R.; Romero, C. Multi-source and multimodal data fusion for predicting academic performance in blended learning university courses. *Comput. Electr. Eng.* **2021**, *89*, 106908. [CrossRef]
27. Mahasseni, B.; Lam, M.; Todorovic, S. Unsupervised video summarization with adversarial lstm networks. In Proceedings of the IEEE conference on Computer Vision and Pattern Recognition, Honolulu, HI, USA, 21–26 July 2017; pp. 202–211.
28. Li, S.; Zhao, Z.; Hu, R.; Li, W.; Liu, T.; Du, X. Analogical reasoning on chinese morphological and semantic relations. *arXiv* **2018**, arXiv:1805.06504.
29. Cao, D.; Miao, L.; Rong, H.; Qin, Z.; Nie, L. Hashtag our stories: Hashtag recommendation for micro-videos via harnessing multiple modalities. *Knowl.-Based Syst.* **2020**, *203*, 106114. [CrossRef]
30. Salampasis, M.; Siomos, T.; Katsalis, A.; Diamantaras, K.; Christantonis, K.; Delianidi, M.; Karaveli, I. Comparison of RNN and Embeddings Methods for Next-item and Last-basket Session-based Recommendations. In Proceedings of the 2021 13th International Conference on Machine Learning and Computing, Shenzhen, China, 26 February–1 March 2021; pp. 477–484.
31. Pradhan, T.; Kumar, P.; Pal, S. CLAVER: An integrated framework of convolutional layer, bidirectional LSTM with attention mechanism based scholarly venue recommendation. *Inf. Sci.* **2021**, *559*, 212–235. [CrossRef]
32. Li, X.; Li, X.; Tang, J.; Wang, T.; Zhang, Y.; Chen, H. Improving Deep Item-Based Collaborative Filtering with Bayesian Personalized Ranking for MOOC Course Recommendation. International Conference on Knowledge Science, Engineering and Management, Hangzhou, China, 28–30 August 2020; pp. 247–258.
33. Sakboonyarat, S.; Tantatsanawong, P. Massive Open Online Courses (MOOCs) Recommendation Modeling using Deep Learning. In Proceedings of the 2019 23rd International Computer Science and Engineering Conference (ICSEC), Phuket, Thailand, 30 October–1 November 2019; pp. 275–280.
34. Volk, N.A.; Rojas, G.; Vitali, M.V. UniNet: Next Term Course Recommendation using Deep Learning. In Proceedings of the 2020 International Conference on Advanced Computer Science and Information Systems (ICACSIS), Depok, Indonesia, 17–18 October 2020; pp. 377–380.
35. Lu, W.; Yu, Y.; Chang, Y.; Wang, Z.; Li, C.; Yuan, B. A dual input-aware factorization machine for CTR prediction. In Proceedings of the Twenty-Ninth International Conference on International Joint Conferences on Artificial Intelligence, Virtual Reality, 19–26 August 2021; pp. 3139–3145.
36. Liu, J.; Zhang, H.; Liu, Z. Research on Online Learning Resource Recommendation Method Based on Wide Deep and Elmo Model. *J. Phys. Conf. Ser.* **2020**, *1437*, 012015. [CrossRef]
37. Trirat, P.; Noree, S.; Yi, M.Y. IntelliMOOC: Intelligent Online Learning Framework for MOOC Platforms. In Proceedings of the 13th International Conference on Educational Data Mining (EDM 2020), Virtual Conference, 10–13 July 2020; pp. 682–685.

sustainability

MDPI

Article

Depth-Wise Separable Convolution Attention Module for Garbage Image Classification

Fucong Liu [1,2,†], Hui Xu [1,3,†], Miao Qi [1,2], Di Liu [1,4], Jianzhong Wang [1,*] and Jun Kong [1,2,*]

1 College of Information Sciences and Technology, Northeast Normal University, Changchun 130117, China; liufc001@nenu.edu.cn (F.L.); xuh504@nenu.edu.cn (H.X.); qim801@nenu.edu.cn (M.Q.); liud737@nenu.edu.cn (D.L.)
2 Institute for Intelligent Elderly Care, Changchun Humanities and Sciences College, Changchun 130117, China
3 Key Laboratory for Applied Statistics of MOE, Northeast Normal University, Changchun 130024, China
4 School of Computer Science, Northeast Electric Power University, Jilin 132000, China
* Correspondence: wangjz019@nenu.edu.cn (J.W.); kongjun@nenu.edu.cn (J.K.)
† These authors contributed equally to this work.

Abstract: Currently, how to deal with the massive garbage produced by various human activities is a hot topic all around the world. In this paper, a preliminary and essential step is to classify the garbage into different categories. However, the mainstream waste classification mode relies heavily on manual work, which consumes a lot of labor and is very inefficient. With the rapid development of deep learning, convolutional neural networks (CNN) have been successfully applied to various application fields. Therefore, some researchers have directly adopted CNNs to classify garbage through their images. However, compared with other images, the garbage images have their own characteristics (such as inter-class similarity, intra-class variance and complex background). Thus, neglecting these characteristics would impair the classification accuracy of CNN. To overcome the limitations of existing garbage image classification methods, a Depth-wise Separable Convolution Attention Module (DSCAM) is proposed in this paper. In DSCAM, the inherent relationships of channels and spatial positions in garbage image features are captured by two attention modules with depth-wise separable convolutions, so that our method could only focus on important information and ignore the interference. Moreover, we also adopt a residual network as the backbone of DSCAM to enhance its discriminative ability. We conduct the experiments on five garbage datasets. The experimental results demonstrate that the proposed method could effectively classify the garbage images and that it outperforms some classical methods.

Keywords: garbage classification; deep learning; attention mechanism; depth-wise separable convolution

Citation: Liu, F.; Xu, H.; Qi, M.; Liu, D.; Wang, J.; Kong, J. Depth-Wise Separable Convolution Attention Module for Garbage Image Classification. *Sustainability* **2022**, *14*, 3099. https://doi.org/10.3390/su14053099

Academic Editor: Miltiadis D. Lytras

Received: 12 January 2022
Accepted: 2 March 2022
Published: 7 March 2022

1. Introduction

Due to the population increase and economic development, the amount of garbage produced every day is growing rapidly, especially in developing countries [1]. If such a large amount of garbage is not treated effectively, it will cause severe environmental pollution and a massive waste of resources. Efficient sorting, recycling, and regeneration treatment are the key and effective means to solve this problem. In recent years, more and more nations have started to explore recycling strategies to improve the environment with the ultimate goal of having a cyclical economy and sustainable development [2]. Scholars have done much research on the garbage classification problem. However, most of their proposed solutions focused on the terminal recycling method [3–8], which is highly dependent on people's cooperation. At present, the most widely used garbage sorting method is based on manual classification. Although manual garbage classification could obtain highly accurate results, it is always time-consuming and requires well-trained operators, which seriously limits the efficiency of garbage classification. Therefore, the automated garbage sorting method is an effective way to solve this problem.

With the development of artificial intelligence (AI) technology, AI-related applications, such as computer vision, speech recognition and natural language processing, have been gaining more and more attention in many industries. Some researchers have also leveraged the artificial intelligence methods for garbage classification and sorting. Wang et al. [9] utilized support vector machine (SVM) [10] and boosting algorithm to classify garbage images. Similarly, Liu et al. [11] also combined multi-class SVM classifier with a speeded up robust feature (SURF) [12] to establish a smart waste sorting system. Though SVM can effectively accomplish the garbage classification task, it belongs to the shallow classification model. Thus, the classification result obtained by SVM may not be optimal. Recently, with the rapid development of deep learning, some deep convolutional neural network-(CNN) based methods were gradually incorporated into the garbage classification task to improve its accuracy. Ozkaya et al. [13] attempted to train different CNNs on a small garbage dataset with transfer learning, and then adopted an SVM to classify the feature obtained by CNNs. Similarly, Fu et al. [14] employed a transfer learning strategy to train a CNN model so that the classification accuracy of garbage images can be improved. Meng et al. [15] designed a network that can reuse and fuse features obtained by different layers of CNN, the multi-scale feature interaction in their method can significantly improve the classification performance. Singh [16] used a modified CNN model called Xception to accomplish the classification task for plastic bags. Besides, CNN has also been combined with some hardware to form various garbage classification systems. Nowakowski and Pamuła [17] proposed a region-based CNN algorithm to classify the garbage images so that the users can recognize the category of garbage by smartphones. Chu et al. [18] deployed a detection system for municipal solid waste recycling, which consists of a high-resolution camera, a bridge sensor, an inductor and a PC. This system takes the pictures of garbage and then utilizes a CNN for classification. In [19], Yu et al. proposed a deep CNN which fuses multiple features in multiple scales for solid garbage detection and sorting. Although Yu's method can improve the detection accuracy by adding multi-view feature, it relies on expensive 3D camera hardware to get the depth information of garbage images. Kokoulin and Tur [20] integrated CNN with IoT hardware and a reverse vending machine (a device that accepts used beverage containers and returns money to the user) to classify and recycle beverage containers. Wang et al. [21] proposed a framework which first classifies the garbage images by a CNN and then monitors the operating state of garbage containers using smart sensors.

From the aforementioned works, it can be seen that classifying the garbage into their corresponding categories is a preliminary and important step in many garbage sorting and recycling tasks. However, the CNN models in most of the existing work (such as Alexnet, VGG and Xception used in [16,22,23]) were originally proposed for classifying natural images rather than garbage images. Hence, the characteristics of garbage images were neglected in them, which limits their performance. Actually, the garbage image classification is much more complicated than some other image classification tasks. For example, the accuracy of garbage image classification often suffers from the inter-class similarity and intra-class variance problem. That is, the appearance of some garbage belonging to different classes is more similar than those from the same class, as shown in Figure 1. Moreover, the various backgrounds around the target garbage will also impair the classification accuracy, as shown in Figure 2. To solve this problem, we introduce a new deep network named Depth-wise Separable Convolution Attention Module (DSCAM) in this paper. Compared with other studies [11–15], the proposed DSCAM is specially designed for garbage image classification in the following aspects. Firstly, in order to suppress the interference of other factors (such as the background) and make our model focus on the target garbage in images, an attention-based module is introduced into the proposed DSCAM. Secondly, unlike the existing work [13–16] which adopts the CNN models with only a small number of convolution layers to extract the feature of garbage image, we employ Resnet [24] as the backbone of our DSCAM. Since Resnet contains more convolution layers, the feature extraction ability of the proposed DSCAM can be enhanced to capture the discriminative features of garbage from different classes. At last, because the attention module and Resnet will bring more parameters into our model, the depth-wise separable convolution technique is used to compensate for the increase of computational burden. The effectiveness of the proposed method is demonstrated by extensive experiments on five garbage image classification datasets.

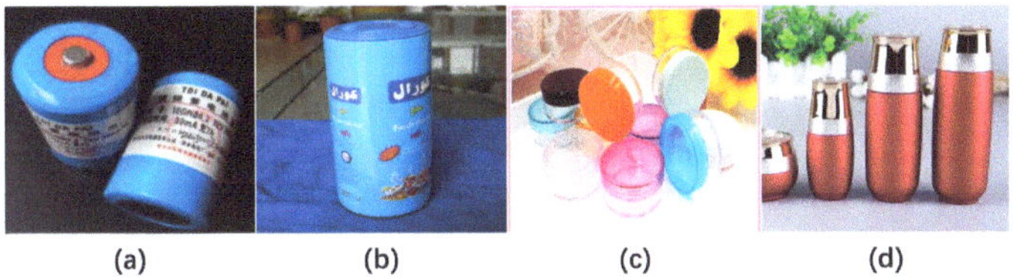

Figure 1. Samples selected from the Baidu-214 dataset. (**a**) belongs to class "Hazardous /Battery" and (**b**) belongs to class "Recyclable/milk powder can", which are from different categories but have similar appearances (inter-class similarity). (**c**,**d**) are from the same class "Cosmetic bottle" but have different appearances (intra-class variance).

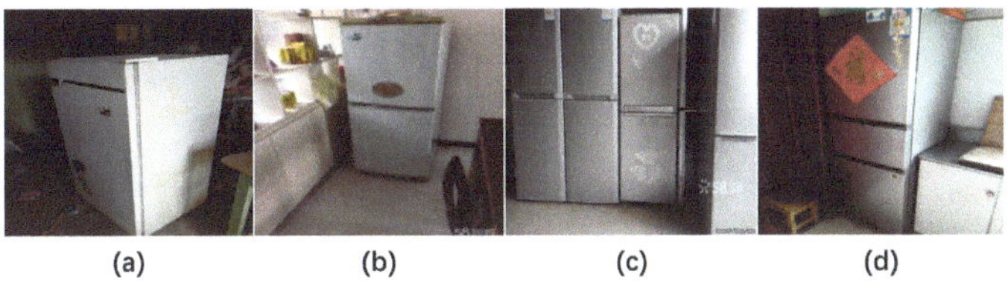

Figure 2. Samples selected from the same class with various backgrounds in the Baidu-214 dataset. (**a**–**d**) belong to the same class "Recyclable/Major Appliances", all their backgrounds have changed significantly.

The rest of this paper is organized as follows: Section 2 briefly reviews some work related to our method. Section 3 presents the proposed DSCAM method. The experimental results on five datasets are shown and analyzed in Section 4. Finally, Section 5 concludes the paper.

2. Related Work

Deep convolutional neural networks (CNNs) have been widely used in the computer vision community and achieved remarkable progress in various tasks, e.g., image classification, object detection and semantic segmentation. Starting from the groundbreaking AlexNet [25] which successfully won the championship of ImageNet image classification in 2012, researchers have realized the importance of CNN for image feature extraction and have committed to further improving its performance [26,27]. The VGG [26] model has proved the importance of network depth in enhancing the effectiveness of CNN model. GoogLeNet [27] designed an Inception module so that the network could capture image features of different scales. Resnet [24] proposed a residual block with skipped connections to construct a deeper network architecture with more layers. Moreover, some other models have also been proposed to improve the performance of CNN in various aspects. For example, DenseNet [28] constructs connections in the CNN network so that the output feature of a layer can be regarded as the input of all its subsequent layers, which can improve the flow of information throughout the network to enhance the feature learning ability.

The attention mechanism has proved to be an effective way to promote the performance of deep CNNs. Thus, the incorporation of attention module into CNN has attracted a lot of interest [29,30]. One of the representative attention-based CNN methods is squeeze-and-excitation networks (SE-Net) [29], which learns channel attention for each convolutional block and brings apparent performance gain for various problems. Subsequently, some other attention modules were developed to enhance the feature aggregation or combine the channel and spatial attention. Specifically, CBAM [30] employed both average and max pooling to aggregate feature and fused channel and spatial attentions into one module, which achieves considerable performance improvements on many computer vision tasks.

To reduce the number of parameters and the computational burden of CNN, some efficient network models were proposed. Among these models, the most widely used are group convolutions [31–33] and depth-wise separable convolutions [34,35]. A group convolution can be viewed as a regular convolution with separable channel convolution kernels, where each kernel corresponds to a partition of channels without connections to other partitions. Xie et al. [32] and Zhang [33] used group convolutions to improve the architecture of the CNN network, which achieves better results while ensuring the number of parameters. Depth-wise separable convolution is an extension of the group convolution. Firstly, depth-wise separable convolution performs group convolution independently over each channel of the input feature, then a point-wise convolution, i.e., an 1×1 convolution, is utilized to project the output of group convolution to a new channel space. The original idea of separable convolution operation comes from the Inception [27] network. Inspired by Inception, Chollet [34] proposed the Xception network, which uses depth-wise separable convolutional to further optimize the module structure and achieves satisfactory performance. The biggest benefit of depth-wise separable convolution is that it allows for significantly increasing the number of convolution units in a deep network without an uncontrolled blow-up in computational complexity. Thus, it has also been adopted in MobileNet [35] and ShuffleNet [36], which are designed for mobile device or embedded vision applications.

3. Proposed Network

The whole structure of the proposed DSCAM is shown Figure 3. As can be seen from this figure, our DSCAM employs the Resnet-50 (i.e., Resnet with 50 layers) as its backbone. Compared with other CNN models with a small number of layers, Resnet introduces the "shortcut connections" that skip several network layers to avoid the vanishing gradient during network training. Thus, it can construct a deeper network with more layers to better extract discriminative features from the garbage images. In our network, the initial feature of input garbage image is first obtained by a shallow convolutional layer with kernel size of 7×7. Then, the shallow feature is inputted into four Resnet blocks for feature refining. In our work, an attention module is embedded in each sub-block of Resnet. The attention module has two sequential processes: channel attention and spatial attention. Each process adopts the depth-wise separable convolution to obtain an attention map which consists of weights to indicate the importance of each channel and spatial position. Through point-wise multiplying the feature with attention maps, our network could adaptively emphasize the informative objects in the garbage image and suppress irrelevant background. At last, a global average pooling layer and cross-entropy loss are employed for image classification. The details of a depth-wise separable convolution and attention module will be described in Sections 3.1 and 3.2.

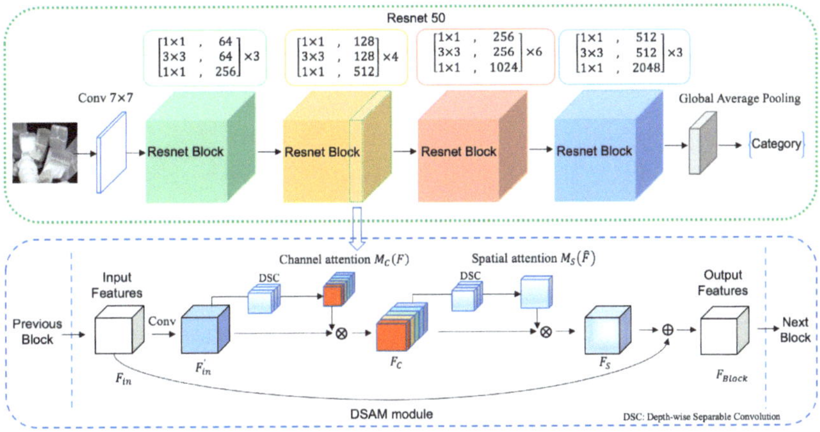

Figure 3. The network architecture of DSCAM.

3.1. Depth-Wise Separable Convolution

Here we briefly introduce how depth-wise separable convolution factorizes a standard convolution (as shown in Figure 4) into a depth convolution and a point-wise convolution (1×1 convolution). Given the input feature $F \in R^{W \times H \times C}$, where W and H denote the spatial dimensions and C is the number of channels, through a standard convolution with kernel size $k \times k \times C \times C'$, we can get the output $F' \in R^{W' \times H' \times C'}$, where W' and H' denote the spatial dimensions of F' and C' is the number of output channels.

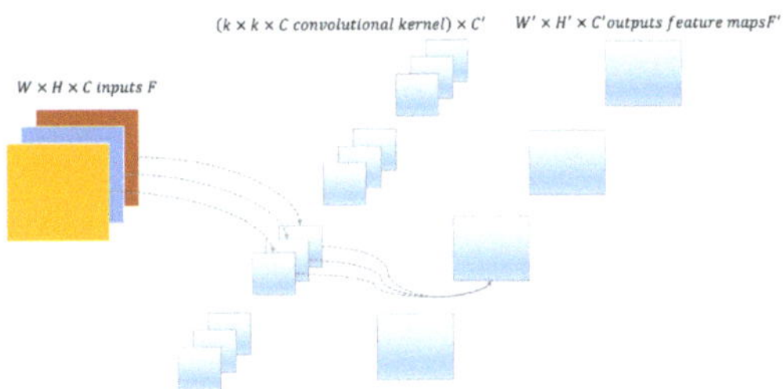

Figure 4. Standard convolution operation.

In order to reduce the computational burden and parameters in standard convolution, depth-wise separable convolution first uses a depth-wise convolution (one filter per input channel) to convolve with the input feature, as shown in Figure 5a, which can be formularized as:

$$\hat{F} = F \otimes K \tag{1}$$

where $F \in R^{W \times H \times C}$ is the input, $K = R^{k \times k \times C}$ is the depth-wise convolution kernel, is the convolution operation. Note that the output $\hat{F}$ has the same number of channels as the input feature. Then, an additional point-wise convolution with a 1×1 convolution kernel is applied to the output of the depth-wise convolution as shown in Figure 5b, which can be formularized as:

$$F' = \hat{F} \otimes \overline{K} \tag{2}$$

where $\hat{F} \in R^{W \times H \times C}$ is the outputs of depth convolution, $\overline{K} = R^{1 \times 1 \times C \times C'}$ is an 1×1 convolution and $F' \in R^{W' \times H' \times C'}$ is the output of whole depth-wise separable convolution.

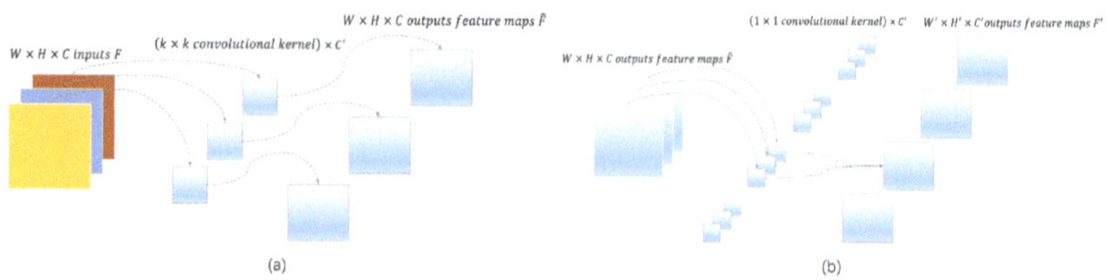

Figure 5. Depth-wise separable convolution operation. (**a**) Depth convolution operation. (**b**) Point-wise convolution operation.

For standard convolution with the kernel size of $k \times k \times C \times C'$, the number of parameters which needs to be optimized is $P_{std} = k^2 \times C \times C'$. On the contrary, a depth-wise separable convolution can reduce the number of parameters to $P_{ds} = k \times k \times C + C \times C' = (k^2 + C') \times C$. Thus, the ratio between them is $\frac{P_{ds}}{P_{std}} = \frac{1}{C'} + \frac{1}{k^2}$. In real-world applications, the number of output channels and kernel size are usually much larger than 1 (i.e., $1 << C'$, $1 << k^2$). Thus, the depth-wise separable convolution can effectively compress the number of parameters and computational burden in a convolutional network. Moreover, some studies have also shown that depth-wise separable convolution could improve the classification performance of a network due to the cross-channel and spatial features being sufficiently decoupled and separately handled in it [34,35].

3.2. Depth-Wise Separable Convolution Attention Module

In this section, we mainly describe how to extract channel and spatial attention weights using depth-wise separable convolution.

First, our purpose is to extract the channel attention weights which model interdependencies between channels. For arbitrary input features, a point-wise convolution is employed to fuse the information of different channels and reduce the dimension of channels to $1/r$. After that, multiple depth convolution layers are used to extract local spatial information, and a max-pooling operation is utilized to squeeze the spatial dimension to 1. Next, a 1×1 convolution kernel is adopted to recover the channel dimension to the input feature dimension and obtain the specific channel attention weights $M_C(F)$. Finally, the input feature is multiplied by the channel attention weights in each channel to generate the weighted feature. The entire process is demonstrated in Figure 6.

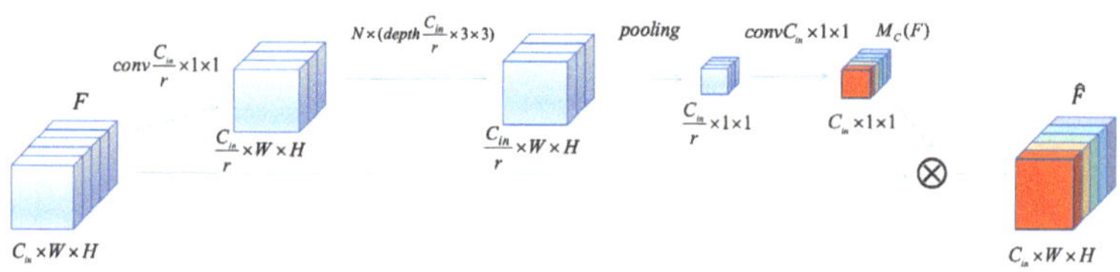

Figure 6. Channel attention with depth-wise separable convolution.

The procedure of obtaining channel attention weights can be formularized as follows:

$$M_C(F) = \overline{W}\left(f^s_{pooling}\left(W_g\left(W_p(F)\right)\right)\right) \quad (3)$$

where $F \in R^{W \times H \times C_{in}}$ is the input feature. W_p is point-wise convolution operation which reduces the input dimension from $W \times H \times C_{in}$ to $\frac{W \times H \times C_{in}}{r}$, where (in Figure 6) represents the reduction ratio. W_g is N times depth convolution. $f^s_{pooling}$ is a spatial max-pooling operation and $\overline{W}$ denotes an 1×1 convolution followed by a Sigmoid activation function, which ensures the channel dimension of the output feature is the same as the input. All convolution operations mentioned above are followed by the ReLU activation function (except the specific one mentioned).

Second, we describe the process for spatial attention weights extraction. To calculate the internal relationship of features in space, multiple depth convolution operations are first performed on the input feature. Then, we utilize a point-wise convolution operation to reduce the channel dimension so that the channel information can be fused. After the pooling operation along the channel axis, we can obtain the spatial attention weights $M_S(F)$. Lastly, the input feature is multiplied by spatial attention weights in a point-wise manner to obtain the weighted feature. These operations are shown in Figure 7.

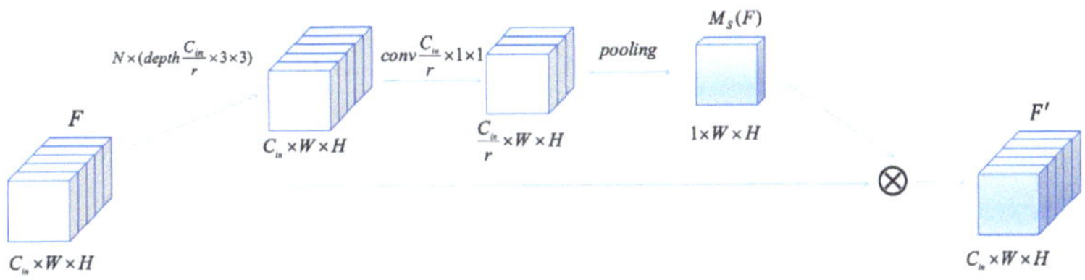

Figure 7. Spatial attention with depth-wise separable convolution.

The procedure of spatial attention weights computation can be formularized as follows:

$$M_S(F) = f^c_{pooling}\left(\left(W_p\left(W_g(F)\right)\right)\right) \tag{4}$$

where W_g denotes N times depth convolution operations, W_P is a 1×1 point-wise convolution, r is the reduction ratio, $f^c_{pooling}$ is the pooling operation along the channel axis. All convolution operations mentioned above are followed by the ReLU activation function, and we use max-pooling for all pooling operations.

3.3. DSCAM Block in Resnet

In this study, we integrate DSCAM into Resnet-50, thus each block in Resnet-50 can be formularized by Equations (5)–(8). In the notation that follows, we take f_{conv} in Equation (5) to be a standard convolution operation, which convolves the input feature F_{in} to F'_{in}. Equations (6) and (7) are used to sequentially compute the weighted channel attention feature F_C and weighted spatial attention feature F_S. In order to avoid the loss of information and make the network converge rapidly, we employ a skip connection to fuse the input F_{in} with the feature F_S by an element-wise summation (denoted by $\oplus$) in Equation (8), so that the final output F_{Block} of the current Resnet block can be obtained.

$$F'_{in} = f_{conv}(F_{in}) \tag{5}$$

$$F_C = M_C(F'_{in}) \otimes F'_{in} \tag{6}$$

$$F_S = M_S(F_C) \otimes F_C \tag{7}$$

$$F_{Block} = F_{in} \oplus F_S \tag{8}$$

3.4. Classification

To classify the feature of input garbage image, we use cross-entropy as the final loss function, which can be defined as:

$$Z = Softmax(f_{Avgpool}(F'_{Block})) \tag{9}$$

$$L = -\frac{1}{T}\sum_{i=1}^{T} y_i \ln Z_i \tag{10}$$

In Equation (9), F'_{Block} denotes the output feature of the last convolution layer in Resnet-50 with attention modules, $f_{Avgpool}$ represents global average pooling operation, and the softmax function is defined as $Softmax(x) = exp(x_k) / \sum_{k=1}^{M} exp(x_k)$, where M is the number of classes and x_k is the k-th element of the output after average pooling. Through Equation (9), the classification result of the feature F'_{Block} can be obtained. For cross-entropy loss in Equation (10), T denotes the total number of samples in training set, Z_i represents the classification result of the i-th sample obtained by Equation (9) and y_i is the true label (i.e., ground truth) of the i-th sample.

4. Experiments

In this section, we evaluate the effectiveness of our proposed method on garbage image datasets and compare its performance with other methods.

4.1. Garbage Datasets

In this study, three publicly available garbage image datasets constructed by Huawei Cloud and Baidu AI Studio are employed to evaluate the performance of our proposed DSCAM.

The Huawei Garbage Classification Challenge Cup dataset (Huawei-40 for short) contains 18,112 images with 40 classes in total (eight types of food waste, 23 types of recyclables, six types of other garbage and three types of hazardous garbage, respectively), which are all common garbage in daily life. The image sizes vary from 113 × 76 to 4000 × 3000 in the Huawei-40 dataset, and the distribution of samples in each category is uneven, ranging from 50 images to 800 images per category.

Baidu's garbage dataset (Baidu-214 for short) has 58,063 images belonging to 214 classes (106 types of recyclables, 53 types of food waste, 36 types of other garbage and 19 types of hazardous garbage, respectively). The minimum and maximum sizes of images in this dataset are 78 × 78 and 6720 × 4480. The number of images in each class ranges from 13 to 1654.

The Baidu recyclable garbage dataset (Baidu-RC for short) has 16,847 images from 21 recyclable garbage classes. The resolution of images in this dataset varies from 78 × 78 to 8150 × 5315, and the distribution of samples in Baidu-RC ranges from 250 images to 1000 images per category.

Some samples of the three datasets mentioned above are shown in Figure 8 (a), (b) and (c), respectively.

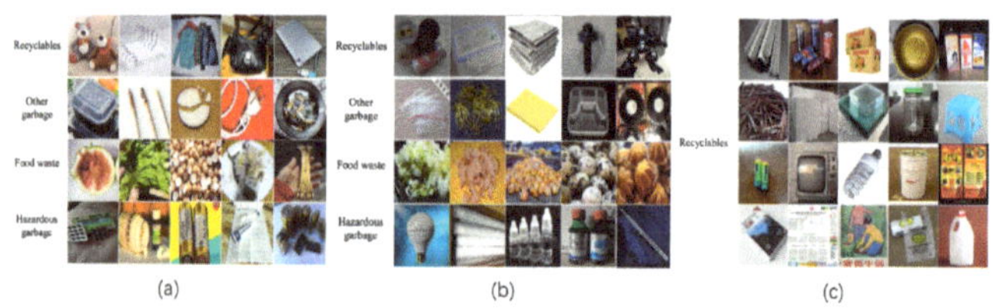

Figure 8. Some samples from (**a**) Huawei-40 dataset, (**b**) Baidu-214 dataset and (**c**) Baidu-RC dataset.

Because recyclable garbage is the most valuable among domestic garbage, we combine the recyclable garbage images from Huawei-40, Baidu-214, and Baidu-RC datasets to generate a new dataset named as BR-124 to further comprehensively evaluate our method's performance. Specifically, we first select the images containing recyclable garbage from Huawei-40 and Baidu-214 datasets. Then, for the selected images whose categories exist in Baidu-RC dataset, we merge them with samples from the same category in Baidu-RC to form their corresponding categories in a new dataset. On the contrary, if the categories of some selected images are not included in Baidu-RC, they are directly put into the new BR-124 dataset. As a result, the BR-124 dataset contains 55,513 images distributed in 124 classes.

The procedure of BR-124 dataset construction is shown in Figure 9. Table 1 summarizes the information of datasets used in this experiment.

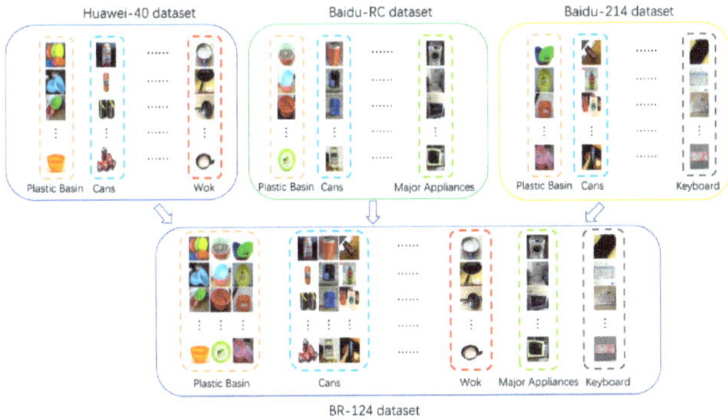

Figure 9. The construction of the new BR-124 dataset.

Table 1. Dataset Attributes.

Dataset	Number of Samples	Number of Categories
Huawei-40 dataset	19,735	40
Baidu-214 dataset	58,063	214
Baidu-RC dataset	16,847	21
BR-124 dataset	55,513	124

For all the above datasets, we convert the class label of each image into a one-hot vector (a vector has the same length as the number of categories, in which the i-th element is set to 1 if the garbage image belongs to the i-th class and the rest are all set to 0). All images are resized to $224 \times 224 \times 3$, where 224 is the image size, 3 is the number of RGB color channels.

4.2. Experimental Setup

In the experiment, all networks were implemented using the Pytorch framework and performed on two NVIDIA GeForce RTX 2080 Ti GPUs. We randomly selected 90% samples from each garbage dataset as the training set and the remaining 10% as the test set. The random selecting process is repeated five times, and then the average classification accuracy is reported. In addition, we use stochastic gradient descent (SGD) as the optimization method during the network training. The learning rate was initially set to 0.01 with a weight decay 0.0005 and Momentum 0.9. Then, we set the learning rate drops to 10% and 5% of its initial value at the 75th and 150th training iterations. Each drop in the learning rate can make the network fine-tuned locally.

4.3. Experimental Results

First, we compared our proposed network with some other widely used CNN architectures in garbage classification, including VGG-19 [23], Xception [34], X-DenseNet [15], MobileNet-V3 [37] and GNet [14]. From the classification accuracy of different methods in Table 2, the following points can be found. The VGG-19 network has the lowest accuracy in all compared methods since it has a simple architecture and shallow layers. Xception adopts depth-wise separable convolution to construct a complex network with more layers. Thus, it achieves higher accuracy than VGG-19. X-DenseNet is an extension of Xception, which uses a dense block to realize feature reuse and fusion. Due to this advantage, X-DenseNet outperforms Xception on all datasets. Nevertheless, the network architecture of Xception and X-DenseNet do not contain any attention module, which leads them to ignore the different importance of extracted feature. For MobileNet-V3 and GNet, since they embed the SE attention module [19] into some layers of their network, we can find that their performance is better than Xception and X-DenseNet. However, SE attention only takes the channel information into account. Hence, the classification results of MobileNet-V3 and GNet are still inferior to the proposed DSCAM, in which both the channel and spatial attention are considered.

Table 2. Classification accuracy (%) ± Standard deviation (%) obtained by different methods.

Methods	Huawei-40	Baidu-214	Baidu-RC	BR-124
VGG-19	84.96 ± 0.63	72.98 ± 0.61	86. 38 ± 0.47	79.13 ± 0.47
Xception	88.45 ± 0.65	79.09 ± 0.55	90.12 ± 0.52	82.88 ± 0.56
X-DenseNet	89.31 ± 0.59	80.38 ± 0.57	91.74 ± 0.46	83.53 ± 0.41
MobileNet-V3	90.83 ± 0.47	82.01 ± 0.63	92.48 ± 0.41	84.70 ± 0.51
GNet	90.99 ± 0.33	82.57 ± 0.43	92.63 ± 0.35	84.89 ± 0.46
DSCAM	**91.20 ± 0.43**	**83.01 ± 0.48**	**92.77 ± 0.45**	**85.35 ± 0.43**

Then, to justify the effectiveness of each component in our DSCAM, we compare the proposed method with three other networks, including Resnet (the original Resnet-50 without attention), SE network and CBAM network. In SE network, we embed SE module in the last convolution layer of each Resnet-50 block. Similarly, the CBAM network sequentially integrates channels and spatial attention modules of CBAM into the same position of Resnet-50 as SE network. Table 3 shows the classification accuracy of these networks in each dataset. First, we can see that Resnet-50 achieves better classification accuracy than some of the networks in Table 2, which indicates more convolution layers could help to enhance the discriminative ability of extracted features. Second, since SE network introduces the squeeze-and-excitation based attention mechanism into Resnet, its performance is superior to Resnet-50. However, since SE network can only learn the channel attention, its classification result is lower than the CBAM network. Third, although the CBAM network takes both the channel and spatial attention into consideration, it merely uses a pooling operation to calculate maximal or average activations along channels or spatial, which may lose some important information. In our proposed DSCAM, the depth-wise separable convolution is adopted to refine the feature before pooling operations, which can improve the capacity of channel and spatial attention maps obtained by our method. Therefore, as can be seen in Table 3, our DSCAM achieves the best classification accuracy.

To further demonstrate the effectiveness of attention in our DSCAM, we visualize the attention maps obtained by our network. From Figure 10, it can be seen that the proposed attention modules can effectively make our method focus on the target object in the image and ignore the interference of background.

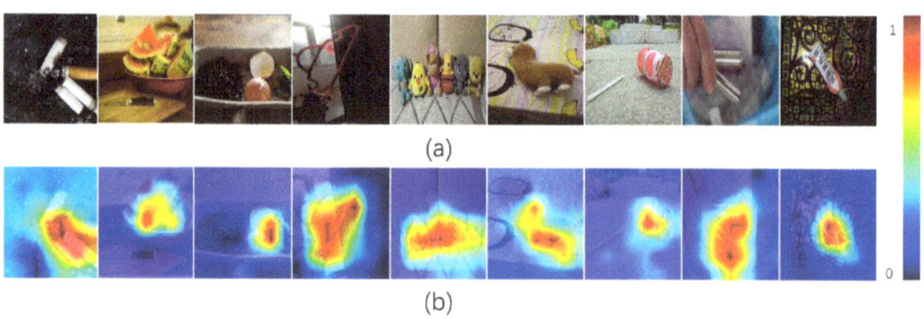

Figure 10. Visualization of the attention maps obtained by our method. (**a**) The origin images randomly selected from Huawei-40 dataset. (**b**) The heatmaps of attention maps, in which redder areas are more important.

Table 3. Classification accuracy (%) ± Standard deviation (%) obtained by each method.

Dataset	Methods	Accuracy
Huawei-40	Resnet	89.47 ± 0.51
	CBAM	90.82 ± 0.48
	SE	90.79 ± 0.41
	DSCAM	**91.20 ± 0.43**
Baidu-214	Resnet	81.67 ± 0.46
	CBAM	82.17 ± 0.57
	SE	82.00 ± 0.48
	DSCAM	**83.01 ± 0.43**
Baidu-RC	Resnet	91.50 ± 0.42
	CBAM	92.25 ± 0.46
	SE	92.26 ± 0.49
	DSCAM	**92.77 ± 0.45**
BR-124	Resnet	83.61 ± 0.51
	CBAM	84.19 ± 0.48
	SE	84.17 ± 0.46
	DSCAM	**85.35 ± 0.43**

Figure 11 shows the accuracy curve of each method from Table 3. In the early stage of iterations, since the parameters in all networks are randomly initialized, their classification results are all very bad. Thus, we only show the accuracy curves between the 25th to 225th training iterations of different methods in this figure. It can be found that the accuracy of our method is lower than other methods when the number of training iterations is small, which may be due to our DSCAM containing more convolution operations. Moreover, since we initially set a larger learning rate to ensure that the networks can quickly achieve a high accuracy at the beginning, the fluctuant phenomena can be clearly seen from the accuracy curves of different methods at the early training stage. Nevertheless, after the learning rate is reduced at the 75th and 150th training iterations, the fluctuation of accuracy curves obtained by all methods become small and the performance of our method improves rapidly and gradually surpasses other approaches with the increasing of training iterations. At last, we can also observe that the average accuracy of all methods becomes nearly stable after 200 iterations, so the classification accuracy at the 200th training iteration is taken as the final experimental result in Table 3.

Next, the confusion matrices of accuracy obtained by various methods on the Baidu-RC dataset are provided for comparison. The confusion matrix can reflect the classification accuracy of each class. The diagonal of the confusion matrix represents the accuracy of each class, and the off-diagonal elements indicate the degree of misclassification. From Figure 12, it can be seen that the confusion matrix obtained by our method is sparser than other methods, which means that our method can extract more discriminative features and classify fewer garbage images into incorrect categories.

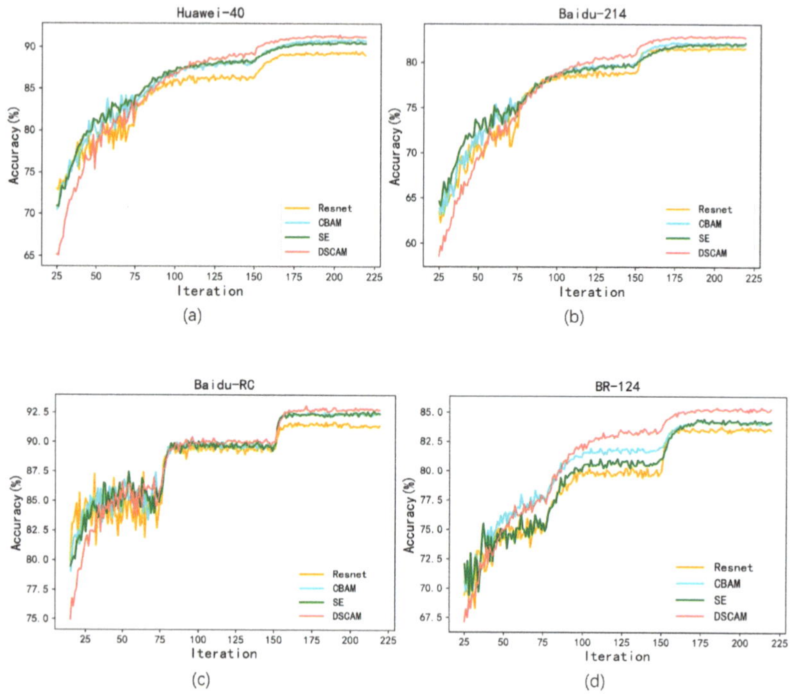

Figure 11. Accuracy curves of different methods on each dataset. (**a**) Huawei-40 dataset; (**b**) Baidu-214 dataset; (**c**) Baidu-RC dataset; (**d**) BR-124 dataset.

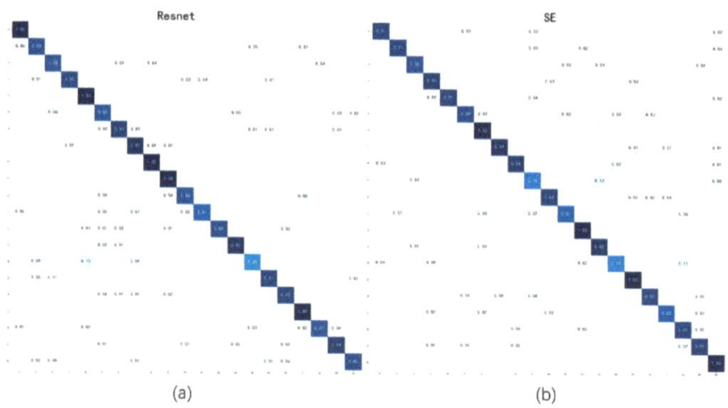

Figure 12. *Cont.*

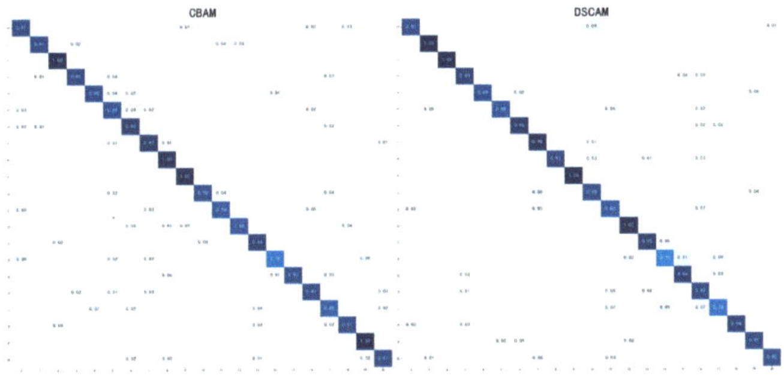

Figure 12. Normalized Confusion matrix of different methods on Baidu-RC dataset. (**a**) Resnet; (**b**) SE; (**c**) CBAM; (**d**) DSCAM. (Zoom in for better view).

To test the statistical significance between different models, the McNemar–Bowker test [38], which can analyze the classification outcome of more than two classes is employed. In our experiment, the significant level is set as 0.05. Table 4 demonstrates the results (p-value) of the McNemar–Bowker test between our DSCAM and compared methods. From these results, we can find that the p-values are all smaller than the significant level. Therefore, through comprehensively considering the results in Table 4, Figures 11 and 12, we can see that the performance of our DSCAM is significantly superior to other methods.

Table 4. McNemar–Bowker test between DSCAM and compared methods.

Dataset	DSCAM & Resnet	DSCAM & CBAM	DSCAM & SE
Huawei-40	$p = 0.0205$	$p = 0.0362$	$p = 0.0275$
Baidu-214	$p = 0.0136$	$p = 0.0188$	$p = 0.0166$
Baidu-RC	$p = 0.0435$	$p = 0.0491$	$p = 0.0494$
BR-124	$p = 0.0065$	$p = 0.0183$	$p = 0.0155$

To illustrate that DSCAM can effectively deal with the inter-class similarity and intra-class variance problem, we demonstrate some images in Figure 13, which misclassified by comparison methods (i.e., Resnet, SE, CBAM) but correctly classified by our method. In this figure, the garbage images in columns (a) and (b) are from different classes. Nevertheless, since the two images in the same column exhibit similar appearances, it is difficult to distinguish them. Contrarily, the garbage images in columns (c) and (d) come from the same class but have quite different appearances. Thus, they are also easily misclassified. However, with the help of the Resnet architecture, the attention module and depth separable convolution, the proposed DSCAM can capture more discriminative deep features and neglect the irrelevant information in the image. Thus, the images in Figure 13 are correctly classified to their corresponding categories by our method. This means that our DSCAM can address the limitations of other methods to some extent.

Figure 13. Samples of misclassification by the comparison methods in the Baidu-214 dataset. (**a**) samples selected from different classes "Recyclables_Mobile phone" and "Recyclables_Portable charger"; (**b**) samples selected from different classes "Recyclables_Plastic toys" and "Recyelables_Plush toys"; (**c**) samples selected from the same class "Recyclables_Shapoo bottle"; (**d**) samples selected from the same class "Kitchen waste_Biscuits".

4.4. Parameter Sensitivity

There exist some parameters whose values may influence the performance of our proposed DSCAM. Thus, we conducted some experiments to test the sensitivity of our method with these parameters.

First, the impact of the parameters (i.e., kernel size and reduction ratio) in depth-wise separable convolution on our DSCAM is evaluated. The performances of our method under various parameter values can be seen in Table 5. From this table, it can be found that DSCAM achieves its best classification accuracy when the kernel size and reduction ratio are set as 3×3 and $r = 16$. The reason may be that a large kernel size will enlarge the receptive field, which overlooks the detailed fine feature in the image. Besides, the purpose of the reduction ratio is to control the degree of information compression along channels. Thus, a larger reduction ratio will lose some useful information, and a smaller reduction ratio will cause information redundancy.

Table 5. Classification accuracy (%) with different parameter values in depth-wise separable convolution.

Depth Convolution Size and Reduction Ratio	Huawei-40	Baidu-214	Baidu-RC	BR-124
3 × 3 (r = 8)	91.02	82.66	92.53	84.78
3 × 3 (r = 16)	**91.20**	**83.01**	**92.77**	**85.35**
3 × 3 (r = 32)	89.90	81.87	91.65	83.86
5 × 5 (r = 8)	90.35	82.47	92.03	83.87
5 × 5 (r = 16)	90.76	82.70	92.36	84.21
5 × 5 (r = 32)	89.63	81.25	91.42	83.01
7 × 7 (r = 8)	89.92	81.86	91.76	83.43
7 × 7 (r = 16)	90.25	82.01	91.95	83.89
7 × 7 (r = 32)	89.03	80.57	91.06	82.62

Since spatial and channel attention have different processes and functions, the order of them may affect the overall performance of our method. Therefore, we compare three different ways of arranging the channel and spatial attention modules: sequential channel+spatial (C + S), sequential spatial+channel (S + C), and parallel use both attention modules (S&C). From the comparison result in Table 6, we can see that sequentially generating the attention maps outperforms the parallel manner. Furthermore, the channel-first order performs slightly better than the spatial-first order. This result justifies the design of our network architecture in Figure 3.

Table 6. Classification accuracy (%) in different architecture combinations of our methods.

Architecture	Huawei-40	Baidu-214	Baidu-RC	BR-124
C + S	91.20	83.01	92.77	85.35
S + C	**90.17**	**81.84**	**92.30**	**84.21**
S&C	89.92	80.24	91.75	83.60

4.5. Ablation Study

The depth of a network (i.e., the number of layers in a network) is an important factor to affect its performance. Generally, more layers will make the network obtain high accuracy. But more layers are also accompanied by more parameters, which would bring longer training and test time. Here, we conduct some experiments to justify the rationality of Resnet-50 as the backbone of our network.

First, we replace the backbone of our network with deeper Resnet-101 and Resnet-152. From the experimental results in Table 7, it can be seen that the backbone network with more layers can slightly improve the performance of our method from 91.20% to 91.27% (Resnet-101) and 91.32% (Resnet-152). However, Resnet-101 and Resnet-152 also greatly increase the number of parameters, training/test time and GPU memory size. Then, we also employ Densenet-121 and Densenet-169 [28] as the backbone of our network. From Table 7, it can also be found that the Densenets with more layers achieve better performance than Resnet-50. Nevertheless, although Densenet-121 and Densenet-169 outperform Resnet-101 and Resnet-152 with fewer parameters, they require larger GPU memory sizes due to the massive concatenation operations in them. Besides, Densenet also needs more training/test time due to it uses many small convolutions in the network, which runs slower on a GPU than large compact convolutions with the same number of GFLOPS [39]. At last, through taking all factors (such as accuracy, time, number of parameters and memory size) into consideration, we choose Resnet-50 as the backbone of our network since it could obtain comparable garbage classification accuracy without very large computation and memory consumption.

Table 7. Comparison of different backbone networks on the Huawei-40 dataset.

Backbone	Accuracy	Training/Test Time Per Image	Params Size (MB)	Memory Size (MB)
DSCAM + Resnet-101	91.27	0.150/0.060	198.90	835.35
DSCAM + Resnet-152	91.32	0.220/0.075	272.71	1183.20
DSCAM + Densenet-121	91.30	0.242/0.088	31.91	5589.45
DSCAM + Densenet-169	91.35	0.278/0.102	53.25	6847.50
DSCAM + Resnet-50	**91.20**	**0.081/0.048**	**109.30**	**523.79**

4.6. Real Scene Application

In order to test the classification result of the proposed DSCAM in real scenes, we construct a simple real garbage dataset. As mentioned before, recyclable garbage is very valuable in real life. Therefore, 400 recyclable garbage images with 20 categories (20 images per category) in real scenes are collected through taking pictures and an online search. Some samples of this dataset are shown in Figure 14. At the same time, to enrich the amount of data, the images in the dataset are randomly clipped, rotated and filled with background for data augmentation. As a result, a total of 2000 recyclable garbage images (100 images per category) were obtained.

Figure 14. Samples of the collected real scene garbage dataset.

In this experiment, we directly inputted the collected garbage images to different networks pre-trained on the BR-124 dataset. Table 8 shows the classification result obtained by each network. First, it can be seen that VGG-19 and Xception obtain the worst accuracy among all methods. Second, the performances of X-DenseNet and Resnet are inferior to other models with an attention mechanism, such as SE, MobileNet-V3, GNet and CBAM. At last, the proposed DSCAM achieved the best results. The observations in Table 8 are consistent with those in the previous experiments, which shows our method has good generalization and can effectively deal with the garbage images in real scenes.

Table 8. Classification accuracy (%) obtained by different methods on the Real Scene dataset.

Methods	Real Scene Dataset
VGG-19	91.0
Xception	92.5
X-DenseNet	95.6
MobileNet-V3	98.5
GNet	98.2
Resnet	97.6
SE	98.0
CBAM	98.7
DSCAM	**98.9**

5. Conclusions

In this paper, we focus on developing a specific deep CNN for garbage image classification problems. To this end, we proposed the attention module DSCAM, which provides a novel mechanism to construct attention weights. Unlike the original attention mechanism, which only uses a pooling layer to infer correlations in channels and spatial, DSCAM utilizes depth-wise separable convolutions to construct the inherent relationship of channel and spatial, which can make the network obtain more discriminative features. Moreover, a Resnet-50 with more convolutional layers is also adopted as the backbone of our method, so that its classification ability can be further improved. Several experiments were conducted to evaluate our method on five garbage datasets. The experimental results illustrate that our method achieves better performance than the compared methods.

In the future, we will embed our attention modules into more recent proposed backbone networks (such as visual transformer [40]) to test their performance on garbage image classification. Furthermore, combining the proposed DSCAM with some hardware and devices (such as a robot chassis, a robotic arm and a camera) to create an automatic garbage sorting system is another interesting topic for future study.

Author Contributions: Conceptualization, F.L. and H.X.; methodology, F.L. and J.W.; software, F.L.; validation H.X.; formal analysis, J.W.; investigation, F.L. and H.X.; resources, J.K.; data curation, M.Q. and D.L., writing—original draft preparation, F.L.; writing—review and editing, J.W. and D.L.; visualization, F.L.; supervision, J.W.; project administration, J.K.; funding acquisition, J.K. and M.Q. All authors have read and agreed to the published version of the manuscript.

Funding: This research was funded in part by the National Natural Science Foundation of China under Grant 61907007, in part by the National Key R&D Program of China under Grant 2020YFA0714102 and in part by the Fund of the Jilin Provincial Science and Technology Department under Grants 20210101187JC.

Institutional Review Board Statement: Not applicable.

Informed Consent Statement: Not applicable.

Data Availability Statement: Baidu AI Studio: https://aistudio.baidu.com/aistudio/index Huawei Cloud: https://pan.baidu.com/s/1NZDZ6LtaDdsdTWoWeVkddw Extraction code: xx0n.

Conflicts of Interest: The authors declare that they have no conflicts of interest.

Real Scene Garbage Dataset: https://pan.baidu.com/s/1h0SOO2jks2O8aP_TK4_OUA Extraction code: rsg2.

References

1. Daniel, H.; Perinaz, B.T. *What a Waste: A Global Review of Solid Waste Management*; World Bank: Washington, DC, USA, 2012.
2. Zhang, D.; Tan, S.K.; Gersberg, R.M. A comparison of municipal solid waste management in Berlin and Singapore. *Waste Manag.* **2010**, *30*, 921–933. [CrossRef] [PubMed]
3. Wang, J.J.; Zhao, N.N.; Li, J.H. Current situation of marine microplastics pollution and prevention proposals in China. *China Environ. Sci.* **2019**, *39*, 3056–3063.
4. Li, W.B.; Ma, G.; Yang, E.Q.; Cai, Y.M.; Chen, Z.; Gao, R.; Pan, E.J. Study on characteristics of electric dust removal fly ash and bag fly ash in circulating fluidized bed waste incineration system. *Proc. CSEE* **2019**, *39*, 1397–1405.
5. Porshnov, D.; Ozols, V.; Klavins, M. Thermogravimetric analysis as express tool for quality assessment of refuse derived fuels used for pyro-gasification. *Environ. Technol.* **2020**, *41*, 29–35. [CrossRef] [PubMed]
6. Pardini, K.; Rodrigues, J.J.; Diallo, O.; Das, A.K.; De Albuquerque, V.H.C.; Kozlov, S.A. A Smart Waste Management Solution Geared towards Citizens. *Sensors* **2020**, *20*, 2380. [CrossRef] [PubMed]
7. Cheng, Y.; Chen, N.; Zhang, H. Coal fly ash as an inducer to study its application in the production of methane gas from domes-tic waste. *Fresenius Environ. Bulletin* **2020**, *29*, 1082–1089.
8. Ren, Y.; Yang, J.; Cao, H.; Zhang, Q.Y.; Liu, Q. All components resourcing system of rural garbage based on post-gather automatic sorting and disposal technology and its application. *Trans. Chin. Soc. Agric. Eng.* **2019**, *35*, 248–254.
9. Wang, W.; Baobao, Z.; ZhiQiang, W.; FangZhi, Z.; Qiang, L. Garbage image recognition and classification based on hog feature and SVM-Boosting. *J. Phys. Conf. Ser.* **2021**, *1966*, 012002.
10. Dong, S. Multi class SVM algorithm with active learning for network traffic classification. *Expert Syst. Appl.* **2021**, *176*, 114885. [CrossRef]
11. Liu, Y.; Fung, K.C.; Ding, W.; Guo, H.; Qu, T.; Xiao, C. Novel Smart Waste Sorting System based on Image Processing Algorithms: SURF-BoW and Multi-class, SVM. *Comput. Inf. Sci.* **2018**, *11*, 35–49. [CrossRef]
12. Bay, H.; Tuytelaars, T.; Van Gool, L. Surf: Speeded up Robust Features. In Proceedings of the European Conference on Computer Vision, Graz, Austria, 7–13 May 2006; pp. 404–417.
13. Ozkaya, U.; Seyfi, L. Fine-tuning models compareds on garbage classification for recyclability. *arXiv* **2019**, arXiv:1908.04393.
14. Fu, B.; Li, S.; Wei, J.; Li, Q.; Wang, Q.; Tu, J. A Novel Intelligent Garbage Classification System Based on Deep Learning and an Embedded Linux System. *IEEE Access* **2021**, *9*, 131134–131146. [CrossRef]
15. Meng, S.; Zhang, N.; Ren, Y. X-DenseNet: Deep learning for garbage classification based on visual images. *J. Phys. Conf. Ser.* **2020**, *1575*, 012139. [CrossRef]
16. Singh, D. Polyth-Net: Classification of Polythene Bags for Garbage Segregation Using Deep Learning. In Proceedings of the 2021 International Conference on Sustainable Energy and Future Electric Transportation (SEFET), Hyderabad, India, 21–23 January 2021; pp. 1–4.
17. Nowakowski, P.; Pamuła, T. Application of deep learning object classifier to improve e-waste collection planning. *Waste Manag.* **2020**, *109*, 1–9. [CrossRef]
18. Chu, Y.; Huang, C.; Xie, X.; Tan, B.; Kamal, S.; Xiong, X. Multilayer hybrid deep-learning method for waste classification and recycling. *Comput. Intell. Neurosci.* **2018**, *2018*, 1–9. [CrossRef] [PubMed]
19. Yu, Y.; Zou, S.; Yin, K. A novel detection fusion network for solid waste sorting. *Int. J. Adv. Robot. Syst.* **2020**, *17*, 1729881420941779. [CrossRef]

20. Kokoulin, A.N.; Tur, A.I.; Yuzhakov, A.A. Convolutional neural networks application in plastic waste recognition and sorting. In Proceedings of the 2018 IEEE Conference of Russian Young Researchers in Electrical and Electronic Engineering (EIConRus), St. Petersburg, Russia, 29 January–1 February 2018; pp. 1094–1098.
21. Wang, C.; Qin, J.; Qu, C.; Ran, X.; Liu, C.; Chen, B. A smart municipal waste management system based on deep-learning and Internet of Things. *Waste Manag.* **2021**, *135*, 20–29. [CrossRef]
22. Yang, M.; Thung, G. *Classification of Trash for Recyclability Status*; CS229 Project Report 2016; Stanford University: Stanford, CA, USA, 2016.
23. Chen, Z.; Hebin, Z.; Yanbo, W.; Yu, L.; Binyan, L. A vision-based robotic grasping system using deep learning for garbage sorting. In Proceedings of the 2017 36th Chinese Control Conference (CCC), Dalian, China, 26–28 July 2017; pp. 11223–11226.
24. He, K.; Zhang, X.; Ren, S.; Sun, J. Deep residual learning for image recognition. In Proceedings of the 2016 IEEE Conference on Computer Vision and Pattern Recognition (CVPR), Las Vegas, NV, USA, 27–30 June 2016; pp. 770–778.
25. Krizhevsky, A.; Sutskever, I.; Hinton, G.E. Imagenet classification with deep convolutional neural networks. *Adv. Neural Inf. Processing Syst.* **2012**, *25*, 1097–1105. [CrossRef]
26. Simonyan, K.; Zisserman, A. Very deep convolutional networks for large-scale image recognition. *arXiv* **2014**, arXiv:1409.1556.
27. Szegedy, C.; Liu, W.; Jia, Y.; Sermanet, P.; Reed, S.; Anguelov, D.; Erhan, D.; Vanhoucke, V.; Rabinovich, A. Going deeper with convolutions. In Proceedings of the 2015 IEEE Conference on Computer Vision and Pattern Recognition (CVPR), Boston, MA, USA, 7–12 June 2015; pp. 1–9.
28. Huang, G.; Liu, Z.; Van Der Maaten, L.; Weinberger, K.Q. Densely connected convolutional networks. In Proceedings of the IEEE Conference on Computer Vision and Pattern Recognition, Honolulu, HI, USA, 21–26 July 2017; pp. 4700–4708.
29. Hu, J.; Shen, L.; Sun, G. Squeeze-and-excitation networks. In Proceedings of the 2018 IEEE Conference on Computer Vision and Pattern Recognition (CVPR), Salt Lake City, UT, USA, 18–23 June 2018; pp. 7132–7141.
30. Woo, S.; Park, J.; Lee, J.Y.; Kweon, I.S. Cbam: Convolutional block attention module. In Proceedings of the European Conference on Computer Vision (ECCV), Munich, Germany, 8–14 September 2018; pp. 3–19.
31. Ioannou, Y.; Robertson, D.; Cipolla, R.; Criminisi, A. Deep roots: Improving cnn efficiency with hierarchical filter groups. In Proceedings of the 2017 IEEE Conference on Computer Vision and Pattern Recognition (CVPR), Honolulu, HI, USA, 21–26 July 2017; pp. 1231–1240.
32. Xie, S.; Girshick, R.; Dollár, P.; Tu, Z.; He, K. Aggregated residual transformations for deep neural networks. In Proceedings of the 2017 IEEE Conference on Computer Vision and Pattern Recognition (CVPR), Honolulu, HI, USA, 21–26 July 2017; pp. 1492–1500.
33. Zhang, T.; Qi, G.J.; Xiao, B.; Wang, J. Interleaved Group Convolutions. In Proceedings of the 2017 IEEE International Conference on Computer Vision (ICCV), Venice, Italy, 22–29 October 2017; pp. 4373–4382.
34. Chollet, F. Xception: Deep learning with depthwise separable convolutions. In Proceedings of the IEEE Conference on Computer Vision and Pattern Recognition, Honolulu, HI, USA, 21–26 July 2017; pp. 1251–1258.
35. Howard, A.G.; Zhu, M.; Chen, B.; Kalenichenko, D.; Wang, W.; Weyand, T.; Andreetto, M.; Adam, H. Mobilenets: Efficient convolutional neural networks for mobile vision applications. *arXiv* **2017**, arXiv:1704.04861.
36. Zhang, X.; Zhou, X.; Lin, M.; Sun, J. Shufflenet: An extremely efficient convolutional neural network for mobile devices. In Proceedings of the 2018 IEEE Conference on Computer Vision and Pattern Recognition (CVPR), Salt Lake City, UT, USA, 18–23 June 2018; pp. 6848–6856.
37. Howard, A.; Sandler, M.; Chu, G.; Chen, L.C.; Chen, B.; Tan, M.; Wang, W.; Zhu, Y.; Pang, R.; Vasudevan, V.; et al. Searching for MobileNetV3. In Proceedings of the 2019 IEEE/CVF International Conference on Computer Vision (ICCV), Seoul, Korea, 27 October–2 November 2019; pp. 1314–1324.
38. Filippini, C.; Cardone, D.; Perpetuini, D.; Chiarelli, A.M.; Gualdi, G.; Amerio, P.; Merla, A. Convolutional neural networks for differential diagnosis of raynaud's phenomenon based on hands thermal patterns. *Appl. Sci.* **2021**, *11*, 3614. [CrossRef]
39. Zhang, C.; Benz, P.; Argaw, D.M.; Lee, S.; Kim, J.; Rameau, F.; Bazin, J.C.; Kweon, I.S. ResNet or DenseNet? Introducing Dense Shortcuts to ResNet. In Proceedings of the 2021 IEEE Winter Conference on Applications of Computer Vision (WACV), Waikoloa, HI, USA, 3–8 January 2021; pp. 3549–3558.
40. Wu, B.; Xu, C.; Dai, X.; Wan, A.; Zhang, P.; Yan, Z.; Tomizuka, M.; Gonzalez, J.; Keutzer, K.; Vajda, P. Visual transformers: Token-based image representation and processing for computer vision. *arXiv* **2020**, arXiv:2006.03677.

information

Review

A Survey on Text Classification Algorithms: From Text to Predictions

Andrea Gasparetto [1,*], Matteo Marcuzzo [1], Alessandro Zangari [1] and Andrea Albarelli [2]

[1] Department of Management, Ca' Foscari University, 30123 Venice, Italy; matteo.marcuzzo@unive.it (M.M.); alessandro.zangari@unive.it (A.Z.)
[2] Department of Environmental Sciences, Informatics and Statistics, Ca' Foscari University, 30123 Venice, Italy; albarelli@unive.it
* Correspondence: andrea.gasparetto@unive.it

Abstract: In recent years, the exponential growth of digital documents has been met by rapid progress in text classification techniques. Newly proposed machine learning algorithms leverage the latest advancements in deep learning methods, allowing for the automatic extraction of expressive features. The swift development of these methods has led to a plethora of strategies to encode natural language into machine-interpretable data. The latest language modelling algorithms are used in conjunction with ad hoc preprocessing procedures, of which the description is often omitted in favour of a more detailed explanation of the classification step. This paper offers a concise review of recent text classification models, with emphasis on the flow of data, from raw text to output labels. We highlight the differences between earlier methods and more recent, deep learning-based methods in both their functioning and in how they transform input data. To give a better perspective on the text classification landscape, we provide an overview of datasets for the English language, as well as supplying instructions for the synthesis of two new multilabel datasets, which we found to be particularly scarce in this setting. Finally, we provide an outline of new experimental results and discuss the open research challenges posed by deep learning-based language models.

Keywords: text classification; tokenisation; topic labelling; news classification; transformer; shallow learning; deep learning; multilabel corpora

Citation: Gasparetto, A.; Marcuzzo, M.; Zangari, A.; Albarelli, A. A Survey on Text Classification Algorithms: From Text to Predictions. *Information* **2022**, *13*, 83. https://doi.org/10.3390/info13020083

Academic Editor: Gennady Agre

Received: 10 January 2022
Accepted: 9 February 2022
Published: 11 February 2022

1. Introduction

Text classification (TC) is a task of fundamental importance, and it has been gaining traction thanks to recent developments in the fields of text mining and natural language processing (NLP). Text classification methods share the common goal of designating a predefined label for a given input text, though this denomination can refer to a variety of specialised methods applied to different domains.

Classic examples of TC include information retrieval, topic labelling, sentiment analysis, and news classification. However, TC has practical applications that extend beyond simple categorisation, such as extractive question answering and summarisation systems. In this case, the intuitive notion of "label" is substituted with a choice between candidates (e.g., an answer or a sentence to include in a summary).

The speed at which textual information is currently being created has long outclassed manual solutions to these tasks, meaning that TC methods are not only useful, but also strictly necessary. Accordingly, developing accurate and unbiased TC systems is of paramount importance.

1.1. Text Classification Tasks

A variety of standard definitions for TC tasks exist in the NLP research area, often used as benchmarks to evaluate new methods. We outline the main representatives, approximately following the taxonomy proposed by Li et al. [1]:

- Sentiment analysis (SA): the task of understanding affective states and subjective information contained in a piece of text, often categorised in terms of stirred emotions;
- Topic labelling (TL): the task of recognising one or more themes for a piece of text (i.e., its topics);
- News classification (NC): the task of assigning categories to news pieces, such as topics or user interests;
- Question answering (QA): the task of selecting an answer to a question, selecting from potential candidate sentences (usually extracted from a context document). This task is usually framed as binary or multiclass classification;
- Natural language inference (NLI): the task of determining whether two sentences entail one another (classifying if the entailment occurs in one of the two directions, or neither);
- Named entity recognition (NER): the task of locating named entities within unstructured text, labelling them with predefined categories;
- Syntactic parsing (SP): a series of tasks related to predicting morpho-syntactic properties of words, such as part-of-speech (PoS) tagging, speech dependencies and semantic role labelling.

It is worth noting that these are generic formulations, meaning that specific tasks will present themselves with slight differences as contextualized to a specific domain.

1.2. *Text Representation*

An important step required by any TC procedure is the projection of text features in a chosen feature space. Because of its lack of structure (from a computational point of view), it is necessary to apply a number of operations, in order to gradually transform it into a form that is digestible for a computer. Preprocessing must be mindful of the models that are intended to be used in the later stages of the classification pipeline, since there is no "silver bullet" solution.

Most notably, earlier methods rely heavily on a manual feature engineering step, which necessitates considerate treatment and domain expertise. Later methods based on deep learning, on the other hand, are notably different because of the automatic extraction of features. As we will explore, preprocessing is still important for these methods, though it may be applied differently because of the assumptions that they make.

1.3. *Broad Categorization of Text Classification Methods*

In this work, machine learning methods are divided into the two broad categories of "shallow" and "deep" approaches, which we proceed to define.

1.3.1. Shallow Learning Approaches

Earlier methods are frequently defined as "shallow learning" approaches. However, since this definition is not one that is particularly standardised or agreed upon, we clarify that, with this term, we refer to all those traditional or classical methods related to conventional machine learning. That is, this group encompasses all of those methods preceding neural networks of which the prediction is based on hand-engineered features. In addition, this category also includes neural networks with very few (0–2) hidden layers, which are themselves referred to as "shallow" and which bridge the gap between this group of methods and their deep learning-based successors.

Shallow learning approaches are the successors of rule-based approaches, which they surpassed in both accuracy and stability. Shallow learning methods are still popular in many practical contexts, or as strong baselines. While they do not scale well to large amounts of data, they shine when resources are too scarce for deep methods to be effective. These classical approaches require a feature engineering step that may be costly, depending on the complexity of the domain. While the computational side of this cost can be significant, the domain knowledge requirements that are necessary for the correct application of appropriate feature extraction techniques may be more difficult to achieve in practice.

1.3.2. Deep Learning Approaches

The advent of deep learning models has affected all fields of artificial intelligence, including text classification. These methods have gained traction because of their ability to model complex features without the necessity of hand engineering them, removing part of the domain knowledge requirement. Instead, work has gone towards the development of neural network architectures able to extract effective representations for textual units. Recent developments have been particularly successful in this, giving birth to semantically meaningful and contextual representations. Automatic feature extraction is particularly advantageous in modelling textual data, as it is capable of leveraging the underlying linguistic structure of a document. This structure is intuitive to us if we understand the language, but is usually incomprehensible to a machine.

1.4. Major Differences and Contributions

Recent publications have explored text classification methods from a generic perspective. Among them, we cite the work by Li et al. [1], which provides a complete investigation of models, ranging from shallow to deep. Kowsari et al.'s [2] survey provides an excellent exploration of preprocessing steps, such as feature extraction and dimensionality reduction. Minaee et al.'s [3] work, on the other hand, focuses solely on a thorough exploration of deep approaches, though it also provides quantitative results for classical methods in its experimental performance analysis.

This work aims to enrich the landscape of text classification surveys by giving an outlook of each step involved in the development of a classifier for textual data. Therefore, we provide a detailed description of the most important data preparation operations utilised in conjunction with text classification algorithms. These passages of the TC pipeline are often overlooked, yet understanding their usage and the motivation behind their choices can prove fundamental in building an effective framework for this task. We continue by summarising information about primary English TC datasets and a general benchmark of state-of-the-art approaches in various sub-tasks. Moreover, we provide results on two newly synthesized multilabel TC datasets, laying out the process to reproduce them. We believe this to be an important contribution, as the sub-tasks that they address (namely multilabel TL and NC) are underrepresented.

In summary, this study's main contributions are as follows:

- We present an analysis of TC procedures, with emphasis on each step of the TC pipeline;
- We showcase a variety of datasets in English, and provide the code for synthesising two multilabel classification datasets;
- We present an overview of quantitative results for main methods in the literature, as well as for the datasets that we presented, complete with an analysis of evaluation metrics.

The rest of the survey is organised as follows. Section 2 explores preprocessing operations in TC pipelines, going through standard operations as well as common tokenisation strategies for more recent, deep learning-based models. Section 3 describes methodologies utilised to project preprocessed data into feature space, starting again from earlier approaches, while also describing word embeddings based on shallow neural networks. Section 4 gives a brief overview of generic classifiers commonly used with such features, while Section 5 goes more in depth into explaining the current landscape of deep learning-based classification methods. We move towards the experimental part of the survey in Section 6, showcasing prominent datasets (as well as two newly synthesised ones) and quantitative results. Finally, we discuss future research directions in Section 7 and conclude in Section 8.

Code and datasets used for experiments in this work are published and available (when legally possible) at https://gitlab.com/distration/dsi-nlp-publib (accessed on 28 December 2021).

2. Preprocessing

Input data for natural language tasks like TC consist of raw, unstructured text. Textual information, unlike other types of data such as images or temporal series, does not possess an intrinsic numerical representation; before feeding it to any classifier, then, it must be projected into an appropriate feature space. Preprocessing procedures are therefore of particular importance, as without them, there is no foundation for feature extraction procedures nor classification algorithms.

In this section, we will provide an overview of the preprocessing operations frequently utilised in TC methods. In the first part, we will cover standard methodologies, most frequently utilised in preparation of manual feature extraction techniques (which are explained in Section 3). We instead reserve the second part of this section to preprocessing operations as contextualized to deep models. By providing an overview of these approaches, we aim to highlight differences and innovations utilised by more recent models, thus offering a more thorough and transparent outlook on how textual data representation is done by state-of-the-art approaches.

2.1. *Standard Preprocessing Operations*

Preprocessing operations clean and normalise input data in order to improve the results of the feature extraction stage, regardless of whether it is manually built (in earlier methods) or automatically learnt (in deep learning methods). In this section, we outline the most prominent preprocessing techniques.

2.1.1. Tokenisation

The most basic preprocessing operation that must be applied to text is that of tokenisation. This procedure is what determines the level of granularity at which we analyse and generate textual data, and may be generally described as the process of breaking a stream of text into smaller chunks (historically called tokens). Until recently, most NLP models have utilised words as their atomic unit of choice, but recent approaches have been decomposing text into smaller units (such as character n-grams or even more maximal forms of decomposition, such as underlying bytes [4]).

Tokenisation [5], while often taken for granted, is a complex matter in itself and has been widely researched. Conventional methods are rule-based, and may be as simple as separating by space, punctuation and contractions. Clearly, much more refined knowledge-based approaches have been developed, still based on linguistic concepts. However, recent developments have moved strongly in the direction of data-driven tokenisers, which yield better results, but in which tokens no longer correspond to the traditional definition of a typographic unit. In other words, the modern meaning of "tokenisation" refers to the task of segmenting a sentence into units that, crucially, do not have a linguistic motivation or explanation. Because some of these methods still utilise traditional tokenisers as an initial step, conventional approaches are now often called "pre-tokenisers".

An in-depth exploration of this topic is outside the scope of this survey, and we point to Mielke et al.'s work [6] for an excellent historical review of the evolution of tokenisers over recent years. As both tokenisation and classification approaches evolved in parallel, it is more common to associate conventional methods with pre-tokenisers. On the other hand, deep learning methods tend to approach tokenisation with the more recent and sophisticated methods, as will be outlined in Section 2.2.

2.1.2. Stopword and Noise Removal

The set of tokens produced by the tokenisation procedure may contain unnecessary or misleading elements. Textual noise such as special characters or superfluous symbols should be removed. It can be useful to remove stopwords [7], i.e., non-informative words that appear in large numbers but carry no semantic importance. Other normalisation procedures, such as lowercasing, misspelling correction, and the standardisation of slang words and abbreviations, may be useful in reducing the number of different elements in the feature space.

These processes, however, may cause interpretation issues (e.g., in English, "US" could be used as an acronym for the "United States", but lowercasing would make it indistinguishable from "us" as a pronoun) [2]. Furthermore, the removal of stopwords and other pieces of text can be detrimental to more recent approaches. Deep models are trained to understand natural language; elements that constitute the syntax of a sentence may be fundamental to the model for it to understand the context of a piece of text. As such, destructive operations should not be applied indiscriminately.

2.1.3. Further Standardisation of Text

On the other hand, classical approaches may benefit from a further simplification of the feature space. Such methods are unable to capture significant semantic information about words; two inflections of the same word, then, cannot be distinguished (e.g., child vs. children), and will be treated as completely different pieces of information. Because of this, simplifying words by reducing inflections to a common form may be beneficial. This is achieved through the process of stemming (deriving a root form for a word) [8] or lemmatisation (deriving the lemma/canonical form for a word) [9]. These procedures can improve the performance of the overall classification, but at the same time, are not devoid of issues. Most notably, words with different meanings might have the same root or lemma, and these approaches will make them indistinguishable.

There are other useful operations that could help towards the improvement of a TC algorithm. An example of this is part-of-speech tagging; a "part of speech" is a category of words with similar grammatical properties, such as noun, verb, and adjective. This is sometimes included as a preprocessing step whenever the task could benefit from limiting the vocabulary to one or a subset of these parts of speech (for example, one might be interested in only nouns and adjectives). PoS tagging is a classification task in itself, and it is easy to recognise how misclassifications can result in worse overall performance by the models employing this procedure. An error of this kind can either introduce unwanted words or seclude important ones by mistake.

2.2. *Preprocessing for Deep Models*

As briefly mentioned, preprocessing operations should not be applied without consideration of the method being used. Recent models based on deep neural network architectures usually include similar steps for removing special characters, lowercasing and stripping letters of accents and apostrophes. Additionally, tokenised documents are also truncated or padded to a specified number of tokens to ensure that the model receives input samples of uniform size (i.e., with the same number of tokens).

Most of the attention, however, is shifted towards the effective tokenisation of text. Modern tokenisation approaches must strike a balance between a sufficiently large and expressive vocabulary and one that is too expensive in terms of memory. Therefore, different approaches have been proposed, of which we highlight the most prominent. As a side note, it is fairly common for modern tokenisers to also apply normalisation operations within their procedures [6].

2.2.1. Tokenisation in Deep Models

Deep models learn a vectorial representation for each token seen in training data. Learnt vectors are stored inside an embedding matrix, a data structure that maps recognised vocabulary terms to corresponding embeddings. The matrix size is hence dependent on the number of vocabulary terms and the embedding dimensionality. As mentioned, limited time and available memory imply the inability to deal with arbitrarily large vocabularies. These considerations motivate the development of different tokenisation techniques aimed at reducing the vocabulary size, minimising, at the same time, the number of unrepresented words. The latter are denoted as out-of-vocabulary (OOV) words, and are the central weakness of word-level models; these lead to unknown tokens at test time, which are not acceptable for most NLP tasks.

In pre-tokenisers, due to the fact that a representation is learnt for every word in the training corpus, related words like derivations and inflections (e.g., "snowboarding" and "snowboard") are considered distinct, forcing the model to learn redundant encodings for each of them. Considering that the number of learnable parameters of such models is linearly dependent on the number of distinct words in the corpus, this issue may lead to unmanageable space and time requirements. Conversely, character-level tokenisation would imply a vocabulary set composed of all single characters in the corpus; this results in a minimal vocabulary, at least for alphabetical languages. While this strategy reduces memory and time complexity with respect to word-level tokenisation, it makes it much harder for the model to learn meaningful token representations. For instance, it is much harder to learn meaningful representation for the character "s" than it is for the word "snow".

Since both of these simple strategies are not entirely satisfactory, most models employ hybrid techniques that segment words in sub-words. The general principle is that frequently used words should not be split into smaller words, but rare words should be decomposed into more meaningful segments. Table 1 summarises the main features of tokenisers commonly paired with deep neural models, which are presented in the next sections.

Table 1. Overview of modern tokenisation approaches. SentencePiece is included, but should not be mistaken as a separate algorithm.

Tokeniser	Pre-Tokenisation	Inference Procedure	Language Support
BPE	Yes	Merge incrementally, keeping merge if in vocabulary	Whitespaced only
WordPiece	Yes	Decrementally search for longest first substring of words within vocabulary	Whitespaced only
UnigramLM	No (instead, creates common sub-strings)	Uses Viterbi Algorithm to find which substrings maximise likelihood	All languages
SentencePiece (sw package)	Depends on method utilised	Fast optimised methods for internal algorithms	All languages

2.2.2. Byte Pair Encoding

The strategy that is now considered the breakthrough for sub-word tokenisation is Byte Pair Encoding (BPE). Originally proposed as a data compression algorithm [10], it was later adapted for sub-word segmentation (e.g., segmenting "snowboarding" as "snow"-"board"-"ing") [11]. When learning a tokenisation, this algorithm iteratively computes the occurrences of consecutive pairs of vocabulary terms and merges the most frequent one into a new vocabulary word. When presented with unseen text to tokenise, the same merging procedure is performed by executing all recorded merges in the order in which they were performed during training. BPE's vocabulary is initialised with the set of characters that appears in the training corpus.

A recent variation of this strategy is used with the GPT-2 [12] and RoBERTa [13] language models (see Section 5.4.4). In particular, these models utilise byte-level BPE [14], which applies this procedure to raw bytes rather than characters.

2.2.3. WordPiece

The WordPiece tokeniser [15], initially devised as a solution to Japanese text segmentation problems, implements a data-driven approach for splitting words into sub-words. It relies on the creation of n-gram-based language models (see Section 3.2) to recognise recurring syllables, prefixes and word segments in a corpus. Referring to the previous "snowboarding" example, the tokeniser must learn that "-ing" is a common ending for verbs, and a word should be likely split before and not after it. Hence, the goal of the model is the creation of a vocabulary of sub-words, such that the size of the training corpus is minimal when segmented according to the selected vocabulary. A greedy algorithm is used to solve this optimisation problem. Again, the initial vocabulary is composed of all single characters appearing in the input documents. Every successive iteration increases the vocabulary size by one, selecting the pair of sub-words that maximises the language model likelihood (hence the greedy nature) and merging them into a new vocabulary term.

The algorithm is stopped when the expected likelihood given by merging falls below a predefined threshold, or the maximum vocabulary size is reached. Recent approaches such as BERT [16] use tokenisers based on WordPiece; BERT, in particular, utilises a vocabulary of 30,000 tokens calibrated on English text.

2.2.4. UnigramLM

Similar to WordPiece, UnigramLM [17] is also based on the usage of language models to judge sub-word candidates (the name is derived from the fact that it utilises a simple unigram LM). Differently from Wordpiece, however, UnigramLM goes in the opposite direction, by initialising to a vocabulary size that is much larger than the count of sub-words desired (such as all pre-tokenised words and common sub-strings), and proceeds to iteratively remove them. In every iteration, the expectation–maximization algorithm is used to prune the lowest probability items, cycling until the vocabulary has reached the desired size.

Most interestingly, this sort of probabilistic setup results in different possible segmentations, all of which are consistent with the given strings. While the algorithm chooses the most likely segmentation in practice, it allows a sampling of different tokenisations based on their probabilities (which are recorded during training), allowing for what is defined as "sub-word regularisation", which has been shown to improve results on some tasks. While UnigramLM is not commonly utilised by itself, it has seen use as a component of SentencePiece, which we now describe.

2.2.5. SentencePiece

Kudo and Richardson [18] propose some improvements over the previously described algorithms in a software package called SentencePiece, which contains optimised versions of the above approaches (though it is often mistakenly assumed to be a separate algorithm). In particular, all the tokenisers described so far depend on knowing which characters act as word separators in the corpus; these characters can be language-dependent, and specific pre-tokenisation procedures can be used to create rules for recognising word boundaries. SentencePiece removes the dependence on this step by considering text as a raw stream of characters, including word separators. Additionally, it implements sub-word regularisation techniques, mainly to improve segmentation for machine translation tasks. XLNet [19] and XLM-R [20] use SentencePiece to tokenise input data.

3. Projecting into Feature Space

Generally speaking, preprocessing pipelines transform bodies of text into lists of separated, standardised tokens. From a technical perspective, words are mapped to an index-based vocabulary, such as to simplify the internal representation of the tokens. However, tokens (or their index) must still be represented in a form that is digestible by a machine, i.e., a vectorial form. Many have been proposed over the years; this section provides an overview of the most popular and effective ones.

3.1. Bag-of-Words

The most intuitive representation can be found in the bag-of-words (BoW). This approach simplifies bodies of text by considering them as unordered collections of words. Clearly, this has the disadvantage of ignoring sentence structure and semantic relationships between sentence elements (as if shuffled inside of a "bag" of words). Nonetheless, despite its strong assumptions, it has been shown to obtain good results and has seen wide use. In general, efficient feature extraction can usually lead to strong performances, even when discarding important but difficult to encapsulate information, which is why this approach has been utilised extensively in NLP as well as other machine learning fields (where it is referred to as "bag-of-features") [21–23].

The original idea behind BoW models is to represent each word as a one-hot-encoded vector with the same size as the vocabulary. It is immediate to see that this may lead to size issues, since the vocabulary itself may have a cardinality in the order of millions. As such, BoW models are usually implemented in conjunction with feature extraction techniques based on the multiplicity of words, which allows one to maintain a single vector per document rather than one for each word. For this purpose, term frequency (TF) counts the number of times that a word occurs within a body of text. When applied to a corpus of texts, rather than the explicit count, it is common practice to utilise the relative frequency of a term in the text in relation to other documents. One may also observe that, in very large corpora, common words, in particular, are inherently less useful (as they appear in all documents, hence do not help in distinguishing them). Consequently, TF is often weighted by inverse document frequency (IDF) [24], which lessens the effect of common words by penalising their overall score (and boosting the one of rarer words).

Depending on the size of the vocabulary, the size of TF-IDF representations may still be excessively large. To alleviate issues related to time complexity and memory consumption, it is possible to set a limit to the maximum number of features in the vectors (effectively pruning low-score words from the vocabulary). Alternatively, it is also possible to utilise a dimensionality reduction algorithm on the full-sized representations. Although a detailed description is outside the scope of this review, the general aim of these algorithms is to find a mapping between these representations and a lower-dimensional, compressed feature space. Popular methods include Principal Components Analysis (PCA) [25], Linear Discriminant Analysis (LDA) [26], and non-negative matrix factorization (NMF) [27]. See Kowsari et al.'s [2] survey for an introduction of these methods.

3.2. Language Models

Language modelling is the task of predicting the likelihood of a string given a sequence of preceding or surrounding context words—at its simplest, guess the next word in a sentence. Language models play an important part in more recent, deep learning-based developments, but their inception much precedes neural networks. N-gram models [28] are some of the earliest implementations of language models, and work by assigning probabilities to sequences of words (i.e., sentences). The specific interpretation of this probability value is task-dependent; intuitively, a higher score is associated with a better-structured sentence (e.g., a good translation).

While the goal is to assign probabilities to whole sentences, the task is related to the computation of the probability of an upcoming word, and is framed as such. These models usually make a simplifying assumption (the Markov assumption): it is assumed that the probability of an upcoming word only depends on the last n words before it. The probability value itself can be computed with methods such as Maximum Likelihood Estimation (MLE) [29], a robust estimation approach used thoroughly in probabilistic approaches. Notably, the BoW model can be seen as an n-Gram model with $n = 1$.

3.3. Word Embeddings

While previous methods have focused on capturing the syntactic representations of words and, in some cases, a small subset of the syntactic relationships that tie them together in sentences, they critically still lack the capability of capturing their semantic meaning. A classic example of this issue is represented by word synonyms; though they are—semantically speaking—the same, these models cannot capture their similarity. When looking at the feature space, this translates into representations that are orthogonal to each other, meaning that they are seen as completely different and separate.

In the last decade, researchers have proposed word embeddings as a solution to this problem. Intuitively, this self-supervised feature learning technique is aimed at learning a mapping between each piece of text (most commonly words, hence the name) to a n-dimensional vector of real numbers. An embedding is therefore a vectorial representation, which is digestible by a machine, but also encodes part of the underlying meaning of the words. These approaches are based on neural networks, which learn these mappings through different learning procedures; in general, they are based on the assumption that a word's meaning can be extracted from its surrounding words in a sentence (similarly to language modelling, but focused on the embeddings as individual entities).

Earlier word embeddings are often informally defined as "static", which can be attributed to how, in their basic form, they encode words outside of context. Practically speaking, this means that they do not model polysemy [30] (where an individual word can have different meanings). The embedding for a word is just one, regardless of how many meanings it could have; if a word token is particularly polysemous, it is likely that its embedding will be a combination of its multiple senses.

As an example, consider the word "sound"; as a noun, we may associate this with something that can be heard, whereas, as an adjective, we can associate it with a description of something in good condition. However, these are just two of the almost 50 different meanings that this word can have; based on this consideration, it appears obvious that a single representation can hardly be effective at representing them all at once. As will be further discussed in Section 5.4.7, recent developments have proposed to utilise surrounding context words in order to differentiate a word's meaning.

3.3.1. Word2Vec

One of the first popular families of word embedding architectures is to be attributed to Mikolov et al.'s Word2Vec [31,32]. Their approach utilises shallow neural networks in order to create a high dimensional vector for each word (though they are much smaller and denser than, for example, TF-IDF vectors). Word2Vec was first proposed with two architectural variants: the Continuous-bag-of-words (CBOW) and the continuous skip-gram.

The way in which the CBOW approach learns word representations is to try to predict a middle word based on its surrounding context words. The reference to bag-of-words derives from the fact that the order of the context elements is not actually taken into account. The continuous skip-gram model instead flips the task on its head, as it attempts to predict the neighbours of a word, given the word itself.

These tasks are clearly hard, and the model is not meant to learn perfectly how to guess these words; the true aim is not that of predicting the words correctly, but to create meaningful mappings for words to embeddings [31].

3.3.2. GloVe

Global Vectors for Word Representations (GloVe) [33] represent another popular word embedding technique. The approach is similar to Word2Vec, though it differs fundamentally by being a count-based model, whereas standard Word2Vec is a predictive model. While predictive models learn word vectors by minimising the loss between target and prediction given context words and vector representations, a count-based model essentially learns semantic similarity between words by explicitly probing the underlying statistics of the corpus, such as words co-occurrence [34]. The key difference is that Word2Vec only leverages local information (the context of each word) to obtain word vectors, while GloVe embeddings are trained by also considering global co-occurrence statistics. Something that should also be noted is that GloVe models utilise a dimensionality reduction step in order to handle the large dimensions of the word co-occurrence matrix that it uses in its calculations. Although compressing representations can arguably lead to a more robust representation (as it theoretically forces the model to try to preserve the most significant pieces of information), a bigger advantage comes from the fact that this approach is more suitable for parallelisation, making it easier to train on more data.

3.3.3. FastText

In the context of static word embeddings, FastText is one of the most novel techniques, developed by Bojanowski et al. [35]. The main concern addressed by this method is the fact that its predecessors ignore the morphology of words by assigning a distinct vector to each word. In FastText, each word is represented instead by a "bag-of-characters n-gram". For example, the word "where", with $n = 3$ would be represented as:

$$< wh, whe, her, ere, re >$$

as well as the complete word as a special sequence.

In terms of architecture, FastText embeddings are trained using a skip-gram architecture. Due to the way in which words are represented, however, the final vector for a word will be constituted by the sum of its character n-grams. This is beneficial, as it allows it to generate good word embeddings for rare words—their n-grams will also be shared by more common words. Most importantly, this also means that FastText is able to handle OOV words, as long as it has seen its composing n-grams during training. Both GloVe and Word2Vec are instead unable to handle the case of OOV words.

4. Overview of Shallow Learning Classification Methods

Shallow learning models were, up until recently, the go-to approach for text classification. In terms of actual classification algorithms, these methods mostly rely on general-purpose classifiers that are not specific to this context. The particular challenges presented by textual data are somewhat "offloaded" to the preceding steps of the TC pipeline (Figure 1), which consist in the extraction of machine-interpretable features and representations from documents (i.e., text interpretation).

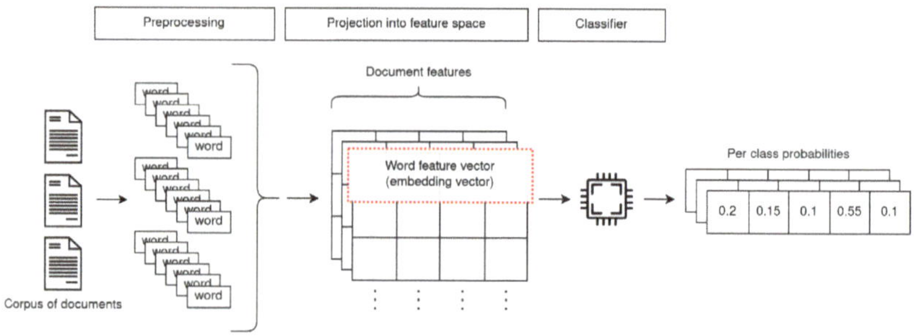

Figure 1. An overview of the two-step procedure adopted by shallow learning methods.

In this section, we provide a brief overview of TC classification algorithms, which are usually categorised as shallow learning-based. These methods are built on generic classification approaches; as highlighted, a large amount of focus is placed on scrupulous data preparation and feature engineering in order to obtain competitive results. As such, we only provide high-level concepts for these approaches, and refer to Kowsari et al.'s work [2] for a more in-depth exploration. Table 2 provides an overview of these methods.

Table 2. Conventional shallow classification techniques.

Model	Advantages	Disadvantages
PGM	• Methods like Naïve Bayes are easy to implement and train, and faster than most conventional methods • Methods like CRFs increase flexibility and expressive power	• Naïve Bayes has strong assumptions that do not always work out • CRFs have higher computational complexity as well as issues with online learning
K-NN	• Non-parametric and fast under the right conditions • Formulation easily adapts to multiclass	• Scales unfavorably with high-dimensional spaces • The value of k is hard to choose • Distance function is hard to define for text
SVM	• Can model non-linear decision boundaries effectively • Effective in high dimensional spaces • Robust against overfitting	• Does not produce probabilities, which must be obtained with expensive cross-validation procedures • High number of dimension causes loss of transparency • High memory complexity • Choosing a kernel function is difficult
DT	• Easily models categorical features • Particularly effective if the decision boundaries are parallel to the feature axis • Fast and easily interpretable	• Extremely susceptible to noise • Very easily overfits • Has issues with diagonal decision boundaries
LR	• Easy to implement and train • Does not necessitate re-scaling of features or fine-tuning	• Can only solve linear problems • Strong assumption of data points as independent
RF	• Fast in comparison to other ensembles • Reduces the variance of single decision trees	• Still prone to overfitting • Difficult to interpret • Slower inference when compared to single trees
Ensembles	• Improve robustness and accuracy • Reduce overfitting	• Training multiple classifiers is expensive • Difficult to interpret • Requires careful fine-tuning

4.1. Probabilistic Classification

Probabilistic graphical models (PGMs) have seen broad use thanks to their effectiveness. Models like Naïve Bayes (NB) [36] are especially popular because of their simplicity, both in structure and calculation process. The simplicity is derived from its independence assumption—that is, each feature is assumed to have no influence on the others. The essence of the Naïve Bayes approach stands in utilising the prior probability of a class given the features—as observed in the training set—to calculate its posterior probability. Other PGM models such as hidden Markov models [37] (HMMs) and conditional random fields (CRFs) [38] have been proposed to encapsulate the sequential nature of textual data (ignored by the independence assumption of NB).

4.2. K-NN-Based Classification

Text classification based on k-nearest neighbours (k-NN) algorithms [39,40] tackles the problem differently, by finding the k most similar labelled instances and, in its simplest iteration, assigning the most common category to the unlabelled instance being classified. This non-parametric method can be quite fast, since it only needs to calculate the distances between data points. However, its performance greatly depends on the chosen distance function, and different functions or even approximations may be needed to deal with large datasets, where k-NN-based methods' performance may suffer [41]. Furthermore, the crucial assumption that similar instances are close in the feature space gradually falters as the number of dimensions in this space becomes larger and larger (i.e., the infamous "curse of dimensionality" [42]).

4.3. Support Vector Machines

Support vector machines (SVMs) [43,44] have been historically robust prediction methods, and have seen success by turning TC tasks into (possibly multiple) binary classification tasks. In short, SVMs construct an optimal hyperplane in a simple one-dimensional feature space, such as to separate the two categories belonging to the binary classification sub-task. To extend to multi-dimensional, non-linear classification, SVMs map their inputs to a higher-dimensional space, so that they may be able to better separate training categories. This procedure is referred to as the kernel trick, since the function that maps to this higher-dimensional space is called a kernel function. Choosing its form and parameters is fundamental to achieving good performance.

4.4. Decision Trees and Random Forests

Decision trees [45] are some of the earliest and most popular classifiers, based on an intuitive tree structure learning method that hierarchically decomposes the data space. The desired result is a tree in which each internal node serves to examine a feature, sending unlabelled instances towards one of its children, depending on the feature value found. Leaf nodes will represent specific categories. Although very intuitive, their basic form can be sensitive to small perturbations in data and overfitting. Random forests [46,47], collections of decision trees trained using random subsets of features, have achieved much better performance and are more used in practice.

4.5. Logistic Regression

One of the earliest methods for classification worth citing is that of logistic regression (LR) [48]. LR is a linear classifier, which predicts probabilities over classes by attempting to discern which features are most useful to discriminate examples. Its base formulation is best suited to binary classification tasks, but it can be extended to the multinomial case [49] by utilizing a formulation that includes (usually) the softmax function, or by creating an ensemble of multiple binary classifiers in a one-vs.-rest scheme.

4.6. Ensemble Learning

Integration-based (or ensemble learning) methods are also another noteworthy mention; these approaches aggregate the results of multiple algorithms, such as to obtain better performance and interpretation. These include various subcategories, the most popular of which consist of bagging and boosting. Bagging [50] (also referred to as bootstrap aggregation methods) averages the output of multiple classifiers without strong dependencies, training each of them separately on a subset of the training data (sampling with replacement). This procedure improves accuracy and stability, and random forests are the most common example of such an approach. Boosting [51] instead sequentially trains weak classifiers in order to obtain a strong ensemble. Each weak classifier "re-weights" the data points based on its own accuracy, giving greater weight to misclassified examples and lower ones to correct predictions. Thus, its successors will know to focus on the "hard" data points that were previously wrongly classified. A classical example of boosting is given by AdaBoost [52].

4.7. Neural-Based Methods

Even though, with regards to neural networks, deep learning methods are undoubtedly the most popular approaches in the NLP field as of now, shallow architectures have been used for TC tasks with competitive results. As an example, the FastText classifier is based on one such architecture; this method extracts n-gram features (based on the embeddings extracted as previously explained in Section 3.3.3), averages them, and then feeds them to a linear classifier (e.g., with a softmax output, if we are in a multiclass scenario) [53]. Another domain-specific shallow neural method is GHS-NET [54], which combines a CNN and GRU layer to classify biomedical texts into a predefined set of disease codes. The model achieved competitive results on several multilabel biomedical benchmark datasets.

4.8. Summary

While it covers the most popular approaches, the list of methods showcased in the previous subsections is not an exhaustive representation of all existing conventional classification models. What is perhaps most notable about these methods, however, is that they have arguably received the most attention in terms of improvements and updated iterations even in recent years, which is why we chose to highlight them. In particular, ensemble meta-algorithms provide an excellent way of utilising "weaker" classifiers and still obtain good results.

5. Deep Learning Methods

One of the limiting factors of classical models is their reliance on explicit feature extraction procedures. Good feature engineering requires extensive domain knowledge, which in turn, makes it difficult to generalise approaches to new tasks. Furthermore, manual feature crafting does not utilise the full potential of large amounts of training data because of how features are predefined rather than discovered.

As a consequence, the development of word embeddings marks the beginning of a paradigm shift towards approaches able to leverage vast amounts of data. Deep learning approaches have gained popularity thanks to their ability to capture meaningful latent information in the training data (depicted in Figure 2); in the case of NLP, such architectures are able to create semantically meaningful representations of text. Recently developed contextualized versions of word embeddings have obtained outstanding results in classification, even when paired with very simple classifiers. An informal yet intuitive explanation of this result is that understanding the content of a body of text is the most important step in the classification pipeline, much like a person would likely be able to label a piece of text if they understood what it meant.

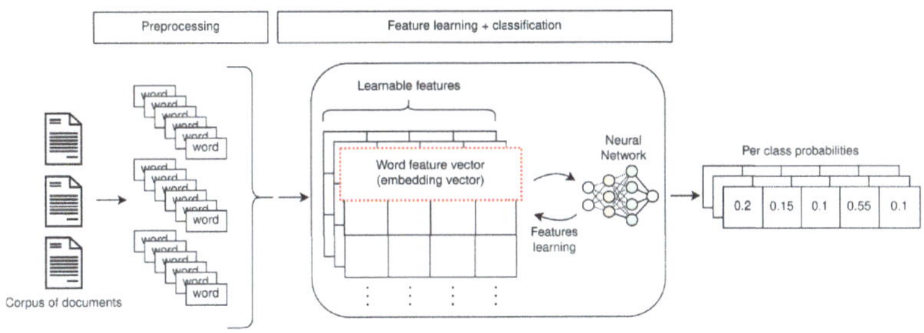

Figure 2. Overview of the training procedure used with deep learning methods.

5.1. Multilayer Perceptrons

Simpler architectures, such as multilayer perceptrons (MLPs), bridge the gap between shallow and deep methods. These are some of the most basic neural network designs, yet they are the foundation of the first word embedding techniques, while also obtaining excellent results as stand-alone classifiers. MLP models such as these usually treat input text as an unordered bag-of-words, where input words are represented through some feature extraction technique (like TF-IDF or word embeddings). For example, deep averaging networks (DAN) [55] are able to perform comparably or better than much more sophisticated methods [3], despite ignoring the syntactic ordering of words. Paragraph-Vec [56] tries instead to incorporate such ordering and the contextual information of paragraphs by utilising an approach that is comparable with CBOW [57], leading to better performances than previous methods.

5.2. Recurrent Neural Networks

The most influential architectures of this period of time were, however, recurrent neural networks (RNNs). RNNs are a popular choice for any type of sequential data; these architectures are aimed at extracting information regarding the structure of sentences, thus capturing latent relationships between context words. The input of an RNN model is usually a sentence represented by a sequence of word embeddings, with its words entering the model one at a time. Long short-term memory networks (LSTMs) are the most popular variant of RNNs, as they address the gradient vanishing or exploding issues faced by standard RNN architectures [58]. Many TC approaches using LSTMs have been proposed, and we mention a few. Tree-LSTM [59] extends LSTMs to tree structures, rather than sequential chain structures, arguing that trees provide a more suitable representation for phrases. TopicRNN [60] integrates the capabilities of latent topic models to overcome the difficulties of RNNs on long-range dependencies. Both of these approaches exhibit improvements on baselines when applied to TC tasks (in particular, sentiment analysis). Howard and Ruder [61] propose the Universal Language Model Fine-tuning (ULMFiT), a recurrent architecture based on an LSTM network trained using discriminative fine-tuning, which allows the tuning of LSTMs using different learning rates in each layer. The Disconnected Recurrent Neural Network (DRNN), introduced by Wang [62], demonstrates the benefits of boosting the feature extraction capabilities of RNNs by incorporating position–invariance—a property attributed to CNNs—into a network based on gated recurrent units (GRU) [63]. In both cases, the proposed enhancements allow surpassing state-of-the-art results on several benchmark datasets. The introduction of bidirectionality in RNNs has also been proven beneficial [64], and has been applied to LSTMs, with notable results such as ELMo [65], a language modelling approach that relies on bidirectional LSTMs, and is one of the first milestones in the development of contextualized word embeddings.

5.3. Convolutional Neural Networks

Convolutional neural networks (CNNs) are commonly associated with computer vision applications, yet they have also seen applications in the context of NLP and TC specifically [66]. The easiest way to understand these approaches in this context is to examine their input, which also relies on word embeddings. While, generally speaking, RNNs input the words of a sentence sequentially, sentences in CNNs are instead presented as a matrix in which each row is the embedding of a word (therefore, the number of columns corresponds to the size of the embeddings). To make a comparison, in image-related tasks, convolutional filters usually slide over local patches of an image across two directions. Instead, filters in text-related tasks are most commonly made as wide as the embedding size, so that this operation only moves in directions that make sense sentence-wise, always considering the entire embedding for each word. In general, the main upside associated with CNNs is their speed and how efficient their latent representations are. Conversely, other properties that could be exploited while working with images, such as location invariance and local compositionality [67,68], make little sense when analysing text. Many approaches have been proposed, one of the most popular being TextCNN [69], a comparatively simple CNN-based model with a one layer convolution structure that is placed on top of word embeddings. More recently, the interest in CNN applications to TC tasks has been renewed thanks to the introduction of temporal convolutional networks (TCN) [70–72], which enhances CNNs with the ability to capture high-level temporal information. For instance, Conneau et al. [73] propose the Very Deep CNN classifier, a 29-layer CNN with skip connections, alternating temporal convolutional blocks to max pooling operations. They achieve state-of-the-art performance on reported TL and SA datasets. Duque et al. [74] modify the VDCNN architecture to lessen the resource requirements and fit mobile platform constraints, while retaining most of its performance.

5.4. Deep Language Models for Classification

In this section, we outline the theoretical notions that lead to the development of current deep language model-based approaches. Then, we proceed to discuss some of these models in more detail. Currently, very few other models are able to compete performance-wise with such classifiers (among them, we discuss the particularly interesting graph-based models).

5.4.1. RNN Encoder–Decoders

For many years, networks based on RNN-like architectures have dominated sequence transduction approaches. Researchers started pushing the boundaries of TC through recurrent language models (evolutions of classic word embedding techniques) and RNN-based encoder–decoder architectures [63,75] (Figure 3).

RNN encoder–decoder architectures are particularly important in this regard. As an example, consider a machine translation task; the input sequence may be a sentence in English, and the output sequences its translation in a different language. In this case, each word of the input sentence is fed to the encoder sequentially, such that at time step t, the model receives the new input word and the hidden state at time step $t-1$. The input is hence consumed sequentially, and the dependence on the output of the previous step should, in theory, allow RNNs to learn long- and short-term dependencies between words. The output of the encoder is the compressed representation of the input sequence, called "context". The decoder then proceeds to interpret the context and produce a new sequence of words (e.g., a translation in a different language) in a sequential manner, where every word depends on the output of the previous time step. With encoder–decoder networks, semantically and syntactically meaningful information is implicitly captured during the encoding phase (the context), which can then potentially be used for tasks such as text classification.

The main issue with this approach is that the encoder needs to compress all relevant information in a fixed-length vector. This was found to be problematic, especially for longer sentences, and it was reported that the performance of a basic encoder–decoder deteriorates rapidly as the length of an input sentence increases [76]. In addition, recurrent models have inherent limitations due to their sequential nature. Sequentiality precludes parallelisation, which means higher computational complexity. Longer sentences can run into memory constraints and, more crucially, are seen as RNNs' true bottleneck because of how the network tends to forget earlier parts of the sequence, making for an incomplete representation (an issue linked to the vanishing gradient problem) [77].

One of the solutions devised to solve the limitations of recurrent architectures was that of the attention mechanism [76,78]. This mechanism would eventually become an integral part of the Transformer architecture, which represents one of the most important milestones in the field of NLP. We denote the particularly effective language models of this era as "Deep" (deep language models, DLM for short), such as to distinguish them from earlier approaches and highlight their reliance on deep architectures. As a matter of fact, depth in these architectures has been proven to be extremely beneficial to their performance, while LSTM-based models were unable to gain much from a large increase in their size [79].

5.4.2. The Attention Mechanism

Attention was initially utilised as an enhancement to various architectures, and was meant to allow the learning process to focus on more relevant parts of input sentences (i.e., give them "attention"). As previously stated, seq2seq tasks have been traditionally approached with RNN-based encoder–decoder architectures, where both the encoder and the decoder are stacked RNN layers.

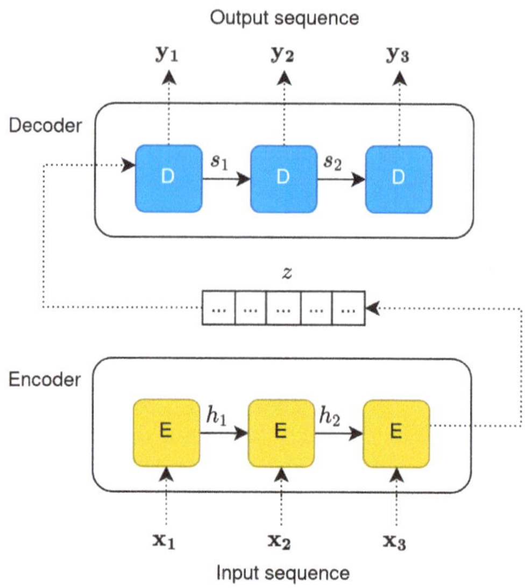

Figure 3. Unfolding of encoder–decoder architecture based on RNN.

The concept of attention was introduced by Bahdanau et al. [76] to mitigate this problem in neural machine translation tasks. The authors argued that, by propagating information about the complete input sequence, the decoder can discern between input words and learn which of them are relevant for generating the next target word. Formally, the attention mechanism relies on enriching the input context (z_i) of each decoder unit, with the hidden state of the encoder h_i (also called "annotation") that carries information on the whole input sequence. Figure 4 highlights the architecture of a simple encoder–decoder recurrent model enhanced with the attention mechanism. Equation (1) describes the computation of the attention score between the input word at position j and the generated word i. This score can be seen as an alignment score, measuring how well the two words match. In the equation, s_{i-1} denotes the previously generated word.

$$\alpha_{ij} = softmax(attention(s_{i-1}, h_j)) \tag{1}$$

The context of each decoder unit is then computed as in Equation (2), by multiplying the annotations of every word position (h_j) with the attention score generated by the attention model.

$$z_i = \sum_{j=1}^{N} \alpha_{ij} h_j \tag{2}$$

The mechanism described here is generally addressed as "additive attention". While there are different approaches to the integration of the attention mechanism in seq2seq architectures (content-based, dot-product, etc.), the goal remains to learn an alignment score that measures the relative importance between words in the input and output sequence.

Applications of attentive neural networks are now widespread, even beyond the boundaries of natural language processing, where attention initially proved its value. Notable examples of application in the text classification domain are hierarchical attention networks (HAN) [71,80]. These methods rely on the application of attention both at the word level, while encoding document sentences, and at the sentence level, roughly encoding the importance of each sentence with respect to the target sequence. Recently, Abreu et al. [71] have proposed a hybrid model (HAHNN) resorting to attention, along with temporal convolutional layers and GRUs.

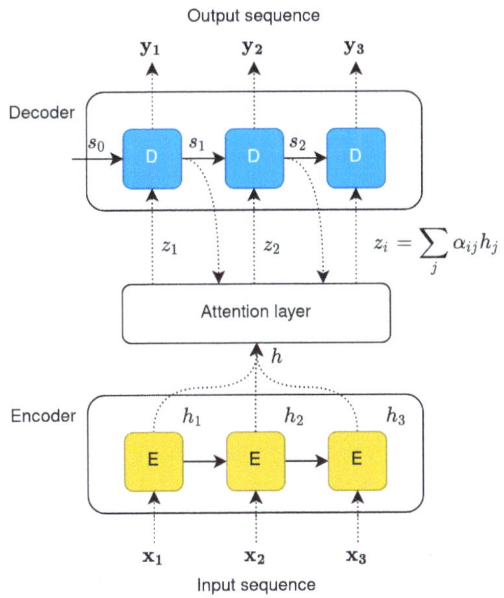

Figure 4. Unfolding of encoder–decoder architecture with attention mechanism.

However, attention has been since used as a strong foundational basis rather than just an added augmentation. The Transformer architecture is based on this idea, retaining a well-known encoder–decoder structure, but making no use of recurrence; instead, dependencies are drawn between input and output through the attention mechanism alone [81]. Transformers have been shown to lead to better results, while also gaining much in speed because of them being highly parallelisable.

5.4.3. The Transformer Architecture

Vaswani et al. [81] introduced the Transformer architecture, a novel encoder–decoder model that allows one to process all input tokens (e.g., words) simultaneously, rather than sequentially. Input sequences in Transformers are presented as a bag of tokens, without any notion of order. To learn dependencies between tokens, the Transformer relies on what is defined as a "self-attention" mechanism. Additionally, a special encoding step performed before the first layer of the encoder ensures that the embeddings for the same word appearing at a different position in the sentence will have a different representation. This step is called positional encoding, and its purpose is injecting information about the relative positioning between words, which would otherwise be lost.

The key component of this architecture is the self-attention layer, which intuitively allows the encoder to look at other words in the input sentence whenever processing one of its words. Stacking multiple layers of this type creates a multi-head attention (MHA) layer, as shown in Figure 5. The individual outputs are then condensed in a single matrix by concatenating the head outputs and passing the result through a linear layer.

Encoder

In the encoding part, the input embeddings are multiplied by three separate weight matrices (Equation (3)) to generate different word representations, which the authors name Q (queries), K (keys) and V (values), following the naming convention used in information retrieval.

$$Q = X \cdot W_Q, \qquad K = X \cdot W_K, \qquad V = X \cdot W_V \tag{3}$$

$W_Q, W_K, W_V \in \mathbb{R}^{dim \times d_k}$ are the learned weight matrices, $X \in \mathbb{R}^{N \times dim}$ are the word embeddings for the input sequence and $Q, V, K \in \mathbb{R}^{N \times d_k}$ contain the word representations. The product of the query and key matrices (Q and K) is taken to compute the self-attention score. With respect to the general attention mechanism introduced in the previous section, this can be seen as the alignment score between each word and the other word in the sentence, which depends on the relative importance between them. Moreover, the authors decided to use scaled dot-product attention instead of additive attention, mainly for efficiency reasons. In order to improve gradient stability, the score is divided by the square root of the representation length for each word, and a softmax is used to obtain a probability distribution. Eventually, the representation of each word is obtained by multiplying the scaled term with the V matrix containing the input representation. This operation is defined in Equation (4).

$$Z = \text{Attention}(Q, K, V) = softmax\left(\frac{Q \cdot K^T}{\sqrt{d_k}}\right) \cdot V \tag{4}$$

The Transformer multi-head self-attention layer performs these operations multiple times in parallel. According to the authors, this was done to extend the diversity of representation sub-spaces that the model can attend to. The output of the attention heads is concatenated and passed through a linear layer to obtain the final representation that condenses information from all the attention heads. This representation is then summed with residual input, normalised and passed to a feed-forward linear layer.

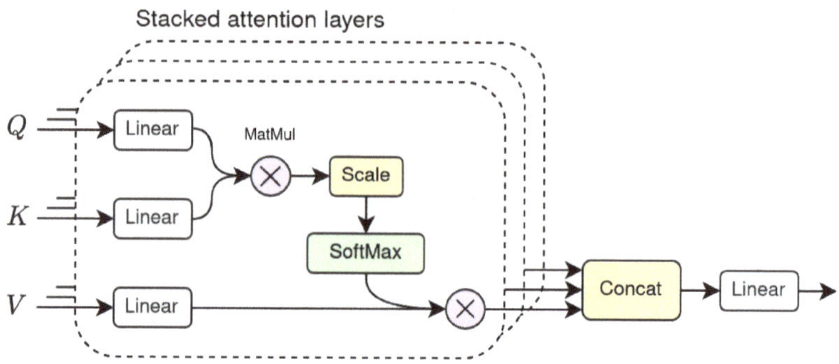

Figure 5. The multi-head attention layer used in the Transformer architecture.

Decoder

During the decoding phase, every decoder layer receives the output of the encoder (the K and V matrices) and the output of the previous decoder layer (or the target sequence for the first one). The key difference between the encoder and decoder layers is the presence, in the latter, of an additional MHA layer, applied to the K and V matrices from the encoder. Additionally, self-attention layers are modified into what are defined as "Masked" self-attention layers. During training, the Transformer decoder is initially fed the real target sequence—this is done to remove the necessity of having to be trained in an autoregressive fashion, hence becoming subject to the same bottleneck as RNNs (a procedure called "teacher forcing"). On the other hand, at inference time, the decoder is indeed autoregressive, since there are no target data and the generation of the next word necessarily depends only on the previously generated sequence. While predicting the token at position i, the masked MH self-attention layer ensures that only the self-attention scores between words at position $[1, i-1]$ are used. This is done by adding a factor M to the word embeddings in Equation (5). M is set to -inf for masked positions, and 0 otherwise. The exponential in the softmax operation will zero the attention scores for masked tokens.

$$Z = \text{MaskedAttention}(Q, K, V) = softmax\left(\frac{Q \cdot K^T + M}{\sqrt{d_k}}\right) \cdot V \tag{5}$$

This allows one to train the model in a parallel fashion, while ensuring that the model behaves consistently between training and inference time (without "cheating" at training time by looking ahead). The additional MHA layer of the decoder works like the one previously described, with the only exception being that the query matrix Q is the output of the preceding masked MHA layer. Residual connections and layer normalisation are also used in the decoder after each attention and feed-forward layer.

5.4.4. BERT and GPT

Recent DLMs are built upon and expand the Transformer architecture. The performance achievable on various NLP tasks, such as TC subtypes, has been boosted by what is considered the new standard in NLP transfer learning. This consists of utilising a pre-trained language model developed on generic textual data, which are subsequently specialised for the task at hand through a "fine-tuning" process.

The fine-tuning process consists of the adaptation of a model to a downstream task (i.e., classification). Classifiers are often just a small segment of the overall algorithm, yet its training procedure usually affects the pre-trained representation of the DLM as well. This is meant to specialise the overarching pipeline on the domain of the task, and usually leads to better results. Technically speaking, the internal representation of the pre-trained language model receives minimal changes from the fine-tuning process, which is a desirable outcome, as otherwise the DLM would incur an excessive loss of generality.

GPT

Although the original Transformer architecture is well suited for language modelling, researchers have argued that, for some tasks, it may learn redundant information in its encoding and decoding phases. It has been suggested that limiting the architecture to either encoders or decoders may result in equivalent performance and lighter models [82]. According to this principle, the Generative Pre-trained Transformer (GPT) [83] utilises a decoder-only architecture, which stacks multiple Transformer–decoder layers. The decoder block of a Transformer is autoregressive (AR), as it defines the conditional probability distribution of a target sequence, given the previous one. Simply put, GPT's pre-training task is that of next word prediction. Precisely because it is an autoregressive model, the GPT architecture produces a language model of which the predictions are only conditioned by either its left or right context (GPT's original architecture is left-to-right). This means that it is not bidirectional (a property that, for instance, BiLSTMs possess).

The decoder-only architecture has the additional benefit of performing better than traditional encoder–decoder models when dealing with longer sequences of text, making it particularly suitable for generative tasks (e.g., abstractive text summarisation and generative question answering). The authors of GPT provide adaptations for it to be used in a variety of downstream tasks, including classification.

BERT

Bidirectional Encoder Representations from Transformers (BERT) [16] closely followed the GPT model, and traded its auto-regressive properties in exchange for bidirectional conditioning. BERT's architecture is, in contrast with GPT, a multi-layer bidirectional Transformer encoder. BERT's training objective is also different, as it follows a "masked language model" (MLM) procedure. A percentage of the input is masked and, rather than predicting the next word, the network tries to predict masked tokens. This training strategy is crucial to achieving bidirectional conditioning, and allows the model to learn the relationship between masked words and their left and right context [84]. BERT is also trained on a secondary task, namely next sentence prediction (NSP), in which it is presented with two sentences and must guess in a binary fashion whether the second sentence follows the first. This task is meant to allow the model to better learn sentence relationships.

Most interestingly, the adaptation of BERT to downstream tasks is very simple. In fact, outstanding results have been obtained in classification by simply fine-tuning a model that passes the representations obtained by the encoders through a single-layer, feed-forward neural network [16].

5.4.5. Recent Transformer Language Models

Since the NLP landscape is constantly evolving, we deem it useful to provide a selection of a few recently proposed language models, and outline their main contributions and differences with previous methods. Most recent models build on the Transformer architecture or introduce variations on the BERT and GPT training approaches. Many are general language models jointly trained on multiple NLP tasks, and are evaluated on the GLUE [85] multi-task benchmark.

Direct Successors

BERT and GPT were just the starting point for the development of many variations and improvements, of which we cite the most influential. Robustly Optimised BERT Pretraining Approach (RoBERTa) [13] was introduced by Liu et al. as a successor of BERT. It improves the training procedure by introducing dynamic masking (masked tokens are changed during training epochs), while also removing the next sentence prediction task. The GPT model has also had successors, the most popular being the GPT-2 [12]. The development of GPT-2 models was mostly in terms of data utilised and the scale of the models (a trend that continued in GPT-3 [86]).

XLNet

As with GPT, XLNet [19] is an autoregressive model. More specifically, it is a generalised autoregressive pre-training approach, which calculates the probability of a word token conditioned on all permutations of word tokens in a sentence. To maintain information about the original relative ordering of words in the midst of all the permutations, XLNet borrows the idea of attention masking, by which, during context calculations, it masks tokens outside of the current context. As an example, consider an input sequence with four tokens and its permutation $P([1,2,3,4]) = [3,2,4,1]$. The first element has no contextual knowledge, and, to reflect this, the attention mask should forbid looking at the other sequence positions (producing mask $[0,0,0,0]$), while the second element should know about the first (namely, that it should be in position 3, giving mask $[0,0,1,0]$). XLNet also introduces a two-stream self-attention schema to allow position-aware word prediction. In other words, this grants the model the ability for token probabilities to be conditioned on their position index without trivially knowing if the word should be part of the sentence or not. XLNet's approach allows it to overcome the lack of bidirectionality of previous autoregressive models. Furthermore, this approach addresses the downsides of masked language modelling (namely, the discrepancy between pre-training and fine-tuning tasks), as it does not rely on data corruption. XLNet has achieved state-of-the-art performance on many TC benchmarks in English, as will be showcased in Section 6.3.

T5

The Text-To-Text Transfer Transformer (T5) [87] is a unified text-to-text model trained to solve a variety of NLP tasks. While general multi-task language models are mostly trained using task-specific architectural components and loss functions, T5's authors build a unified learning framework that casts every NLP problem as a text-to-text problem. This allows them to use the exact same model, loss function and hyperparameters to produce a single, unified multi-task model. The architecture of T5 closely follows the original Transformer architecture. Differences from the original implementation are limited to the layer normalisation (simplified by not using additive bias), the usage of dropout on the linear and attention layers, and a relative positional encoding strategy. Here, word embeddings are altered depending on the alignment between the "key" and "query" matrices computed by the multi-head attention layer, instead of using a fixed embedding for every position. T5 is pre-trained on the C4 English corpus—the Colossal Clean Crawled Corpus, derived from the Common Crawl—using a masked language modelling objective (similarly to BERT) that masks 15% of the tokens in every sentence. Fine-tuning is performed on all tasks at once: all datasets—including the ones for translation, question answering, and classification—are concatenated in a single dataset using a sampling strategy to ensure a balance between samples from each tasks. Differently from BERT, which has an encoder-only architecture, T5 uses the same causal masking strategy used in the Transformer decoder, where the attention output is masked to prevent the model from attending to subsequent positions. T5 has had a recent successor, T0 [88], which utilises the same model, but pushes the boundaries of transfer learning towards zero-shot generalisation further.

DeBERTa

The Decoding-Enhanced BERT with Disentangled Attention [89] is another language model based on the Transformer architecture. As with BERT, the model is pre-trained using a MLM objective, and it is fine-tuned using adversarial training. This strategy improves its generalisation ability and its robustness to adversarial examples. The architecture introduces a disentangled attention layer. Unlike the Transformer, where positional encoding is added to the word content embeddings to create the input representation, DeBERTa encodes words and positions separately. Attention scores for every position are computed using disentangled matrices and explicitly depend on the word content and the word relative position in the sentence. The decoder is also modified to consider word absolute positions when predicting a masked token.

ByT5

ByT5 is an adaptation of the T5 model able to process raw bytes of text, instead of tokens. Models like BERT rely on a separate tokenisation step to chunk documents into a vocabulary of sub-words. This translates into heightened memory constraints, since large vocabularies require massive embedding matrices with a large number of parameters. Additionally, the authors argue that sub-word tokenisation is not robust in handling typos and lexical variants, and reducing all unknown words to the same out-of-vocabulary token prevents the model from distinguishing between different OOV situations. Instead of processing words or sub-words, ByT5 is fed UTF-8 bytes without any preprocessing operation. To represent byte-level embeddings, the model only needs a dense matrix of 256 embeddings with additional embeddings for the special tokens used to pad sequences and to signal the end of a sentence. In this case, there is no need to handle the OOV case, since all possible bytes are represented. Furthermore, removing the dependency on tokenisation allows for the simpler training of the model on multilingual corpora, since there is no need for language-specific tokenisation strategies. ByT5 uses a slightly modified T5 architecture with a heavier encoder, which is three times the size of the decoder. The authors empirically find that a bigger encoder works better for byte-level language models. The model is pre-trained with a "span corruption" self-supervised objective, where the model learns to reconstruct sequences of 20 bytes that are masked in the input sentences.

ERNIE 3.0

ERNIE [90] enhances the language models pre-training phase integrating knowledge-graph information. As with other language models, ERNIE is trained using different unsupervised tasks, including MLM, sentence reordering, and sentence distance. The latter is a variation of the NSP task, where the model is asked to predict whether two sentences are adjacent, within the same document (but not adjacent), or from different documents altogether. In order to integrate knowledge graph information in the training phase, the tasks of knowledge masked language modelling and universal knowledge–text prediction are added. The former objective is used to let the model learn higher-level entity representations, for instance by masking the name of a character (e.g., "Harry Potter"). The latter strategy extends the former by explicitly embedding a knowledge graph. The model is provided with a knowledge graph representation of the sentence (in the shape of a triple) and a sentence, both with masked tokens, and must use this information to fill in the blanks. According to the authors, these tasks allow the model to gain a lexical understanding of words, while traditional language models only capture more global syntactical and semantic knowledge. ERNIE is composed of two modules: a universal representation module, meant to capture shared word representations, and a task-specific module pair, one for natural language understanding tasks, and the other for natural language generation tasks. Both modules use the Transformer-XL architecture [91], which differs from the original Transformer, mainly for the addition of an auxiliary recurrence module to aid in the modelling of long sequences.

FLAN

The Fine-tuned Language Net (FLAN) [92] explores the usage of instruction fine-tuning to enhance the zero-shot generalisation ability of a general pre-trained language model (similarly to T0). Zero-shot learning aims at giving models the ability to solve novel tasks—unseen during training—at inference time. To do so, the authors gathered data from 62 NLP datasets and 12 different tasks—including language inference, translation, question answering, text classification—and aggregated them in a mixed multi-task dataset. A maximum of 30,000 samples are selected from each dataset in order to limit task imbalance. From the final dataset, samples are enriched with instruction templates, defined as descriptions of the task that the model should solve expressed in natural language. For instance, the sentence "classify this movie review as positive or negative" can be used to ask the model to perform binary classification. To increase diversity, 10 different templates are manually created for each dataset, some of them asking the model to perform collateral-derived tasks (e.g., asking for a movie classification from an SA task). FLAN uses a Transformer decoder-only architecture, similarly to GPT, with 137 billion parameters. The model is pre-trained on a collection of English documents that includes a wide variety of textual data, ranging from Wikipedia articles to computer code. This pre-trained model is then used for the instruction-tuning procedure.

5.4.6. Graph Neural Networks

Recent developments in the field of AI are exploring the application of neural networks to graph representations [93]. In this context, graph neural networks (GNNs) [94,95] are architectures that utilise such structures to capture the dependencies and relations between nodes. Well-established approaches are being generalised to arbitrarily structured graphs, most notably CNNs [96,97].

Successful Approaches

Representations are based on heterogeneous graphs in which nodes are both words and documents have seen wide success; TC is thus cast as a node classification task for document nodes. TextGCN [98] utilises a graph convolutional network (GCN) on text mapped onto a graph structure. Words are connected to other words as well as documents, but there are no inter-document relations (documents are able to indirectly exchange information through multiple convolutional layers). Weights of document–word edges are based on word occurrence measurements (usually TF-IDF), while word–word edges are weighted on word co-occurrence in the whole corpus (usually variations of point-wise mutual information [99]). The training procedure learns word and document embeddings, and is therefore connected to text embedding techniques (such embeddings are learnt simultaneously in this case). On this account, it is easy to see how graph architectures can also be integrated with deep language models. BertGCN [100], for example, trains a GCN jointly with a BERT-like model, in order to leverage the advantages of both pre-trained language models and graph-based approaches. Document nodes are initialised through BERT-style embeddings and updated iteratively by the GCN layers. Another example is the MPAD (Message Passing Attention Network for Document Classification) [101], which proposes the application of the message passing framework to TC. The nodes represent unique words and the directed edges encode the text flow and word co-occurrence statistics. Nodes and edges information is iteratively aggregated and combined using GRUs to update word representations.

Transductive Nature

Many GNN architectures include unlabelled test document nodes in the training procedure, making them inherently transductive. Transductive learning [102] can be useful because of how it can allow label influence to be propagated to unlabelled test data during training, notably removing the need for modelling the relation between data and target labels. Often, the data required for comparable performances are fewer than that for traditional inductive learning approaches. However, this also has the downside of not being able to quickly generate predictions for unseen test documents that were not included in the training procedure (i.e., online testing). Furthermore, building a GNN for a large-scale text corpus can be costly due to memory limitations [103].

Weaknesses and Solutions

Reducing the modelling cost has been an important topic of discussion. Huang et al. [103], for example, change the model training strategy, opting to build graphs for each input text and utilizing global parameter sharing to reduce the burden, while maintaining global information. Methods such as SGC [104] work instead on reducing model complexity; in particular, SGC repeatedly removes nonlinearities between consecutive layers in deep GCNs, collapsing the resulting function into a single linear transformation. The authors argue that the expressive power of GCNs originates from graph propagation rather than nonlinear feature extraction, and exhibit comparable or even superior performances to other methods. Related to such nonlinearities is the phenomenon of over-smoothing, which suggests that increasing the number of layers in GCNs causes the node representations to converge to a same value (and become hence indistinguishable) [105]. While removing such non-linearities may certainly help in this regard, recent research has developed techniques aimed directly at combatting this phenomenon. For example, GCNII [106] extends the vanilla GCN model with initial residual and identity mapping. Initial residual connections are related to the idea of residual connections in residual networks, but modify this concept by adding a skip connection to the input layer instead (i.e., the initial representation), while identity mapping (also borrowing from residual networks) adds an identity matrix to the weight matrix. They show great improvements on the over-smoothing problems and state-of-the-art results in TC tasks. SSGC [107] also addresses this problem, by combining techniques from previous works—namely, SGC and a specialised graph convolution filter [108,109].

Summary

While this section is merely a brief introduction to GNNs with a few notable examples, they are among the few architectures able to compete with deep language models in recent years. They are of particular interest both, because of their excellent performance—sometimes even achieving state-of-the-art results—but also because of how they can perform quite well in low label rate datasets [98], a characteristic which can be attributed to their transductive nature.

5.4.7. Contextualisation of Word Embeddings

As anticipated previously, recent language models have the extremely valuable capability of incorporating context into their representations of text. This means that these representations, in principle, are effectively able to understand and distinguish polysemous words by looking at the body of text that they appear in, as well as containing complex characteristics of word use (its syntax and semantics). A practical difference with the earlier word embeddings described in Section 3.3 is that, while older embeddings were "static" and could therefore be extracted as individual entities, contextualised embeddings require their originating language model to be generated. Such language models are only able to understand words in context—they are not meant or suited for isolated words.

Contextualised embeddings have been proven to be "very contextual", meaning that they can adapt well to the different semantic significance of words [110]. Furthermore, researchers have extracted word representations from the lower hidden states of BERT-like language models, and managed to create static embeddings of much higher quality than the ones created by static embedding procedures, further proving the benefits of this learning process.

Contextualised embeddings were not a novelty introduced by Transformers; multiple research contributions pushed on their development [111,112]. As a prime example, ELMo [65] introduced a bidirectional LSTM architecture trained for language modelling (i.e., next word prediction) that also produced contextualised embeddings. Due to its recurrent structure, however, it struggled with long-term dependencies.

5.4.8. Challenges of Language Models

One of the main challenges posed by very deep language models is their size. This depth incurs a huge number of parameters that must be loaded into memory, an operation that is not exclusive to the training phase. In practice, this might be an issue when pre-trained models are used for inference under low-latency constraints, or fine-tuning in lower resource settings.

Solutions to this problem have been devised in order to create more compact models (i.e., with fewer parameters). A successful approach to this end is knowledge distillation [113]. In this process, the larger, original model is utilised as a "teacher" in the creation of the more manageable "student" sub-model. The key idea of distillation is that it should be easier to train a student model to mimic the output distribution (i.e., the knowledge) of a bigger teacher model, with enough capacity to learn a concise knowledge representation from raw data. Thus, the probability scores assigned by the teacher to input samples are used as predictive targets for the student, allowing it to encode the teacher knowledge in a compressed form, without having to learn it from scratch. DistilBERT [114] leverages this to reduce the size of BERT models by up to 40%, while retaining 97% of its effectiveness. Much in the same way, TinyBERT [115] is also based on knowledge distillation, but extends it to the task-specific learning stage. This model is also able to retain most of its teacher's performance.

On the other hand, seeing that such approaches still require a teacher model, there have been methods that propose tackling the issue from a different angle. ALBERT [116], for example, introduces parameter reduction techniques to reduce memory consumption and increase the training speed of BERT models. ELECTRA [117] introduces token detection, a more sample-efficient pre-training task, in place of masked language modelling. This task corrupts the input by replacing tokens with plausible alternatives rather than masks, and changing the learning task to a discriminative one, where the model must identify whether each token in the corrupted input had been artificially replaced or not. In particular, this reframing has proven to be effective in models with fewer parameters, increasing their viability.

In summary, downsized models such as these are essential in practical applications of DLMs. It is worth noting that, whenever these alternatives are not available, it may be possible to perform a fine-tuning procedure without involving the language model in the learning process (i.e., "freezing" the base model's weights). This would be similar to utilising the word embeddings contained in the LM in their agnostic state (contextualised, but not specialised). The resource requirements are hence reduced, though this shrinks the learning capacity of the system.

6. Experimental Performance Analysis

We report, in this section, the performance metrics of a selection of machine learning models on a number of TC tasks among those introduced in Section 1.1. Moreover, we introduce two new multilabel datasets and showcase results for a selection of TC approaches on them. To test models on a TL task, we introduce a new public dataset based on Wikipedia articles labelled with their topics. We also perform tests on the RCV1 dataset as a NC representative, which is available freely for research purposes upon request.

6.1. Datasets

We researched the most popular TC datasets utilised in recent works. We focused on the document-level tasks; hence, we excluded from our search the NLI, SP, and NER tasks. The list of datasets is shown in Table 3, and we refer to the reviews by Li et al. [1], Minaee et al. [3] for a description of each one.

Most datasets with document-level annotations have a single target label for each sample, and are hence used in binary or multiclass classification tasks. For the development of models for TL and NC tasks, we found a limiting unavailability of multilabel datasets. Despite this, real-world applications of supervised topic extraction methods for documents and news are likely to require the assignment of multiple topics to pieces of text, rather than just one. For the evaluation of methods on TL tasks, Wikipedia articles provide a source of semi-structured text labelled with multiple categories (topics) assigned by contributors. Such data can be extracted by accessing the DBpedia https://www.dbpedia.org (accessed on 28 December 2021) project, or from the articles' raw files available in Wikipedia dumps https://dumps.wikimedia.org (accessed on 28 December 2021). Although these corpora have already seen use in various works [118,119], there is no consistent set of annotations to be used as reference for a multilabel formulation. For news classification, we synthesised a new multilabel dataset from RCV1 [120], a collection of English news articles from the Reuters agency. Statistics for both datasets are appended to Table 3.

6.1.1. New Datasets Distillation

EnWiki-100

The synthesised EnWiki-100 dataset contains the text of more than 300,000 English Wikipedia pages, along with a variable number of topics related to the page content. While categories assigned to Wikipedia pages by contributors are often used as predictive targets, we found that these frequently contain too generic or too specific categories that are not informative of the page's main topics. Therefore, extracted articles are annotated with the name of their respective Wikipedia portals. In support of this decision, one should consider that English Wikipedia contains roughly 500 portals, while categories are more than 500,000. Portals themselves are stated to be "entry points" for articles belonging to the same broad subject [121]. Consequently, they are better suited as targets for a TL task.

Articles are extracted from the September 2021 Wikipedia dump, using a customised version of the WikiExtractor https://github.com/attardi/wikiextractor (accessed on 28 December 2021) tool, modified for the extraction of the portal names from the page metadata. Only the 100 most populous portals are kept for the final dataset. In addition, article frequency has been limited to a maximum of 50,000 per label, such as to reduce the size of the already large dataset, as well as to limit class imbalance.

RCV1-57

The RCV1-57 dataset is extracted from the RCV1 (version 1) collection of English news. Articles are labelled with a variety of tags that describe their content in different manners. The most consistent and appropriate were topic codes, which describe the general subject of the news piece. Such codes have a hierarchical structure, with the higher levels being more abstract, and the lowest being the most specific. We decide to cut off the codes at the second level of specificity, since they were the most complete and descriptive overall. We clean up labels by stripping tags from other levels, and discard topics with fewer than 500 representatives. If articles contain only scrapped topics, they are also discarded.

Table 3. English TC datasets.

Name	Task	Classes	Multilabel	Samples
AG News [122]	NC	4	✗	127,600
20 News (20 Newsgroup) [123]	NC	20	✗	18,846
R52 [124]	NC	52	✗	9100
R8 [124]	NC	8	✗	7674
Yelp2 (Yelp Polarity) [125,126]	SA	2	✗	598,000
Yelp5 (Yelp Full) [125,126]	SA	5	✗	700,000
Amz2 (Amazon Polarity) [125]	SA	2	✗	4,000,000
Amz5 (Amazon Full) [125]	SA	5	✗	3,650,000
IMDb [127]	SA	2	✗	50,000
IMDb10 [127]	SA	10	✗	50,000
MR (Movie Review) [128]	SA	2	✗	10,662
Yah!A (Yahoo! Answers) [125]	TL	10	✗	1,450,000
TREC [129]	TL	6	✗	5952
Ohs (Ohsumed) [130,131]	TL	23	✓	13,929
DBP (DBpedia-14) [125]	TL	14	✗	630,000
EnWiki-100 [1]	TL	100	✓	329,600
RCV1-57 [1]	NC	57	✓	758,100

[1] Datasets presented in this paper.

6.2. Evaluation Metrics

Accuracy is the most adopted evaluation metric for TC tasks. This metric is a simple and interpretable way of measuring the overall fraction of correct predictions. It is defined as the number of correct predictions over the total number of samples (N), as in Equation (6). The equations use standard notations for evaluation metrics, namely true positive (TP), true negative (TN), false positive (FP), and false negative (FN).

$$\text{Accuracy} = \frac{\text{TP} + \text{TN}}{\text{N}} \tag{6}$$

In a multilabel setting, the definition of this metric can change depending on the definition of true positives and negatives. A prediction may be considered correct when the predicted labels exactly match the ground truth (referred to as "subset accuracy"). Alternatively, predictions can be flattened and reduced to a single-label task before the accuracy computation.

Precision and recall are other popular metrics, especially for multilabel classification. The former is the fraction of correct predictions among all the ones that have been predicted as true (TP + TN), while the latter is the fraction of correct predictions over all the ones that should have been predicted (TP + FN). *F-score*, and in particular F_1*-score*, is a popular combination of these two; it is defined as the harmonic mean of precision and recall, as in Equation (7).

$$\text{Precision} = \frac{\text{TP}}{\text{TP} + \text{FP}} \qquad \text{Recall} = \frac{\text{TP}}{\text{TP} + \text{FN}} \qquad F_1 = 2 \cdot \frac{\text{Precision} \cdot \text{Recall}}{\text{Precision} + \text{Recall}} \tag{7}$$

For multilabel and multiclass problems, these metrics can be computed separately for each class and then averaged, obtaining the macro-averaged metrics. In such a case, every class contributes equally to the final score, and hence provides a more challenging metric for unbalanced datasets. On the other hand, a micro-average reduction strategy is used when computing the metric globally with no weighting.

6.3. Quantitative Results

The accuracy score for several previously introduced algorithms on prominent benchmarks are shown in Tables 4 and 5. We report only accuracy, since it is the most consistently used metric across different studies. GLUE and RACE benchmarks are popular choices for the evaluation of NLP methods on a variety of tasks, hence we report results on these datasets in Table 6, whenever available. We evaluated a selection of deep and shallow methods on our new synthesised datasets. Since our datasets are multilabel, we report the F_1-*score* with macro-averaging across all categories, along with the subset accuracy, computed on the test set. Tables 4 and 5 showcases our results, averaged on four different evaluation splits. The following section provides additional details on the training procedure.

Table 4. Test accuracy score (%) on SA benchmark datasets (best results in bold).

Model	Yelp2	Yelp5	Amz2	Amz5	IMDb	IMDb10	MR
XLNet [19]	**98.63**	72.95	**97.89**	**68.33**	96.80	-	-
BERT-base [100,132,133]	98.13	70.80	-	-	95.63	54.20	85.70
BERT-large [132–134]	98.19	71.38	97.37	65.83	95.79	**55.60**	-
L-MIXED [135]	-	-	-	-	95.68	-	-
DNC + CUW [136]	96.40	65.60	-	-	91.30	-	-
SVDCNN [74]	95.26	63.20	-	-	-	-	-
ULMFiT (small-data) [137]	97.10	67.60	96.10	64.10	-	-	-
BERT-large-UDA (ft) [134]	97.95	67.92	96.50	62.88	95.80	-	-
USE_T + CNN [138]	-	-	-	-	-	-	81.59
MPAD [101]	-	66.80	-	-	91.87	-	-
STM + TSED + PT + 2L [139]	-	-	-	-	-	-	80.09
VLAWE [140]	-	-	-	-	-	-	93.30
TM [141]	-	-	-	-	-	-	77.51
ErnieDoc-large [142]	-	-	-	-	97.10	-	-
BigBird [143]	-	72.16	-	-	95.20	-	-
NB-weight-BON + CS [144]	-	-	-	-	**97.42**	-	-
HAHNN (CNN) [71]	-	**73.28**	-	-	92.26	-	-
HAHNN (TCN) [71]	-	72.63	-	-	95.17	-	-
RoBERTa + ODPT [145]	-	-	-	-	96.60	-	-
HAN [80]	-	71.00	-	63.60	-	49.40	-
SSGC [107]	-	-	-	-	-	-	76.70
SGCN [104]	-	-	-	-	-	-	75.90
TextGCN [98]	-	-	-	-	-	-	76.74
GCN [98]	-	-	-	-	-	-	76.30
RMDL (15 m) [146]	-	-	-	-	90.79	-	-
GraphStar [147]	-	-	-	-	-	-	76.60

Table 4. *Cont.*

Model	Yelp2	Yelp5	Amz2	Amz5	IMDb	IMDb10	MR
BertGCN [100]	-	-	-	-	-	-	86.00
RoBERTaGCN [100]	-	-	-	-	-	-	89.70
DRNN [62]	97.27	69.15	96.49	64.43	-	-	-
ULMFiT [61]	97.84	70.02	-	-	95.40	-	-
DPCNN [148]	97.36	69.42	96.68	65.19	-	-	-
LSTM-reg [149]	-	68.70	-	-	-	52.80	-
KD-LSTM-reg [133]	-	69.40	-	-	-	53.70	-
CCCapsNet [150]	96.48	65.85	94.96	60.95	-	-	-
CharCNN [125]	95.40	62.05	95.07	59.57	-	-	-
EFL [151]	-	64.90	-	-	96.10	-	**92.50**
byte mLSTM [152]	-	-	-	-	92.20	-	86.80
BLSTM-2DCNN [153]	-	-	-	-	-	-	82.30
oh-2LSTMp [154]	-	-	-	-	91.86	-	-
VDCNN [73]	95.72	64.72	95.72	63.00	-	-	-
RoBERTa [100,143]	-	71.75	-	-	95.00	-	89.40

Table 5. Test accuracy score (%) on NC and TL benchmark datasets (best results in bold).

	NC				TL			
Model	**AG**	**20N**	**R52**	**R8**	**Yah!A**	**TREC**	**Ohs**	**DBP**
XLNet [19]	**95.55**	-	-	-	-	-	-	**99.40**
BERT-base [100,132]	95.20	85.30	96.40	97.80	**78.14**	**98.20**	70.50	99.35
BERT-large [132]	95.34	-	-	-	-	-	-	99.39
L-MIXED [135]	95.05	-	-	-	-	-	-	99.30
DNC + CUW [136]	93.90	-	-	-	74.30	-	-	99.00
SVDCNN [74]	90.55	-	-	-	-	-	-	-
ULMFiT (small-data) [137]	93.70	-	-	-	74.30	-	-	99.20
USE_T + CNN [138]	-	-	-	-	-	97.44	-	-
MPAD [101]	-	-	-	97.57	-	95.60	-	-
STM + TSED + PT + 2L [139]	-	-	-	-	-	93.48	-	-
DELTA [155]	-	-	-	-	75.10	92.20	-	
VLAWE [140]	-	-	-	89.30	-	94.20	-	-
TM [141]	-	-	89.14	97.50	-	90.04	-	-
HAN [80]	-	-	-	-	75.80	-	-	-
SSGC [107]	-	88.60	94.50	97.40	-	-	68.50	-
SGCN [104]	-	88.50	94.00	97.20	-	-	68.50	-
TextGCN [98]	-	86.34	93.56	97.07	-	-	68.36	-
GCN [98]	-	87.90	93.80	97.00	-	-	68.20	-
NABoE-full	-	86.80	-	97.10	-	-	-	-
RMDL (15 m) [146]	-	87.91	-	-	-	-	-	-

Table 5. *Cont.*

	NC					TL		
Model	**AG**	**20N**	**R52**	**R8**	**Yah!A**	**TREC**	**Ohs**	**DBP**
GraphStar [147]	-	86.90	95.00	97.40	-	-	64.20	-
SCDV-MS [156]	-	86.19	-	-	-	-	-	-
STC-Q [157]	-	87.30	-	-	-	-	-	-
BertGCN [100]	-	89.30	**96.60**	98.10	-	-	**72.80**	-
RoBERTaGCN [100]	-	**89.50**	96.10	**98.20**	-	-	**72.80**	-
RepSet	-	77.02	-	96.85	-	-	66.12	-
Rel-RWMD kNN	-	74.78	-	95.61	-	-	58.74	-
DRNN [62]	94.47	-	-	-	-	-	-	99.19
ULMFiT [61]	94.99	-	-	-	-	96.40	-	99.20
DPCNN [148]	93.13	-	-	-	76.10	-	-	99.12
CCCapsNet [150]	92.39	-	-	-	73.85	-	-	98.72
CharCNN [125]	91.45	-	-	-	71.20	-	-	98.63
EFL [151]	86.10	-	-	-	-	80.90	-	-
byte mLSTM [152]	-	-	-	-	-	90.40	-	-
BLSTM-2DCNN [153]	-	-	-	-	-	96.10	-	-
oh-2LSTMp [154]	-	86.68	-	-	-	-	-	-
VDCNN [73]	91.33	-	-	-	73.43	-	-	98.71
RoBERTa [100]	-	83.80	96.20	97.80	-	-	70.70	-

Table 6. GLUE and RACE score updated to latest results (best ones in bold).

	GLUE		RACE
Model	**Reported**	**Latest**	**Latest**
BERT (base) [16]	79.60	-	65.00
BERT (large)	82.10	80.50	67.90
RoBERTa [13]	88.50	88.10	83.20
XLNet (large) [13,19]	88.40	88.28	81.75
ALBERT (ensemble) [116]	89.40	88.16	**89.40**
ELECTRA (large) [117]	89.40	89.40	-
ELECTRA (base) [117]	85.10	-	-
T5 [87]	90.30	90.30	87.10
GPT [83]	72.80	-	59.00
DeBERTa (large) [89]	**90.00**	90.80	-
ERNIE [90]	-	**91.10**	-

6.3.1. Custom Experimental Setup

We selected seven algorithms (Table 7) that either were recently, or have been, state-of-the-art approaches in TC tasks, and we test their performances on the EnWiki-100 and RCV1-57 datasets to showcase their performances on new multilabel datasets. As strong baseline representatives for classical methods, we present the results of Naïve Bayes and linear support vector classifiers, both using TF-IDF features. These methods are utilised in a one-vs.-rest ensemble fashion, as is common in multilabel applications. We then present the results of FastText [53], a re-implementation of XML-CNN [118,133] and a BiLSTM-based classifier as representatives of those methods bridging the gap between earlier methods and DLMs. Lastly, we fine-tuned and tested Transformer-based language models utilising the pre-trained open-source models, published in the Hugging Face library [158].

Datasets are split into training, validation, and test set: 40% of the data are reserved for testing, and 20% of the remaining samples are used for validation. Splits are produced in a way to preserve the distribution of labels, through a stratification strategy [159,160]. Final metrics reported in Table 7 are obtained by averaging results on the test set over all runs (four runs of each method, except for XLM-R, due to computational complexity). Hyperparameters for the Naïve Bayes and SVM classifiers are tuned using a grid search with 10 fold cross-validation. FastText hyperparameters are selected using the auto-tuning procedure provided in the Python package. Due to resource constraints, only a limited tuning of the most impactful parameters is performed on the bidirectional LSTM and Transformer-based models.

Table 7. Test set Macro-F_1 and Subset Accuracy (%) on synthesized datasets (best results in bold).

	F_1-*Score*		Accuracy	
Model	**EnWiki-100**	**RCV1-57**	**EnWiki-100**	**RCV1-57**
Naïve Bayes (OVA)	63.64	56.30	39.24	47.27
Linear SVM (OVA)	80.29	71.73	66.93	67.67
FastText Classifier	74.45	69.65	68.23	67.02
BiLSTM (GloVe)	81.22	76.62	68.05	72.48
XML-CNN (FastText)	78.19	73.02	66.64	70.98
BERT (base)	**85.52**	**78.07**	**75.28**	73.48
XLM-R (base)	84.60	77.19	74.25	**74.01**

6.3.2. Discussion on Results

Tables 4 and 5 show the dominance of Transformer-based language models in all TC tasks. Specifically, BERT and XLNet reach state-of-the-art performances on most datasets. Moreover, TC using graph representation for documents, paired with graph convolutional networks, proves to be an effective approach for the extraction of useful features, especially when node embeddings are initialized with contextual representations generated by BERT-like models (BertGCN, RoBERTaGCN). Notably, Thongtan and Phienthrakul [144] propose to learn document-level embeddings by minimising cosine similarity, and combining features extracted from a Naïve Bayes model before the classification step. The model architecture is very simple in comparison with BERT and other deeper language models, but manages to achieve the best accuracy on a binary SA dataset. The authors argue that, for this kind of task, tailored document representations are more important than the choice of classifier. Abreu et al. [71] achieve state-of-the-art results on the Yelp5 dataset, by using a CNN to extract features from word vectors and using the attention mechanism twice, first at the word level, to generate a sentence representation, and then at the sentence level, to weight the importance of sentences in the document embedding used for classification. This model makes use of recurrent blocks (GRU) and reiterates the importance of the attention mechanism to enhance seq2seq architectures.

The latest trend for the evaluation of deep language models is the leveraging of general multi-task benchmarks. Table 6 showcases the results of two of them. These benchmarks measure the model ability to generalise to multiple tasks, and are not specific for TC. The reported score averages the performance metrics of the model across all tasks.

Finally, Table 7 lists the results that we obtained on the new synthesised multilabel datasets. Since these are new datasets, we use them to test the classification performance of a few notable language models, within the limits of our resources. Unsurprisingly, BERT and XML-R achieved the best results over the two new corpora, with accuracy values aligned with the respective scores of similar multilabel TL datasets (e.g., Ohsumed). We clarify that, since these datasets are different, metrics should not be directly compared, but we expected classification accuracy to fall in a range of values that are reasonably comparable to the other TL datasets. Interestingly, XML-R performed only slightly worse than BERT, despite being a multilingual model which was not specifically trained to understand the English language. On the one side, this is interesting, as the curse of multilinguality—which states that the per-language capacity decreases as the number of languages grows—suggests that BERT should be able to understand much better the single language that it was trained on. On the other hand, one should consider that the smaller XML-R model has more than 2.5 times the parameters of the BERT base, and the two models are pre-trained on different corpora. XML-R is pre-trained on the Common Crawl in 100 languages, while our BERT model is pre-trained on the combination of BookCorpus and Wikipedia dumps. While the former corpus amounts to 2.5 TB of data, the latter does not exceed one hundred GB. The BiLSTM also performed better than other non-Transformer methods, as expected, since its architecture makes it suitable to model dependencies between adjacent chunks of text.

7. Future Research Directions

Despite achieving state-of-the-art results throughout NLP literature, large-scale language models are not infallible. For example, recent studies [161,162] reported that BERT models show considerable performance pitfalls when faced with adversarial text examples generated by adversarial networks. These text sequences are subtly altered with word-level replacements or sentence rephrasings that do not change their meaning (as far as human understanding is concerned), but are enough to fool these language models. This phenomenon raises questions about the ability of these models to make decisions based on the actual meaning carried by a portion of text. Particularly interesting is therefore the development of robustness benchmarks such as AdvGLUE [163], which aim to quantitatively and thoroughly explore the vulnerabilities of modern large-scale language models. Such benchmarks are carefully filtered and validated by human annotators, such as to ensure that adversarial examples are valid and unambiguous. Benchmarks like these will be fundamental in the development of more sophisticated adversarial attacks, as well as more robust language models able to withstand them.

The current trend of proposing deeper and deeper models with an ever-increasing number of parameters has led some scientists to warn against the creation of so-called "stochastic parrots" that effectively emulate language understanding by memorising large training datasets, but with very limited actual generalisation ability, let alone the ability to comprehend human language. Furthermore, the black-box nature of deep neural networks adds to these concerns, as learnt features are hardly interpretable, while some computational linguists also bring legitimate concern about harmful and dangerous biases learnt from the data [79]. As a consequence, future research should put the robustness of large-scale language models under mindful scrutiny and provide tools to lessen the interpretability issues which currently afflict deep learning. The former problem is addressed by Wang et al. [164], who propose a framework for a more robust fine-tuning of BERT language models.

Finally, recent developments have gone against the trend of scaling larger and larger models, proposing that smaller generative LMs can perform competitively with massively larger models when augmented with search/query information from a retrieval database [165,166]. For example, Retrieval-Enhanced Transformer (RETRO) performs on par with GPT-3, despite being 4% its size. Pursuing more reasonably sized models is certainly a research direction that will be worth exploring.

8. Conclusions

In this manuscript, we provide an overview of existing models for text classification (TC), highlighting the steps necessary to encode textual data into a feature space for classification. We presented an analysis of preprocessing operations for both shallow and deep learning methods, exploring procedures that are often taken for granted. We outlined the development in the approaches to NLP tasks and TC in particular, showcasing how they have evolved, up to the very recent Transformer-based methods. We provided an overview of English datasets, while also laying out instructions to synthesise two new multilabel classification datasets. We move towards the conclusion by presenting experimental results for methods as researched in the literature, as well as new results on our datasets for a choice of methods. Finally, we explore future research challenges, mainly tied to the robustness of DLMs and what is being done to address their weaknesses.

Author Contributions: Conceptualization, A.G., A.Z. and M.M.; methodology, A.Z.; software, A.Z. and M.M.; validation, A.G., A.Z. and M.M.; formal analysis, M.M.; investigation, A.Z. and M.M.; resources, A.Z.; data curation, A.Z.; writing—original draft preparation, A.Z. and M.M.; writing—review and editing, A.G., A.Z. and M.M.; visualization, A.Z. and M.M.; supervision, A.G. and A.A.; project administration, A.G. and A.A. All authors have read and agreed to the published version of the manuscript.

Funding: This research received no external funding.

Institutional Review Board Statement: Not applicable.

Informed Consent Statement: Not applicable.

Data Availability Statement: Publicly available datasets were analysed in this study. These data, along with all the code produced, can be found here: https://gitlab.com/distration/dsi-nlp-publib (accessed on 28 December 2021).

Acknowledgments: We would like to thank NIST for allowing us to utilise the RCV1 dataset in our experiments.

Conflicts of Interest: The authors declare no conflict of interest.

Abbreviations

The following abbreviations are used in this manuscript:

TC	Text Classification
NLP	Natural Language Processing
DLM	Deep Language Model
CNN	Convolutional Neural Network
RNN	Recurrent Neural Network
LSTM	Long Short-Term Memory (Network)
GRU	Gated Recurrent Units
OOV	Out-of-Vocabulary (word)
BERT	Bidirectional Encoder Representations from Transformers
GPT	Generative Pre-trained Transformer
GNN	Graph Neural Network
MHA	Multi-Head Attention

References

1. Li, Q.; Peng, H.; Li, J.; Xia, C.; Yang, R.; Sun, L.; Yu, P.S.; He, L. A Survey on Text Classification: From Shallow to Deep Learning. *arXiv* **2020**, arXiv:2008.00364.
2. Kowsari, K.; Jafari Meimandi, K.; Heidarysafa, M.; Mendu, S.; Barnes, L.; Brown, D. Text Classification Algorithms: A Survey. *Information* **2019**, *10*, 150. [CrossRef]
3. Minaee, S.; Kalchbrenner, N.; Cambria, E.; Nikzad, N.; Chenaghlu, M.; Gao, J. Deep Learning–Based Text Classification: A Comprehensive Review. *Acm Comput. Surv.* **2021**, *54*, 1–40. [CrossRef]
4. Graves, A. Generating Sequences With Recurrent Neural Networks. *arXiv* **2013**, arXiv:1308.0850.
5. Manning, C.D.; Raghavan, P.; Schütze, H. *Introduction to Information Retrieval*; Cambridge University Press: Cambridge, UK, 2008. [CrossRef]
6. Mielke, S.J.; Alyafeai, Z.; Salesky, E.; Raffel, C.; Dey, M.; Gallé, M.; Raja, A.; Si, C.; Lee, W.Y.; Sagot, B.; et al. Between words and characters: A Brief History of Open-Vocabulary Modeling and Tokenization in NLP. *arXiv* **2021**, arXiv:2112.10508.

7. Saif, H.; Fernandez, M.; He, Y.; Alani, H. On Stopwords, Filtering and Data Sparsity for Sentiment Analysis of Twitter. In Proceedings of the Ninth International Conference on Language Resources and Evaluation (LREC'14), Reykjavik, Iceland, 26–31 May 2014; pp. 810–817.
8. Jivani, A. A Comparative Study of Stemming Algorithms. *Int. J. Comput. Technol. Appl.* **2011**, *2*, 1930–1938.
9. Plisson, J.; Lavrac, N.; Mladenic, D. A rule based approach to word lemmatization. In Proceedings of the 7th International Multiconference on Information Society (IS04), Ljubljana, Slovenia, 11–15 October 2004.
10. Gage, P. A New Algorithm for Data Compression. *C Users J.* **1994**, *12*, 23–38.
11. Sennrich, R.; Haddow, B.; Birch, A. Neural Machine Translation of Rare Words with Subword Units. In Proceedings of the 54th Annual Meeting of the Association for Computational Linguistics, Berlin, Germany, 7–12 August 2016; pp. 1715–1725. [CrossRef]
12. Radford, A.; Wu, J.; Child, R.; Luan, D.; Amodei, D.; Sutskever, I. Language Models are Unsupervised Multitask Learners. *OpenAI Blog* **2019**, *1*, 9.
13. Liu, Y.; Ott, M.; Goyal, N.; Du, J.; Joshi, M.; Chen, D.; Levy, O.; Lewis, M.; Zettlemoyer, L.; Stoyanov, V. RoBERTa: A Robustly Optimized BERT Pretraining Approach. *arXiv* **2019**, arXiv:1907.11692.
14. Wang, C.; Cho, K.; Gu, J. Neural Machine Translation with Byte-Level Subwords. *Proc. AAAI Conf. Artif. Intell.* **2020**, *34*, 9154–9160. [CrossRef]
15. Schuster, M.; Nakajima, K. Japanese and Korean voice search. In Proceedings of the 2012 IEEE International Conference on Acoustics, Speech and Signal Processing (ICASSP), Kyoto, Japan, 25–30 March 2012; pp. 5149–5152.
16. Devlin, J.; Chang, M.W.; Lee, K.; Toutanova, K. BERT: Pre-training of Deep Bidirectional Transformers for Language Understanding. In Proceedings of the 2019 Conference of the North American Chapter of the Association for Computational Linguistics: Human Language Technologies, Minnesota, MN, USA, 2–7 June 2019; pp. 4171–4186.
17. Kudo, T. Subword Regularization: Improving Neural Network Translation Models with Multiple Subword Candidates. In Proceedings of the 56th Annual Meeting of the Association for Computational Linguistics (Volume 1: Long Papers), Melbourne, Australia, 15–20 July 2018; pp. 66–75. [CrossRef]
18. Kudo, T.; Richardson, J. SentencePiece: A simple and language independent subword tokenizer and detokenizer for Neural Text Processing. In Proceedings of the 2018 Conference on Empirical Methods in Natural Language Processing: System Demonstrations, Brussels, Belgium, 31 October–4 November 2018; pp. 66–71. [CrossRef]
19. Yang, Z.; Dai, Z.; Yang, Y.; Carbonell, J.; Salakhutdinov, R.; Le, Q.V. XLNet: Generalized Autoregressive Pretraining for Language Understanding. In Proceedings of the 33rd International Conference on Neural Information Processing Systems, Vancouver, BC, Canada, 8–14 December 2019.
20. Conneau, A.; Khandelwal, K.; Goyal, N.; Chaudhary, V.; Wenzek, G.; Guzmán, F.; Grave, E.; Ott, M.; Zettlemoyer, L.; Stoyanov, V. Unsupervised Cross-lingual Representation Learning at Scale. In Proceedings of the 58th Annual Meeting of the Association for Computational Linguistics, Online event, 5–10 July 2020; pp. 8440–8451. [CrossRef]
21. Rodolà, E.; Cosmo, L.; Litany, O.; Bronstein, M.M.; Bronstein, A.M.; Audebert, N.; Hamza, A.B.; Boulch, A.; Castellani, U.; Do, M.N.; et al. Deformable Shape Retrieval with Missing Parts. In Proceedings of the Eurographics Workshop on 3D Object Retrieval, Lyon, France, 23–24 April 2017; Pratikakis, I., Dupont, F., Ovsjanikov, M., Eds.; Eurographics Association: Geneva, Switzerland, 2017. [CrossRef]
22. Gasparetto, A.; Minello, G.; Torsello, A. Non-parametric Spectral Model for Shape Retrieval. In Proceedings of the 2015 International Conference on 3D Vision, Lyon, France, 19–22 October 2015; pp. 344–352. [CrossRef]
23. Pistellato, M.; Bergamasco, F.; Albarelli, A.; Cosmo, L.; Gasparetto, A.; Torsello, A. Robust phase unwrapping by probabilistic consensus. *Opt. Lasers Eng.* **2019**, *121*, 428–440. [CrossRef]
24. Jones, K.S. A statistical interpretation of term specificity and its application in retrieval. *J. Doc.* **1972**, *28*, 11–21. [CrossRef]
25. Maćkiewicz, A.; Ratajczak, W. Principal components analysis (PCA). *Comput. Geosci.* **1993**, *19*, 303–342. [CrossRef]
26. Tharwat, A.; Gaber, T.; Ibrahim, A.; Hassanien, A.E. Linear discriminant analysis: A detailed tutorial. *Ai Commun.* **2017**, *30*, 169–190. [CrossRef]
27. Tsuge, S.; Shishibori, M.; Kuroiwa, S.; Kita, K. Dimensionality reduction using non-negative matrix factorization for information retrieval. In Proceedings of the 2001 IEEE International Conference on Systems, Man and Cybernetics. e-Systems and e-Man for Cybernetics in Cyberspace (Cat.No.01CH37236), Tucson, AZ, USA, 7–10 October 2001; Volume 2, pp. 960–965. [CrossRef]
28. Rosenfeld, R. Two decades of statistical language modeling: Where do we go from here? *Proc. IEEE* **2000**, *88*, 1270–1278. [CrossRef]
29. Jurafsky, D.; Martin, J. *Speech and Language Processing: An Introduction to Natural Language Processing, Computational Linguistics, and Speech Recognition*, 3rd ed. (draft) Unpublished. 2021; pp. 30–35. Available online: https://web.stanford.edu/~jurafsky/slp3/ed3book.pdf (accessed on 28 December 2021).
30. Huang, E.H.; Socher, R.; Manning, C.D.; Ng, A.Y. Improving Word Representations via Global Context and Multiple Word Prototypes. In Proceedings of the 50th Annual Meeting of the Association for Computational Linguistics, Jeju Island, Korea, 8–14 July 2012; Volume 1, pp. 873–882.
31. Mikolov, T.; Chen, K.; Corrado, G.; Dean, J. Efficient Estimation of Word Representations in Vector Space. In Proceedings of the 1st International Conference on Learning Representations (ICLR 2013), Scottsdale, AZ, USA, 2–4 May 2013.

32. Mikolov, T.; Sutskever, I.; Chen, K.; Corrado, G.S.; Dean, J. Distributed Representations of Words and Phrases and their Compositionality. In Proceedings of the 26th International Conference on Neural Information Processing Systems, Lake Tahoe, NV, USA, 5–10 December 2013; Volume 2, pp. 3111–3119.
33. Pennington, J.; Socher, R.; Manning, C. GloVe: Global Vectors for Word Representation. In Proceedings of the 2014 Conference on Empirical Methods in Natural Language Processing (EMNLP), Doha, Qatar, 25–29 October 2014; pp. 1532–1543. [CrossRef]
34. Baroni, M.; Dinu, G.; Kruszewski, G. Don't count, predict! A systematic comparison of context-counting vs. context-predicting semantic vectors. In Proceedings of the 52nd Annual Meeting of the Association for Computational Linguistics, Baltimore, MD, USA, 22–27 June 2014; Volume 1, pp. 238–247. [CrossRef]
35. Bojanowski, P.; Grave, E.; Joulin, A.; Mikolov, T. Enriching Word Vectors with Subword Information. *Trans. Assoc. Comput. Linguist.* **2017**, *5*, 135–146. [CrossRef]
36. Xu, S.; Li, Y.; Wang, Z. Bayesian Multinomial Naïve Bayes Classifier to Text Classification. In *Advanced Multimedia and Ubiquitous Engineering*; Springer: Singapore, 2017; pp. 347–352. [CrossRef]
37. van den Bosch, A. Hidden Markov Models. In *Encyclopedia of Machine Learning and Data Mining*; Springer: Boston, MA, USA, 2017; pp. 609–611. [CrossRef]
38. Sutton, C.; McCallum, A. An Introduction to Conditional Random Fields. *Found. Trends® Mach. Learn.* **2012**, *4*, 267–373. [CrossRef]
39. Cover, T.; Hart, P. Nearest Neighbor pattern classification. *IEEE Trans. Inf. Theory* **1967**, *13*, 21–27. [CrossRef]
40. Li, B.; Yu, S.; Lu, Q. An Improved k-Nearest Neighbor Algorithm for Text Categorization. *arXiv* **2003**, arXiv:cs/0306099.
41. Ali, N.; Neagu, D.; Trundle, P. Evaluation of k-nearest neighbour classifier performance for heterogeneous data sets. *SN Appl. Sci.* **2019**, *1*, 1–15. [CrossRef]
42. Bellman, R. Dynamic Programming. *Science* **1966**, *153*, 34–37. [CrossRef] [PubMed]
43. Cortes, C.; Vapnik, V.N. Support-Vector Networks. *Mach. Learn.* **1995**, *20*, 273–297. [CrossRef]
44. Boser, B.E.; Guyon, I.M.; Vapnik, V.N. A Training Algorithm for Optimal Margin Classifiers. In Proceedings of the Fifth Annual Workshop on Computational Learning Theory, Pittsburgh, PA, USA, 27–29 July 1992; pp. 144–152. [CrossRef]
45. Safavian, S.; Landgrebe, D. A survey of decision tree classifier methodology. *IEEE Trans. Syst. Man Cybern.* **1991**, *21*, 660–674. [CrossRef]
46. Ho, T.K. Random decision forests. In Proceedings of the 3rd International Conference on Document Analysis and Recognition, Montreal, QC, Canada, 14–15 August 1995; Volume 1, pp. 278–282. [CrossRef]
47. Islam, M.Z.; Liu, J.; Li, J.; Liu, L.; Kang, W. A Semantics Aware Random Forest for Text Classification. In Proceedings of the 28th ACM International Conference on Information and Knowledge Management (CIKM '19), Beijing, China, 3–7 November 2019; pp. 1061–1070. [CrossRef]
48. Genkin, A.; Lewis, D.; Madigan, D. Large-Scale Bayesian Logistic Regression for Text Categorization. *Technometrics* **2007**, *49*, 291–304. [CrossRef]
49. Krishnapuram, B.; Carin, L.; Figueiredo, M.; Hartemink, A. Sparse multinomial logistic regression: Fast algorithms and generalization bounds. *IEEE Trans. Pattern Anal. Mach. Intell.* **2005**, *27*, 957–968. [CrossRef]
50. Breiman, L. Bagging predictors. *Mach. Learn.* **2004**, *24*, 123–140. [CrossRef]
51. Schapire, R.E. The Strength of Weak Learnability. *Mach. Learn.* **1990**, *5*, 197–227. [CrossRef]
52. Freund, Y.; Schapire, R.E. A Decision-Theoretic Generalization of On-Line Learning and an Application to Boosting. *J. Comput. Syst. Sci.* **1997**, *55*, 119–139. [CrossRef]
53. Joulin, A.; Grave, E.; Bojanowski, P.; Mikolov, T. Bag of Tricks for Efficient Text Classification. In Proceedings of the 15th Conference of the European Chapter of the Association for Computational Linguistics, Valencia, Spain, 3–7 April 2017; Volume 2, pp. 427–431.
54. Ibrahim, M.A.; Ghani Khan, M.U.; Mehmood, F.; Asim, M.N.; Mahmood, W. GHS-NET a generic hybridized shallow neural network for multi-label biomedical text classification. *J. Biomed. Inform.* **2021**, *116*, 103699. [CrossRef]
55. Iyyer, M.; Manjunatha, V.; Boyd-Graber, J.; Daumé, H., III Deep Unordered Composition Rivals Syntactic Methods for Text Classification. In Proceedings of the 53rd Annual Meeting of the Association for Computational Linguistics and the 7th International Joint Conference on Natural Language Processing, Beijing, China, 27–31 July 2015; Volume 1, pp. 1681–1691. [CrossRef]
56. Le, Q.; Mikolov, T. Distributed Representations of Sentences and Documents. In Proceedings of the 31st International Conference on Machine Learning, Beijing, China, 21–26 June 2014; Volume 32, pp. 1188–1196.
57. Mikolov, T.; Le, Q.V.; Sutskever, I. Exploiting Similarities among Languages for Machine Translation. *arXiv* **2013**, arXiv:1309.4168.
58. Hochreiter, S.; Schmidhuber, J. Long Short-term Memory. *Neural Comput.* **1997**, *9*, 1735–1780. [CrossRef] [PubMed]
59. Tai, K.S.; Socher, R.; Manning, C.D. Improved Semantic Representations From Tree-Structured Long Short-Term Memory Networks. In Proceedings of the 53rd Annual Meeting of the Association for Computational Linguistics and the 7th International Joint Conference on Natural Language Processing, Beijing, China, 27–31 July 2015; Volume 1, pp. 1556–1566. [CrossRef]
60. Dieng, A.B.; Wang, C.; Gao, J.; Paisley, J. TopicRNN: A Recurrent Neural Network with Long-Range Semantic Dependency. *arXiv* **2016**, arXiv:1611.01702.
61. Howard, J.; Ruder, S. Universal Language Model Fine-tuning for Text Classification. In Proceedings of the 56th Annual Meeting of the Association for Computational Linguistics, Melbourne, Australia, 15–20 July 2018; Volume 1, pp. 328–339. [CrossRef]

62. Wang, B. Disconnected Recurrent Neural Networks for Text Categorization. In Proceedings of the 56th Annual Meeting of the Association for Computational Linguistics, Melbourne, Australia, 15–20 July 2018; Volume 1, pp. 2311–2320. [CrossRef]
63. Cho, K.; van Merriënboer, B.; Gulcehre, C.; Bahdanau, D.; Bougares, F.; Schwenk, H.; Bengio, Y. Learning Phrase Representations using RNN Encoder–Decoder for Statistical Machine Translation. In Proceedings of the 2014 Conference on Empirical Methods in Natural Language Processing (EMNLP), Doha, Qatar, 25–29 October 2014; pp. 1724–1734. [CrossRef]
64. Schuster, M.; Paliwal, K. Bidirectional recurrent neural networks. *IEEE Trans. Signal Process.* **1997**, *45*, 2673–2681. [CrossRef]
65. Peters, M.E.; Neumann, M.; Iyyer, M.; Gardner, M.; Clark, C.; Lee, K.; Zettlemoyer, L. Deep Contextualized Word Representations. In Proceedings of the 2018 Conference of the North American Chapter of the Association for Computational Linguistics: Human Language Technologies, New Orleans, LA, USA, 1–6 June 2018; Volume 1, pp. 2227–2237. [CrossRef]
66. Zhang, Y.; Wallace, B.C. A Sensitivity Analysis of (and Practitioners' Guide to) Convolutional Neural Networks for Sentence Classification. In Proceedings of the Eighth International Joint Conference on Natural Language Processing (IJCNLP), Taipei, Taiwan, 27–30 November 2017; Volume 1, pp. 253–263.
67. Stone, A.; Wang, H.; Stark, M.; Liu, Y.; Phoenix, D.; George, D. Teaching Compositionality to CNNs. In Proceedings of the 2017 IEEE Conference on Computer Vision and Pattern Recognition (CVPR), Honolulu, HI, USA, 21–26 July 2017; pp. 732–741. [CrossRef]
68. Pistellato, M.; Cosmo, L.; Bergamasco, F.; Gasparetto, A.; Albarelli, A. Adaptive Albedo Compensation for Accurate Phase-Shift Coding. In Proceedings of the 2018 24th International Conference on Pattern Recognition (ICPR), Beijing, China, 20–24 August 2018; pp. 2450–2455. [CrossRef]
69. Kim, Y. Convolutional Neural Networks for Sentence Classification. In Proceedings of the 2014 Conference on Empirical Methods in Natural Language Processing (EMNLP), Doha, Qatar, 25–29 October 2014; pp. 1746–1751. [CrossRef]
70. Lea, C.; Flynn, M.D.; Vidal, R.; Reiter, A.; Hager, G.D. Temporal Convolutional Networks for Action Segmentation and Detection. In Proceedings of the 2017 IEEE Conference on Computer Vision and Pattern Recognition (CVPR), Honolulu, HI, USA, 21–26 July 2017; pp. 1003–1012. [CrossRef]
71. Abreu, J.; Fred, L.; Macêdo, D.; Zanchettin, C. Hierarchical Attentional Hybrid Neural Networks for Document Classification. In *Artificial Neural Networks and Machine Learning—ICANN 2019: Workshop and Special Sessions*; Springer International Publishing: Cham, Switzerland, 2019; pp. 396–402. [CrossRef]
72. Yan, J.; Mu, L.; Wang, L.; Ranjan, R.; Zomaya, A.Y. Temporal Convolutional Networks for the Advance Prediction of ENSO. *Sci. Rep.* **2020**, *10*, 8055. [CrossRef]
73. Conneau, A.; Schwenk, H.; Barrault, L.; Lecun, Y. Very Deep Convolutional Networks for Text Classification. In Proceedings of the 15th Conference of the European Chapter of the Association for Computational Linguistics, Valencia, Spain, 3–7 April 2017; Volume 1, pp. 1107–1116.
74. Duque, A.B.; Santos, L.L.J.; Macêdo, D.; Zanchettin, C. Squeezed Very Deep Convolutional Neural Networks for Text Classification. In *Artificial Neural Networks and Machine Learning—ICANN 2019: Theoretical Neural Computation*; Springer International Publishing: Cham, Switzerland, 2019; pp. 193–207. [CrossRef]
75. Sutskever, I.; Vinyals, O.; Le, Q.V. Sequence to Sequence Learning with Neural Networks. In Proceedings of the 27th International Conference on Neural Information Processing Systems, Montreal, QC, Canada, 8–13 December 2014; MIT Press: Cambridge, MA, USA, 2014; Volume 2, pp. 3104–3112.
76. Bahdanau, D.; Cho, K.; Bengio, Y. Neural Machine Translation by Jointly Learning to Align and Translate. *arXiv* **2015**, arXiv:1409.0473.
77. Pascanu, R.; Mikolov, T.; Bengio, Y. On the difficulty of training recurrent neural networks. In Proceedings of the 30th International Conference on Machine Learning, Atlanta, GA, USA, 16–21 June 2013; Dasgupta, S., McAllester, D., Eds.; PMLR: Atlanta, GA, USA, 2013; Volume 28, pp. 1310–1318.
78. Luong, T.; Pham, H.; Manning, C.D. Effective Approaches to Attention-based Neural Machine Translation. In Proceedings of the 2015 Conference on Empirical Methods in Natural Language Processing, Lisbon, Portugal, 17–21 September 2015; pp. 1412–1421. [CrossRef]
79. Bender, E.M.; Gebru, T.; McMillan-Major, A.; Shmitchell, S. On the Dangers of Stochastic Parrots: Can Language Models Be Too Big? In Proceedings of the 2021 ACM Conference on Fairness, Accountability, and Transparency, Online event, 3–10 March 2021; pp. 610–623. [CrossRef]
80. Yang, Z.; Yang, D.; Dyer, C.; He, X.; Smola, A.; Hovy, E. Hierarchical Attention Networks for Document Classification. In Proceedings of the 2016 Conference of the North American Chapter of the Association for Computational Linguistics: Human Language Technologies, San Diego, CA, USA, 12–17 June 2016; pp. 1480–1489. [CrossRef]
81. Vaswani, A.; Shazeer, N.; Parmar, N.; Uszkoreit, J.; Jones, L.; Gomez, A.N.; Kaiser, L.; Polosukhin, I. Attention is All You Need. In Proceedings of the 31st International Conference on Neural Information Processing Systems, Long Beach, CA, USA, 4–9 December 2017; pp. 6000–6010.
82. Liu, P.J.; Saleh, M.; Pot, E.; Goodrich, B.; Sepassi, R.; Kaiser, L.; Shazeer, N. Generating Wikipedia by Summarizing Long Sequences. *arXiv* **2018**, arXiv:1801.10198.
83. Radford, A.; Narasimhan, K. Improving Language Understanding by Generative Pre-Training. *OpenAI Blog* **2018**. Available online: https://www.cs.ubc.ca/~amuham01/LING530/papers/radford2018improving.pdf (accessed on 28 December 2021).

84. Von Platen, P. Transformers-Based Encoder-Decoder Models. Available online: https://huggingface.co/blog/encoder-decoder (accessed on 28 December 2021).
85. Wang, A.; Singh, A.; Michael, J.; Hill, F.; Levy, O.; Bowman, S. GLUE: A Multi-Task Benchmark and Analysis Platform for Natural Language Understanding. In Proceedings of the 2018 EMNLP Workshop BlackboxNLP: Analyzing and Interpreting Neural Networks for NLP, Brussels, Belgium, 31 October–4 November 2018; pp. 353–355. [CrossRef]
86. Brown, T.; Mann, B.; Ryder, N.; Subbiah, M.; Kaplan, J.D.; Dhariwal, P.; Neelakantan, A.; Shyam, P.; Sastry, G.; Askell, A.; et al. Language Models are Few-Shot Learners. In Proceedings of the 34th Annual Conference on Neural Information Processing Systems, Online event, 6–12 December 2020; Volume 33, pp. 1877–1901.
87. Raffel, C.; Shazeer, N.; Roberts, A.; Lee, K.; Narang, S.; Matena, M.; Zhou, Y.; Li, W.; Liu, P.J. Exploring the Limits of Transfer Learning with a Unified Text-to-Text Transformer. *J. Mach. Learn. Res.* **2020**, *21*, 1–67.
88. Sanh, V.; Webson, A.; Raffel, C.; Bach, S.H.; Sutawika, L.; Alyafeai, Z.; Chaffin, A.; Stiegler, A.; Scao, T.L.; Raja, A.; et al. Multitask Prompted Training Enables Zero-Shot Task Generalization. *arXiv* **2021**, arXiv:2110.08207.
89. He, P.; Liu, X.; Gao, J.; Chen, W. DeBERTa: Decoding-Enhanced BERT with Disentangled Attention. In Proceedings of the 2021 International Conference on Learning Representations (ICLR 2021), Vienna, Austria, 4–8 May 2021.
90. Sun, Y.; Wang, S.; Feng, S.; Ding, S.; Pang, C.; Shang, J.; Liu, J.; Chen, X.; Zhao, Y.; Lu, Y.; et al. ERNIE 3.0: Large-scale Knowledge Enhanced Pre-training for Language Understanding and Generation. *arXiv* **2021**, arXiv:2107.02137.
91. Dai, Z.; Yang, Z.; Yang, Y.; Carbonell, J.; Le, Q.; Salakhutdinov, R. Transformer-XL: Attentive Language Models beyond a Fixed-Length Context. In Proceedings of the 57th Annual Meeting of the Association for Computational Linguistics, Florence, Italy, 28 July–2 August 2019; pp. 2978–2988. [CrossRef]
92. Wei, J.; Bosma, M.; Zhao, V.Y.; Guu, K.; Yu, A.W.; Lester, B.; Du, N.; Dai, A.M.; Le, Q.V. Finetuned Language Models Are Zero-Shot Learners. *arXiv* **2021**, arXiv:2109.01652.
93. Schiavinato, M.; Gasparetto, A.; Torsello, A. Transitive assignment kernels for structural classification. *Lect. Notes Comput. Sci.* **2015**, *9370*, 146–159. [CrossRef]
94. Cai, H.; Zheng, V.W.; Chang, K. A Comprehensive Survey of Graph Embedding: Problems, Techniques, and Applications. *IEEE Trans. Knowl. Data Eng.* **2018**, *30*, 1616–1637. [CrossRef]
95. Battaglia, P.W.; Hamrick, J.B.; Bapst, V.; Sanchez-Gonzalez, A.; Zambaldi, V.F.; Malinowski, M.; Tacchetti, A.; Raposo, D.; Santoro, A.; Faulkner, R.; et al. Relational inductive biases, deep learning, and graph networks. *arXiv* **2018**, arXiv:1806.01261.
96. Bruna, J.; Zaremba, W.; Szlam, A.; Lecun, Y. Spectral networks and locally connected networks on graphs. In Proceedings of the International Conference on Learning Representations (ICLR 2014), Banff, AB, Canada, 14–16 April 2014.
97. Torsello, A.; Gasparetto, A.; Rossi, L.; Bai, L.; Hancock, E. Transitive State Alignment for the Quantum Jensen-Shannon Kernel. *Lect. Notes Comput. Sci.* **2014**, *8621*, 22–31. [CrossRef]
98. Yao, L.; Mao, C.; Luo, Y. Graph Convolutional Networks for Text Classification. *Proc. AAAI Conf. Artif. Intell.* **2019**, *33*, 7370–7377. [CrossRef]
99. Church, K.W.; Hanks, P. Word Association Norms, Mutual Information, and Lexicography. In Proceedings of the 27th Annual Meeting on Association for Computational Linguistics, Vancouver, BC, Canada, 26–29 June 1989; pp. 76–83. [CrossRef]
100. Lin, Y.; Meng, Y.; Sun, X.; Han, Q.; Kuang, K.; Li, J.; Wu, F. BertGCN: Transductive Text Classification by Combining GNN and BERT. In *Findings of the Association for Computational Linguistics: ACL-IJCNLP 2021*; Association for Computational Linguistics: Stroudsburg, PA, USA, 2021; pp. 1456–1462. [CrossRef]
101. Nikolentzos, G.; Tixier, A.; Vazirgiannis, M. Message Passing Attention Networks for Document Understanding. *Proc. AAAI Conf. Artif. Intell.* **2020**, *34*, 8544–8551. [CrossRef]
102. Gammerman, A.; Vovk, V.; Vapnik, V. Learning by Transduction. In Proceedings of the Fourteenth Conference on Uncertainty in Artificial Intelligence, Madison, WI, USA, 24–26 July 1998; pp. 148–155.
103. Huang, L.; Ma, D.; Li, S.; Zhang, X.; Wang, H. Text Level Graph Neural Network for Text Classification. In Proceedings of the 2019 Conference on Empirical Methods in Natural Language Processing and the 9th International Joint Conference on Natural Language Processing (EMNLP-IJCNLP), Hong Kong, China, 3–7 November 2019; pp. 3444–3450. [CrossRef]
104. Wu, F.; Souza, A.; Zhang, T.; Fifty, C.; Yu, T.; Weinberger, K. Simplifying Graph Convolutional Networks. In Proceedings of the 36th International Conference on Machine Learning, Irvine, CA, USA, 13–16 November 2019; Volume 9, pp. 6861–6871.
105. Li, Q.; Han, Z.; Wu, X.M. Deeper Insights into Graph Convolutional Networks for Semi-Supervised Learning. In Proceedings of the Thirty-Second AAAI Conference on Artificial Intelligence, New Orleans, LA, USA, 2–7 February 2018.
106. Chen, M.; Wei, Z.; Huang, Z.; Ding, B.; Li, Y. Simple and Deep Graph Convolutional Networks. In Proceedings of the 37th International Conference on Machine Learning, Online event, 13–18 July 2020; Volume 119, pp. 1725–1735.
107. Zhu, H.; Koniusz, P. Simple Spectral Graph Convolution. In Proceedings of the 2021 International Conference on Learning Representations (ICLR 2021), Vienna, Austria, 4–8 May 2021.
108. Klicpera, J.; Bojchevski, A.; Günnemann, S. Predict then Propagate: Graph Neural Networks meet Personalized PageRank. *arXiv*, **2019**, arXiv:1810.05997.
109. Gasparetto, A.; Cosmo, L.; Rodola, E.; Bronstein, M.; Torsello, A. Spatial Maps: From low rank spectral to sparse spatial functional representations. In Proceedings of the 2017 International Conference on 3D Vision (3DV), Qingdao, China, 10–12 October 2017; pp. 477–485. [CrossRef]

110. Ethayarajh, K. How Contextual are Contextualized Word Representations? Comparing the Geometry of BERT, ELMo, and GPT-2 Embeddings. In Proceedings of the 2019 Conference on Empirical Methods in Natural Language Processing and the 9th International Joint Conference on Natural Language Processing (EMNLP-IJCNLP), Hong Kong, China, 3–7 November 2019; pp. 55–65. [CrossRef]
111. Peters, M.E.; Ammar, W.; Bhagavatula, C.; Power, R. Semi-supervised sequence tagging with bidirectional language models. In Proceedings of the 55th Annual Meeting of the Association for Computational Linguistics, Vancouver, BC, Canada, 31 July–4 August 2017; Volume 1, pp. 1756–1765. [CrossRef]
112. McCann, B.; Bradbury, J.; Xiong, C.; Socher, R. Learned in Translation: Contextualized Word Vectors. In Proceedings of the 31st International Conference on Neural Information Processing Systems, Long Beach, CA, USA, 4–9 December 2017; Curran Associates Inc.: Red Hook, NY, USA, 2017; pp. 6297–6308.
113. Hinton, G.; Vinyals, O.; Dean, J. Distilling the Knowledge in a Neural Network. *arXiv* **2015**, arXiv:1503.02531.
114. Sanh, V.; Debut, L.; Chaumond, J.; Wolf, T. DistilBERT, a distilled version of BERT: Smaller, faster, cheaper and lighter. *arXiv* **2019**, arXiv:1910.01108.
115. Jiao, X.; Yin, Y.; Shang, L.; Jiang, X.; Chen, X.; Li, L.; Wang, F.; Liu, Q. TinyBERT: Distilling BERT for Natural Language Understanding. In Proceedings of the Findings of the Association for Computational Linguistics: EMNLP 2020, Online event, 16–20 November 2020; pp. 4163–4174. [CrossRef]
116. Lan, Z.; Chen, M.; Goodman, S.; Gimpel, K.; Sharma, P.; Soricut, R. ALBERT: A Lite BERT for Self-supervised Learning of Language Representations. *arXiv* **2019**, arXiv:1909.11942.
117. Clark, K.; Luong, M.T.; Le, Q.V.; Manning, C.D. ELECTRA: Pre-training Text Encoders as Discriminators Rather Than Generators. In Proceedings of the ICLR 2020: Eighth International Conference on Learning Representations, Online event, 26 April–1 May 2020.
118. Liu, J.; Chang, W.C.; Wu, Y.; Yang, Y. Deep Learning for Extreme Multi-Label Text Classification. In Proceedings of the 40th International ACM SIGIR Conference on Research and Development in Information Retrieval, Shinjuku Tokyo, Japan, 7–11 August 2017; SIGIR '17, pp. 115–124. [CrossRef]
119. Zhang, W.; Yan, J.; Wang, X.; Zha, H. Deep Extreme Multi-Label Learning. In Proceedings of the 2018 ACM on International Conference on Multimedia Retrieval, Yokohama, Japan, 11–14 June 2018; pp. 100–107. [CrossRef]
120. Lewis, D.D.; Yang, Y.; Rose, T.G.; Li, F. RCV1: A New Benchmark Collection for Text Categorization Research. *J. Mach. Learn. Res.* **2004**, *5*, 361–397.
121. Wikipedia:Portal. Available online: https://en.wikipedia.org/wiki/Wikipedia:Portal (accessed on 28 December 2021).
122. AG's Corpus of News Articles. Available online: http://groups.di.unipi.it/~gulli/AG_corpus_of_news_articles.html (accessed on 28 December 2021).
123. The 20 Newsgroups Data Set. Available online: http://qwone.com/~jason/20Newsgroups (accessed on 28 December 2021).
124. Ohsumed-R8-R52. Available online: https://www.kaggle.com/weipengfei/ohr8r52 (accessed on 28 December 2021).
125. Zhang, X.; Zhao, J.; LeCun, Y. Character-Level Convolutional Networks for Text Classification. In Proceedings of the 28th International Conference on Neural Information Processing Systems, Montreal, QC, Canada, 7–12 December 2015; Volume 1, pp. 649–657.
126. Yelp Open Dataset: An all-Purpose Dataset for Learning. Available online: https://www.yelp.com/dataset (accessed on 28 December 2021).
127. Maas, A.L.; Daly, R.E.; Pham, P.T.; Huang, D.; Ng, A.Y.; Potts, C. Learning Word Vectors for Sentiment Analysis. In Proceedings of the 49th Annual Meeting of the Association for Computational Linguistics: Human Language Technologies, Portland, OR, USA, 19–24 June 2011; pp. 142–150.
128. Pang, B.; Lee, L.; Vaithyanathan, S. Thumbs up? Sentiment Classification Using Machine Learning Techniques. In Proceedings of the ACL-02 Conference on Empirical Methods in Natural Language Processing, Philadelphia, PA, USA, 6–7 July 2002; Volume 10, pp. 79–86. [CrossRef]
129. Li, X.; Roth, D. Learning Question Classifiers. In Proceedings of the 19th International Conference on Computational Linguistics, Taipei, Taiwan, 24 August–1 September 2002; Volume 1, pp. 1–7. [CrossRef]
130. Joachims, T. Text Categorization with Support Vector Machines: Learning with Many Relevant Features. In Proceedings of the 10th European Conference on Machine Learning, Chemnitz, Germany, 21–23 April 1998; pp. 137–142. [CrossRef]
131. Text Categorization Corpora. Available online: https://disi.unitn.it/moschitti/corpora.htm (accessed on 28 December 2021).
132. Sun, C.; Qiu, X.; Xu, Y.; Huang, X. How to Fine-Tune BERT for Text Classification? In *Chinese Computational Linguistics*; Springer International Publishing: Cham, Switzerland, 2019; pp. 194–206. [CrossRef]
133. Adhikari, A.; Ram, A.; Tang, R.; Lin, J. DocBERT: BERT for Document Classification. *arXiv* **2019**, arXiv:1904.08398.
134. Xie, Q.; Dai, Z.; Hovy, E.; Luong, M.T.; Le, Q.V. Unsupervised Data Augmentation for Consistency Training. *arXiv* **2020**, arXiv:1904.12848.
135. Sachan, D.S.; Zaheer, M.; Salakhutdinov, R. Revisiting LSTM Networks for Semi-Supervised Text Classification via Mixed Objective Function. *Proc. AAAI Conf. Artif. Intell.* **2019**, *33*, 6940–6948. [CrossRef]
136. Le, H.; Tran, T.; Venkatesh, S. Learning to Remember More with Less Memorization. *arXiv* **2019**, arXiv:1901.01347.

137. Prabhu, A.; Dognin, C.; Singh, M. Sampling Bias in Deep Active Classification: An Empirical Study. In Proceedings of the 2019 Conference on Empirical Methods in Natural Language Processing and the 9th International Joint Conference on Natural Language Processing (EMNLP-IJCNLP), Hong Kong, China, 3–7 November 2019; pp. 4058–4068. [CrossRef]
138. Cer, D.; Yang, Y.; Kong, S.-y.; Hua, N.; Limtiaco, N.; John, R.S.; Constant, N.; Guajardo-Cespedes, M.; Yuan, S.; Tar, C.; et al. Universal Sentence Encoder. *arXiv* **2018**, arXiv:1803.11175.
139. Shin, B.; Yang, H.; Choi, J.D. The Pupil Has Become the Master: Teacher-Student Model-Based Word Embedding Distillation with Ensemble Learning. In Proceedings of the Twenty-Eighth International Joint Conference on Artificial Intelligence, IJCAI-19, Macao, China, 10–16 August 2019; pp. 3439–3445. [CrossRef]
140. Ionescu, R.T.; Butnaru, A. Vector of Locally-Aggregated Word Embeddings (VLAWE): A Novel Document-level Representation. In Proceedings of the 2019 Conference of the North American Chapter of the Association for Computational Linguistics: Human Language Technologies, Minneapolis, MN, USA, 2–7 June 2019; Volume 1, pp. 363–369. [CrossRef]
141. Yadav, R.K.; Jiao, L.; Granmo, O.C.; Goodwin, M. Enhancing Interpretable Clauses Semantically using Pretrained Word Representation. In Proceedings of the Fourth BlackboxNLP Workshop on Analyzing and Interpreting Neural Networks for NLP, Punta Cana, Dominican Republic, 11 November 2021; pp. 265–274.
142. Ding, S.; Shang, J.; Wang, S.; Sun, Y.; Tian, H.; Wu, H.; Wang, H. ERNIE-Doc: A Retrospective Long-Document Modeling Transformer. In Proceedings of the 59th Annual Meeting of the Association for Computational Linguistics and the 11th International Joint Conference on Natural Language Processing (Volume 1: Long Papers), Online event, 1–6 August 2021; Volume 1, pp. 2914–2927. [CrossRef]
143. Zaheer, M.; Guruganesh, G.; Dubey, A.; Ainslie, J.; Alberti, C.; Ontanon, S.; Pham, P.; Ravula, A.; Wang, Q.; Yang, L.; et al. Big Bird: Transformers for Longer Sequences. *arXiv* **2020**, arXiv:2007.14062.
144. Thongtan, T.; Phienthrakul, T. Sentiment Classification Using Document Embeddings Trained with Cosine Similarity. In Proceedings of the 57th Annual Meeting of the Association for Computational Linguistics: Student Research Workshop, Florence, Italy, 28 July–2 August 2019; pp. 407–414. [CrossRef]
145. Sun, Z.; Fan, C.; Sun, X.; Meng, Y.; Wu, F.; Li, J. Neural Semi-supervised Learning for Text Classification Under Large-Scale Pretraining. *arXiv* **2020**, arXiv:2011.08626.
146. Kowsari, K.; Heidarysafa, M.; Brown, D.E.; Meimandi, K.J.; Barnes, L.E. RMDL: Random Multimodel Deep Learning for Classification. In Proceedings of the 2nd International Conference on Information System and Data Mining, Lakeland, FL, USA, 9–11 April 2018; pp. 19–28. [CrossRef]
147. Lu, H.; Huang, S.H.; Ye, T.; Guo, X. Graph Star Net for Generalized Multi-Task Learning. *arXiv* **2019**, arXiv:1906.12330.
148. Johnson, R.; Zhang, T. Deep Pyramid Convolutional Neural Networks for Text Categorization. In Proceedings of the 55th Annual Meeting of the Association for Computational Linguistics, Vancouver, BC, Canada, 31 July–4 August 2017; Volume 1, pp. 562–570. [CrossRef]
149. Adhikari, A.; Ram, A.; Tang, R.; Lin, J. Rethinking Complex Neural Network Architectures for Document Classification. In Proceedings of the 2019 Conference of the North American Chapter of the Association for Computational Linguistics: Human Language Technologies, Minneapolis, MN, USA, 2–7 June 2019; Volume 1, pp. 4046–4051. [CrossRef]
150. Ren, H.; Lu, H. Compositional coding capsule network with k-means routing for text classification. *arXiv* **2018**, arXiv:1810.09177.
151. Wang, S.; Fang, H.; Khabsa, M.; Mao, H.; Ma, H. Entailment as Few-Shot Learner. *arXiv* **2021**, arXiv:2104.14690.
152. Khodak, M.; Saunshi, N.; Liang, Y.; Ma, T.; Stewart, B.M.; Arora, S. A La Carte Embedding: Cheap but Effective Induction of Semantic Feature Vectors. In Proceedings of the 56th Annual Meeting of the Association for Computational Linguistics, Melbourne, Australia, 15–20 July 2018; Volume 1, pp. 12–22. [CrossRef]
153. Zhou, P.; Qi, Z.; Zheng, S.; Xu, J.; Bao, H.; Xu, B. Text Classification Improved by Integrating Bidirectional LSTM with Two-dimensional Max Pooling. In Proceedings of the COLING 2016, the 26th International Conference on Computational: Technical Papers, Osaka, Japan, 11–16 December; Technical Papers. The COLING 2016 Organizing Committee: Osaka, Japan, 2016; pp. 3485–3495.
154. Johnson, R.; Zhang, T. Supervised and Semi-Supervised Text Categorization Using LSTM for Region Embeddings. In Proceedings of the 33rd International Conference on International Conference on Machine Learning, New York, NY, USA, 19–24 June 2016; Volume 48, pp. 526–534.
155. Han, K.; Chen, J.; Zhang, H.; Xu, H.; Peng, Y.; Wang, Y.; Ding, N.; Deng, H.; Gao, Y.; Guo, T.; et al. DELTA: A DEep learning based Language Technology plAtform. *arXiv* **2019**, arXiv:1908.01853.
156. Gupta, V.; Kumar, A.; Nokhiz, P.; Gupta, H.; Talukdar, P.P. Improving Document Classification with Multi-Sense Embeddings. *Front. Artif. Intell. Appl.* **2020**, *325*, 2030–2037. [CrossRef]
157. Guidotti, E.; Ferrara, A. An Explainable Probabilistic Classifier for Categorical Data Inspired to Quantum Physics. *arXiv* **2021**, arXiv:cs.LG/2105.13988.
158. Wolf, T.; Debut, L.; Sanh, V.; Chaumond, J.; Delangue, C.; Moi, A.; Cistac, P.; Rault, T.; Louf, R.; Funtowicz, M.; et al. Transformers: State-of-the-Art Natural Language Processing. In Proceedings of the 2020 Conference on Empirical Methods in Natural Language Processing: System Demonstrations, Online event, 16–20 November 2020; pp. 38–45. [CrossRef]
159. Sechidis, K.; Tsoumakas, G.; Vlahavas, I. On the Stratification of Multi-label Data. In *Machine Learning and Knowledge Discovery in Databases*; Springer: Berlin/Heidelberg, Germany, 2011; Volume 6913, pp. 145–158. [CrossRef]

160. Szymański, P.; Kajdanowicz, T. A Network Perspective on Stratification of Multi-Label Data. In Proceedings of the First International Workshop on Learning with Imbalanced Domains: Theory and Applications, Skopje, Macedonia, 22 September 2017; Luís Torgo, P.B., Moniz, N., Eds.; Volume 74, pp. 22–35.
161. Jin, D.; Jin, Z.; Zhou, J.T.; Szolovits, P. Is BERT Really Robust? A Strong Baseline for Natural Language Attack on Text Classification and Entailment. *Proc. AAAI Conf. Artif. Intell.* **2020**, *34*, 8018–8025. [CrossRef]
162. Wang, B.; Pan, B.; Li, X.; Li, B. Towards Evaluating the Robustness of Chinese BERT Classifiers. *arXiv* **2020**, arXiv:cs.CL/2004.03742.
163. Wang, B.; Xu, C.; Wang, S.; Gan, Z.; Cheng, Y.; Gao, J.; Awadallah, A.H.; Li, B. Adversarial GLUE: A Multi-Task Benchmark for Robustness Evaluation of Language Models. In Proceedings of the 35th Annual Conference on Neural Information Processing System (NeurIPS 2021), Online event, 6–14 December 2021.
164. Wang, B.; Wang, S.; Cheng, Y.; Gan, Z.; Jia, R.; Li, B.; Liu, J. InfoBERT: Improving Robustness of Language Models from an Information Theoretic Perspective. *arXiv* **2020**, arXiv:2010.02329.
165. Borgeaud, S.; Mensch, A.; Hoffmann, J.; Cai, T.; Rutherford, E.; Millican, K.; van den Driessche, G.; Lespiau, J.; Damoc, B.; Clark, A.; et al. Improving language models by retrieving from trillions of tokens. *arXiv* **2021**, arXiv:2112.04426.
166. Nakano, R.; Hilton, J.; Balaji, S.; Wu, J.; Ouyang, L.; Kim, C.; Hesse, C.; Jain, S.; Kosaraju, V.; Saunders, W.; et al. WebGPT: Browser-assisted question-answering with human feedback. *arXiv* **2021**, arXiv:2112.09332.

remote sensing

MDPI

Article

SR-Net: Saliency Region Representation Network for Vehicle Detection in Remote Sensing Images

Fanfan Liu [1,2,3,4], Wenzhe Zhao [1,2,3,4,*], Guangyao Zhou [1,2], Liangjin Zhao [1,2] and Haoran Wei [1,2,3,4]

[1] Aerospace Information Research Institute, Chinese Academy of Sciences, Beijing 100190, China; liufanfan19@mails.ucas.ac.cn (F.L.); zhougy@aircas.ac.cn (G.Z.); zhaolj004896@aircas.ac.cn (L.Z.); weihaoran18@mails.ucas.ac.cn (H.W.)
[2] Key Laboratory of Network Information System Technology (NIST), Aerospace Information Research Institute, Chinese Academy of Sciences, Beijing 100190, China
[3] University of Chinese Academy of Sciences, Beijing 100190, China
[4] School of Electronic, Electrical and Communication Engineering, University of Chinese Academy of Sciences, Beijing 100190, China
* Correspondence: zwz@mail.ie.ac.cn

Abstract: Vehicle detection in remote sensing imagery is a challenging task because of its inherent attributes, e.g., dense parking, small sizes, various angles, etc. Prevalent vehicle detectors adopt an oriented/rotated bounding box as a basic representation, which needs to apply a distance regression of height, width, and angles of objects. These distance-regression-based detectors suffer from two challenges: (1) the periodicity of the angle causes a discontinuity of regression values, and (2) small regression deviations may also cause objects to be missed. To this end, in this paper, we propose a new vehicle modeling strategy, i.e., regarding each vehicle-rotated bounding box as a saliency area. Based on the new representation, we propose SR-Net (saliency region representation network), which transforms the vehicle detection task into a saliency object detection task. The proposed SR-Net, running in a distance (e.g., height, width, and angle)-regression-free way, can generate more accurate detection results. Experiments show that SR-Net outperforms prevalent detectors on multiple benchmark datasets. Specifically, our model yields 52.30%, 62.44%, 68.25%, and 55.81% in terms of AP on DOTA, UCAS-AOD, DLR 3*K* Munich, and VEDAI, respectively.

Keywords: vehicle detection; distance-regression-free; remote sensing imagery

Citation: Liu, F.; Zhao, W.; Zhou, G.; Zhao, L.; Wei, H. SR-Net: Saliency Region Representation Network for Vehicle Detection in Remote Sensing Images. *Remote Sens.* **2022**, *14*, 1313. https://doi.org/10.3390/rs14061313

Academic Editor: Andrzej Stateczny

Received: 7 January 2022
Accepted: 5 March 2022
Published: 9 March 2022

1. Introduction

Vehicle detection plays a significant role in optics remote sensing (RS) imagery interpretation, supporting numerous downstream applications, e.g., surveillance, defense, traffic planning, etc. [1–5]. However, vehicle detection is an extremely challenging task because vehicles are usually small with varied appearances and directions, and are parked densely in many scenarios.

In the beginning, researchers used traditional methods for vehicle detection in remote sensing images. These conventional methods mainly utilized low-level features and some manual features designed with prior human knowledge. These features include a histogram of oriented gradients (HOG) [6], scale-invariant feature transform (SIFT) [7], color histogram, texture feature, etc. Specifically, Shao et al. incorporate multiple visual features, a local binary pattern (LBP) [8], HOG, and an opponent histogram for vehicle detection in high-resolution RS images. Moranduzzo et al. [9] first use SIFT to detect interest points of vehicles, and then train a support vector machine (SVM) to classify these interest points into the vehicle target or not into the vehicle target. They later present an approach [10] that performs filtering operations in horizontal and vertical directions to extract HOG features and yields vehicle detection after the computation of a similarity measure, using a catalog of vehicles as a reference. Liu and Mattyus [11] used the AdaBoost classifier and similar Haar features to achieve a more robust and fast vehicle detection. Besides, ElMikaty et al. [12] use a sliding window framework consisting of four stages, namely, window evaluation; the extraction and encoding of features; classification; and post-processing, to detect cars in complex urban environments by using a combined feature of the local distributions of gradients, colors, and texture. In addition, Zhou et al. [13] rotate the sliding window and then classify the rotated sliding window to accomplish vehicle detection in any direction, but the detection speed of this method is relatively slow. Later, Kalantar et al. [14] utilize a region-matching approach to detect vehicles in remote sensing video frames captured by UAVs.

For almost a decade, neural networks have been experiencing a Renaissance, since Alexnet [15] won an image classification contest in 2012. Subsequently, ZFnet [16], VGGnet [16], and Resnet [17] appeared one after another, and the convolutional neural network (CNN) keeps refreshing the records of the classification task [18–20]. Ross and Kaiming introduce CNN into the detection task and propose a series of effective algorithms, such as R-CNN [21], Fast R-CNN [22], Faster R-CNN [23], and an improved detector based on Faster R-CNN [24–26], whose results reach an unprecedented high level.

Of course, the above methods also promote the development of RS vehicle detection. Recently, a series of advanced vehicle detectors [27,28] place more emphasis on how to output high-quality rotated bounding boxes due to the flexible representation. Among these methods, detectors based on anchor and regression occupy the mainstream, and offer the representation of multi-oriented vehicles by a rotated bounding box (bbox) or quadrangles. Although these oriented detectors have achieved promising results, they still suffer from some fundamental problems. Specifically, in the anchor-based detector, if the vehicle has a large aspect ratio, some issues, as shown in Figure 1a, may occur when the anchor matches with the ground truth (GT). The intersection-over-union (IoU) between a detected predicted box and the ground truth box is large enough. However, it is unreasonable to define the detected predicted box as a positive sample, as it contains only a tiny amount of object information. If using this method to match the positive and negative samples, the performance and stability of the models will be seriously affected. RRPN [29] tries to solve this problem by introducing six angles for each generated anchor on the vanilla basis, as shown in Figure 1b. In this way, RRPN is more suitable for the oriented object detection task. However, it introduces multiple anchors, which leads to a large decrease in terms of speed. More seriously, these angle-guided distance-regression-based detectors suffer from two problems: (1) discontinuous regression values, which are directly caused by angular periodicity, and (2) the difficulty of regressing small vehicles. For small objects, even tiny regression deviations (height, width, or angle) can result in an object miss.

Accordingly, the accuracy of angle prediction is critical for vehicle detectors that use rotating bounding boxes. Small angular deviations can lead to severe (IoU) degradation, which, in turn, leads to inaccurate vehicle detection, especially in the case of vehicles with large aspect ratios. Some recent works try to solve these problems; for example, Yang et al. propose the IoU-smooth L1 loss [30], which eliminated the mutation of angular regression by adding boundary constraints. Later, they proposed the circular smooth label (CSL) [31] to change the angular regression problem to an angular classification problem in order to avoid the discontinuity of numerical regression. Based on this, they also proposed densely coded labels (DCL) [32] to use dense angle labels instead of CSL to speed up the angle classification branch. Zhou et al. [33] migrated the targets in the right-angle coordinate system to the polar coordinate system using two polar angles and one polar diameter to represent the targets. Wei et al. [27] modeled a remotely sensed target as two intersecting centerlines, and Chang et al. [34] modeled a remotely sensed target as nine on the target. These methods fundamentally solve the regression inaccuracy problem, but such methods are still not applicable due to the small and intensive nature of vehicle detection. Obviously, the above methods still employ the distance regression and still suffer from the problems mentioned above.

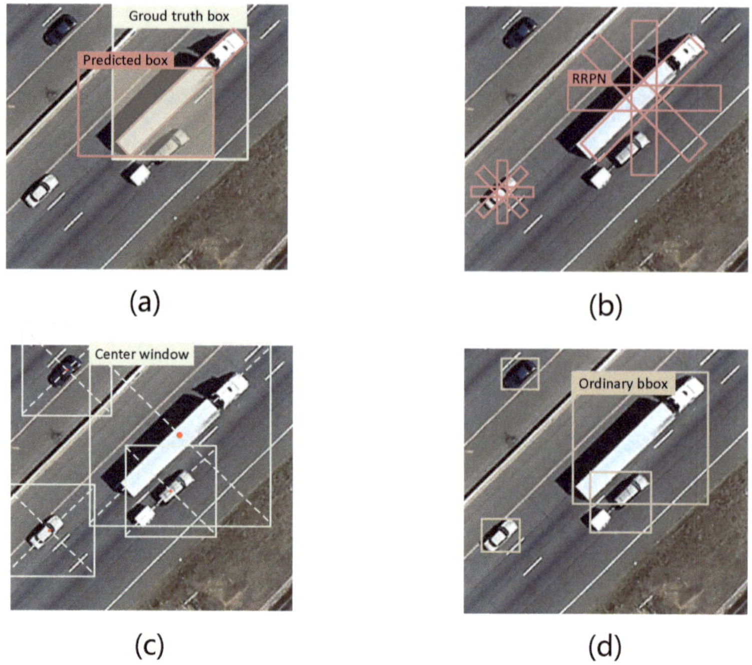

Figure 1. (**a**) The traditional anchor and ground truth matching. (**b**) The multi-angle anchor in RRPN. (**c**) The SR-Net-proposed center window. (**d**) The ordinary bbox.

To fundamentally avoid distance regression and considering the above shortcomings of current models in the vehicle detection task, in this paper, we propose a novel vehicle instance modeling method that regards each vehicle box as a saliency area. Based on the new representation, we propose SR-Net, which transforms the vehicle detection task into a saliency object detection task. This design, without any distance regression, allows our model to fundamentally avoid the aforementioned problems. Specifically, we first introduce the concept of the center window for each vehicle object. We define the center window as a square region and the vehicle's center point as the center point of the center window. We divide the center window into two sizes, i.e., large and small. As shown in Figure 1c, we define all pixels in the oriented vehicle box located in the center of the center window as the saliency area (salient object). It is worth noting that our saliency region is very different from binary segmentation, as shown in Figure 2. Besides, we design two networks to estimate both the center window and the saliency area. We make the center window detection network focus on global information to estimate all center windows and make the saliency region detection network only focus on local information to better predict the saliency area. The overall flow of our approach is shown in the Figure 3. To obtain a more robust saliency region in the training stage, we add supervision to each side path of the saliency region estimator network. Finally, an edge capture module is devised to enhance the boundary information in order to obtain a more accurate saliency region boundary. In the inference stage, we only take the maximum value on the heatmap of the center window detection network as the target center window to avoid the post-processing of complex rotating non-maximum suppression (NMS). Experiments show that the oriented boxes decoded by the saliency region output by our SR-Net are more accurate. They do not suffer from the problems of inaccurate and discontinuous angular regression, as shown in Figure 4. In summary, the main contributions of this paper are four-fold.

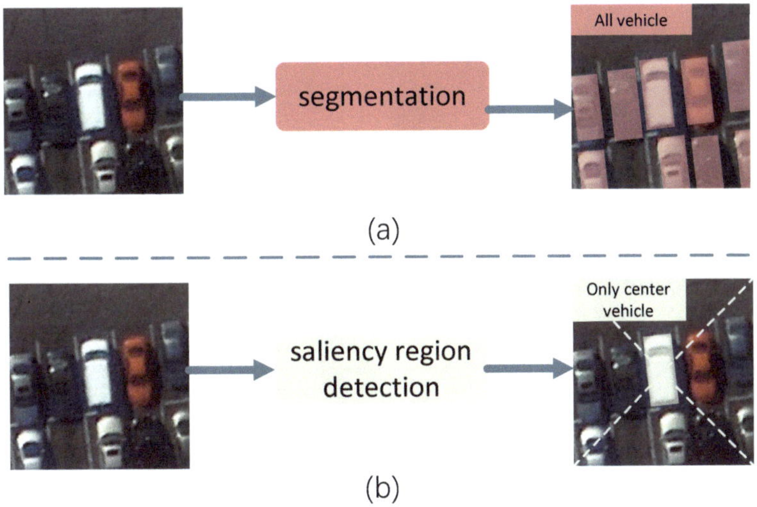

Figure 2. Difference between our saliency area (**b**) and binary segmentation (**a**). In our proposed SR-Net, we define only the vehicles at the center as the saliency area, i.e., the foreground. The saliency region is properly defined as the core of our SR-Net. Based on this definition, our results will not have the problem of sticking adjacent vehicle pixels. In contrast, the binary segmentation defines all vehicles as foreground.

1. We propose a new model, SR-Net, to exploit saliency-area-based representation and localize the vehicle objects in remote sensing images. To the best of our knowledge, we are the first to detect vehicles via saliency object detection;
2. Our model can handle the discontinuous problem of angular regression by replacing vanilla oriented-box-based representation with the proposed distance-regression-free saliency-area-based representation;
3. To obtain a more accurate boundary of the saliency region and enhance the edge information of the saliency area, we design a contour-aware module to capture the object's edge;
4. To eliminate the divergence of feature construction, we propose a new pipeline that divides the localization networks and saliency-based representation networks into two paths.

Extensive ablation studies are conducted to verify the effectiveness of our developments. Specifically, our method is compared with state-of-the-art (SOTA) methods and achieves a competitive performance in vehicle detection. Specifically, for the DOTA, UCAS-AOD, DLR 3*K* Munich, and VEDAI datasets, the average AP boosting of our model is 2.32 compared to the SOTA.

In the rest of the article, the method is described in detail in Section 1. Then, we conducted experiments to compare with some of the state-of-the-art methods in Section 2. Section 3 presents the ablation study, analysis, and discussion. Finally, Section 4 provides a summary of our approach and an outlook for future work.

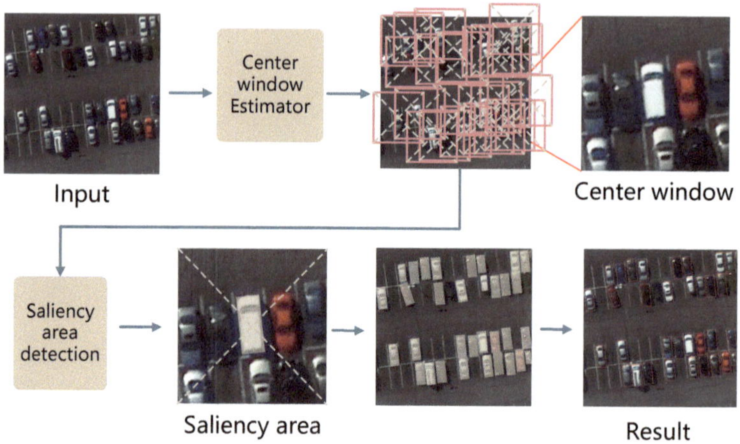

Figure 3. The pipeline of our algorithm. The center window estimator module obtains the center window of each vehicle. Then, the saliency region detection module detects the saliency region detection for each center window and decodes final predicted bboxes.

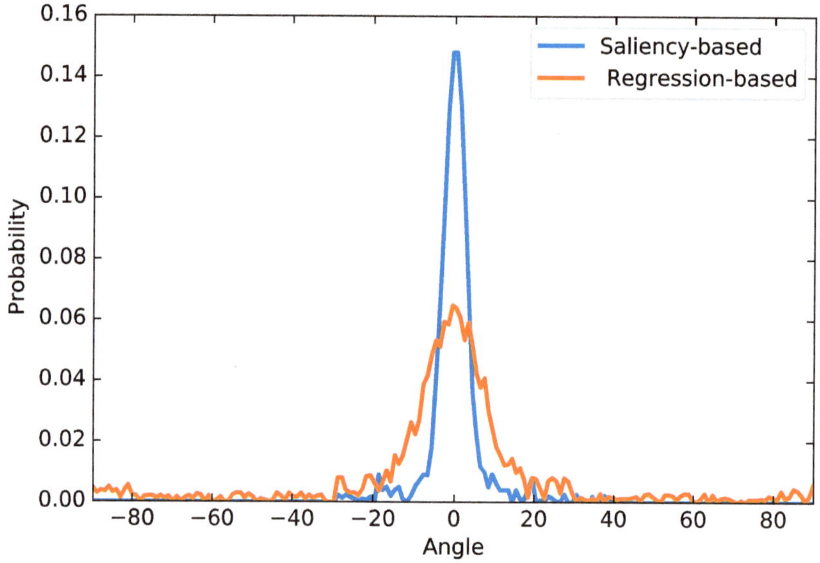

Figure 4. In the UCAS-AOD dataset, the errors of the angles are statistically obtained in the figure for the oriented bboxes generated based on regression and the oriented bboxes generated based on significance, respectively.

2. Method

We first provide an overview of the approach. As shown in Figure 5, SR-Net consists of two modules, namely, the center window estimator module and the saliency region detection module. We first generated the center window on the center window estimator pipeline. With all of the center windows, we then introduced a saliency region detection algorithm to pursue a high-quality representation of the vehicle and decoded final predicted bboxes. In this section, we first introduce the detailed structure of the center window estimator module and the design of loss functions. Then, the saliency region detection module and edge capture module are introduced to generate the saliency region with high precision.

2.1. Center Window Estimator Module

In the center window estimator network, a vehicle is defined as a center point with a fixed-size window. We used the convolutional neural network (CNN) to predict a series of heat maps to represent the center of the vehicle. Meanwhile, the network also predicts the window size (large or small) corresponding to the center of the vehicle and slightly adjusts the position of the center point by predicting the offset. Upon the predicted vehicle center point, center window category, and offset, we adopted simple post-processing to obtain the corresponding center window of each vehicle. Figure 5 provides an overview of the center window estimator network. An hourglass network [35] was used as the backbone network, followed by a prediction module. Unlike other detectors, we did not use features from different scales to solve the scale discrepancy problem.

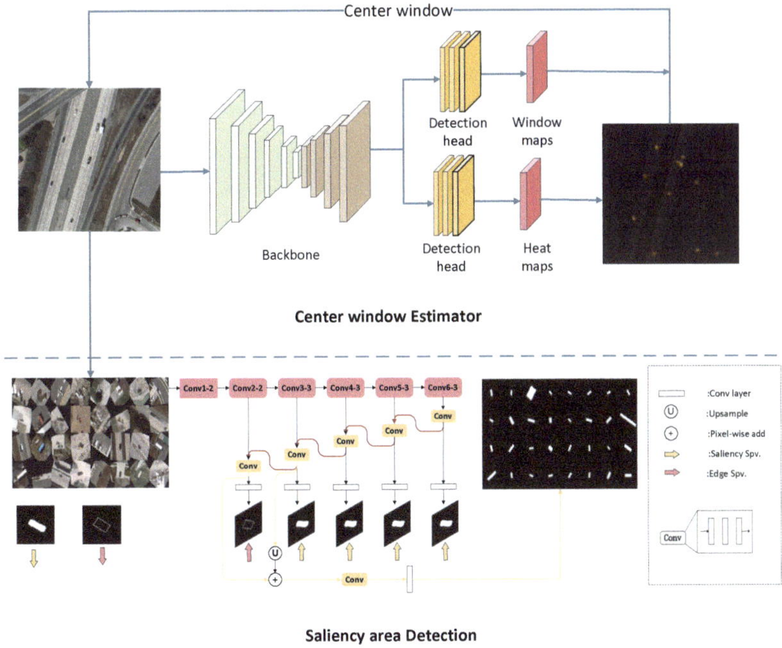

Figure 5. An overview of SR-Net. The model is divided into two main parts: the center window estimator module and the saliency region detection module. The saliency region detection module is cascaded behind the center window estimator module. The output of the center window estimator module is used as the input of the saliency region detection module.

2.1.1. Center Window Estimator

As shown in Figure 5, we refer to the design of Center-Net [36]. The center window estimator network is a pure convolutional network, and an hourglass network [35] that introduces introducing channel attention was adopted as the backbone network for the center point estimator. The detection head consists of two prediction modules to predict the center point of each vehicle, and the category of the center window. Each prediction head consists of three layers of convolution. The first two convolution layers use the ReLU activation function, and the three convolution layers use convolution kernels of sizes $3 \times 3 \times 128$, $3 \times 3 \times 128$, and $1 \times 1 \times 1$. Here, the last channel means that there is only one output category of SR-Net, i.e., vehicles. In the last output layer, the sigmoid function was selected for activation in order to map output values to the range of 0 to 1, as shown in Figure 6.

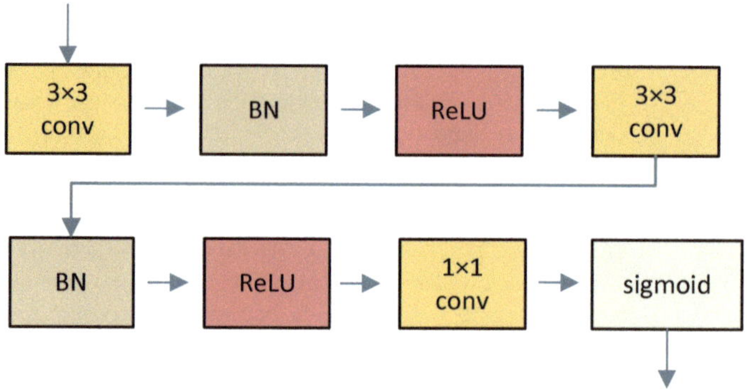

Figure 6. Schematic diagram of the structure of the detection head, where "3 × 3 conv" represents a convolutional layer with a convolutional kernel size of 3 × 3, "BN" represents the batch normalization layer, "ReLU" represents the ReLU activation function, and "sigmoid" is the sigmoid activation layer.

2.1.2. Center Estimator Loss Function

Heat map channels were used to predict central points, among which is a binary mask indicating the locations of the center for an object. For each center, there is one ground truth positive location, and all other locations are negative. In the training process, not all negative positions are equivalently penalized, but the penalty for negative positions within the radius of the positive position is reduced.

The false center detections that are close to the ground truth locations can still produce the locating centers accurately. Specifically, the radius depends on the size of the vehicle, and the points within the radius are ensured at least inside the target ground truth bbox. Meanwhile, the extent of penalty reduction is given by an un-normalized 2D Gaussian according to the radius, $e^{\frac{x^2+y^2}{2\sigma^2}}$, in which, σ^2 is $1/2$ of the radius. p_{ij} is denoted as the score at location (i,j) for object in the predicted heatmaps and y_{ij} is denoted as the "groundtruth" heatmap augmented with the unnormalized Gaussians. Based on the above analysis, we designed a variation of focal loss :

$$L_{det} = \frac{1}{N}\sum_{i=1}^{H}\sum_{j=1}^{W}\begin{cases} (1-p_{ij})^{\alpha}\log(p_{ij}) & if\ y_{ij}=1 \\ (1-y_{ij})^{\beta}(p_{ij})^{\alpha}\log(1-p_{ij}) & otherwise \end{cases} \tag{1}$$

where N denotes the number of vehicles in the current image, α is the hyperparameter that controls the weight of difficult samples, and β is the hyperparameter that controls the imbalance between positive and negative samples. We set α to 2 and β to 3 in all experiments. The $(1-y_{ij})$ term reduces the penalty around the ground truth locations with the Gaussian bumps encoded in y_{ij}. Downsampling layers are involved in our networks to gather global information and reduce memory consumption. The size of the output is usually smaller than the image when fully convolutionally applying them to an image. Therefore, the position in the original image (x,y) corresponds to the position in the heat map $(\frac{x}{n},\frac{y}{n})$, where n is the coefficient of downsampling of the heat map with respect to the original image. Some precision may have been lost when we remapped the locations from the heatmaps to the input image, which can seriously affect the center-locating accuracy of small vehicles. To solve this problem, we predicted the offset of the centroid position and slightly adjusted the centroid position before generating the center window based on the centroid.

$$o_k = (\frac{x_k}{n} - [\frac{x_k}{n}], \frac{y_k}{n} - [\frac{y_k}{n}]) \tag{2}$$

where o_k is the offset and x_k and y_k are the x and y coordinates for center k. Particularly, we predicted a set of offsets shared by the centers of all categories. In the training stage, the $smooth_{L1}$ loss (Girshick, 2015) was applied at the ground truth center locations:

$$L_{off} = \frac{1}{N}\sum_{k=1}^{N} Smooth_{L1}(o_k, \hat{o}_k) \tag{3}$$

2.1.3. Center Window

Different from other networks, after we obtained the center point of each vehicle, we did not directly regress the length, width, and angle of the vehicle We predicted a vehicle size category (large and small) for the center point of each vehicle. The unique center window was jointly determined by the the center point and size of each vehicle. In this way, we ensured that the vehicle was not only inside the center window but also in the center of the center window. Our central window category prediction head was also composed of three convolutional layers, with convolutional kernels of size $3 \times 3 \times 128$, $3 \times 3 \times 128$, and $1 \times 1 \times 2$ in each convolutional layer. Here, the last channel implies the classification result. In the final output layer, we used the softmax function to activate it. For training, we applied the crossentropy loss.

$$L_{class}(p_i, c_i) = \frac{1}{N}\sum_{i=1}^{i=N} \begin{cases} -\log(p_i(c_i)) & if \;\; c_i = large \\ -\delta\log(p_i(c_i)) & else \end{cases} \tag{4}$$

where δ is a parameter used to balance the imbalance of the size vehicle data. We set it to 0.3 In the center window estimator stage, our loss function could be expressed as:

$$L_{CE} = \gamma L_{class} + (L_{off} + L_{det}) \tag{5}$$

where L_{CE} denotes the total loss in the center window estimator stage, and γ is the coefficient.

2.2. *Saliency Region Detection*

The role of the saliency region detection network is to accurately represent a coarse representation of the vehicle (center window) as an accurate representation at the pixel level. The design of our saliency region detection network took into account some existing saliency detection models [37–39]. In the saliency region detection network, we cut the center window on the input image according to the center window generated by the center window estimator network. For each center window area, we used the saliency region detection network to detect its saliency region, which is defined as the central object in this area. The specific structure of our saliency region detection network is shown in Figure 5. The effect is shown in Figure 7. In the saliency region detection network, we obtained five features with high resolution and rich semantic information through multi-level feature fusion. We supervised each feature map, where four of them were used to obtain the region of significance and the other to obtain the edge of the area. Finally, we fused the results from the edge capture module's features and the saliency region detection module's features and obtained a accurate saliency detection.

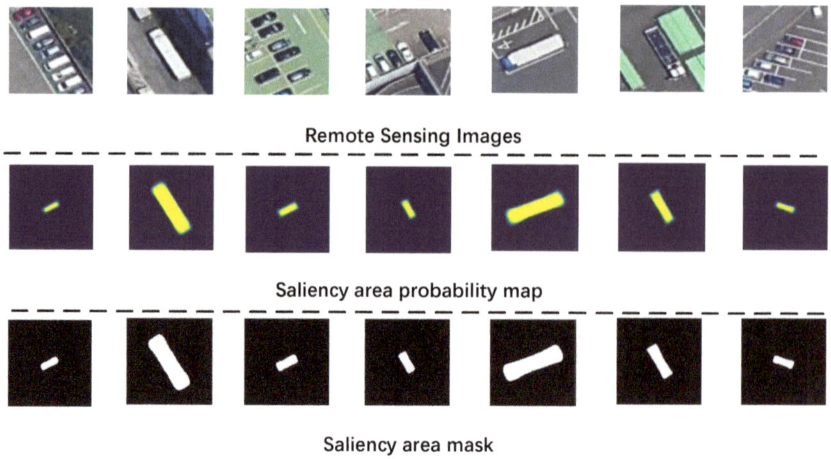

Figure 7. The figure shows the visualization of the output probability map of the saliency region detection stage and the visualization of the saliency region generated from the probability map.

2.2.1. Complementary Information Modeling

Our proposed network architecture is independent of the feature extraction backbone; here, the architecture is based on VGGNet, like many previous CNN-based methods. The architecture we proposed can be found in the saliency region detection part of Figure 5. Following DSS, we discarded the full connection layer in the original network. Six features called Conv1-2, Conv2-2, Conv3-3, Conv4-3, Conv5-3, and Conv6-3 were generated from VGGNet; we denoted them as $C^{(1)}$, $C^{(2)}$, $C^{(3)}$, $C^{(4)}$, $C^{(5)}$, and $C^{(6)}$ for simplicity. Then, we used side paths to obtain corresponding side outputs, except $C^{(1)}$, because the number of convolutional layers spaced between $C^{(1)}$ and the inputs was too low and the perceptual field was too small. Five side paths are denoted by a path set S:

$$S = \left\{ S^{(2)}, S^{(3)}, S^{(4)}, S^{(5)}, S^{(6)} \right\} \tag{6}$$

Conv2-2 has the highest resolution of these features while retaining the complete edge information due to its proximity to the input. We leveraged $S^{(2)}$ to obtain edge output and other outputs to obtain saliency object output in the later network.

2.2.2. Saliency Object Features Extraction

In order to leverage rich semantic and high-resolution information in multi-level side outputs, we imitated widely used architecture U-Net [35] for feature fusion. Different from the original U-Net [35], we added three extra convolutional and ReLU layers (orange Conv block in Figure 5) on each side path to obtain more robust fused features. For simplicity, we used H (Table 1) to denote these convolutional layers and ReLU layers in the side path. Then, we used a convolutional layer which was denoted as O (Table 1) to transit the feature maps to the single-channel prediction masks. The details of convolutional layers in the side path can be found in Table 1.

Table 1. The detailed convolutional structure of the prediction head. H denotes the feature enhancement module in each prediction header, each prediction header contains three convolutional layers—H_1, H_2, H_3—and each convolutional layer is followed by a ReLu activation function. The table shows the parameters of each convolutional layer; for example, 3, 1, 64 indicates a convolutional layer with convolutional kernel size of 3, padding of 1, and channel number of 64. O represents the output layer, which transforms the multi-channel features into a single-channel activation map.

S		H_1			H_2			H_3			O	
2	3	1	64	3	1	64	3	1	64	3	1	1
3	3	1	128	3	1	128	3	1	128	3	1	1
4	3	1	256	3	1	256	3	2	256	3	1	1
5	5	2	256	5	2	256	5	2	256	3	1	1
6	5	2	256	5	2	256	5	2	256	3	1	1

2.2.3. Saliency Edge and Region Masks Extraction

In this section, our aim is to model the saliency edge information while obtaining saliency edge features. As mentioned in the previous section, Conv2-2 has the highest resolution and retains the most complete edge information. Therefore, our local edge information was extracted from Conv2-2. However, saliency edge features cannot be obtained with only local edge information. We also needed high-level semantic information. When the semantic information at the top layer is passed back to the bottom layer from the top layer, like the U-Net structure, the semantic information at the top layer will be diluted. Therefore, we designed top-down layer-by-layer information propagation, in which, supervision is added at each layer.

Our high-level semantic information was propagated to the side path $S^{(2)}$ via a top-down path to suppress non-saliency edges. The fused saliency edge features C_E can be expressed as:

$$C_E = C^{(2)} + U(ReLU(Conv(\widehat{\Lambda}^{(6)}; \varphi)); C^{(2)}) \tag{7}$$

where $Conv(*; \varphi)$ is a convolutional layer with trainable weight φ and a convolutional kernel size of $1 * 1$. It was intended to have the same number of channels for the features. The $ReLU$ denotes the $ReLU$ activation function and $U(*; C^{(2)})$ denotes the size of the $*$ upsampled to $C^{(2)}$ resolution (using a bilinear interpolation method). $\widehat{\Lambda}^{(6)}$ is the fused feature in the side path. $\widehat{\Lambda}^{(6)}$ can be denoted as $E(C^{(6)}; W_T^{(6)})$ and the fused feature $\widehat{\Lambda}^{(3)}$, $\widehat{\Lambda}^{(4)}$, $\widehat{\Lambda}^{(5)}$ in the same side paths S^3, S^4, S^5 can be denoted as

$$\widehat{\Lambda}^{(i)} = E(C^{(i)} + U(ReLU(Conv(\widehat{\Lambda}^{(i+1)}; \theta)); W_H^{(i)}) \tag{8}$$

where $E(*; W_H^{(i)})$ denotes the convolution layer and activation function with parameters $W_H^{(i)}$. Saliency edge features F_E in $S^{(2)}$ could be computed as $E(\overline{C}^{(2)}; W_H^{(2)})$. The configuration details could be found in Table 1. We supervised the saliency edge and the saliency region prediction maps, respectively. For saliency edge supervision, we used the cross-entropy loss, which could be defined as:

$$\begin{aligned} L_{edge}(F_E; W_O^{(2)}) = &- \sum_{j \in G_+} \log P_r(y_i = 1 | F_E; W_O^{(2)}) - \\ &\sum_{j \in G_-} \log P_r(y_i = 0 | F_E; W_O^{(2)}) \end{aligned} \tag{9}$$

where G_+ and G_- denote the pixels in the saliency edge and the pixels in the background region, respectively. W_O denotes the parameters of the prediction layer, as shown in Table 1, and $P_r(y_i = 1 | F_E; W_O^{(2)})$ denotes the confidence level that the pixel has a saliency edge. In addition, for saliency region supervision, we used the cross-entropy loss, which could be defined as:

$$L_{area}(\widehat{\Lambda}^{(i)}; W_O^{(i)}) = -\sum_{j\in A_+} \log P_r(y_i = 1|\widehat{\Lambda}^{(i)}; W_O^{(i)}) - \sum_{j\in A_-} \log P_r(y_i = 0|\widehat{\Lambda}^{(i)}; W_O^{(i)}) \quad (10)$$

where A_+ and A_- denote the pixels in the saliency region and the pixels in the background area, respectively. The total relay supervision L_{Relay} can be expressed as:

$$L_{Relay} = L_{edge}(F_E; W_O^{(2)}) + \sum_{i=3}^{6} L_{area}(\widehat{\Lambda}^{(i)}; W_O^{(i)}) \quad (11)$$

2.2.4. Saliency Object Detection

After obtaining the saliency edge and saliency areas outputs, our goal is to use edge output to obtain better segmentation results. Then, we can obtain the detection boxes by calculating the minimum external rectangle of these segmentation results. Specifically, we fused F_E with edge features and $\widehat{F}^{(3)}$ with regional features to obtain the complete feature F_W.

$$F_W = \widehat{F}^{(3)} + Up(ReLU(Conv(\widehat{F}^{(3)}; \theta)); F_E) \quad (12)$$

We also supervised the final predicted saliency region, with the loss function expressed as the following equation.

$$L_{Saliency}(F_W; W_E) = -\sum_{j\in Y_+} \log P_r(y_i = 1|F_W; W_E) - \sum_{j\in Y_-} \log P_r(y_i = 0|F_W; W_E) \quad (13)$$

The supervision losses of all parts were combined as follows:

$$L_t = \eta L_{Relay} + L_{Saliency} \quad (14)$$

where η denotes the weight of the relay supervision loss to the total loss, and the effect of η on the final result is discussed in the next section.

3. Result

In this section, we conduct experiments to evaluate the performance of SR-Net. First, we introduce some training details of SR-Net, such as dataset, hyperparameters, etc. Then, we introduce the evaluation metrics. Finally, we present the ablation experiments with some parameters and structures of our SR-Net and compare them with current state-of-the-art detectors to derive the effectiveness of our method.

3.1. Data Set

(1) DOTA [40]: DOTA is a standard benchmark for the detection of objects in remote sensing images.It contains two detection tasks: detection with oriented bounding boxes and detection with horizontal bounding boxes. Only detection with oriented bounding boxes is used in our experiments. The DOTA contains 2806 remote sensing images. DOTA is labelled with 15 categories (e.g., aircraft, small vehicle, large vehicle), and, in this paper, we only use two categories: small vehicles and large vehicles. In training, we crop the large-scale images in DOTA into multi-scale remote sensing image slices of size 384 × 384, 512 × 512, and 800 × 800.

(2) UCAS-AOD [41]: The UCAS-AOD dataset contains 510 large-scale remote sensing images with two types of targets (aircraft and vehicles) in the UCAS-AOD dataset containing 7114 vehicle targets. We randomly select 400 images for training and the other images for testing. Similarly, we slice the large-scale remote sensing images to obtain multi-scale slices of size 384 × 384, 512 × 512, and 800 × 800 for training.

(3) DLR 3K Munich [11]: In the DLR 3K Munich, vehicles are accurately labeled by rotatable rectangular boxes. This dataset consists of three categories (car, bus, and truck), and we consider the three categories as one class.The dataset contains a total of 8268 vehicles, of which, the training set contains 5214 vehicles and the test set contains 3054 vehicles.

(4) VEDAI [42]: VEDAI contains two modal images, RGB and NIR, and four classes (car, pickup, truck, van), which, again, we still consider as one class. The dataset contains a total of 5494 vehicles, with a training set of 2792 and a test set of 2702.

3.2. Evaluation Metric

In the detection task, the output of the model can be represented as a rotatable rectangular bounding boxes and its corresponding class. For our vehicle detection, since we use single-category detection, our output is only rotatable rectangular bounding boxes. We evaluate the accuracy of the detected boxes by using the intersection over union (IoU) metric. The IoU denotes the number of detected boxes and the overlap rate between detected boxes and ground truth boxes. The formula for IoU is defined as:

$$IOU = (S_{bbox} \cap S_{gt}) / (S_{bbox} \cup S_{gt}) \tag{15}$$

where S_{box} denotes the region of the detection box and S_{gt} denotes the region of the ground truth box. We use average precision (AP) to evaluate the merit of the vehicle detection method. In order to calculate the AP value, we need to define and calculate the following values based on the detection results true positives (TPs), false positives (FPs), true negatives (TNs) , and false negatives (FNs).

In our vehicle detection task, we set the threshold of IoU between the detection box and the ground truth box. If the IoU between a detection box and ground truth box is greater than the threshold, this detection box is considered as TP. Conversely, if the IoU is less than the threshold, this detection box is considered FP (also called false alarm). If a ground truth box is not detected, this ground truth box is called FN (also called miss alarm). Based on the above definition, we define the equation of the precision and recall metrics as:

$$Precision = \frac{TP}{(TP + FP)} \tag{16}$$

$$Recall = \frac{TP}{(TP + FN)} \tag{17}$$

The AP is calculated by integrating the $P(Precision)$-$R(Recall)$ curve. The P-R curve can be obtained by calculating the recall values corresponding to different precision values based on the detection results. Based on this definition, it is known that, the higher the AP value, the better the performance of the model. The AP value can be defined as:

$$AP = \int_0^1 P(R)d(R) \tag{18}$$

3.3. Training Details

We implement SR-Net in PyTorch (Paszke et al., 2017). During the center window forecasting phase, our center window estimator section is trained on an image with an input resolution of 512 × 512, corresponding to our output of a heat map with a resolution of 128 × 128. To ensure a diversity of vehicle sizes, our training set was produced with multiple scales (384 × 384, 512 × 512, and 800 × 800), so we uniformly resized it to a 512 × 512 size using bilinear interpolation, which, in addition, enhanced the robustness of the model and reduced the risk of overfitting. We also used data enhancement methods, such as rotation, color transformation, etc. Finally, Adam [43] was used to optimize the overall objective.

We train our SR-Net with a batch-size of 32 (on two GPUs), and we train the model from scratch to 60,000 iterations in order to achieve the best results. The convergence curve of the model is shown in Figure 8. During the saliency region detection phase, since the data are labeled in the form of rotating boxes, which do not meet the supervision information required, we first need to generate the center window for each vehicle based on the definition of our saliency area. We need to transform the rotating frame annotation into a pixel-level segmentation annotation by defining the pixels in the target region as saliency areas. To enhance stability enhancements were also made as shown in Figure 9. In order to train our edge information enhancement module, we need to label the saliency region to extract the edges using the edge extraction operator (we use the Canny operator). During training, we resize each center window to a uniform size of 32 × 32. Again, to ensure sample diversity and eliminate the risk of overfitting, we use the same data enhancement as in the center window estimator stage. Again, we use Adam to optimize the parameters. In this stage, we only need 10,000 iterations to obtain the best results.

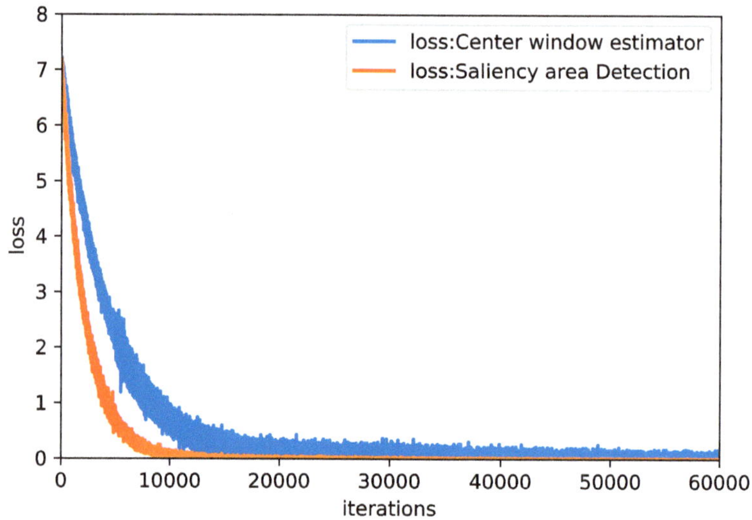

Figure 8. Training loss of SR-Net. The loss of saliency region detection has converged at 10,000 iterations. The loss of the center window estimator requires 60,000 iterations to fully converge.

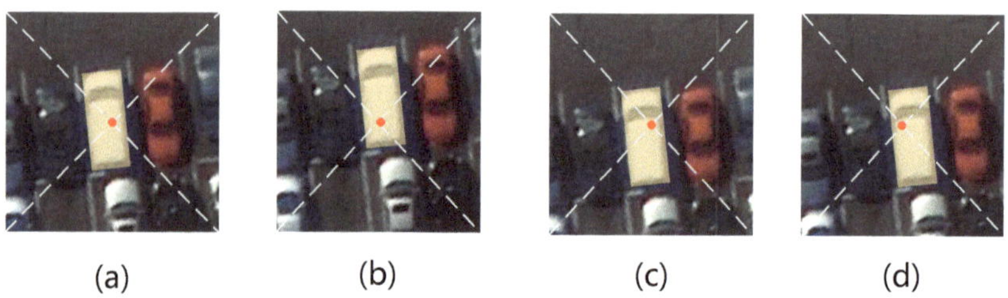

Figure 9. Figure (**a**) shows the standard saliency region. In order to increase the robustness of saliency region detection, we translated the saliency region, as shown in (**b**–**d**), in the training phase, and the translation range ensures that the center of the center window is within the saliency area.

3.4. Comparisons with State-of-the-Art Detectors

In this section, we compare SR-Net with other state-of-the-art detectors. We evaluate our model and other models in a vehicle detection task on four datasets. For SR-Net, we use two backbone networks (104-Hourglass, ResNet-101). For the other models, we follow the authors' previous backbone networks. We also test the difference in inference speed between our model and the other models to verify the advantages of our model. The detection results of our model on DLR 3K Munich and VEDAI datasets are shown in Figure 10, and the test results on DOTA and UCAS-AOD are shown in Figure 11. In addition, to verify the generalization ability of our model, we trained our SR-Net on the ensemble of four datasets and tested our model on a panoramic map of a residential region from Google Maps. The results are shown in Figure 12.

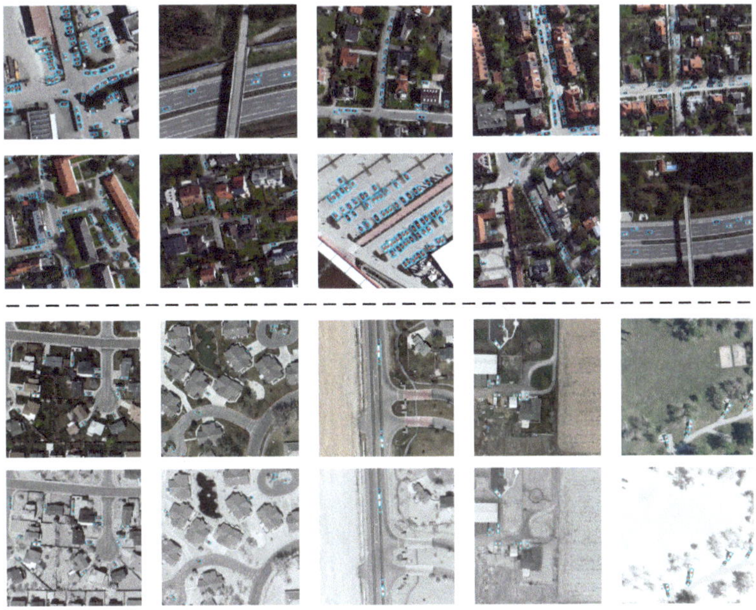

Figure 10. The figure shows some examples of our proposed SR-Net for vehicle detection in any direction. The first two rows visualize the test results in the DLR 3K Munich dataset. The second two rows visualize the test results in the VEDAI dataset (two image modes: RGB and NIR). The effect is better when viewed in a larger size.

3.5. Qualitative Quantity Analysis

As shown in Table 2, our proposed SR-Net achieved the highest results for AP values on the DOTA, UCAS-AOD, and DLR 3K Munich datasets, and AP^{75} still achieved the best results on the VEDAI dataset despite the fact that AP did not achieve the best results, again illustrating the advantages of our proposed method in terms of accuracy. The average results on the four datasets (AP^{m}) showed an improvement of 2.32 compared to the previous method. In terms of speed, since we do not need rotational non-maximal suppression (RNMS), anchor, etc., our model achieves the highest AP value, while our speed is approximately twice that of R^3-Net.

Table 2. State-of-the-art comparisons. AP is the average of 10 values of AP taken between $iou = 50$ and $iou = 95$. AP^1 denotes the AP value obtained in the DOTA validation set, AP^2 denotes the AP value obtained in the UCAS-AOD test set, AP^3 denotes the AP value obtained in the DLR 3K Munich dataset test set, and AP^4 denotes the AP value obtained in the VEDAI test set. AP^m denotes the average of the AP values obtained in the four datasets. AP_{50} is the AP value obtained in the $IoU = 50$, and AP_{75} is the AP value obtained at $IoU = 75$.

Models	FPS	AP^1	AP^1_{50}	AP^1_{75}	AP^2	AP^2_{50}	AP^2_{75}	AP^3	AP^3_{50}	AP^3_{75}	AP^4	AP^4_{50}	AP^4_{75}	AP^m
YOLOv2(O) [44]	15.31	4.43	6.70	4.73	11.29	18.21	12.74	12.84	20.32	14.30	12.25	19.80	14.36	10.20
R^2CNN [45]	3.81	34.94	57.20	41.64	59.83	87.95	61.88	59.65	87.22	64.54	60.79	64.59	46.31	53.80
RRPN [29]	5.25	34.09	53.19	39.25	59.83	86.17	62.50	58.17	85.43	61.93	51.85	62.34	44.88	50.99
R-DFPN [46]	5.84	34.22	54.33	39.17	60.41	87.37	61.72	55.08	84.35	59.21	54.91	60.57	43.73	51.15
ICN [47]	6.54	45.64	65.77	48.20	58.24	89.24	64.93	54.34	86.12	63.81	56.15	63.22	44.94	53.59
RetinaNet(O) [25]	7.34	42.08	68.26	50.99	62.54	89.65	64.60	54.54	86.30	63.08	53.16	65.41	47.74	53.08
Roi-Transformer [48]	3.92	43.77	70.27	50.31	60.90	90.56	66.69	55.86	88.53	65.15	56.12	66.52	48.82	54.16
P-RSDet [33]	7.82	46.55	71.52	51.70	57.93	90.87	66.19	54.75	88.75	64.96	61.94	70.33	49.58	55.29
SCRDet [30]	6.37	40.17	65.97	47.56	57.27	90.25	64.40	56.58	89.67	63.21	55.77	74.65	54.34	52.44
O^2-DNet [28]	7.62	47.09	72.45	51.43	59.42	91.06	67.15	63.08	90.64	67.88	56.01	71.34	52.07	56.40
R^3-Net [49]	3.23	49.58	73.24	52.07	60.19	91.22	66.21	63.12	91.48	67.69	56.26	69.23	50.39	57.28
R^3-Det [50]	6.53	46.49	75.24	55.15	61.09	91.63	64.85	62.01	90.14	66.34	56.42	70.58	51.02	56.50
SR-Net(ResNet-101)	7.64	50.27	76.40	57.75	61.18	92.03	69.55	64.47	90.68	69.82	53.42	70.53	56.54	57.33
SR-Net(104-Hourglass)	7.43	52.30	80.09	59.03	62.44	93.24	70.55	68.25	91.01	71.89	55.81	72.53	58.16	59.60

Figure 11. The figure shows some examples of our proposed SR-Net for vehicle detection in any direction. The first two rows visualize the test results of UCAS-AOD in the UCAS-AOD dataset. The last two rows visualize the test results in the DOTA dataset. As can be seen, vehicles of multiple sizes parked in multiple directions can be detected, and accurate bboxes are obtained. The effect is better when viewed in a larger size.

Figure 12. In order to verify the robustness of the model, we took a remote sensing panorama from Google Maps, and the selected region contains several scenes, such as densely parked parking lots, intersections, streets, private garages, etc. We used SR-Net to perform the detection, and this figure shows the detection results. It is best when viewed zoomed in.

4. Discussion

In this section, we perform detailed ablation experiments on the parameters and components of SR-Net. Besides, the advantages and disadvantages of the proposed method are discussed. All ablation experiments are performed using ResNet-101 as the backbone network and are trained and evaluated on the UCAS-AOD dataset.

Loss weight. We analyze the effect of the γ parameter in Equation (5) on the SR-Net. As shown in Table 3, the best results were achieved at a γ value of 2. In addition, we found that a small selection of γ values did not have a great impact on the final results.

This indicates that the results of the center window prediction are more dependent on the accuracy of the center point prediction. It also shows that the network can easily predict the category of the center window. Similarly, we analyze the effect of the loss function of the relay supervision, i.e., the effect of the size of the parameter η in Equation (14) on the final results. As shown in Table 4, the best results are obtained when the value of t is 1.2, while it can be observed that the final result decreases significantly if no relay supervision is added, and that the final results are affected if the weight of relay supervision is too large.

Table 3. The AP values of the results were obtained by comparing different γ values on the UCAS-AOD dataset. (AP is the average of 10 values of AP taken between $IoU = 50$ and $IoU = 95$).

γ	**0.1**	**1**	**2**	**3**	**5**	**7**
AP	0.602	0.608	0.612	0.610	0.609	0.593

Table 4. The AP values of the results were obtained by comparing different η values on the UCAS-AOD dataset. (AP is the average of 10 values of AP taken between $IoU = 50$ and $IoU = 95$).

η	**0**	**0.1**	**0.4**	**0.8**	**1.2**	**1.4**	**1.6**
AP	0.574	0.586	0.604	0.609	0.612	0.610	0.608

Edge supervision. Since our method models vehicles at the pixel level, the accuracy of the saliency region edges directly affects the accuracy of the oriented bboxs generated from the saliency region decoding. In addition, in remote sensing images, the edges of vehicles are relatively clear, based on which, we design the edge supervision module to make the edges of saliency regions more precise. To verify the effectiveness of our edge supervision module, we perform ablation experiments. As shown in Table 5, equipped with the area supervision module with the edge supervision module, the resulting AP value gains a 0.9% improvement, AP_{50} remains basically the same, and AP_{75} yields a 1.2% improvement, proving that the devised edge supervision can achieve finer boundaries.

Table 5. A comparison of the effect of adding edge supervision module and region supervision module on the results on UCAS-AOD dataset. (AP is the average of 10 values of AP taken between $IoU = 50$ and $IoU = 95$. AP_{50} is the AP value obtained in the $IoU = 50$, and AP_{75} is the AP value obtained at $IoU = 75$).

Edge Supervision	Area Supervision	AP	AP_{50}	AP_{75}
	✓	0.613	0.919	0.683
✓		0.612	0.942	0.695

Saliency-based. In this section, we will discuss the advantages of saliency-based over regression-based detection methods in detail. In the previous regression-based detection methods, the vehicle detection results are represented by the vehicle center, length, width, and angle, where length, width, and angle are obtained by distance regression. The objective function in distance regression is generally L1 loss, L2 loss, or a deformation based on them, and, in the distance regression, the network needs a clear objective value, yet, in the definition of an angle, it is periodic without a clear objective value, e.g., 0 degrees and 360 degrees. In the proposed saliency-based detection method, the detection result of the vehicle is represented by the center point of the vehicle and the corresponding saliency region. The saliency region is obtained by modeling the probability that each pixel belongs to the foreground, and modeling each pixel in the instance foreground is a binary classification problem, i.e., a logistic regression problem in which the loss function is the cross-entropy loss, as shown in Equation (14). Therefore, the saliency-based method does not have distance regression values in principle, and there is no problem with inaccurate regression values.

To further illustrate the problem, we compared the saliency-based method and the regression-based method through an experiment. In order to eliminate the influence of other components on the final results, we only compare the accuracy of the generated bbox. In particular, we obtain the labeled center directly in order to ensure the consistency of the vehicle center, assuming that all vehicle centroids are detected correctly. Then, for each vehicle centroid, regression is used to regress the vehicle's length, width, and angle. Similarly, we use a saliency-based approach to generate the saliency region of the vehicle while keeping the centroids unchanged and decoding the vehicle's aspect and angle based on the saliency area. For the angles of the oriented bboxes obtained by the two methods, we obtain Figure 4. This statistical plot clearly shows the angular error of oriented bboxes obtained by the two methods. The angle errors obtained by our saliency-based method are concentrated between -10 degrees and 10 degrees, whereas the angle errors obtained by the traditional regulation-based method are concentrated between -20 degrees and 20 degrees, and the angle deviations of some samples reach 90 degrees due to the unclear angle definitions caused by the angle periodicity. In summary, the bboxes obtained by the saliency-based method are more accurate than those obtained by the traditional distance-regression-based method.

Summary. For vehicle detection challenges, such as dense parking, small sizes, and various angles, our SR-Net can be a good solution. Conventional anchor-based detectors set a fixed step of anchor arrangement, which may lead to missed vehicle detection once the density of vehicles is greater than the density of anchors. We estimate the center point of each vehicle and obtain the corresponding center window to ensure that the dense arrangement of vehicles is not missed. As described in the **Edge supervision** section, our edge-enhanced supervision makes the vehicle edges more accurate and avoids sticking between neighboring vehicles. The **Saliency-based** section describes the traditional regression-based approach facing the problem of inaccurate angle regression due to the periodic nature of angles when facing the problem of variable angles of vehicles. By comparison, our saliency-based approach fundamentally solves this problem of various angles. However, for the problem of small sizes in vehicle detection, our model is not deliberately designed because this problem has been largely solved in the center-based detector. Our center window estimator module is designed precisely based on the center-based detector. Of course, our SR-Net still has some shortcomings; for example, the pipeline is not neat enough, which may face challenges in future industrial deployment.

5. Conclusions

In this paper, we transform vehicle detection into a saliency region detection task for the first time and propose SR-Net (saliency region representation network for vehicle detection in remote sensing images). Our method works in an anchor-free, NMS-free, and regression-free way via modeling a vehicle instance as a saliency region without the problems of complex positive and negative sample matching and angular regression discontinuity faced by vanilla models. Besides, compared with existing state-of-the-art detectors on public vehicle detection datasets, the proposed method achieves competitive results.

We apply our model to other categories of objects, e.g., ships, aircraft, etc. Finally, we suggest that researchers should think about migrating our detector from vehicle detection to full class target detection. Of course, our model has its own shortcomings; for example, the pipeline is complicated, and we will continue to optimize (streamline) our network in the future.

With the development of remote sensing technology, remote sensing images have gradually developed from single-frame images to remote sensing videos. In order to meet the demand of real-time, the speed of vehicle detectors has also been challenged. In the future, not only is the detection of vehicles needed, but the tracking of vehicles also needs to be realized.

Author Contributions: Conceptualization, F.L., W.Z., G.Z., L.Z., H.W.; methodology, F.L., W.Z.; software, F.L., H.W.; validation, F.L., W.Z., G.Z., L.Z., H.W.; formal analysis, F.L., W.Z.; data curation, L.Z.; writing—original draft preparation, F.L.; writing—review and editing, F.L., W.Z., H.W.; visualization, F.L.; project administration, W.Z.; All authors have read and agreed to the published version of the manuscript.

Funding: Key Laboratory of Network Information System Technology (NIST), Aerospace Information Research Institute, Chinese Academy of Sciences.

Institutional Review Board Statement: Not applicable.

Informed Consent Statement: Not applicable.

Data Availability Statement: Not applicable.

Conflicts of Interest: The authors declare no conflict of interest.

References

1. Zhang, F.; Du, B.; Zhang, L.; Xu, M. Weakly Supervised Learning Based on Coupled Convolutional Neural Networks for Aircraft Detection. *IEEE Trans. Geosci. Remote Sens.* **2016**, *54*, 5553–5563. [CrossRef]
2. Mou, L.; Zhu, X.X. Spatiotemporal scene interpretation of space videos via deep neural network and tracklet analysis. In Proceedings of the 2016 IEEE International Geoscience and Remote Sensing Symposium, Beijing, China, 10–15 July 2016; pp. 1823–1826. [CrossRef]
3. Mou, L.; Zhu, X.; Vakalopoulou, M.; Karantzalos, K.; Paragios, N.; Saux, B.L.; Moser, G.; Tuia, D. Multitemporal Very High Resolution From Space: Outcome of the 2016 IEEE GRSS Data Fusion Contest. *IEEE J. Sel. Top. Appl. Earth Obs. Remote Sens.* **2017**, *10*, 3435–3447. [CrossRef]
4. Audebert, N.; Saux, B.L.; Lefèvre, S. Segment-before-Detect: Vehicle Detection and Classification through Semantic Segmentation of Aerial Images. *Remote Sens.* **2017**, *9*, 368. [CrossRef]
5. Li, Q.; Mou, L.; Liu, Q.; Wang, Y.; Zhu, X.X. HSF-Net: Multiscale Deep Feature Embedding for Ship Detection in Optical Remote Sensing Imagery. *IEEE Trans. Geosci. Remote Sens.* **2018**, *56*, 7147–7161. [CrossRef]
6. Dalal, N.; Triggs, B. Histograms of Oriented Gradients for Human Detection. In Proceedings of the 2005 IEEE Computer Society Conference on Computer Vision and Pattern Recognition (CVPR 2005), San Diego, CA, USA, 20–26 June 2005; pp. 886–893. [CrossRef]
7. Lowe, D.G. Object Recognition from Local Scale-Invariant Features. In Proceedings of the International Conference on Computer Vision, Kerkyra, Corfu, Greece, 20–25 September 1999; pp. 1150–1157. [CrossRef]
8. Ojala, T.; Pietikäinen, M.; Mäenpää, T. Multiresolution Gray-Scale and Rotation Invariant Texture Classification with Local Binary Patterns. *IEEE Trans. Pattern Anal. Mach. Intell.* **2002**, *24*, 971–987.[CrossRef]
9. Moranduzzo, T.; Melgani, F. Automatic Car Counting Method for Unmanned Aerial Vehicle Images. *IEEE Trans. Geosci. Remote Sens.* **2014**, *52*, 1635–1647. [CrossRef]
10. Moranduzzo, T.; Melgani, F. Detecting Cars in UAV Images With a Catalog-Based Approach. *IEEE Trans. Geosci. Remote Sens.* **2014**, *52*, 6356–6367. [CrossRef]
11. Liu, K.; Máttyus, G. Fast Multiclass Vehicle Detection on Aerial Images. *IEEE Geosci. Remote Sens. Lett.* **2015**, *12*, 1938–1942. [CrossRef]
12. Elmikaty, M.; Stathaki, T. Detection of Cars in High-Resolution Aerial Images of Complex Urban Environments. *IEEE Trans. Geosci. Remote Sens.* **2017**, *55*, 5913–5924. [CrossRef]
13. Zhou, H.; Wei, L.; Lim, C.P.; Creighton, D.C.; Nahavandi, S. Robust Vehicle Detection in Aerial Images Using Bag-of-Words and Orientation Aware Scanning. *IEEE Trans. Geosci. Remote Sens.* **2018**, *56*, 7074–7085. [CrossRef]
14. Kalantar, B.; Mansor, S.; Halin, A.A.; Shafri, H.Z.M.; Zand, M. Multiple Moving Object Detection From UAV Videos Using Trajectories of Matched Regional Adjacency Graphs. *IEEE Trans. Geosci. Remote Sens.* **2017**, *55*, 5198–5213.[CrossRef]
15. Krizhevsky, A.; Sutskever, I.; Hinton, G.E. ImageNet Classification with Deep Convolutional Neural Networks. In *Advances in Neural Information Processing Systems 25, Proceedings of the 26th Annual Conference on Neural Information Processing Systems, Lake Tahoe, NV, USA, 3–6 December 2012*; Morgan Kaufmann Publishers, Inc.: Lake Tahoe, NV, USA, 2012; pp. 1106–1114.
16. Zeiler, M.D.; Fergus, R. Visualizing and Understanding Convolutional Networks. *arXiv* **2013**, arXiv:1311.2901. Available online: http://xxx.lanl.gov/abs/1311.2901 (accessed on 8 May 2012).
17. He, K.; Zhang, X.; Ren, S.; Sun, J. Deep Residual Learning for Image Recognition. *CoRR* **2015**, *abs/1512.03385*. Available online: http://xxx.lanl.gov/abs/1512.03385 (accessed on 8 May 2012).
18. Huang, G.; Liu, Z.; van der Maaten, L.; Weinberger, K.Q. Densely Connected Convolutional Networks. In Proceedings of the IEEE Conference on Computer Vision and Pattern Recognition (CVPR), Honolulu, HA, USA, 21–26 July 2017; pp. 2261–2269.
19. Xie, S.; Girshick, R.B.; Dollár, P.; Tu, Z.; He, K. Aggregated Residual Transformations for Deep Neural Networks. In Proceedings of the IEEE Conference on Computer Vision and Pattern Recognition (CVPR), Honolulu, HA, USA, 21–26 July 2017; pp. 5987–5995.

20. Howard, A.G.; Zhu, M.; Chen, B.; Kalenichenko, D.; Wang, W.; Weyand, T.; Andreetto, M.; Adam, H. MobileNets: Efficient Convolutional Neural Networks for Mobile Vision Applications. *arXiv* **2017**, arXiv:1704.04861.
21. Girshick, R.B.; Donahue, J.; Darrell, T.; Malik, J. Rich Feature Hierarchies for Accurate Object Detection and Semantic Segmentation. In Proceedings of the 2014 IEEE Conference on Computer Vision and Pattern Recognition, CVPR 2014, Columbus, OH, USA, 23–28 June 2014; pp. 580–587. [CrossRef]
22. Girshick, R.B. Fast R-CNN. InProceedings of the Proceedings of the IEEE International Conference on Computer Vision (ICCV), Santiago, Chile, 7–13 December 2015. Available online: http://xxx.lanl.gov/abs/1504.08083 (accessed on 8 May 2012).
23. Ren, S.; He, K.; Girshick, R.B.; Sun, J. Faster R-CNN: Towards Real-Time Object Detection with Region Proposal Networks. In Proceedings of the 28th International Conference on Neural Information Processing Systems, Montreal, QC, Canada, 7–12 December 2015. Available online: http://xxx.lanl.gov/abs/1506.01497 (accessed on 8 May 2012).
24. Cai, Z.; Vasconcelos, N. Cascade R-CNN: Delving Into High Quality Object Detection. In Proceedings of the IEEE Conference on Computer Vision and Pattern Recognition (CVPR), Salt Lake City, UT, USA, 18–22 June 2018; pp. 6154–6162.
25. Lin, T.; Goyal, P.; Girshick, R.B.; He, K.; Dollár, P. Focal Loss for Dense Object Detection. In Proceedings of the IEEE International Conference on Computer Vision, Venice, Italy, 22–29 October 2017; pp. 2999–3007. [CrossRef]
26. Tian, Z.; Shen, C.; Chen, H.; He, T. FCOS: Fully Convolutional One-Stage Object Detection. In Proceedings of the IEEE/CVF International Conference on Computer Vision (ICCV), Seoul, Korea, 27–28 October 2019; pp. 9626–9635.
27. Wei, H.; Zhang, Y.; Wang, B.; Yang, Y.; Li, H.; Wang, H. X-LineNet: Detecting Aircraft in Remote Sensing Images by a Pair of Intersecting Line Segments. *IEEE Trans. Geosci. Remote Sens.* **2021**, *59*, 1645–1659. [CrossRef]
28. Wei, H.; Zhou, L.; Zhang, Y.; Li, H.; Guo, R.; Wang, H. Oriented Objects as pairs of Middle Lines. *ISPRS J. Photogramm. Remote Sens.* **2020**, *169*, 268–279. [CrossRef]
29. Ma, J.; Shao, W.; Ye, H.; Wang, L.; Wang, H.; Zheng, Y.; Xue, X. Arbitrary-Oriented Scene Text Detection via Rotation Proposals. *IEEE Trans. Multimed.* **2017**, *20*, 3111–3122. Available online: http://xxx.lanl.gov/abs/1703.01086 (accessed on 8 May 2012). [CrossRef]
30. Yang, X.; Yang, J.; Yan, J.; Zhang, Y.; Zhang, T.; Guo, Z.; Sun, X.; Fu, K. SCRDet: Towards More Robust Detection for Small, Cluttered and Rotated Objects. In Proceedings of the 2019 IEEE/CVF International Conference on Computer Vision, ICCV 2019, Seoul, Korea, 27 October–2 November 2019; pp. 8231–8240. [CrossRef]
31. Yang, X.; Yan, J. Arbitrary-Oriented Object Detection with Circular Smooth Label. In Lecture Notes in Computer Science; Springer:Berlin/Heidelberg, Germany, 2020; Volume 12353, pp. 677–694.
32. Yang, X.; Hou, L.; Zhou, Y.; Wang, W.; Yan, J. Dense Label Encoding for Boundary Discontinuity Free Rotation Detection. In Proceedings of the IEEE/CVF Conference on Computer Vision and Pattern Recognition (CVPR), Nashville, TN, USA, 20–25 June 2021; pp. 15819–15829.
33. Zhou, L.; Wei, H.; Li, H.; Zhao, W.; Zhang, Y. Objects detection for remote sensing images based on polar coordinates. *arXiv* **2020**, arXiv:2001.02988. Available online: http://xxx.lanl.gov/abs/2001.02988 (accessed on 8 May 2012).
34. Fu, K.; Chang, Z.; Zhang, Y.; Sun, X. Point-Based Estimator for Arbitrary-Oriented Object Detection in Aerial Images. *IEEE Trans. Geosci. Remote Sens.* **2021**, *59*, 4370–4387. [CrossRef]
35. Newell, A.; Yang, K.; Deng, J. Stacked Hourglass Networks for Human Pose Estimation. In Proceedings of the Computer Vision—ECCV 2016—14th European Conference, Amsterdam, The Netherlands, 11–14 October 2016; Part VIII; pp. 483–499. [CrossRef]
36. Zhou, X.; Wang, D.; Krähenbühl, P. Objects as Points. *arXiv* **2019**, arXiv:1904.07850.
37. Zhao, J.; Liu, J.; Fan, D.; Cao, Y.; Yang, J.; Cheng, M. EGNet: Edge Guidance Network for Salient Object Detection. In Proceedings of the IEEE/CVF International Conference on Computer Vision (ICCV), Seoul, Korea, 27–28 October 2019.
38. Kümmerer, M.; Theis, L.; Bethge, M. Deep Gaze I: Boosting Saliency Prediction with Feature Maps Trained on ImageNet. *arXiv* **2015**, arXiv:1411.1045.
39. Pan, J.; McGuinness, K.; Sayrol, E.; O'Connor, N.E.; Giró-i-Nieto, X. Shallow and Deep Convolutional Networks for Saliency Prediction. In Proceedings of the IEEE Conference on Computer Vision and Pattern Recognition (CVPR), Honolulu, HA, USA, 21–26 July 2016.
40. Xia, G.; Bai, X.; Ding, J.; Zhu, Z.; Belongie, S.J.; Luo, J.; Datcu, M.; Pelillo, M.; Zhang, L. DOTA: A Large-Scale Dataset for Object Detection in Aerial Images. In Proceedings of the 2018 IEEE Conference on Computer Vision and Pattern Recognition, CVPR 2018, Salt Lake City, UT, USA, 18–22 June 2018; pp. 3974–3983.[CrossRef]
41. Zhu, H.; Chen, X.; Dai, W.; Fu, K.; Ye, Q.; Jiao, J. Orientation robust object detection in aerial images using deep convolutional neural network. In Proceedings of the 2015 IEEE International Conference on Image Processing, ICIP 2015, Quebec City, QC, Canada, 27–30 September 2015; pp. 3735–3739. [CrossRef]
42. Razakarivony, S.; Jurie, F. Vehicle detection in aerial imagery: A small target detection benchmark. *J. Vis. Commun. Image Represent.* **2016**, *34*, 187–203. [CrossRef]
43. Kingma, D.P.; Ba, J. Adam: A Method for Stochastic Optimization. *arXiv* **2014**, arXiv:1412.6980.
44. Redmon, J.; Divvala, S.K.; Girshick, R.B.; Farhadi, A. You Only Look Once: Unified, Real-Time Object Detection. In Proceedings of the 2016 IEEE Conference on Computer Vision and Pattern Recognition, CVPR 2016, Las Vegas, NV, USA, 27–30 June 2016; pp. 779–788. [CrossRef]

45. Jiang, Y.; Zhu, X.; Wang, X.; Yang, S.; Li, W.; Wang, H.; Fu, P.; Luo, Z. R2CNN: Rotational Region CNN for Orientation Robust Scene Text Detection. *arXiv* **2017**, arXiv:1706.09579. Available online: http://xxx.lanl.gov/abs/1706.09579 (accessed on 8 May 2012).
46. Yang, X.; Sun, H.; Fu, K.; Yang, J.; Sun, X.; Yan, M.; Guo, Z. Automatic Ship Detection in Remote Sensing Images from Google Earth of Complex Scenes Based on Multiscale Rotation Dense Feature Pyramid Networks. *Remote Sens.* **2018**, *10*, 132. [CrossRef]
47. Azimi, S.M.; Vig, E.; Bahmanyar, R.; Körner, M.; Reinartz, P. Towards Multi-class Object Detection in Unconstrained Remote Sensing Imagery. In Proceedings of the Computer Vision—ACCV 2018—14th Asian Conference on Computer Vision, Perth, Australia, 2–6 December 2018; Revised Selected Papers, Part III; pp. 150–165. [CrossRef]
48. Ding, J.; Xue, N.; Long, Y.; Xia, G.; Lu, Q. Learning RoI Transformer for Oriented Object Detection in Aerial Images. In Proceedings of the IEEE Conference on Computer Vision and Pattern Recognition, CVPR 2019, Long Beach, CA, USA, 16–20 June 2019; pp. 2849–2858. [CrossRef]
49. Li, Q.; Mou, L.; Xu, Q.; Zhang, Y.; Zhu, X.X. R^3-Net: A Deep Network for Multioriented Vehicle Detection in Aerial Images and Videos. *IEEE Trans. Geosci. Remote Sens.* **2019**, *57*, 5028–5042. [CrossRef]
50. Yang, X.; Liu, Q.; Yan, J.; Li, A. R3Det: Refined Single-Stage Detector with Feature Refinement for Rotating Object. *arXiv* **2019**, arXiv:1908.05612. Available online: http://xxx.lanl.gov/abs/1908.05612 (accessed on 8 May 2012).

remote sensing

Article

Electromagnetic Signal Classification Based on Class Exemplar Selection and Multi-Objective Linear Programming

Huaji Zhou [1,2], Jing Bai [1,*], Linchun Niu [1], Jie Xu [2], Zhu Xiao [3], Shilian Zheng [2], Licheng Jiao [1] and Xiaoniu Yang [2]

1 School of Artificial Intelligence, Xidian University, Xi'an 710071, China; 19172110640@stu.xidian.edu.cn (H.Z.); lcniu_1@stu.xidian.edu.cn (L.N.); lchjiao@mail.xidian.edu.cn (L.J.)
2 Science and Technology on Communication Information Security Control Laboratory, Jiaxing 314033, China; dayjay151@126.com (J.X.); lianshizheng@126.com (S.Z.); yxn2117@126.com (X.Y.)
3 College of Computer Science and Electronic Engineering, Hunan University, Changsha 410082, China; zhxiao@hnu.edu.cn
* Correspondence: baijing@mail.xidian.edu.cn

Abstract: In the increasingly complex electromagnetic environment, a variety of new signal types are appearing; however, existing electromagnetic signal classification (ESC) models cannot handle new signal types. In this context, the emergence of class-incremental learning aims to incrementally update the classification model as new categories emerge. In this paper, an electromagnetic signal classification framework based on class exemplar selection and a multi-objective linear programming classifier (CES-MOLPC) is proposed in order to continuously learn new classes in an incremental manner. Specifically, our approach involves the adaptive selection of class exemplars considering normalized mutual information and a multi-objective linear programming classifier. The former is used to maintain the classification capability of the model for previous categories by selecting key samples, while the latter is used to allow the model to adapt quickly to new categories. Meanwhile, a weighted loss function based on cross-entropy and distillation loss is presented in order to fine-tune the model. We demonstrate the effectiveness of the proposed CES-MOLPC method through extensive experiments on the public RML2016.04c data set and the large-scale real-world ACARS signal data set. The results of the comparative experiments demonstrate that our method can achieve significant improvements over state-of-the-art methods.

Keywords: class incremental learning; electromagnetic signal classification; deep learning; multi-objective linear programming; class exemplar selection; normalized mutual information

Citation: Zhou, H.; Bai, J.; Niu, L.; Xu, J.; Xiao, Z.; Zheng, S.; Jiao, L.; Yang, X. Electromagnetic Signal Classification Based on Class Exemplar Selection and Multi-Objective Linear Programming. *Remote Sens.* **2022**, *14*, 1177. https://doi.org/10.3390/rs14051177

Academic Editor: Danilo Orlando

Received: 5 January 2022
Accepted: 24 February 2022
Published: 27 February 2022

1. Introduction

Electromagnetic signal classification (ESC) is a key technology in the field of information processing, which forms the basis for non-cooperative communications [1], electronic counter-measures [2], smart antennas [3], software radio [4], and wireless spectrum management [5]. With the widespread use of various types of radio equipment, the types and numbers of radiation sources in the electromagnetic environment are increasing, and the electromagnetic spectrum is becoming more and more congested. More importantly, electromagnetic signal data are acquired faster than before. Under these circumstances, electromagnetic signal data have applications in many fields, such as electromagnetic spectrum monitoring [6], cognitive radio [7], and cyberspace security [8]. ESC is an important requirement for these applications. ESC includes electromagnetic signal type classification and specific emitter identification (SEI). The main task of the former is to exploit the characteristics of electromagnetic signals to distinguish different signal types [8], while the latter involves measuring the characteristics of the received electromagnetic signals and determining the individual radiation sources that generate the signals, based on available prior information [9].

Existing methods based on deep learning have recently achieved remarkable success in the field of signal classification. Many correlative algorithms [10,11] have been proposed and have achieved excellent classification results. For instance, in [12], a graph convolutional modulation recognition framework has been proposed to identify the modulation type of the signal. In [13], a framework based on the capsule network has been proposed to solve the small sample modulation-type detection problem. In [14], a sequence-based network has been proposed. In [15], a convolutional neural network for radio modulation classification has been presented, which is not only suitable for the complex time domain of radio signals but also can achieve excellent results under low signal-to-noise ratio (SNR) conditions. A residual neural network [16] has been used in [17] to perform signal classification over a range of configurations and channel impairments, providing reliable statistics. To solve the task of modulation signal classification, AlexNet and GoogleNet have been used in [18] and achieved great results, proving the effectiveness of deep learning methods in this area. In [10], the authors proposed a complex-valued convolutional neural network, which was used to study the intrinsic properties of the radar interference signal. The authors in [19] have developed a small sample signal modulation recognition framework using an attention relation network. In [20], a convolutional long short-term deep neural network has been used to fully exploit the temporal properties of electromagnetic signals and achieve excellent performance. In [21], a multitasking-based generalized automatic modulation classification framework has been proposed, which was more robust than conventional methods. Moreover, in [22], the authors proved the effectiveness of a deep model in identifying interference sources using frequency band, SNR, and sample selection to optimize the training time. A generative adversarial network has been proposed, in [8], to address the problem of insufficiently labeled electromagnetic signal samples.

Although researchers have conducted many explorations on deep learning-based ESC methods and achieved good results, there is a pre-condition for such deep learning algorithms, which is that the number of signal categories that these algorithms can recognize is fixed. When multiple new categories appear, a natural approach to incremental learning is to simply fine-tune a pre-trained model with the training data of new categories. However, a serious challenge to this approach is catastrophic forgetting [23]; specifically, fine-tuning a model on new data usually results in a significant drop in performance on previous categories. The model has to be trained with the training data of all the previously observed categories in order to recognize the new and the old classes simultaneously. This leads to high computational effort and memory requirements.

With the increasing number of signal processing tasks, electromagnetic signal data have shown an explosive growth trend. In particular, we are faced with the constant emergence of new categories of electromagnetic signal data. For example, in the field of coded communication and adaptive modulation, the receiver recognizes the coding and the modulation method used by the transmitter, and then uses the corresponding decoding and demodulation algorithm to decode and demodulate the acquired signal. When the transmitter changes the modulation mode, the signal received by the receiver will be a new class. When managing the spectrum, cognitive radio can monitor whether illegal signals are interfering with the user's communications by detecting radio signals in the perceived frequency band. If there are multiple categories of illegal signals, they should be identified immediately. Moreover, in realistic applications, new categories are expected to keep appearing over time. If the classification accuracy on the previous categories of the classification model decreases after the learning task of the new class is completed at each stage, there will theoretically be an infinite number of new classes to learn, thus creating an infinite loop.

Class incremental learning is a suitable learning mode for this scenario, where the data increase gradually. Class-incremental learning algorithms can learn a set of gradually evolving classification tasks, and the categories in each task have no overlap with the categories in other tasks. In recent years, driven by practical applications, class incremental learning has gradually become an important research topic. Existing deep learning-based algorithms can be mainly divided into the following three categories: regularization-based, rehearsal-based, and bias-correction-based. The core idea of regularization-based approaches is to estimate the importance of each parameter in the network [23]. The regularization element was added, in [24,25], in order to penalize parameters that vary widely. To constrain important parameters, ref. [26] have introduced regularization terms in the loss function. Furthermore, the authors of [27] have attempted to use the idea of data regularization to prevent activation drift. MAS [28] uses the gradient of the loss function to measure the importance of neurons. The work in [29] preserved old knowledge by freezing the weights of the last layer and penalizing differences between activations before the classification layer. Rehearsal-based methods can mitigate the catastrophic forgetting phenomenon by storing a small number of old samples [30,31], or using a generation network to generate fake samples [32,33]. In [30], the distillation loss and an exemplar set of old categories were used to train the model for the first time. In [34], a semantic drift compensation algorithm was used to estimate the center of each category and perform the prediction by the NCM classifier [30]. In [35], a supervised contrast loss was used to bring samples from the same category closer together in the embedding space, while samples of different classes were further apart in the embedding space. Bias-correction-based methods are dedicated to addressing the problem of imbalance between new and old tasks [31,36,37]. To alleviate such an imbalance, cosine normalization, less-forget constraint, and inter-class separation have been proposed in [36]. To address the problem that the norm of the classifier weight vector of new categories is larger than that of old categories, the weight aligning [37] method has been proposed, which cooperates better with the distillation loss. Another effective algorithm has been proposed in [31], which corrects the biases of different tasks by adding an additional linear layer.

In real ESC applications, an incremental learning algorithm may be able to use part of the previous samples. What an incremental learning method based on the class exemplar selection should do is retain the previous classification capabilities in the form of exemplars of the old categories. The samples selected by the rehearsal-based methods described above can only maintain the classification performance of the model for samples in the center of the class. In order to maintain the classification ability of the model for samples far from the class center at the same time, more efficient samples should be selected. Many exemplar selection-based strategies have recently been proposed [38]. They can be broadly divided into two categories: model output-based and training data distribution-based. Model output-based methods use some output indicators of the classification model as the exemplar selection criteria. In [39], samples were selected based on the lowest confidence level, but the sample set selected in this way was redundant. In [40], the output probability of the model was used as the selection criterion. The entropy-based method was proposed in [41]. Studies have also considered the prediction of errors [42] and interface-distance-based sampling [23]. Training data distribution-based methods use the characteristics of the data itself as the selection criterion. In [30], the average feature vector was extracted. Then, selection was performed according to the distance between the feature vector of each sample and the center of the class feature. In [43], clustering information was used to select the effective samples. Some samples were randomly selected from the whole data set as exemplars. In general, methods based on random sampling are used for comparison with other methods [23,44]. However, the samples selected by the abovementioned methods can only represent the class very well, while discrimination between classes is not sufficient for the models to achieve good classification results for old classes.

It is not an easy task to develop a good classification model which is suitable for new classes. The model must be able to classify new classes while maintaining its ability to recognize old categories. Nevertheless, existing deep learning approaches are not suitable for the above situations. In summary, the challenges facing ESC can be expressed as follows:

1. How to maintain the classification performance of the model when new categories of electromagnetic signal data emerge. As the number of different electromagnetic devices increases and the amount of collected signal data continues to grow, various

new types of signals may be mixed with the existing data. Furthermore, data are updated rapidly, making it an unrealistic task to save all signal data for further training when memory is limited. The emergence of multiple categories and the updating of large data require the deep model to rapidly update its classification capabilities; however, existing state-of-the-art deep learning approaches are not able to learn new categories efficiently; and

2. How to maintain the model's ability to classify old classes after learning the new class. A model trained on old categories may suffer from catastrophic forgetting [23] after learning new categories. If all of the data from the previously observed classes are used to train the new model, significant computational overhead and time costs will be incurred. Mainstream methods based on exemplar selection are usually based on prototypes [30]. These methods use the mean value of the extracted features as the selection criterion. However, samples selected in this way can only represent the class very well, while discrimination between classes is not sufficient for the model to achieve good classification results on the old classes.

To overcome the aforementioned problems, we present a multi-class incremental learning framework, CES-MOLPC, for ESC, which not only can learn new classes rapidly but also ensures that its classification accuracy on old classes is maintained at a high value. Specifically, we develop a multi-objective linear programming (MOLP) incremental classifier based on the average feature vectors extracted by the deep extractor. Inspired by [45], the proposed MOLP classifier is obtained by solving the MOLP problem, where the whole process only requires forward propagation of the old model without using a back-propagation algorithm to train the model. In addition, inspired by [46–48], we develop an adaptive class exemplar selection approach, in view of the normalized mutual information, in order to maintain the accuracy of new model on previous categories. We also design a weighted loss function based on cross-entropy and distillation loss, which is used to fine-tune the incremental model, thus making the model more robust. The presented method requires only a small amount of computational and storage resources. In summary, the main innovative contributions of this article are as follows:

1. The presented MOLP classifier learns multiple new categories simultaneously by increasing multiple weight vectors, which can be obtained by solving the MOLP problem based on the average feature vector of each class;
2. The presented adaptive class exemplar selection approach, in view of normalized mutual information, preserves the classification ability of the model for previous categories by selecting key samples; and
3. A weighted loss function based on cross-entropy and distillation loss is presented to fine-tune the new model using both the new category data and the old exemplar set.

The remainder of this article is organized as follows: in Section 2, the proposed ESIC framework is presented. Section 3 shows the numerical examples to verify the classification performance of our method. In Section 4, we discuss the proposed method and experimental results. Finally, in Section 5, we summarize the main results of this paper.

2. Methodology

According to the electromagnetic signal increment classification (ESIC) framework, shown in Figure 1, the proposed framework is mainly composed of the following three modules:

(1) Adaptive class exemplar selection;
(2) MOLP incremental classifier; and
(3) Incremental learning.

The specific description, with respect to the proposed algorithm, is given in the following. Assuming that there are N categories of signal data, a model $\mathcal{M}(\mathcal{F}_{old}, \psi_{old}(N; W_{old}))$ trained using these data has the ability to accurately classify N old categories, where $\mathcal{F}_{old}$ is a feature extractor, $\psi_{old}(N; W_{old})$ is an N-class classifier, and W_{old} is the weight matrix of the classifier. In our model, the bias of the classifier is set to 0. In the ESIC task, the signal data constantly appear in a stream. In the case of limited memory size, only a small part of the training data can be saved. Therefore, the exemplar selection approach, in view of normalized mutual information, is presented for the selection of key samples. When T unknown categories are used as input to the model, the model must correctly identify the N old categories while also being able to identify the T new categories well. At this time, the output layer of the model must add T output nodes; that is, the weight matrix of the classifier needs to add T weight columns. Thus, the MOLP classifier $\psi_{new}(N+T; W_{new})$ is proposed to learn new classes by adding T weight columns. W_{new} is obtained by solving the MOLP problem. Finally, in the incremental learning stage, the exemplars of the old classes and the data of the new class are fused to fine-tune the model $\mathcal{M}(\mathcal{F}_{old}, \psi_{new}(N+T; W_{new}))$. At the same time, cross-entropy loss and distillation loss are adopted for fine-tuning. Our goal is to obtain the incremental model $\mathcal{M}(\mathcal{F}_{new}, \psi_{new}(N+T; W_{new}))$.

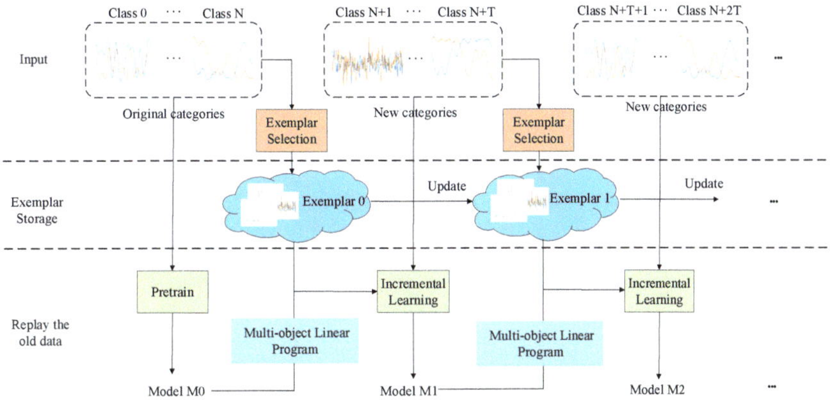

Figure 1. Overview of the proposed framework, CES-MOLPC.

2.1. Adaptive Exemplar Selection

The idea behind the proposed exemplar selection approach is to calculate the normalized mutual information between the one-hot vector corresponding to the true label of the sample and the probability vector predicted by model for the sample. The one-hot vector and the class vector predicted by the model for the sample can be regarded as two clusters, where the resemblance between the two clusters is measured by the normalized mutual information. We assume that the harder a sample is to identify as belonging to the correct category, the closer it is to the decision boundary. Specifically, on the premise that a sample is correctly classified, if there is less normalized mutual information between the ground-truth label and the probability of the sample predicted by model, the more difficult it is for the sample to be recognized as the correct category, which means that we can select a small number of boundary samples to describe the sample distribution of the entire category.

As shown in Algorithm 1, given the data of N old categories $\mathcal{X}^N = \{X^1, X^2, \ldots, X^N\}$, the exemplar is selected from $\mathcal{X}^N$. An initial model, $\mathcal{M}$, is trained to recognize N classes using all data from $\mathcal{X}^N$. When T new classes appear, they are integrated with the exemplars of old categories to train a new model. The normalized mutual information between the one-hot ground-truth label $\mathcal{Y}$ and predicted label $\hat{\mathcal{Y}}$ is shown in (1).

$$\mathrm{NMI}(\mathcal{Y}, \hat{\mathcal{Y}}) = \frac{2\mathrm{I}(\mathcal{Y}, \hat{\mathcal{Y}})}{\mathrm{H}(\mathcal{Y}) + \mathrm{H}(\hat{\mathcal{Y}})}, \quad (1)$$

where $\mathrm{I}(\mathcal{Y}, \hat{\mathcal{Y}})$ refers to the mutual information between $\mathcal{Y}$ and $\hat{\mathcal{Y}}$, as shown in (2); $H(\mathcal{Y})$ refers to the entropy of $\mathcal{Y}$, as shown in (3); and $H(\hat{\mathcal{Y}})$ refers to the entropy of $\hat{\mathcal{Y}}$, as shown in (4).

$$\mathrm{I}(\mathcal{Y},\hat{\mathcal{Y}}) = \sum_{y,\hat{y}} p(y,\hat{y}) log \frac{p(y,\hat{y})}{p(y)p(\hat{y})}. \tag{2}$$

$$\mathrm{H}(\mathcal{Y}) = -\sum_{y} p(y) log p(y). \tag{3}$$

$$\mathrm{H}(\hat{\mathcal{Y}}) = -\sum_{\hat{y}} p(\hat{y}) log p(\hat{y}). \tag{4}$$

Algorithm 1 Adaptive class exemplar selection

Input:

$\mathcal{X}^N = \{X^1, X^2, \ldots, X^N\}$: all training data, $X^i = \{(x_1^i, y_1^i), (x_2^i, y_2^i), \ldots, (x_{n_i}^i, y_{n_i}^i)\}$ are the training data of the ith category;

$\mathcal{K}$: size of memory;

$\mathcal{M}(\mathcal{F}_{old}, \psi_{old}(N; W_{old}))$: pre-trained model;

$\{n_1, n_2, \ldots, n_N\}$: the number of training samples for each old class;

Output:

Class exemplar set.

1. For each training sample x_1^i in class i:
   i. The sample is input into the model $\mathcal{M}$ to obtain its predicted probability $\hat{y}_j^i$;
   ii. Calculate the normalized mutual information between $\hat{y}_j^i$ and y_j^i, according to Equation (1);
2. For each class i:
   i. Obtain the classification accuracy OA_i of ith category through forward propagation;
   ii. Obtain the average feature vector f of all samples through forward propagation and the average feature vector f_{mean}^i;
   iii. Calculate the weight factor η_i of the ith category according to Equation (5);
   iv. Calculate the number of samples m_i that need to be saved for the ith category;
   v. Find m_i samples P_i with the smallest mutual information;
3. Obtain the exemplar set $\{\mathcal{P}_1, \mathcal{P}_2, \ldots, \mathcal{P}_N\}$.

Previous algorithms based on sample playback select the same number of samples in each category. In this work, an adaptive exemplar selection method, which is weighted based on the number of samples, is proposed to maintain the classification ability on old categories. As the classification difficulty for each category is different, it is not necessary for each category to preserve the same number of samples. For the categories that are difficult to distinguish, more samples should be retained, while samples that are easy to distinguish need only retain a small number of samples. When selecting the exemplar set, the recognition accuracy of the old model on the old training data is used as feedback. If the recognition rate for a certain old category is higher, then fewer samples should be saved, accordingly. In addition, the degree of dispersion of the category distribution is also used as a measure. We use the mean value of the distance between the feature vector of all samples in a category and the average feature vector to measure the distribution of samples in the category. The more concentrated the sample distribution of the category, the higher the similarity within the category, such that fewer samples are required to describe the category. The weight factor is shown in (5):

$$\eta_i = -log\mathrm{OA}_i + \frac{1}{n_i}\sum_{k=1}^{n_i} \|f_k - f_{mean}^i\|_2, \tag{5}$$

where OA_i is the classification accuracy on the ith class, n_i refers to the number of training samples in the ith class, f_k refers to the feature vector of the kth sample, and f_{mean}^i refers to the average feature vector of the ith class.

As the memory size is fixed, considering the number of samples in each category and then normalizing the weights of all old categories to between 0 and 1, the number of samples that need to be saved for each category can be obtained, as shown in (6):

$$m_i = \mathcal{K} \cdot \frac{exp(\eta_i)}{\sum_{k=1}^{N} exp(\eta_k)}, \tag{6}$$

where m_i represents the number of samples that need to be saved in the signal data of the first type in the old data.

2.2. MOLP Incremental Classifier

As shown in Figure 2, when the model needs to learn T new categories, the classifier must add T output nodes; that is, the weight matrix needs to add T columns. According to the classification principle of the linear classifier, when the new category data are input to the CNN model, the feature vector is first extracted through the feature extractor, following which the feature is used as input to the linear classification layer to obtain a vector. Each component in the vector denotes the possibility that the sample belongs to certain category. Thus, the component value corresponding to the new class should be maximized.

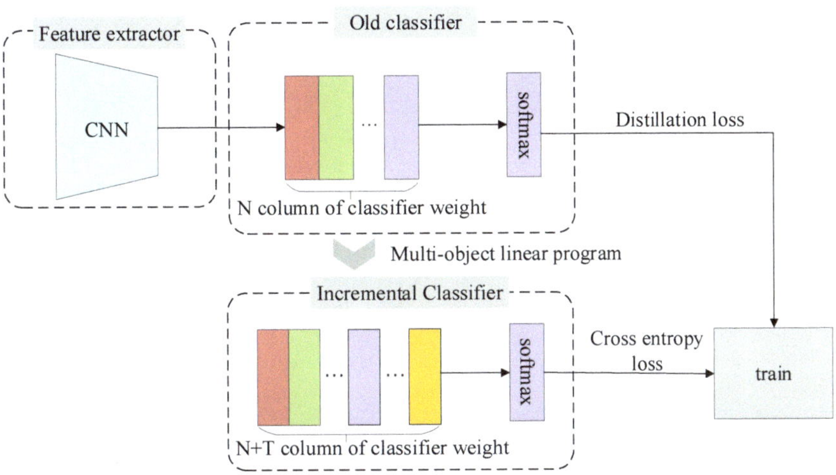

Figure 2. MOLP incremental classifier and training of the model.

Given the model $\mathcal{M}(\mathcal{F}_{old}, \psi_{old}(N; W_{old}))$ trained on old categories and a sample x of the ith category, as well as the eigenvector f of x extracted by the feature extractor $\mathcal{F}_{old}$, the existing classifier $\psi_{old}(N; W_{old})$ gives the score by classifier dot multiplying f with W_{old}^i. Then, the softmax activation function is used to obtain the probability of x belonging to the ith category, as shown in (7):

$$p_i(x) = \frac{exp(f \cdot W_{old}^i)}{\sum_k exp(f \cdot W_{old}^j)}, \tag{7}$$

where $\mathcal{F}_{old}$, W_{old} is $L \times N$ dimensional, as shown in (8); W^i_{old} refers to the ith column of the weight matrix; and L refers to the dimension of the eigenvector. When T new categories arrive, the new classifier corresponds to the weight matrix W_{new}, as shown in (9):

$$W_{old} = \begin{pmatrix} W_1^1 & W_2^1 & \cdots & W_N^1 \\ \vdots & \vdots & & \vdots \\ W_1^l & W_2^l & \cdots & W_N^l \\ \vdots & \vdots & & \vdots \\ W_1^L & W_2^L & \cdots & W_N^L \end{pmatrix}. \quad (8)$$

$$W_{new} = \begin{pmatrix} W_1^1 & \cdots & W_N^1 & W_{N+1}^1 & \cdots & W_{N+T}^1 \\ \vdots & & \vdots & \vdots & & \vdots \\ W_1^l & \cdots & W_N^l & W_{N+1}^l & \cdots & W_{N+T}^l \\ \vdots & & \vdots & \vdots & & \vdots \\ W_1^L & \cdots & W_N^L & W_{N+1}^L & \cdots & W_{N+T}^L \end{pmatrix}. \quad (9)$$

The exemplar set of old categories and new categories of data are used as input to the model. The class mean eigenvector of each class is obtained by averaging each eigenvector extracted by the sophisticated extractor $\mathcal{F}_{old}$:

$$f = \{f_1, f_2, \ldots, f_N, f_{N+1}, \ldots, f_{N+T}\}, \quad (10)$$

where $f_1, f_2, \ldots, f_N$ refers to the mean feature vector of old categories, and $f_{N+1}, \ldots, f_{N+T}$ refers to the mean feature vector of new categories at a time. When the mean feature vector of the ith old category is input to the classification layer, it will predict a score Score_i by multiplying the ith class feature vector by the new weight matrix, as shown in (11). $\text{Score}_i[i]$ should be maximized in Score_i; thus, we can formulate it as $f_i \cdot W_i > f_i \cdot W_j$, $i \in [1, N]$, $j \in [N+1, N+T]$, which leads to the constraint of the MOLP problem.

$$\text{Score}_i = \begin{pmatrix} f_i \cdot W_1 \\ f_i \cdot W_2 \\ \vdots \\ f_i \cdot W_N \\ f_i \cdot W_{N+1} \\ \vdots \\ f_i \cdot W_{N+T} \end{pmatrix}. \quad (11)$$

If the feature vector of the jth new category is input into the classifier, it will predict a score Score_j by multiplying the jth class feature vector with the new weight matrix, as shown in (12). $\text{Score}_j[j]$ should be maximized in Score_j; thus, we can formulate it as $f_j \cdot W_j > f_i \cdot W_i$, $f_j \cdot W_j > f_i \cdot W_k$, $k \in [N+1, N+T]$, $k \neq j$, which leads to the constraint of the MOLP problem. We take $\text{Score}_j[j]$ as the maximization goal of the MOLP.

$$\text{Score}_j = \begin{pmatrix} f_j \cdot W_1 \\ f_j \cdot W_2 \\ \vdots \\ f_j \cdot W_N \\ f_j \cdot W_{N+1} \\ \vdots \\ f_j \cdot W_{N+T} \end{pmatrix}. \quad (12)$$

In summary, we can express the MOLP problem as follows:

$$\begin{aligned} max \quad & f_{N+1} \cdot W_{N+1}, \ldots, f_{N+T} \cdot W_{N+T} \\ s.t. \quad & f_i \cdot W_i > f_i \cdot W_j \\ & f_j \cdot W_j > f_j \cdot W_i \\ & f_j \cdot W_j > f_j \cdot W_k \\ -max\{|W_{old}|\} < & W_j^l < max\{|W_{old}|\} \end{aligned} , \tag{13}$$

where f_j is the mean feature vector of the jth new category, f_i is the mean feature vector of the ith old category, W_j denotes the jth column of the new weight matrix, W_k denotes the kth column of the new weight matrix, and $\max\{|W_{old}|\}$ denotes the largest value of the old weight matrix. By solving the above MOLP problem, we can obtain the new classifier $\psi_{new}(N+T; W_{new})$.

2.3. Incremental Learning

After completing the exemplar $\{\mathcal{P}_1, \mathcal{P}_2, \ldots, \mathcal{P}_N\}$ selection of old classes and obtaining the incremental classifier $\psi_{new}(N+T; W_{new})$, the last step is incremental learning. The specific algorithm flow is detailed in Algorithm 2. The old exemplar set and data of new categories are integrated to fine-tune the model $\mathcal{M}(\mathcal{F}_{old}, \psi_{new}(N+T; W_{new}))$. In the incremental learning phase, both the cross-entropy loss and distillation loss are used when fine-tuning the new incremental model. The cross-entropy loss is designed to guide the model to better distinguish signal categories, while distillation loss is used to help the new model not forget the knowledge of the old model as much as possible. Our loss function can be expressed as follows:

$$L = \lambda L^c + (1-\lambda) L^d, \tag{14}$$

where λ is equal to $\frac{N}{N+T}$, N is the number of old classes, and T is the number of new classes. The standard cross-entropy loss L^c and the distillation loss L^d are as follows:

$$L^c = -\sum_{i=1}^{|C|} \delta_{i=y} log(p_i), \tag{15}$$

$$L^d = -\sum_{i=1}^{|C|} \varphi_i(\hat{p}_i) log(\varphi_i(p_i)), \tag{16}$$

where C collects the old exemplar set and all the data of the new classes, σ is the indicator function, p_i is the probability that the input signal belongs to the ith category obtained by the new model, $\hat{p}$ is the soft label of the input signal generated by old model $\mathcal{M}(\mathcal{F}_{old}, \psi_{old}(N+T; W_{old}))$ on all observed categories and y is the ground-truth label, and $\varphi_i(q) = q_i^{1/\tau} / \sum_k q_k^{1/\tau}$ is a rescaling function, where τ is scaling factor ($\tau = 2$ in our experiments) which increases the weights of small values.

Algorithm 2 Incremental learning

Input:
$\{\mathcal{P}_1, \mathcal{P}_2, \ldots, \mathcal{P}_N\}$: the exemplar set of old categories;
$\mathcal{X}^{new} = \{X^{N+1}, \ldots, X^{N+T}\}$: new training data, $X^j = \{(x_1^j, y_1^j), (x_2^j, y_2^j), \ldots, (x_{n_j}^j, y_{n_j}^j)\}$ is the training data of jth category;
$\mathcal{M}(\mathcal{F}_{old}, \psi_{old}(N; W_{old}))$: old model;
$\mathcal{M}(\mathcal{F}_{old}, \psi_{new}(N+T; W_{new}))$: incremental model;
Output:
$\mathcal{M}(\mathcal{F}_{new}, \psi_{new}(N+T; W_{new}))$: new model.
// form combined training set:
1. $\mathcal{D} \leftarrow \{\mathcal{P}_1, \mathcal{P}_2, \ldots, \mathcal{P}_N\} \bigcup \{X^{N+1}, \ldots, X^{N+T}\}$
2. For each true label $y \in [1, N+T]$:
   i. Obtain the according probability $\hat{q}_i^y$ through forward propagation of model $\mathcal{M}(\mathcal{F}_{old}, \psi_{old}(N; W_{old}))$, for all $(x_i^y, \cdot) \in \mathcal{D}$;
   ii. Obtain the according probability q_i^y through forward propagation of model $\mathcal{M}(\mathcal{F}_{old}, \psi_{new}(N+T; W_{new}))$, for all $(x_i^y, \cdot) \in \mathcal{D}$;
3. Calculate the total loss according to Equations (14)–(16), as follows:
$loss = -\sum_{y=1}^{y=N+T} \sum_{(x_i^y, y_i^y) \in \mathcal{D}} [\lambda \sigma_{y=y_i} + (1-\lambda)\hat{q}_i^y log(q_i^y)]$
4. Run the fine-tuning and update the parameters of model $\mathcal{M}(\mathcal{F}_{old}, \psi_{new}(N+T; W_{new}))$

3. Results

3.1. Data Set

To test the effectiveness of the presented framework for signal classification (CES-MOLPC), we performed comparative experiments on both the public RML2016.04c data set and the large-scale real-world ACARS signal data set.

The signal-to-noise ratio of the RML2016.04c dataset ranges from −20 to 18 dB, with an interval of 2 dB. The corresponding in-phase and quadrature signal (IQ) for each signal-to-noise ratio includes 11 modulation types, of which 8 are digital modulations and the remaining 3 are analog modulations. We used the 12 dB data as the experimental data set. The total number of training samples was 8103, but the numbers of samples for each category differed. The dimensions of each signal sample were 2×128.

As a digital data transmission system, ACARS is generally used to transmit messages between aircraft and ground stations by radio or satellite. At present, ACARS is used in many civil aviation systems. The data set in this article was collected using our own experimental system, whose antenna was placed at a specific place in Jiaxing, Zhejiang Province, China. There were 10,000 ACARS signals from 20 aircraft overall, such that the data set had a total of 20 categories, with each category containing 500 samples. In our experiment, the training samples and test samples each accounted for half of the samples. The dimensions of each IQ signal sample were 2×4096.

3.2. Experimental Setup

3.2.1. Parameters

In the following experiment, the memory size $\mathcal{K}$ was limited to 200, the Adam optimizer was used when fine-tuning the model, the learning rate was set to 0.001 throughout the training of the old classes and the fine-tuning phases, the weight decay was 0.00001 in the whole process, the batch size in the whole process was 32, and the number of iterations was 500 for the training of the old classes and 80 for the fine-tuning phase.

3.2.2. Network Structure

We created two convolutional neural networks for incremental classification on the above two data sets. The network structure corresponding to the RML2016.04c data set is given in Table 1, which mainly consisted of three convolutional layers, three pooling layers, and three regularization layers. A dropout layer was used to reduce overfitting.

The network structure corresponding to the ACARS data set is shown in Table 2. This network mainly consisted of seven convolutional layers, seven pooling layers, and seven regularization layers.

Table 1. The neural network structure (RML2016.04c).

Layer_Name	Output_Shape	Kernel_Size	Stride
Conv2d-1	(−1,128,126,1)	(3,2)	(1,1)
BatchNorm2d-2	(−1,128,126,1)	N/A	N/A
MaxPool2d-3	(−1,128,42,1)	(3,1)	(3,1)
Conv2d-4	(−1,256,40,1)	(3,1)	(1,1)
BatchNorm2d-5	(−1,256,40,1)	N/A	N/A
MaxPool2d-6	(−1,256,13,1)	(3,1)	(3,1)
Conv2d-7	(−1,512,11,1)	(3,1)	(1,1)
BatchNorm2d-8	(−1,512,11,1)	N/A	N/A
MaxPool2d-9	(−1,512,3,1)	(3,1)	(3,1)
Linear-10	(−1512)	N/A	N/A
Dropout-11	(−1512)	N/A	N/A
Linear-12	(−1,N)	N/A	N/A

Table 2. The neural network structure (ACARS).

Layer_Name	Output_Shape	Kernel_Size	Stride
Conv2d-1	(−1,150,4082,1)	(15,2)	(1,1)
BatchNorm2d-2	(−1,150,4082,1)	N/A	N/A
MaxPool2d-3	(−1,150,2041,1)	(2,1)	(2,1)
Conv2d-4	(−1,150,2035,1)	(7,1)	(1,1)
BatchNorm2d-5	(−1,150,2035,1)	N/A	N/A
MaxPool2d-6	(−1,150,1017,1)	(2,1)	(2,1)
Conv2d-7	(−1,150,1013,1)	(5,1)	(1,1)
BatchNorm2d-8	(−1,150,1013,1)	N/A	N/A
MaxPool2d-9	(−1,150,506,1)	(2,1)	(2,1)
Conv2d-10	(−1,150,502,1)	(5,1)	(1,1)
BatchNorm2d-11	(−1,150,502,1)	N/A	N/A
MaxPool2d-12	(−1,150,251,1)	(2,1)	(2,1)
Conv2d-13	(−1,150,249,1)	(3,1)	(1,1)
BatchNorm2d-14	(−1,150,249,1)	N/A	N/A
MaxPool2d-15	(−1,150,124,1)	(2,1)	(2,1)
Conv2d-16	(−1,150,120,1)	(5,1)	(1,1)
BatchNorm2d-17	(−1,150,120,1)	N/A	N/A
MaxPool2d-18	(−1,150,60,1)	(2,1)	(2,1)
Conv2d-19	(−1,150,56,1)	(5,1)	(1,1)
BatchNorm2d-20	(−1,150,56,1)	N/A	N/A
MaxPool2d-21	(−1,150,28,1)	(2,1)	(2,1)
Linear-22	(−1512)	N/A	N/A
Dropout-23	(−1512)	N/A	N/A
Linear-24	(−1,N)	N/A	N/A

3.2.3. Evaluation Metrics

We defined the evaluation metrics Old OA, New OA, and Total OA, where New OA refers to the ratio of the number of correctly classified samples in the test samples of all the new classes to the total number of test samples in the new class; Old OA refers to the ratio of the number of correctly classified samples in all the test samples of the old class to the total number of test samples in the old class; and Total OA refers to the ratio of the number of correctly classified samples in the test samples of all the categories to the total number of test samples in all the categories.

3.2.4. Environment

Our experiment was conducted on a computer with an Inter(R) Xeon(R) Gold 6240 CPU with 128 GB RAM at 2.60 GHz and an NVIDIA GeForce RTX 3090 GPU with 24 GB RAM. The software environment was the Ubuntu 16.04 64 bit system, and the deep learning frameworks of PyTorch were used.

3.3. *Result and Analysis of Experiments*

3.3.1. Multi-Class Recognition Performance

To validate the advantages of our presented overall framework, CES-MOLPC, we performed multi-class recognition experiments and compared the results with those of several competing or representative methods, including Incremental Classifier and Representation Learning (ICARL) [30], Bias Correction (BIC) [31], and Weight Aligning (WA) [37]. These methods all use a portion of the old class signal samples to maintain the incremental model's ability to classify the old classes after learning a new class. To ensure the fairness of the experiment, the size of the exemplar set $\mathcal{K}$ of our approach matched that used in the above approaches, and the network structure and hyper-parameters of the comparison methods were also consistent with those of our method. The experiments considering all of the abovementioned methods were performed five times, and the average value was taken as the final result.

For the RML2016.04c data set, we set the number of new categories to 1, 2, or 3 (T = 1, 2, 3). Table 3 shows the experimental results on the RML2016.04c data set. When $T = 1$, the proposed approach (CES-MOLPC) achieved the highest values of Old OA and Total OA, while the difference in New OA between our method and the best was only 0.26. The WA method does not learn any new knowledge at all, and its New OA was not as high as that of our method. In contrast, the BIC method is biased towards new classes. The possible reason for this is that the weighted loss function is not balanced. The ICARL method performed worse than the BIC method. When $T = 2$, the CES-MOLPC method also achieved the best Old OA and Total OA, but its New OA value was low. Meanwhile, the BIC method had the best New OA, but it obviously forgot the old knowledge. The Total OA of our method was 0.3 higher than that of the second-highest method. When $T = 3$, CES-MOLPC performed best on old categories. The OA of the second-best method, WA, was only 0.21 lower than our method, but its New OA was poor. The BIC approach was still biased towards new classes. In the above three cases, CES-MOLPC achieved the highest Total OA. Obviously, the proposed method achieved the best classification results, when compared to the other methods.

For the ACARS data set, we set the number of new categories to 2, 4, or 6 (T = 2, 4, 6). Table 4 shows the experimental results on the ACARS data set. When $T = 2$, the proposed approach, CES-MOLPC, achieved the best Old OA and New OA, and its Total OA was 2% higher than that of the runner-up. The Old OA value of the WA method was only 0.04% lower than that of the proposed method, but its New OA was very poor. When $T = 4$, CES-MOLPC had the highest New OA and Total OA, while the Old OA of the proposed method was lower than that of WA; however, the New OA value of WA is only 0.59%, which means that the method had basically not learned the knowledge of the new classes at all. This is probably due to this method scaling the weights of the new classifiers too much, such that the new classifiers tend too much to the old classes. When $T = 6$, CES-MOLPC performed best for the overall category. The ICARL method had the highest New OA, while the WA method had the highest Old OA. In the above three cases, the proposed CES-MOLPC outperformed all of the compared methods, in terms of the Total OA. This means that our approach was more robust.

Table 3. Results of experiments on the RML2016.04c data set ($T = 1, 2, 3$).

Class	$T=1$				$T=2$				$T=3$			
	ICARL	BIC	WA	CES-MOLPC	ICARL	BIC	WA	CES-MOLPC	ICARL	BIC	WA	CES-MOLPC
1	57.00	80.19	64.25	57.00	28.02	14.01	80.68	45.89	32.85	2.42	93.72	77.78
2	92.63	82.72	100.00	87.25	92.35	86.97	100.0	97.17	94.90	97.73	100.00	97.45
3	100.00	100.00	100.00	100.00	100.00	100.00	100.00	100.00	100.00	100.00	100.00	100.0
4	91.35	98.24	67.79	98.56	90.22	98.40	99.20	97.44	98.88	97.28	99.52	98.72
5	90.87	100.00	92.15	100.00	96.79	98.88	93.11	100.00	98.72	92.63	100.00	100.00
6	68.75	78.21	93.27	94.87	93.75	80.29	89.10	95.19	86.38	78.04	93.27	97.60
7	99.04	998.71	99.36	98.39	99.36	98.07	99.36	99.04	96.78	97.75	94.86	98.71
8	92.90	88.39	87.10	94.84	96.13	95.48	90.32	97.42	89.68	90.32	78.71	98.71
9	91.26	88.35	98.06	81.55	96.12	61.17	93.20	72.82	97.09	95.15	0.0	86.41
10	61.41	68.81	69.77	89.07	79.42	97.75	0.32	81.99	90.68	98.71	0.0	49.20
11	94.62	99.74	0.00	99.49	88.97	98.97	10.00	95.13	93.59	97.69	0.0	91.79
Old OA [1]	84.61	89.63	86.79	**93.42**	90.82	87.21	94.63	**94.64**	90.82	86.71	96.71	**96.92**
New OA [2]	94.61	**99.74**	0.00	99.48	84.74	**98.72**	5.71	89.30	84.74	**97.76**	0.0	88.73
Total OA [3]	85.57	90.60	78.44	**94.01**	89.77	89.20	79.26	**93.46**	89.77	88.90	77.53	**93.32**

[1] refers to the ratio of the number of correctly classified samples in all test samples of the old class to the total number of test samples in the old class. [2] refers to the ratio of the number of correctly classified samples in the test samples of all new classes to the total number of test samples in the new class. [3] refers to the ratio of the number of correctly classified samples in the test samples of all categories to the total number of test samples in all categories.

In order to further observe the effectiveness of our method in alleviating catastrophic forgetting problems, we set up several continuous incremental processes. For the RML2016.04c data set, set the number of old classes to 5 and the number of new classes to 6, adding one class each time. For the ACARS data set, we set the number of old and new categories to 10, and added two categories each time. The experimental results are shown in Figures 3 and 4. The abscissa in the figure represents the number of classes learned by the model, and the ordinate represents the classification accuracy of the current model on the test data of all learned classes.

From the figures, we can observe the advantages of our proposed CES-MOLPC: the accuracy of this method decreased very slowly in all the categories. The BIC method performed well on the RML2016.04c data set, but it was still not as good as our method. When the number of categories learned was 9, the decrease was even more significant. The ICARL method performed well on the ACARS data set, but was second to our method. The WA method performed poorly on both data sets, with significant variation and low accuracy. It can be seen that our proposed method was not only effective in learning new categories, but also retained the classification ability of the model for old categories.

Table 4. Results of experiments on the ACARS data set ($T = 2, 4, 6$).

Class	$T=2$				$T=4$				$T=6$			
	ICARL	BIC	WA	CES-MOLPC	ICARL	BIC	WA	CES-MOLPC	ICARL	BIC	WA	CES-MOLPC
1	100.00	100.00	100.00	100.00	100.00	100.00	100.00	100.00	100.00	100.00	100.00	100.00
2	81.01	79.35	80.17	80.17	81.01	79.35	80.17	80.17	80.59	79.35	78.90	80.17
3	95.42	79.92	95.83	95.42	95.42	79.92	95.83	95.42	95.00	79.92	94.58	95.42
4	95.98	83.27	93.57	95.18	95.98	83.27	93.57	95.18	93.57	83.27	96.39	95.98
5	96.72	91.98	96.72	95.49	96.72	91.98	96.72	95.49	96.72	91.98	97.95	96.31
6	96.61	75.58	94.12	95.76	96.61	75.58	94.12	95.76	95.34	75.58	97.03	96.61
7	92.65	89.29	93.48	94.49	92.65	89.29	93.48	94.49	94.12	89.29	94.12	94.12
8	93.87	88.66	97.84	95.02	93.87	88.66	97.84	95.02	93.87	88.66	93.87	96.93
9	96.54	87.92	92.06	96.97	96.54	87.92	92.06	96.97	97.40	87.92	98.70	98.27
10	88.49	85.60	97.58	93.65	88.49	85.60	97.58	93.65	87.70	85.60	96.03	96.03
11	97.18	83.40	92.99	96.77	97.18	83.40	92.99	96.77	94.76	83.40	97.98	97.58
12	92.61	83.27	94.94	93.77	92.61	83.27	94.94	93.77	88.33	83.27	92.22	93.39

Table 4. *Cont.*

Class	T = 2				T = 4				T = 6			
	ICARL	**BIC**	**WA**	**CES-MOLPC**	**ICARL**	**BIC**	**WA**	**CES-MOLPC**	**ICARL**	**BIC**	**WA**	**CES-MOLPC**
13	94.94	79.27	97.24	96.50	94.94	79.27	97.24	96.50	92.61	79.27	96.50	96.50
14	92.91	93.00	95.25	95.67	92.91	93.00	95.25	95.67	96.46	93.00	97.64	97.24
15	94.86	79.51	95.88	96.05	94.86	79.51	95.88	96.05	96.84	79.51	0.04	83.79
16	95.13	74.02	98.43	96.63	95.13	74.02	98.43	96.63	95.13	74.02	0.00	91.01
17	94.53	8.54	96.33	98.44	94.53	8.54	96.33	98.44	96.48	8.54	0.00	78.52
18	88.57	25.40	93.46	93.06	88.57	25.40	93.46	93.06	92.24	25.40	0.02	86.94
19	92.64	97.97	0.0	95.74	92.65	97.97	0.0	95.74	94.57	97.97	0.00	81.01
20	97.55	93.17	0.0	96.73	96.98	93.17	0.0	96.73	100.00	93.17	0.00	87.35
Old OA [1]	93.77	77.16	94.93	**94.97**	92.89	92.36	**93.93**	93.87	80.59	92.16	**95.13**	94.34
New OA [2]	95.03	95.56	0.00	**96.22**	94.52	94.66	0.00	**95.12**	**95.00**	93.23	0.59	93.88
Total OA [3]	93.90	78.98	85.38	**95.10**	93.22	92.78	84.68	**94.20**	93.57	92.56	66.32	**94.13**

[1] refers to the ratio of the number of correctly classified samples in all test samples of the old class to the total number of test samples in the old class. [2] refers to the ratio of the number of correctly classified samples in the test samples of all new classes to the total number of test samples in the new class. [3] refers to the ratio of the number of correctly classified samples in the test samples of all categories to the total number of test samples in all categories.

3.3.2. Comparison to Other Class Exemplar Selection Methods

To determine the additional benefits of class exemplar selection, we compared the proposed CES-MOLPC with other class exemplar selection approaches. The other algorithms included Random, Entropy [41], and Prototype [30]. The Random method randomly selects old exemplars from the old data. The Prototype approach selects the signal samples whose feature vectors are close to the average feature vector of its own category. All experiments were performed five times and the average was taken as the final result.

Table 5 shows the comparative experimental results of the class exemplar selection methods on the RML2016.04c data set. It can be seen that the proposed CES-MOLPC achieved the highest classification accuracy. The Entropy exemplar selection method performed better than the Prototype method. Both Old OA and Total OA of the Random method outperformed the other two methods, and were only slightly behind those of our method. The New OA of these methods was low, mainly as the old classes contained similar categories to the new classes. Table 6 lists the comparative experimental results of class exemplar selection methods on the ACARS data set. We can see that the presented method achieved the highest classification accuracy in all three indicators (i.e., Old OA, New OA, and Total OA). As the Prototype method only considers representativeness when selecting samples but ignores the distinction between classes, the Old OA of this method was not as good as that of the Entropy and Random methods.

Table 5. Comparison of class exemplar selection methods on the RML2016.04c data set ($T = 2$).

Method	Old OA	New OA	Total OA
Entropy	93.71	87.73	92.92
Random	93.11	88.49	91.68
Prototype	90.19	88.12	89.38
CES-MOLPC	**95.49**	**89.58**	**93.91**

Table 6. Comparison of class exemplar selection methods on the ACARS data set ($T = 2$).

Method	Old OA	New OA	Total OA
Entropy	94.13	93.29	93.78
Random	94.02	93.63	93.98
Prototype	92.10	94.10	92.84
CES-MOLPC	**95.81**	**96.82**	**95.82**

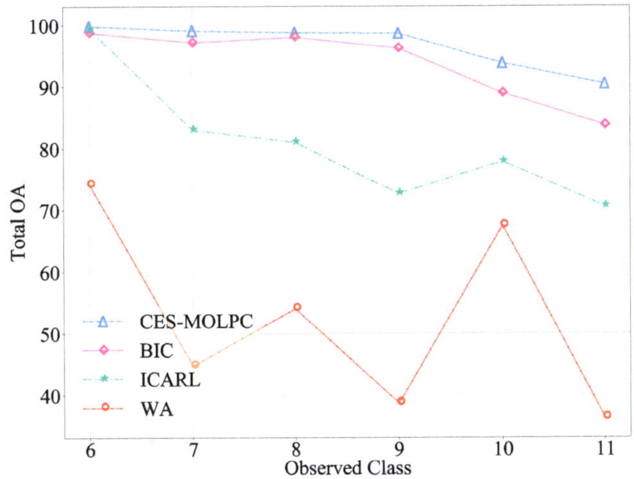

Figure 3. The Total OA of multiple incremental classification on the RML2016.04c data set ($T = 1$).

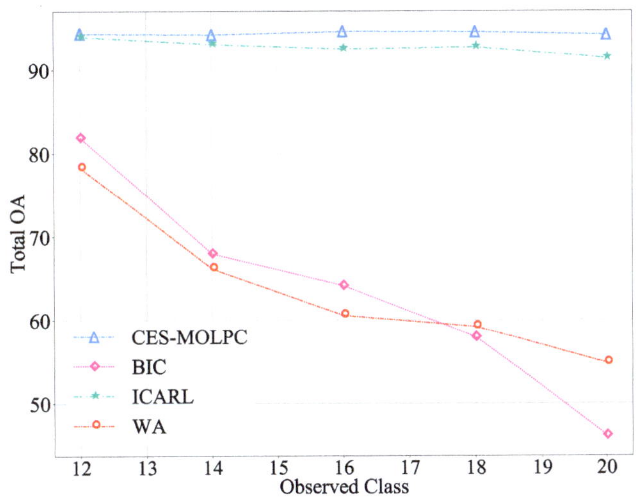

Figure 4. The Total OA of multiple incremental classification on the ACARS data set ($T = 2$).

4. Discussion

In terms of computational complexity, the proposed algorithms mainly included multi-objective linear programming, adaptive class example selection, and incremental learning. The interior-point method was used to solve linear programming problems, such that the time complexity was $O(n^{3.5}L^2)$. During the process of adaptive class exemplar selection, the normalized mutual information of each sample and the Euclidean distance between each sample and its class feature center had to be computed, and so its computational complexity was $O(n) + O(n^2)$. During the process of incremental learning, the deep model was fine-tuned. Therefore, we introduce the concept of floating point operations (FLOPs) to illustrate the computational complexity. The FLOPs of the model corresponding to the RML2016.04c data set were 1.28 M, while the FLOPs of the model corresponding to the ACARS data set were 2.83 M.

To further illustrate the computational complexity of the proposed method, we performed a comparative experiment considering the runtime of the algorithm, setting $T = 2$

on both the RML2016.04c data set and the ACARS data set. Tables 7 and 8 show the running times of the proposed method and other comparable methods for learning new classes. We can see that the running time of the proposed CES-MOLPC was much shorter than those of the other methods. The proposed method, CES-MOLPC, took only 1.64 min to complete learning of a new class on the RML2016.04c data set; however, that on the ACARS data set took a longer time, due to the high dimensionality of the data. Taken together, the other methods took more than three times as much time as the proposed method, CES-MOLPC.

From the experimental results shown in Tables 3 and 4, the proposed method showed good generalization performance for the public data set RML2016.04c, as well as for the signal data set ACARS recorded in a real environment. The Total OA corresponding to the proposed method surpassed that of other comparison methods; however, the New OA and Old OA values were different. In fact, as the distillation loss was used to fine-tune the model, it was difficult for New OA and Old OA to reach a relatively balanced condition simultaneously. We privately performed a more extreme experiment, in order to adjust the ratio of cross-entropy loss and distillation loss. When the cross-entropy loss was much larger than the distillation loss, the resulting New OA was higher and the Old OA was lower. Meanwhile, if the distillation loss was much greater than the cross-entropy loss, then New OA would be higher and Old OA would be lower. In the WA method, it was assumed that the L2-norms of the weight vectors of the classifiers corresponding to the new and old classes were different, such that the weight alignment method was used to balance the new and old classes, for which distillation loss was also used. However, in the experimental results, the Old OA corresponding to the WA method was almost 0 in any case and, so, this method is not suitable for the problem considered in this paper. Therefore, the advantages of the proposed method can be seen from the experimental results. In incremental learning, both the Old OA and the New OA must be considered. Taken together, the proposed method outperformed the comparison methods. The common feature of the BIC, ICARL, and WA methods was that they all used the Prototype exemplar selection method. From Tables 5 and 6, it can be seen that the adaptive sample selection method based on normalized mutual information proposed in this paper had better performance than other exemplar selection methods. This indicates that the proposed exemplar selection method can select more valuable samples from the old categories. The results shown in Figures 3 and 4 demonstrate that the proposed method outperformed the other methods, regardless of how many new classes were learned.

Table 7. Algorithm running time comparison on RML2016.04c data set ($T = 2$).

Method	ICARL	BIC	WA	CES-MOLPC
Time (min)	5.67	6.76	5.86	1.64

Table 8. Algorithm running time comparison on ACARS data set ($T = 2$).

Method	ICARL	BIC	WA	CES-MOLPC
Time (min)	56.88	59.29	57.94	16.73

In all of the experiments conducted in this paper, the total exemplar size $\mathcal{K}$ of the old classes was set to a fixed value of 200, which is relatively small, compared to the total number of training samples. This allows the proposed algorithm to guarantee its robustness under small sample conditions. However, in a real-world ESC application, when the number of classes for which the model has learned exceeds 200, the number of samples stored in each previous class is less than 1. At this point, the value of $\mathcal{K}$ must be increased, in order to make the algorithm suitable for the learning of new classes. Obviously, the larger the value of $\mathcal{K}$, the more training samples are retained from each old class, and the classification accuracy of the model after incremental learning will be higher. When the value of $\mathcal{K}$ exceeds the sum of all training samples of all old classes, the trained model reaches the upper bound of performance. Assuming that only one sample is reserved for each category as the exemplar of the category, it may be difficult for this sample to represent the feature distribution of the whole category. In this case, if there are too many training samples in the new class, there will be an extreme imbalance between the old and new classes. This will result in the New OA being significantly higher than Old OA; however, the distillation loss may offset the effects of the imbalance between old and new classes. A key problem researchers face in incremental learning tasks is catastrophic forgetting, and so the metric researchers should pay more attention to is Old OA. In practical applications, the signals that we are interested in may belong to a new class or an old class. Therefore, New OA and Old OA should be considered equally important. If the model can both recognize the new classes and store the data features of the old classes only by learning the new class without storing samples from the old classes, it will not be constrained by memory resources. Another feasible idea is to retain the key features of the previous categories, then replay them when learning new classes. This can save memory, compared to storing the training samples directly.

5. Conclusions

In this paper, the CES-MOLPC method was proposed in order to learn new categories from a continuous data stream. This approach is capable of handling multiple new categories of electromagnetic signals. The class exemplars of the old categories are selected based on the normalized mutual information, and we obtain the classifier weights of the new categories by solving the MOLP problem.

To demonstrate the effectiveness of our proposed ESC framework CES-MOLPC, we conducted several comparative experiments on the public RML2016.04c data set and the large-scale ACARS data set. The experimental results demonstrated that our approach performs better than other representative incremental learning approaches under limited memory conditions. The proposed CES-MOLPC was able to converge quickly and achieved the highest Total OA. To illustrate the power of proposed class exemplar selection method, we compared our method with different exemplar selection methods. Extensive experiments revealed that our approach has the highest accuracy.

As the number of various applications using electromagnetic signals keeps increasing and new signal categories keep appearing, the associated deep learning algorithms need to be able to handle new signal categories while maintaining their capabilities on old ones. In this work, a more realistic problem is considered, and the proposed method provides a new idea to address the problems pertaining to electromagnetic signal classification. At the same time, it also provides a solution to the problem of incremental learning. However, a limitation is that it is difficult to balance the performance of the model on new and old classes. In future research, the class exemplar selection method deserves further exploration. Very few key samples may be needed to reconstruct the distribution of the entire training data set.

Author Contributions: Formal analysis, J.X.; Funding acquisition, L.J. and X.Y.; Methodology, H.Z.; Supervision, J.B., L.J. and X.Y.; Writing—original draft, H.Z. and L.N.; Writing—review & editing, J.B., Z.X. and S.Z. All authors have read and agreed to the published version of the manuscript.

Funding: This work is supported by the National Natural Science Foundation of China (No. 61772401,61871398). This paper is also supported by the Science and Technology on Communication Information Security Control Laboratory.

Conflicts of Interest: The authors declare no conflict of interest.

References

1. Ahmed, A.; Zhang, Y.D.; Hassanien, A. Joint Radar–Communications Exploiting Optimized OFDM Waveforms. *Remote Sens.* **2021**, *13*, 4376. [CrossRef]
2. Huang, L.; Zhang, Y.; Pan, W.; Chen, J.; Qian, L.P.; Wu, Y. Visualizing deep learning-based radio modulation classifier. *IEEE Trans. Cognit. Commun. Netw.* **2020**, *7*, 47–58. [CrossRef]

3. Turhan-Sayan, G. Real time electromagnetic target classification using a novel feature extraction technique with PCA-based fusion. *IEEE Trans. Antennas Propag.* **2005**, *53*, 766–776. [CrossRef]
4. Zhuang, Y.; Li, X.; Ji, H.; Zhang, H.; Leung, V.C. Optimal Resource Allocation for RF-Powered Underlay Cognitive Radio Networks With Ambient Backscatter Communication. *IEEE Trans. Veh. Technol.* **2020**, *69*, 15216–15228. [CrossRef]
5. Cai, P.; Zhang, Y. Intelligent cognitive spectrum collaboration: Convergence of spectrum sensing, spectrum access, and coding technology. *Intell. Converg. Netw.* **2020**, *1*, 79–98. [CrossRef]
6. Zheng, S.; Chen, S.; Qi, P.; Zhou, H.; Yang, X. Spectrum sensing based on deep learning classification for cognitive radios. *China Commun.* **2020**, *17*, 138–148. [CrossRef]
7. Lin, Y.; Wang, C.; Wang, J.; Dou, Z. A novel dynamic spectrum access framework based on reinforcement learning for cognitive radio sensor networks. *Sensors* **2016**, *16*, 1675. [CrossRef]
8. Zhou, H.; Jiao, L.; Zheng, S.; Yang, L.; Shen, W.; Yang, X. Generative adversarial network-based electromagnetic signal classification: A semi-supervised learning framework. *China Commun.* **2020**, *17*, 157–169. [CrossRef]
9. Dudczyk, J.; Wnuk, M. The utilization of unintentional radiation for identification of the radiation sources. In Proceedings of the 34th European Microwave Conference, Amsterdam, The Netherlands, 12–14 October 2004; Volume 2, pp. 777–780.
10. Zhang, H.; Yu, L.; Chen, Y.; Wei, Y. Fast Complex-Valued CNN for Radar Jamming Signal Recognition. *Remote Sens.* **2021**, *13*, 2867. [CrossRef]
11. Huang, H.; Song, Y.; Yang, J.; Gui, G.; Adachi, F. Deep-learning-based millimeter-wave massive MIMO for hybrid precoding. *IEEE Trans. Veh. Technol.* **2019**, *68*, 3027–3032. [CrossRef]
12. Liu, Y.; Liu, Y.; Yang, C. Modulation recognition with graph convolutional network. *IEEE Wirel. Commun. Lett.* **2020**, *9*, 624–627. [CrossRef]
13. Li, L.; Huang, J.; Cheng, Q.; Meng, H.; Han, Z. Automatic Modulation Recognition: A Few-Shot Learning Method Based on the Capsule Network. *IEEE Wirel. Commun. Lett.* **2020**, *10*, 474–477. [CrossRef]
14. Wei, S.; Qu, Q.; Zeng, X.; Liang, J.; Shi, J.; Zhang, X. Self-Attention Bi-LSTM Networks for Radar Signal Modulation Recognition. *IEEE Trans. Microw. Theory Tech.* **2021**, *69*, 5160–5172. [CrossRef]
15. O'Shea, T.J.; Corgan, J.; Clancy, T.C. Convolutional radio modulation recognition networks. In Proceedings of the International Conference on Engineering Applications of Neural Networks, Aberdeen, UK, 2–5 September 2016; pp. 213–226.
16. He, K.; Zhang, X.; Ren, S.; Sun, J. Deep residual learning for image recognition. In Proceedings of the IEEE Conference on Computer Vision and Pattern Recognition, Las Vegas, NV, USA, 27–30 June 2016; pp. 770–778.
17. O'Shea, T.J.; Roy, T.; Clancy, T.C. Over-the-air deep learning based radio signal classification. *IEEE J. Sel. Top. Signal Process.* **2018**, *12*, 168–179. [CrossRef]
18. Peng, S.; Jiang, H.; Wang, H.; Alwageed, H.; Zhou, Y.; Sebdani, M.M.; Yao, Y.D. Modulation classification based on signal constellation diagrams and deep learning. *IEEE Trans. Neural Netw. Learn. Syst.* **2018**, *30*, 718–727. [CrossRef]
19. Zhang, Z.; Li, Y.; Gao, M. Few-Shot Learning of Signal Modulation Recognition based on Attention Relation Network. In Proceedings of the 2020 28th European Signal Processing Conference (EUSIPCO), Amsterdam, The Netherlands, 18–21 January 2021; pp. 1372–1376.
20. Liu, X.; Yang, D.; El Gamal, A. Deep neural network architectures for modulation classification. In Proceedings of the 2017 51st Asilomar Conference on Signals, Systems, and Computers, Pacific Grove, CA, USA, 29 October–1 November 2017; pp. 915–919.
21. Wang, Y.; Gui, G.; Ohtsuki, T.; Adachi, F. Multi-task learning for generalized automatic modulation classification under non-Gaussian noise with varying SNR conditions. *IEEE Trans. Wirel. Commun.* **2021**, *20*, 3587–3596. [CrossRef]
22. Zhang, X.; Seyfi, T.; Ju, S.; Ramjee, S.; El Gamal, A.; Eldar, Y.C. Deep learning for interference identification: Band, training SNR, and sample selection. In Proceedings of the 2019 IEEE 20th International Workshop on Signal Processing Advances in Wireless Communications (SPAWC), Cannes, France, 2–5 July 2019; pp. 1–5.
23. Chaudhry, A.; Dokania, P.K.; Ajanthan, T.; Torr, P.H. Riemannian walk for incremental learning: Understanding forgetting and intransigence. In Proceedings of the European Conference on Computer Vision (ECCV), Munich, Germany, 8–14 September 2018; pp. 532–547.
24. Lee, S.W.; Kim, J.H.; Jun, J.; Ha, J.W.; Zhang, B.T. Overcoming catastrophic forgetting by incremental moment matching. *arXiv* **2017**, arXiv:1703.08475.
25. Zenke, F.; Poole, B.; Ganguli, S. Continual learning through synaptic intelligence. In Proceedings of the International Conference on Machine Learning, Sydney, NSW, Australia, 6–11 August 2017; pp. 3987–3995.
26. Farajtabar, M.; Azizan, N.; Mott, A.; Li, A. Orthogonal gradient descent for continual learning. In Proceedings of the International Conference on Artificial Intelligence and Statistics, Sicily, Italy, 26–28 August 2020; pp. 3762–3773.
27. Hinton, G.; Vinyals, O.; Dean, J. Distilling the knowledge in a neural network. *arXiv* **2015**, arXiv:1503.02531.
28. Aljundi, R.; Babiloni, F.; Elhoseiny, M.; Rohrbach, M.; Tuytelaars, T. Memory aware synapses: Learning what (not) to forget. In Proceedings of the European Conference on Computer Vision (ECCV), Munich, Germany, 8–14 September 2018; pp. 139–154.
29. Jung, H.; Ju, J.; Jung, M.; Kim, J. Less-forgetting learning in deep neural networks. *arXiv* **2016**, arXiv:1607.00122.
30. Rebuffi, S.A.; Kolesnikov, A.; Sperl, G.; Lampert, C.H. icarl: Incremental classifier and representation learning. In Proceedings of the IEEE conference on Computer Vision and Pattern Recognition, Honolulu, HI, USA, 21–26 July 2017; pp. 2001–2010.
31. Wu, Y.; Chen, Y.; Wang, L.; Ye, Y.; Liu, Z.; Guo, Y.; Fu, Y. Large scale incremental learning. In Proceedings of the IEEE/CVF Conference on Computer Vision and Pattern Recognition, Long Beach, CA, USA, 15–20 June 2019; pp. 374–382.

32. Shin, H.; Lee, J.K.; Kim, J.; Kim, J. Continual learning with deep generative replay. *arXiv* **2017**, arXiv:1705.08690.
33. Wu, C.; Herranz, L.; Liu, X.; Wang, Y.; Van de Weijer, J.; Raducanu, B. Memory replay gans: Learning to generate images from new categories without forgetting. *arXiv* **2018**, arXiv:1809.02058.
34. Yu, L.; Twardowski, B.; Liu, X.; Herranz, L.; Wang, K.; Cheng, Y.; Jui, S.; Weijer, J.V.D. Semantic drift compensation for class-incremental learning. In Proceedings of the IEEE/CVF Conference on Computer Vision and Pattern Recognition, Seattle, WA, USA, 13–19 June 2020; pp. 6982–6991.
35. Mai, Z.; Li, R.; Kim, H.; Sanner, S. Supervised Contrastive Replay: Revisiting the Nearest Class Mean Classifier in Online Class-Incremental Continual Learning. In Proceedings of the IEEE/CVF Conference on Computer Vision and Pattern Recognition, Nashville, TN, USA, 19–25 June 2021; pp. 3589–3599.
36. Hou, S.; Pan, X.; Loy, C.C.; Wang, Z.; Lin, D. Learning a unified classifier incrementally via rebalancing. In Proceedings of the IEEE/CVF Conference on Computer Vision and Pattern Recognition, Long Beach, CA, USA, 15–20 June 2019; pp. 831–839.
37. Zhao, B.; Xiao, X.; Gan, G.; Zhang, B.; Xia, S.T. Maintaining discrimination and fairness in class incremental learning. In Proceedings of the IEEE/CVF Conference on Computer Vision and Pattern Recognition, Seattle, WA, USA, 14–19 June 2020; pp. 13208–13217.
38. Parisi, G.I.; Kemker, R.; Part, J.L.; Kanan, C.; Wermter, S. Continual lifelong learning with neural networks: A review. *Neural Netw.* **2019**, *113*, 54–71. [CrossRef] [PubMed]
39. Cui, Z.; Tang, C.; Cao, Z.; Dang, S. SAR unlabeled target recognition based on updating CNN with assistant decision. *IEEE Geosci. Remote Sens. Lett.* **2018**, *15*, 1585–1589. [CrossRef]
40. Rudd, E.M.; Jain, L.P.; Scheirer, W.J.; Boult, T.E. The extreme value machine. *IEEE Trans. Pattern Anal. Mach. Intell.* **2017**, *40*, 762–768. [CrossRef]
41. Zhou, Z.; Shin, J.; Zhang, L.; Gurudu, S.; Gotway, M.; Liang, J. Fine-tuning convolutional neural networks for biomedical image analysis: Actively and incrementally. In Proceedings of the IEEE Conference on Computer Vision and Pattern Recognition, Honolulu, HI, USA, 21–26 July 2017; pp. 7340–7351.
42. Shao, J.; Huang, F.; Yang, Q.; Luo, G. Robust prototype-based learning on data streams. *IEEE Trans. Knowl. Data Eng.* **2017**, *30*, 978–991. [CrossRef]
43. Shadloo, M.; Beigy, H.; Haghiri, S. Exploiting structural information of data in active learning. In Proceedings of the International Conference on Artificial Intelligence and Soft Computing, Zakopane, Poland, 1–5 June 2014; pp. 796–808.
44. Dang, S.; Cao, Z.; Cui, Z.; Pi, Y.; Liu, N. Open set incremental learning for automatic target recognition. *IEEE Trans. Geosci. Remote Sens.* **2019**, *57*, 4445–4456. [CrossRef]
45. Bai, J.; Yuan, A.; Xiao, Z.; Zhou, H.; Wang, D.; Jiang, H.; Jiao, L. Class incremental learning with few-shots based on linear programming for hyperspectral image classification. *IEEE Trans. Cybern.* **2020**, 1–12. [CrossRef]
46. Rybak, Ł.; Dudczyk, J. Variant of Data Particle Geometrical Divide for Imbalanced Data Sets Classification by the Example of Occupancy Detection. *Appl. Sci.* **2021**, *11*, 4970. [CrossRef]
47. Bang, J.; Kim, H.; Yoo, Y.; Ha, J.W.; Choi, J. Rainbow Memory: Continual Learning with a Memory of Diverse Samples. In Proceedings of the IEEE/CVF Conference on Computer Vision and Pattern Recognition, Nashville, TN, USA, 20–25 June 2021; pp. 8218–8227.
48. Rybak, Ł.; Dudczyk, J. A geometrical divide of data particle in gravitational classification of moons and circles data sets. *Entropy* **2020**, *22*, 1088. [CrossRef]

future internet

Article

A Hybrid Robust-Learning Architecture for Medical Image Segmentation with Noisy Labels

Jialin Shi *, Chenyi Guo and Ji Wu

The Department of Electronic Engineering, Tsinghua University, Beijing 100089, China; guochy@mail.tsinghua.edu.cn (C.G.); wuji_ee@mail.tsinghua.edu.cn (J.W.)
* Correspondence: shi-jl16@mails.tsinghua.edu.cn

Abstract: Deep-learning models require large amounts of accurately labeled data. However, for medical image segmentation, high-quality labels rely on expert experience, and less-experienced operators provide noisy labels. How one might mitigate the negative effects caused by noisy labels for 3D medical image segmentation has not been fully investigated. In this paper, our purpose is to propose a novel hybrid robust-learning architecture to combat noisy labels for 3D medical image segmentation. Our method consists of three components. First, we focus on the noisy annotations of slices and propose a slice-level label-quality awareness method, which automatically generates label-quality scores for slices in a set. Second, we propose a shape-awareness regularization loss based on distance transform maps to introduce prior shape information and provide extra performance gains. Third, based on a re-weighting strategy, we propose an end-to-end hybrid robust-learning architecture to weaken the negative effects caused by noisy labels. Extensive experiments are performed on two representative datasets (i.e., liver segmentation and multi-organ segmentation). Our hybrid noise-robust architecture has shown competitive performance, compared to other methods. Ablation studies also demonstrate the effectiveness of slice-level label-quality awareness and a shape-awareness regularization loss for combating noisy labels.

Keywords: deep learning; noisy labels; medical image segmentation; robust learning

Citation: Shi, J.; Guo, C.; Wu, J. A Hybrid Robust-Learning Architecture for Medical Image Segmentation with Noisy Labels. *Future Internet* **2022**, *14*, 41. https://doi.org/10.3390/fi14020041

Academic Editor: Daniel Gutiérrez Reina

Received: 30 December 2021
Accepted: 25 January 2022
Published: 26 January 2022

1. Introduction

Medical image segmentation is of importance for clinical diagnosis, pathology research, and the development of treatment plans. A typical situation in the medical image field is the usage of 3D whole-volume data (e.g., CT or MRI data), where the whole volume is a set consisting of tens or hundreds of slices for every patient. With the advent of deep learning, numerous methods based on deep neural networks (DNNs) have been developed and show promising performance for medical image segmentation [1–3]. The success relies on the existence of an abundance of correctly labeled data. However, this assumption often does not hold for real applications. The label quality heavily depends on the experience and preferences of operators, and less-experienced annotators provide relative noisy labels. Especially for 3D medical images, the data is annotated in 2D and slice-by-slice, resulting in heavy labeling work and inconsistent quality among slices [4]. Many works have proven that low-quality annotations will degrade the performance of models and lead to some misunderstandings for subsequent computer-aided diagnoses. In this study, we aim to develop noise-robust methods for 3D medical image segmentation with noisy labels.

Recent methods for learning from noisy labels can be roughly divided as follows. First, a label transition matrix based on pre-defined knowledge is proposed to correct corresponding loss functions [5–8]. The matrix is used to learn latent noise patterns between noisy and clean annotations. However, they regard categories with clear meanings at states, which is suitable for classification tasks but cannot be directly applied to segmentation tasks. Second, some methods resort to redesigning robust loss functions instead of the conventional loss functions, such as generalized cross-entropy loss [9] and symmetric cross-entropy loss [10]. However, the effectiveness has not been verified for medical image segmentation tasks [11]. Third, many methods are proposed based on the loss re-weighting strategy, which is a very popular way to assign different weights to different samples. They make the relatively clean samples contribute more, while the relatively noisy samples contribute less for the training. The core is designing reliable criteria to estimate and select samples. For example, Mirikharaji et al. [12] proposed deploying a meta-learning framework for automatically learning the weights of pixels based on gradient directions, which was an extension of the "learning to re-weight" [13] method, from classification tasks to segmentation tasks. However, this method requires extra, trusted subsets of samples for training. Zhu et al. [14] introduced a pick-and-learn method with label-quality evaluation and an over-fitting control module. The important weights were assigned for dealing with 2D noisy-labeled image segmentation. Zhang et al. [15] proposed a tri-network by extending co-teaching [16], where they maintained three networks simultaneously for sample selection. This method can be regarded as hard-weighting. However, both the pick-and-learn and tri-network methods have been developed for 2D medical images, and cannot be directly applied for resource-intensive 3D medical image segmentation. Presently, the related investigations for 3D biomedical segmentation with low-quality noisy labels have not aroused enough attention.

In this paper, we concentrate on the problem of noisy-labeled medical image segmentation, especially for 3D medical volumetric data with tens or hundreds of slices in every whole volume. Specially, we propose a novel hybrid robust-learning architecture to combat noisy labels. First, we propose a novel slice-level label-quality awareness module (LQAM), which is jointly trained with a conventional segmentation module to predict label quality scores for each slice. Inspired by the idea of set-to-set recognition, we regard the whole volume of patients as a set of slices. By incorporating a 3D image segmentation module and a 2D LQAM, our hybrid architecture could predict label quality scores of the whole-volume slices in one shot. Second, we employ shape-awareness regularization to encourage the introduction of prior shape information. To be concrete, we apply the Hausdorff loss, based on distance transform maps, to boost the segmentation performance. Third, we construct our hybrid architecture based on the inspiration of the re-weighting strategy, which could adjust the contributions of relatively clean and noisy slices to improve the noise tolerance of the trained method. Finally, we experiment with two representative, publicly available datasets, i.e., liver segmentation and abdominal multi-organ segmentation. The results have shown the competitive performance of our noise-robust hybrid architecture. The main contributions are as follows:

(1) Different from previous studies, we target the challenging problem of 3D medical image segmentation with noisy labels, especially the inconsistent noisy label qualities among different slices. To address this problem, we propose a novel end-to-end hybrid robust-learning architecture to combat noisy labels from the perspective of slice-level label-quality awareness;
(2) We propose a novel slice-level label-quality awareness method, which automatically generates quality scores for each slice in a set without knowing the prior noise distribution. With the help of re-weighting, our method can alleviate the negative effect of noisy labels. The design is particularly effective for 3D medical image segmentation by satisfying the constraints of noise tolerance and the capacity limitations of GPUs;
(3) We propose a shape-awareness regularization loss to introduce prior shape information to provide extra performance gains. In the presence of noisy labels, we regard it as an auxiliary loss instead of the main learning targets and, further, it benefits the model training together with slice-level label-quality awareness. To our knowledge, this is the first attempt to apply prior shape information for the problem of learning with noisy labels.

2. Related Works

We review some related works for developing noise-robust methods. The representative methods can be roughly divided as follows. A label transition matrix based on pre-defined knowledge is proposed to correct the corresponding loss function. The matrix is used to learn latent noise patterns between noisy and clean annotations [17]. The key is how to accurately estimate the label transition matrix. For example, Goldberger et al. added a linear noise layer at the end of a backbone convolutional neural network to implicitly estimate the matrix [18]. Patrini et al. estimated a transition matrix by forward and backward loss-correction, which depended on strong assumptions to stack the maximum of each class [5]. Hendrycks et al. proposed gold loss-correction via mean prediction, which relied on extra, trusted subsets of data to accurately estimate the corrected matrix [6]. Wang et al. adapted a meta-learning method to directly optimize a noise transition matrix with the help of a large noisy dataset and a small trusted subset [7]. Han et al. proposed the introduction of prior knowledge for estimating a noise transition matrix [19]. As for medical applications, Dgani et al. followed the line of a noise-adaptation layer to represent the transition matrix and checked its applicability for breast microcalcification classification [8]. However, the above methods regard categories with clear meanings at states, which is suitable for classification tasks but cannot be directly applied to segmentation tasks.

Some methods focus on re-designing a robust loss function to combat noisy labels, which means they design a new robust loss function as the alternative to the conventional loss function in the presence of noisy labels. For example, Zhang and Sabuncu et al. proposed generalized cross-entropy loss [9], which combined the advantages of categorical cross-entropy and mean absolute error [20]. Wang et al. proposed a popular symmetric cross-entropy loss, which was the combination of cross-entropy and reverse cross-entropy [10]. Menon et al. leveraged gradient clipping for designing new loss functions [21]. However, these robust loss functions are developed based on their specific constraints, and the effectiveness may be reduced when encountering relatively complex medical segmentation data [11].

Some regularization methods have also been proposed to reduce the overfitting effect based on explicit or implicit forms. The former applies an explicit form to modify the training loss, such as early-learning regularization [22] or trace regularization [23]. The latter means they have the similar effect of regularization without the explicit form [24–27]. However, this kind of method should be designed with specific characteristics or it should introduce sensitive model-dependent hyperparameters.

Some methods are proposed based on the idea of loss re-weighting. Here, they make the relatively clean samples contribute more, while the relatively noisy samples contribute less for the training. The core of this method is how to design the accurate criterion to estimate and select samples. For example, Jiang et al. trained a mentor network to guide a student network by assigning weights to samples [28]. Arazo et al. calculated sample weights by modeling losses per-sample with a mixture model [29]. Han et al. used multiple class prototypes to assign attention weights to data samples [30]. Lee et al. designed an additional network to decide whether a label was noisy or not and produced the weight of each sample to reduce the influence of noisy labels [31]. Ren et al. proposed re-weighting samples based on their gradient directions and a meta-learning framework [13]. Under this investigation, Shu et al. proposed an explicit mapping method for sample weighting based on a meta-weight net [32]. However, the mentioned methods are all used for classification tasks of a natural-image domain; they cannot be directly applied to segmentation tasks, especially for 3D medical image segmentation tasks.

In fact, the research on medical image segmentation tasks with noisy labels is in its infancy. For example, Mirikharaji et al. [12] proposed deploying a meta-learning framework for automatically learning the weights of pixels based on gradient directions, which was an extension of the "learning to re-weight" [13] method, from classification tasks to segmentation tasks. However, this method requires extra, trusted subsets of samples for training. Zhu et al. introduced a pick-and-learn method with label-quality evaluation and an over-fitting control module [14]. Then, important weights were assigned for dealing with 2D noise-labeled biomedical segmentation. Zhang et al. proposed a tri-network [15] by extending co-teaching [16], where they maintained three networks simultaneously for sample selection. This method can be regarded as hard-weighting. However, both the pick-and-learn and tri-network methods have been developed for 2D medical images, and they cannot be directly applied for resource-intensive 3D medical volumetric image segmentation. From the perspective of sample selection, Min et al. [33] proposed their two-stream networks, based on pixel-wise sample selection. They selected samples from the disagreement area, where two predictions coming from two networks were different [34]. However, this method selected pixel-wise samples and was not suitable for estimating the quality of every slice. Presently, the related investigations for 3D biomedical segmentation with low-quality noisy labels has not aroused enough attention. In this paper, we target the problem of noisy labeled 3D medical organ segmentation. In particular, we focus on the inconsistent noisy labeled slices among the whole volume. After obtaining the quality-awareness scores of slices, we further propose our hybrid noise-tolerance architecture with the help ofa re-weighting strategy, which is different from previous works.

3. Methods

We aim to develop a hybrid architecture to weaken the negative effect of noisy labels. This architecture could automatically estimate the label quality of slices without extra clean labels. As shown in Figure 1, the novel hybrid framework contains four major modules: (1) a 3D segmentation module, which generates whole-volume segmentation probability maps in one shot; (2) a 2D slice-level label-quality awareness module, which predicts label quality scores for each slice; (3) a shape-awareness regularization loss for introducing prior information; (4) a final re-weighting module, which assigns different weights for different slices to construct the final loss.

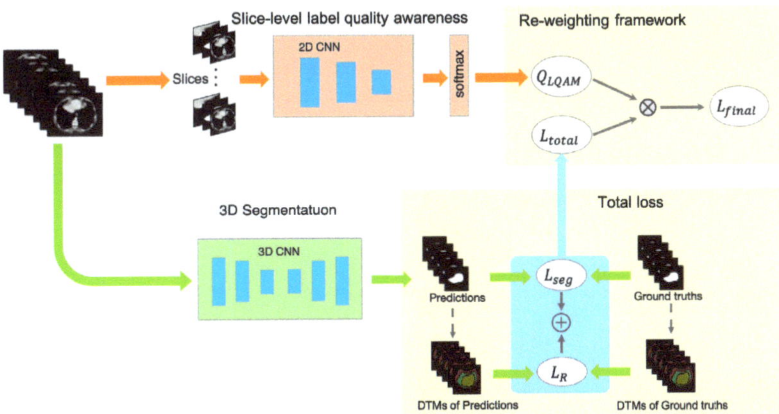

Figure 1. The architecture of the proposed hybrid noise-robust architecture. It includes a 3D segmentation module, a 2D label-quality awareness module, a shape-awareness regularization loss, and a loss re-weighting module. The input is whole-volume 3D medical images, which consist of multiple slices (e.g., CT and MRI data). For the 2D label-quality awareness module, the input is the concatenation of the image, and the foreground and background for every slice. The outputs of the LQAM are the score set of Q_{LQAM}. The shape-awareness regularization loss L_R is constructed with distance transform maps (DTMs). The final loss L_{final} is constructed with the idea of a re-weighting strategy. L_{seg} means the segmentation loss, and L_{total} is the weighted sum of L_{seg} and L_R. The ground truths have noisy information.

3.1. Segmentation Module

Considering the complementary relationship among slices of CT (or MRI) data, we prefer to choose the 3D neural network as the basic segmentation module to leverage the spatial relationship between slices, instead of the 2D segmentation model. Considering the promising performance of U-net, we select 3D U-net as the backbone [35], which has proven its effectiveness for 3D volumetric images. Further, we extend our backbone with two important modules. First, we adopt a residual skip connection for each convolutional block to aid the total training [36]. Second, we introduce 3D squeeze-and-excitation (SE) blocks for better feature representation. Inspired by the recent proposed 2D SE module for image classification [37], we introduce the 3D extension for our image segmentation.

3.2. Label-Quality Awareness Module

A label-quality awareness module (LQAM) is proposed to automatically learn the quality of labels among slices, thus reducing the impact of noisy labels with the re-weighting strategy. Considering that the slices that belong to the whole volume are annotated slice-by-slice and in 2D by less-experienced operators, the label quality of slices may be noisy and inconsistent. We regard the whole-volume data as the set of slices, and the quality scores of slices as another set.

In this section, we design our label-quality awareness module to estimate label quality of slices. The set $S = \{I_1, I_2, \ldots, I_N\}$ is used to denote the set of slices with different label qualities, which can be compared with each other. We use I_i to denote the information of the i-th slice and use N to represent the number of slices belonging to one volume. Let $Q(S)$ denote the set of quality scores belonging to the whole -volume, satisfying:

$$Q(S) = F(I_1, I_2, \ldots, I_N), \tag{1}$$

where $F(\cdot)$ is the quality-awareness function. The challenge is to find an optimized $F(\cdot)$, which aggregates features from the whole volumetric set to obtain different quality scores. As neural networks could approximate various functions, we obtain the quality scores via neural network learning [30,38]. The notion is that images with higher-quality labeled information are easier to recognize, while lower-quality labeled information are hard to fit. This is related to the memorization of neural networks, and has been widely used for the recognition of noisy labels in existing natural image classification tasks with noisy labels [39]. We also use this trick to estimate the label quality of the slice set. More specifically, the slices with noisy labels have relatively higher losses, while the slices with well-annotated labels have relatively smaller losses during the training process. Therefore, with the help of losses coming from the main segmentation network, the LQAM could assess the relative quality of every slice.

We construct our label-quality awareness module using deep neural networks, which are trained together with a main segmentation network. We focus on learning with noisy labels where the images are intact and the labels are noisy. As for the input information to the LQAM, we use the set $S = \{I_1, I_2, \ldots, I_N\}$. In order to deal with noisy labels, the concatenations of image, foreground, and background are used as the whole input information I_i of the i-th slice. We use these three components to contain enough information about noisy labels, which is used for subsequent label-quality awareness. Moreover, the LQAM module is jointly trained with the main segmentation. We only use this kind of concatenation as the input to keep the number of parameters constant, regardless of the number of organs. This construction is very helpful for satisfying the memory limitations of GPUs [40], and is appliable for resource-intensive volumetric segmentation. After obtaining the label quality scores of slices, the segmentation loss could be adjusted to combat noisy labels. With the main segmentation network and the label-quality awareness network, we construct our hybrid noise-robust architecture in Section 3.4.

3.3. Shape-Awareness Regularization Loss

We attempt to investigate the method for considering prior shape information to provide extra performance gains. Instead of using the complex techniques that require computationally expensive optimizations, we resort to the distance transform maps (DTMs), which have been investigated for introducing shape information for segmentation tasks [41,42]. In their studies, the segmentation labels are clean and provide accurate supervision information. However, in our study, the labels are noisy and the supervision signals are confusing. If directly applying the DTMs as the learning targets, as with the representative works [41,42], it will lead to unstable training or severe overfitting problems. There are no related studies to apply DTMs for the tasks of learning with noisy labels. In order to benefit from the shape information and exclude the negative effect of noisy labels, we propose shape-awareness regularization loss as the auxiliary loss, which is very different from previous studies.

Inspired by the studies of [41,42], our regularization loss L_R is formulated as:

$$L_R = \frac{1}{|\Omega|} \sum_{\Omega} (P - G)^2 \circ (G_{DTM}^2 + P_{DTM}^2), \tag{2}$$

where G denotes the original labels and contains noise. P means the outputs of predictions. G_{DTM} and P_{DTM} denote the DTMs of original labels G and predictions P. Let Ω denote the grid on which the slice is defined. Formally, the DTM of G is written as:

$$G_{DTM} = \begin{cases} \inf_{y \in \partial G} ||v - y||_2, & v \in G_{in}; \\ 0, & others \end{cases}, \tag{3}$$

where $||v - y||_2$ is the Euclidian distance between voxels v and y, and G_{in} denotes the inside of the object. G_{DTM} computes the distance transformation of the foreground and the presentation for P_{DTM} is similar. We further normalize the G_{DTM} to be in the range $(-1, 1)$ by dividing the maximum value. As an implicit shape representation, DTMs embed contours in a higher dimensional space. With the aid of slice-level label-quality awareness, the regularization loss is useful for boosting the final segmentation performance.

3.4. The Final Framework

We apply the idea of re-weighting to develop the final framework, which makes relatively clean samples contribute more to the training and compresses the influence of relatively noisy samples. We construct our total loss consisting of two parts. The first one L_{seg} is the cross-entropy loss, which is calculated with prediction maps and the original labels. The second one L_R is defined as the shape-awareness regularization loss. The final re-weighted loss L_{final} is:

$$L_{final} = Q_{LQAM}(L_{seg} + \lambda L_R), \tag{4}$$

where Q_{LQAM} are the quality weights originating from the LQAM. λ is the hyper-parameter. Our method is independent of an auxiliary clean dataset or prior information about noise distribution, which is more applicable for 3D biomedical segmentation tasks.

4. Experiments and Results

4.1. Data and Implementation Details

We experiment with the publicly available medical dataset "Chaos" to justify the performance of our hybrid robust framework. The Chaos challenges have been held in The IEEE international Symposium on Biomedical Imaging (ISBI) 2019 and contain CT and MRI datasets. More details about the original data descriptions and data acquisition can be found in [43]. We choose CT data for liver segmentation and MRI data for abdominal multi-organ segmentation (i.e., liver, right kidney, left kidney and spleen). Each volume in these two datasets corresponds to a series of DICOM images belonging to a single patient. The CT dataset consists of 20 different patients and is divided into training (12 patients) and testing (8 patients) in experiments. For the MRI dataset, we choose T2-SPIR (20 patients). We divide the 12 cases for training and the remaining 8 cases for testing. We resize the CT images as $64 \times 196 \times 196$ and resize the MRI images as $32 \times 196 \times 196$. We clip all values larger than 2000 as 2000, and then pre-process the data by normalization.

As both the CT dataset and the MRI dataset have clean labels, we first generate noisy labeled images under different noise settings. Following the previous works of [14,15], we set the number of slices selected for noise generation as the noise rate, and set the morphological changes within pixels for every selected slice as the noise level. For example, we set the noise rate to 50% and noise level to 1–8 pixels, which means we randomly select 50% of the slices from the whole volume and further erode or dilate every selected slice with 1–8 pixels.

We utilize PyTorch for realizing our experiments. The learning rate is set empirically to 0.001. The Adam optimizer is used for optimization and the Betas of the optimizer are set to 0.9 and 0.999. We set the maximum epoch as 10,000 in the experiments. We repeat the experiments three times and report the average for the final performance. During training, we select 3D SE ResUnet as the backbone with four residual and SE blocks. The backbone consists of a contracting path and an expanding path. We prefer the contracting path as the LQAM module. During training, the batch size for the 3D segmentation model is set to 1, which means we obtain all predictions belonging to a patient in one shot. We use the widely used metric of dice score to measure the segmentation performance. λ is set to 0.01 for the experiments.

As for the network architecture, we use 3D Unet as the basic backbone, with residual blocks and SE blocks. We expand the 2D squeeze, excitation, scale, and convolutional functions to obtain the 3D SE counterparts. The encoder consists of four residual modules with 16, 32, 32, and 64 output channels, followed by a down-sampling layer. The decoder shares the symmetric structure, but with up-sampling layers. For each residual module, we use two $3 \times 3 \times 3$ convolution blocks, two SE modules, and corresponding up-sampling or down-sampling layers. The down-sampling layer is achieved via 3D max-pooling with stride 2 and the up-sampling is achieved with the strategy of nearest neighbor. Each convolution block consists of convolutional layers, an ELU activation function, and a 3D batch-normalization layer. We prefer the contracting path as the LQAM module.

4.2. Comparisons on Liver Segmentation Dataset

We conduct experiments on a liver segmentation benchmark to demonstrate the effectiveness of our hybrid robust architecture. The dataset has 3D CT images. We train our network on the noisy-labeled data under different noise settings and test the network with clean-labeled data. Specifically, we set two noise levels (noise level 1 of 1–8 pixels and noise level 2 of 5–18 pixels) and three noise rates (noise rate = 25%, 50%, and 75%). Following the generation process of noisy labels [14,15], we obtain the data with simulated noisy labels.

The comparing baselines include: (1) Plain [35]: the conventional training based on a 3D segmentation model with noisy labels. We use 3D U-net with the additions of residual blocks and 3D SE blocks as the backbone. (2) Pick-and-learn [14]: a representative method for addressing 2D medical image segmentation with noisy labels. We regard every slice as a 2D datum and further adopt the pick-and-learn strategy for the corresponding robust learning. (3) Disagreement [33]: a noise-robust method from the perspective of sample selection, which selects pixels from the disagreement area to obtain informative samples. (4) Area-aware method proposed by [44]: the area aware-factor is calculated as the ratio between the area of the foreground and the area of the background, and is then introduced by a simple multiplication strategy. (5) INT [45]: this method aims to distill effective supervision information from image-level data to develop a noise-tolerance algorithm.

The experimental results on the liver segmentation dataset with simulated noisy labels are illustrated in Table 1. On the clean-annotated dataset, we have the performance upper as the dice value of 88.55. However, the liver segmentation performance for the Plain baseline decreases sharply when adding different noise rates (from no noise to 25%, 50%, and 75%) and noise levels (from noise level 1 of 1–8 pixels to noise level 2 of 5–18 pixels). Comparing with other baselines, we could observe that our proposed hybrid architecture has a consistent performance improvement. Even in the hardest case, with noise level 2 and a noise rate of 75%, our proposed method shows its robustness to noisy labels.

Table 1. Results on the CT dataset for liver segmentation under different noise settings. We report the mean value (±std) of the dice score (%) with 3 runs.

Method	Noise Level 1			Noise Level 2		
	25%	50%	75%	25%	50%	75%
Plain [35]	70.94 ± 0.96	68.97 ± 0.43	65.24 ± 1.32	59.61 ± 0.41	56.82 ± 0.19	48.07 ± 2.38
Pick-and-learn [14]	65.67 ± 0.44	59.36 ± 0.73	54.10 ± 0.86	50.91 ± 0.59	47.31 ± 0.26	40.32 ± 0.45
Disagreement [33]	71.43 ± 1.10	69.88 ± 0.24	67.58 ± 0.28	66.64 ± 0.07	55.72 ± 0.19	47.23 ± 0.25
INT [45]	77.34 ± 0.14	75.33 ± 0.08	71.67 ± 0.06	70.38 ± 0.81	60.80 ± 0.64	53.56 ± 0.67
Area-aware [44]	76.62 ± 1.75	74.92 ± 1.27	69.45 ± 1.96	70.63 ± 1.32	60.44 ± 1.98	54.82 ± 1.56
Ours	78.31 ± 0.46	76.29 ± 0.63	72.78 ± 0.60	71.72 ± 0.18	64.05 ± 0.30	56.99 ± 0.66

Comparing with pick-and-learn, our method achieves performance gains, which demonstrates that adopting a 3D network as a backbone is more appropriate because of the implicit complementary information among slices. Comparing with the disagreement method, our superior performance shows that our hybrid architecture could mine more effective information, while disagreement only selects samples from a disagreement area, resulting few samples during training. For image-level noise-robust learning, we compare an INT (image-level noise-tolerance) method with our method. The results show the improvement, which verifies the effectiveness of our method from the perspective of slice-level noise tolerance. We further observe the results of our method and the area-aware strategy, which also verifies that our slice-level label-quality awareness and regularization loss show better performance advantages.

4.3. Comparisons on Multi-Organ Segmentation Dataset

We demonstrate the experiments on an abdomen multi-organ MRI-T2SPIR dataset, and the corresponding organs include the liver, right kidney, left kidney, and spleen. Table 2 depicts the comparative performance of multiple methods with noise ratios of 25%, 50%, and 75% and the noise level of 1–5 pixels. The noisy labels are generated in the same way as the liver segmentation. We report the average dice value (%) of these four organs. The upper bound with the average dice value of 74.48 is acquired by training the 3D segmentation model on a clean-labeled dataset. Other results are obtained by training the model on a dataset with corrupted labels. As illustrated, our proposed hybrid architecture outperforms other methods, which demonstrates the effectiveness of our hybrid noise-robust architecture.

Table 2. Experimental results (dice %) for multi-organ segmentation on abdomen MRI-T2SPIR. We report the performances of the liver, right kidney, left kidney, and spleen, as well as the average values.

Noise Rates	Method	Liver	Right Kidney	Left Kidney	Spleen	Average
No noise	Plain [35]	84.20	75.13	64.93	73.66	74.48
25%	Plain [35]	79.36	55.09	42.88	52.51	57.46
	Pick-and-learn [14]	80.31	45.79	38.33	38.44	50.72
	Disagreement [33]	73.46	49.79	44.66	52.09	55.00
	INT [45]	78.93	56.63	45.66	59.14	60.09
	Area-aware [44]	75.05	52.83	54.17	51.11	58.29
	Ours	78.00	60.72	49.62	60.95	62.32
50%	Plain [35]	76.86	52.43	42.75	54.40	56.61
	Pick-and-learn [14]	70.36	48.87	41.26	48.55	52.26
	Disagreement [33]	71.37	49.87	41.26	49.55	53.01
	INT [45]	79.10	55.10	46.97	56.29	59.37
	Area-aware [44]	75.47	51.80	46.66	54.19	57.03
	Ours	80.27	54.07	47.54	60.49	61.60
75%	Plain [35]	75.18	56.29	41.75	52.10	56.33
	Pick-and-learn [14]	70.79	48.13	36.83	44.22	49.99
	Disagreement [33]	72.99	48.49	40.41	46.42	52.08
	INT [45]	76.66	50.69	47.25	57.63	58.06
	Area-aware [44]	76.47	48.99	45.12	57.20	56.94
	Ours	78.62	52.08	51.84	59.62	60.54

4.4. Ablation Study and Visualization

We also show the ablation study in Figure 2. We use "*Plain*" to denote the basic baseline, where the model is trained without any noise-robust strategy. We use "$LQAM$" to represent our hybrid architecture with only slice-level label-quality awareness. We use "$LQAM + L_R$" to denote our proposed hybrid architecture with $LQAM$ and regularization loss L_R. The experiments are conducted on a liver segmentation dataset with different noise rates and noise levels. Comparing $LQAM$ and *Plain*, we can observe that adding a label-quality awareness module is necessary for developing noise-robust methods. The comparisons between $LQAM$ and $LQAM + L_R$ have shown the effectiveness of introducing shape-awareness regularization loss. This may be because the shape-awareness loss will benefit from learning some boundary information. Together with the re-weighted framework, the regularization loss helps to further improve the contribution of relatively clean samples.

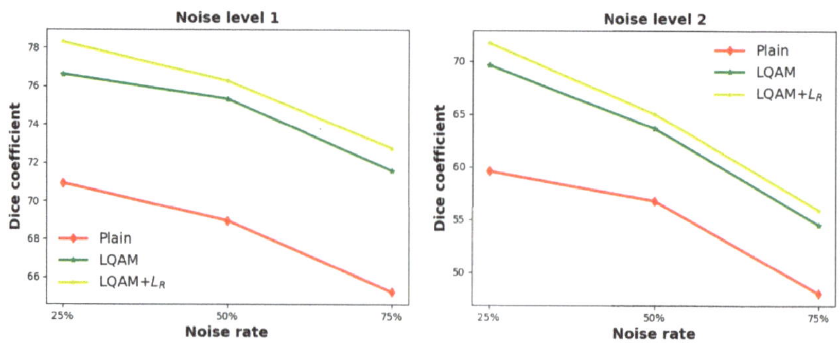

Figure 2. The results of ablation study. The experiments are conducted on the liver segmentation dataset under different noise settings.

We visualize some slices to show the effectiveness of our hybrid noise-robust architecture in Figure 3. For the liver segmentation dataset, we randomly select slices with noise levels of 5–18 pixels and a noise rate of 25%. For multi-organ segmentation, we show the slices with a noise level of 1–5 pixels and a noise rate of 25%.

We also show some bad examples in Figure 4, where the slices are selected from multi-organ segmentation on the abdomen MRI-T2SPIR dataset. Note that the 3D whole-volume data is a set consisting of tens or hundreds of slices for every patient. The size variances of organs among different slices are very obvious, and sometimes the organs present a relatively small size on some slices. The bad case, on the left, is the example. The organ size is relatively small in this slice, and the performance is poor. We should investigate the segmentation of small organs to further improve the performance of learning with noisy labels, which is the future work. When we observe the bad case on the right, the results show that there are some overlaps among different organs based on our method. The main reason is that the boundaries of organs in medical images are usually unclear, making it more difficult to distinguish them in the presence of noisy labels. A possible solution is to introduce prior information about the relationship among organs, which has the potential to improve the segmentation performance.

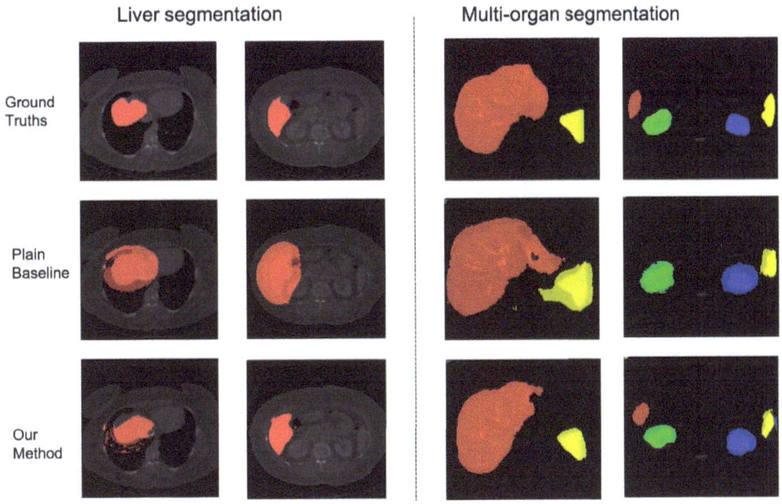

Figure 3. Some visualization results. The left panel is the selected slices of liver segmentation and the right panel shows some slices of multi-organ segmentation. Three rows denote the ground truths, the plain baseline, and our method, respectively.

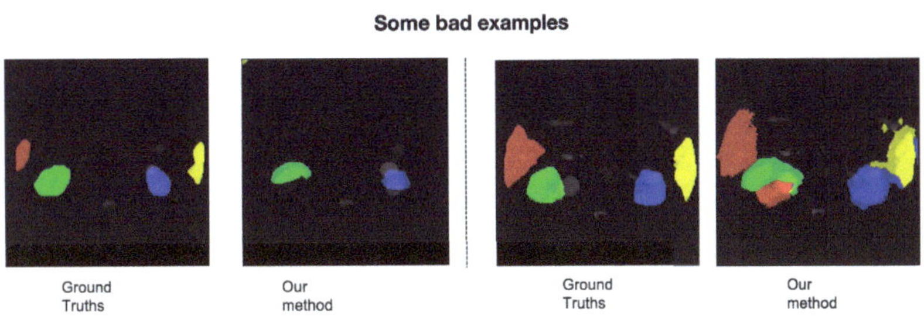

Figure 4. Some visualization results for demonstrating bad examples. The selected slices are from multi-organ segmentation on the abdomen MRI-T2SPIR dataset.

5. Conclusions

We concentrate on the problem of 3D medical image segmentation with noisy labels. We present a novel noise-robust hybrid architecture to combat noisy labels. Specifically, we propose a slice-level label-quality awareness module, which is jointly trained with a conventional segmentation module to predict label quality scores for each slice. We further introduce regularization loss by distance transform maps to boost the segmentation performance. With the calculated quality scores, we apply a re-weighting strategy on the total loss to distill effective supervision information from relatively clean samples. Our architecture can be trained without knowing the prior noise distribution or the availability of an extra trusted subset. Experimental results on the publicly available medical datasets (CT for liver segmentation and MRI-T2SPIR for abdomen multi-organ segmentation) demonstrate the effectiveness of our hybrid noise-robust architecture. In the future, we will investigate the relationships of multiple organs and hard sample mining for learning with noisy labels.

Author Contributions: Formal analysis, J.S.; Funding acquisition, J.W.; Investigation, J.S.; Methodology, J.S.; Project administration, J.W.; Supervision, C.G. and J.W.; Validation, J.S.; Writing—original draft, J.S.; Writing—review & editing, J.S. and C.G. All authors have read and agreed to the published version of the manuscript.

Funding: This research was funded by the Beijing Municipal Natural Science Foundation (No. L192026) and Tsinghua–Foshan Innovation Special Fund (TFISF) (No. 2020THFS0111).

Institutional Review Board Statement: Not applicable.

Informed Consent Statement: Not applicable.

Data Availability Statement: Not applicable.

Conflicts of Interest: The authors declare no conflict of interest.

References

1. Tang, H.; Liu, X.; Sun, S.; Yan, X.; Xie, X. Recurrent mask refinement for few-shot medical image segmentation. *arXiv* **2021**, arXiv:2108.00622.
2. Liu, L.; Cheng, J.; Quan, Q.; Wu, F.-X.; Wang, Y.-Y.; Wang, J. A survey on U-shaped networks in medical image segmentations. *Neurocomputing* **2020**, *409*, 244–258. [CrossRef]
3. Gao, Y.; Zhou, M.; Metaxas, D.N. UTNet: A hybrid transformer architecture for medical image segmentation. In Proceedings of the International Conference on Medical Image Computing and Computer-Assisted Intervention (MICCAI), Strasbourg, France, 27 September–1 October 2021.
4. Xue, Y.; Tang, H.; Qiao, Z.; Gong, G.; Yin, Y.; Qian, Z.; Huang, C.; Fan, W.; Huang, X. Shape-aware organ segmentation by predicting signed distance maps. In Proceedings of the 34th AAAI Conference on Artificial Intelligence, New York, NY, USA, 7–12 February 2020; Volume 34, pp. 12565–12572.
5. Patrini, G.; Rozza, A.; Krishna Menon, A.; Nock, R.; Qu, L. Making Deep Neural Networks Robust to Label Noise: A Loss Correction Approach. *arXiv* **2017**, arXiv:1609.03683.
6. Hendrycks, D.; Mazeika, M.; Wilson, D.; Gimpel, K. Using trusted data to train deep networks on labels corrupted by severe noise. *arXiv* **2018**, arXiv:1802.05300.
7. Wang, Z.; Hu, G.; Hu, Q. Training noise-robust deep neural networks via meta-learning. In Proceedings of the IEEE/CVF Conference on Computer Vision and Pattern Recognition, Seattle, WA, USA, 14–19 June 2020.
8. Dgani, Y.; Greenspan, H.; Goldberger, J. Training a neural network based on unreliable human annotation of medical images. In Proceedings of the 2018 IEEE 15th International Symposium on Biomedical Imaging (ISBI), Washington, DC, USA, 4–7 April 2018, pp. 39–42.
9. Zhang, Z.; Sabuncu, M.R. Generalized cross entropy loss for training deep neural networks with noisy labels. In Proceedings of the 32nd Conference on Neural Information Processing Systems (NeurIPS), Montréal, QC, Canada, 3–8 December 2018.
10. Wang, Y.; Ma, X.; Chen, Z.; Luo, Y.; Yi, J.; Bailey, J. Symmetric cross entropy for robust learning with noisy labels. In Proceedings of the IEEE/CVF International Conference on Computer Vision, Seoul, Korea, 27 October–2 November 2019. pp. 322–330.
11. Karimi, D.; Dou, H.; Warfield, S.K.; Gholipour, A. Deep learning with noisy labels: Exploring techniques and remedies in medical image analysis. *Med. Image Anal.* **2020**, *65*, 101759. [CrossRef] [PubMed]
12. Mirikharaji, Z.; Yan, Y.; Hamarneh, G. Learning to segment skin lesions from noisy annotations. *arXiv* **2019**, arXiv:1906.03815.
13. Ren, M.; Zeng, W.; Yang, B.; Urtasun, R. Learning to reweight examples for robust deep learning. In Proceedings of the International Conference on Machine Learning, Stockholm, Sweden, 10–15 July 2018; pp. 4334–4343.

14. Zhu, H.; Shi, J.; Wu, J. Pick-and-learn: Automatic quality evaluation for noisy-labeled image segmentation. *arXiv* **2019**, arXiv:1907.11835v1.
15. Zhang, T.; Yu, L.; Hu, N.; Lv, S.; Gu, S. Robust medical image segmentation from non-expert annotations with tri-network. In Proceedings of the Medical Image Computing and Computer Assisted Intervention—MICCAI 2020, 23rd International Conference, Lima, Peru, 4–8 October 2020; pp. 249–258.
16. Han, B.; Yao, Q.; Yu, X.; Niu, G.; Xu, M.; Hu, W.; Tsang, I.; Sugiyama, M. Co-teaching: Robust training of deep neural networks with extremely noisy label. In Proceedings of the Conference on Neural Information Processing Systems (NeurIPS), Montréal, QC, Canada, 3–8 December 2018,
17. Song, H.; Kim, M.; Park, D.; Shin, Y.; Lee, J.-G. Learning from noisy labels with deep neural networks: A survey. *arXiv* **2020**, arXiv:2007.08199.
18. Goldberger, J.; Ben-Reuven, E. Training deep neural-networks using a noise adaptation layer. In Proceedings of the 4th International Conference on Learning Representations (ICLR 2016), San Juan, Puerto Rico, 2–4 May 2016.
19. Han, B.; Yao, J.; Niu, G.; Zhou, M.; Tsang, I.; Zhang, Y.; Sugiyama, M. Masking: A new perspective of noisy supervision. In Proceedings of the 32nd Conference on Neural Information Processing Systems (NeurIPS), Montréal, QC, Canada, 3-8 December 2018.
20. Ghosh, A.; Kumar, H.; Sastry, P. Robust loss functions under label noise for deep neural networks. In Proceedings of the 31st AAAI Conference on Artificial Intelligence, San Francisco, CA, USA, 4–9 February 2017.
21. Menon, A.K.; Rawat, A.S.; Reddi, S.J.; Kumar, S. Can gradient clipping mitigate label noise? In Proceedings of the International Conference on Learning Representations, New Orleans, LA, USA, 6–9 May 2019.
22. Liu, S.; Niles-Weed, J.; Razavian, N.; Fernandez-Granda, C. Early-learning regularization prevents memorization of noisy labels. In Proceedings of the Advances in Neural Information Processing Systems 33: Annual Conference on Neural Information Processing Systems (NeurIPS 2020), Virtual, 6–12 December 2020.
23. Tanno, R.; Saeedi, A.; Sankaranarayanan, S.; Alexander, D.C.; Silberman, N. Learning from noisy labels by regularized estimation of annotator confusion. In Proceedings of the IEEE/CVF Conference on Computer Vision and Pattern Recognition, Long Beach, CA, USA, 16–20 June 2019.
24. Jindal, I.; Nokleby, M.; Chen, X. Learning deep networks from noisy labels with dropout regularization. In Proceedings of the 2016 IEEE 16th International Conference on Data Mining (ICDM), Barcelona, Spain, 12–15 December 2016.
25. Zhang, H.; Cisse, M.; Dauphin, Y.N.; Lopez-Paz, D. Mixup: Beyond empirical risk minimization. In Proceedings of the 5th International Conference on Learning Representations (ICLR), Toulon, France, 24–26 April 2017.
26. Nishi, K.; Ding, Y.; Rich, A.; Höllerer, T. Augmentation strategies for learning with noisy labels. In Proceedings of the IEEE/CVF Conference on Computer Vision and Pattern Recognition, Virtual, 19–25 June 2021.
27. Zheltonozhskii, E.; Baskin, C.; Mendelson, A.; Bronstein, A.M.; Litany, O. Contrast to divide: Self-supervised pre-training for learning with noisy labels. In Proceedings of the IEEE/CVF Winter Conference on Applications of Computer Vision, Waikoloa, HI, USA, 4–8 January 2022.
28. Jiang, L.; Zhou, Z.; Leung, T.; Li, L.-J.; Li, F.-F. Mentornet: Learning data-driven curriculum for very deep neural networks on corrupted labels. In Proceedings of the International Conference on Machine Learning, Stockholm, Sweden, 10–15 July 2018;
29. Arazo, E.; Ortego, D.; Albert, P.; O'Connor, N.E.; McGuinness, K. Unsupervised label noise modeling and loss correction. In Proceedings of the 36th International Conference on Machine Learning, Long Beach, CA, USA, 10-15 June 2019.
30. Han, J.; Luo, P.; Wang, X. Deep self-learning from noisy labels. In Proceedings of the IEEE/CVF International Conference on Computer Vision, Seoul, Korea, 27 October–2 November 2019.
31. Lee, K.-H.; He, X.; Zhang, L.; Yang, L. Cleannet: Transfer learning for scalable image classifier training with label noise. In Proceedings of the IEEE Conference on Computer Vision and Pattern Recognition, Salt Lake City, UT, USA, 18-23 June 2018.
32. Shu, J.; Xie, Q.; Yi, L.; Zhao, Q.; Zhou, S.; Xu, Z.; Meng, D. Meta-weight-net: Learning an explicit mapping for sample weighting. In Proceedings of the Conference on Neural Information Processing Systems (NeurIPS), Vancouver, BC, Canada, 8–14 December 2019.
33. Min, S.; Chen, X.; Zha, Z.J.; Wu, F.; Zhang, Y. A two-stream mutual attention network for semi-supervised biomedical segmentation with noisy labels. In Proceedings of the AAAI Conference on Artificial Intelligence, Honolulu, HI, USA, 27 January–1 February 2019; Volume 33, pp. 4578–4585.
34. Malach, E.; Shalev-Shwartz, S. Decoupling "when to update" from "how to update". In Proceedings of the 31st Conference on Neural Information Processing Systems (NeurIPS), Long Beach, CA, USA, 4–9 December 2017.
35. Çiçek, Ö.; Abdulkadir, A.; Lienkamp, S.S.; Brox, T.; Ronneberger, O. 3D U-Net: learning dense volumetric segmentation from sparse annotation. In Proceedings of the 19th International Conference on Medical Image Computing and Computer-Assisted Intervention, Athens, Greece, 17–21 October 2016; pp. 424–432.
36. He K.; Zhang, X.; Ren, S.; Sun, J. Deep residual learning for image recognition. In Proceedings of the IEEE Conference on Computer Vision and Pattern Recognition, Las Vegas, NV, USA, 27–30 June 2016; pp. 770–778.
37. Hu, J.; Shen, L.; Sun, G. Squeeze-and-excitation networks. In Proceedings of the IEEE Conference on Computer Vision and Pattern Recognition, Salt Lake City, UT, USA, 18–23 June 2018; pp. 7132–7141.
38. Xu, Y.; Zhu, L.; Jiang, L.; Yang, Y. Faster meta update strategy for noise-robust deep learning. In Proceedings of the IEEE/CVF Conference on Computer Vision and Pattern Recognition, Nashville, TN, USA, 19–25 June 2021; pp. 144–153.

39. Han, B.; Yao, Q.; Liu, T.; Niu, G.; Tsang, I.W.; Kwok, J.T.; Sugiyama, M. A survey of label-noise representation learning: Past, present and future. *arXiv* **2020**, arXiv:2011.04406.
40. Zheng, H.; Zhang, Y.; Yang, L.; Liang, P.; Zhao, Z.; Wang, C.; Chen, D.Z. A new ensemble learning framework for 3D biomedical image segmentation. In Proceedings of the AAAI Conference on Artificial Intelligence, Honolulu, HI, USA, 27 January–1 February 2019; pp. 5909–5916.
41. Ma, J.; Wei, Z.; Zhang, Y.; Wang, Y.; Lv, R.; Zhu, C.; Gaoxiang, C.; Liu, J.; Peng, C.; Wang, L.; Wang, Y.; Chen, J. How distance transform maps boost segmentation CNNs: An empirical study. In Proceedings of the Medical Imaging with Deep Learning, Montreal, QC, Canada, 6–9 July 2020; pp. 479–492.
42. Karimi, D.; Salcudean, S.E. Reducing the hausdorff distance in medical image segmentation with convolutional neural networks. *IEEE Trans. Med. Imaging* **2019**, *39*, 499–513. [CrossRef] [PubMed]
43. Chaos Challenge. Available online: https://doi.org/10.5281/zenodo.3431873 (accessed on 10 January 2022).
44. Shi, J.; Ding, X.; Liu, X.; Li, Y.; Liang, W.; Wu, J. Automatic clinical target volume delineation for cervical cancer in CT images using deep learning. *Med. Phys.* **2021**, *48*, 3968–3981 [CrossRef] [PubMed]
45. Shi, J.; Wu, J. Distilling effective supervision for robust medical image segmentation with noisy labels. In Proceedings of the International Conference on Medical Image Computing and Computer-Assisted Intervention, Strasbourg, France, 27 September–1 October 2021.

remote sensing

Article

HyperLiteNet: Extremely Lightweight Non-Deep Parallel Network for Hyperspectral Image Classification

Jianing Wang [1,*,†], Runhu Huang [1,†], Siying Guo [1], Linhao Li [1], Zhao Pei [2] and Bo Liu [1]

1 Key Laboratory of Intelligent Perception and Image Understanding of Ministry of Education of China, School of Artificial Intelligence, Xidian University, No. 2 South TaiBai Road, Xi'an 710071, China; 20171213745@stu.xidian.edu.cn (R.H.); 20171213779@stu.xidian.edu.cn (S.G.); lhli_7@stu.xidian.edu.cn (L.L.); liub@xidian.edu.cn (B.L.)
2 School of Computer Science, Shaanxi Normal University, Xi'an 710062, China; zpei@snnu.edu.cn
* Correspondence: jnwang@xidian.edu.cn
† These authors contributed equally to this work.

Abstract: Deep learning (DL) is widely applied in the field of hyperspectral image (HSI) classification and has proved to be an extremely promising research technique. However, the deployment of DL-based HSI classification algorithms in mobile and embedded vision applications tends to be limited by massive parameters, high memory costs, and the complex networks of DL models. In this article, we propose a novel, extremely lightweight, non-deep parallel network (HyperLiteNet) to address these issues. Based on the development trends of hardware devices, the proposed HyperLiteNet replaces the deep network by the parallel structure in terms of fewer sequential computations and lower latency. The parallel structure can extract and optimize the diverse and divergent spatial and spectral features independently. Meanwhile, an elaborately designed feature-interaction module is constructed to acquire and fuse generalized abstract spectral and spatial features in different parallel layers. The lightweight dynamic convolution further compresses the memory of the network to realize flexible spatial feature extraction. Experiments on several real HSI datasets confirm that the proposed HyperLiteNet can efficiently decrease the number of parameters and the execution time as well as achieve better classification performance compared to several recent state-of-the-art algorithms.

Keywords: lightweight; dynamic convolution; parallel structure; hyperspectral image; classification

Citation: Wang, J.; Huang, R.; Guo, S.; Li, L.; Pei, Z.; Liu, B. HyperLiteNet: Extremely Lightweight Non-Deep Parallel Network for Hyperspectral Image Classification. *Remote Sens.* **2022**, *14*, 866. https://doi.org/10.3390/rs14040866

Academic Editors: Miltiadis D. Lytras and Andreea Claudia Serban

Received: 30 December 2021
Accepted: 7 February 2022
Published: 11 February 2022

1. Introduction

With continuous improvement to the accuracy of image acquisition equipment, the spectral dimension of hyperspectral images (HSIs) has expanded greatly, which provides great discrimination ability for the component analysis of materials and land covers, etc. Therefore, the availability of abundant spectral–spatial information has allowed HSIs to be widely used in various applications, e.g., Earth observation [1], military reconnaissance [2], environmental protection [3], and resource management [4].

Numerous handcrafted features and traditional classifiers have been proposed over the past decades. However, HSI data suffers from the "curse of dimensionality" problem, due to the extremely limited hyperspectral training samples [5]. Therefore, many linear and nonlinear dimension reduction algorithms, such as band selection [6], principal component analysis [7], independent component analysis [8], and the maximum noise fraction [9] have been proposed to address this problem. Such algorithms employ feature optimization to remove redundant features and improve classification performance. Traditional classifiers can be mainly divided into two categories, according to their feature extraction characteristics: (1) spectral-based classification algorithms, such as support vector machines (SVM) [10], K-nearest neighbor [11], multinomial logistic regression [12], and random forest [13]; (2) spectral–spatial-based classification algorithms, e.g., SVM with composite kernels (SVMCK) [14], sparse representation-based classifiers [15], spatial–spectral derivative-aided kernel joint sparse representation [16], and adaptive nonlocal spatial–spectral kernels [17]. However, above-mentioned algorithms usually adopt handcrafted features in terms of the requirements of professional domain knowledge in the HSI classification area.

In contrast, deep learning (DL)-based algorithms demonstrate automatic representative and abstract feature extraction due to the rapid iteration and GPU-based parallel implementation. In the remote sensing field, DL-based algorithms also demonstrate extremely promising results. For instance, the stacked autoencoder [18] demonstrates the effectiveness of DL models in abstract high-level semantic feature extraction. The restricted Boltzmann machine and deep belief network [19] and unsupervised greedy learning algorithms realize the DL of objects. With the development of DL, convolutional neural networks (CNN) are widely used in HSI classification tasks. For example, Zhong et al. proposed the spectral–spatial CNN-based algorithm (SSRN) [20] by combining 3D convolutions and residual connections to realize robust HSI classification results. To refine the feature fusion process, Gao et al. [21] performed multi-scale feature extraction and dense feature fusion to expand the features of different levels and scales. Li et al. [22] proposed a hierarchical homogeneity–attention network to consolidate the position supervision of object regions and accelerate computation by reducing the number of similar operations. In terms of the visual invariance of the CNN, input image visual transformation causes drastic fluctuations in network performance. To address the issue of insufficient spectral–spatial feature extraction, Zhang et al. [23] proposed a spectral–spatial attention fusion with a deformable convolution residual network. The capsule network (CapsNet) was proposed [24] to improve network robustness and preserve more efficient position information that is omitted in traditional CNNs. Then, the deep convolutional capsule neural network (DC-CapsNet) [25] introduced CapsNet in 3D convolution to enhance the robustness of the learned spectral–spatial features. The generative adversarial network (GAN) has recently demonstrated satisfying performance for alleviating the problem caused by an insufficient number of labeled samples of CapsNet. However, it is difficult for GAN to model and preserve the relative positions between features accurately. To eliminate the mode collapse and gradient disappearance problem caused by traditional GANs, Wang et al. [26] developed a dual-channel spectral–spatial fusion capsule GAN (DcCapsGAN) by integrating CapsNet with a GAN. The progress of DL-based algorithms also exposed some challenging issues. With the expansion in complexity of DL networks, though the performance improved dramatically, the requirement of training samples and depths increase proportionally with the complexity of the model and make it difficult to deploy on edge devices. Therefore, realizations of an efficient DL structure with lower time costs and fewer training samples are more requisite for improving the HSI classification performance; it is a more crucial and applicable research task for mobile devices and in-orbit aerospace applications.

Recent research has revealed that neural networks with deep structures typically exhibit higher nonlinearity capabilities and stronger function approximations when automatically learning more abstract features [27]. However, deeper network structures also give rise to the vanishing and exploding gradient problems. Recent studies have indicated that a class of widened ResNets is far superior than their deep counterparts [28]. Shallow network structures with high parallelizations tend to provide fast responses and low latency. In addition, Wu et al. proved that a well-designed shallow neural network can outperform many deep neural networks [29], and recent empirical work has further indicated that obtaining the best accuracy requires a balance of depth and width [30]. However, most of these previous studies only focused on networks with linear and sequential structures. Conversely, in [31], it is found that the performance of a shallow network with a parallel structure is similar to that of the deep network. Therefore, compared to deep networks with more sequential processing and higher latency, the parallel shallow networks are more suitable for the hardware applications and deployment [31].

On the other hand, compressing a well-trained complex model or learning a lightweight model with carefully designed network structures are the current mainstream methods to handle the network efficiency problem caused by CNN-based algorithms. Network operating efficiency is primarily affected by the following four components: the number of parameters, the number of computations, memory access, and memory usage. The main design principle of a lightweight network architecture model can be mainly divided into the convolutional level, the convolution operator level, and the computation method:

(1) In the convolutional level: Redesigning the convolutional level primarily involves small-size convolution kernels, a bottleneck layer, and channel concatenation. For example, SqueezeNet [32] replaces 3×3 convolution with 1×1 convolution, reduces the number of input channels in the squeeze layer, and concatenates two sets of convolution output channels to obtain the target number of output channels to reduce network complexity.

(2) In the convolution operator level: Standard convolution has the inherent properties for global spacial and channel feature extraction. ShuffleNet V1 [33] limits the convolution operation to each group and performs a channel shuffle to promote the flow of information between groups. This is equivalent to decomposing the global channel feature extraction process into a local channel feature extraction and channel shuffle process. Furthermore, the depthwise separable (DW) convolution completely separates the space and channel feature extractions. From the implementation perspective, DW convolution can be considered as a special case of group convolution. The MobileNet family has successfully applied DW convolution and recently made a series of improvements [34–36]. Subsequently, different from DW convolution, Ghost [37] performs a one-to-many mapping output of a single feature map. In addition to channel decomposition, standard convolution can be further decomposed in spatial dimensions. For example, Zhang et al. [38] developed a compact convolution module to divide the input feature maps into groups and perform cyclic recursive feature extraction in each group. Compared to the channel-specific and spatially-agnostic nature of convolution, Involution [39] involves a spatially-specific and channel-agnostic operation that is distinct in the spatial extent but shared across channels. However, extremely low computational costs are likely to cause significant performance degradation. To address this problem, MicroNet [40] integrates sparse connectivity into convolution and constructs a dynamic activation function, i.e., dynamic shift max.

(3) In the computation level: Current CNNs primarily rely on multiplication operations, which are more computationally complex than addition operations. To address this issue, AdderNets [41] replaces massive number of multiplication operations in CNNs with less-expensive addition operations to reduce computational costs. Note that most of the aforementioned algorithms are based on the parameter and computation as model design indicators, and the actual acceleration effect is inferior to the theoretical analysis value. Shufflenet V2 [42] deeply analyzes the influence of different elements, e.g., memory access cost. Therefore, in addition to theoretical indicators, more attention should be paid to the actual time costs on equal hardware. The superiority of the lightweight structures has also been confirmed in the HSI classification tasks. Zhang et al. [43] proposed 3-D-LWNet, which has a deeper network structure and exhibits better classification performance than conventional 3D-CNN models. In addition, LWCNN [44] employs spatial–spectral Schroedinger eigenmap (SSSE)-based feature extraction and a dual-scale convolution module. The former reduces dimensions, and the latter address the SSSE features from a 1D vector viewpoint to reduce the number of parameters. Based on the conclusions of a neural architecture search [45], LMAFN is proposed to combine Ghost and ECA modules for further achieving efficient feature extraction performance [46]. LiteDepthwiseNet [47] used 3D depthwise convolution to reduce the number of parameters as well as remove the ReLU layer and batch normalization layer in the original 3D depthwise convolution to improve the overfitting phenomenon of the model on small-sized datasets.

Considering that compact lightweight models still achieved comparable or even better performance with fewer parameters, a lightweight model indicates more effective representative knowledge capacity than traditional DL models. However, blindly reducing model size can easily lead to knowledge degeneration, which contradicts the purpose of the lightweight design. To address this issue, CondConv [48] embeds multiple sets of convolution kernels in the convolutional layer and adaptively combines the weights according to the input, which is equivalent to learning specialized convolutional kernels for each example in terms of increasing network capacity while maintaining inference efficiency. To reorganize multiple sets of static convolution kernels according to the input, DYNet [49] performs a low-redundancy and efficient dynamic convolution to avoid redundant calculations and similar feature generation while maintaining a lightweight structure and efficient inference ability. Moreover, the study finds that dynamic networks can address domain conflicts by aggregating residual matrices and a static convolution matrix [50].

In the HSI classification area, the limited availability of labeled samples hinders the classification accuracy and optimization performance of DL. Therefore, recently, few-shot learning [51], domain adaptation [52], transfer learning [53], and meta learning [54] have been gradually proposed to address this issue. For example, Li et al. [55] proposed DCFSL, which utilizes a conditional adversarial domain-adaptation strategy to overcome domain shift and achieve domain distribution alignment. In addition, AMF-FSL [56] transfers the learned classification capabilities from multiple source data to the target data. To realize the promising performance with limited labeled samples, the above DL methods are generally based on complicated network structures and learning strategies. Therefore, an efficient network model that can obtain effective classification results with limited labeled samples is required. Intuitively, a lightweight model can also mitigate the requirements for the quantity of labeled samples to a certain extent.

Based on the above-mentioned knowledge, and aiming at realizing high efficiency and performance at lower computation cost, a HyperLiteNet is proposed and designed in this paper, which is an extremely lightweight non-deep parallel network. In the proposed HyperLiteNet, a parallel shallow narrow network is constructed to reduce network complexity and execution cost, where a parallel dual-branch structure is employed to enhance the nonlinearity of network and the capability of feature extraction. In addition, the spatial features and spectral features are extracted via a parallel dual-branch structure to, respectively, promote the diversity and divergence of the extracted features. As the main components of the dual-branch structure, lightweight pointwise convolution units and lightweight dynamic convolution units are constructed to further reduce the network's complexity. In addition, the introduction of dynamic modules enables more elastic mapping with the input sample to generate compact and representative features. The experimental results confirm that the proposed HyperLiteNet outperforms other comparison algorithms with an extremely low number of parameters and high execution efficiency.

The remainder of this paper is organized as follows. Section 2 presents the proposed HyperLiteNet in detail. Experimental validation of the proposed HyperLiteNet compared to existing algorithms, including parameter analysis, classification results, classification efficiency, and accuracies, are presented in Section 3. Some experimental results are discussed in Section 4. Finally, conclusions are presented in Section 5.

2. Proposed Methods

The overall framework of the proposed HyperLiteNet is shown in Figure 1. The proposed HyperLiteNet can be roughly divided into five main components: (1) the parallel interconnection module (PIM); (2) the pointwise convolution branch (PCB) to extract spectral features; (3) the dynamic convolution branch (DCB) to extract spatial features; (4) the feature interconnection module (FIM); (5) the classification module (CM). In summary, the specific spectral and spatial features are extracted independently by the PCB and the DCB, respectively. Then, the spectral–spatial features are fused by the FIM and classified by the CM module. The details of different components are illustrated in the following parts.

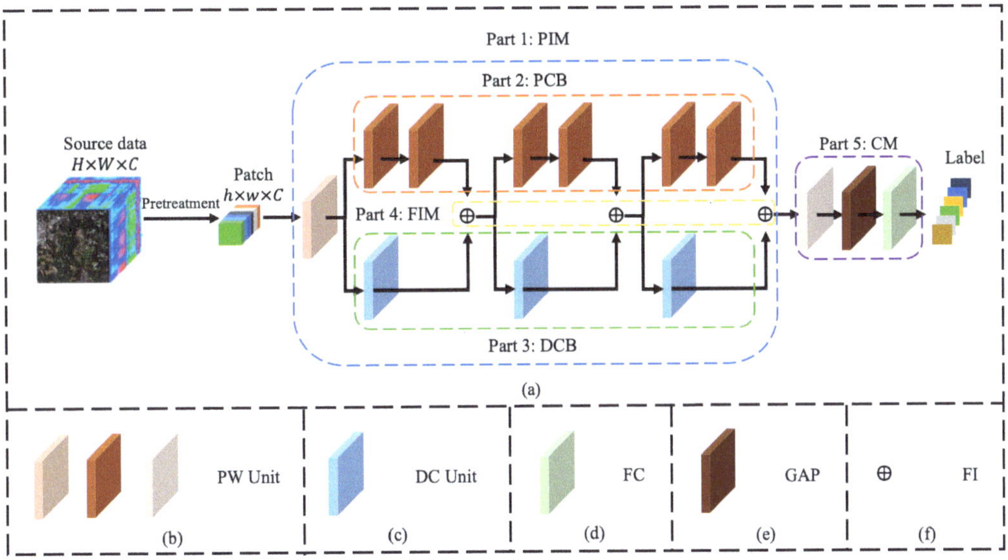

Figure 1. (**a**) The overall flowchart of HyperLiteNet. The five main parts: parallel interconnection module (PIM), pointwise convolution branch (PCB), dynamic convolution branch (DCB), feature interconnection module (FIM), and classification module (CM); (**b**) pointwise convolution (PW) units with different numbers of convolution kernels; (**c**) dynamic convolution (DC) unit; (**d**) fully connected layer (FC); (**e**) global average pooling (GAP); (**f**) feature interconnect (FI) structure.

2.1. Parallel Interconnection Module (PIM)

The PIM is the primary component of the proposed HyperLiteNet, where the parallel structure contains more nonlinear activation layers than a serial structure at the same depth. The paralleled shallow structure realizes an efficient nonlinear activation ability to extract abstract high-level features through utilizing PCB and DCB structures. With the reduced network depth, the vanishing and exploding gradient problems are mitigated in the PIM. This dual-branch structure is adopted to extract abstract spectral and spatial features independently and in a parallel manner through an extremely shallow network structure. As a result, due to the task decomposition of the PCB and DCB, fewer channels are established by the PIM in each branch than traditional spectral–spatial combination networks. Therefore, the network structure can be efficiently compressed in a compact, parallel manner with a lighter structure and better feature extraction. The structure of the PIM is mainly composed of three stages ($stage_1, stage_2, stage_3$), which extract shallow pixel-level features, mid-level features, and deep high-level features, respectively. The features extracted in each stage are fused at the tail of each stage and then input to the next stage to further increase the discriminativeness of the extracted features. Meanwhile, each stage contains two pointwise convolution (PW) units and one depthwise separable dynamic convolution (DC) unit. In the feature extraction process, the PCB and DCB comprise lightweight PW and DC units, respectively. Therefore, the spectral and spacial feature can be extracted in a parallel manner to accelerate execution efficiency. Further, for combining and fusing the spectral and spatial features at different stages, the FIM is designed to connect each stage to the subsequent stage. The entire structure provides a compact and efficient parallel method to extract and fuse the spectral and spatial features with a small number of parameters and low computational cost.

2.2. Pointwise Convolution Branch (PCB) and Dynamic Convolution Branch (DCB)

2.2.1. Pointwise Convolution Branch (PCB)

As shown in Figure 2a, lightweight PW units are employed in the PCB to extract spectral features. Here, only continuous spectral features are processed at a certain point in the feature map. The PCB is concentrated to effectively extract spectral distinguishing features. The pointwise (PW) convolution is calculated as follows:

$$o_{c,ij} = \mathbf{w}_c \otimes \mathbf{x}_{ij} \quad (1)$$

where $\mathbf{x}_{ij} \in \mathbb{R}^{1\times 1\times C_{in}}$ is the spectral vector at the (i,j) of the input features. $\mathbf{w}_c \in \mathbb{R}^{1\times 1\times C_{in}}$ is the c-th PW convolution to convolve with $\mathbf{x}_{ij}$ for obtaining the output $o_{c,ij} \in \mathbb{R}^1$, where $o_{c,ij}$ is the value at (i,j) in the c-th channel of the output. It can be seen from the calculation process that the PW convolution is essentially a weighted summation of the spectral vectors of the input features. The procedure intensifies the channels that contributed more discriminant information with a dynamic learning process, which can indicate more satisfying effects and efficiency than traditional band-selection algorithms.

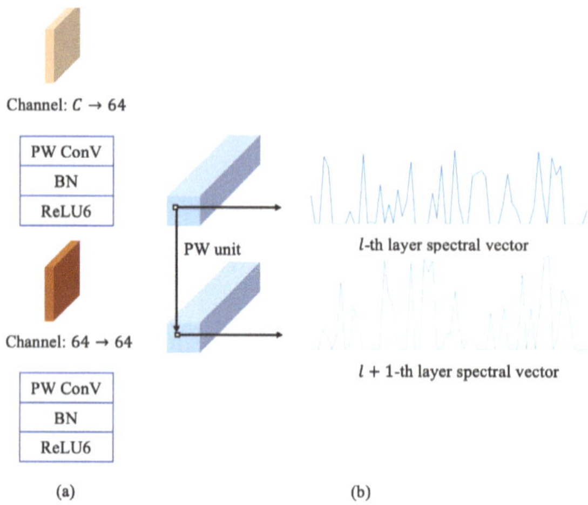

Figure 2. (**a**) PW units with the pointwise convolution, BN, and ReLU6. Convert the number of x channels from input channels to 64 (upside); convert the number of input features channels from 64 to 64 (underside). (**b**) Spectral vectors at a certain point in the l-th layer and the $l+1$-th layer.

2.2.2. Dynamic Convolution Branch (DCB)

In traditional convolution, all input samples are treated equally to the convolution kernel, and the parameters of the convolution kernel are determined after the training process in a static network manner. Differing from static networks, dynamic networks attempt to adjust the network structure adaptively and dynamically according to the varied inputs. Therefore, for different inputs, a dynamic network can deform the network structure corresponding to the local areas, which efficiently improves the robustness and knowledge capacity of the network with low latency. The introduction of a dynamic network enables a more elastic mapping. Define the network convolutions in Figure 3a as follows:

$$\tilde{\mathbf{W}} = \mathbf{W} + \hat{\mathbf{W}} \quad (2)$$

where $\mathbf{W} \in \mathbb{R}^{k\times k\times C_{in}\times C_{out}}$ and $\hat{\mathbf{W}} \in \mathbb{R}^{k\times k\times C_{in}\times C_{out}}$ are the static convolution kernel and the dynamic convolution kernel, respectively. C_{in} and C_{out} are the number of input channels and the number of output channels. If $\mathbf{W}$ is ignored, $\tilde{\mathbf{W}}$ is equivalent to $\hat{\mathbf{W}}$.

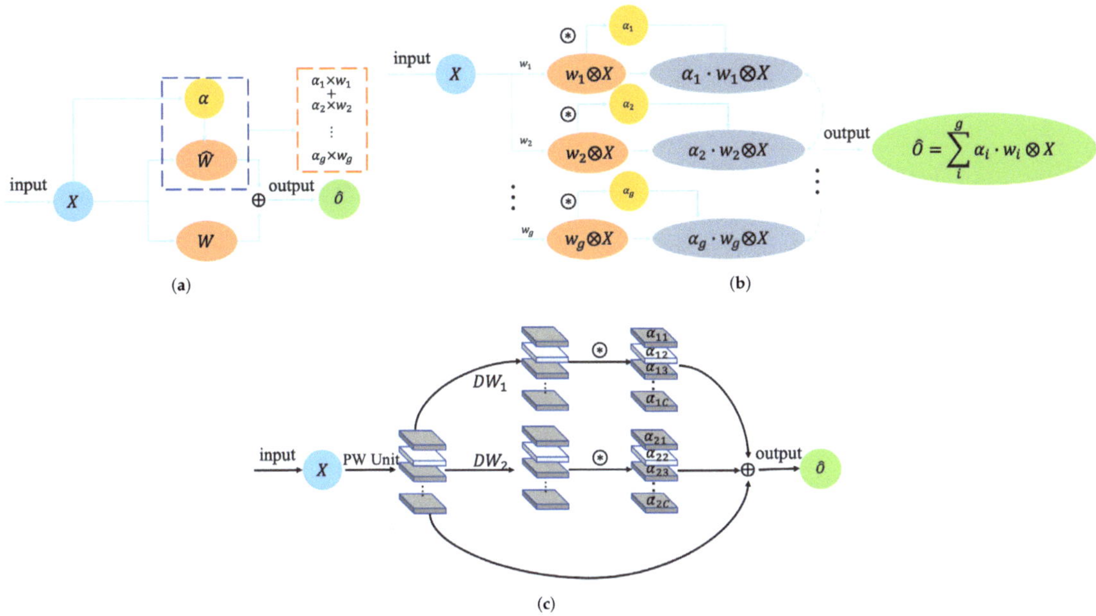

Figure 3. Adaptive dynamic convolution. (**a**) Dynamic module composed of the static convolution kernel $\mathbf{W}$ and the dynamic convolution kernel $\hat{\mathbf{W}}$. $\hat{\mathbf{W}}$ realizes subspace routing based on the input adaptive α. If $\mathbf{W}$ is ignored, $\bar{\mathbf{W}}$ is equivalent to $\hat{\mathbf{W}}$. (**b**) Standard dynamic convolution $\hat{\mathbf{W}}$ implementation. $\otimes$ refers to the convolution operation. $\circledast$ is the input adaptation. The adaptive importance α multiplied by the features is equivalent to path selection based on different input samples. (**c**) DW dynamic convolution implementation. (Both BN and ReLU6 are ignored.)

The structure of the standard dynamic convolution is shown in Figure 3b, where the multi-group parallel standard convolutions extract features independently. Then, the extracted features are adjusted dynamically through each attention dynamic unit. The multi-group features are dynamically fused by a dynamic fusion unit. Essentially, the dynamic fusion unit reorganizes multi-group standard dynamic convolutions with different weights. The standard dynamic kernel can be generated by the following formulas:

$$\hat{\mathbf{w}}_c = \sum_i^g \alpha_c^i \cdot \mathbf{w}_c^i \tag{3}$$

where $c = 0, \ldots, C_{out}$ and $i = 0, \ldots, g$ are the output channels and number of static convolution kernel groups, respectively. α represents the importance weights of different sets of static convolution kernels based on the input. $\mathbf{w}_c^i \in \mathbb{R}^{k \times k \times C_{in}}$ indicates the c-th convolution kernel of the i-th static convolution. Based on the linear combination characteristic of convolution, the dynamic convolution output can be mathematically represented as:

$$\begin{aligned}\hat{\mathbf{O}}_c = \hat{\mathbf{w}}_c \otimes \mathbf{X} = \sum_i^g (\alpha_c^i \cdot \mathbf{w}_c^i) \otimes \mathbf{X} = \sum_i^g (\alpha_c^i \cdot \mathbf{w}_c^i \otimes \mathbf{X}) \\ = \sum_i^g (\alpha_c^i \cdot (\mathbf{w}_c^i \otimes \mathbf{X})) = \sum_i^g (\alpha_c^i \cdot \mathbf{o}_c^i)\end{aligned} \tag{4}$$

where the output $\hat{\mathbf{O}}_c$ of the dynamic convolution kernel can be expressed as a linear combination of the output $\mathbf{O}_c^i$ of the static convolution kernel.

Further, for realizing the design of the lightweight structure, we adopt the lightweight DW dynamic convolution as the primary component of the DC unit in Figure 3c. Specifically, the parameter definition is different in the realization of the DW dynamic convolution. $\mathbf{w}_c^i \in \mathbb{R}^{k \times k \times 1}$ indicates the c-th channel of the i-th static DW convolution due to the characteristic of DW convolution. Compared with the standard dynamic convolution, DW dynamic convolution realizes the single-channel-level dynamic fusion to further decompose and refine the dynamic convolution. In addition, we further analyze the DW dynamic convolution from the perspective of parameters and calculations. Assume the standard convolution $\mathbf{W}_S \in \mathbb{R}^{k \times k \times C_{in} \times C_{out}}$ and the DW dynamic convolution $\mathbf{W}_{DW} \in \mathbb{R}^{k \times k \times C_{out} \times C_{out} \times g}$. The PW convolution adjusts the number of channels in the DW dynamic convolution. Ignoring the input adaptive correlation calculation, the reduced computation and parameters effect can be written as follows:

$$\begin{aligned} &\frac{k \times k \times C_{out} \times s \times s \times g + C_{in} \times C_{out} \times s \times s}{k \times k \times C_{in} \times C_{out} \times s \times s} \\ &= \frac{g}{C_{in}} + \frac{1}{k^2} \approx \frac{1}{k^2} (C_{in} \gg g) \end{aligned} \tag{5}$$

where s is the space size of the feature map.

Compared to the traditional standard convolution, the diversity and spatial differences of the spatial characteristics can be extracted flexibly and enhanced by the dynamic module. In addition, the DCB realizes the potential for reducing the number of parameters and computations, which improves the efficiency of the network operations.

2.3. Feature Interconnection Module (FIM) and Classification Module (CM)

The function of FIM is to fuse the spectral and spatial features of the two branches and feed the fused features to the next stage of the network. Here, suppose the input spectral and spatial features in $stage_3$ are defined as $\mathbf{X}^{P_6}$ and $\mathbf{X}^{D_3}$, respectively. The calculation process of FIM is determined by the following formula:

$$\mathbf{X}_{FIM_3} = \mathbf{X}^{P_6} + \mathbf{X}^{D_3} \tag{6}$$

For the CM, the PW unit progressively reduces the number of channels and prevents the sharp reduction in the number of channels for avoiding the feature information loss. In this module, we utilize cross entropy as the loss function for HSI classification. The cross entropy loss can be represented as:

$$Loss = -\frac{1}{N} \sum_{n=1}^{N} \sum_{c=1}^{C} y_c^n \log(\hat{y}_c^n) \tag{7}$$

where C indicates the number of classes, y_c^n and $\hat{y}_c^n$ are the truth and predicted labels, respectively, and N is the number of samples in a minibatch.

3. Experiments and Analysis

In this section, we mainly evaluate our proposed frameworks on three real HSI datasets. The experimental settings and results, including parameters analysis and evaluation metrics are illustrated. Finally, the comparison result with the existing algorithms are analyzed and discussed in this section.

3.1. Datasets Descriptions

The Indian Pines (IN) dataset was captured in 1992 over the Indian Pines agriculture experimental area by the Airborne Visible/Infrared Imaging Spectrometer (AVIRIS) sensor. The IN dataset has 220 bands with a wavelength range from 0.4 to 2.5 μm, and with

145 × 145 pixels. Note that 20 bands of noisy or water-absorption regions are removed in our experiments. The IN dataset contains 16 different ground-truth classes. The three-band false-color composite of the IN images and the corresponding ground-truth data are shown in Figure 4.

The University of Pavia (UP) dataset was collected in 2001 over the Pavia university campus through the Reflective Optics System Imaging Spectrometer (ROSIS). The UP dataset has 115 bands with a wavelength range from 0.43 to 0.86 μm, and with 610 × 340 pixels. Note that 12 bands of noisy or water-absorption regions are removed in our experiments. The UP dataset contains nine different ground-truth classes. The three-band false-color composite of the IN images and the corresponding ground-truth data are shown in Figure 5.

The Salinas (SA) dataset was gathered over the Salinas Valley in California by the AVIRIS sensor. The SA dataset has 220 bands with a wavelength range from 0.36 to 0.5 μm, and with 512 × 217 pixels. Note that 20 bands of noisy or water-absorption regions are removed in our experiments. The SA dataset contains 16 different ground-truth classes. The three-band false-color composite of IN images and the corresponding ground-truth data are shown in Figure 6.

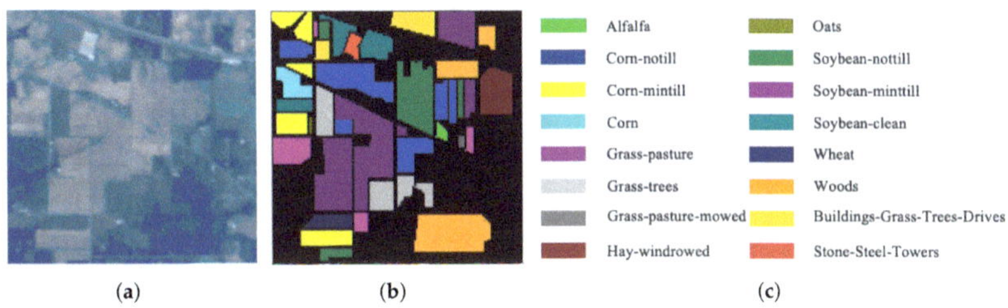

Figure 4. IN. (**a**) False-color image. (**b**) Ground-truth map. (**c**) Ground-truth classes.

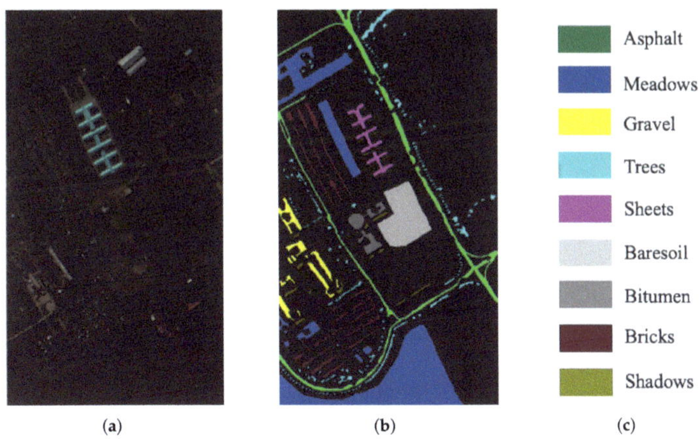

Figure 5. UP. (**a**) False-color image. (**b**) Ground-truth map. (**c**) Ground-truth classes.

3.2. *Experimental Configuration and Parameter Analysis*

3.2.1. Experimental Configuration

The experimental configuration is as follows: an Intel Xeon W-2123 CPU at 3.60 GHz with 64-GB RAM and an NVIDIA GeForce GTX 2080 Ti graphical processing unit (GPU) with 11-GB RAM were used. The software environment includes a 64-bit Windows 10 system and the Pytorch 1.6.0 DL frameworks. Here, the learning rate is set to 0.0001, and the Adam optimizer is used with a batch size of 72. In addition, five labeled samples are randomly selected from each class as a training set for each HSI. To avoid the inclination in random sampling, the evaluated mean value is reported for parameter analysis and classification result comparison with ten random repetitions.

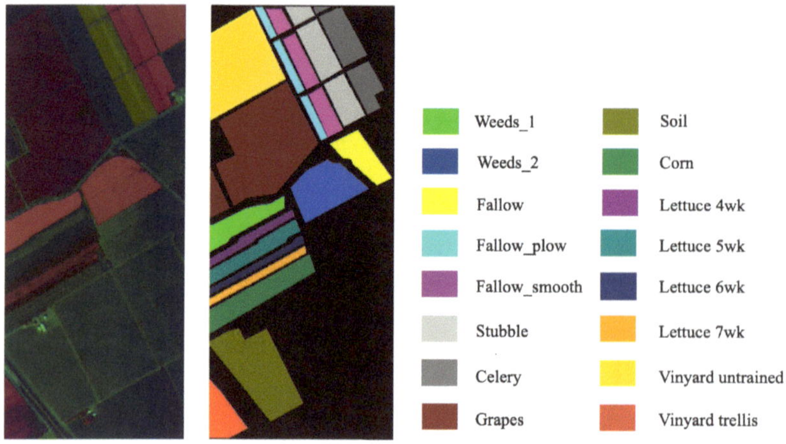

(a) (b) (c)

Figure 6. SA. (**a**) False-color image. (**b**) Ground-truth map. (**c**) Ground-truth classes.

3.2.2. Experimental Parameter Analysis

To evaluate the effectiveness of the proposed HyperLiteNet, we mainly evaluate the effect of the number of kernels (extracted channels), the quantity of parameters, and the training epoch. We first analyze the effect for the number of kernels in each conventional layer. The number of kernels in each layer is selected from $\{16, 32, 48, 64, 72, 96, 128\}$ for each dataset. The effect analysis results are illustrated in Figure 7a,b; we can notice that the number of 16 kernels, corresponding to the least number of parameters, presents the worst performance for all experimental datasets. The overall accuracy (OA) gradually increased with the increasing number of channels. Compared to 128 channels, we observe from Figure 7a that 16 channels obtain lower OA with the same number of training epochs (4000). It is notable that the increasing number of channels can improve the classification accuracy significantly. However, when the number of convolution kernels exceeds a certain threshold, the performance gradually increases to a stable value. The excessive number of convolution kernels brings additional an parameter burden without beneficial effects on performance. The aforementioned data analysis indicates that the limited model performance and difficult model optimization phenomenon exist in low-parameter models. Therefore, the optimal parameters acquired from the above-mentioned analysis are utilized in the following comparison experiments.

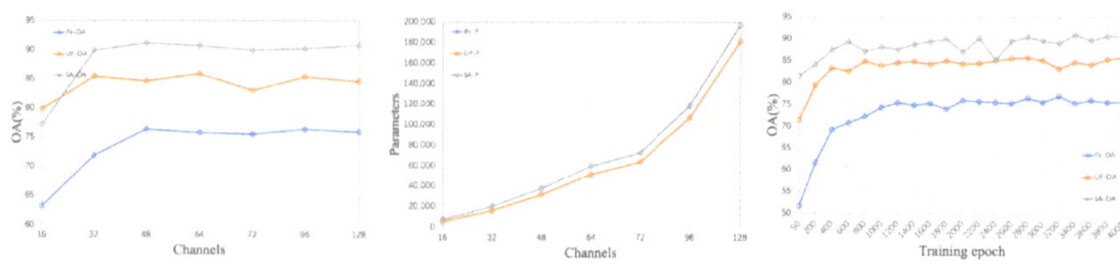

(a) (b) (c)

Figure 7. Effect analysis for the number of convolution kernels, the number of parameters, and the training epoch. (**a**) OA corresponding to different channel numbers (4000 epochs). (**b**) Parameters corresponding to different channel numbers. (**c**) OA corresponding to different training epochs (64 channels).

Because of the correlation between time costs and model convergence, we evaluate OA under different numbers of training epochs to achieve a balance between time costs and performance, where training epochs are a varied set from 50 to 4000 with step 200. The analysis results are presented in Figure 7c. We can notice that the proposed model reaches a stable and exceptional OA after 2000 epochs. Moreover, MiniNet achieves acceptable results in 1000 epochs. Considering the necessity of adequate training, 4000 is chosen as the number of training epochs in the following experiment.

3.3. Ablation Study for DW Dynamic Convolution

To verify the proposed effectiveness of the DW dynamic convolution (DC), we compare the proposed DC with standard convolution (SC) (3×3 convolution kernel is applied), and multi-scale convolution (MSC) (3×3 and 5×5 convolution kernels are applied). The comparison results are illustrated in Table 1. It is obvious that the DC improves the classification performance while effectively compressing the model size. In addition to quantitative analysis, the t-SNE feature visualization technology is further exhibited to qualitatively analyze the differences between the three convolutions in Figure 8. The blue dotted line indicates the comparison between DC and SC, and the red dotted line represents the comparison between DC and MSC. From Figure 8a–c, we can observe that the dotted line in Figure 8a has a higher concentration and purity ratio. This phenomenon also appeared in Figure 8g–i. The red dotted area in Figure 8d,g exhibits better concentration than Figure 8f,i. Moreover, it is roughly observed that the distinction between the red and green cluster in the blue dotted rectangle region for Figure 8d is inferior to Figure 8e. Meanwhile, in real classification investigations, there are many misclassified samples in these two classes for Figure 8d,f, which further indicates that these two classes are more indistinguishable in real classification tasks. Overall, from the t-SNE feature visualization, as illustrated in Figure 8, the DC achieves a more distinguishable and recognizable feature extraction capability.

Table 1. Model performance analysis (OA) and parameters (P) with different convulution types. (1) DC: depthwise separable dynamic convolution; (2) SC: standard convolution; (3) MSC: multi-scale convolution.

Type \ Metric	OA	P	OA	P	OA	P
DC	75.75 ± 3.25	59,658	85.89 ± 4.60	51,939	90.70 ± 1.76	59,914
SC	75.19 ± 3.16	153,840	83.96 ± 4.88	146,121	90.35 ± 2.07	154,096
MSC	75.34 ± 3.62	461,616	83.11 ± 3.89	453,897	88.72 ± 4.21	461,872
Dataset	IN		UP		SA	

3.4. Classification Accuracy and Performance

In this part, we compare the proposed HyperLiteNet with six comparable methods: two traditional methods, SVM and SVMCK, and four deep learning-based methods, SSRN, LMAFN, DcCapsGAN, and DCFSL. The SVM only extracts spectral features, and the SVMCK further introduces spatial features for classification. The SSRN is a classic method using 3D convolution for HSI classification. The DcCapsGAN combines the capsule network and GAN to enhance the model robustness. Based on the lightweight characteristics of the proposed HyperLiteNet, we also compared it with the LMAFN. In order to test the effectiveness of proposed HyperLiteNet under the condition of limited labeled samples, we further compared the proposed HyperLiteNet with the few-shot learning method DCFSL. The quantitative experimental results of different methods are shown in Tables 2–4. The metrics include overall accuracy (OA(%)), average accuracy (AA(%)), kappa (KP(%)), the number of parameters (PA), floating point operations (FLOPs (M)), training time (Train(s)), and test time (Test(s)).

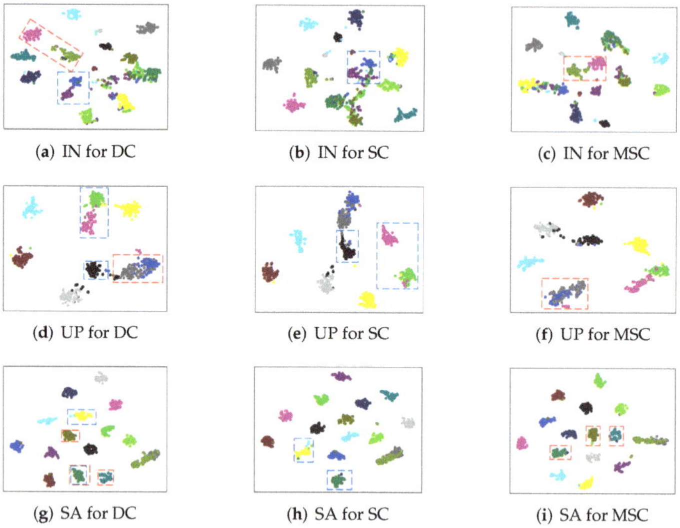

Figure 8. 2D t-SNE feature visualization for different convolution types on three datasets. The corresponding colored dotted lines are used in the figure to indicate representative differences. (**a**) IN for DC; (**b**) IN for SC; (**c**) IN for MSC; (**d**) UP for DC; (**e**) UP for SC; (**f**) UP for MSC; (**g**) SA for DC; (**h**) SA for SC; (**i**) SA for MSC.

Table 2. Classification results (%) on the IN dataset (five labeled samples per class). The best performance values are marked in bold.

Class	SVM	SVMCK	SSRN	LMAFN	DcCapsGAN	DCFSL	HyperLiteNet
1	83.17 ± 10.32	94.63 ± 3.75	45.60 ± 21.99	99.51 ± 0.98	**100 ± 0.00**	93.17 ± 9.05	99.76 ± 0.73
2	28.09 ± 9.09	51.33 ± 11.37	**64.03 ± 11.93**	55.85 ± 11.61	50.50 ± 0.29	43.25 ± 9.14	61.33 ± 15.75
3	37.53 ± 10.10	51.25 ± 6.99	63.89 ± 14.14	67.82 ± 9.63	44.36 ± 0.10	53.82 ± 4.95	**69.95 ± 7.97**
4	59.18 ± 11.14	88.71 ± 11.50	44.11 ± 15.82	90.60 ± 8.90	46.70 ± 0.54	78.19 ± 18.67	**93.41 ± 8.93**
5	67.91 ± 9.21	64.16 ± 12.82	78.58 ± 12.99	80.94 ± 7.47	65.06 ± 0.17	72.53 ± 6.42	**81.72 ± 8.07**
6	72.43 ± 13.84	63.03 ± 15.13	90.83 ± 7.92	94.52 ± 3.56	83.77 ± 0.07	85.31 ± 9.64	**96.40 ± 2.53**
7	90.00 ± 5.52	98.70 ± 1.99	31.90 ± 14.60	**100 ± 0.00**	**100 ± 0.00**	98.70 ± 3.91	**100 ± 0.00**
8	67.51 ± 15.56	96.62 ± 3.86	**97.58 ± 3.86**	93.78 ± 9.86	80.20 ± 0.20	86.89 ± 11.06	95.88 ± 7.87
9	91.33 ± 13.35	96.00 ± 8.00	25.50 ± 15.94	**100 ± 0.00**	**100 ± 0.00**	99.33 ± 2.00	**100 ± 0.00**
10	53.58 ± 14.38	57.61 ± 6.27	57.57 ± 9.91	**69.51 ± 8.70**	62.36 ± 0.22	62.53 ± 7.68	**71.98 ± 7.57**
11	36.12 ± 15.10	53.05 ± 14.20	69.27 ± 7.80	62.73 ± 11.90	65.02 ± 0.13	58.45 ± 8.90	64.12 ± 12.27
12	34.32 ± 9.08	**69.46 ± 12.03**	51.80 ± 16.10	69.37 ± 13.83	48.58 ± 0.32	45.90 ± 7.84	69.12 ± 15.96
13	88.00 ± 6.34	95.20 ± 7.96	88.56 ± 15.34	**99.05 ± 1.60**	99.01 ± 0.00	98.40 ± 3.49	98.50 ± 1.75
14	61.27 ± 12.99	71.52 ± 12.69	**95.24 ± 5.47**	88.13 ± 10.50	94.89 ± 0.19	82.02 ± 9.45	89.61 ± 9.64
15	28.69 ± 7.84	73.78 ± 10.15	65.58 ± 14.58	76.93 ± 17.21	35.61 ± 0.75	76.09 ± 11.54	**86.25 ± 12.16**
16	89.20 ± 6.37	98.52 ± 2.16	70.83 ± 18.10	**99.32 ± 1.04**	98.86 ± 0.24	97.95 ± 4.09	98.75 ± 2.00
OA	47.45 ± 5.32	62.74 ± 5.60	67.10 ± 3.24	73.41 ± 3.95	65.57 ± 0.03	64.96 ± 2.03	**75.75 ± 3.25**
AA	61.77 ± 3.20	77.01 ± 4.00	65.05 ± 2.96	84.26 ± 1.91	73.43 ± 0.04	77.03 ± 0.86	**86.05 ± 1.50**
KP	41.67 ± 5.38	58.46 ± 6.05	63.00 ± 3.49	70.08 ± 4.37	60.94 ± 0.03	60.65 ± 2.08	**72.71 ± 3.50**
PA	–	–	346,784	161,451	33,521,328	4,270,121	**59,658**
Flops (M)	–	–	110.245	7.369	433.220	–	**2.920**
Train(s)	–	–	436.20	281.83	2394.88	2629.22	**80.84**
Test(s)	–	–	31.38	1.43	11.51	2.21	**0.24**

Table 3. Classification results (%) on the UP dataset (five labeled samples per class). The best performance values are marked in bold.

Class	SVM	SVMCK	SSRN	LMAFN	DcCapsGAN	DCFSL	HyperLiteNet
1	49.41 ± 13.47	55.21 ± 11.76	**96.30 ± 2.20**	75.68 ± 11.73	36.40 ± 4.74	74.94 ± 17.10	86.64 ± 6.63
2	44.45 ± 14.50	56.70 ± 11.93	**96.42 ± 3.57**	77.49 ± 14.00	74.92 ± 0.52	85.96 ± 8.44	80.62 ± 12.13
3	62.34 ± 18.77	71.77 ± 8.74	61.35 ± 8.69	68.64 ± 15.32	77.94 ± 1.21	62.43 ± 12.45	**87.40 ± 7.65**
4	86.33 ± 11.81	93.46 ± 5.50	76.77 ± 18.53	96.21 ± 5.48	**98.29 ± 0.11**	91.62 ± 5.01	92.94 ± 6.24
5	96.19 ± 6.99	99.57 ± 0.22	99.19 ± 1.20	99.45 ± 0.87	**100 ± 0.00**	99.25 ± 0.95	99.57 ± 0.79
6	52.54 ± 13.28	71.86 ± 15.90	62.92 ± 12.21	79.36 ± 16.01	54.15 ± 0.38	71.07 ± 10.49	**88.07 ± 9.58**
7	73.50 ± 21.41	94.42 ± 2.07	77.03 ± 15.96	92.33 ± 7.52	98.04 ± 0.43	77.69 ± 8.81	**98.29 ± 2.36**
8	49.04 ± 15.44	72.46 ± 12.68	74.35 ± 12.96	80.18 ± 20.70	69.86 ± 2.33	68.16 ± 11.62	**89.12 ± 11.37**
9	93.68 ± 7.54	86.69 ± 7.78	99.25 ± 0.69	97.07 ± 5.68	**100 ± 0.00**	99.03 ± 1.41	97.29 ± 4.14
OA	54.05 ± 6.35	66.15 ± 5.95	82.03 ± 3.31	80.15 ± 6.04	69.94 ± 0.90	80.66 ± 4.02	**85.89 ± 4.60**
AA	67.50 ± 3.37	78.97 ± 3.36	82.62 ± 3.07	85.16 ± 3.29	78.84 ± 0.65	81.13 ± 2.70	**91.11 ± 2.25**
KP	44.90 ± 5.99	58.52 ± 6.41	77.24 ± 3.76	74.87 ± 6.97	62.29 ± 1.07	74.92 ± 4.81	**82.03 ± 5.38**
PA	–	–	199,153	153,060	21,468,326	4,259,294	**51,939**
Flops (M)	–	–	54.178	7.001	342.850	–	**2.553**
Train(s)	–	–	139.12	216.60	766.54	1593.04	**79.47**
Test(s)	–	–	73.67	4.78	22.17	7.11	**0.70**

Table 4. Classification results (%) on the SA dataset (five labeled samples per class). The best performance values are marked in bold.

Class	SVM	SVMCK	SSRN	LMAFN	DcCapsGAN	DCFSL	HyperLiteNet
1	97.45 ± 1.70	98.73 ± 0.87	98.49 ± 4.51	97.74 ± 2.37	97.72 ± 1.11	**99.25 ± 1.05**	98.65 ± 2.48
2	95.89 ± 2.99	93.76 ± 7.00	99.44 ± 0.88	90.74 ± 12.57	**99.95 ± 0.00**	99.62 ± 0.30	95.92 ± 10.14
3	69.86 ± 17.07	71.02 ± 17.71	96.29 ± 1.53	85.40 ± 13.74	**100 ± 0.00**	89.48 ± 13.06	91.16 ± 10.05
4	98.42 ± 2.40	99.63 ± 0.18	95.86 ± 3.51	99.83 ± 0.19	**99.86 ± 0.00**	99.29 ± 0.48	99.73 ± 0.40
5	94.36 ± 5.00	96.61 ± 2.09	**97.08 ± 2.28**	94.54 ± 3.57	95.57 ± 0.11	91.55 ± 2.85	93.03 ± 11.65
6	99.21 ± 0.29	99.50 ± 0.12	99.78 ± 0.34	99.63 ± 0.67	**100 ± 0.00**	99.20 ± 0.79	99.57 ± 0.59
7	98.36 ± 1.50	97.59 ± 2.03	99.29 ± 1.04	99.86 ± 0.10	99.83 ± 0.00	99.13 ± 1.14	**99.96 ± 0.07**
8	58.39 ± 13.63	61.67 ± 13.65	**82.17 ± 5.25**	63.25 ± 14.20	61.66 ± 0.52	74.96 ± 11.32	78.79 ± 9.25
9	96.28 ± 0.46	95.61 ± 3.00	99.10 ± 0.68	99.57 ± 0.70	**99.87 ± 0.00**	99.44 ± 0.73	99.67 ± 0.68
10	76.97 ± 6.46	83.15 ± 6.71	**90.97 ± 3.72**	85.47 ± 13.79	79.71 ± 0.16	84.36 ± 5.98	90.82 ± 4.19
11	93.88 ± 3.43	91.83 ± 8.65	93.08 ± 2.84	99.10 ± 0.95	98.28 ± 0.09	97.38 ± 2.21	**99.20 ± 0.77**
12	82.39 ± 19.68	93.11 ± 9.79	97.65 ± 3.94	99.53 ± 0.75	**99.90 ± 0.04**	98.55 ± 1.82	98.69 ± 1.29
13	97.65 ± 0.75	97.86 ± 0.79	92.87 ± 7.42	98.97 ± 2.27	**99.74 ± 0.05**	99.20 ± 0.47	99.58 ± 0.38
14	89.92 ± 5.11	91.61 ± 3.66	92.73 ± 10.92	98.07 ± 1.23	**99.41 ± 0.04**	98.48 ± 1.78	97.29 ± 5.03
15	59.13 ± 16.60	66.76 ± 16.26	64.45 ± 7.39	69.89 ± 17.11	38.65 ± 1.63	74.50 ± 6.53	**78.39 ± 14.47**
16	85.03 ± 6.05	83.17 ± 5.96	**99.55 ± 0.63**	93.88 ± 6.07	97.71 ± 0.07	87.56 ± 8.17	93.53 ± 6.70
OA	80.55 ± 2.29	82.95 ± 2.08	88.95 ± 2.03	85.51 ± 2.47	82.07 ± 0.19	88.79 ± 2.48	**90.70 ± 1.76**
AA	87.04 ± 2.43	88.85 ± 1.89	93.67 ± 0.83	92.22 ± 2.03	91.74 ± 0.14	93.25 ± 1.66	**94.62 ± 1.13**
KP	78.43 ± 2.53	81.09 ± 2.30	87.74 ± 2.23	83.93 ± 2.71	80.17 ± 0.20	87.55 ± 2.73	**89.67 ± 1.96**
PA	–	–	352,928	161,707	34,627,364	4,270,521	**59,914**
Flops (M)	–	–	112.178	7.382	1274.300	–	**2.933**
Train(s)	–	–	398.63	206.90	2585.84	2574.94	**79.98**
Test(s)	–	–	178.85	5.50	13.59	10.29	**1.20**

From Tables 2–4, we can observe that, in most cases, the accuracy of SVM and SVMCK are inferior to DL-based methods. In terms of the spectral–spatial information involved in SVMCK, which presents higher classification accuracy than SVM, data-driven characteristics and nonlinearity fitting of CNNs can approximate any function, which facilitates the extraction of discriminant features. Therefore, most of the DL-based models in this paper perform better than traditional SVM and SVMCK methods. The implementation of 3D convolution is consistent with the continuity of the HSI spectral curve. Therefore, the SSRN obtains relatively stable classification performance on most datasets. Our previous research work presents LMAFN based on a lightweight design, which achieves the best operating efficiency among all comparable DL-based methods. The ingenious combination of the GAN and capsule network promotes the robustness of the DcCapsGAN. However, the huge number of parameters in the capsule network also becomes a significant burden for DcCapsGAN. The DCFSL obtains acceptable numerical indexes with low efficiency based on conditional adversarial domain daptation strategy. Compared with LMAFN, we can observe from Tables 2–4 that the proposed HyperLiteNet further compresses the quantity of parameters and greatly improves network efficiency, which is beneficial to the deployment of the HyperLiteNet on edge devices. Meanwhile, the feature extraction capabilities of the HyperLiteNet support satisfactory classification results with limited labeled samples. Generally, the proposed HyperLiteNet achieves the optimal classification accuracy among all comparison methods with the lowest number of parameters and execution costs.

The qualitative comparisons of the classification maps for different algorithms are illustrated in Figures 9–11. In our expectation, due to the lack of spatial information, SVM indicates the most random misclassified noise labels on the three datasets. In contrast, other methods that exploit spectral–spatial combination information present smoother classification maps than SVM. In addition, the SSRN and DcCapsGAN have more distinguishable classification boundaries based on more elaborately designed structures involved in the structure, whereas the number of parameters also increased dramatically. We can further observe that the HyperLiteNet produces a smoother classification area compared with other comparison algorithms in the red dotted-line area in Figure 11i. These results further demonstrate that the proposed HyperLiteNet can obtain more satisfactory results with fewer parameters and more efficient execution times, even under a limited training sample size.

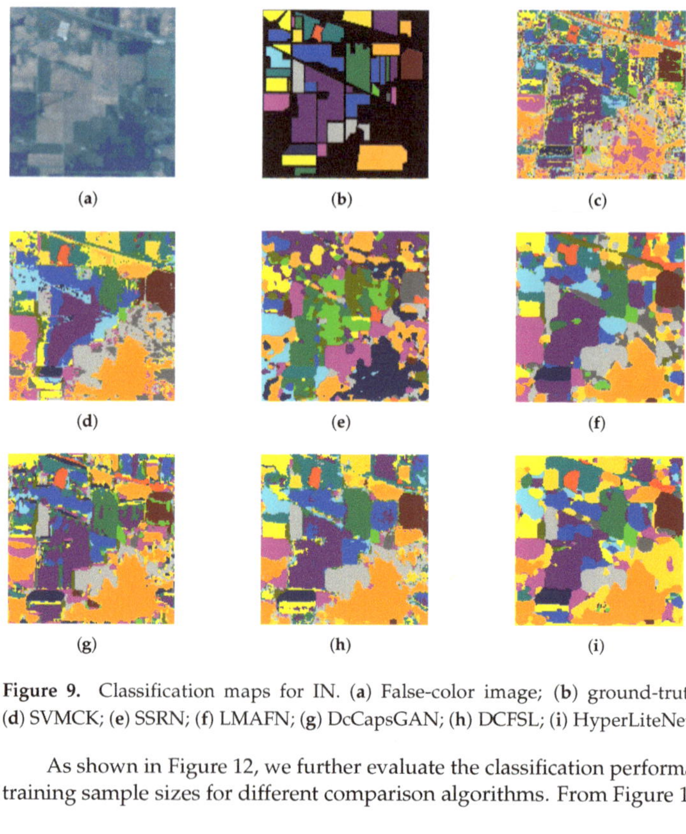

Figure 9. Classification maps for IN. (**a**) False-color image; (**b**) ground-truth map; (**c**) SVM; (**d**) SVMCK; (**e**) SSRN; (**f**) LMAFN; (**g**) DcCapsGAN; (**h**) DCFSL; (**i**) HyperLiteNet.

As shown in Figure 12, we further evaluate the classification performance with varied training sample sizes for different comparison algorithms. From Figure 12, we can notice that the classification accuracy gradually improves as the number of training samples increases. It is easy to find that the proposed HyperLiteNet exhibits acceptable satisfactory classification results with different sample sizes.

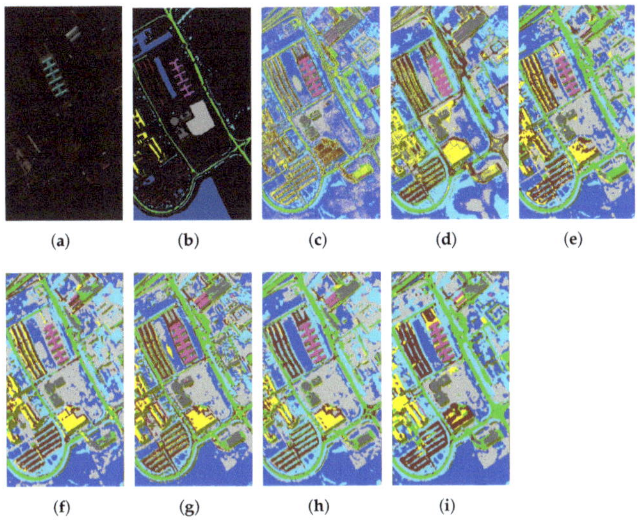

Figure 10. Classification maps for UP. (a) False-color image; (b) ground-truth map; (c) SVM; (d) SVMCK; (e) SSRN; (f) LMAFN; (g) DcCapsGAN; (h) DCFSL; (i) HyperLiteNet.

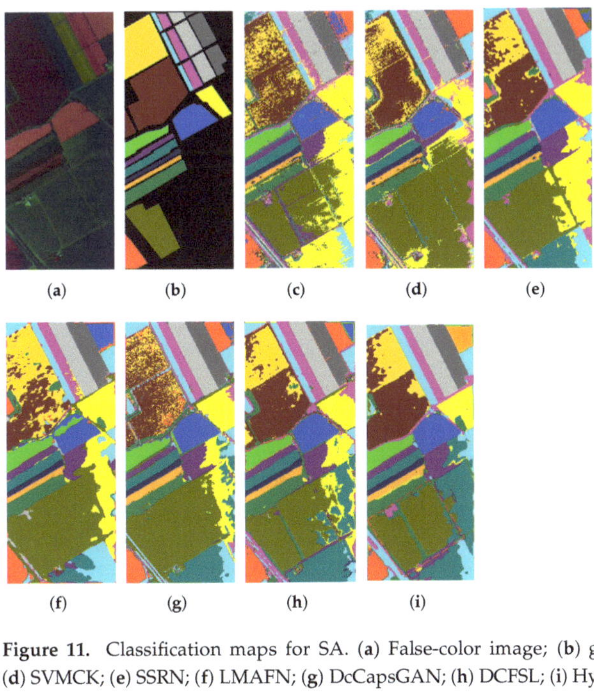

Figure 11. Classification maps for SA. (a) False-color image; (b) ground-truth map; (c) SVM; (d) SVMCK; (e) SSRN; (f) LMAFN; (g) DcCapsGAN; (h) DCFSL; (i) HyperLiteNet.

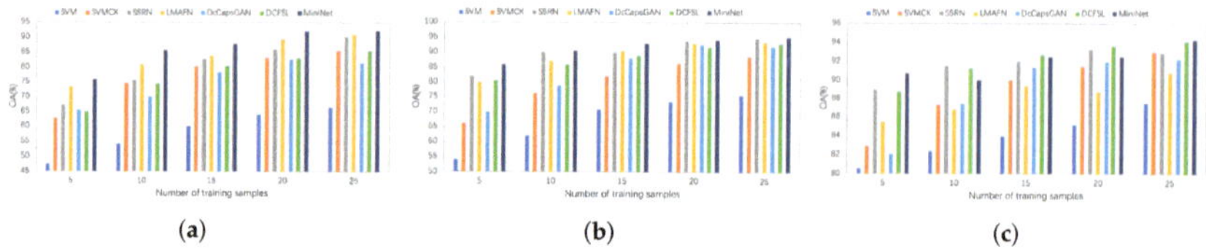

Figure 12. Overall Accuracy (OA) % with different training samples with 5, 10, 15, 20, and 25 for (a) Indian Pines (IN), (b) Pavia University (UP), and (c) Salinas (SA).

4. Discussion

4.1. Assessment of the Model Design

According to the experimental results of the different numbers of training epochs (Figure 7c), the HyperLiteNet has the characteristic of fast convergence for different datasets during the training process. In addition, Figure 7a,b confirm that the model complexity and model convergence ability are negatively correlated. The smaller the model, the more difficult it is to find the appropriate parameters. Therefore, the key is to achieve a suitable balance between the efficiency and the model size. Many practical applications only focus on inference costs. Therefore, more training time in exchange for less inference time is feasible for small algorithm models, which can facilitate online leaning in open-scenario circumstances. On the other hand, as can be seen from Table 1, the dynamic convolution on the three datasets improves model performance (from 0.35% to 2.78%) while reducing the number of parameters over 60% and 80% when compared to the SC and MSC. Meanwhile, the dynamic characteristic support the stability of the model on different datasets. This conclusion is further demonstrated through the t-SNE visualization in Figure 8.

4.2. Assessment of the Model Performances

As illustrated in Tables 2–4, the HyperLiteNet achieves the highest classification OA—75.75%, 85.89%, and 90.70%—with gains of 2.34%, 3.86%, and 1.75% over the optimal results in the comparison algorithms on the three experimental datasets under extremely small training sizes. Through an elaborate, lightweight design, the HyperLiteNet reduces the number of parameters by 66%–98% and Flops by 60%–99%, compared to other DL-based comparison algorithms. In terms of training time and testing time, even compared with the fastest LMAFN, the HyperLiteNet still achieves 2.6-times and 4.6-times efficiency improvements. Experiments on three representative datasets indicate that the proposed HyperLiteNet outperforms these state-of-the-art DL-based comparison algorithms with extremely high efficiency and a low number of parameters. However, Figures 9–11 show that the classification maps generated by the HyperLiteNet presents oversmoothed results. This phenomenon illustrates that HyperLiteNet achieves efficient classification performance, whereas the lightweight structure of spectral information is apt to result in oversmooth boundaries of classes.

5. Conclusions

At present, small, portable smart devices require low energy consumption and fast response times. This limits the efficiency and size of the DL models. Therefore, it is not advisable to blindly pursue the high complexity of the model in exchange for the improvement of the performance. In this article, an extremely lightweight, non-deep MiniNet model is proposed for HSI classification. The proposed model employs dynamic convolution and a feature interaction module to realize the flexible extraction and interaction of abstract features in different stages, which can further improve the discriminant information of different inputs. Meanwhile, the dual-branch PW and dynamic convolutional structure plays an important role in extracting spectral and spatial features in a hardware-friendly manner through the shallow parallel lightweight structure design, with extremely low numbers of parameters and execution costs.

Author Contributions: Conceptualization, J.W. and R.H.; methodology, J.W. and R.H.; validation, S.G., B.L. and L.L.; investigation, J.W., R.H. and S.G.; writing—original draft preparation, J.W. and R.H.; writing—review and editing, J.W., R.H., Z.P. and B.L. All authors have read and agreed to the published version of the manuscript.

Funding: This work was supported in part by the National Natural Science Foundation of China under grant numbers 61801353 and 61971273, in part by the Key Research and Development Program in Shaanxi Province of China under grant number 2021GY-032, in part by GHfund B under grant number 202107020822, and in part by the Project Supported by the China Postdoctoral Science Foundation funded project under grant number 2018M633474.

Institutional Review Board Statement: Not applicable.

Informed Consent Statement: Not applicable.

Data Availability Statement: Not applicable.

Conflicts of Interest: The authors declare no conflict of interest. The funders had no role in the design of the study; in the collection, analyses, or interpretation of data; in the writing of the manuscript, or in the decision to publish the results.

References

1. Xue, Z.; Li, J.; Cheng, L.; Du, P. Spectral–spatial classification of hyperspectral data via morphological component analysis-based image separation. *IEEE Trans. Geosci. Remote Sens.* **2014**, *53*, 70–84.
2. Zhang, L.; Zhang, L.; Tao, D.; Huang, X.; Du, B. Hyperspectral remote sensing image subpixel target detection based on supervised metric learning. *IEEE Trans. Geosci. Remote Sens.* **2013**, *52*, 4955–4965. [CrossRef]
3. Wan, Y.; Hu, X.; Zhong, Y.; Ma, A.; Wei, L.; Zhang, L. Tailings reservoir disaster and environmental monitoring using the UAV-ground hyperspectral joint observation and processing: A case of study in Xinjiang, the belt and road. In Proceedings of the IGARSS 2019-2019 IEEE International Geoscience and Remote Sensing Symposium, Yokohama, Japan, 28 July–2 August 2019; pp. 9713–9716.
4. Coppin, P.; Jonckheere, I.; Nackaerts, K.; Muys, B.; Lambin, E. Digital change detection methods in ecosystem monitoring: A review. *Int. J. Remote Sens.* **2004**, *25*, 1565–1596. [CrossRef]
5. Hughes, G. On the mean accuracy of statistical pattern recognizers. *IEEE Trans. Inf. Theory* **1968**, *14*, 55–63. [CrossRef]
6. Wang, L.; Li, H.C.; Xue, B.; Chang, C.I. Constrained band subset selection for hyperspectral imagery. *IEEE Geosci. Remote Sens. Lett.* **2017**, *14*, 2032–2036. [CrossRef]
7. Palsson, F.; Sveinsson, J.R.; Ulfarsson, M.O.; Benediktsson, J.A. Model-based fusion of multi-and hyperspectral images using PCA and wavelets. *IEEE Trans. Geosci. Remote Sens.* **2014**, *53*, 2652–2663. [CrossRef]
8. Falco, N.; Benediktsson, J.A.; Bruzzone, L. A study on the effectiveness of different independent component analysis algorithms for hyperspectral image classification. *IEEE J. Sel. Top. Appl. Earth Obs. Remote Sens.* **2014**, *7*, 2183–2199. [CrossRef]
9. Gordon, C. A generalization of the maximum noise fraction transform. *IEEE Trans. Geosci. Remote Sens.* **2000**, *38*, 608–610. [CrossRef]
10. Tarabalka, Y.; Fauvel, M.; Chanussot, J.; Benediktsson, J.A. SVM-and MRF-based method for accurate classification of hyperspectral images. *IEEE Geosci. Remote Sens. Lett.* **2010**, *7*, 736–740. [CrossRef]
11. Song, W.; Li, S.; Kang, X.; Huang, K. Hyperspectral image classification based on KNN sparse representation. In Proceedings of the 2016 IEEE International Geoscience and Remote Sensing Symposium (IGARSS), Beijing, China, 10–15 July 2016; pp. 2411–2414.
12. Li, J.; Bioucas-Dias, J.M.; Plaza, A. Spectral–spatial hyperspectral image segmentation using subspace multinomial logistic regression and Markov random fields. *IEEE Trans. Geosci. Remote Sens.* **2011**, *50*, 809–823. [CrossRef]
13. Joelsson, S.R.; Benediktsson, J.A.; Sveinsson, J.R. Random forest classifiers for hyperspectral data. In Proceedings of the 2005 IEEE International Geoscience and Remote Sensing Symposium, Seoul, Korea, 29 July 2005; Volume 1, pp. 4–8.
14. Camps-Valls, G.; Gomez-Chova, L.; Muñoz-Marí, J.; Vila-Francés, J.; Calpe-Maravilla, J. Composite kernels for hyperspectral image classification. *IEEE Geosci. Remote Sens. Lett.* **2006**, *3*, 93–97. [CrossRef]
15. Chen, Y.; Nasrabadi, N.M.; Tran, T.D. Classification for hyperspectral imagery based on sparse representation. In Proceedings of the 2010 2nd Workshop on Hyperspectral Image and Signal Processing: Evolution in Remote Sensing, Reykjavik, Iceland, 14–16 June 2010; pp. 1–4.
16. Wang, J.; Jiao, L.; Liu, H.; Yang, S.; Liu, F. Hyperspectral image classification by spatial–spectral derivative-aided kernel joint sparse representation. *IEEE J. Sel. Top. Appl. Earth Obs. Remote Sens.* **2015**, *8*, 2485–2500. [CrossRef]
17. Wang, J.; Jiao, L.; Wang, S.; Hou, B.; Liu, F. Adaptive nonlocal spatial–spectral kernel for hyperspectral imagery classification. *IEEE J. Sel. Top. Appl. Earth Obs. Remote Sens.* **2016**, *9*, 4086–4101. [CrossRef]
18. Zhang, L.; Zhang, L.; Du, B. Deep learning for remote sensing data: A technical tutorial on the state of the art. *IEEE Geosci. Remote Sens. Mag.* **2016**, *4*, 22–40. [CrossRef]

19. Chen, Y.; Zhao, X.; Jia, X. Spectral–spatial classification of hyperspectral data based on deep belief network. *IEEE J. Sel. Top. Appl. Earth Obs. Remote Sens.* **2015**, *8*, 2381–2392. [CrossRef]
20. Zhong, Z.; Li, J.; Luo, Z.; Chapman, M. Spectral–spatial residual network for hyperspectral image classification: A 3-D deep learning framework. *IEEE Trans. Geosci. Remote Sens.* **2017**, *56*, 847–858. [CrossRef]
21. Gao, H.; Zhang, Y.; Chen, Z.; Li, C. A Multiscale Dual-Branch Feature Fusion and Attention Network for Hyperspectral Images Classification. *IEEE J. Sel. Top. Appl. Earth Obs. Remote Sens.* **2021**, *14*, 8180–8192. [CrossRef]
22. Li, S.; Luo, X.; Wang, Q.; Li, L.; Yin, J. H2AN: Hierarchical Homogeneity-Attention Network for Hyperspectral Image Classification. *IEEE Trans. Geosci. Remote Sens.* **2021**, 1–16. [CrossRef]
23. Zhang, T.; Shi, C.; Liao, D.; Wang, L. A Spectral Spatial Attention Fusion with Deformable Convolutional Residual Network for Hyperspectral Image Classification. *Remote Sens.* **2021**, *13*, 3590. [CrossRef]
24. Sabour, S.; Frosst, N.; Hinton, G.E. Dynamic Routing Between Capsules. *arXiv* **2017**, arXiv:1710.09829.
25. Lei, R.; Zhang, C.; Liu, W.; Zhang, L.; Zhang, X.; Yang, Y.; Huang, J.; Li, Z.; Zhou, Z. Hyperspectral Remote Sensing Image Classification Using Deep Convolutional Capsule Network. *IEEE J. Sel. Top. Appl. Earth Obs. Remote Sens.* **2021**, *14*, 8297–8315. [CrossRef]
26. Wang, J.; Guo, S.; Huang, R.; Li, L.; Zhang, X.; Jiao, L. Dual-Channel Capsule Generation Adversarial Network for Hyperspectral Image Classification. *IEEE Trans. Geosci. Remote Sens.* **2022**, *60*, 1–16. [CrossRef]
27. Bianchini, M.; Scarselli, F. On the complexity of neural network classifiers: A comparison between shallow and deep architectures. *IEEE Trans. Neural Netw. Learn. Syst.* **2014**, *25*, 1553–1565. [CrossRef] [PubMed]
28. Zagoruyko, S.; Komodakis, N. Wide Residual Networks. *arXiv* **2016**, arXiv:1605.07146.
29. Wu, Z.; Shen, C.; Van Den Hengel, A. Wider or deeper: Revisiting the resnet model for visual recognition. *Pattern Recognit.* **2019**, *90*, 119–133. [CrossRef]
30. Tan, M.; Le, Q. Efficientnet: Rethinking model scaling for convolutional neural networks. In *International Conference on Machine Learning*; PMLR: Long Beach, CA, USA, 2019; pp. 6105–6114.
31. Goyal, A.; Bochkovskiy, A.; Deng, J.; Koltun, V. Non-deep Networks. *arXiv* **2021**, arXiv:2110.07641.
32. Iandola, F.N.; Moskewicz, M.W.; Ashraf, K.; Han, S.; Dally, W.J.; Keutzer, K. SqueezeNet: AlexNet-level accuracy with 50x fewer parameters and <1 MB model size. *arXiv* **2016**, arXiv:1602.07360.
33. Zhang, X.; Zhou, X.; Lin, M.; Sun, J. Shufflenet: An extremely efficient convolutional neural network for mobile devices. In Proceedings of the IEEE Conference on Computer Vision and Pattern Recognition, Salt Lake City, UT, USA, 18–23 June 2018; pp. 6848–6856.
34. Howard, A.G.; Zhu, M.; Chen, B.; Kalenichenko, D.; Wang, W.; Weyand, T.; Andreetto, M.; Adam, H. MobileNets: Efficient Convolutional Neural Networks for Mobile Vision Applications. *arXiv* **2017**, arXiv:1704.04861.
35. Sandler, M.; Howard, A.; Zhu, M.; Zhmoginov, A.; Chen, L.C. Mobilenetv2: Inverted residuals and linear bottlenecks. In Proceedings of the IEEE Conference on Computer Vision and Pattern Recognition, Salt Lake City, UT, USA, 18–23 June 2018; pp. 4510–4520.
36. Howard, A.; Sandler, M.; Chu, G.; Chen, L.C.; Chen, B.; Tan, M.; Wang, W.; Zhu, Y.; Pang, R.; Vasudevan, V.; et al. Searching for mobilenetv3. In Proceedings of the IEEE/CVF International Conference on Computer Vision, Long Beach, CA, USA, 9 May 2019; pp. 1314–1324.
37. Han, K.; Wang, Y.; Tian, Q.; Guo, J.; Xu, C.; Xu, C. Ghostnet: More features from cheap operations. In Proceedings of the IEEE/CVF Conference on Computer Vision and Pattern Recognition, Seattle, WA, USA, 13–19 June 2020; pp. 1580–1589.
38. Zhang, C.; Xu, Y.; Shen, Y. CompConv: A Compact Convolution Module for Efficient Feature Learning. In Proceedings of the IEEE/CVF Conference on Computer Vision and Pattern Recognition, Nashville, TN, USA, 20–25 June 2021; pp. 3012–3021.
39. Li, D.; Hu, J.; Wang, C.; Li, X.; She, Q.; Zhu, L.; Zhang, T.; Chen, Q. Involution: Inverting the inherence of convolution for visual recognition. In Proceedings of the IEEE/CVF Conference on Computer Vision and Pattern Recognition, Nashville, TN, USA, 10 March 2021; pp. 12321–12330.
40. Li, Y.; Chen, Y.; Dai, X.; Chen, D.; Liu, M.; Yuan, L.; Liu, Z.; Zhang, L.; Vasconcelos, N. MicroNet: Towards Image Recognition with Extremely Low FLOPs. *arXiv* **2020**, arXiv:2011.12289.
41. Chen, H.; Wang, Y.; Xu, C.; Shi, B.; Xu, C.; Tian, Q.; Xu, C. AdderNet: Do we really need multiplications in deep learning? In Proceedings of the IEEE/CVF Conference on Computer Vision and Pattern Recognition, Seattle, WA, USA, 13–19 June 2020; pp. 1468–1477.
42. Ma, N.; Zhang, X.; Zheng, H.T.; Sun, J. Shufflenet v2: Practical guidelines for efficient cnn architecture design. In Proceedings of the European Conference on Computer Vision (ECCV), Munich, Germany, 8–14 September 2018; pp. 116–131.
43. Zhang, H.; Li, Y.; Jiang, Y.; Wang, P.; Shen, Q.; Shen, C. Hyperspectral classification based on lightweight 3-D-CNN with transfer learning. *IEEE Trans. Geosci. Remote Sens.* **2019**, *57*, 5813–5828. [CrossRef]
44. Jia, S.; Lin, Z.; Xu, M.; Huang, Q.; Zhou, J.; Jia, X.; Li, Q. A lightweight convolutional neural network for hyperspectral image classification. *IEEE Trans. Geosci. Remote Sens.* **2020**, *59*, 4150–4163. [CrossRef]
45. Radosavovic, I.; Kosaraju, R.P.; Girshick, R.; He, K.; Dollár, P. Designing network design spaces. In Proceedings of the IEEE/CVF Conference on Computer Vision and Pattern Recognition, Seattle, WA, USA, 14–19 June 2020; pp. 10428–10436.
46. Wang, J.; Huang, R.; Guo, S.; Li, L.; Zhu, M.; Yang, S.; Jiao, L. NAS-Guided Lightweight Multiscale Attention Fusion Network for Hyperspectral Image Classification. *IEEE Trans. Geosci. Remote Sens.* **2021**, *59*, 8754–8767. [CrossRef]

47. Cui, B.; Dong, X.M.; Zhan, Q.; Peng, J.; Sun, W. LiteDepthwiseNet: A Lightweight Network for Hyperspectral Image Classification. *IEEE Trans. Geosci. Remote Sens.* **2022**, *60*, 5502915. [CrossRef]
48. Yang, B.; Bender, G.; Le, Q.V.; Ngiam, J. Soft Conditional Computation. *arXiv* **2019**, arXiv:1904.04971.
49. Zhang, Y.; Zhang, J.; Wang, Q.; Zhong, Z. DyNet: Dynamic Convolution for Accelerating Convolutional Neural Networks. *arXiv* **2020**, arXiv:2004.10694.
50. Li, Y.; Yuan, L.; Chen, Y.; Wang, P.; Vasconcelos, N. Dynamic Transfer for Multi-Source Domain Adaptation. In Proceedings of the IEEE/CVF Conference on Computer Vision and Pattern Recognition, Nashville, TN, USA, 20–25 June 2021; pp. 10998–11007.
51. Liu, B.; Yu, X.; Yu, A.; Zhang, P.; Wan, G.; Wang, R. Deep few-shot learning for hyperspectral image classification. *IEEE Trans. Geosci. Remote Sens.* **2018**, *57*, 2290–2304. [CrossRef]
52. Li, S.; Zhang, J.; Ma, W.; Liu, C.H.; Li, W. Dynamic Domain Adaptation for Efficient Inference. In Proceedings of the IEEE/CVF Conference on Computer Vision and Pattern Recognition, Nashville, TN, USA, 10 March 2021; pp. 7832–7841.
53. Zhong, C.; Zhang, J.; Wu, S.; Zhang, Y. Cross-Scene Deep Transfer Learning With Spectral Feature Adaptation for Hyperspectral Image Classification. *IEEE J. Sel. Top. Appl. Earth Obs. Remote Sens.* **2020**, *13*, 2861–2873. [CrossRef]
54. Hospedales, T.M.; Antoniou, A.; Micaelli, P.; Storkey, A.J. Meta-Learning in Neural Networks: A Survey. *arXiv* **2020**, arXiv:2004.05439.
55. Li, Z.; Liu, M.; Chen, Y.; Xu, Y.; Li, W.; Du, Q. Deep Cross-Domain Few-Shot Learning for Hyperspectral Image Classification. *IEEE Trans. Geosci. Remote Sens.* **2022**, *60*, 5501618. [CrossRef]
56. Liang, X.; Zhang, Y.; Zhang, J. Attention Multisource Fusion-Based Deep Few-Shot Learning for Hyperspectral Image Classification. *IEEE J. Sel. Top. Appl. Earth Obs. Remote Sens.* **2021**, *14*, 8773–8788. [CrossRef]

sustainability

Article

Machine Learning for Pan Evaporation Modeling in Different Agroclimatic Zones of the Slovak Republic (Macro-Regions)

Beáta Novotná [1,*], Ľuboš Jurík [1], Ján Čimo [1], Jozef Palkovič [2], Branislav Chvíla [3] and Vladimír Kišš [4]

[1] Institute of the Landscape Engineering, Faculty of Horticulture and Landscape Engineering, Slovak University of Agriculture, 949 76 Nitra, Slovakia; Lubos.Jurik@uniag.sk (Ľ.J.); Jan.Cimo@uniag.sk (J.Č.)
[2] Institute of Statistics, Operation Research and Mathematics, Faculty of Economics and Management, Slovak University of Agriculture, 949 76 Nitra, Slovakia; Jozef.Palkovic@uniag.sk
[3] Meteorological and Climatological Monitoring, Network of Ground Synoptic Stations, Slovak Hydrometeorological Institute, 833 15 Bratislava, Slovakia; branislav.chvila@shmu.sk
[4] AgroBioTech Research Centre, Slovak University of Agriculture, 949 76 Nitra, Slovakia; vladimir.kiss@uniag.sk
* Correspondence: Beata.Novotna@uniag.sk

Abstract: Global climate change is likely to influence evapotranspiration (ET); as a result, many ET calculation methods may not give accurate results under different climatic conditions. The main objective of this study is to verify the suitability of machine learning (ML) models as calculation methods for pan evaporation modeling on the macro-regional scale. The most significant PE changes in the different agroclimatic zones of the Slovak Republic were compared, and their considerable impacts were analyzed. On the basis of the agroclimatic zones, 35 meteorological stations distributed across Slovakia were classified into six macro-regions. For each of the meteorological stations, 11 variables were applied during the vegetation period in the years from 2010 to 2020 with a daily time step. The performance of eight different ML models—the neural network (NN) model, the autoneural network (AN) model, the decision tree (DT) model, the Dmine regression (DR) model, the DM neural network (DM NN) model, the gradient boosting (GB) model, the least angle regression (LARS) model, and the ensemble model (EM)—was employed to predict PE. It was found that the different models had diverse prediction accuracies in various geographical locations. In this study, the results of the values predicted by the individual models are compared.

Keywords: pan evaporation; agroclimatic zone; macro-region; climatic characteristic; machine learning

Citation: Novotná, B.; Jurík, Ľ.; Čimo, J.; Palkovič, J.; Chvíla, B.; Kišš, V. Machine Learning for Pan Evaporation Modeling in Different Agroclimatic Zones of the Slovak Republic (Macro-Regions). *Sustainability* **2022**, *14*, 3475. https://doi.org/10.3390/su14063475

Academic Editors: Miltiadis D. Lytras and Andreea Claudia Serban

Received: 11 December 2021
Accepted: 15 February 2022
Published: 16 March 2022

1. Introduction

Changes in climatic factors, such as in temperature, wind speed, sunshine hours, humidity, and solar radiation, can have a significant impact on the evapotranspiration (ET) process [1]. ET is the combination of two separate water loss processes: water's evaporation from the soil and plant surfaces and plant transpiration, through which water escapes from a plant's body into the ambient air through its stomata in the form of steam [2,3]. Evaporation is an important component of the hydrological cycle [4]. Pan evaporation (PE) is extensively used in developing irrigation projects and provincial water resources [5,6]. PE is an important climatic variable for developing efficient water resource management strategies [7]. Generally, two approaches—(i) direct (i.e., pan evaporimeter) and (ii) indirect (i.e., empirical or semi-empirical equations)—are used to measure evaporation [8,9]. The direct estimation of PE using a class A pan evaporimeter has limited spatial coverage because of practical and instrumental problems [9–11]. In contrast, the application of the indirect PE estimation method based on the relationship of ET with various climatic parameters is often restricted due to data availability and climate variability [9,12,13]. Considering the limitations of both the methods, machine learning (ML) techniques have been used in recent years as an alternative [9]. In the past, many ML models for PE modeling using different combinations of available climatic variables have been reported in the literature [13]. Owing to the rapid development of ML techniques, a series of powerful tools has been proposed over the last two decades, allowing the scientific community to obtain new insights into the patterns of ET on different spatial scales, ranging from ecosystem to global. Elucidating the biophysical mechanisms governing the exchange of water vapor between land and the atmosphere is particularly crucial for addressing water scarcity under climate change [14]. Some applications of ML models for PE estimation are described in [1,9,13,15,16]. In the last decade, advances in computation have led to the introduction of ML methodologies for referencing evapotranspiration calculations, and the high accuracy of their results has been proven by using different approaches [17]. In the last few decades, ML techniques have been increasingly utilized to estimate hydrological variables [18–22], ecological variables [23], and renewable energy variables [24], as described in [14]. Because ML techniques solve the non-linear relationships between input and output variables [14], many ML techniques have been proposed to estimate ET for hydrological applications [25], such as k-nearest neighbors [26], support vector machines [27], random forests [26], and artificial neural networks (ANNs) [28]. Previously, most studies applied ML approaches to in situ measurements; however, many recent studies have also applied ML approaches to remote sensing data [29–31], as shown in [32], as well as [33,34].

Potential evapotranspiration represents the upper limit of ET when this process is not limited by water deficit in the soil [35]. Potential evapotranspiration in Slovakia is estimated according to empirical or semi-empirical relationships based on measurements of other meteorological elements, as described by the authors of [36], who compared the daily reference crop's (grass cover) potential evapotranspiration results which were calculated with two modifications of the Penman–Monteith equation. The authors of [37] studied drought occurrence using the Standardized Precipitation and Evapotranspiration Index (SPEI) and the Standardized Precipitation Index (SPI). Maps were constructed based on data calculated using the Budyko–Zubenokova method in 31 Slovak climatological stations [35].

Several local-scale studies have been conducted in the Slovak Republic, e.g., [35,38,39]; however, an overall PE estimation with ML techniques and by using all of the available pan vapor data for the whole territory of the country is missing, and no similar works have been carried out by using these kinds of methods. In order to obtain an overview about the present state of research in this field, a review of related research articles was carried out, and details of some recently published articles are listed in Table 1.

Table 1. Example of ML applications for PE estimation.

Author	Country	Recommended Model for PE Estimation	Input Climatic Data Variables *	Statistical Indices **	Number of the Meteorological Stations
Majhi and Naidu (2021) [13]	India	functional link artificial neural network (FLANN)	PE, T_{max}, T_{min}, RH1, RH11	RMSE = 0.85; MAE = 0.63; EF = 0.70	3
Kisi, O. (2015) [40]	Mediterranean Region of Turkey	multivariate adaptive regression splines (MARS), M5 Model Tree (M5Tree)	PE, T_{aver} SR, RH, Us	RMSE = 0.189	2
Zounemat-Kermani et al. (2021) [16]	Turkey	Levenberg–Marquardt (MLP-LM)	PE, T_{max}, T_{min}, SR, S, RH, Us	MAE = 0.492; d = 0.981	2
Malik et al. (2021) [9]	Northern India	Slap Swarm Algorithm (SVR-SSA)	PE, T_{max}, T_{min}, RH_{max}, Rhmin, SR, Us	MAE = 0.697; RMSE = 1.1; IOS = 0.250; NSE = 0.861; PCC = 0.929; IOA = 0.960	3

Table 1. *Cont.*

Author	Country	Recommended Model for PE Estimation	Input Climatic Data Variables *	Statistical Indices **	Number of the Meteorological Stations
Abed et al. (2021) [41]	Malaysia	Long Short-Term Memory Neural Network (LSTM)	PE, T_{aver}, T_{max}, T_{min}, RH, SR, Us	R^2 = 0.970; MAE = 0.135; MSE = 0.027; RMSE = 0.166; RAE = 0.173; RSE = 0.029	2
Ferreira et al. (2019) [42]	Brazil	multivariate adaptive regression splines (MARS)	Etr, T_{aver}, SR, Us, G, es, ea, Δ, y	R^2 (0.79–0.85); RMSE (0.41–0.54); MAE (0.34–0.46)	8
Al-Mukhtar (2021) [1]	middle, south, and north of Iraq	weighted K-nearest neighbor (KKNN)	PE, T_{max}, T_{min}, T, RH, Us	R^2 = 0.98; RMSE = 26.39; MAE = 18.62; NSE = 0.97; PBIAS = 3.8	3
Wang et al. (2017) [11]	China	multiple linear regression (MLR), Stephens and Stewart model (SS)	PE,T_{aver}, SR, S, RH, Us	R^2 = 0.988; RSME (0.314–0.405)	8
Sattari et al. (2021) [43]	Northwest Iran	M5 tree model (M5Tree)	PE, T_{aver}, RH, Us, P	RMSE (0.0042–0.0058); R^2(0.9916–0.9952); *t*-test (0.722–0.96); NSE (0.989 to 0.994)	4
Adnan et al. (2017) [44]	Pakistan	principal component analysis (PCA)	PE, T_{max}, T_{min}, T_{aver} RH, SR, Us, P	R = 0.83426	1
Simon-Gáspár et al. (2021) [45]	Hungary	multiple stepwise regression (MLR)	PE, T_{aver}, T_{max}, T_{min}, RH, Us, Rs	RMSE = 0.834; MAE = 0.660; S = 0.217	1

* Pan evaporation (PE); reference evapotranspiration (Etr); average, maximum, minimum, morning, and afternoon relative humidity (RH, Rh_{max}, Rh_{min}, RH1, and RH11); average, maximum, and minimum air temperature (T_{aver}, T_{min}, and T_{max}); wind speed (Us); relative sunshine duration (S); solar radiation (SR); soil heat flux (G); saturation vapor pressure (es); actual vapor pressure (ea); slope of the saturation vapor pressure function (Δ); psychometric constant (y); precipitation (P); global radiation (Rs). ** Root mean square error (RMSE); mean absolute error (MAE); efficiency factor (EF); Willmott index (d); absolute error (MSE); determination coefficient (R2); regression value (R); relative absolute error (RAE); relative squared error (RSE); percentage bias (PBias); unpaired two-sample *t*-test (*t*-test); Nash–Sutcliffe efficiency index (NSE); scatter index (SI).

The North Atlantic Stream, as part of the Gulf Stream, has a favorable effect on the climatic conditions of Western and Northwestern Europe. These effects are mainly reflected in the reduction in temperature fluctuations throughout the year; this means that, for example, during the winter period, the air temperatures do not decrease below the freezing point [46]. However, the weather in Slovakia also affects the flows of air from the Atlantic Ocean and from the Adriatic Sea. The individual weather forecasting models used in Slovakia have also contradicted each other in certain time periods. An example is the year 2021, when weather differed from most of the years of the second decade of the 20th century. Frequent changes in weather were typical for most of the year 2021. This has occurred since the end of summer and during autumn, and it was also characteristic for the month of December. At the beginning of this month, it warmed up, then there was a cooling period; however, this only occurred in some parts of Slovakia, for example in the west and southwest, and it was accompanied by heavy snowfall. In the middle of the month of December, it warmed noticeably, and it lasted until the end of the year [47]. The climate of a particular area is also affected by microclimatic factors, especially the shape of the relief (convex or concave), the orientation of the relief towards the world and the prevailing flow, relative altitude fragmentation, vegetation, and anthropogenic influences. The aim of this study is to obtain information about the areas that are the most affected by PE changes, which is closely related to the manifestations of global climate change.

For the purpose of this study, eight newly explored ML models for estimating PE losses in six macro-regional scales were assessed. The macro-regions were classified according to individual agroclimatic zones. Using available climatic data from 35 different meteorological stations across the Slovak Republic, we performed the estimation of the PE in the vegetation period applying the datasets selected for the years 2010 to 2020, including their different elevations above sea level and geographical locations.

As a result of the above-mentioned literature resources, ML is an important tool for the goal of leveraging technologies around artificial intelligence. ML is, in part, based on a model of brain cell interaction. Although the model was created in 1949 by Donald Hebb (1949) [48] in a book titled *The Organization of Behavior* (Keith, 2021) [49], the applications of ML techniques to evapotranspiration estimation problems are currently limited and the knowledge on the topic is still partial and fragmented [50].

The results of this study are unique because: (1) there is no comparative study for the whole of the Slovak Republic, where all available climate data for the last climate decade for PE estimation are applied; (2) the new progressive ML method was used for PE assessment; and (3) besides the selection of the applicability and predictability of eight different ML models, the macro-regions with the most pronounced manifestations of global climate change are identified.

2. Materials and Methods

2.1. Study Location and Climatic Data Collection

In connection with global climate change, the gradual modification of the energy balance globally, and especially locally, was observed. Therefore, the assessment of PE changes and their trends in the Slovak Republic has its justification. The climatic data from the climatological network of Slovak stations were provided by the Slovak Hydrometeorological Institute (SHMI) in Bratislava. The available PE dataset for the vegetation periods of the years from 2010 to 2020 was analyzed according to individual regions of the Slovak Republic.

The daily data of PE during the main vegetation period (from April to October) from 35 meteorological stations distributed across Slovakia (see Figure 1) were used in this study. There were evaluated data mostly from the ten consecutive years (2010–2020). All available PE data measured in the Slovak Republic by SHMI during the observed periods were applied in this study. The reason for this is not only the availability of the data, but also the application of the same instrument for their measurement, i.e., a GGI—3000 evaporator (see principle in [51]).

Slovak Republic Zoning Criteria

The zoning criteria are based on the agroclimatic division of Slovakia [52]. For the agroclimatic division of Slovakia, three basic agroclimatic indicators were used, according to the area divided into agroclimatic macro-areas, sub-areas, and districts.

(a) The agroclimatic temperature indicator (TS10) is the sum of average daily air temperatures during the period with an average daily temperature of ≥10.0 °C. According to this indicator, the territory of Slovakia is divided into three agroclimatic macro-areas and eight agroclimatic areas:

1. Warm agroclimatic macro-area with TS10 from 3100 to 2400 °C;
2. Slightly warm agroclimatic macro-region with TS10 from 2400 to 2000 °C; and
3. Cold agroclimatic macro-region with TS10 from 2000 to 1600 °C.

(b) The agroclimatic moisture indicator ($K_{VI–VIII}$) is given by the difference of potential evaporation (Eo) and precipitation (P) in the summer months (June–August (VI–VIII)):

$$K_{VI–VIII} = E_o - P \text{ (mm)} \quad (1)$$

E_o values were calculated for the territory of Slovakia by [53]. The climatic indicator "K" expresses the moisture balance of the area well (expressed in millimeters). Positive values $E_o - P$ are characterized by the water deficiency and negative values by the moisture excess.

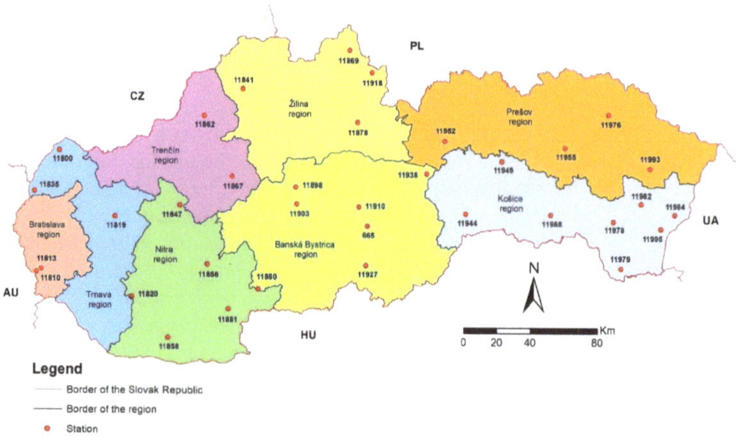

Figure 1. Geographical location of the used climatological stations in the Slovak Republic divided into eight regions and their macro-regional classification: northwest (NW): Trenčín region; southwest (SW): Trnava region, Bratislava region, Nitra region; north-central (NC): Žilina region; south-central (SC): Banská Bystrica region; northeast (NE): Prešov region; southeastern (SE): Košice region.

According to the climatic indicator ($K_{VI-VIII}$), there are seven sub-areas in Slovakia, in which agricultural crops have differently ensured their moisture requirements:

1. Subarea with $K_{VI-VIII} \geq 150$ mm—very dry
2. Subarea with $K_{VI-VIII}$ 150 to 100 mm—mostly dry
3. Subarea with $K_{VI-VIII}$ 100 to 50 mm—slightly dry
4. Subarea with $K_{VI-VIII}$ 50 to 0 mm—slightly humid
5. Subarea with $K_{VI-VIII}$ 0 to -50 mm—mostly humid
6. Subarea with $K_{VI-VIII}$ -50 to -100 mm—humid
7. Subarea with $K_{VI-VIII}$ -100 mm—very humid

(c) The agroclimatic wintering indicator (T_{min}) represents the average of the annual absolute temperature minima. This characteristic effectively describes the climatic conditions during the winter. Absolute temperature minima are important factors for the cultivation of winter crops and fruit trees. They effectively express the critical freezing temperatures. According to the conditions of wintering, there are five districts:

1. Agroclimatic district of mostly mild winter with $T_{min} \geq -18$ °C;
2. Agroclimatic district of relatively mild winters with T_{min} from -18.0 °C to -20.0 °C;
3. Agroclimatic district of mildly cold winter with T_{min} from -20.0 °C to -22.0 °C;
4. Agroclimatic district of mostly cold winter with T_{min} from -22.0 to -24.0 °C;
5. Agroclimatic district of cold winter with $T_{min} \leq -24.0$ °C.

The average annual vertical temperature gradient in Slovakia is 0.61 °C per 100 m of height. During the summer months, however, its value increases to 0.76 °C and, in the winter months, it decreases to 0.33 °C. The territory of the Slovak Republic was divided into eight regions (see Figure 1). The classification of the eight Slovak regions into six macro-regions with 35 individual meteorological stations across Slovakia is shown in Table 2.

According to Figure 1 and Table 2, the division of eight regions into six macro-regions according to the different agroclimatic zones is as follows:

- Northwest (NW): Trenčín region;
- Southwest (SW): Trnava region, Bratislava region, and Nitra region;

- North-central (NC): Žilina region;
- South-central (SC): Banská Bystrica region;
- Northeast (NE): Prešov region;
- Southeastern (SE): Košice region.

The characteristics of the agroclimatic zones were applied to classify eight regions into six macro-regions, which are:

(1) Northwestern Slovakia (NW): Trenčín region—From an agroclimatic point of view, the NW area is assigned to the macro-area of a mildly warm, agroclimatic area of a lightly warm subarea that is moderately humid to mostly humid. Moreover, this area is assigned to agroclimatic precincts with a mild/cold winter to a mostly cold winter. This area also transitions north to the macro-area with a cold, agroclimatic area with a moderately cold sub-area that is mostly humid to humid, and a precinct with mostly cold winters.

(2) Southwestern Slovakia (SW): Bratislava region, Trnava region, and Nitra region—From an agroclimatic point of view, the SW area is assigned to the macro-area of a warm and agroclimatic area that is very warm, a sub-area that is very dry, and a predominantly dry and agroclimatic precinct that has mainly mild winters.

(3) North-central Slovakia (NC): Žilina region—From an agroclimatic point of view, the NC area is assigned to the macro-area of a slightly warm to cold agroclimatic area from a slightly to moderate/mildly warm sub-area up to a slightly cold sub-area. This area is also assigned to slightly dry, moderately humid, mostly humid, and agroclimatic precincts that are mildly cold to mostly cold in the winter. This area transitions north to a macro-area of a cold, agroclimatic area that is mostly cold, a sub-area that is mostly humid to humid, and a precinct that is mostly cold/cold in the winter.

(4) South-central Slovakia (SC): Banská Bystrica region—From an agroclimatic point of view, the SC area at the southernmost part of the state border is assigned to a warm macro-area, a very warm agroclimatic area, a very dry and predominantly dry sub-area, and an agroclimatic precinct with a predominantly mild winter. This area transitions to a macro-area with a moderately warm, an agroclimatic area that is relatively mild/warm, a sub-area that is slightly humid to mostly humid, and a precinct that is slightly cold in the winter.

(5) Northeastern Slovakia (NE): Prešov region—From an agroclimatic point of view, this area is the most diverse. In the southern part, it is considered a warm macro-area, with an agroclimatic area that is sufficiently warm to relatively/moderately warm, sub-areas that are predominantly dry to moderately dry, and agroclimatic precincts that have relatively mild winters to mild cold winters. This area transitions north to a macro-area that is warm or moderately warm to cold, an agroclimatic area that is relatively mild/warm to slightly cold, sub-areas that are moderately dry or slightly humid, and a precinct that is slightly cold in the winter to mostly cold in the winter.

(6) Southeastern Slovakia (SE): Košice region—From an agroclimatic point of view, the SE area is assigned as the macro-area of a warm and agroclimatic area that is very warm, a sub-area that is very dry and a predominantly dry, and a agroclimatic precinct with a predominantly mild winter. This area transitions to an agroclimatic area that is mostly warm, a sub-area that is very dry, and a precinct that has a relatively mild winter.

Table 2. Identification of the individual regions across the territory of the Slovak Republic and their integration according to macro-regions.

Number	Identification	Station Name	Pan Evaporation Measurement		Classification	Region Identification	Classification	Region Identification	Macro-region Classification		
			Setting date	Ending date							
1	665	ĎUBÁKOVO	1 May 2011	30 June 2016	BB (Ban-skobystrický)	SC (south of central Slovakia)	NW (north of western Slovakia)	TN (Trenčiansky)	NW	NC	NE
2	11800	HOLÍČ	1 April 2011	30 June 2015	TT (Trnavský)	SW (south of western Slovakia)	SW (south of western Slovakia)	TT (Trnavský)	SW	SC	SE
3	11810	BRATISLAVA–M. DOLINA	1 April 2011	31 October 2020	BA (Bratislavský)	SW (south of western Slovakia)		BA (Bratislavský)			
4	11813	BRATISLAVA-KOLIBA	1 April 2011	31 October 2011	BA (Bratislavský)	SW (south of western Slovakia)		NR (Nitriansky)			
5	11819	JASLOVSKÉ BOHUNICE	1 April 2011	31 July 2011	TT (Trnavský)	SW (south of western Slovakia)	NC (north of central Slovakia)	ZA (Žilinský)			
6	11820	ŽIHÁREC	1 April 2011	31 October 2020	NR (Nitriansky)	SW (south of western Slovakia)	SC (south of central Slovakia)	BB (Ban-skobystrický)			
7	11835	MORAVSKÝ SVÄTÝ JÁN	1 April 2011	31 May 2016	TT (Trnavský)	SW (south of western Slovakia)	NE (north of eastern Slovakia)	PO (Prešovský)			
8	11841	DOLNÝ HRIČOV	1 April 2011	30 September 2011	ZA (Žilinský)	NC (north of central Slovakia)	SE (south of eastern Slovakia)	KE (Košický)			
9	11847	TOPOĽČANY	1 April 2011	31 October 2020	NR (Nitriansky)	SW (south of western Slovakia)					
10	11856	MOCHOVCE	1 April 2011	31 July 2011	NR (Nitriansky)	SW (south of western Slovakia)					
11	11858	HURBANOVO	1 April 2011	31 October 2020	NR (Nitriansky)	SW (south of western Slovakia)					

Table 2. *Cont.*

Number	Identification	Station Name	Pan Evaporation	Measurement	Classification	Region Identification	Classification	Region Identification	Macro-region Classification
12	11862	BELUŠA	1 April 2011	31 August 2015	TN (Trenčiansky)	NW (north of western Slovakia)			
13	11867	PRIEVIDZA	1 April 2011	31 October 2020	TN (Trenčiansky)	NW (north of western Slovakia)			
14	11869	RABČA	1 May 2011	31 October 2020	ZA (Žilinský)	NC (north of central Slovakia)			
15	11878	LIPTOVSKÝ MIKULÁŠ	1 May 2011	31 October 2020	ZA (Žilinský)	NC (north of central Slovakia)			
16	11880	DUDINCE	1 April 2011	31 October 2020	BB (Ban-skobystrický)	SC (south of central Slovakia)			
17	11881	ŽELIEZOVCE	1 June 2011	31 October 2014	NR (Nitriansky)	SW (south of western Slovakia)			
18	11898	BANSKÁ BYSTRICA	1 May 2011	31 October 2020	BB (Ban-skobystrický)	SC (south of central Slovakia)			
19	11903	SLIAČ	1 April 2011	31 October 2020	BB (Ban-skobystrický)	SC (south of central Slovakia)			
20	11910	LOM NAD RIMAVICOU	1 May 2011	31 October 2020	BB (Ban-skobystrický)	SC (south of central Slovakia)			
21	11918	LIESEK	1 April 2011	31 October 2020	ZA (Žilinský)	NC (north of central Slovakia)			
22	11927	BOĽKOVCE	1 April 2011	31 October 2020	BB (Ban-skobystrický)	SC (south of central Slovakia)			
23	11938	TELGÁRT	1 May 2011	12 October 2020	BB (Ban-skobystrický)	SC (south of central Slovakia)			
24	11944	ROŽŇAVA	1 April 2011	30 October 2017	KE (Košický)	SE (south of eastern Slovakia)			
25	11949	SPIŠSKÉ VLACHY	1 May 2011	31 October 2020	KE (Košický)	SE (south of eastern Slovakia)			

Table 2. *Cont.*

Number	Identification	Station Name	Pan Evaporation	Measurement	Classification	Region Identification	Classification	Region Identification	Macro-region Classification
26	11952	GÁNOVCE	20 April 2011	30 October 2020	PO (Prešovský)	NE (north of eastern Slovakia)			
27	11955	PREŠOV-VOJSKO	1 May 2011	31 October 2020	PO (Prešovský)	NE (north of eastern Slovakia)			
28	11968	KOŠICE, LETISKO	1 April 2011	31 October 2020	KE (Košický)	SE (south of eastern Slovakia)			
29	11976	TISINEC	1 April 2011	31 October 2020	PO (Prešovský)	NE (north of eastern Slovakia)			
30	11978	TREBIŠOV, MILHOSTOV	1 April 2011	30 October 2020	KE (Košický)	SE (south of eastern Slovakia)			
31	11979	SOMOTOR	1 April 2011	31 October 2014	KE (Košický)	SE (south of eastern Slovakia)			
32	11982	MICHALOVCE	1 April 2011	31 October 2020	KE (Košický)	SE (south of eastern Slovakia)			
33	11984	ORECHOVÁ	1 April 2011	31 October 2020	KE (Košický)	SE (south of eastern Slovakia)			
34	11993	KAMENICA N. CIROCHOU	1 April 2011	31 October 2020	PO (Prešovský)	NE (north of eastern Slovakia)			
35	11995	VYSOKÁ NAD UHOM	1 April 2011	31 October 2012	KE (Košický)	SE (south of eastern Slovakia)			

Extreme temperatures are considered a limitation for the production process of field and garden crops and often for all agricultural production. The absolute maximum temperature in Slovakia was measured on 20 July 2017 in Hurbanovo at 40.3 °C and the absolute minimum temperature was measured on 11 February 1929 in Vígľaš-Pstruša at −41.0 °C.

2.2. Case Study and Data Description

In this study, the following measured meteorological data were applied on a daily basis: (1) daily pan evaporation in millimeters (PE); (2) minimum temperature in degrees Celsius (T_{min}); (3) maximum temperature in degrees Celsius (T_{max}); (4) average temperature in degrees Celsius (T_{aver}); (5) relative humidity as a percentage (RH); (6) average wind speed in meters per second (Us) with (7) wind speed measured at 7, 14, and 21 h, as well as (8) prevailing wind directions measured at 7, 14, and 21 h; (9) the sum of precipitation in millimeters (P); and (10) vapor pressure in hectopascals (E). For each station, the (11) elevation above sea level in meters (elevation) and (12) geographical location (coordinates) were applied.

2.3. ML Models and Evaluation Criteria

The data collected daily from 35 meteorological stations were processed into an extensive database (including about 176,000 observation records) in the form of panel data. Each observation was labelled by time and cross-sectional characteristics. The database was divided into a ratio of 40:30:30 into data for model estimation (train), for model validation (Validate) and for model testing (Test). The analysis was performed in SAS Enterprise Miner 15.1. As part of the workflow, the most used modeling techniques for large-scale data analysis were applied. As the target variable, pan evaporation was selected. All the other variables were used as the input variables, including geographical location. Based on the input data, the following models were estimated, validated, and tested.

2.3.1. Neural Network (NN)

The neural network with multilayer perception (MLP) model architecture was tested. Artificial neurons are similar to biological neurons in terms of learning based on experience and generalizations of previous experiences and applying acquired experience to new data. MLP is a neural network inspired by the structure of the real brain [54]. It consists of layers that contain neurons as processing units. Each unit in a layer is associated with all units in the previous layer. The knowledge of the whole network is encoded in the weights of these connections. Such a neural network can accurately predict the outputs of any unknown function. The network uses three nodes (input layer, hidden layer, and output layer), which are neurons with a nonlinear activation function [55] (see Figure 2).

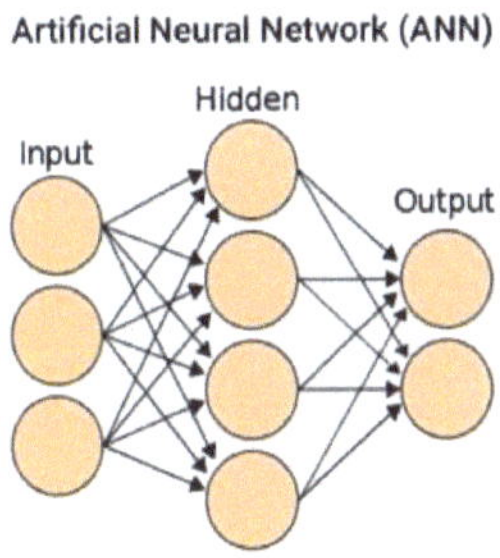

Figure 2. Architecture of a convolutional neural network according to [56].

2.3.2. AutoNeural Network (AN)

An autoneural network is a modeling tool in which the neural network architecture is automatically configured by the software to achieve the best result. The main advantage of a neural network is its ability to detect complex nonlinear relationship between dependent and independent variables and all possible interactions between them [57]. Another advantage of neural networks is the availability of multiple training algorithms which may be applied without deeper statistical knowledge. As disadvantages could be considered a higher computational burden, it has the empirical nature of model development and the tendency towards overfitting. Another disadvantage is the "black box" nature of neural networks, which complicates the further modification and deeper interpretation of model results compared to other types of models.

2.3.3. Decision Tree (DT)

An empirical decision tree represents a segmentation of data that is created based on a series of simple decision rules. The tree is created based on an algorithm that identifies different ways of dividing the dataset into individual segments similar to branches. They can be used for the analysis of categorical as well as quantitative variables. The rule for forming the branches of the decision tree is based on the method of extracting the relationship between the analysis object and one or more inputs to create the segments [58].

2.3.4. Dmine Regression (DR)

Dmine regression is an estimation method using a sequentially applied least squares method. In each step, an independent variable is selected that contributes most to the explained model's variability. In the estimation, bidirectional interactions between classification variables are considered, a possible non-linear relationship between the input variables and the target variable is identified, and group variables are used to reduce the number of the classification variable levels.

2.3.5. DM Neural Network (DM NN)

A DM neural network is the method used to estimate the additive nonlinear model, whereby the main components are used to predict the target variable. In contrast to the neural network, it overcomes the problem of nonlinear estimation and the identification of the global optimal solution with a significantly shorter computational time. In the first step, the method of the main components is applied to the training data, from which a smaller number of components is subsequently selected for use in the next modeling process. At each stage of the modeling process, eight training functions are estimated separately in the training dataset, from which the variant that achieves the best results is selected.

2.3.6. Gradient Boosting (GB)

Gradient boosting is the application of the algorithm published by [59]. The algorithm searches for the optimal partition of data defined in terms of the values of one variable. The optimality of the criterion depends on how the target variable is divided into the individually defined partitions. The more similar the values of the target variable within a segment, the higher the value of the partition. The majority of the similar algorithms work through a process called recursive partitioning. The partitions are then combined to create a predictive model. The accuracy of the model is evaluated by standard indicators based on a comparison with the values of the target variable. As with decision trees, there is no assumption about the distribution of variables that need to comply with the gradient boosting. For interval input variables, the model only works with variable sequences. For analyzed data, where decision trees have achieved good results, gradient boosting can often further improve this result.

2.3.7. Least Angle Regression (LARS)

This type of regression was first published by Efron et al. (2004) [60]. Similar to the forward selection that is often used in regression models, the algorithm produces a sequence of regression models. One additional parameter is added at each step. The sequential procedure is completed after all parameters have entered the model. The algorithm begins by centering the input and output variables. The input variable scales are adjusted to the same number of corrected squares. The initial level of the coefficients is zero, as is the predicted effect on the output variable. Subsequently, the predictor that is most correlated with the residues is identified and a step in the direction of this predictor follows. The step length determines the predictor coefficient. The step length is determined so that any of the other predictors and the current predicted effect have the same correlation with the residues. At this point, the predicted effect moves in a direction that is rectangular between the two predictors. This will ensure that already mentioned predictors have the same correlation with residues. The predicted effect moves in the same direction until the third predictor also has the same correlation with residues as the previous two predictors previously in the model. The new direction is designed to be rectangular between the three predictors. The predicted effect moves again in this direction until a fourth predictor that has the same correlation with residues enters the model. In this way, the process continues until all predictors enter the model.

2.3.8. Least Ensemble Model (EM)

This type of model is one of the methods that combines posterior probabilities and two or more predictive models to create a potentially more accurate model [61]. In the case of the analysis performed, the ensemble model was based on the results of all the above models.

The modeling process is shown in Figure 3. In the initial step, data were collected and processed into the input database. The database was divided into train, validate, and test datasets. In the next step, all the modeling techniques described above in the methodology were applied in the train dataset. Estimated models were validated and tested using validate and test datasets. In the final step, the best model was selected based on the smallest value of average squared error achieved in validation and testing.

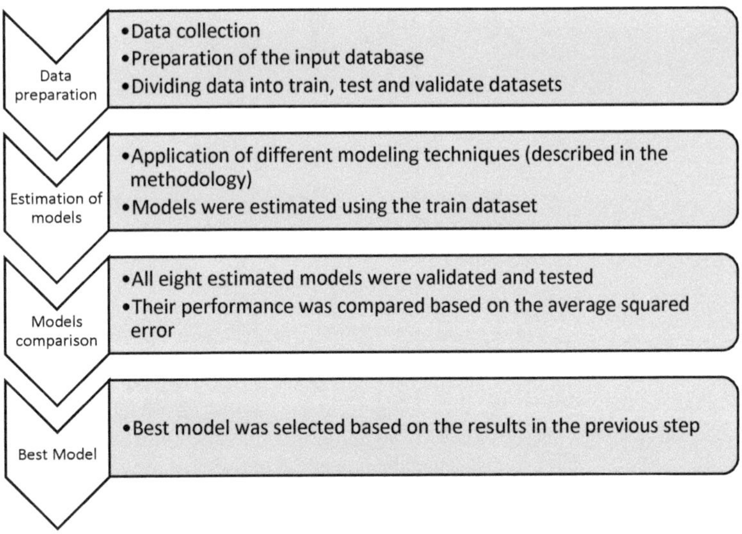

Figure 3. Workflow of the modeling process.

3. Results

3.1. PE Changes at the Macro-Regional Level

Table 3 shows the basic characteristics of PE for individual regions. On average, the lowest PE values were recorded in the NC area, and the highest were recorded in the SE and SW regions. As for the largest change over the period, it could be expressed using the variation range in the last column. It follows that the largest change in terms of PE occurred in the SE region, followed by SC and SW. The smallest change in terms of evaporation was in the NC and NW regions. Similar conclusions can be taken from Figure 4.

Table 3. Overview of the individually analyzed regions of the Slovak Republic with the statistic assessment of their evaluated elements.

Analysis Variable: Pan Evaporation (PE) (mm)						
Region Orientation	**Observation Numbers**	**Mean**	**Median**	**Standard Deviation**	**Coefficient of Variation**	**Range**
SW	23,723	2.55	2.40	1.43	56.27	14.00
SC	18,557	2.28	2.20	1.20	52.76	15.20
NC	11,128	1.96	1.90	1.10	56.18	7.00
SE	22,256	2.55	2.40	1.51	59.35	18.50
NE	11,128	2.32	2.20	1.23	53.02	12.60
NW	5381	2.30	2.20	1.37	59.60	8.50

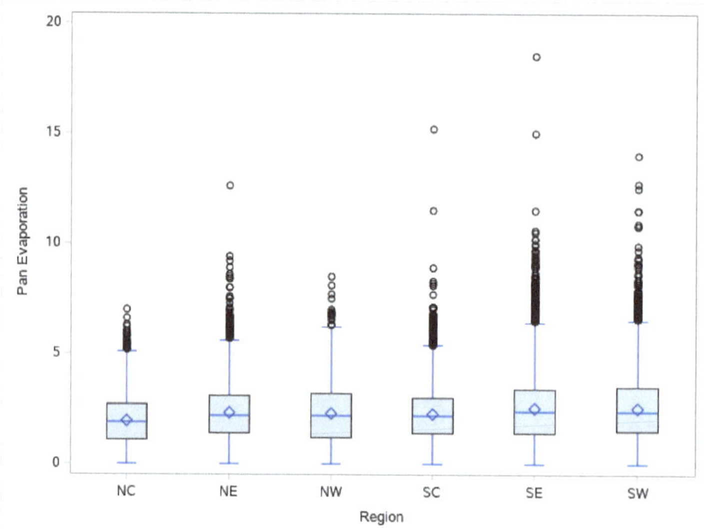

Figure 4. This figure shows the comparison of the individual analyzed regions of the Slovak Republic with the minimum and maximum values of the PE (mm). The box is bounded by the lower and upper quartile, the line in the middle is the median, and the diamond means the average value.

Figure 4 compares the measured PE values in the individual regions. For each region, Figure 4 shows the minimum and maximum values, the box is bounded by the lower and upper quartile, the line in the middle is the median, and the diamond means the average value. Values that are displayed as points above the graph are considered extreme values. The values are in accordance with Table 4, where the lowest value on average is in NC, and the highest values are in SW and SE. The displayed values represent the measurements taken during the vegetation season. In terms of the graphical range of values, similar results were also achieved, either with or without the consideration of extreme values. The highest margin is in SE and SW, or SC, where the highest PE changes were found. If we assess the differences from a statistical point of view, Table 4 shows the differences between areas that can be considered statistically significant.

Table 4. Overview of the statistically significant differences in PE according to six macro-regions: northwest (NW), southwest (SW), north-central (NC), south-central (SC), northeast (NE), and southeast (SE).

Region Comparison	Difference between Means	Simultaneous 95% Confidence Limits		
SE—SW	0.00009	−0.04029	0.04047	
SE—NE	0.22104	0.17463	0.26744	***
SE—NW	0.24533	0.18232	0.30834	***
SE—SC	0.26869	0.22804	0.30933	***
SE—NC	0.58997	0.54258	0.63735	***
SW—SE	−0.00009	−0.04047	0.04029	
SW—NE	0.22094	0.17405	0.26783	***
SW—NW	0.24523	0.18186	0.30860	***
SW—SC	0.26859	0.22739	0.30980	***
SW—NC	0.58987	0.54201	0.63773	***
NE—SE	−0.22104	−0.26744	−0.17463	***
NE—SW	−0.22094	−0.26783	−0.17405	***
NE—NW	0.02429	−0.04308	0.09166	
NE—SC	0.04765	0.00053	0.09477	***
NE—NC	0.36893	0.31589	0.42197	***
NW—SE	−0.24533	−0.30834	−0.18232	***
NW—SW	−0.24523	−0.30860	−0.18186	***
NW—NE	−0.02429	−0.09166	0.04308	
NW—SC	0.02336	−0.04018	0.08690	
NW—NC	0.34464	0.27659	0.41269	***
SC—SE	−0.26869	−0.30933	−0.22804	***
SC—SW	−0.26859	−0.30980	−0.22739	***
SC—NE	−0.04765	−0.09477	−0.00053	***
SC—NW	−0.02336	−0.08690	0.04018	
SC—NC	0.32128	0.27319	0.36936	***
NC—SE	−0.58997	−0.63735	−0.54258	***
NC—SW	−0.58987	−0.63773	−0.54201	***
NC—NE	−0.36893	−0.42197	−0.31589	***
NC—NW	−0.34464	−0.41269	−0.27659	***
NC—NC	−0.32128	−0.36936	−0.27319	***

Comparisons that are Significant at the 0.05 Level Are Indicated by ***.

3.2. ML Models' Accuracy Evaluation

The indicator for comparing the accuracy of forecasts of individual models was the average squared error (ASE) indicator. The model with the lowest value of this indicator was evaluated as the most suitable model for prediction.

Table 5 compares the prediction accuracy results of all estimated models, according to which the ASE chooses the model with the smallest average squared error value. The comparison is made using the average squared error of the estimate, which is calculated as the square of the difference between the actual and predicted value of the target variable. The lowest value of the criterion is considered as the best one, because it is the most accurate estimation. The second column shows the accuracy that was achieved when estimating the model (Train). In this case, higher accuracy is usually always achieved in the case of model validation and training. The first column shows the value of the average squared error achieved when predicting the validation sample, and the last column of Table 5 shows the error values achieved when training the model on another data sample as large as the validation sample. These values are slightly higher than in the model estimate. The model that achieved the lowest average square error in PE prediction was Dmine regression. Both models of the neural network were evaluated as the least suitable models for the prediction of PE values.

Table 5. Comparison and evaluation of the individual machine learning models ranking from best fit to least suitable. The average squared error (ASE) chooses the model with the smallest average squared error value.

ML Model	Valid Average Squared Error	Train Average Squared Error	Test Average Squared Error
Dmine Regression	0.78819	0.77826	0.78094
Gradient Boosting	0.79695	0.78867	0.79537
Decision Tree	0.84862	0.81904	0.84500
Ensemble model	0.93492	0.93011	0.93512
DM Neural	1.13691	1.14073	1.14558
LARS	1.38220	1.37880	1.38476
AutoNeural	1.62204	1.61178	1.63309
Neural Network	1.62526	1.61762	1.63568

Figure 5 compares the results of the average predicted values according to the individual models. With the three best models (Dmine regression, gradient boosting, and decision tree), a very similar development of predicted values can be observed. The most significant differences in the predicted values were recorded at higher values; on the contrary, at low predicted values, the resulting predictions of all models were very similar.

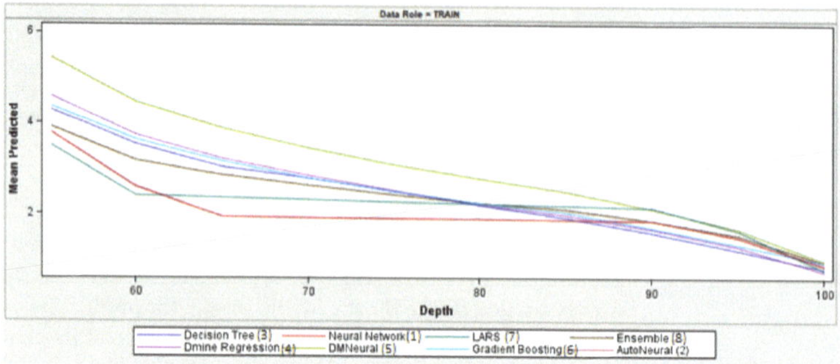

Figure 5. Comparison of the individual machine learning models: (1) neural network, (2) autoneural network, (3) decision tree, (4) Dmine regression, (5) DM neural network, (6) gradient boosting, (7) least angle regression, and (8) ensemble model.

Figure 6 indicates the importance of the individual input variables for the PE value according to the most accurate model (Dmine regression). The results show that humidity and temperature have the highest effect on the PE results. All other factors are considered to be far less important. The term OOV16 means that there are variables through which some numeric variables, such as binned variables, are expressed, i.e., categorical variables are created according to numerical variable values, for example. The RH shows a significant difference in the impact of the different expressions.

For comparison, the results according to the second best model are also presented (gradient boosting). The order of the importance of the variables in this case was almost identical to the results obtained with the decision tree. Even in this case, humidity and temperature have the greatest influence (although there is a difference in the importance of the individual variables related to temperature, compared to the first model). In the third case, the elevation above sea level is in third place. Significant variables selected in the Dmine regression model, which was considered superior, are shown in Table 6. The table includes information about the explanatory ability of each variable, the sum of squares, and its significance expressed by the F-value and *p*-value. Variables denoted as AOV16 represent the categorical expression of quantitative variables. They were used in cases in which this kind of variable achieved better results than the original one. According to the results in Table 6, it can be concluded that the most significant variables influencing pan evaporation were average humidity and minimum temperature. Other variables were also significant, but with much smaller influence on the target variable.

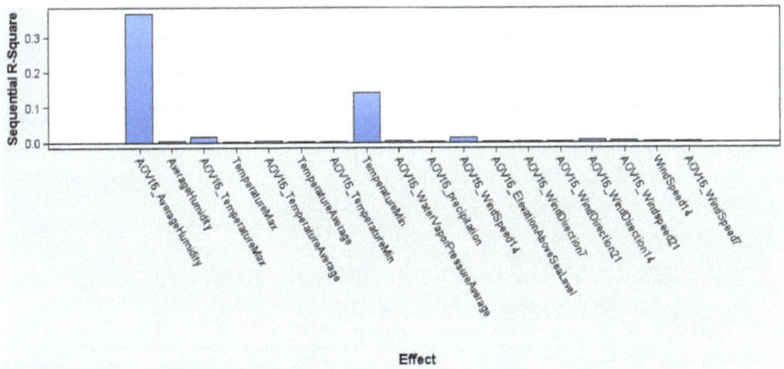

Figure 6. Overview of the individual input variables' importance for the PE value assessment according to the most accurate model, Dmine regression.

Table 6. The results according to the best model, Dmine regression.

Effect	DF	R-Square	F Value	*p*-Value	Sum of Squares
AOV16: Average Humidity	15	0.367893	1290.510136	<0.0001	22095
Var: Temperature Min	1	0.142492	9679.358776	<0.0001	8557.782572
AOV16: Temperature Max	10	0.015878	111.436181	<0.0001	953.574047
AOV16: WindSpeed14	15	0.012232	58.721617	<0.0001	734.603369
AOV16: WindDirection14	15	0.006584	32.049942	<0.0001	395.401311
AOV16: Windspeed21	14	0.003448	18.115361	<0.0001	207.096720
AOV16: Temperature Average	9	0.002904	23.875931	<0.0001	174.387981
AOV16: Water Vapor Pressure—Average	14	0.002906	15.453458	<0.0001	174.513166
Var: Average Humidity	1	0.002524	189.014114	<0.0001	151.605476
AOV16: WindDirection21	15	0.002285	11.461461	<0.0001	137.246991
AOV16: Elevation Above Sea Level	10	0.001970	14.880498	<0.0001	118.297418
Var: Temperature Max	1	0.001751	132.805014	<0.0001	105.159677
AOV16: Wind Direction7	15	0.001362	6.905653	<0.0001	81.803511
AOV16: Temperature Min	9	0.001215	10.295476	<0.0001	72.991107
AOV16: Wind Speed7	15	0.000957	4.873271	<0.0001	57.482001
AOV16: Precipitation	15	0.000854	4.355927	<0.0001	51.301760
Var: Wind Speed14	1	0.000800	61.284456	<0.0001	48.030878
Var: Temperature—Average	1	0.000740	56.825843	<0.0001	44.461511

This study's results are in accordance with other studies, e.g., [39,62–64], which concluded that ML may be a powerful tool for the prediction of actual evapotranspiration when a time series of a few years is available. Starting from the measurements of a sufficient number of climatic parameters, it is possible to obtain forecasting models characterized by very high accuracy and precision.

The final model was selected as the best one, from all common data mining techniques usually used in the modeling of large-scale data. The dataset was, in this case, compiled from measurements in different places and included a large number of observations. The overall explanatory ability of this model measured by R-squared analysis is slightly smaller than the value presented by other authors; however, this could be explained by a very large number of observations used in the presented analysis, which was unique especially due to the extent of the data (both in time and space). The presented model can be used as the universal model for the prediction of pan evaporation in the Slovak Republic.

3.3. Relationship between PE and Variables

Concerning the relationship between PE, temperature and elevation above sea level, the correlation analysis was simplified; only T_{min} and T_{max} were included instead of T_{aver}. Table 7 shows that there is a significant yet weak relationship between PE and elevation above sea level. On the other hand, there is a moderately strong relationship between PE and T_{aver}, which is a moderate relationship. Comparing T_{min} and T_{max} clearly shows that T_{max} is a slightly more important factor.

Table 7. Comparison of the strongest relationship variables and PE on the basis of the Pearson correlation coefficient.

Pearson Correlation Coefficients	
Prob > |r| under H0: Rho = 0	
Number of Observations	
	PE
T_{max} (°C) **T_max (°C)**	0.60209 <0.0001 77,531
T_{min} (°C) **T_min (°C)**	0.43345 <0.0001 77,531
T_{aver} **T_{aver} (°C)**	0.58680 <0.0001 77,525
Elevation above sea level (m)	−0.11332 <0.0001 77,534

4. Discussion

The results of this study are built on existing evidence of ML models and statistical techniques as useful frameworks for making predictions of complex climatological indices, such as the hydrological PE, as concluded by numerous authors (see Table 1). Investigating outcomes indicate that the ML models perform well in predicting PE at different climatical regions [1,9,13,16,40–42], which is also consistent with this case study. The results according to [11] showed that the ML models have different accuracies in different climates.

ML models based on multiple regression, such as DR in this study, have the best accuracy of results, which is in line with the hypotheses of the previous studies [11,40,42,45]. However, the results do not fit with the theory according to Zounemat-Kermani (2021) [40], where the kriging model, as well as the support vector regression (SVR), radial basis function neural network (RBFNN), and = Levenberg–Marquardt (MLP-ML) models, performed better compared to the RSM (the modified response surface method) and M5Tree (M5 model tree).

Contrary to this study's hypothesized association, in which AN and NN were evaluated as the least suitable models for the prediction of PE, the study of Sudheer (2002) [50] proved that PE values could be reasonably estimated using temperature data only through the ANN technique. The results of this study also contradict the claims of the research [13], in which the estimation results obtained with the functional link artificial neural network (FLANN) model are compared with those obtained by multi-layer artificial neural networks (MLANN) and two empirical methods using the same raw data and corresponding features. In summary, the proposed FLANN models provide improved estimates, and they are able to model the daily evaporation process more accurately compared to MLANN. Out of MLANN and FLANN, the FLANN is less complex and better in terms of accuracy. However, among the three agroclimatic zones, the improvement in the plains zone (Raipur) is more visible, as compared to the plateau (Jagdalpur) and hill (Ambikapur) zones of the study area [13].

The study [13] concludes that appropriate FLANN models can also be selected depending upon the availability of climatic parameters. A comparison according to [40] indicates that the models, without local climatic data, provide more scattered estimates. The studies that used more input climate data variables [9,40,44] have low predicted values; on the contrary, the most significant relationships in the predicted values were recorded in studies with less input climatic data variables [11,40,43]. As confirmed by [11], there is a slight difference between RH and T_{aver}, as they were much more effective at modeling PE than the other variables in their study. A different study [42] concluded that besides T_{aver}, SR had the best performance, followed by models that used RH and, finally, Us. However, the temperature-based models showed the worst performance in their study. From a study conducted by Al-Mukhtar et al. [1], it is evident that the temperatures (T_{max}, T_{min}, and T_{aver}), RH and Us were all significantly associated at the 0.05 level with the PE, justifying that the PE rates at these stations can be nicely modelled by these elements. The results of this study are consistent with the above-mentioned studies and proved the strong relationship between PE and T_{aver}.

The results of this study match with the study of Majhi and Naidu (2021) [13], where they stated that the evaporation rate increases with an increase in the radiation and temperature of the evaporating zone. Similarly, wind speed also helps to remove the water from water bodies to some extent and triggers the evaporation process. Bright sunshine and low humidity and atmospheric pressure are the other climatic variables that significantly contribute to the evaporation process [13]. The study used the long-term climatic data on T_{max} (°C), T_{min} (°C), and RH for morning and afternoon hours (RH1 and RH11 (%)), and Up, SR, and PE were collected from the certified observatories located in selected stations in India [13].

The different studies used different numbers of meteorological stations for their calculations, from one [44,45] or two [9,16,40] up to eight [11,42]. However, in this case study, 35 meteorological stations were used to improve the calculation accuracy, which provides new insight and is an advantage of this study.

In a study from the same country as this case study (Košice, Slovak Republic) [35], the soft computing techniques for estimating daily PE were investigated, using daily SR, RH, T, and Us as the meteorological variables for modeling. These model results show that the ML model performs better than another study that used soft computing techniques [35]; this was the main motivation of this study. However, in this case study, the applicability of the newly explored ML models was investigated, with the best accuracy obtained by DR. For future research, it can be compared with the best performing models of ML from Table 1. The macro-regions with the greatest change in PE will be used for further research applying PE with other variables (SE and SW regions of the Slovak Republic) with the uniform climatic characteristics.

In general terms, no one ML algorithm is the best for all problems [50]. The performance of different machine learning algorithms strongly depends on the size and structure of the available data. As concluded in the study by Granata (2019) [50], this study also infers that the individual ML models have been chosen because they usually achieve high performance and they are very good at learning complex, highly non-linear relationships. However, due to the large dispersion of climatic, geographical, and local conditions, it is necessary to test individual ML models independently for each site.

5. Conclusions

This study explored the abilities of eight different ML models—NN, AN, DT, DR, DM NN, GB, LARS, and EM—in modeling PE, utilizing the meteorological dataset combinations of the variables T_{min}, T_{max}, T_{aver}, RH, and Us with wind speed in the directions at 7, 14, and 21 degrees; prevailing wind directions at 7, 14, and 21 degrees; PE; elevation above sea level; and geographical location. The climatic datasets obtained from 35 stations in different geographical locations were used as inputs for training and testing. The datasets were taken from eight regions in the Slovak Republic during the vegetation periods of the years from 2010 to 2020.

The eight regions were classified into six macro-regions—NW, SW, NC, SC, NE, and SE—according to the agroclimatic zones of the Slovak Republic. The results of this study could be applied practically in the field of regional PE estimation. According to results of the performed analysis, the best method for the modeling of pan evaporation was Dmine regression. Both neural network models were the worst performing models from all applied methods. This suggests that the character of the modelled relationships could be explained better by other methods, despite the ability of neural networks to explain complex relationships. On the other hand, Dmine regression allows us to use a standard modeling approach and offers better insight into the modelled relationships, which is not the case of the "black box" neural network.

From this study, we conclude the following:

(a) The lowest PE values were recorded in the NC area, and the highest were recorded in the SE and SW regions. The largest PE change over the observed period (expressed by using the variation range) occurred in the SE region, followed by SC and SW. The smallest PE change was in the NC and NW regions.

(b) The best accuracy of the ML models was obtained by DR (TASE = 0.78819), followed by GB (TASE = 0.77826) and DT (TASE = 0.78094), though it is possible to see very similar results of the predicted values. Both neural network models, AN and NN, were evaluated as the least suitable models for the prediction of PE.

(c) There is a significant but weak relationship between PE and elevation above sea level. However, there is a moderately strong relationship between PE and T_{aver}. A comparison between T_{min} and T_{max} shows that T_{max} is a slightly more important factor.

Because ET is one of the main parts of the hydrology cycle, it is crucial for estimating PE, especially when there are limited data available. Based on the results, the use of the best three evaluated ML models is recommended (DR, GB, and DT) in the region with the largest PE change and with uniform climatic characteristics. The previous studies that used more input climate data variables have low predicted values; on the contrary, the most significant relationships in the predicted values were recorded in studies with less input climatic data variables.

The results of this study can be used in water resources management to make sustainable irrigation plans, design sustainable water supply systems, or carry out sustainable reservoir management.

Author Contributions: Conceptualization: L'.J., B.N. and J.Č.; methodology, L'.J., B.N., J.Č. and J.P.; investigation, L'.J., B.N., J.Č., J.P. and B.C.; resources, B.N., J.Č. and J.P.; data curation: J.P., B.C. and V.K.; writing—original draft preparation: B.N. and J.P.; writing—review and editing: L'.J., B.N., J.Č., J.P. and B.C.; supervision: L'.J.; project administration, B.N. and J.Č.; funding acquisition B.N. and J.Č. All authors have read and agreed to the published version of the manuscript.

Funding: This publication was funded by the Cultural and Educational Grant Agency KEGA Project no. 29SPU-4/2020: Interactive learning as a tool for analysis and solutions design for the water modeling training in the landscape, and no. 026SPU-4/2020: Climate change and its impact on temperature conditions in Slovakia. At the same time, it is the result of the project implementation no. ITMS2014+313011W580: Scientific support of climate change adaptation in agriculture and mitigation of soil degradation supported by the Integrated Infrastructure Operational Programme funded by the ERDF.

Institutional Review Board Statement: This study used secondary data and ethical approval was not required.

Informed Consent Statement: Not applicable.

Data Availability Statement: The data generated during and/or analyzed during the current study are available from the corresponding author on reasonable request.

Conflicts of Interest: The authors declare no conflict of interest.

References

1. Al-Mukhtar, M. Modeling of pan evaporation based on the development of machine learning methods. *Arch. Meteorol. Geophys. Bioclimatol. Ser. B* **2021**, *146*, 961–979. [CrossRef]
2. Ghorbani, M.A.; Kazempour, R.; Ahmadi, M.; Shamshirband, S.; Ghazvinei, P.T. Forecasting pan evaporation with an integrated artificial neural network quantum-behaved particle swarm optimization model: A case study in Talesh, Northern Iran. *Eng. Appl. Comput. Fluid Mech.* **2018**, *12*, 724–737. [CrossRef]
3. Moazenzadeh, R.; Mohammadi, B.; Shamshirband, S.; Ahmadi, M. Coupling a firefly algorithm with support vector regression to predict evaporation in northern Iran. *Eng. Appl. Comput. Fluid Mech.* **2018**, *12*, 584–597. [CrossRef]
4. Wu, L.; Peng, Y.; Fan, J.; Wang, Y. Machine learning models for the estimation of monthly mean daily reference evapotranspiration based on cross-station and synthetic data. *Water Policy* **2019**, *50*, 1730–1750. [CrossRef]
5. Azorin-Molina, C.; Vicente-Serrano, S.M.; Sanchez-Lorenzo, A.; McVicar, T.R.; Morán-Tejeda, E.; Revuelto, J.; El Kenawy, A.; Martín-Hernández, N.; Tomas-Burguera, M. Atmospheric evaporative demand observations, estimates and driving factors in Spain (1961–2011). *J. Hydrol.* **2015**, *523*, 262–277. [CrossRef]
6. Burn, D.H.; Hesch, N.M. Trends in evaporation for the Canadian Prairies. *J. Hydrol.* **2007**, *336*, 61–73. [CrossRef]
7. Majidi, M.; Alizadeh, A.; Farid, A.; Vazifedoust, M. Estimating Evaporation from Lakes and Reservoirs under Limited Data Condition in a Semi-Arid Region. *Water Resour. Manag.* **2015**, *29*, 3711–3733. [CrossRef]
8. Malik, A.; Rai, P.; Heddam, S.; Kisi, O.; Sharafati, A.; Salih, S.; Al-Ansari, N.; Yaseen, Z. Pan Evaporation Estimation in Uttarakhand and Uttar Pradesh States, India: Validity of an Integrative Data Intelligence Model. *Atmosphere* **2020**, *11*, 553. [CrossRef]
9. Malik, A.; Tikhamarine, Y.; Al-Ansari, N.; Shahid, S.; Sekhon, H.S.; Pal, R.K.; Rai, P.; Pandey, K.; Singh, P.; Elbeltagi, A.; et al. Daily pan-evaporation estimation in different agro-climatic zones using novel hybrid support vector regression optimized by Salp swarm algorithm in conjunction with gamma test. *Eng. Appl. Comput. Fluid Mech.* **2021**, *15*, 1075–1094. [CrossRef]
10. Shiri, J.; Martí, P.; Singh, V.P. Evaluation of gene expression programming approaches for estimating daily evaporation through spatial and temporal data scanning. *Hydrol. Process.* **2012**, *28*, 1215–1225. [CrossRef]
11. Wang, L.; Kisi, O.; Zounemat-Kermani, M.; Li, H. Pan evaporation modeling using six different heuristic computing methods in different climates of China. *J. Hydrol.* **2017**, *544*, 407–427. [CrossRef]
12. Ghaemi, A.; Rezaie-Balf, M.; Adamowski, J.; Kisi, O.; Quilty, J. On the applicability of maximum overlap discrete wavelet transform integrated with MARS and M5 model tree for monthly pan evaporation prediction. *Agric. For. Meteorol.* **2019**, *278*, 107647. [CrossRef]
13. Majhi, B.; Naidu, D. Pan evaporation modeling in different agroclimatic zones using functional link artificial neural network. *Inf. Process. Agric.* **2021**, *8*, 134–147. [CrossRef]
14. Dou, X.; Yang, Y. Evapotranspiration estimation using four different machine learning approaches in different terrestrial ecosystems. *Comput. Electron. Agric.* **2018**, *148*, 95–106. [CrossRef]
15. Ghumman, A.; Jamaan, M.; Ahmad, A.; Shafiquzzaman, M.; Haider, H.; Al Salamah, I.; Ghazaw, Y. Simulation of Pan-Evaporation Using Penman and Hamon Equations and Artificial Intelligence Techniques. *Water* **2021**, *13*, 793. [CrossRef]
16. Zounemat-Kermani, M.; Keshtegar, B.; Kisi, O.; Scholz, M. Towards a Comprehensive Assessment of Statistical versus Soft Computing Models in Hydrology: Application to Monthly Pan Evaporation Prediction. *Water* **2021**, *13*, 2451. [CrossRef]
17. Bellido-Jiménez, J.A.; Estévez, J.; García-Marín, A.P. New machine learning approaches to improve reference evapotranspiration estimates using intra-daily temperature-based variables in a semi-arid region of Spain. *Agric. Water Manag.* **2021**, *245*, 106558. [CrossRef]
18. Abrahart, R.J.; Anctil, F.; Coulibaly, P.; Dawson, C.; Mount, N.J.; See, L.M.; Shamseldin, A.; Solomatine, D.P.; Toth, E.; Wilby, R.L. Two decades of anarchy? Emerging themes and outstanding challenges for neural network river forecasting. *Prog. Phys. Geogr. Earth Environ.* **2012**, *36*, 480–513. [CrossRef]
19. Keshtegar, B.; Piri, J.; Kisi, O. A nonlinear mathematical modeling of daily pan evaporation based on conjugate gradient method. *Comput. Electron. Agric.* **2016**, *127*, 120–130. [CrossRef]
20. Kousari, M.R.; Hosseini, M.E.; Ahani, H.; Hakimelahi, H. Introducing an operational method to forecast long-term regional drought based on the application of artificial intelligence capabilities. *Arch. Meteorol. Geophys. Bioclimatol. Ser. B* **2017**, *127*, 361–380. [CrossRef]
21. Moghaddamnia, A.; Gousheh, M.G.; Piri, J.; Amin, S.; Han, D. Evaporation estimation using artificial neural networks and adaptive neuro-fuzzy inference system techniques. *Adv. Water Resour.* **2009**, *32*, 88–97. [CrossRef]

22. Park, S.; Im, J.; Jang, E.; Rhee, J. Drought assessment and monitoring through blending of multi-sensor indices using machine learning approaches for different climate regions. *Agric. For. Meteorol.* **2016**, *216*, 157–169. [CrossRef]
23. Crisci, C.; Ghattas, B.; Perera, G. A review of supervised machine learning algorithms and their applications to ecological data. *Ecol. Model.* **2012**, *240*, 113–122. [CrossRef]
24. Voyant, C.; Notton, G.; Kalogirou, S.; Nivet, M.-L.; Paoli, C.; Motte, F.; Fouilloy, A. Machine learning methods for solar radiation forecasting: A review. *Renew. Energy* **2017**, *105*, 569–582. [CrossRef]
25. Wilby, R.L.; Yu, D. Rainfall and temperature estimation for a data sparse region. *Hydrol. Earth Syst. Sci.* **2013**, *17*, 3937–3955. [CrossRef]
26. Courault, D.; Seguin, B.; Olioso, A. Review on estimation of evapotranspiration from remote sensing data: From empirical to numerical modeling approaches. *Irrig. Drain. Syst.* **2005**, *19*, 223–249. [CrossRef]
27. Mu, Q.; Heinsch, F.A.; Zhao, M.; Running, S.W. Development of a global evapotranspiration algorithm based on MODIS and global meteorology data. *Remote Sens. Environ.* **2007**, *111*, 519–536. [CrossRef]
28. McCabe, M.F.; Wood, E.F. Scale influences on the remote estimation of evapotranspiration using multiple satellite sensors. *Remote Sens. Environ.* **2006**, *105*, 271–285. [CrossRef]
29. Long, D.; Longuevergne, L.; Scanlon, B.R. Uncertainty in evapotranspiration from land surface modeling, remote sensing, and GRACE satellites. *Water Resour. Res.* **2014**, *50*, 1131–1151. [CrossRef]
30. Gomis-Cebolla, J.; Jimenez, J.C.; Sobrino, J.A.; Corbari, C.; Mancini, M. Intercomparison of remote-sensing based evapotranspiration algorithms over amazonian forests. *Int. J. Appl. Earth Obs. Geoinf.* **2019**, *80*, 280–294. [CrossRef]
31. Chia, M.Y.; Huang, Y.F.; Koo, C.H.; Fung, K.F. Recent Advances in Evapotranspiration Estimation Using Artificial Intelligence Approaches with a Focus on Hybridization Techniques—A Review. *Agronomy* **2020**, *10*, 101. [CrossRef]
32. Jang, J.-C.; Sohn, E.-H.; Park, K.-H.; Lee, S. Estimation of Daily Potential Evapotranspiration in Real-Time from GK2A/AMI Data Using Artificial Neural Network for the Korean Peninsula. *Hydrology* **2021**, *8*, 129. [CrossRef]
33. Zhang, L.; Lemeur, R. Evaluation of daily evapotranspiration estimates from instantaneous measurements. *Agric. For. Meteorol.* **1995**, *74*, 139–154. [CrossRef]
34. Bae, H.; Ji, H.; Lim, Y.-J.; Ryu, Y.; Kim, M.-H.; Kim, B.-J. Characteristics of drought propagation in South Korea: Relationship between meteorological, agricultural, and hydrological droughts. *Nat. Hazards* **2019**, *99*, 1–16. [CrossRef]
35. Kaya, Y.Z.; Zelenakova, M.; Üneş, F.; Demirci, M.; Hlavata, H.; Mesaros, P. Estimation of daily evapotranspiration in Košice City (Slovakia) using several soft computing techniques. *Theor. Appl. Clim.* **2021**, *144*, 287–298. [CrossRef]
36. Hlaváčiková, H.; Novák, V. Comparison of daily potential evapotranspiration calculated by two procedures based on Penman-Monteith type equation. *J. Hydrol. Hydromech.* **2013**, *61*, 173–176. [CrossRef]
37. Fendeková, M.; Gauster, T.; Labudová, L.; Vrablíková, D.; Danáčová, Z.; Fendek, M.; Pekarova, P. Analysing 21st century meteorological and hydrological drought events in Slovakia. *J. Hydrol. Hydromech.* **2018**, *66*, 393–403. [CrossRef]
38. Parajka, J.; Szolgay, J.; Meszaros, I.; Kostka, Z. Grid-based mapping of the long-term mean annual potential and actual evapotranspiration in upper Hron River basin. *J. Hydrol. Hydromech. ÚH SAV* **2004**, *4*, 239–254.
39. Kubiak-Wójcicka, K.; Nagy, P.; Zeleňáková, M.; Hlavatá, H.; Abd-Elhamid, H. Identification of Extreme Weather Events Using Meteorological and Hydrological Indicators in the Laborec River Catchment, Slovakia. *Water* **2021**, *13*, 1413. [CrossRef]
40. Kisi, O. Pan evaporation modeling using least square support vector machine, multivariate adaptive regression splines and M5 model tree. *J. Hydrol.* **2015**, *528*, 312–320. [CrossRef]
41. Abed, M.; Alam Imteaz, M.; Ahmed, A.N.; Huang, Y.F. Application of long short-term memory neural network technique for predicting monthly pan evaporation. *Sci. Rep.* **2021**, *11*, 20742. [CrossRef] [PubMed]
42. Ferreira, L.B.; Duarte, A.B.; Da Cunha, F.F.; Filho, E.I.F. Multivariate adaptive regression splines (MARS) applied to daily reference evapotranspiration modeling with limited weather data. *Acta Sci. Agron.* **2018**, *41*, 39880. [CrossRef]
43. Sattari, M.T.; Ahamadifar, V.; Delirhasannia, R.; Apaydin, H. Estimation of pan evaporation coefficient in cold and dry climate conditions with a decision-tree model. *Atmósfera* **2020**, *34*, 289–300. [CrossRef]
44. Adnan, M.; Latif, M.A.; Rehman, A.-U.; Nazir, M. Estimating Evapotranspiration using Machine Learning Techniques. *Int. J. Adv. Comput. Sci. Appl.* **2017**, *8*, 108–113. [CrossRef]
45. Simon-Gáspár, B.; Soós, G.; Anda, A. Estimation Standard and Seeded Pan Evaporation Using Modelling Approach. *Hydrol. Earth Syst. Sci. Discuss.* **2021**, *in review*, preprint. [CrossRef]
46. Pecho, J. The Gulf Stream Is not Weakening Due to Climate Change. 2010. Available online: https://www.shmu.sk/sk/?page=2049&id=159 (accessed on 6 December 2021).
47. Kopcsay, M. Weather Weather Forecast: Three Weather Scenarios for Christmas. 2021. Available online: https://www.teraz.sk/pocasie/velka-predpoved-pocasia-tri-scenar/598631-clanok.html (accessed on 10 December 2021).
48. Morris, R.D.O. Hebb: The Organization of Behavior, Wiley: New York; 1949. *Brain Res. Bull.* **1999**, *50*, 437. [CrossRef]
49. Keith, D. A Brief History of Machine Learning. 2021. Available online: https://www.dataversity.net/a-brief-history-of-machine-learning/# (accessed on 11 December 2021).
50. Granata, F. Evapotranspiration evaluation models based on machine learning algorithms—A comparative study. *Agric. Water Manag.* **2019**, *217*, 303–315. [CrossRef]
51. World Meteorological Organization Guide to Hydrological Practice. *Hydrology—From Measurement to Hydrological Information*; World Meteorological Organization: Geneva, Switzerland, 2008; Volume I, ISBN 9789263101686.

52. Kurpelová, M.; Coufal, L.; Čulík, J. *Agroklimatické podmienky ČSSR*; Hydrometeorologický Ústav: Bratislava, Slovakia, 1975.
53. Tomlain, J. Climate change impacts on evapotranspiration from the forest on the territory of Slovakia. *Acta Meteorol. Univ. Comen.* **2000**, *29*, 1–8.
54. Rojas, R. *Neural Networks: A Systematic Introduction*; Springer Science and Business Media: Berlin, Germany, 2013; Volume 46.
55. Assi, K.J.; Nahiduzzaman, K.M.; Ratrout, N.T.; Aldosary, A.S. Mode choice behavior of high school goers: Evaluating logistic regression and MLP neural networks. *Case Stud. Transp. Policy* **2018**, *6*, 225–230. [CrossRef]
56. Desai, M.; Shah, M. An anatomization on breast cancer detection and diagnosis employing multi-layer perceptron neural network (MLP) and Convolutional neural network (CNN). *Clin. eHealth* **2021**, *4*, 1–11. [CrossRef]
57. Tu, J. Advantages and disadvantages of using artificial neural networks versus logistic regression for predicting medical outcomes. *J. Clin. Epidemiol.* **1996**, *49*, 1225–1231. [CrossRef]
58. deVille, B.; Padraic, N. *Decision Trees for Analytics Using SAS® Enterprise Miner™*; SAS Institute Inc.: Cary, NC, USA, 2013.
59. Friedman, J.H. Greedy function approximation: A gradient boosting machine. *Ann. Stat.* **2001**, *45*, 1189–1232. [CrossRef]
60. Avila, J.; Hauck, T. Least angle regression. *Ann. Stat.* **2004**, *32*, 407–499. [CrossRef]
61. Czika, W.; Maldonado, M.; Liu, Y. *Ensemble Modeling: Recent Advances and Applications, SAS3120-2016*; Proceedings of the SAS® Global Forum 2016 Conference; SAS Institute Inc.: Cary, NC, USA, 2016; Available online: https://support.sas.com/resources/papers/proceedings16/SAS3120-2016.pdf (accessed on 11 December 2021).
62. Ravindran, S.M.; Bhaskaran, S.K.M.; Ambat, S.K.N. A Deep Neural Network Architecture to Model Reference Evapotranspiration Using a Single Input Meteorological Parameter. *Environ. Process.* **2021**, *8*, 1567–1599. [CrossRef]
63. Čimo, J.; Aydın, E.; Šinka, K.; Tárník, A.; Kišš, V.; Halaj, P.; Toková, L.; Kotuš, T. Change in the Length of the Vegetation Period of Tomato (*Solanum lycopersicum* L.), White Cabbage (*Brassica oleracea* L. var. *capitata*) and Carrot (*Daucus carota* L.) Due to Climate Change in Slovakia. *Agronomy* **2020**, *10*, 1110. [CrossRef]
64. Sudheer, K.P.; Gosain, A.K.; Rangan, D.M.; Saheb, S.M. Modelling evaporation using an artificial neural network algorithm. *Hydrol. Process.* **2002**, *16*, 3189–3202. [CrossRef]

remote sensing

MDPI

Article

Comparison of the Novel Probabilistic Self-Optimizing Vectorized Earth Observation Retrieval Classifier with Common Machine Learning Algorithms

Jan Pawel Musial * and Jedrzej Stanislaw Bojanowski

Institute of Geodesy and Cartography, Remote Sensing Centre, PL02-679 Warsaw, Poland; igik@igik.edu.pl
* Correspondence: jan.musial@igik.edu.pl

Abstract: The Vectorized Earth Observation Retrieval (VEOR) algorithm is a novel algorithm suited to the efficient supervised classification of large Earth Observation (EO) datasets. VEOR addresses shortcomings in well-established machine learning methods with an emphasis on numerical performance. Its characteristics include (1) derivation of classification probability; (2) objective selection of classification features that maximize Cohen's kappa coefficient (κ) derived from iterative "leave-one-out" cross-validation; (3) reduced sensitivity of the classification results to imbalanced classes; (4) smoothing of the classification probability field to reduce noise/mislabeling; (5) numerically efficient retrieval based on a pre-computed look-up vector (LUV); and (6) separate parametrization of the algorithm for each discrete feature class (e.g., land cover). Within this study, the performance of the VEOR classifier was compared to other commonly used machine learning algorithms: K-nearest neighbors, support vector machines, Gaussian process, decision trees, random forest, artificial neural networks, AdaBoost, Naive Bayes and Quadratic Discriminant Analysis. Firstly, the comparison was performed using synthetic 2D (two-dimensional) datasets featuring different sample sizes, levels of noise (i.e., mislabeling) and class imbalance. Secondly, the same experiments were repeated for 7D datasets consisting of informative, redundant and insignificant features. Ultimately, the benchmarking of the classifiers involved cloud discrimination using MODIS satellite spectral measurements and a reference cloud mask derived from combined CALIOP lidar and CPR radar data. The results revealed that the proposed VEOR algorithm accurately discriminated cloud cover using MODIS data and accurately classified large synthetic datasets with low or moderate levels of noise and class imbalance. On the contrary, VEOR did not feature good classification skills for significantly distorted or for small datasets. Nevertheless, the comparisons performed proved that VEOR was within the 3–4 most accurate classifiers and that it can be applied to large Earth Observation datasets.

Keywords: Vectorized Earth Observation Retrieval (VEOR); machine learning; artificial intelligence; classification; support vector machines; Gaussian process; random forest; artificial neural networks; Naive Bayes

Citation: Musial, J.P.; Bojanowski, J.S. Comparison of the Novel Probabilistic Self-Optimizing Vectorized Earth Observation Retrieval Classifier with Common Machine Learning Algorithms. *Remote Sens.* **2022**, *14*, 378. https://doi.org/10.3390/rs14020378

Academic Editors: Andreea Claudia Serban and Miltiadis D. Lytras

Received: 3 December 2021
Accepted: 9 January 2022
Published: 14 January 2022

1. Introduction

The importance of image classification acquired by remote sensing sensors was recognized early on by [1] with the launch of satellites from the Landsat series. Numerous algorithms have been developed since then to partially or fully automatize the classification processes used in various environmental applications, such as land cover mapping [2–5], crop type identification [6–9], wetland monitoring [10–13], water body mapping [14–17], forestry [18–21], ice/snow cover monitoring [22,23], urban planning [24–26] and monitoring of weather phenomena [27–31]. Nowadays, the abundance of remote sensing datasets featuring multi-spectral, high-spatio-temporal-resolution imagery hinders manual image interpretation, and necessitates an efficient machine learning method easily adaptable to a broad suite of sensors. Furthermore, the results of image classification should be complemented by uncertainty estimates that contribute to the computation of error statistics of consecutive (higher-level) products.

In general, image classification algorithms can be categorized as: (1) supervised and unsupervised; (2) parametric and non-parametric; (3) hard and soft (fuzzy) classifiers; (4) per-pixel, subpixel and per-field/per-object [32]. The supervised classifiers, as opposed to unsupervised ones, require a training dataset composed of spectral samples labeled with categorical classes. The labeling may originate from in situ measurements/observations, pixels or image segments labeled by a human interpreter, or from other remote sensing products. The parametric classifiers assume a Gaussian distribution when describing input features (e.g., maximum likelihood, Quadratic Discriminant Analysis) that allows for the derivation of statistical properties such as mean vector or covariance matrix describing the distribution of input features. Non-parametric classifiers do not assume statistical distributions and are suitable for the incorporation of ancillary (non-spectral) input features. The "hard" classifiers assign a single class to a pixel/object, while soft/fuzzy classifiers may assign multiple classes to a pixel/object along with the multiple corresponding classification probabilities. Special types of classifier include the sub-pixel methods that assume the heterogeneous spectral signature of a pixel composed of a linear or non-linear mix of spectral signatures originating from homogeneous endmembers (e.g., land cover classes). This is particularly true for coarse-spatial-resolution imagery and allows for the estimation of the fractional membership of each pixel to each endmember. These types of classifiers have strong and weak points, which were well described by [32]. Nonetheless, a robust classifier should be accurate, self-optimizing (capable of increasing classification accuracy by internally optimizing its hyper-parameters), numerically efficient (applicable to large datasets), non-parametric (easily applicable and transferable) and should provide uncertainty estimates for a generated classification.

In this article, a fast, robust, non-parametric, probabilistic Vectorized Earth Observation Retrieval (VEOR) image classifier is introduced that is capable of feature selection and is applicable to the large multidimensional datasets frequently occurring in the remote sensing discipline. The VEOR classifier was evaluated together with other popular machine learning algorithms: K-nearest neighbors (KNN), support vector machines (SVM), Gaussian process (GP), decision trees (DT), random forest (RF), artificial neural networks (ANN), AdaBoost (ADA), Naive Bayes (NB) and Quadratic Discriminant Analysis (QDA). The comparison of the classifiers was firstly performed using two synthetic datasets: a simplified 2D (two-dimensional) case, and a more complex 7D case for which some of the features (dimensions) were redundant or irrelevant for the classification. The synthetic datasets featured different sample sizes, levels of noise (i.e., mislabeling), shapes of class separation boundaries (linear, non-linear, circular) and class imbalance in order to test the performance of the selected classifiers under a wide range of disturbing factors. Evaluation of algorithms by means of the synthetic data is a common practice [33–35] as it allows to analyze, in a controlled manner, the response of a model to a particular disturbing factor or a combination of factors. Secondly, the classifiers were inter-compared within real case scenarios where cloudiness over water and ice/snow surfaces was discriminated using spectral measurements of the Moderate Resolution Imaging Spectroradiometer (MODIS). Ultimately, the numerical performance of the classifiers was evaluated with respect to training and retrieval processes.

The following sections of this article include: the most important challenges to the image classification process (Section 2); a detailed description of the proposed VEOR classifier (Section 3); brief descriptions of other selected classifiers, together with their numerical implementations in the Python scikit-learn library (Section 4); a description of the experimental design, including the generation of synthetic and real case datasets, as well as the methodology of the evaluation of classifiers (Section 5); the presentation of acquired classification results (Section 6); discussion of the results (Section 7); the final conclusions drawn from the study (Section 8); and a list of optimal features (Appendix A).

2. The Challenges of Image Classification

Classification of remote sensing imagery is usually applied to a large multidimensional dataset that often consists of many cross-correlated measurements in multiple spectral bands acquired at different times. Furthermore, the neighboring pixels within an image time series feature spatio-temporal relationships that represent the evolution (in time and space) of the phenomenon to be classified (e.g., land cover, cloud types). This allows for the exploitation of additional features, apart from spectral information, such as image filters, topology of objects, edge filters, principal components and other numerical combinations of spectral bands [32]. However, the images within a time series are often distorted by a number of factors, such as variable illumination/acquisition geometry, anisotropy of reflectance, atmospheric disturbances, topographic effects, geolocation shifts, temporal degradation of imaging instrument, data gaps (e.g., missing scan lines), residual cloud/snow cover, variable surface moisture, radio-frequency interference, thermal noise and many more. This turns image classification into a complex problem, which may be difficult to solve even for an intuitively simple task such as cloud detection on satellite imagery.

The classification of physical phenomena on remote sensing imagery is a complex process that can be divided into the following steps: (1) selection of suitable remote sensing sensor(s) featuring appropriate spatial, temporal, radiometric and spectral resolutions; (2) selection of training samples; (3) image pre-processing; (4) feature extraction; (5) feature selection; (6) selection of suitable classification algorithm(s); (7) image classification; (8) image post-classification (optional filtering of classification results, e.g., clustering small groups of pixels to fulfil minimum mapping unit criteria); (9) accuracy assessment. Steps 1–4 require expert knowledge for a particular classification problem and can be regarded as experimental design. Steps 5–9 can be generalized and automated to some extent in order to facilitate the image classification process and to reduce the human workload. There are several common challenges to supervised image classification related to sparse, mislabeled and imbalanced datasets featuring classes that may be inseparable, and which are discussed in the following subsections.

2.1. Training Sample Selection

Supervised image classification requires a training dataset composed of pixels featuring distinct spectral signatures labeled within a single class, or more in the case of fuzzy classifiers. The labeling might originate from the visual interpretation of the imagery to be classified or from other referential datasets derived from more sensitive sensors or observed/measured in situ. The visual image interpretation is usually performed by geographic information system (GIS) software and it involves manual delineation and labeling of polygons covering homogeneous areas. The usage of other referential remote sensing datasets often requires spatio-temporal collocation between two sensors by means of interpolation or nearest neighbor matching. Both labeling approaches should follow common rules to produce the optimal training dataset:

1. The reference dataset should be large enough to populate the feature space with information sufficient to correctly infer labels for new test cases that have not appeared in the training dataset. This implies that the new samples should be very similar to the ones used for classifier training. Such a situation occurs only rarely for simple classification tasks based on a few features. The multi-temporal and multi-spectral remote sensing imagery composed of heterogeneous pixels with mixed classes requires an enormous training dataset to represent all of the potential relationships between the features. The training datasets are usually affected by sampling issues related to an uneven distribution of in situ observations due to limited accessibility (e.g., remote areas, swamps, high mountains), temporal and spatial constraints while collocating different data sources and the subjective selection of training polygons. Consequently, a training dataset may be sparse and under-represented. Therefore, a robust image classifier should be able to generalize available a priori information to unseen cases without overfitting the training dataset. The overfitting implies that

a classifier matches the training dataset too well (including noisy cases instead of only the "true" signal). On the contrary, underfitting implies that a classifier performs poorly on the training data, presumably due to crude assumptions or incorrectly selected hyper-parameters or classification features.

2. The training dataset should feature low spatial and temporal autocorrelation in order to maximize information within the feature space and to objectively assess the quality of the generated classification. If pixels are highly correlated, then their spectral signatures are mostly redundant. Consequently, a classifier has little information to learn from and to generalize to new test cases. Such a situation often occurs in the case of visual image interpretation, where an operator marks a few large polygons covering the same class. Even though a large number of pixels are selected, very limited spectral information is actually used to train a classifier. Moreover, if a subset of those pixels is used for the evaluation of classification quality, then the derived statistics are inflated and unreliable. Thus, as suggested by [36], the most robust solution to this problem is a random selection of pixels forming training and validation datasets. Furthermore, it is advantageous to temporarily decouple both datasets so that the training dataset is acquired at a different time (e.g., a different year) than the validation dataset.
3. The reference dataset should reflect the same proportion of classes as the one occurring in the classified image. This is especially important for Bayesian classifiers, which rely on the a priori probability of class occurrence [37]. Similar conclusions were reported by [36] in the case of decision tree algorithms.
4. The reference data should be of high quality, with as few mislabeled pixels as possible. This obvious statement is difficult to implement in reality, where the reference data may originate from: (1) other classification products featuring their own uncertainty; (2) in situ observations altered by instrumental malfunction or typographical errors; (3) crowdsourcing data featuring many subjective observations; and (4) poorly delineated training polygons. Furthermore, in the case of low-resolution imagery covering a heterogeneous landscape, pixels are usually composed of mixed classes, which ultimately have to be classified into a single class. Alternatively, some sub-pixel classification techniques [38–40] can be applied to infer the fractional proportion of classes within such pixels. Nevertheless, from the perspective of an image classifier, the mislabeled pixels introduce noise, which has to be dissected from the true spectral signature associated with a certain class. A limited number of studies have elaborated on the sensitivity of the image classifiers to mislabeled input data. In [34], it is found that the random forest algorithm is less sensitive to class mislabeling than support vector machines (SVMs) thanks to the bootstrap and random split operations incorporated into the random forest method. The robust multi-class AdaBoost (Rob_MulAda) classifier [41] was proposed to overcome the class overfitting problem of the standard AdaBoost algorithm, where there are mislabeled noisy training datasets. In [42], the Naive Bayes classifier was found to be less sensitive to class mislabeling as compared to the decision tree classifier [43], IBk instance-based learner [44] and the Sequential Minimal Optimization (SMO) support vector machine classifier [45]. Recently, in [46], a novel CMR-NLD method has been proposed to filter out the mislabeled samples by the thresholding of a covariance matrix calculated between sample dimensions. Ultimately, it has to be emphasized that a reliable training dataset is a prerequisite for the high accuracy of supervised image classification. Thus, it is advisable to perform data screening prior to classification instead of relying on a classifier to cope with noise in the input data.

2.2. Feature Selection

Classification of remote sensing imagery is based on a set of features that might be spectrally redundant or irrelevant from the perspective of class recognition. Moreover, the spectral variability of the Earth's surface further complicates the classification process.

In this respect, certain image features might be important only over certain kinds of land cover. For instance, visible spectral channels are useful for the detection of bright clouds over a dark water surface but are ineffective over a bright ice surface. This implies that it might be advantageous to separately select image features for some discrete classes (e.g., land cover or soil types). The selection of an optimal feature subset from the input feature vector leads to dimensionality reduction and consequently to: (1) improvement in classification accuracy; (2) lower computational demand; and (3) better understanding of the relationship between features and classification results [47]. Many techniques for feature selection have been developed and these can be grouped essentially into three approaches. The first one, called filtering, involves the ranking of the input features with respect to the output class according to some correlation, dependency or distance metrics [48]—for example, correlation coefficient, mutual information, symmetric uncertainty (SU), information distance or Fisher score. The second approach involves validation with the bootstrap of a machine learning algorithm trained against different subsets of input features. Consequently, the quality metrics of derived classifications are used to rank each subset of input features in order to select the optimal one. Such a technique depends on the numerical robustness of a classifier iteratively applied within a wrapper algorithm, and usually it is applicable to a limited number of feature combinations. For large sets of features, the number of possible subsets is enormous and computationally too demanding, which turns feature selection into a so-called NP (non-deterministic polynomial-time) hardness problem [49]. The subsets of features can be derived either through forward selection or through backward elimination. Within forward selection, the variables are progressively added to form larger subsets, while, on the contrary, within the backward elimination technique, the least promising features are removed from a larger feature subset. From the perspective of numerical performance, it is argued that forward selection algorithms are less computationally demanding, whereas backward elimination techniques are more accurate [48]. The third approach towards feature selection involves methods embedded within a classifier training process. An example of such a technique could be Optimal Brain Damage (OBD), used for node pruning within neural network classifiers [50]. Ultimately, the filter and wrapper feature selection techniques can be combined in order to firstly reduce the number of features by means of the selection technique, and then to apply the wrapper method. Other approaches [51] incorporate variations of the Principal Component Analysis (PCA) to decompose an initial set of features into new orthogonal features, which are further used for classification. Nonetheless, reduction of the feature vector improves class separability and classification accuracy and this has been reported by numerous studies [52–54].

2.3. Imbalanced Distribution of Classes

This common classification problem is related to an imbalanced class distribution, where one class occurs more frequently than other(s). Consequently, different classifiers tend to overestimate the frequency of the dominant class. This improves some of the classification quality statistics (e.g., accuracy) but leads to a severe misclassification of the rare classes. This may turn the classification task based on input features into a class assignment based on a priori information about class frequency distribution. To mitigate this problem, two approaches are possible. The first is to balance the training dataset by oversampling the rare classes or undersampling the prevalent classes. Some more sophisticated methods, such as the Synthetic Minority Oversampling Technique (SMOTE) [55], may be used to combine both sampling techniques. The second approach to the class imbalance problem involves different corrections applied to classifiers, such as cost-sensitive learning to more severely penalize the misclassification of rare classes. This can be implemented by the utilization of specific quality measures during classifier optimization, such as the Hansen–Kuipers Skill score [56], F-score [57], G-mean [58], the area under the Receiver Operating Characteristics (ROC) curve [59], Cohen kappa coefficient [60] and average mutual information [61], which are less sensitive to the class imbalance problem as opposed to overall classification accuracy. Comprehensive reviews of the deficiencies of common classification algorithms in the presence of the imbalanced distribution of classes were reported by [41,62,63]. In this section, only the most important aspects related to the class imbalance problem are discussed.

In the case of the KNN classifier, it is quite intuitive that the neighborhood of a pixel within a feature space more likely consists of a prevalent class than infrequent classes. Thus, infrequent classes are more likely to be incorrectly classified. This effect can be partially diminished by balancing the class distribution within the training dataset. In the case of the decision tree classifier, an imbalanced training dataset results in many tests (branches within the trees) to separate infrequent classes from prevalent classes. Usually, too few tests are implemented to correctly delineate small classes or they are removed during the tree pruning procedure due to not being significant. Some solutions to this problem involve adjustment of the scores generated by each test [64] or the involvement of more advanced tree pruning techniques [65]. SVM classifiers are also reported to be sensitive to the class imbalance problem [66]. In such a case, the class-separating hyperplanes become more skewed towards the prevalent class as the number of support vectors is higher for this class. Consequently, the nearest neighborhood of a test sample (especially when it is located near the separating hyperplane) is likely to be dominated by the support vectors of the dominant class, which may lead to a misclassification of the infrequent class. To overcome this problem, [66] proposed the class boundary alignment algorithm, which adjusts the class boundary either by transforming the kernel function, where the training data can be represented in a vector space, or by modifying the kernel matrix when the data do not have vector space representation. For the neural network algorithm, the imbalanced distribution of classes alters the weights of connections between neurons during the iterative optimization of a network. Consequently, the classification error decreases for the prevalent class and at the same time increases for the infrequent classes. One of the potential solutions to this problem involves undersampling the prevalent class by eliminating highly correlated cases [67]. In the case of the Naive Bayes classifier, the imbalanced training dataset may lead to an overfitting of the learned parameters [68] and to poor classification performance.

2.4. Classifier Numerical Performance

The numerical performance of classifiers expressed in terms of fast processing time and low RAM (random-access memory) usage is becoming a problem for the classification of the multi/hyperspectral remote sensing imagery featuring petabytes of data volume per year. In this respect, an accurate but slow classifier may not be applicable or its application may require significant financial resources and may not be easy to reprocess. Therefore, the aspect of numerical efficiency should be considered while selecting a classifier for a large remote sensing dataset.

The numerical performance may differ significantly between classifier training and retrieval processes and as such should be evaluated separately. Classifier (re)training is performed rarely with an updated/corrected training dataset or after the improvement of a classifier (i.e., a new version). This implies that algorithm training may be computationally expensive in order to generate the best classification model. On the contrary, the classifier retrieval time should be as limited as possible in order to classify a large set of images. Ultimately, it has to be emphasized that the numerical performance of a classifier depends not only on the algorithm itself, but also on its implementation within a particular programming language (which may feature its own numerical limitations) and the hardware employed. Consequently, it might be difficult to objectively assess the numerical performance of several classifiers implemented in different programming languages and compiled with different (more/less aggressive) compiler flags used within the process of building computer software.

3. The Vectorized Earth Observation Retrieval Classifier

The Vectorized Earth Observation Retrieval (VEOR) classifier stems from the Probabilistic Cloud Mask (PCM) algorithm proposed by [69], which was further reimplemented for the detection of low stratiform clouds [70]. Following this, the VEOR prototype was used for the derivation of cloud climatology over Central Europe from Advanced Very High Resolution Radiometer (AVHRR) imagery [71]. Its usability was also assessed for the generation of the EUMETSAT Satellite Application Facility on Climate Monitoring (CM SAF) climate data records [72]. Ultimately, the VEOR methodology was improved within the current study, especially concerning classification accuracy, self-optimization and numerical performance. There are two common concepts behind all of these algorithm versions. The first is related to the computation of the multidimensional classification probability histogram, which is flattened to a vector form during algorithm training (this is the origin of the vectorized adjective in the VEOR name), and the second concept is related to the computation of an index that assigns a unique identifier (ID) to each element in the flattened probability vector. The vector with ID is generated by placing values of input classification features transformed to 8-bit integers on consecutive bit ranges, within a 64-bit integer, by means of bitwise operators (bitwise_shift and bitwise_or). This second vector is analogous to the quality flag products usually associated with Earth observation (EO) products (e.g., MYD35 cloud mask), where different ancillary information is provided. Thus, after VEOR training, two look-up vectors (LUV) are derived: one with a classification probability and the second with sorted ID. The classification of new input data involves the computation of a new ID vector that is itself located within the sorted LUV by means of a fast binary search algorithm [73] in order to retrieve the classification probability. Comprehensive explanation of the VEOR algorithm for a simplified two-dimensional (2D) synthetic dataset is provided in the following subsections.

3.1. Derivation of Look-Up Vectors (LUV)

The simplified derivation of LUV for the 2D case, with binary labels and two artificial features, is presented in Figures 1 and 2. Red points indicate label 1, and blue points label 0. The gray lines denote percentiles of feature values ranging from 0 to 100% with 10% intervals forming an irregular grid, where the probability Px for a cell x (Equation (1)) is computed as a ratio between the number of samples labeled with 1 (red points) and the total number of samples N within cell x (red + blue points).

$$P_x = \frac{\sum_{i=1}^{N} c_i}{N} \quad (1)$$

where:

P_x—classification probability within cell x;
N—number of samples within cell x;
c_i—binary label of a sample within cell x.

However, if there is a significant imbalance between the number of red and blue points, the computed probability is biased towards the more frequent class. Therefore, the initial probability is recomputed after class balancing in which every class is oversampled relative to the total number of samples; this effectively doubles the initial number of samples. The oversampling is performed while randomly assuming the initial probability distribution of each class within each cell of the percentile grid. Incorporation of percentiles as the grid coordinates instead of varying magnitudes of features (corresponding to different physical units) suppresses the need for feature scaling. Within the standard VEOR implementation, the grid is composed of 254 percentiles ranging from 0 to 100, which normalize values for each feature to 2^8 (8-bit) integers. Further, this allows for the computation of a unique ID for each cell consisting of 8-bit values stacked together (by means of bitwise_or and bitwise_left_shift operators for little-endian machines) on consecutive bit ranges within a 64-bit value. Consequently, the VEOR can analyze up to eight features with 256 degrees of freedom each. The percentile grid significantly reduces the number of samples relative to the number of cells, which gives great flexibility in terms of the size of the training sample and the incorporation of new samples. This feature of the VEOR algorithm is especially important for frequently updated training datasets that require frequent adjustment of LUV with classification probabilities.

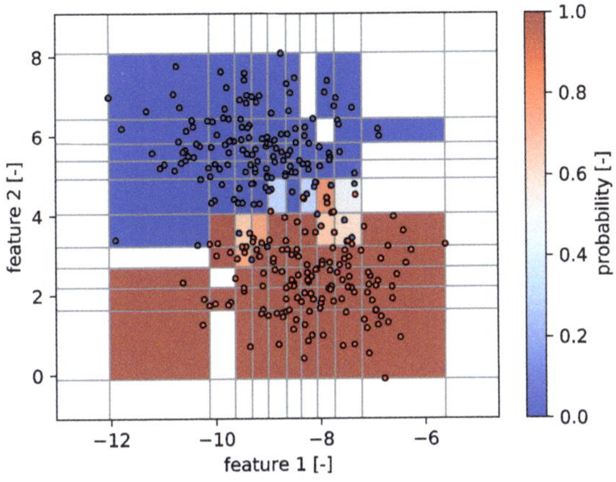

Figure 1. The derivation of classification probability for two clearly separable classes. The gray grid denotes the 0%, 10%, 20%, 30%, . . . , 90%, 100% percentiles for each feature.

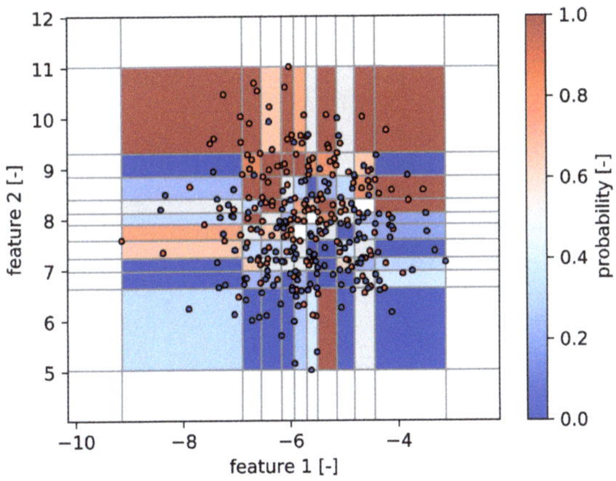

Figure 2. The derivation of classification probability for two weakly separable classes. The gray grid denotes the 0%, 10%, 20%, 30%, . . . , 90%, 100% percentiles for each feature.

3.2. Feature Selection

The percentile grid should include features with strong classification skills. This minimizes the number of cells with the mixed classes, as depicted in Figure 1. On the contrary, if the features have weak classification skills, many grid cells contain mixed classes and the derived probability field is not decisive, as depicted in Figure 2. In order to determine the optimal feature subset, VEOR applies an iterative forward selection procedure (see Figure 3 and Algorithm A2 in Appendix B) where the features are progressively added whenever the kappa score is improved. Forward selection starts with a single feature for which the percentile grid is simply a 1D histogram with classification probabilities. Then, for each grid cell (i.e., histogram bin), the K-nearest cells and Euclidean distances d_k are determined using a fast cKDTree nearest neighbor search method [74]. The selected K-nearest cells are used to reconstruct the classification probability P'_x for cell x (Equation (2)). This is achieved by dividing two sums weighted by inverse Euclidean distances between cell coordinates expressed as percentile numbers (8-bit values). The first sum is computed as the weighted total number of samples labeled with 1 within the K cells. The second sum is the weighted total number of samples N_k within the K cells. The reconstructed probability value describes each sample within cell x and is ultimately rounded to a reconstructed binary label c'_i for all samples within cell x.

$$P'_x = \frac{\sum_{k=1}^{K} \frac{1}{d_k} \sum_{i=1}^{N_k} c_i}{\sum_{k=1}^{K} \frac{1}{d_k} N_k} \tag{2}$$

where:

P'_x—reconstructed probability value for cell x;
K—number of nearest neighbouring cells in the vicinity of cell x;
d_k—Euclidean distance between cell k and cell x expressed as a percentile number;
N_k—number of samples within cell k;
c_i—binary label of a sample within cell k.

This allows for the computation of a confusion matrix between the original c and reconstructed c' binary labels, and consequently for a derivation of the kappa score for a particular feature or combination of features.

The next parameter to be optimized is the number of nearest neighbors (K) used to reconstruct the probability value. The higher the K, the smoother the reconstructed classification probability field, which, on the one hand, reduces the impact of mislabeled samples (i.e., reduces noise effects), but, on the other, decreases the accuracy of correctly labeled data. Optimal K is derived iteratively for each feature combination by repeating the probability reconstruction procedure with increasing even numbers $K \in \{2, 4, 6, \ldots\}$. It was found that, for small odd K values, asymmetry is introduced around the class separation border. In particular, if $K = 3$ and the main cell is exactly at the class separation border, then two cells from one class and only one cell from the other class are used for reconstruction of the probability value. This obviously introduces bias into the reconstructed value. The optimal K number is derived whenever adding more neighbors starts to decrease the kappa score.

After the kappa score is computed for all single features, then the feature with the highest kappa is selected, and within the following iterations, features are added to it. After the best two-feature subset is selected, single features are added again, and so on. Whenever adding new features does not bring any improvement to the kappa value, the forward feature selection procedure converges and the feature subset with the highest class separability skills is returned. The sequence in which the optimal features are selected further corresponds to the 8-bit positions within the 64-bit ID value. Features derived at the beginning of the forward feature selection procedure occupy more significant bit ranges than features derived later. Furthermore, VEOR ranks not only the separation skill of single features but also the separation skill of feature combinations, which greatly facilitates the understanding of the acquired classification results.

The last part of the VEOR algorithm training involves the oversampling and compaction of the LUV derived for the optimal feature subset. Firstly, oversampling involves reconstruction of the classification probability using the optimal K-nearest grid cell number for the user-defined number (default 10^6) of random cells located around the grid cells filled with samples within an Euclidean distance selected from the normal distribution with the user-defined scale (default 10). Secondly, the oversampling involves reconstruction of the classification probability using the 5^D kernel around each grid cell filled with samples, where D is the number of dimensions (features). Consequently, classification probability is predicted for grid cells not populated with samples during algorithm training (marked in white in Figures 1 and 2). This is achieved by means of the inverse distance interpolation between the grid cells populated with data, which essentially extrapolates a classification probability to grid cells without any data. Such a prediction of classification probability within a multidimensional percentile grid is performed only once, which significantly increases the numerical performance of the VEOR classification.

Ultimately, compaction of LUV involves the removal of entries that are redundant from the perspective of a binary search algorithm. The LUV with ID is sorted, and therefore if several consecutive ID have exactly the same classification probability (which happens often if the classes are clearly separable), then only the IDs at the sequence edges are required for the binary search algorithm. This procedure reduces the number of entries within LUV several times, which further reduces the computational effort of the VEOR classification process.

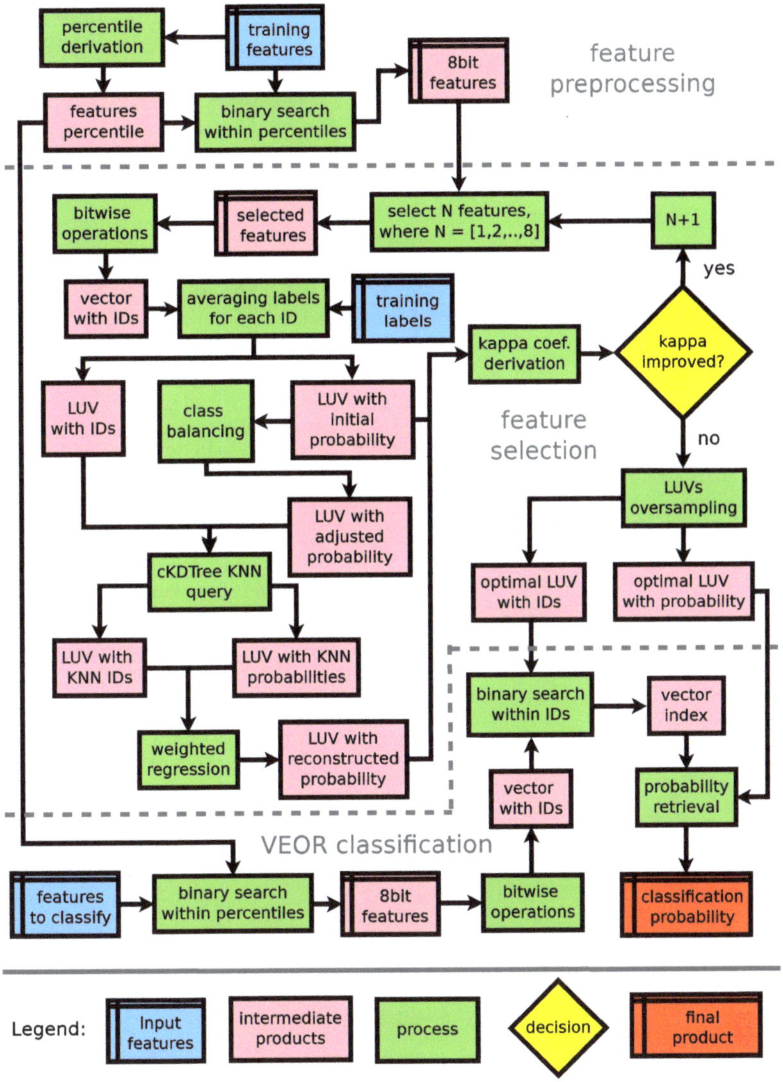

Figure 3. The VEOR algorithm workflow.

3.3. Inclusion of Discrete Features

The features discussed so far with respect to VEOR training were of the continuous type, which allows for the computation of Euclidean distances and percentiles. In the case of discrete/categorical features (e.g., land cover), VEOR training is performed separately for each discrete class, which results in several separate LUV. Further, the integer labels describing classes are placed on the most significant bit ranges within each 64-bit ID value. This, in turn, allows the combination of all of the LUV using a bitwise_or operator, as each LUV with ID occupies a different bit range within the merged LUV. Obviously, several discrete features can be placed on the most significant bit ranges, giving great flexibility for the adjustment of VEOR settings to a wide range of environmental conditions. In any case, the limiting factor is the 64-bit value of the ID, which should encompass both continuous and discrete features.

3.4. Classification of New Data

The classification of new samples by means of the VEOR algorithm only requires the extraction of probability values from LUV using a fast binary search technique. This search algorithm is based on the partitioning of sorted data without derivation of distance metrics at all. In the worst-case scenario, this reduces the computation time to O(log N) operations, where N is the number of LUV entries. In order to extract values from LUV, the input features have to first be transformed to an 8-bit integer using binary search within the percentiles derived during VEOR training. Further 8-bit values are stacked into 64-bit values by means of bitwise operations. The derived 64-bit values are located within the LUV with sorted ID by means of the binary search technique. This results in a vector of indexes (positions), which is finally used to extract values from the LUV with the classification probabilities derived during algorithm training (see Figure 3 and Algorithm A1 in Appendix B).

4. Description of Selected Classifiers Implemented in the Python Scikit-Learn Library

The VEOR classifier proposed for this study was evaluated together with ten machine learning algorithms implemented within the Python scikit-learn (former sklearn) library [75] version 0.24.2, released in April 2021 (Table 1).

Table 1. The list of scikit-learn classification routines used for benchmarking with the VEOR algorithm.

Algorithm	Module/Routine
Multi-layer perceptron classifier that optimizes the log-loss function using the limited memory Broyden–Fletcher–Goldfarb–Shanno (LBFGS) gradient descent algorithm [76,77]	sklearn.neural_network/MLPClassifier
K-nearest neighbor classifier [78]	sklearn.neighbors/KneighborsClassifier
Support vector classification with a linear kernel [79]	sklearn.svm/LinearSVC
Support vector classification with a Radial Basis Function (RBF) kernel [79]	sklearn.svm/SVC
Gaussian process classification (GPC) based on the Laplace approximation [80]	sklearn.gaussian_process/ GaussianProcessClassifier
Decision tree classifier [43]	sklearn.tree/DecisionTreeClassifier
Random forest classifier [81]	sklearn.ensemble/RandomForestClassifier
AdaBoost-SAMME classifier [82,83]	sklearn.ensemble/AdaBoostClassifier
Gaussian Naive Bayes classifier [84]	sklearn.naive_bayes/GaussianNB
Quadratic Discriminant Analysis [85]	sklearn.discriminant_analysis/ QuadraticDiscriminantAnalysis

The scikit-learn library is very well established within the machine learning community (developed for >10 years). However, it has to be emphasized that the numerical implementation of some classifiers might still be suboptimal or erroneous. Moreover, this library makes external calls to other libraries implemented in Python (i.e., Numpy version 1.21.1) or C (i.e., BLAS version 3.7.1), which adds another source of potential software bugs. Nevertheless, debugging of the scikit-learn classification routines is beyond the scope of this work and the library was used as it was. Besides classifiers, the scikit-learn library comes with a set of routines used within this study for (1) generation of synthetic datasets, (2) computation of classification performance metrics, (3) data pre-processing and standardization, (4) hyper-parameter optimization, (5) feature selection and (6) visualization of classifiers' performance (plot_classifier_comparison.py). In the following subsections, a short summary of each classifier is given in order to facilitate further understanding and interpretation of the results.

4.1. Nearest Neighbor Classifier

The nearest neighbor algorithm [78] categorizes a sample based on a simple majority vote taken from a subset of adjacent training samples located in the multidimensional feature space. Subset selection can be performed based on the number of samples (KNN: K-nearest neighbor method) or the radius around the analyzed sample. Moreover, majority voting may take into account various distance metrics to privilege samples located closer to the analyzed sample. The main drawback of the nearest neighbor classifier is related to the handling of the entire training dataset in order to classify a single sample and to the computational overhead of the extensive search. On the positive side, the classifier is robust and accurate for relatively small data samples.

4.2. Decision Tree (DT) Classifier

A decision tree [43] is a suite of non-parametric multi-stage learning methods that break down a complex decision-making process into a collection of simpler decisions. At each stage (a tree node), the algorithm partitions an input training dataset (a tree root), by means of simple thresholding tests, into smaller data subsets, forming consecutive branches of a decision tree. These tests are constructed statistically in a variety of ways to increase the homogeneity of the resulting subgroups. At the last stage within a branch (a tree leaf), the residual data subset is assigned to one of the final classification labels. Usually, the tree grows until all of the training observations are correctly classified (i.e., classification accuracy = 100%). This leads to overfitting of the decision tree and results in a poor classification of samples unseen in the training dataset. Thus, the ultimate step of a DT classifier should involve tree pruning to remove sections of a tree with limited classification significance and to improve its predictive skills. Approaches towards the construction of a DT can be categorized into three groups: (1) univariate (decisions at a single stage made using a single feature), (2) multivariate (decisions at a single stage made using multiple features), (3) hybrid (fusion of univariate and multivariate trees within a single tree). The process of tree building can be optimized within the DecisionTreeClassifier routine by means of several hyper-parameters that control the number of leaves, tree depth, minimum number of points to split a node or a criterion of splitting. The advantages of a DT classifier are related to: (1) direct interpretation of a classification result depicted as a graph, and (2) handling of numeric and categorical variables featuring non-linear relationships with the classes. On the downside, DT have a tendency to overfit training data and do not perform well for high-dimensional data [86].

4.3. Random Forest (RF) Classifier

The random forest method [81] is an ensemble (bagged) predictor composed of a set of classifications generated by the DT algorithm, called the Classification and Regression Tree (CART). The classifier is trained against randomly sampled (with replacement) data subsets using randomly selected features from a feature vector. For each DT, an input dataset is split into training (referred to as in-bag samples) and validation subsets (referred to as out-of-bag samples) using a 67/33% proportion. This allows for internal cross-validation of derived classification and computation of goodness-of-fit statistics called an out-of-bag (OOB) error, defined as an error rate (number of misclassifications divided by the total number of cases). Decision trees are not pruned and the final classification is computed by majority voting across results acquired from all of the generated trees. The main benefit of the RF classifier over the DT method is insensitivity to overfitting and to some level of noise in a training dataset. On the negative side, RF classifiers may be computationally expensive in the case of large training datasets.

4.4. AdaBoost Classifier

The AdaBoost classifier, known as Adaptive Boosting [82], is conceptually similar to the RF algorithm in the way that it combines many weak learners (such as DT classifiers) to develop a strong predictor. Contrary to the RF method, where DTs are generated independently/parallelly, AdaBoost constructs a forest of DTs sequentially based on the classification skills of previous DTs. In fact, DT learners within the AdaBoost classifier are called "stumps", simply because they have one node and two leaves. The stumps alone may feature predictive skills only slightly better than a random guess, but a group of stumps, with each one progressively learning from the misclassified cases of its predecessor, "boosts" the final classification accuracy. In practice, initially, AdaBoost assigns equal weights to all training samples and selects the first decision stump with the highest classification skill expressed by a logistic loss function. Further, the weights of the misclassified samples from the first stump are increased so the second stump is more focused on those samples. This procedure ends whenever all of the samples are correctly classified or the maximum number of stumps has been reached. Ultimately, for each sample, the predictive skills of positive stumps (with a "yes" label) and negative stumps (with a "no" label) are summed up separately, and the final label is assigned according to a larger sum. AdaBoost is suited for imbalanced datasets as it is insensitive to overfitting but underperforms for noisy datasets [41].

4.5. Support Vector Machine (SVM) Classifier

SVMs [79] are supervised learning algorithms that construct an optimal hyperplane or a set of hyperplanes to linearly discriminate two groups of samples. The optimal hyperplane maximizes a distance between two planes (referred to as support vectors) that are a part of the convex hulls (margins) of those groups in a feature space. SVM learning is an iterative procedure to select a classifier featuring the optimal hyperplane that minimizes misclassifications. If samples are incorrectly labeled in a training dataset and consequently fall onto the wrong side of a hyperplane, then they are down-weighted to reduce their influence. In the case of a non-linear relationship between classes and samples in the feature space, these samples are projected by means of kernel techniques to a (usually) higher-dimensional space where the separation of classes becomes linear. The choice of optimal type (e.g., linear, Radial Basis Function (RBF), etc.) and the parameters of a kernel greatly influence the classification quality. SVMs perform a multi-class classification by fitting binary sub-classifiers to each class and selecting the one that occurs in the most cases. From a numerical perspective, optimization of SVMs is a quadratic problem not suitable for large datasets. The advantage of this method is related to reliable classification results in the case of sparse and noisy training datasets [87].

4.6. Artificial Neural Network (ANN) Classifier

ANNs [76,77] are a collection of machine learning algorithms inspired by the decision-making processes occurring in a human brain. ANNs consist of many interconnected artificial neurons (also called nodes) forming a network called the perceptron, composed of one or more layers of neurons. An artificial neuron is a thresholding processing unit that generates a response (binary or continuous) if a weighted sum of inputs from other neurons exceeds a certain activation level (threshold value). The weights determine the connection strength between neurons, which may express an excitatory (positive) or inhibitory (negative) connection. Through supervised learning techniques, the weights are iteratively adjusted to minimize log-loss cost function. The layers of neurons can be consecutively connected, forming a feed-forward network, or can be arranged in loops (reversely connected), forming a recurrent network. The neurons can be divided into three types: input units, which receive data from outside the neural network; output units, which send data out of the neural network; and hidden units, whose input and output signals remain within the neural network. The distributed structure of an ANN allows for massive computational parallelization provided by modern GPU (graphical processing unit) processors. Optimal parametrization of an ANN might be difficult; however, once achieved the algorithm provides results superior to other machine learning methods for large datasets. The main disadvantage of an ANN is the "black box" nature of the derived optimal network configuration, which is not human-interpretable. Consequently, it is difficult to track and justify why a particular output was generated by the algorithm.

4.7. Gaussian Process (GP) Classifier

GP is a stochastic model used within the machine learning discipline for classification and regression analyses [80]. GP can be considered an infinite collection of random variables that feature multivariate Gaussian (normal) distributions for any finite subset of this collection. The random variables are described by an infinite set of functions, which are further constrained by means of observations to derive posterior distributions over functions. In this way, GP does not make any assumptions about the parameters of the function featuring a "best-fit" to observed values, but instead, a priori, it places directly on the space of functions. The closer to training samples a new sample is within this space, the higher probability that the retrieved value will be correct. GP allows for the quantification of this probability. It is characterized by its mean and covariance functions, which facilitate analytical manipulations due to the simple mathematical solutions originating from the properties of a multivariate Gaussian distribution. The mean and covariance functions have to be selected a priori, which may be difficult for a small noisy training dataset. In such a case, the mean function is often set to zero for the sake of numerical convenience, since it is always possible to center a data distribution around zero. Derivation of an appropriate covariance function is more problematic since it is defined in terms of hyper-parameters, which have to be inferred by the GP itself. The definition of hyper-parameters allows for the incorporation of expert knowledge, e.g., about the smoothness or periodicity of the modeled phenomenon. The main advantages of the GP method are related to the derivation of classification probability and to the incorporation of a priori knowledge about the classified phenomena. The main disadvantage of GP is the computational complexity, which quickly scales up with the number of samples; thus, sparse approximation techniques are required for large training datasets.

4.8. Naive Bayes Classifier

Bayesian theory (formulated by Thomas Bayes in 1763) describes the a posteriori probability of an event as a function of conditional probabilities and the a priori probabilities of other events. From the perspective of an image classification, the a posteriori probability is defined as the likelihood of occurrence of a certain class given a certain feature vector (e.g., spectral bands). The conditional probabilities are the likelihoods of occurrence of a particular feature vector for a particular class. The a priori probabilities are the overall likelihood of occurrence of a particular class and the overall likelihood of occurrence of a particular feature vector. Conditional probabilities induce many correlations between features and as such are numerically expensive to derive for a large feature vector with many degrees of freedom. The solution to this problem assumes an independence of features and is the foundation of the Naive Bayes method. Such a simplification allows for the reduction of a multidimensional probability density matrix to one-dimensional density estimation. Although independence of features is a weak assumption, in practice, the Naive Bayes method was found to be robust, especially for large training datasets with many features. The performance of the Naive Bayes classifier under the violated independence assumption was studied by [88], where it was proven that the classification results are unaffected if the dependencies between features are distributed evenly between the classes or if the dependencies cancel each other out. The main advantages of the Naive Bayes method are related to the quantification of classification uncertainty and numerical efficiency for large datasets. The main disadvantages are related to the assumption of independent features and to the "zero frequency problem", where one of the conditional probabilities is zero and consequently the entire a posteriori probability equals zero. In such a case, a Laplace smoothing technique can be used that involves adding a single row (one count) while computing conditional probabilities. This solution eliminates the "zero frequency problem" without any significant influence on the classification results.

4.9. Quadratic Discriminant Analysis (QDA) Classifier

The QDA classifier [85] defines discriminant function(s) that separate(s) classes that are supposed to feature Gaussian (normal) probability density distributions characterized by mean values and covariance matrices. The discriminant function(s) determines points within the feature space where probability density estimates for each class are equal (i.e., the classification likelihood is 50%/50%). The Gaussian parameters for each class are derived from a training dataset using the maximum likelihood (ML) method, which is convenient for small, under-represented classes. A simplified version of the QDA called the Linear Discriminant Analysis (LDA) assumes that the covariance matrices for each class are equal, which results in linear discriminant functions. Nevertheless, both classifiers require that the number of samples is larger than the number of features. The advantages of QDA and LDA are related to their simplicity and accurate classification in the case of classes that feature Gaussian probability distributions. If this assumption is not met, then the classification results might be poor.

5. Experimental Design

The presented experiments aimed at evaluating the performance of the selected classifiers using synthetic datasets featuring various characteristics/malformations that are challenging from the perspective of class separation, as well as real cases involving cloud discrimination using MODIS spectral measurements. All of the datasets were labeled with a binary categorical variable. Datasets labeled with several classes were not considered, as every multiclass classification can be divided into several binary classifications and combined again using the majority voting scheme. Lastly, the numerical performance of the classifiers was evaluated. The following subsections provide methodological details about these experiments.

5.1. Generation of Synthetic Datasets

Synthetic datasets with binary labels, generated by the scikit-learn library, were divided into three simplified two-dimensional (2D) cases and one complex seven-dimensional (7D) case. Further, within each case, the dataset was (re)generated for a varying number of samples (100, 250, 500, 1000), with different fractions of mislabeled samples referred to as noise level (10%, 20%, 30%), and with different proportions of class balance (30%/70%, 40%/60%, 50%/50%). Each synthetic dataset was split into training and validation datasets according to the 70%/30% proportion. The datasets were grouped into three categories with respect to classification difficulty:

1. *easy*—perfectly balanced with 10% noise;
2. *moderate*—40%/60% class imbalance with 20% noise;
3. *difficult*—30%/70% class imbalance with 30% noise.

For the generation of the 2D cases, the following scikit-learn routines were used:

1. *linear dataset*—adapted from the "classifier comparison" script with a linear boundary between groups;
2. *moons dataset*—adapted from the routine "make_moons" from the "sklearn.datasets" module with a non-linear boundary between groups;
3. *circles dataset*—adapted from the routine "make_circles" from the "sklearn.datasets" module with one group forming a ring surrounding another.

For the generation of the 7D case, the "make_classification" routine from the "sklearn.datasets" module was used with the following settings: seven features, three informative features, two redundant features, two irrelevant features, one cluster per class and two classes (binary labeling). The informative features consist of Gaussian clusters each located around the vertices of a hypercube. For each cluster, informative features are drawn independently from N(0, 1) and then randomly linearly combined within each cluster in order to add covariance. The clusters are then placed on the vertices of the hypercube. Redundant features are random linear combinations of the informative features and irrelevant features are merely random samples from the univariate "normal" (Gaussian) distribution. The "make_classification" routine was adapted from [89], where it was used to generate the non-linear "Madelon" dataset used during the Neural Information Processing Systems (NIPS) workshop on feature extraction in 2003. The parameters describing the synthetic datasets are given in Table 2.

Table 2. Parameters describing synthetic datasets used for the evaluation of selected classifiers.

number of features	2, 7
number of samples	100, 250, 500, 1000
fraction of mislabeled samples (in %)	10, 20, 30
class imbalance (in %)	30/70, 40/60, 50/50

5.2. Real Case Datasets

The real case datasets were composed of spectral measurements from the MODIS sensor mounted aboard the AQUA satellite, which were spatially and temporally collocated with the reference cloud mask originating from the Cloud-Aerosol Lidar with Orthogonal Polarization (CALIOP) lidar data acquired by the CALIPSO satellite, as well as the Cloud Profiling Radar (CPR) data acquired by the CloudSat satellite. The AQUA, CALIPSO and CloudSat platforms followed each other closely as part of the A-Train satellite constellation, crossing the equator in an ascending (northbound) direction at approximately 1:30 p.m. local solar time. CloudSat and CALIPSO lagged AQUA by 1 to 2 min and were separated from each other by 10 to 15 s [90]. This allowed for constant spatio-temporal collocations of AQUA's MODIS measurements with CALIPSO's CALIOP and CloudSat's 94 GHz CPR vertical profiles in order to jointly analyze the rapidly changing cloud cover. In this respect, the 2B-GEOPROF-LIDAR product version R05 [90] used in this study includes a number of hydrometeor layers in the vertical column of the CloudSat profile. To separate cloud layers, at least 960 m of hydrometeor free space, as defined by either radar or lidar, must be present. The cloud layers detected by the lidar are computed from the 5 km horizontally averaged Vertical Feature Mask product featuring a native resolution of 333 m. The horizontally averaged product was found to be in best agreement with the MODIS AQUA (MYD35) cloud mask product [90]. Within this study, the reference binary cloud mask used for the training and validation of selected classifiers was generated assuming cloud cover whenever at least one cloud layer was detected either by the CALIOP or CPR instruments.

Along with the 2B-GEOPROF-LIDAR, the MODIS-AUX product [91] is generated, which consists of the Collection 6 AQUA MODIS products that overlap and surround each CPR footprint, namely (1) radiance (MYD02_1KM_L1B product), (2) cloud mask (MYD35_L2 product) and (3) geodetic 1-kilometer resolution latitude and longitude (MYD03). Within the MODIS-AUX product, all MODIS datasets are collocated with the CPR ray using the great-circle nearest neighbor scheme, forming a 3-pixel across-track by 5-pixel along-track spatial subset overlapping each CPR footprint. The maximum allowed collocation distance between a MODIS 1 km pixel and a CPR footprint was set to 0.95 km. As the A-Train constellation satellites follow each other along very similar orbits, most of the collocations are close to the nadir view (mean MODIS satellite zenith angle 15.15° +/− 2.84°). Consequently, the dataset used to train classifiers featured very little angular variability across-track and, in this respect, it was not representative.

Within the presented study, the 2B-GEOPROF-LIDAR and the MODIS-AUX products were further matched by selecting a pixel from the 3-pixel across-track by 5-pixel along-track spatial subset of MODIS products with the smallest distance to the corresponding CPR footprint. Spectral reflectances, brightness temperatures and Sun/sensor geometries derived from the embedded MYD02_1KM_L1B product were used as explanatory variables for the training of selected classifiers and the retrieval of classifications. All explanatory variables were standardized (centered around the mean value with a unit standard deviation) in order to improve the reliability of distance computation required by some classifiers (e.g., KNN, SVM). For the sake of comparison, the MODIS cloud mask was also derived from the embedded MYD35 product and transformed to a binary mask assuming that "confident clear" and "probably clear" pixels are cloud-free whereas "confident cloudy" and "probably cloudy" pixels are filled with clouds. The 2B-GEOPROF-LIDAR and the MODIS-AUX products were split into two independent, temporally decoupled validation and training datasets acquired in 2006 and in 2007, respectively. The analyzed time period was selected due to the earliest possible joint operation time of the AQUA (launched in 2002), CALIPSO (launched in 2006) and CloudSat (launched together with CALIPSO) satellites. At that time, the instruments had experienced very small degradations apart from the MODIS AQUA 1.6 μm channel. The full list of spectral and ancillary features used by the classifiers is presented in Table 3.

The MODIS collocations with CALIOP and CPR were divided into two real cases. The first simple one consisted of daytime data (Sun zenith angle < 85°) over water, away from sun glint areas and excluding polar regions (60°–90° latitudes). The second, more difficult case consisted of daytime data (Sun zenith angle < 85°) over ice/snow. Information about sun glint areas as well as the land/water and snow masks was extracted from the MYD35 product. Each real case dataset was randomly subsampled into 100,000 training samples and 30,000 validation (hold-out) samples.

Table 3. Spectral and angular features used for cloud discrimination over water surfaces using MODIS imagery.

Abbreviation	Full Name
B1_645nm	Band 1 reflectance at 645 nm
B2_858nm	Band 2 reflectance at 858 nm
B3_469nm	Band 3 reflectance at 469 nm
B4_555nm	Band 4 reflectance at 555 nm
B5_1240nm	Band 5 reflectance at 1.240 μm
B7_2130nm	Band 7 reflectance at 2.130 μm
B17_905nm	Band 17 reflectance at 905 nm
B18_936nm	Band 18 reflectance at 936 nm
B19_940nm	Band 19 reflectance at 940 nm
B26_1375nm	Band 26 reflectance at 1.375 μm
B20_3750nm	Band 20 brightness temperature at 3.750 μm
B27_6715nm	Band 27 brightness temperature at 6.715 μm

Table 3. *Cont.*

Abbreviation	Full Name
B28_7325nm	Band 28 brightness temperature at 7.325 µm
B29_8550nm	Band 29 brightness temperature at 8.550 µm
B30_9730nm	Band 30 brightness temperature at 9.730 µm
B31_11030nm	Band 31 brightness temperature at 11.030 µm
B32_12020nm	Band 32 brightness temperature at 12.020 µm
B33_13335nm	Band 33 brightness temperature at 13.335 µm
B34_13635nm	Band 34 brightness temperature at 13.635 µm
B35_13935nm	Band 35 brightness temperature at 13.935 µm
B36_14235nm	Band 36 brightness temperature at 14.235 µm
B1_645nm_B2_858nm_ratio	Reflectance ratio between Bands 1 and 2
B31_11030nm_B20_3750nm_diff	Brightness temperature difference between Bands 31 and 20
B32_12020nm_B20_3750nm_diff	Brightness temperature difference between Bands 32 and 20
B31_11030nm_B32_12020nm_diff	Brightness temperature difference between Bands 31 and 32
B28_7325nm_B31_11030nm_diff	Brightness temperature difference between Bands 28 and 31
B29_8550nm_B31_11030nm_diff	Brightness temperature difference between Bands 29 and 31
B29_8550nm_B28_7325nm_diff	Brightness temperature difference between Bands 29 and 28
sunz	Sun zenith angle
satz	Satellite zenith angle
suna	Sun azimuth angle
sata	Satellite azimuth angle
razi	Relative azimuth angle
scat	Scattering angle
NDVI	Normalized Difference Vegetation Index $\mathrm{NDVI} = \frac{\mathrm{B2\_858nm} - \mathrm{B1\_645nm}}{\mathrm{B2\_858nm} + \mathrm{B1\_645nm}}$

5.3. Optimization of Classifiers: Forward Feature Selection and Hyper-Parameter Tuning

The classifiers implemented within the scikit-learn library come within a set of default hyper-parameters that should perform well for a wide range of datasets. Nevertheless, this flexibility comes at the price of suboptimal accuracy, which can be further improved by the tuning of hyper-parameters. In this respect, the scikit-learn library version 0.24.2 offers a suite of useful routines, amongst which the *HalvingGridSearchCV* routine from the sklearn.model_selection module was used within this study. This routine starts by evaluating all combinations of hyper-parameters with a small amount of resources and iteratively selects the best ones, using more and more resources. The term "resources" typically denotes a number of training samples, but it can also be an arbitrary numeric parameter such as a number of trees in a random forest. The optimal combination of hyper-parameters maximizes some classification quality metric (Cohen's kappa score in this study), which is derived internally by the *HalvingGridSearchCV* routine, which divides the input samples into the training and held-out samples. It has to be emphasized that the input samples used for the forward feature selection and hyper-parameter tuning consisted of training datasets generated within the experiments with the 2D datasets and MODIS data. The optimal set of hyper-parameters varies for different combinations of input features. Therefore, feature selection was performed, prior to the hyper-parameter tuning, by means of the *SequentialFeatureSelector* routine from the sklearn.feature_selection module. This routine performs forward feature selection using 5-fold cross-validation, which maximizes some classification quality metric (Cohen's kappa score in this study). The feature selection procedure was performed only for the real case datasets (summary of the selected features is given in Appendix A). Within the experiments with synthetic 7D datasets with redundant and irrelevant features, the feature selection was not performed on purpose, in order to evaluate the ability of the classifiers to deal with such situations. On the contrary, the hyper-parameter tuning was always performed—separately for every classification process.

5.4. Evaluation of Classifiers

The selected classifiers were assessed in the same consistent way across experiments, using accuracy (ACC) and Cohen's kappa (κ) scores to measure the agreement between classified and reference data. The ACC score, defined as the fraction of correctly labeled samples, is widely used; however, it is difficult to interpret for imbalanced datasets (see Section 2.3). In such a case, Cohen's kappa score [60] is more robust as it takes into account the possibility of agreement occurring by chance. Its values range from -1 to 1, where 1 is perfect agreement between predicted and reference labels, 0 denotes (dis)agreement that would be obtained by a completely random classifier, whereas negative values denote (dis)agreement that is worse than random. The formula for Cohen's kappa score is given by Equation (3).

$$\kappa = \frac{P_o - P_e}{1 - P_e} \tag{3}$$

where:

P_o—empirical probability of agreement between predicted and reference labels (same as ACC);

P_e—hypothetical probability of agreement between randomly assigned and reference labels.

The randomly generated synthetic datasets introduced stochastic variations of derived classification quality metrics. Therefore, the experiments with the synthetic datasets were repeated 30 times for each combination of three variables: number of samples, class imbalance and noise level. In the case of the daytime cloud detection over water surfaces using the MODIS/CALIPSO/CPR collocations, the analysis was run once with the training dataset collected in 2007 and the validation dataset from 2006.

Ultimately, the numerical performance of the classifiers in terms of the execution times of the training and retrieval processes separately was investigated. The machine used for this analysis was based on the AMD EPYC 7401 24-Core Processor with 256 GB RAM, running the Ubuntu version 18.04.5 operating system with a Python 3.9.6 environment containing the NumPy version 1.21.1, Scipy version 1.7.14. and scikit-learn version 0.24.2 libraries. For the sake of comparability, the multicore/hyper-threading support for the selected classifiers (including the VEOR) was deactivated in order to measure the single-core numerical performance. For this analysis, the linear synthetic dataset with an increasing number of samples (10^2, 10^3, 10^4, 10^5, 10^6) was used, and each analysis was repeated five times to generate mean execution times. In the case of the Gaussian process classifier, training of more than 10^4 samples was computationally too expensive and the execution times for those datasets were not measured.

6. Results

6.1. Classification of Synthetic 2D Datasets

Initially, the evaluation of the selected algorithms involved a binary classification of three types of randomly generated synthetic datasets composed of two features separated by linear, non-linear and circular borders (see rows in Figure 4), which were further grouped according to a classification in terms of difficulty: easy, moderate and difficult (see Section 5.1). An exemplary experiment with an easy dataset is presented in Figure 4, where blue points are labeled with 0 and red with 1. Circles denote training points and triangles denote validation points, which were split according to a 70%/30% proportion, respectively. In the background, the classification probability or decision function, scaled to a 0–1 range for SVM, is presented for every sample within the 2D grid. The accuracy (ACC) and kappa (κ) values are also reported, although they are not conclusive due to the randomness of synthetic datasets, which introduces great variability between algorithms, especially for datasets with a low number of samples (see standard deviation whiskers in Figure 5). Nevertheless, some interesting conclusions can be drawn regarding the inability of some classifiers to fit the samples. In this respect, it is not surprising that linear SVM is unable to fit datasets with non-linear and circular borders. However, it is more interesting that the Naive Bayes classifier is unable to fit the simplest linear dataset, most likely because, in this case, the feature independence assumption is strongly violated. Likewise, for the moons dataset, the RBF SVM, Naive Bayes and QDA classifiers performed poorly, but the same methods performed well for the circles dataset (presumably the most difficult to classify). The classification probability/scaled decision function presented in the background of Figure 4 reveals interesting patterns. Algorithms based on data partitioning, such as VEOR, decision tree, random forest and AdaBoost, produced jagged results, whereas algorithms that model/fit the probability field (i.e., SVM, Gaussian process, neural net, Naive Bayes, QDA) produce smoother results. A quantitative analysis of the classification results for the linear, moons and circles datasets, grouped together according to classification difficulty (see Section 5.1), is presented in Figure 5. The general pattern apparent from this analysis reveals that the number of samples is directly proportional to the classification accuracy and inversely proportional to the variability of derived classification metrics. This is particularly noticeable for the VEOR algorithm, where a small training dataset leads to a low number of samples within the percentile grid and consequently to variable classification probability estimates that are difficult to reconstruct by the KNN weighted average. The same applies to noisy training datasets, where the reconstruction of probability is challenging even for a large number of nearest neighbors, resulting in significant probability smoothing. In such a case, the algorithms that model/fit to the classification probability field, such as RBF SVM, Gaussian process and neural net, perform best. On the contrary, the linear SVM, decision tree and Naive Bayes classifiers seem to be the least accurate and they do not "learn" much from the larger training datasets (see the median line marked on the boxes in Figure 5). This implies that if the assumptions behind these algorithms are violated (e.g., linearity, feature independence), they will not improve the classification skills given more training samples. Ultimately, it has to be emphasized that the experiment with the 2D datasets was an idealized case where only two important/meaningful features were used. In reality, the classification is usually based on more features, including some irrelevant or redundant ones. This aspect was investigated in the experiment with the 7D dataset, described in the next subsection.

6.2. Classification of Synthetic 7D Datasets

The experiment with the synthetic 7D datasets was performed analogously to the evaluation of the 2D datasets, apart from the feature vector, which consisted of three informative/meaningful features, two redundant features and two irrelevant features. No feature selection was applied during the optimization of classifiers within this experiment in order to test the sensitivity of each method to irrelevant and redundant features. Evaluation of the classifiers' performance used for the 7D dataset (Figure 6) revealed a similar pattern to the experiment with the 2D datasets, except for the linear SVM and Naive Bayes classifiers, which performed well this time. The least accurate was the decision tree classifier, which is very sensitive to the irrelevant and noisy information that forms spurious branches within a tree structure. This in turn deteriorates the ability of a classifier to generalize to new/unseen cases. Thus, a tree pruning procedure should be applied, which was optimized within this study by means of the *ccp_alpha* hyper-parameter that controls the Minimal Cost-Complexity Pruning method [43]. Surprisingly, for a small number of points, the random forest classifier performed better than the AdaBoost method, which is supposed to incorporate the decision tree classifications more efficiently than the former algorithm. Across selected classifiers, the VEOR algorithm performed well for the easy and moderate datasets, while, for the difficult datasets, the classification accuracy was slightly worse than other methods.

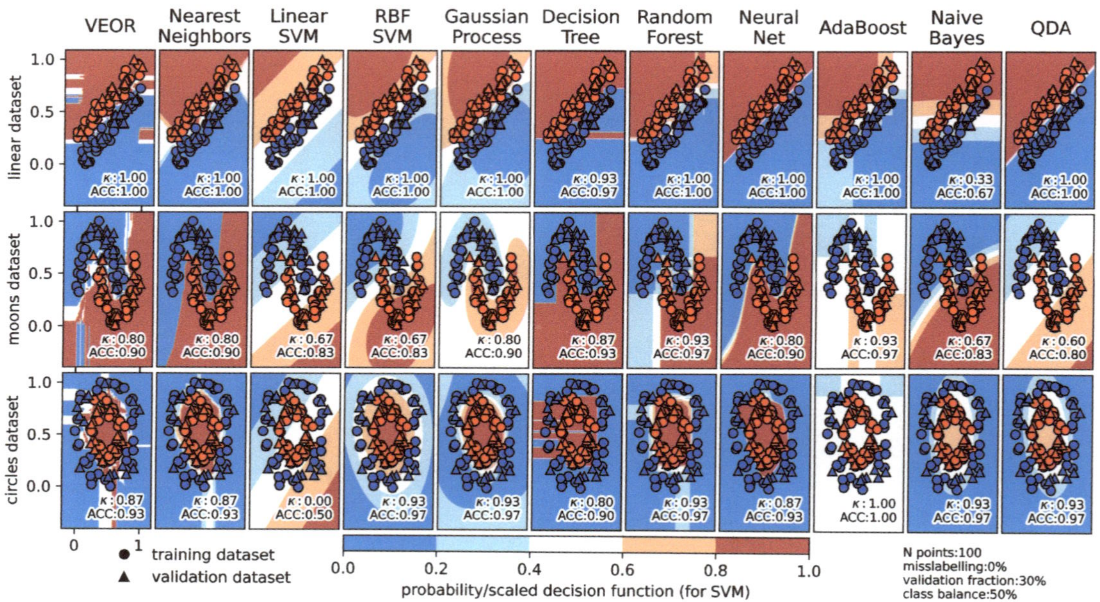

Figure 4. Comparison of classifier performance for a single experiment with synthetic 2D datasets with a 50%/50% class balance and without mislabeling.

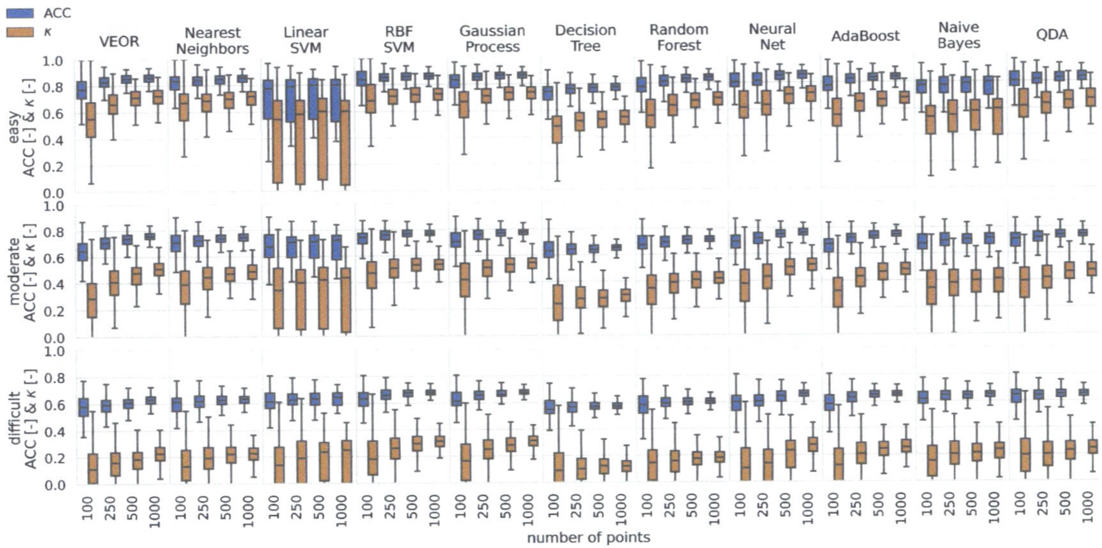

Figure 5. Summary of experiments with combined synthetic linear, moons and circles 2D datasets grouped according to classification difficulty. Easy datasets feature 10% of mislabeled samples and a 50%/50% class balance; moderate datasets feature 20% of mislabeled samples and a 40%/60% class imbalance; difficult datasets feature 30% of mislabeled samples and a 30%/70% class imbalance. Each experiment was repeated 30 times with randomly generated data.

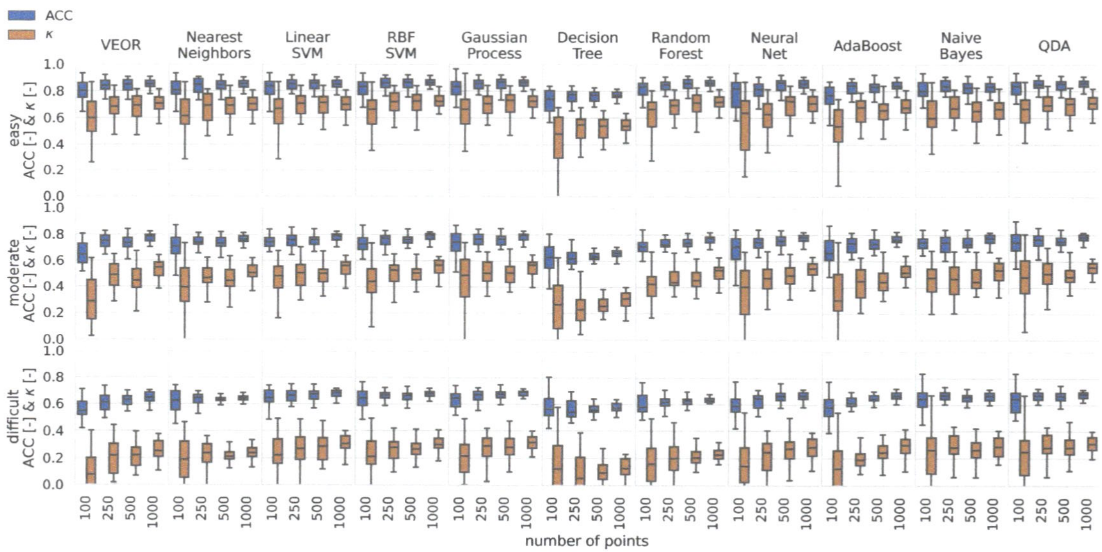

Figure 6. Summary of experiments with 7D datasets grouped according to classification difficulty. Easy datasets feature 10% of mislabeled samples and a 50%/50% class balance; moderate datasets feature 20% of mislabeled samples and a 40%/60% class imbalance; difficult datasets feature 30% of mislabeled samples and a 30%/70% class imbalance. Each experiment was repeated 30 times with randomly generated data.

6.3. Cloud Detection Using Daytime MODIS Spectral Measurements and a Reference Cloud Mask Derived from Combined CALIOP and CPR Data

6.3.1. Cloud Detection over Water Surfaces

Within the real case experiment involving cloud discrimination using MODIS satellite spectral measurements (Figure 7), the reference quality metrics (ACC = 0.879, κ = 0.723) for the selected classifiers were derived from the Collection 6 AQUA MODIS cloud mask product (MYD35). In this respect, it has to be emphasized that the MYD35 product is globally applicable for all satellite zenith angles, whereas the classifiers selected are applicable close to the satellite nadir (due to an unrepresentative training dataset). Despite this fact, the decision tree classifier produced inferior results to the MYD35 product (ACC = 0.840, κ = 0.627), whereas all other methods with optimized hyper-parameters generated very similar and accurate classifications. As compared to the VEOR algorithm (ACC = 0.888, κ = 0.748), the Naive Bayes (ACC = 0.880, κ = 0.734), Gaussian process (ACC = 0.882, κ = 0.736), QDA (ACC = 0.883, κ = 0.736), nearest neighbor (ACC = 0.884, κ = 0.738), linear SVM (ACC = 0.885, κ = 0.739) and RBF SVM (ACC = 0.886, κ = 0.745) classifiers were slightly less accurate, while the random forest (ACC = 0.889, κ = 0.747), neural net (ACC = 0.890, κ = 0.752) and AdaBoost (ACC = 0.891, κ = 0.753) classifiers were slightly more accurate. It could be noticed that many classifiers discriminated cloud cover over water surfaces with almost the same classification quality metrics. This may indicate that several methods reached almost maximum classification accuracy given the radiometric sensitivity of the MODIS instrument and the collocation inaccuracies with the CALIOP and CPR sensors.

Regardless of the very similar classification results, the forward feature selection procedure applied to each classifier resulted in different optimal combinations of input features over water surfaces (see Appendix A). In this respect, the VEOR algorithm selected four optimal features ordered according to decreasing importance: B18_936nm, B32_12020nm, sunz, B30_9730nm (for abbreviations, see Table 3). The first two features (the most important) were also selected by the nearest neighbor, neural net, decision tree and random forest classifiers, while linear SVM, Naive Bayes and QDA did not select even one of them. Other features that were often selected were: B29_8550nm_B31_11030nm_diff (6 times), B29_8550nm_B31_11030nm_diff (3 times), B29_8550nm_B31_11030nm_diff (3 times), NDVI (3 times). All of the algorithms apart from the linear SVM required some angular feature, but the type of this information varied significantly. The Sun zenith angle was required only by the VEOR algorithm. The Sun azimuth angle was optimal for the nearest neighbor, decision tree and Naive Bayes classifiers. The scattering angle was selected by the neural network, random forest, AdaBoost, Naive Bayes and QDA algorithms, while the relative azimuth angle was required by the RBF SVM and Gaussian process classifiers. The satellite zenith angle was not selected by any method because collocations of the CALIPSO and CPR profiles with the AQUA MODIS radiometer were only available close to the satellite nadir. The inconsistency in the selected angular features is not easily explainable as it is a combination of physics and data sampling. Based on the VEOR output, the angular information itself is a weak classification feature for cloud discrimination and it slightly improves the overall accuracy over water surfaces close to the satellite nadir. Thus, in the case of the stochastic classifiers, different features (including angular ones) might be selected between consecutive algorithm runs with the same input data.

Cloud discrimination over water surfaces during daytime is a fairly simple classification task. Therefore, to fully evaluate the performance of the selected classifiers, an analogous experiment over snow/ice surfaces was performed (see next subsection).

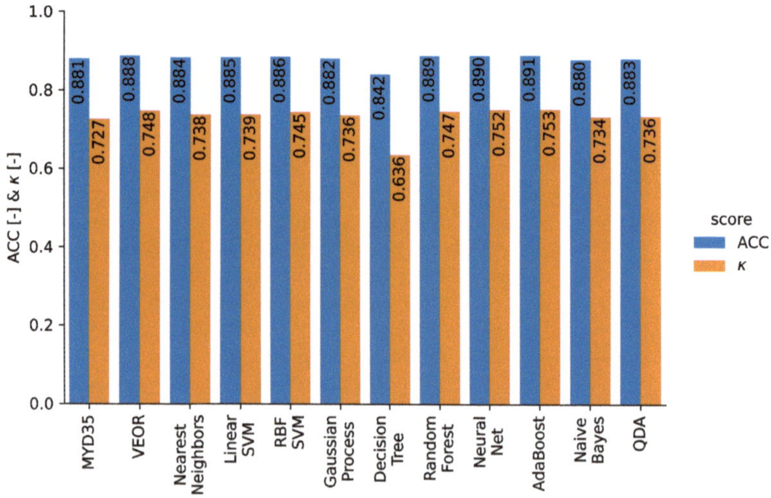

Figure 7. Comparison of classifier performance for cloud discrimination over water surfaces using MODIS spectral measurements collocated with the CALIOP/CPR reference cloud mask. MYD35 denotes AQUA MODIS collection 6 cloud mask product.

6.3.2. Cloud Detection over Ice/Snow Surfaces

Cloud discrimination over ice/snow surfaces during daytime is more challenging than over water surfaces due to the low optical and thermal contrasts between the surface and clouds (see Section 2.2). This is particularly apparent in the case of the MYD35 product (Figure 8), which, over ice/snow, is 21% less accurate than over water (ACC = 0.671 vs. ACC = 0.881), and the Cohens' kappa is 56% lower (ACC = 0.727, κ = 0.317). Amongst the selected classifiers, only the Naive Bayes method produced worse results than the MYD35 product (ACC = 0.659, κ = 0.296). As compared to the VEOR algorithm (ACC = 0.809, κ = 0.611), the QDA (ACC = 0.742, κ = 0.467) and decision tree (ACC = 0.764, κ = 0.526) methods were less accurate, the nearest neighbor (ACC = 0.809, κ = 0.614) and linear SVM (ACC = 0.808, κ = 0.613) classifiers were equally accurate, while the AdaBoost (ACC = 0.816, κ = 0.629), random forest (ACC = 0.818, κ = 0.632), Gaussian process (ACC = 0.822, κ = 0.639), RBF SVM (ACC = 0.886, κ = 0.745) and neural net (ACC = 0.890, κ = 0.752) algorithms were more accurate. Regarding the optimal feature combination for cloud discrimination over ice/snow (see Appendix A), the internal VEOR feature selection procedure selected the following features (ordered by decreasing significance): B7_2130nm, B18_936nm, B26_1375nm, B3_469nm. In this respect, MODIS reflectance at 2130 nm was selected by all of the classifiers apart from the Naive Bayes classifier, while the reflectance at 936 nm was selected by the nearest neighbor, neural network, RBF SVM, random forest and AdaBoost methods. Other features that were often selected were: B29_8550nm_B28_7325nm_diff (5 times), B27_6715nm (4 times), B32_12020nm (3 times), B28_7325nm (3 times). As opposed to the cloud discrimination over water surfaces, only the linear SVM classifier required angular information (i.e., Sun zenith angle) to discriminate cloud cover over ice/snow surfaces.

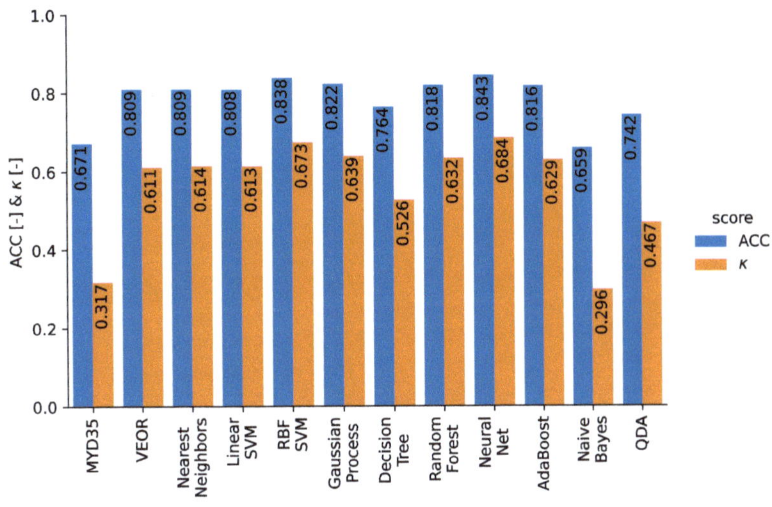

Figure 8. Comparison of classifier performance for cloud discrimination over ice/snow surfaces using MODIS spectral measurements collocated with the CALIOP/CPR reference cloud mask. MYD35 denotes AQUA MODIS collection 6 cloud mask product.

6.4. Numerical Performance Evaluation

The evaluation of numerical performance revealed great differences between the single-core computational requirements of the selected classifiers. The most demanding from the perspective of classifier training is the Gaussian process algorithm, which was not able to be applied for 2D datasets with more than 10^5 samples; thus, only results for 10^2, 10^3, 10^4 are reported in Figure 9. However, the retrieval time for this classifier is in the order of tens of seconds for a dataset with 10^6 samples (Figure 10), which might be applicable to moderate-resolution imagery. Another numerically expensive algorithm is the RBF SVM, which takes hours for training and tens of minutes to classify a dataset with 10^6 samples. Thus, this method might be applicable for the classification of a single image. On the contrary, the linear SVM classifier is numerically robust, with a retrieval time a few orders of magnitude faster than other techniques. Interestingly, the RBF SVM routine can be executed with a linear kernel (instead of the RBF kernel), which is supposed to be conceptually the same as the linear SVM method. However, due to the different libraries used by these two routines (liblinear vs. libsvm), the reduction in processing time is from hours to seconds for algorithm training, and from minutes to milliseconds for classification retrieval. Thus, it has to be emphasized again that the presented results depend heavily on scikit-learn implementation, and for other machine learning libraries, the numerical performance of selected classifiers may differ significantly. The least numerically demanding methods for both training and retrieval are the Naive Bayes and QDA algorithms. Other methods, such as random forest, neural net and AdaBoost, which provided accurate classification results (see Figures 5–7), are also numerically efficient, which might be the reason for their popularity amongst the machine learning community.

The processing time for the VEOR classifier scales efficiently with the number of samples due to the utilization of the fast cKDTree method, which performs worse than a "brute-force" search for a small number of samples but is very robust for large datasets. The largest improvement in processing time for the VEOR algorithm over the nearest neighbor classifier (which also involves searching) is noticeable for classification retrieval. This is due to the utilization of a fast binary search algorithm applied over a vector with sorted ID. Consequently, almost two orders of magnitude of speed improvement are gained over the cKDTree method, which additionally requires the computation of distance metrics in multiple dimensions. Finally, it has to be stated that the reported relationships between the processing times of the selected classifiers may change with other datasets featuring more dimensions, more noise or different class separability. Thus, the presented results are more qualitative than quantitative and could provide a general idea about the numerical performance of the selected classifiers implemented in the scikit-learn library version 0.24.2.

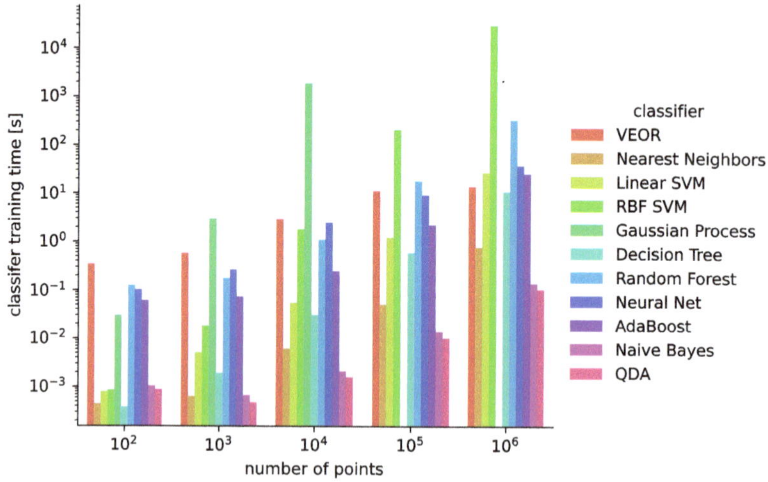

Figure 9. Comparison of single-core algorithm training processing times for selected classifiers.

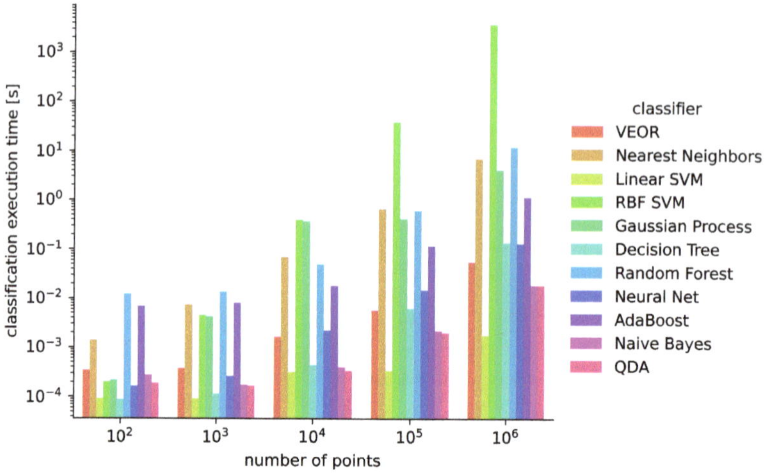

Figure 10. Comparison of single-core algorithm classification processing times for selected classifiers.

7. Discussion

The experiments with the classification of synthetic and satellite datasets revealed that the VEOR algorithm provides similar or better classification results compared with other widely used machine learning algorithms with fine-tuned hyper-parameters and optimized input features. In this respect, it performed among the top 3–5 most accurate algorithms for synthetic datasets with low and moderate levels of noise and class imbalance, as well as for cloud detection using MODIS/CALIOP/CPR collocations. For small, noisy (30% of mislabeling) and imbalanced (30%/70%) datasets, the VEOR provided moderate results comparable to the random forest classifier. This is due to the fact that a significantly disturbed classification probability field is difficult to reconstruct by means of a simple inverse distance weighted average, and thus has to be modeled. Therefore, the classifiers that involve more advanced modeling approaches, such as SVM or Gaussian process, perform better for such datasets but at the price of large computational costs. Nevertheless, classification of an extensive volume of remote sensing data requires a method that is both numerically efficient and accurate. Such a combination is reached by the neural network classifier, which provides accurate classification results, but its internal configuration is difficult to interpret. In this respect, the VEOR method ranks single features and combinations of features according to Cohen's kappa score. This greatly facilitates the understanding of the feature selection process.

Conceptually simple methods such as nearest neighbor, Naive Bayes or QDA do not fit well with all datasets, especially with multispectral MODIS measurements over ice/snow surfaces. Interestingly, for some difficult synthetic datasets, these methods perform well in comparison to the most robust classifiers. This may indicate that if some assumptions behind such methods are met (e.g., Gaussian distribution describing each class and/or linear separability of classes), then they are able to provide accurate classification results with great numerical efficiency. Another type of classifier is the so-called "weak learners", such as the decision tree algorithm, which generally do not feature good classification skills (see Figures 5–8) but are extremely fast and simple to apply. Therefore, they can be executed multiple or hundreds of times with different settings and their stochastic results can be further post-processed to derive an accurate ensemble classifier such as the random forest or AdaBoost. The latter algorithms were found to be robust especially for the classification of multispectral MODIS data and synthetic datasets with a low level of noise and small class imbalance. The reiteration of a stochastic classifier several times with the same input data increases the representability of the classification results, as some initial settings within an algorithm optimization are set by a random guess. Consequently, without repetition, the stochastic classifiers (e.g., decision tree) may provide significantly different results for the same input data. Another source of algorithm instability may be caused by a convergence problem when an iterative algorithm optimization procedure fails due to an unrealistic assumption (e.g., linearity in the case of the linear SVM) or a small number of iterations. Apart from the classifier itself, an additional source of variability in the performed experiments originates from the stochastic wrapper procedures to optimize the hyper-parameters (HalvingGridSearchCV—scikit-learn library routine) and to select the most optimal features (SequentialFeatureSelector—scikit-learn library routine). Therefore, it has to be noted that the presented evaluation results and ranking of the classifiers are prone to small changes related to the random state/random seed of the pseudo-random number generator embedded within the scikit-learn library.

Application of a classifier to large EO datasets requires numerical efficiency, especially concerning the retrieval processing time. In this respect, the classification of a 2D dataset composed of 10^6 samples took longest for the RBF SVM method (1668 s), while, for the linear SVM, it took the shortest time (0.002 s). The main reason for this drastic reduction in processing time originates from the different libraries, and not from algorithm differences per se (retrieval time for the SVM routine with linear instead of an RBF kernel also took several minutes). This implies that before rejecting a classifier due to numerical inefficiency, it is advisable to test its different implementations (if possible). The VEOR implementation in the Python programming language appeared to be numerically robust due to the implementation of the fast cKDTree algorithm for finding the nearest cells within the percentile grid and a binary search algorithm to extract the classification probabilities from the LUV. Thus, the VEOR retrieval is one of the fastest amongst the selected classifiers and therefore can be applied to extremely large EO datasets. On the contrary, the scikit-learn implementation of the RBF SVM and Gaussian process classifiers is not suitable for the processing of large datasets. The analyses performed revealed that there is no single classifier that would outperform all the other methods for all analyzed datasets. However, a few algorithms performed well in almost all cases. In this respect, the neural network method accurately classified synthetic datasets, which were significantly disturbed by noise and class imbalance. It also discriminated cloud cover over water surfaces using MODIS spectral measurements well, with accuracy comparable to the random forest and AdaBoost methods, which provided the best results in this case ($\sim$1% better than the MYD35 product). However, more noticeable differences were revealed by the experiment with a daytime cloud discrimination over ice/snow, where the neural network classifier performed the best—17% more accurate than the MYD35 product and 0.5% better than the second best RBF SVM classifier (Figure 8). Almost all classifiers, apart from the Naive Bayes, outperformed the MYD35 collection 6 product over snow, which implies that there is still great potential for improvement of the MODIS cloud mask product. However, it has to be again strongly emphasized that the acquired results refer only to low satellite zenith angles as the AQUA MODIS collocations with the CALIOP and CPR profiles are only available close to the satellite nadir. This is also a reason that the satellite zenith angle was not selected either by the VEOR or by the *SequentialFeatureSelector* routine applied to other classifiers (see Appendix A). In this respect, there is interesting inconsistency in combinations of the MODIS spectral and angular features optimized for every classifier, especially for the relatively easy experiment with the daytime cloud discrimination over water surfaces, where almost all of the classifiers performed almost equally well. This may indicate that several combinations of spectral and angular features may lead to the same optimal results. On the other hand, within a more difficult experiment with the daytime cloud discrimination over ice/snow, the performance of the evaluated classifiers varied more significantly along with the combinations of selected features. This may imply that the feature selection procedure may produce incorrect results if the classification skills of the classifier being optimized are low (see results of the Naive Bayes classifier in Figure 8 and the corresponding optimal feature combination listed in Appendix A). Nevertheless, the experiments conducted within this study and the extensive literature [47,48] lead to the conclusion that the feature selection significantly improves the classification results.

Ultimately, it has to be stated that the VEOR classifier performed well for datasets with many samples, including the experiment with MODIS data, where it proved that the combination of four features allows for more accurate (by $\sim$1% over water surfaces and by $\sim$14% over ice/snow) cloud discrimination than the MYD35 product. The feature selection procedure implemented in VEOR reports Cohen's kappa coefficient for every analyzed combination of features. This is computationally feasible due to the simple reconstruction of the classification probability field by means of the weighted inverse distance K-nearest neighbor average. Nevertheless, such a solution is not accurate for small datasets, especially with significant noise and class imbalance. However, the premise for VEOR development was the derivation of an accurate probabilistic classification for large EO datasets requiring some methodological simplifications. Several experiments conducted within this study have proven that the VEOR classifier is an interesting alternative to common machine learning techniques.

8. Conclusions

The main objective of this work was to evaluate the performance of the novel self-optimizing probabilistic Vectorized Earth Observation Retrieval (VEOR) classifier with respect to common machine learning algorithms with fine-tuned hyper-parameters and optimized input features. This was achieved within a series of experiments with synthetic datasets, as well as with daytime cloud discrimination using AQUA MODIS spectral measurements and a reference cloud mask derived from combined CALIOP lidar and CPR radar data. The classification results proved that the VEOR method is able to accurately classify large datasets commonly occurring within various Earth Observation (EO) disciplines. However, it is not suitable for small and noisy datasets due to the imprecise reconstruction of the multidimensional classification probability field by means of a weighted inverse distance average when the number of samples is limited. Nevertheless, comparison with other classifiers revealed that the VEOR algorithm is within the 3–5 most robust techniques for large (i.e., >=1000 samples) synthetic datasets with low or moderate levels of noise and class imbalance. More importantly, it performed well for the daytime cloud masking over water surfaces using MODIS spectral measurements (ACC = 0.888, κ = 0.748), where it was better than the reference MODIS cloud mask collection 6 product (ACC = 0.881, κ = 0.727), but slightly worse than the random forest (ACC = 0.889, κ = 0.747), neural net (ACC = 0.890, κ = 0.752) and AdaBoost (ACC = 0.891, κ = 0.753) classifiers. An analogous experiment over ice/snow revealed that the VEOR algorithm is again more accurate (ACC = 0.809, κ = 0.611) than the MODIS cloud mask (ACC = 0.671, κ = 0.317) and less accurate than the AdaBoost (ACC = 0.816, κ = 0.629), random forest (ACC = 0.818, κ = 0.632), Gaussian process (ACC = 0.822, κ = 0.639), RBF SVM (ACC = 0.886, κ = 0.745) and neural net (ACC = 0.890, κ = 0.752) classifiers. Interestingly, the forward feature selection method applied to each classifier resulted in a wide range of optimal combinations of features. This is particularly surprising for the daytime cloud discrimination over water surfaces, where different combinations of input features led to almost the same results. Within a more difficult experiment with daytime cloud discrimination over ice/snow, the feature selection was not effective for classifiers featuring limited classification skills over ice/snow (i.e., Naive Bayes method). Finally, it has to be emphasized that the acquired results refer only to low satellite zenith angles as the AQUA MODIS collocations with the CALIOP and CPR profiles are only available close to the satellite nadir.

The performed experiments revealed that the decision tree, linear SVM and Naive Bayes produced variable results for the synthetic datasets, provided that some internal assumptions were met or not. The Gaussian process and RBF SVM (based on the libsvm library) classifiers were found to be accurate even for small datasets with a significant noise level. However, their complexity and current implementation within the scikit-learn library do not allow for their application to large EO datasets. Ultimately, the VEOR algorithm features some unique characteristics amongst analyzed classifiers, such as:

- numerically efficient probabilistic classification of large, multidimensional datasets;
- an internal feature selection method that allows the minimal feature subset required for optimal classification results to be identified;
- the ranking of classification skills for analyzed feature combinations according to Cohen's kappa coefficient;
- intuitive interpretation of derived optimal algorithm settings (including selected features);
- easily expandable LUV whenever new data have to be added to a training dataset.

These characteristics, supported by the analytical results in this study, lead to the final conclusion that the VEOR classifier is an interesting alternative to existing machine learning methods, especially in terms of the classification of large multidimensional EO datasets. Further development of the VEOR algorithm will involve adaptive reconstruction of the multidimensional classification probability field, where the number of nearest neighboring cells K (see Equation (2)) will not be constant for all cells, but it will be further optimized to a specific location within the probability field. Moreover, a more sophisticated inverse distant weighting scheme used within the reconstruction of the multidimensional classification probability field will be proposed. These adaptations should further improve the accuracy of the VEOR classifier and increase its competitiveness with other machine learning algorithms.

Author Contributions: Conceptualization, J.P.M. and J.S.B.; methodology, J.P.M.; validation, J.P.M.; formal analysis, J.P.M.; investigation, J.P.M.; writing—original draft preparation, J.P.M.; writing—review and editing, J.P.M. and J.S.B.; visualization, J.P.M.; funding acquisition, J.P.M. All authors have read and agreed to the published version of the manuscript.

Funding: This study was supported by the National Science Centre of Poland (grant number DEC-2013/11/D/ST10/03487).

Institutional Review Board Statement: Not applicable.

Informed Consent Statement: Not applicable.

Data Availability Statement: Not applicable.

Conflicts of Interest: The authors declare no conflict of interest.

Appendix A

Table A1. Optimized features by means of a forward selection method for the experiments with MODIS cloud screening over water and snow.

Classifier	MODIS Water	MODIS Snow
VEOR	B18_936nm, B32_12020nm, sunz, B30_9730nm	B7_2130nm, B18_936nm, B26_1375nm, B3_469nm
Multi-layer perceptron	scat, B18_936nm, B32_12020nm, B29_8550nm_B31_11030nm_diff	B7_2130nm, B18_936nm, B27_6715nm, B32_12020nm
K-nearest neighbor classifier	suna, B18_936nm, B26_1375nm, B32_12020nm	B7_2130nm, B18_936nm, B28_7325nm, B29_8550nm_B28_7325nm_diff
Support vector classification with a linear kernel	B7_2130nm, B33_13335nm, NDVI, B29_8550nm_B31_11030nm_diff	sunz, B7_2130nm, B29_8550nm_B31_11030nm_diff, B29_8550nm_B28_7325nm_diff
Support vector classification with a Radial Basis Function (RBF) kernel	razi, B18_936nm, B31_11030nm_B20_3750nm_diff, B29_8550nm_B31_11030nm_diff	B7_2130nm, B18_936nm, B27_6715nm, B33_13335nm
Gaussian process classification	razi, B18_936nm, B29_8550nm_B31_11030nm_diff, B29_8550nm_B28_7325nm_diff	B3_469nm, B7_2130nm, B19_940nm, B27_6715nm
Decision tree classifier	suna, B18_936nm, B32_12020nm, B31_11030nm_B20_3750nm_diff	B7_2130nm, B19_940nm, B31_11030nm, B29_8550nm_B28_7325nm_diff
Random forest classifier	scat, B18_936nm, B26_1375nm, B32_12020nm_B20_3750nm_diff	B7_2130nm, B18_936nm, B20_3750nm, B28_7325nm
AdaBoost-SAMME classifier	scat, B18_936nm, B26_1375nm, B31_11030nm_B20_3750nm_diff	B7_2130nm, B18_936nm, B27_6715nm, B28_7325nm_B31_11030nm_diff
Gaussian Naive Bayes classifier	suna, scat, NDVI, B29_8550nm_B31_11030nm_diff	B5_1240nm, B28_7325nm, B32_12020nm_B20_3750nm_diff, B29_8550nm_B28_7325nm_diff
Quadratic Discriminant Analysis	scat, B27_6715nm, NDVI, B29_8550nm_B31_11030nm_diff	B7_2130nm, B32_12020nm_B20_3750nm_diff, B28_7325nm_B31_11030nm_diff, B29_8550nm_B28_7325nm_diff

Appendix B

Algorithm A1: Pseudocode for the VEOR classification

```
Function Standardize (input feature vectors):
    Data: input feature vectors with training samples
    Result: standardized feature vectors to 8 bit representation
    for feature vector in input feature vectors do
        declare empty percentile vector;
        for percentile in {0, 0.5%, 1%, 1.5%, ..., 100%} do
            derive percentile value for a feature vector;
            append percentile value to a percentile vector;
        end
        derive standardized feature vector with integer vector indices by locating a
          feature vector within a percentile vector using a binary search algorithm;
    end
    return standardized feature vectors to 8 bit representation;
end
Data: optimal feature combination
Data: look-up vectors LUV_ID and LUV_prob for optimal feature combination
Data: standardized feature vectors using Standardize ()
Result: classification probabilities for input feature vectors
declare empty IDs vector;
declare i = 0;
for standardized feature vector in optimal standardized feature vectors do
    bitwise_shift standardized feature vector by i × 8 bits;
    bitwise_or shifted standardized feature vector with IDs vector;
    i=i+1;
end
derive index vector by locating IDs vector within the sorted LUV_ID vector using a
  binary search algorithm;
use the derived index vector to extract classification probabilities from the LUV_prob
  vector;
```

Algorithm A2: Pseudocode for the VEOR classifier training

```
Data: standardized optimal feature vectors using Standardize ()
Data: vector with training binary labels
Result: LUV with IDs and LUV with classification probabilities derived for optimal
        feature combination
declare initial number of features N = 1;
while Cohen's kappa improved do
    for N standardized feature vector(s) in standardized feature vectors do
        declare empty IDs vector;
        declare i = 0;
        for standardized feature vector in N standardized feature vector(s) do
            bitwise_shift standardized feature vector by i × 8 bits;
            bitwise_or shifted standardized feature vector with IDs vector;
            i = i + 1;
        end
        sort ascending IDs and apply the same sorting to binary training labels;
        declare empty vector for initial classification probabilities P;
        for ID in IDs vector do
            select training binary labels where IDs = ID;
            average selected binary labels to derive initial classification probability;
            append initial classification probability to a vector P;
        end
        balance classes within training binary labels and IDs vectors assuming the
          P probability distribution;
        declare empty vector for initial classification probabilities P_b;
        declare empty LUV_IDs vector for IDs;
        for ID in balanced IDs vector do
            select balanced binary labels where balanced IDs = ID;
            average selected balanced binary labels to derive P_b x probability;
            append P_b x probability to a P_b vector;
            append ID to a LUV_IDs vector;
        end
        declare initial number of neighbors K=2;
        while Cohen's kappa improved do
            declare empty vector for reconstructed classification probabilities P';
            for ID in LUV_IDs do
                select K nearest IDs to the ID value using cKDTree KNN method;
                reconstruct P'_x using the IDW interpolation between K nearest P_b
                  probabilities;
                append reconstructed probability P'_x to the P' vector;
            end
            compute Cohen's kappa between Round (P') and Round (P);
            K = K + 2;
        end
        N = N + 1;
    end
end
oversample the optimal LUVs to IDs values unseen in the training feature vectors;
save optimal LUV_IDs, LUV_prob and optimal feature combination;
```

References

1. Anderson, J.R. *A Land Use and Land Cover Classification System for Use with Remote Sensor Data*; US Government Printing Office: Washington, DC, USA, 1976; Volume 964.
2. Friedl, M.A.; Sulla-Menashe, D.; Tan, B.; Schneider, A.; Ramankutty, N.; Sibley, A.; Huang, X. MODIS Collection 5 global land cover: Algorithm refinements and characterization of new datasets. *Remote Sens. Environ.* **2010**, *114*, 168–182. [CrossRef]
3. Hansen, M.C.; Loveland, T.R. A review of large area monitoring of land cover change using Landsat data. *Remote Sens. Environ.* **2012**, *122*, 66–74. [CrossRef]
4. Inglada, J.; Vincent, A.; Arias, M.; Tardy, B.; Morin, D.; Rodes, I. Operational high resolution land cover map production at the country scale using satellite image time series. *Remote Sens.* **2017**, *9*, 95. [CrossRef]
5. Talukdar, S.; Singha, P.; Mahato, S.; Pal, S.; Liou, Y.A.; Rahman, A. Land-use land-cover classification by machine learning classifiers for satellite observations—A review. *Remote Sens.* **2020**, *12*, 1135. [CrossRef]
6. Defourny, P.; Bontemps, S.; Bellemans, N.; Cara, C.; Dedieu, G.; Guzzonato, E.; Hagolle, O.; Inglada, J.; Nicola, L.; Rabaute, T.; et al. Near real-time agriculture monitoring at national scale at parcel resolution: Performance assessment of the Sen2-Agri automated system in various cropping systems around the world. *Remote Sens. Environ.* **2019**, *221*, 551–568. [CrossRef]
7. Kussul, N.; Lavreniuk, M.; Skakun, S.; Shelestov, A. Deep learning classification of land cover and crop types using remote sensing data. *IEEE Geosci. Remote Sens. Lett.* **2017**, *14*, 778–782. [CrossRef]
8. Li, R.; Xu, M.; Chen, Z.; Gao, B.; Cai, J.; Shen, F.; He, X.; Zhuang, Y.; Chen, D. Phenology-based classification of crop species and rotation types using fused MODIS and Landsat data: The comparison of a random-forest-based model and a decision-rule-based model. *Soil Tillage Res.* **2021**, *206*, 104838. [CrossRef]
9. Sonobe, R.; Tani, H.; Wang, X.; Kobayashi, N.; Shimamura, H. Random forest classification of crop type using multi-temporal TerraSAR-X dual-polarimetric data. *Remote Sens. Lett.* **2014**, *5*, 157–164. [CrossRef]
10. Adam, E.; Mutanga, O.; Rugege, D. Multispectral and hyperspectral remote sensing for identification and mapping of wetland vegetation: A review. *Wetl. Ecol. Manag.* **2010**, *18*, 281–296. [CrossRef]
11. Bourgeau-Chavez, L.; Kasischke, E.; Brunzell, S.; Mudd, J.; Smith, K.; Frick, A. Analysis of space-borne SAR data for wetland mapping in Virginia riparian ecosystems. *Int. J. Remote Sens.* **2001**, *22*, 3665–3687. [CrossRef]
12. DeLancey, E.R.; Simms, J.F.; Mahdianpari, M.; Brisco, B.; Mahoney, C.; Kariyeva, J. Comparing deep learning and shallow learning for large-scale wetland classification in Alberta, Canada. *Remote Sens.* **2020**, *12*, 2. [CrossRef]
13. McCarthy, M.J.; Radabaugh, K.R.; Moyer, R.P.; Muller-Karger, F.E. Enabling efficient, large-scale high-spatial resolution wetland mapping using satellites. *Remote Sens. Environ.* **2018**, *208*, 189–201. [CrossRef]
14. Carroll, M.L.; Townshend, J.R.; DiMiceli, C.M.; Noojipady, P.; Sohlberg, R.A. A new global raster water mask at 250 m resolution. *Int. J. Digit. Earth* **2009**, *2*, 291–308. [CrossRef]
15. Donchyts, G.; Baart, F.; Winsemius, H.; Gorelick, N.; Kwadijk, J.; Van De Giesen, N. Earth's surface water change over the past 30 years. *Nat. Clim. Chang.* **2016**, *6*, 810–813. [CrossRef]
16. Feng, M.; Sexton, J.O.; Channan, S.; Townshend, J.R. A global, high-resolution (30-m) inland water body dataset for 2000: First results of a topographic–spectral classification algorithm. *Int. J. Digit. Earth* **2016**, *9*, 113–133. [CrossRef]
17. Wang, G.; Wu, M.; Wei, X.; Song, H. Water identification from high-resolution remote sensing images based on multidimensional densely connected convolutional neural networks. *Remote Sens.* **2020**, *12*, 795. [CrossRef]
18. Martin, M.; Newman, S.; Aber, J.; Congalton, R. Determining forest species composition using high spectral resolution remote sensing data. *Remote Sens. Environ.* **1998**, *65*, 249–254. [CrossRef]
19. Hościło, A.; Lewandowska, A. Mapping forest type and tree species on a regional scale using multi-temporal Sentinel-2 data. *Remote Sens.* **2019**, *11*, 929. [CrossRef]
20. Sheykhmousa, M.; Mahdianpari, M.; Ghanbari, H.; Mohammadimanesh, F.; Ghamisi, P.; Homayouni, S. Support vector machine vs. random forest for remote sensing image classification: A meta-analysis and systematic review. *IEEE J. Sel. Top. Appl. Earth Obs. Remote Sens.* **2020**, *13*, 6308–6325. [CrossRef]
21. Tang, L.; Shao, G. Drone remote sensing for forestry research and practices. *J. For. Res.* **2015**, *26*, 791–797. [CrossRef]
22. Hall, D.K.; Riggs, G.A.; Salomonson, V.V. MODIS snow and sea ice products. In *Earth Science Satellite Remote Sensing*; Springer: Berlin/Heidelberg, Germany, 2006; pp. 154–181.
23. Cannistra, A.F.; Shean, D.E.; Cristea, N.C. High-resolution CubeSat imagery and machine learning for detailed snow-covered area. *Remote Sens. Environ.* **2021**, *258*, 112399. [CrossRef]
24. Schneider, A.; Friedl, M.A.; Potere, D. A new map of global urban extent from MODIS satellite data. *Environ. Res. Lett.* **2009**, *4*, 044003. [CrossRef]
25. Wang, L.; Li, C.; Ying, Q.; Cheng, X.; Wang, X.; Li, X.; Hu, L.; Liang, L.; Yu, L.; Huang, H.; et al. China's urban expansion from 1990 to 2010 determined with satellite remote sensing. *Chin. Sci. Bull.* **2012**, *57*, 2802–2812. [CrossRef]
26. Zhang, Y.; Qin, K.; Bi, Q.; Cui, W.; Li, G. Landscape Patterns and Building Functions for Urban Land-Use Classification from Remote Sensing Images at the Block Level: A Case Study of Wuchang District, Wuhan, China. *Remote Sens.* **2020**, *12*, 1831. [CrossRef]
27. Heidinger, A.K.; Evan, A.T.; Foster, M.J.; Walther, A. A naive Bayesian cloud-detection scheme derived from CALIPSO and applied within PATMOS-x. *J. Appl. Meteorol. Climatol.* **2012**, *51*, 1129–1144. [CrossRef]

28. Karlsson, K.G.; Johansson, E.; Devasthale, A. Advancing the uncertainty characterisation of cloud masking in passive satellite imagery: Probabilistic formulations for NOAA AVHRR data. *Remote Sens. Environ.* **2015**, *158*, 126–139. [CrossRef]
29. Marchant, B.; Platnick, S.; Meyer, K.; Arnold, G.T.; Riedi, J. MODIS Collection 6 shortwave-derived cloud phase classification algorithm and comparisons with CALIOP. *Atmos. Meas. Tech.* **2016**, *9*, 1587–1599. [CrossRef] [PubMed]
30. Håkansson, N.; Adok, C.; Thoss, A.; Scheirer, R.; Hörnquist, S. Neural network cloud top pressure and height for MODIS. *Atmos. Meas. Tech.* **2018**, *11*, 3177–3196. [CrossRef]
31. Wang, C.; Platnick, S.; Meyer, K.; Zhang, Z.; Zhou, Y. A machine-learning-based cloud detection and thermodynamic-phase classification algorithm using passive spectral observations. *Atmos. Meas. Tech.* **2020**, *13*, 2257–2277. [CrossRef]
32. Lu, D.; Weng, Q. A survey of image classification methods and techniques for improving classification performance. *Int. J. Remote Sens.* **2007**, *28*, 823–870. [CrossRef]
33. Mantero, P.; Moser, G.; Serpico, S.B. Partially supervised classification of remote sensing images through SVM-based probability density estimation. *IEEE Trans. Geosci. Remote Sens.* **2005**, *43*, 559–570. [CrossRef]
34. Pelletier, C.; Valero, S.; Inglada, J.; Champion, N.; Marais Sicre, C.; Dedieu, G. Effect of training class label noise on classification performances for land cover mapping with satellite image time series. *Remote Sens.* **2017**, *9*, 173. [CrossRef]
35. Musial, J.P.; Verstraete, M.M.; Gobron, N. Comparing the effectiveness of recent algorithms to fill and smooth incomplete and noisy time series. *Atmos. Chem. Phys.* **2011**, *11*, 7905–7923. [CrossRef]
36. Millard, K.; Richardson, M. On the importance of training data sample selection in random forest image classification: A case study in peatland ecosystem mapping. *Remote Sens.* **2015**, *7*, 8489–8515. [CrossRef]
37. Langley, P.; Sage, S. Induction of selective Bayesian classifiers. In *Uncertainty Proceedings 1994*; Elsevier: Amsterdam, The Netherlands, 1994; pp. 399–406.
38. Foody, G.M. The continuum of classification fuzziness in thematic mapping. *Photogramm. Eng. Remote Sens.* **1999**, *65*, 443–452.
39. Benz, U.C.; Hofmann, P.; Willhauck, G.; Lingenfelder, I.; Heynen, M. Multi-resolution, object-oriented fuzzy analysis of remote sensing data for GIS-ready information. *ISPRS J. Photogramm. Remote Sens.* **2004**, *58*, 239–258. [CrossRef]
40. Li, Y.; Lu, T.; Li, S. Subpixel-pixel-superpixel-based multiview active learning for hyperspectral images classification. *IEEE Trans. Geosci. Remote Sens.* **2020**, *58*, 4976–4988. [CrossRef]
41. Sun, B.; Chen, S.; Wang, J.; Chen, H. A robust multi-class AdaBoost algorithm for mislabeled noisy data. *Knowl.-Based Syst.* **2016**, *102*, 87–102. [CrossRef]
42. Nettleton, D.F.; Orriols-Puig, A.; Fornells, A. A study of the effect of different types of noise on the precision of supervised learning techniques. *Artif. Intell. Rev.* **2010**, *33*, 275–306. [CrossRef]
43. Breiman, L.; Friedman, J.H.; Olshen, R.A.; Stone, C.J. *Classification and Regression Trees*; Routledge: London, UK, 2017.
44. Aha, D.W.; Kibler, D.; Albert, M.K. Instance-based learning algorithms. *Mach. Learn.* **1991**, *6*, 37–66. [CrossRef]
45. Platt, J. *Sequential Minimal Optimization: A Fast Algorithm for Training Support Vector Machines*; Technical Report; Microsoft: Redmond, WA, USA, 1997.
46. Tu, B.; Kuang, W.; He, W.; Zhang, G.; Peng, Y. Robust Learning of Mislabeled Training Samples for Remote Sensing Image Scene Classification. *IEEE J. Sel. Top. Appl. Earth Obs. Remote Sens.* **2020**, *13*, 5623–5639. [CrossRef]
47. Guyon, I.; Elisseeff, A. An introduction to variable and feature selection. *J. Mach. Learn. Res.* **2003**, *3*, 1157–1182.
48. Cai, J.; Luo, J.; Wang, S.; Yang, S. Feature selection in machine learning: A new perspective. *Neurocomputing* **2018**, *300*, 70–79. [CrossRef]
49. Amaldi, E.; Kann, V. On the approximability of minimizing nonzero variables or unsatisfied relations in linear systems. *Theor. Comput. Sci.* **1998**, *209*, 237–260. [CrossRef]
50. LeCun, Y.; Denker, J.; Solla, S. Optimal brain damage. *Adv. Neural Inf. Process. Syst.* **1989**, *2*, 598–605.
51. Uddin, M.P.; Mamun, M.A.; Hossain, M.A. PCA-based feature reduction for hyperspectral remote sensing image classification. *IETE Tech. Rev.* **2021**, *38*, 377–396. [CrossRef]
52. Hughes, G. On the mean accuracy of statistical pattern recognizers. *IEEE Trans. Inf. Theory* **1968**, *14*, 55–63. [CrossRef]
53. Harsanyi, J.C.; Chang, C.I. Hyperspectral image classification and dimensionality reduction: An orthogonal subspace projection approach. *IEEE Trans. Geosci. Remote Sens.* **1994**, *32*, 779–785. [CrossRef]
54. Krishnaiah, P.R.; Kanal, L.N. *Classification Pattern Recognition and Reduction of Dimensionality*; Elsevier: Amsterdam, The Netherlands, 1982; Volume 2.
55. Chawla, N.V.; Bowyer, K.W.; Hall, L.O.; Kegelmeyer, W.P. SMOTE: Synthetic minority over-sampling technique. *J. Artif. Intell. Res.* **2002**, *16*, 321–357. [CrossRef]
56. Hanssen, A.; Kuipers, W. *On the Relationship between the Frequency of Rain and Various Meteorological Parameters (with Reference to the Problem of Objective Forecasting)*; Koninklijk Nederlands Meteorologisch Instituut: De Bilt, The Netherlands, 1965.
57. Van Rijsbergen, C.J. A theoretical basis for the use of co-occurrence data in information retrieval. *J. Doc.* **1977**, *33*, 106–119. [CrossRef]
58. Kubat, M.; Holte, R.C.; Matwin, S. Machine learning for the detection of oil spills in satellite radar images. *Mach. Learn.* **1998**, *30*, 195–215. [CrossRef]
59. Hanley, J.A.; McNeil, B.J. The meaning and use of the area under a receiver operating characteristic (ROC) curve. *Radiology* **1982**, *143*, 29–36. [CrossRef]
60. Cohen, J. A coefficient of agreement for nominal scales. *Educ. Psychol. Meas.* **1960**, *20*, 37–46. [CrossRef]

61. Finn, J.T. Use of the average mutual information index in evaluating classification error and consistency. *Int. J. Geogr. Inf. Sci.* **1993**, *7*, 349–366. [CrossRef]
62. López, V.; Fernández, A.; Moreno-Torres, J.G.; Herrera, F. Analysis of preprocessing vs. cost-sensitive learning for imbalanced classification. Open problems on intrinsic data characteristics. *Expert Syst. Appl.* **2012**, *39*, 6585–6608. [CrossRef]
63. Thabtah, F.; Hammoud, S.; Kamalov, F.; Gonsalves, A. Data imbalance in classification: Experimental evaluation. *Inf. Sci.* **2020**, *513*, 429–441. [CrossRef]
64. Quinlan, J.R. Improved estimates for the accuracy of small disjuncts. *Mach. Learn.* **1991**, *6*, 93–98. [CrossRef]
65. Zadrozny, B.; Elkan, C. Learning and making decisions when costs and probabilities are both unknown. In Proceedings of the Seventh ACM SIGKDD International Conference on Knowledge Discovery and Data Mining, San Francisco, CA, USA, 26–29 August 2001; pp. 204–213.
66. Wu, G.; Chang, E.Y. Class-boundary alignment for imbalanced dataset learning. In Proceedings of the ICML 2003 Workshop on Learning from Imbalanced Data Sets II, Washington, DC, USA, 21 August 2003; pp. 49–56.
67. Carvajal, K.; Chacón, M.; Mery, D.; Acuna, G. Neural network method for failure detection with skewed class distribution. *Insight-Non Test. Cond. Monit.* **2004**, *46*, 399–402. [CrossRef]
68. Bermejo, P.; Gámez, J.A.; Puerta, J.M. Improving the performance of Naive Bayes multinomial in e-mail foldering by introducing distribution-based balance of datasets. *Expert Syst. Appl.* **2011**, *38*, 2072–2080. [CrossRef]
69. Musial, J.P.; Hüsler, F.; Sütterlin, M.; Neuhaus, C.; Wunderle, S. Probabilistic approach to cloud and snow detection on Advanced Very High Resolution Radiometer (AVHRR) imagery. *Atmos. Meas. Tech.* **2014**, *7*, 799–822. [CrossRef]
70. Musial, J.P.; Hüsler, F.; Sütterlin, M.; Neuhaus, C.; Wunderle, S. Daytime low stratiform cloud detection on AVHRR imagery. *Remote Sens.* **2014**, *6*, 5124–5150. [CrossRef]
71. Musiał, J.P.; Bojanowski, J.S. AVHRR LAC satellite cloud climatology over Central Europe derived by the Vectorized Earth Observation Retrieval (VEOR) method and PyLAC software. *Geoinf. Issues* **2017**, *9*, 39–51.
72. Musial, J. *CM SAF Visiting Scientist Activity CM_VS18_01 Report: Assessing the VEOR Technique for Bayesian Cloud Detection for the Generation of CM SAF Cloud Climate Data Records*; Technical Report; The Satellite Application Facility on Climate Monitoring; Darmstadt, Germany, 2018.
73. Knuth, D.E. *The Art of Computer Programming*; Pearson Education: London, UK, 1997; Volume 3.
74. Maneewongvatana, S.; Mount, D.M. It's okay to be skinny, if your friends are fat. In Proceedings of the Center for Geometric Computing 4th Annual Workshop on Computational Geometry, San Francisco, CA, USA, 28–30 May 1999; Volume 2, pp. 1–8.
75. Pedregosa, F.; Varoquaux, G.; Gramfort, A.; Michel, V.; Thirion, B.; Grisel, O.; Blondel, M.; Prettenhofer, P.; Weiss, R.; Dubourg, V.; et al. Scikit-learn: Machine learning in Python. *J. Mach. Learn. Res.* **2011**, *12*, 2825–2830.
76. McCulloch, W.S.; Pitts, W. A logical calculus of the ideas immanent in nervous activity. *Bull. Math. Biophys.* **1943**, *5*, 115–133. [CrossRef]
77. Rosenblatt, F. The perceptron: A probabilistic model for information storage and organization in the brain. *Psychol. Rev.* **1958**, *65*, 386. [CrossRef]
78. Goldberger, J.; Hinton, G.E.; Roweis, S.T.; Salakhutdinov, R.R. Neighbourhood components analysis. In Proceedings of the Advances in Neural Information Processing Systems, Vancouver, BC, Canada, 5–8 December 2005; pp. 513–520.
79. Cortes, C.; Vapnik, V. Support-vector networks. *Mach. Learn.* **1995**, *20*, 273–297. [CrossRef]
80. Bernardo, J.; Berger, J.; Dawid, A.; Smith, A. Regression and classification using Gaussian process priors. *Bayesian Stat.* **1998**, *6*, 475.
81. Breiman, L. Random forests. *Mach. Learn.* **2001**, *45*, 5–32. [CrossRef]
82. Freund, Y.; Schapire, R.E. A decision-theoretic generalization of on-line learning and an application to boosting. *J. Comput. Syst. Sci.* **1997**, *55*, 119–139. [CrossRef]
83. Hastie, T.; Rosset, S.; Zhu, J.; Zou, H. Multi-class adaboost. *Stat. Its Interface* **2009**, *2*, 349–360. [CrossRef]
84. Chan, T.F.; Golub, G.H.; LeVeque, R.J. Updating formulae and a pairwise algorithm for computing sample variances. In *COMPSTAT 1982 5th Symposium Held at Toulouse 1982*; Springer: Berlin/Heidelberg, Germany, 1982; pp. 30–41.
85. Tharwat, A. Linear vs. quadratic discriminant analysis classifier: A tutorial. *Int. J. Appl. Pattern Recognit.* **2016**, *3*, 145–180. [CrossRef]
86. Pal, M.; Mather, P.M. An assessment of the effectiveness of decision tree methods for land cover classification. *Remote Sens. Environ.* **2003**, *86*, 554–565. [CrossRef]
87. Mountrakis, G.; Im, J.; Ogole, C. Support vector machines in remote sensing: A review. *ISPRS J. Photogramm. Remote Sens.* **2011**, *66*, 247–259. [CrossRef]
88. Zhang, H. The optimality of naive Bayes. *AA* **2004**, *1*, 3.
89. Guyon, I. Design of experiments of the NIPS 2003 variable selection benchmark. In Proceedings of the NIPS 2003 Workshop on Feature Extraction and Feature Selection, Whistler, BC, Canada, 11–13 December 2003; Volume 253.
90. Mace, G.G.; Zhang, Q. The CloudSat radar-lidar geometrical profile product (RL-GeoProf): Updates, improvements, and selected results. *J. Geophys. Res. Atmos.* **2014**, *119*, 9441–9462. [CrossRef]
91. Partain, P. *CloudSat MODIS-AUX Auxiliary Data Process Description and Interface Control Document*; Cooperative Institute for Research in the Atmosphere, Colorado State University: Fort Collins, CO, USA, 2007; p. 23.

remote sensing

MDPI

Article

TAE-Net: Task-Adaptive Embedding Network for Few-Shot Remote Sensing Scene Classification

Wendong Huang [1], Zhengwu Yuan [1], Aixia Yang [2,*], Chan Tang [1] and Xiaobo Luo [1]

[1] Chongqing Engineering Research Center for Spatial Big Data Intelligent Technology, School of Computer Science and Technology, Chongqing University of Posts and Telecommunications, Chongqing 400065, China; s190201036@stu.cqupt.edu.cn (W.H.); yuanzw@cqupt.edu.cn (Z.Y.); s200231016@stu.cqupt.edu.cn (C.T.); luoxb@cqupt.edu.cn (X.L.)

[2] State Key Laboratory of Remote Sensing Science, Aerospace Information Research Institute, Chinese Academy of Sciences, Beijing 100101, China

* Correspondence: yangax@radi.ac.cn; Tel.: +86-10-6480-6256

Abstract: Recently, approaches based on deep learning are quite prevalent in the area of remote sensing scene classification. Though significant success has been achieved, these approaches are still subject to an excess of parameters and extremely dependent on a large quantity of labeled data. In this study, few-shot learning is used for remote sensing scene classification tasks. The goal of few-shot learning is to recognize unseen scene categories given extremely limited labeled samples. For this purpose, a novel task-adaptive embedding network is proposed to facilitate few-shot scene classification of remote sensing images, referred to as TAE-Net. A feature encoder is first trained on the base set to learn embedding features of input images in the pre-training phase. Then in the meta-training phase, a new task-adaptive attention module is designed to yield the task-specific attention, which can adaptively select informative embedding features among the whole task. In the end, in the meta-testing phase, the query image derived from the novel set is predicted by the meta-trained model with limited support images. Extensive experiments are carried out on three public remote sensing scene datasets: UC Merced, WHU-RS19, and NWPU-RESISC45. The experimental results illustrate that our proposed TAE-Net achieves new state-of-the-art performance for few-shot remote sensing scene classification.

Keywords: scene classification; few-shot learning; meta-learning; pre-training; task-adaptive attention

Citation: Huang, W.; Yuan, Z.; Yang, A.; Tang, C.; Luo, X. TAE-Net: Task-Adaptive Embedding Network for Few-Shot Remote Sensing Scene Classification. *Remote Sens.* **2022**, *14*, 111. https://doi.org/10.3390/rs14010111

Academic Editor: Emanuele Frontoni

Received: 2 November 2021
Accepted: 23 December 2021
Published: 28 December 2021

1. Introduction

Remote sensing scene images taken by satellite consist of abundant semantic information of land-cover objects, which have been widely used in various fields, containing urban planning [1], traffic control [2], disaster detection [3], agricultural and environmental modeling [4,5], and other areas [6–9]. For remote sensing images, scene classification aims to divide unseen samples into corresponding scene classes according to the contained semantic information, which has recently attracted growing attention. By contrast, few-shot scene classification of remote sensing images is more inclined to divide remote sensing images into corresponding scene classes with only few labeled samples [10], which has quite vital significance in those areas where only few labeled samples are available. Few-shot scene classification has enormous potential in various fields, containing ecological monitoring, environmental monitoring, road detection, and so on, which can significantly decrease the burden of data collection and manual labeling.

In recent years, deep learning makes great success in traditional remote sensing image recognition [11–15], which typically acquires large amounts of labeled data in the training procedure. To achieve excellent performance, methods based on deep learning tend to contain a great deal of prior knowledge and sophisticated network structure [16,17]. Nevertheless, when training samples are insufficient, the models based on deep learning are susceptible to overfitting and deviation of feature representation [18]. Specifically, in real remote sensing scenarios, collecting and labeling remote sensing data is pretty time-consuming and laboursome [19]. Moreover, massive computing resources are consumed in the training procedure of the model due to millions of learnable parameters, which further increases the burden on researchers.

Inspired by the way humans classify an object into one of several novel categories, few-shot learning is presented to address a series of few-shot tasks, which aims to make models quickly adapt unseen classes with only limited labeled samples [10]. In general, there are three lines of thinking to resolve the issue of few-shot learning: metric learning [20], meta-learning [21], and transfer learning [22]. In terms of metric learning, the similarity between the query sample and support sample is calculated by a distance function (e.g., Euclidean distance or cosine similarity) and then is compared to judge whether they are from the same class, where the higher the similarity between samples, the greater the probability of deriving from the same class. Recently, various approaches based on metric learning are proposed [23–27], where diverse similarity metrics are employed to address problems of few-shot classification. The goal of meta-learning is to learn how to learn [28], which plays a role in guiding the model to learn how quickly adapt to unseen tasks. For example, Zhai et al. [29] develop a scene classification model on the basis of meta-learning for remote sensing images, called LLSR, which aims to rapidly recognize unseen categories with few labeled samples. Hence, few-shot learning is generally regarded as a special case of meta-learning. As for transfer learning, the line of thinking is to train a model on similar tasks, typically known as pre-training, and then apply the pre-trained model with massive prior knowledge into new tasks.

In view of big scales and unique attributes of remote sensing images, some well-designed neural networks [30] are employed to encode images to acquire deep features with abundant semantic information in earlier work, such as AlexNet, GoogLeNet, VGG16, and so on. Nevertheless, when labeled samples are insufficient, the trained deep neural network typically suffers from overfitting, which is particularly obvious in the few-shot scenarios. Recently, a majority of proposed approaches for few-shot learning have focused more on improving feature representation. In RS-MetaNet, Li et al. [31] propose a training scheme based on meta tasks for few-shot classification of remote sensing images, which can prompt the model to learn how to capture a task-level distribution. In DLA-MatchNet, Li et al. [32] introduce spatial attention and channel attention into the feature extractor to learn robust feature representations, and a learnable matcher is developed to adaptively calculate similarity scores between the samples. In addition, Jiang et al. [33] present a multi-scale metric learning method based on the feature pyramid construction, called MSML, which aims to incorporate multi-scale features to improve feature representation of samples. These methods have been verified to improve the scene classification performance of remote sensing images, but in the case of extremely few available samples for each category, the improvement is not obvious.

The methods mentioned above focus more on employing image-level or class-level features for few-shot image classification, which does not fully excavate discriminative information from the perspective of the entire task. In terms of problems of scene classification, the model may not properly distinguish the correct category of the image when the feature extractor can not capture discriminative features [34]. In particular, in the tasks of few-shot scene classification, there are plenty of similar areas between the images from different categories [35], which further increases the difficulty of few-shot scene classification. Figure 1 presents the scene classification task in the 6-way 1-shot scenario. It is observed from Figure 1 that images in the first row have analogical texture features, and images in the second row have nearly approximate backgrounds, which shows that there are many similar feature areas between various categories. That is to say, the shared semantic features among all categories are not really critical for classifying an unseen sample. Furthermore, a remote sensing image may contain multiple different scene objects, which also has an enormous influence on the scene classification performance of the model.

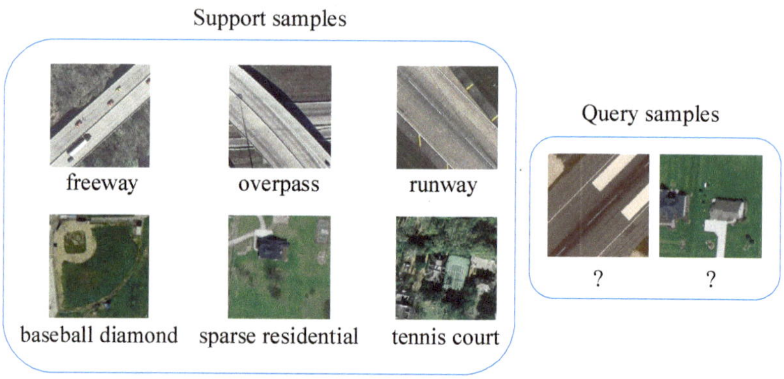

Figure 1. Illustration of few-shot remote sensing scene classification with only one labeled image.

To address the aforementioned issues, a task-adaptive embedding network based on meta-learning is proposed for few-shot remote sensing scene classification, which targets to learn task-relevant semantic features to enhance the discrimination between different scenes in few-shot tasks. In our proposed architecture, a pre-training scheme is employed to learn an effective feature encoder, which aims to make models learn generic feature representations for various scenes. Then, in meta-training stage, a task-adaptive attention module is employed to explore discriminative semantic features from the perspective of the whole task, which enhances weights of salient semantic features and reduces weights of common semantic features shared by diverse scenes. Finally, in the meta-testing stage, new scene categories can be predicted by the trained model. Experimental results on three challenging scene datasets verify state-of-the-art performance of our TAE-Net in few-shot settings.

The primary contributions of this paper are overall summarized as follows:

- A task-adaptive embedding network based on meta-learning, called TAE-Net, is proposed to enhance the generalization performance of the model for unseen remote sensing images in few-shot settings. The proposed TAE-Net can learn generic feature representations by combining pre-training with meta-training, which can allow models to quickly adapt to new categories with extremely few labeled samples.
- A task-adaptive attention module is developed to capture task-specific information, which can remove the effect of task-irrelevant noises. The proposed task-adaptive attention module aims to adaptively and dynamically pick out discriminative semantic features for diverse tasks.
- Comprehensive experiments on three public remote sensing scene classification datasets verify effectiveness of our proposed model, which exceeds existing state-of-the-art few-shot scene classification approaches and acquires new state-of-the-art performance of few-shot scene classification.

2. Related Work

In this section, the work related to our research is reviewed from two perspectives: remote sensing scene classification and few-shot learning.

2.1. Remote Sensing Scene Classification

In the past decades, remote sensing scene classification has been sufficiently researched due to its broad application prospects, which aims to divide images into corresponding scene categories. In recent years, approaches based on deep learning have made considerable progress in the area of remote sensing; particularly, plenty of approaches based on the convolutional neural network (CNN) are proposed [36–38]. To solve the issue of intra-class difference and inter-class similarity in remote sensing images, Cheng et al. [39] design discriminative CNN (D-CNN). Zhang et al. [40] design a CNN-CapsNet for remote sensing scene classification, which consists of CNN for extracting features and CapsNet for classifying features. Wang et al. [35] design an attention recurrent CNN in an end-to-end manner for remote sensing scene classification, which can capture high-level semantic features and remove irrelevant information. Sun et al. [41] develop a gated bidirectional model based on CNN to implement remote sensing scene classification, which can incorporate multilayer convolutional features and eliminate the distraction information. Pires de Lima et al. [42] study the performance of CNN combined with transfer learning on remote sensing scene classification tasks, which confirms that transfer learning from natural images to remote sensing images is effective. To constrain the distribution of the training data belonging to the same category, Xie et al. [43] design a scene classification network with intra-class constraint; besides that, label augmentation is employed to label each augmented sample with a joint label. Considering the considerable complexity of current CNN, Shi et al. [44] construct a lightweight network that combines the attention mechanism and multi-branch feature fusion strategy for remote sensing scene classification.

2.2. Few-Shot Learning

In recent years, methods based on deep learning have made considerable progress in various areas, especially in fields where a large amount of computing resources and labeled data can be obtained. Nevertheless, methods based on deep learning are susceptible to insufficient data. To resolve the issue, few-shot learning is proposed, which has achieved outstanding success. The goal of few-shot learning is to learn to recognize new categories from only limited labeled data, which is typically regarded as a special meta-learning. Subsequently, some representative few-shot learning methods are introduced, which are divided into two main branches: metric-based approaches and optimization-based approaches.

Metric-based approaches: This line of work targets to solve few-shot image classification problems through learning to compare. The line of thinking aims to learn a feature encoder that transforms input images into embedding representations suitable for comparing; when encoded in the feature space, the similarity between the support sample and the query sample can be calculated explicitly through a common pairwise metric, such as cosine similarity or Euclidean distance. Siamese network [23] is proposed to learn generic embedding features, which is employed to achieve the binary classification task. By an attention mechanism, matching network [24] transforms support samples into embedding space, where the query sample is classified through a nearest-neighbor classifier using cosine similarity. Likewise, prototypical network [25] also learns a metric rule to conduct few-shot classification over embeddings, which calculates Euclidean distance between the category-mean embedding and the query embedding. Instead of common metrics, relation network [26] presents a parameterized relation module as a learnable metric. Inspired by prototypical network, Oreshkin et al. [45] introduce a metric-based framework, called TADAM, that integrates three helpful improvements for few-shot learning, that is, task conditioning, metric scale, and auxiliary task co-training. By adding unlabeled data into each few-shot task, Ren et al. [46] carry out few-shot classification of prototypical network in the semi-supervised setting. Three schemes are tried to improve the prototype representation of each category. MetaOptNet [21] indicates that discriminative linear classifiers, such as support vector machine, may be preferable to nearest neighbor classifiers (e.g., cosine similarity or Euclidean distance) in few-shot settings. Compared with nearest neighbor classifiers, linear classifiers can find more appropriate decision boundaries by negative samples.

Optimization-based approaches: Another family of method is typically known as learning to learn (i.e., meta-learning), which aims to learn generic parameter initialization and then make the model quickly adapt to new tasks by a few gradient steps. Finn et al. [47] propose a model-agnostic architecture based on meta-learning, called MAML, which aims at learning a suitable parameter initialization that can be applied to any neural network. In other words, that network can fast adapt to any unseen few-shot task by only a few parameter updating steps. In addition, considering the complexity of the algorithm, the authors also introduce a first-order variant of MAML, which can speed up the model optimization by removing second-order derivatives. Likewise, Reptile [48] is also first-order variant of MAML, which employs a Taylor series expansion to perform first-order gradient optimization. A lot of variants [49–52] of MAML are based on an analogous idea that, given appropriate parameter initialization, the model can quickly adapt to the novel task with a few gradient steps. However, these methods suffer from a key challenge that internal optimization contains as massive parameters as external optimization. Additionally, a critical controversy is whether only an initialization condition guarantees quick adaption for various few-shot tasks. Further, whether the upper limit of model performance is affected by this initialization condition.

Recently, several approaches based on pre-training (i.e., transfer learning) have achieved competitive performance [53–55] as methods based on meta-learning make noteworthy progress in few-shot image classification. Additionally, some methods based the graph neural network are also developed for few-shot learning [56,57]. Our study is more inclined to the first kind of work, that is, employing pre-training to extract good embedding features and learning an appropriate similarity metric by encoding task-specific information.

3. Proposed Method

3.1. Overall Architecture

In this work, a task-adaptive embedding network is proposed to tackle scene classification tasks in few-shot settings, which is depicted in Figure 2. The overall architecture is composed of pre-training, meta-training, and meta-testing. A feature encoder is first pre-trained on the base set D_{base} to map inputs to embedding space suitable for comparison. To be specific, a common neural network model is trained on all base classes by minimizing generalization error. To obtain the trained feature encoder, the fully connected (FC) layer is removed. Next, a meta-learning model H is trained across a cluster of episodes in the meta-training phase. Specifically, different from previous works [54,55], f_φ is not frozen to further fine-tune; instead, it is employed to initialize parameters in the meta-learning model and is optimized by minimizing the cross-entropy loss. For each episode, a relation matrix R is calculated to obtain the embedding relation between the support image and the query image. Then, the task-adaptive attention module g_θ yields the task attention matrix A, which can adaptively pick out the discriminative embedding features for a certain query embedding feature within the support set, like visual recognition of humans. It's worth noting that, instead of independent entity, the task attention centers on the relations between the embedding features. Subsequently, the task attention matrix A is incorporated into the relation matrix R by an element-wise multiplication to eradicate noises, such as the relation formed by the similar embedding features within a task, and strengthen the discriminative embedding features. The predicted scores can be then obtained from the incorporated relation matrix using the sum operation. In the meta-testing phase, the meta-learning model H is evaluated across a cluster of episodes, which are sampled at random from the novel set D_{novel}.

3.2. Problem Formulation

In the few-shot settings, remote sensing scene classification can be referred to as a set of N-way K-shot M-query tasks, which indicates that only K labeled samples are used to recognize M unseen samples from N scene classes. A N-way K-shot M-query task $\mathcal{T}$ is also known as an episode. Therein, M samples without labels and K samples with labels are derived from each class, and accordingly, a support set consists of $N \times K$ labeled samples and a query set consists of $N \times M$ samples. To be specific, when M is set to 1, the quantity of all samples in the query set is equivalent to the total quantity of classes. Figure 3 presents the illustration of 5-way 1-shot classification tasks. The dataset for few-shot scene classification is split into three disjoint subsets, that is, base set D_{base}, validation set D_{val}, and novel set D_{novel}. The overall process contains the meta-training phase, meta-validation phase, and meta-testing phase. The specific details are described below.

Different with standard supervised learning, there are extremely few labeled samples available in few-shot learning, which makes it difficult for the model to learn enough prior knowledge. Therefore, intra-class relationship learning and knowledge transfer become quite vital for few-shot scene classification of remote sensing images. In the meta-training phase, the support set $S = (x_i, y_i)(i = 1, 2, 3, \ldots, N \times K)$ and query set $Q = (x_j, y_j)(j = 1, 2, 3, \ldots, N \times M)$ are sampled from D_{base}, where x_i represents the i-th sample and y_i represents its ground truth label. Furthermore, the parameters in the model are iteratively updated through back propagation. After massive iterations, a well-trained feature encoder can be learned.

The validation phase aims to adjust the hyper-parameters in the model. In the meta-validation phase, D_{val} is also divided into the support set S and the query set Q, where the support set is employed to predict labels of query samples. In addition, only forward propagation is performed in the meta-validation phase, which indicates that the parameters in the model are not updated through back propagation.

In the few-shot scene classification, the categories used in the meta-testing phase are new, which indicates that classes in D_{novel} are distinct from those of D_{base} and D_{val}. Query samples in D_{novel} are predicted by a well-trained model given the support set that only contains N classes and K labeled samples per class. The class corresponding to the largest predicted probability is determined as the label of the query sample.

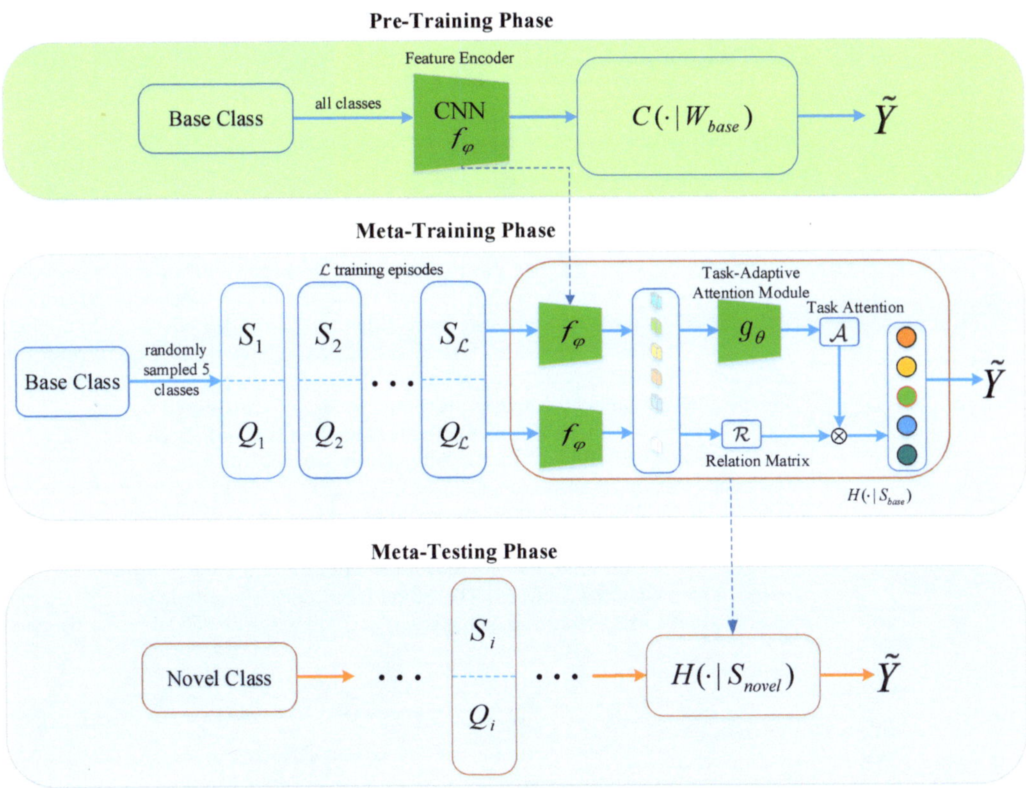

Figure 2. The overall architecture and details of the task-adaptive embedding network (shortened to TAE-Net).

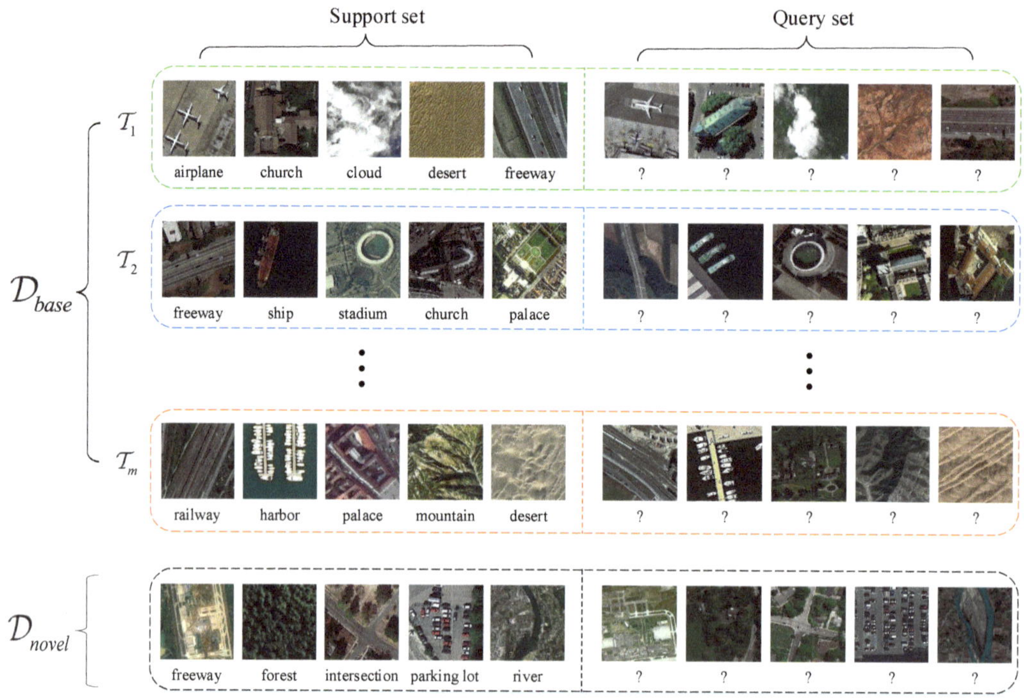

Figure 3. Illustration of data partition of remote sensing scene classification in the 5-way 1-shot setting.

3.3. Pre-Training of Feature Encoder

In the pre-training stage, a feature encoder f_φ with learnable parameters φ is trained on the base set D_{base}, which transforms the input sample into the embedding feature with the shape of $5 \times 5 \times 64$. For a fair comparison, a neural network with 4 convolutional blocks is employed to train a classification model on all base categories. Then, to obtain well-trained f_φ, the fully connected layer and the softmax layer are removed in the neural network. The above process is described in detail as follows. Before starting training, all input samples from D_{base} are adjusted to the size of 84×84. The entire setting of the pre-training model we adopt, presented in Figure 4, is formed from an encoder and a classifier. The encoder contains 4 convolutional blocks (Conv-4), each of which contains 3×3 convolution with 64 kernels, followed by BatchNorm with a momentum value of 0.1, Leaky ReLU activation function, and 2×2 max-pooling. Furthermore, the classifier is made up of a fully connected layer and a softmax layer. To optimize the overall pre-training model, a common cross-entropy loss function is selected as the loss metric in this work. After pre-training, the classifier in the pre-training model is removed to get the encoder for extracting features. Therefore, by inputting a sample into the pre-training model, a embedding feature with the shape of $5 \times 5 \times 64$ can be yielded by the encoder.

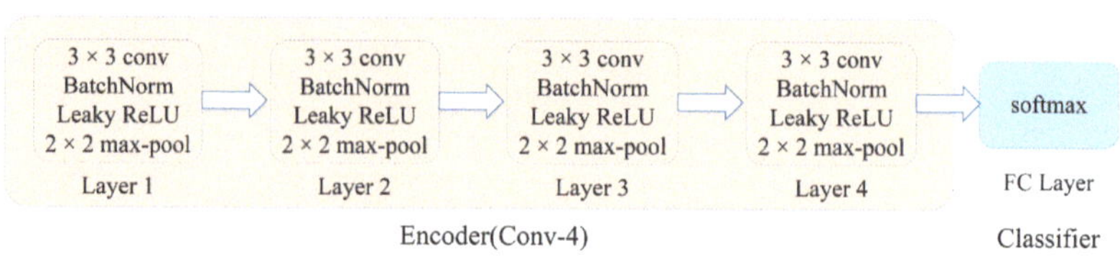

Figure 4. Structure and parameters of the pre-training model.

3.4. Meta-Training with Task-Adaptive Attention Module

The goal of meta-learning is to enhance generalization performance through learning meta-konwledge from multiple tasks, also known as episodes, which is a quite popular method for solving few-shot tasks. In the N-way K-shot setting, a meta-learning model $H(\cdot|S)$ is trained by minimizing the N-way loss. To this end, a set of episodes are sampled at random from the base set. Each episode contains K labeled samples for each class, which means that there are $N \times K$ support samples in total for training and $N \times M$ query samples for testing. Even though each episode only contains a few support samples for training, the model $H(\cdot|S)$ shares the same parameters over multiple episodes. As a result, training $H(\cdot|S)$ across plenty of episodes contributes to decreasing the model demand for data. In the meta-training procedure, the validation set D_{val} is employed to select hyperparameters of $H(\cdot|S)$. Figure 2 presents corresponding meta-training process.

For an episode, a sample x from $S \bigcup Q$ first is inputted into the feature encoder f_φ to attain an embedding representation $f_\varphi(x) \in \mathbb{R}^{W \times H \times C}$. Generally, each input sample is represented as WH C-dimensional embedding features. Thus, the entire support set can be denoted as $NKWH$ C-dimensional support embedding features, that is, $Z^S = f_\varphi(S) \in \mathbb{R}^{NKWH \times C}$ and each query image can be denoted as WH C-dimensional query embedding features, that is, $Z^q = f_\varphi(q) \in \mathbb{R}^{WH \times C}$. Next, the relation matrix R of these embedding features can be formulated as follows:

$$R_{i,j} = s(Z_i^q, Z_j^S) \tag{1}$$

where $i \in \{1,2,3,\ldots,WH\}$, $j \in \{1,2,3,\ldots,NKWH\}$ and $s(\cdot,\cdot)$ is a similarity function. In this paper, Gaussian similarity function is selected to calculate similarity. Different from prior work that is based on class-level [58] or image-level [26] relation, our proposed approach aims to construct task-level relation while sustaining discriminative relations.

Moreover, a convolution block f_v consisting of two 1×1 convolution layers is applied to the embedding features extracted by the feature encoder. Then another relation matrix R' is calculated by the following operation:

$$R'_{i,j} = s(f_v(Z_i^q), f_v(Z_j^S)) \tag{2}$$

where $i \in \{1,2,3,\ldots,WH\}, j \in \{1,2,3,\ldots,NKWH\}$. For relation matrix R', each row denotes the adaptive subspace relation of each space position in the query image to all space positions of all support images. Furthermore, to eradicate the noises (e.g., the irrelevant relations), a threshold λ is applied to the relation matrix R', which can generate task attention A by the following operation:

$$A_{i,j} = \frac{M(R'_{i,j})}{\sum_j M(R'_{i,j})} \tag{3}$$

$$M(x) = \begin{cases} 1, & if\ x > \lambda \\ 0, & otherwise. \end{cases} \tag{4}$$

As shown in Equation (3), the universal embedding features tend to exist in multiple categories in the whole task, which will greatly reduce the attention they receive. Hence, their corresponding attention value will also relatively smaller. Additionally, considering that the impact of each irrelevant noise is small, but the total number is quite large, which still has a considerable effect on the distribution of task attention. To this end, Equation (4) is adopted to filter the task attention, which can select informative relations and eliminate negligible relations.

Considering that the threshold λ is fixed in Equation (4), it can not handle different query features flexibly. For this purpose, a task encoder with the adaptive threshold λ' is proposed, as shown in Figure 5. Unlike Equation (4), a transformer, consisting of three fully connected layers, is employed to adaptively yield the threshold λ' for each C-dimensional embedding feature of the query image. Meanwhile, considering that Equation (4) is indifferentiable, a variant M' of the sigmoid function is adopted to approximate it:

$$M'(x) = 1/(1 + exp^{\beta(\lambda' - x)}) \tag{5}$$

where λ' is the adaptive threshold of x, and x represents one of the elements in A. Conceptually, when β takes a large enough value, M' is equivalent to M.

Subsequently, a weighted relation matrix $A \otimes R$ can be calculated by a Hadamard product between A and R. Then, the predicted score belonging to the c-th category is attained by gathering the weighted relation between the query sample q and the c-th category:

$$Score_c = \frac{1}{WH} \sum_{i=1}^{WH} \sum_{j=r_1^c}^{r_{KWH}^c} (A \otimes R)_{i,j} \tag{6}$$

where r_t^c denotes that the t-th relation of KWH relations belonging to the c-th category in the NKWH relations of all support samples. Hence, the final predicted probability P of the query image q can be formulated as follows:

$$P(y = c|q) = \frac{exp(Score_c)}{\sum_{c'=1}^{N} exp(Score_{c'})} \tag{7}$$

where N represents the total quantity of categories in the N-way K-shot setting.

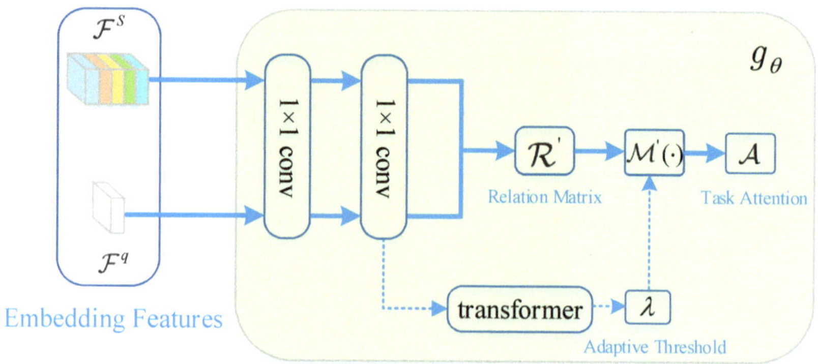

Figure 5. The schematic of task-adaptive attention module.

3.5. Meta-Testing

At the end of meta-training, a novel set D_{novel} is typically utilized to estimate the generalization performance of the trained model $H(\cdot|S_{base})$. It is worth noting that all classes in D_{novel} have never appeared in the meta-training stage. In the meta-testing stage, a set of new episodes are stochastically sampled from D_{novel}, also known as meta-testing set $D_{\mathcal{T}}^{test} = \{\mathcal{T}_i\}_{i=1}^{I}$. Here, $\mathcal{T}$ consists of S_{novel} and Q_{novel}. Then, given the novel support set S_{novel}, categories of samples from Q_{novel} can be predicted by the trained model $H(\cdot|S_{base})$.

4. Results and Discussions

In this section, dataset description and experimental settings are first presented. Then, several state-of-the-art methods for few-shot learning are compared with our proposed approach and the corresponding experimental results are presented below. Moreover, we also conduct a series of ablation experiments to investigate the impact of different components on model performance, containing the pre-training strategy, task-adaptive attention mechanism, and the number of shots.

4.1. Dataset Description

The UC Merced land-use dataset [59] consists of 21 categories with a total of 2100 scene images, each of which corresponds to 100 land-use images containing 256 × 256 pixels. Furthermore, all scene images are of RGB color space. The scene categories contain agricultural, airplane, baseball diamond, beach, buildings, chaparral, and other scenes. This dataset was published by UC Merced Computer Vision Laboratory in 2010, including various urban area imageries from the United States Geological Survey. In this dataset, 10 categories are deemed as the base set, 5 categories are considered as the validation set and the remaining 6 categories are considered as the novel set. In the experiments, the shapes of all images are adjusted to 84 × 84 to fit our proposed feature encoder for feature extraction.

The WHU-RS19 dataset [60] is also composed of remote sensing scenes, which was published by Wuhan University. It consists of 19 scene categories in total, and the quantity of samples for each category is more than or equal to 50 images. A total of 1005 images are contained in this dataset. The entire dataset is divided into three subsets, namely, a base set containing 9 classes, a validation set containing 5 classes, and a novel set containing 5 classes. In order to adapt to our proposed feature encoder, the pixel sizes of all scene images are adjusted to 84 × 84.

The NWPU-RESISC45 is a quite popular scene dataset in the domain of remote sensing, and was published by Cheng et al. [61] in 2017. It contains 45 scene classses and each class has 700 scene images, which consists of a total of 31,500 images, as shown in Figure 6. Each scene image is of 256 × 256 pixels. The scene classes include airplane, beach, circular farmland, dense residential, parking lot, and other scenes. These scene data are gathered by experts from Google Earth, and the spatial resolution of each pixel varies between close to 30 and 0.2. The entire dataset is split into 3 subsets, namely, a base set containing 25 classes, a validation set containing 10 classes, and a novel set containing 10 classes. In the experiments, the pixel sizes of all scene images are adjusted to 84 × 84 , which is to adapt to the designed feature encoder. The segmentation details of the three scene datasets mentioned above are presented in Table 1.

Table 1. The segmentation details of UC Merced dataset, WHU-RS19 dataset, and NWPU-RESISC45 dataset.

Datasets	Base	Validation	Novel
UC Merced	Agricultural; Baseball diamond; Buildings; Parking lot; Harbor; Medium residential; Dense residential; Chaparral; Freeway; Overpass;	Airplane; Forest; Runway; Intersection; Storage tanks;	Beach; River; Golf course; Mobile home park; Sparse residential; Tennis court;
WHU-RS19	Airport; Bridge; Football field; Desert; Mountain; Industrial; Port; Residential; Parking;	Beach; Forest; Farmland; Park; Railway station;	Meadow; Pond; River; Viaduct; Commercial;

Table 1. *Cont.*

Datasets	Base	Validation	Novel
NWPU-RESISC45	Airplane;Church; Baseball diamond; Bridge;Beach; Cloud;Freeway; Desert;Island; Chaparral; Harbor;Lake; Meadow;Mountain; Palace;Ship; Railway;Stadium; Wetland; Golf course; Mobile home park; Sparse residential; Sea ice; Roundabout; Rectangular farmland;	Commercial area; Overpass; Industrial area; Railway station; Snowberg; Runway; Storage tank; Terrace; Thermal power station; Tennis court;	Airport; Dense residential; Basketball court; Circular farmland; Intersection; Forest; Ground track field; Parking lot; Medium residential; River;

Figure 6. Scene images (with 84 × 84 pixels) derived from 45 categories in the NWPU-RESISC45 dataset [61]. Airport, basketball court, circular farmland, dense residential, forest, ground track field, intersection, medium residential, Parking lot, and river are selected as the novel set for performance evaluation.

4.2. Experimental Settings

In this section, the network architecture and hyperparameter setup are described in detail. To make a fair comparison with other work, the same backbone as other methods [24–26], a 4-layer convolutional network (Conv-4), is employed as our encoder to extract embedding features, which consists of 4 convolution modules, as depicted in Figure 4. For each convolution module, there is a convolutional layer with 64 kernels whose size is 3×3, a batch normalization operation, a Leaky ReLU nonlinearity activation function, and a 2×2 max-pooling layer. The padding and stride sizes are both set to 1 in the convolutional layer while for the 2×2 max-pooling, the padding and stride sizes are separately set to 0 and 2. For each batch normalization operation, the momentum factor is set to 0.1 and the value of epsilon ϵ is set to 0.00001. The learning rate η is initialized to 0.001. In addition, all comparison models adopt the cross-entropy loss to evaluate generalization performance. The models are optimized by Adam Optimizer, where $\beta 1$ and $\beta 2$ are separately set to 0.9 and 0.999 to adjust exponential decay rates. The pixel size of each image is adjusted to 84×84. Classification accuracy is employed to evaluate the performance of all models, which is formalized as:

$$\text{Accuracy} = \frac{1}{\mathcal{L}} \sum_{t=1}^{\mathcal{L}} \frac{u^t}{V} \tag{8}$$

where $\mathcal{L}$ denotes the total quantity of tasks, u^t denotes the quantity of samples that are correctly classified in the t-th task, and V denotes the total quantity of samples in the task.

For traditional deep learning, an epoch indicates that the whole dataset passes the neural network by the forward propagation once. For few-shot learning, each episode randomly sampled from the dataset. Though there are only a few labeled images per episode, when the total quantity of episodes is more enough, all samples have been probably sampled in the whole dataset. In our work, an epoch consists of 5000 episodes.

Pre-training is done with D_{base} containing 10, 9, and 25 categories for UC Merced, WHU-RS19, and NWPU-RESISC45 experiments, separately. For the meta-learning (i.e., meta-training and meta-testing) stage, 5 novel classes are randomly chosen from D_{train} per episode, and 1 or 5 samples (i.e., shots) from 5 randomly chosen novel classes makes up the support set. Following the protocol of few-shot learning [25,62], the query set contains 15 samples per class in each episode. The PyTorch framework is adopted to implement our proposed TAE-Net, which is run with four NVIDIA Tesla V100 GPUs (32G $\times$ 4).

4.3. Experimental Results

4.3.1. Experimental Results on UC Merced Dataset

In few-shot scenarios, the classification results on the UC Merced dataset are presented in Table 2, where classification accuracies are acquired by averaging the results of 600 episodes randomly sampled on the novel set with 95% confidence interval. In the experiments, nine other few-shot scene classification approaches of remote sensing images are employed to make a comparison, in which average accuracies with 95% confidence interval are reported. For Table 2, it is observed that our proposed TAE-Net performs best with accuracies of 60.21% and 77.44% in the scenarios of 5-way 1-shot and 5-way 5-shot , outperforming the accuracies of RS-MetaNet with 2.98% and 1.36% improvements, separately. Compared to DLA-MatchNet, the proposed TAE-Net has 6.45% and 14.43% improvements in the scenarios of 5-way 1-shot and 5-way 5-shot, separately. In addition, our proposed TAE-Net exceeds TPN with 6.85% improvement in the 5-way 1-shot scenario and 9.21% improvement in the 5-way 5-shot scenario, separately. The proposed TAE-Net achieves higher classification accuracies than existing state-of-the-art methods. In consequence, it is illustrated that our proposed TAE-Net can take full advantage of a small amount of information from limited samples, which promotes the performance of scene classification in the cases of limited labeled samples.

Table 2. Classification accuracies and standard deviations (%) of 5-way 1-shot and 5-way 5-shot on the UC Merced dataset. The best results per scenario are marked in bold.

Method	5-Way 1-Shot	5-Way 5-Shot
MatchingNet [24]	46.16 ± 0.71	66.73 ± 0.56
Prototypical Network [25]	52.62 ± 0.70	65.93 ± 0.57
MAML [47]	43.65 ± 0.68	58.43 ± 0.64
Meta-SGD [52]	50.52 ± 2.61	60.82 ± 2.00
Relation Network [26]	48.89 ± 0.73	64.10 ± 0.54
TPN [56]	53.36 ± 0.77	68.23 ± 0.52
LLSR [29]	39.47	57.40
RS-MetaNet [31]	57.23 ± 0.56	76.08 ± 0.28
DLA-MatchNet [32]	53.76 ± 0.60	63.01 ± 0.51
TAE-Net (ours)	**60.21 ± 0.72**	**77.44 ± 0.51**

The proposed TAE-Net is capable of learning the task-specific feature representation by task encoder, which eliminates the deviation of embedding features posed by limited labeled samples. Additionally, pre-training scheme is developed for feature embedding, which can provide the feature encoder of the model with better initialization parameters.

4.3.2. Experimental Results on WHU-RS19 Dataset

Comparative experiments on the WHU-RS19 dataset are performed in few-shot cases, and the corresponding experimental results are presented in Table 3. All experimental results are the average of 600 episodes with 95% confidence interval. It can be seen that our proposed few-shot classification approach achieves the accuracies of 73.67% and 88.95% in the 5-way 1-shot scenario and 5-way 5-shot scenario, separately. From Table 3, it is observed that the proposed TAE-Net achieves superior performance over Prototypical Network, with 2.79% improvement in the 5-way 1-shot scenario and 3.33% improvement in the 5-way 5-shot scenario.

Compared with experimental results in the 5-way 1-shot scenario, the performance of scene classification is better than that in the 5-way 1-shot scenario, which illustrates that prior knowledge is more significant. In our proposed TAE-Net, task-adaptive attention module integrates task-specific information into the relationship matrix, which can seek out more informative prior knowledge from extremely limited labeled samples. Hence, the proposed TAE-Net can significantly improve the classification performance of remote sensing scene images in few-shot cases.

Table 3. Classification accuracies and standard deviations (%) of 5-way 1-shot and 5-way 5-shot on the WHU-RS19 dataset. The best results per scenario are marked in bold.

Method	5-Way 1-Shot	5-Way 5-Shot
MatchingNet [24]	60.60 ± 0.68	82.99 ± 0.40
Prototypical Network [25]	70.88 ± 0.65	85.62 ± 0.33
MAML [47]	46.72 ± 0.55	79.88 ± 0.41
Meta-SGD [52]	51.54 ± 2.31	61.74 ± 2.02
Relation Network [26]	60.54 ± 0.71	76.24 ± 0.34
TPN [56]	59.28 ± 0.72	71.20 ± 0.55
LLSR [29]	57.10	70.65
DLA-MatchNet [32]	68.27 ± 1.83	79.89 ± 0.33
TAE-Net (ours)	**73.67 ± 0.74**	**88.95 ± 0.53**

4.3.3. Experimental Results on NWPU-RESISC45 Dataset

Experimental results on the NWPU-RESISC45 dataset are presented in Table 4, where all evaluation results are the average accuracies on 600 episodes. In Table 4, it is distinctly observed that the proposed TAE-Net obtains best performance with classification accuracies of 69.13% and 82.37% in the 5-way 1-shot scenario and 5-way 5-shot scenario, respectively.

Compared with DLA- MatchNet, our proposed TAE-Net achieves more superior performance, with 0.33% and 0.74% improvements in the scenarios of 5-way 1-shot and 5-way 5-shot. Additionally, the proposed TAE-Net exceeds RS-MetaNet with 16.35% improvement in the 5-way 1-shot scenario and 10.88% improvement in the 5-way 5-shot scenario. Besides, the proposed TAE-Net also obtains greater improvements than other methods. The reason for achieving such significant improvements is that our proposed TAE-Net can reduce the interference between similar embedding features while focusing more on enhancing the discrimination between informative embedding features.

Table 4. Classification accuracies and standard deviations (%) of 5-way 1-shot and 5-way 5-shot on the NWPU-RESISC45 dataset. The best results per scenario are marked in bold.

Method	5-Way 1-Shot	5-Way 5-Shot
MatchingNet [24]	54.46 ± 0.77	67.87 ± 0.59
Prototypical Network [25]	50.82 ± 0.84	74.38 ± 0.59
MAML [47]	37.36 ± 0.69	45.94 ± 0.68
Meta-SGD [52]	60.63 ± 0.90	75.75 ± 0.65
Relation Network [26]	58.61 ± 0.83	78.63 ± 0.52
TPN [56]	66.51 ± 0.87	78.50 ± 0.56
LLSR [29]	51.43	72.90
RS-MetaNet [31]	52.78 ± 0.09	71.49 ± 0.81
DLA-MatchNet [32]	68.80 ± 0.70	81.63 ± 0.46
TAE-Net (ours)	**69.13 ± 0.83**	**82.37 ± 0.52**

4.4. Ablation Study

To further validate the effectiveness of our proposed TAE-Net on few-shot scene classification of remote sensing images, a sequence of ablation experiments are performed to analyze the role of each module in our framework and presented as follows.

4.4.1. Effect of Pre-Training Strategy

Before meta-training, pre-training is introduced to improve the feature representation ability of the model in few-shot settings, which can give the model a good initialization. To illustrate the effect of pre-training, experiments with and without pre-training are conducted, in which several cases containing different shots are designed for comparison. Experimental results on UC Merced dataset are depicted in Figure 7.

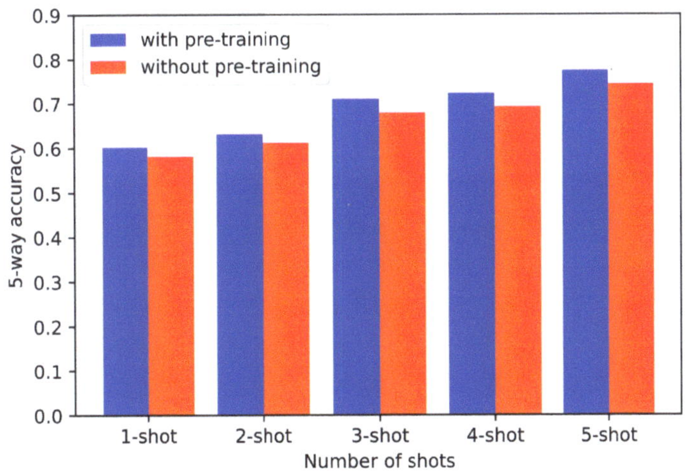

Figure 7. Classification accuracies with different number of shots on UC Merced dataset.

In Figure 7, it can be seen that the classification accuracy without pre-training is only 58.17% in the 5-way 1-shot scenario, which is markedly lower than our proposed TAE-Net by approximately 2%. Likewise, the classification accuracy without pre-training is 74.32% in the 5-way 5-shot scenario, which is obviously lower than our proposed TAE-Net by approximately 3%. Besides, the proposed method with pre-training is superior to that without pre-training in the settings of different shots by a large margin. It can be demonstrated that pre-training before meta-training helps to improve the performance of few-shot scene classification, which can yield a feature encoder with good initialization parameters.

4.4.2. Effect of Task-Adaptive Attention Module

In our proposed approach, a task-adaptive attention module is designed to enhance discriminative embedding features and weaken common embedding features shared by diverse scenes, which can strengthen the discrimination between embedding features from different categories and reduce the distraction of irrelevant information. To verify the effect of the task-adaptive attention module, an ablation experiment is conducted on UC Merced dataset. The results of the ablation experiment for the task-adaptive attention module is depicted in Figure 8. It is observed that when the task-adaptive attention module is ablated, the classification performance of the model notably degenerates in various numbers of shots settings. As the task-adaptive attention module is removed, the model is inclined to treat equally noncritical embedding features and discriminative embedding features, which makes it difficult for the model to pay more attention to informative features. Moreover, it is also noticed that the task-adaptive attention module has a growingly significant contribution to classification performance as the number shots increases. For example, in the case of 5-way 1-shot, the classification accuracy of the model improves by 10% when the task-adaptive attention module is adopted. By contrast, the classification accuracy of the model increases by nearly 13% in the case of 5-way 5-shot when the task-adaptive attention module is adopted. It is verified that our proposed task-adaptive attention module can significantly improve the classification performance of the model in few-shot settings, since it enhances the discriminability between embedding features belonging to diverse categories and reduces the impact of irrelevant information on classification.

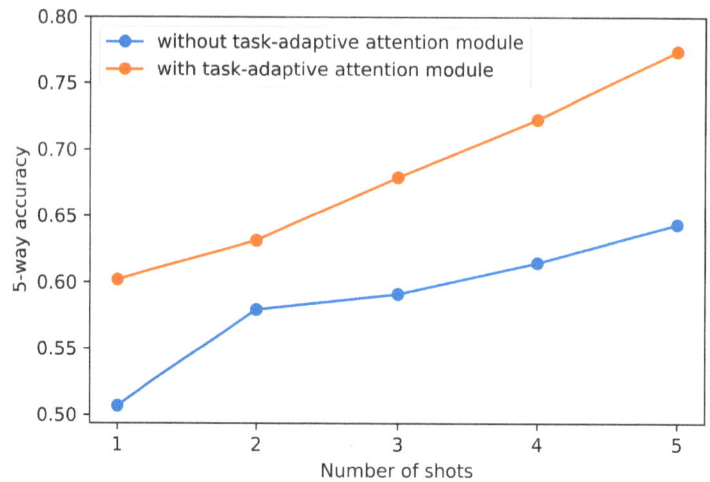

Figure 8. 5-way accuracies of the ablation analysis of task-adaptive attention module. When task-adaptive attention module is removed, the classification performance of the model degenerates dramatically in all scenarios.

To intuitively illustrate the effect of the task-adaptive attention module, 5-way 1-shot classification experiments are carried out on the UC Merced dataset. Figure 9 visualizes corresponding experimental results, including validation accuracies and losses in the cases of with the task-adaptive attention module and without the task-adaptive attention module. The number of epochs ranges from 1 to 350 with the step size of 1. As can be seen in Figure 9, in the meta-training stage, the validation losses with the task-adaptive attention module are smaller than that without the task-adaptive attention module, and the best validation accuracy obtained by our method is higher than that without the task-adaptive attention module.

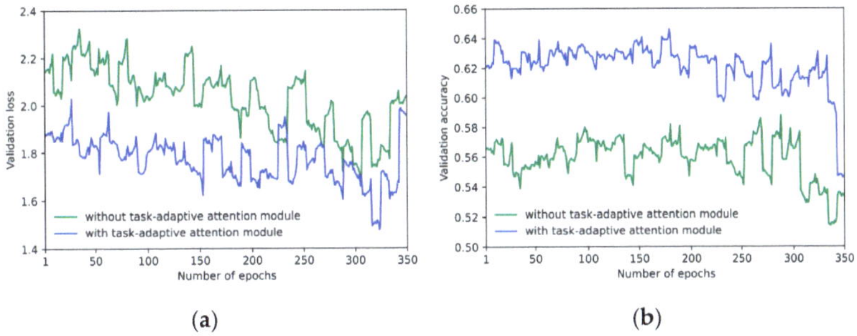

Figure 9. Validation losses and accuracies on the UC Merced dataset in the scenarios of 5-way 1-shot. (**a**) shows the validation losses, and (**b**) shows the validation accuracies.

4.4.3. Effect of Shots

To further validate the 5-way classification performance in the case of different number of shots, a series of experiments are performed on two public datasets, that is, UC Merced and NWPU-RESISC45. Figure 10 presents the corresponding experimental results, where our model is provided with 1, 2, 3, 4, 5, 6, 7, 8, 9, and 10 labeled support images. It is observed from Figure 10 that the classification accuracy significantly improved as the quantity of shots increases from 1 to 5. Nevertheless, the classification performance of the model does not benefit a lot as the number of shots keeps on increasing. It demonstrates that our proposed TAE-Net is well-suited to scene classification tasks in the case of extremely limited labeled samples.

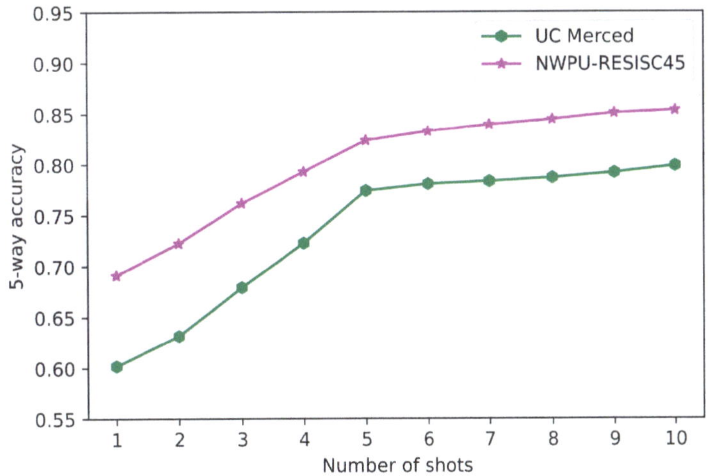

Figure 10. The influence of different shots on 5-way accuracies is presented with 95% confidence interval, in which the experiments are performed on UC Merced and NWPU-RESISC45.

5. Conclusions

In this paper, a task-adaptive embedding network based on meta-learning is proposed for remote sensing scene classification in few-shot settings, which aims to extract more informative embedding features from the perspective of the entire task. Our proposed approach first adopts Conv-4 as an encoder to learn embedding representation from the base set. Next, in the meta-training phase, the whole model is optimized by task-adaptive embeddings. To be specific, a task-adaptive attention mechanism is developed to adaptively filter irrelevant information and retain discriminative information in a certain task, which can avoid distraction from the general features shared by different categories. Experiments on three popular scene datasets illustrate that our proposed approach exceeds existing state-of-the-art few-shot approaches and achieves new state-of-the-art performance. Moreover, a series of ablation experiments are performed to investigate the influence of the pre-training strategy, task-adaptive attention mechanism, and the number of shots.

In the current work, we only use labeled data for few-shot scene classification, while ignoring the impact of unlabeled data on classification performance. Hence, an attractive research focus is the improvement of the performance of few-shot remote sensing scene classification under the semi-supervised setting, which will be the topic of future research.

Author Contributions: Conceptualization, W.H., Z.Y., A.Y. and X.L.; resources, W.H. and Z.Y.; data curation, W.H., Z.Y. and A.Y.; software, W.H. and Z.Y.; investigation, W.H., Z.Y. and A.Y.; methodology, W.H., A.Y. and X.L.; validation, W.H., Z.Y., A.Y. and C.T.; visualization, W.H.; writing—original draft, W.H.; writing—review and editing, W.H. and C.T.; supervision, C.T. and Z.Y. All authors have read and agreed to the published version of the manuscript.

Funding: This research was funded by the Advance Research Project of Civil Space Technology (grant no. D040402) and the National Natural Science Foundation of China (grant no. 41871226).

Institutional Review Board Statement: Not applicable.

Informed Consent Statement: Not applicable.

Data Availability Statement: The data presented in this study are available from the corresponding author.

Conflicts of Interest: The authors declare no conflict of interest.

References

1. Pham, H.M.; Yamaguchi, Y.; Bui, T.Q. A case study on the relation between city planning and urban growth using remote sensing and spatial metrics. *Landsc. Urban Plan.* **2011**, *100*, 223–230. [CrossRef]
2. Zhang, G.; Lu, S.; Zhang, W. CAD-Net: A context-aware detection network for objects in remote sensing imagery. *IEEE Trans. Geosci. Remote Sens.* **2019**, *57*, 10015–10024. [CrossRef]
3. Cheng, G.; Guo, L.; Zhao, T.; Han, J.; Li, H.; Fang, J. Automatic landslide detection from remote-sensing imagery using a scene classification method based on BoVW and pLSA. *Int. J. Remote Sens.* **2013**, *34*, 45–59. [CrossRef]
4. Jahromi, M.N.; Jahromi, M.N.; Pourghasemi, H.R.; Zand-Parsa, S.; Jamshidi, S. Accuracy assessment of forest mapping in MODIS land cover dataset using fuzzy set theory. In *Forest Resources Resilience and Conflicts*; Elsevier: Amsterdam, The Netherlands, 2021; pp. 165–183.
5. Li, Y.; Yang, J. Meta-learning baselines and database for few-shot classification in agriculture. *Comput. Electron. Agric.* **2021**, *182*, 106055. [CrossRef]
6. Li, X.; Shao, G. Object-based urban vegetation mapping with high-resolution aerial photography as a single data source. *Int. J. Remote Sens.* **2013**, *34*, 771–789. [CrossRef]
7. Fang, B.; Li, Y.; Zhang, H.; Chan, J.C.W. Semi-supervised deep learning classification for hyperspectral image based on dual-strategy sample selection. *Remote Sens.* **2018**, *10*, 574. [CrossRef]
8. Tai, X.; Li, M.; Xiang, M.; Ren, P. A mutual guide framework for training hyperspectral image classifiers with small data. *IEEE Trans. Geosci. Remote Sens.* **2021**, 1–17. [CrossRef]
9. Denisova, A.Y.; Kavelenova, L.M.; Korchikov, E.S.; Prokhorova, N.V.; Terentyeva, D.A.; Fedoseev, V.A. Tree species classification for clarification of forest inventory data using Sentinel-2 images. In Proceedings of the Seventh International Conference on Remote Sensing and Geoinformation of the Environment, Paphos, Cyprus, 18–21 March 2019; International Society for Optics and Photonics: Bellingham, WA, USA, 2019; Volume 11174, p. 1117408.

10. Alajaji, D.; Alhichri, H.S.; Ammour, N.; Alajlan, N. Few-shot learning for remote sensing scene classification. In Proceedings of the 2020 Mediterranean and Middle-East Geoscience and Remote Sensing Symposium (M2GARSS), Tunis, Tunisia, 9–11 March 2020; pp. 81–84.
11. Cen, F.; Wang, G. Boosting occluded image classification via subspace decomposition-based estimation of deep features. *IEEE Trans. Cybern.* **2019**, *50*, 3409–3422. [CrossRef]
12. Noothout, J.M.; De Vos, B.D.; Wolterink, J.M.; Postma, E.M.; Smeets, P.A.; Takx, R.A.; Leiner, T.; Viergever, M.A.; Išgum, I. Deep learning-based regression and classification for automatic landmark localization in medical images. *IEEE Trans. Med. Imaging* **2020**, *39*, 4011–4022. [CrossRef]
13. Du, L.; Li, L.; Guo, Y.; Wang, Y.; Ren, K.; Chen, J. Two-Stream Deep Fusion Network Based on VAE and CNN for Synthetic Aperture Radar Target Recognition. *Remote Sens.* **2021**, *13*, 4021. [CrossRef]
14. Andriyanov, N.; Dementiev, V.; Gladkikh, A. Analysis of the Pattern Recognition Efficiency on Non-Optical Images. In Proceedings of the 2021 IEEE Ural Symposium on Biomedical Engineering, Radioelectronics and Information Technology (USBEREIT), Yekaterinburg, Russia, 13–14 May 2021; pp. 319–323.
15. Xu, P.; Li, Q.; Zhang, B.; Wu, F.; Zhao, K.; Du, X.; Yang, C.; Zhong, R. On-Board Real-Time Ship Detection in HISEA-1 SAR Images Based on CFAR and Lightweight Deep Learning. *Remote Sens.* **2021**, *13*, 1995. [CrossRef]
16. Wu, B.; Meng, D.; Zhao, H. Semi-supervised learning for seismic impedance inversion using generative adversarial networks. *Remote Sens.* **2021**, *13*, 909. [CrossRef]
17. Liu, Y.; Zhong, Y.; Fei, F.; Zhang, L. Scene semantic classification based on random-scale stretched convolutional neural network for high-spatial resolution remote sensing imagery. In Proceedings of the IEEE International Geoscience and Remote Sensing Symposium, Beijing, China, 10–15 July 2016; pp. 763–766.
18. Zeng, Q.; Geng, J.; Huang, K.; Jiang, W.; Guo, J. Prototype Calibration with Feature Generation for Few-Shot Remote Sensing Image Scene Classification. *Remote Sens.* **2021**, *13*, 2728. [CrossRef]
19. Geng, J.; Deng, X.; Ma, X.; Jiang, W. Transfer learning for SAR image classification via deep joint distribution adaptation networks. *IEEE Trans. Geosci. Remote Sens.* **2020**, *58*, 5377–5392. [CrossRef]
20. Chang, H.; Yeung, D.Y. Semisupervised metric learning by kernel matrix adaptation. In Proceedings of the International Conference on Machine Learning and Cybernetics, Guangzhou, China, 18–21 August 2005; Volume 5, pp. 3210–3215.
21. Lee, K.; Maji, S.; Ravichandran, A.; Soatto, S. Meta-learning with differentiable convex optimization. In Proceedings of the IEEE Conference on Computer Vision and Pattern Recognition, Long Beach, CA, USA, 16–20 June 2019; pp. 10657–10665.
22. Shao, L.; Zhu, F.; Li, X. Transfer learning for visual categorization: A survey. *IEEE Trans. Neural Netw. Learn. Syst.* **2014**, *26*, 1019–1034. [CrossRef] [PubMed]
23. Koch, G.; Zemel, R.; Salakhutdinov, R. Siamese neural networks for one-shot image recognition. In Proceedings of the International Conference on Machine Learning, Lille, France, 6–11 July 2015; Volume 2.
24. Vinyals, O.; Blundell, C.; Lillicrap, T.; Kavukcuoglu, K.; Wierstra, D. Matching networks for one shot learning. *Proc. Neural Inf. Process. Syst.* **2016**, *29*, 3630–3638.
25. Snell, J.; Swersky, K.; Zemel, R.S. Prototypical networks for few-shot learning. *Proc. Neural Inf. Process. Syst.* **2017**, *30*, 4077–4087.
26. Sung, F.; Yang, Y.; Zhang, L.; Xiang, T.; Torr, P.H.; Hospedales, T.M. Learning to compare: Relation network for few-shot learning. In Proceedings of the IEEE Conference on Computer Vision and Pattern Recognition, Salt Lake City, UT, USA, 18–23 June 2018; pp. 1199–1208.
27. Zhang, C.; Cai, Y.; Lin, G.; Shen, C. DeepEMD: Few-Shot Image Classification With Differentiable Earth Mover's Distance and Structured Classifiers. In Proceedings of the IEEE Conference on Computer Vision and Pattern Recognition, Seattle, WA, USA, 13–19 June 2020; pp. 12203–12213.
28. Xu, Z.; Cao, L.; Chen, X. Learning to learn: Hierarchical meta-critic networks. *IEEE Access* **2019**, *7*, 57069–57077. [CrossRef]
29. Zhai, M.; Liu, H.; Sun, F. Lifelong learning for scene recognition in remote sensing images. *IEEE Geosci. Remote Sens. Lett.* **2019**, *16*, 1472–1476. [CrossRef]
30. Liu, S.; Deng, W. Very deep convolutional neural network based image classification using small training sample size. In Proceedings of the 3rd IAPR Asian Conference on Pattern Recognition, Kuala Lumpur, Malaysia, 3–6 November 2015; pp. 730–734.
31. Li, H.; Cui, Z.; Zhu, Z.; Chen, L.; Zhu, J.; Huang, H.; Tao, C. RS-MetaNet: Deep meta metric learning for few-shot remote sensing scene classification. *arXiv* **2020**, arXiv:2009.13364
32. Li, L.; Han, J.; Yao, X.; Cheng, G.; Guo, L. DLA-MatchNet for few-shot remote sensing image scene classification. *IEEE Trans. Geosci. Remote Sens.* **2020**, *59*, 7844–7853. [CrossRef]
33. Jiang, W.; Huang, K.; Geng, J.; Deng, X. Multi-scale metric learning for few-shot learning. *IEEE Trans. Circuits Syst.* **2020**, *31*, 1091–1102. [CrossRef]
34. Ma, H.; Yang, Y. Two specific multiple-level-set models for high-resolution remote-sensing image classification. *IEEE Geosci. Remote Sens. Lett.* **2009**, *6*, 558–561.
35. Wang, Q.; Liu, S.; Chanussot, J.; Li, X. Scene classification with recurrent attention of VHR remote sensing images. *IEEE Trans. Geosci. Remote Sens.* **2018**, *57*, 1155–1167. [CrossRef]
36. Tang, X.; Ma, Q.; Zhang, X.; Liu, F.; Ma, J.; Jiao, L. Attention consistent network for remote sensing scene classification. *IEEE J. Sel. Top. Appl. Earth Obs. Remote Sens.* **2021**, *14*, 2030–2045. [CrossRef]

37. Cheng, G.; Xie, X.; Han, J.; Guo, L.; Xia, G.S. Remote sensing image scene classification meets deep learning: Challenges, methods, benchmarks, and opportunities. *IEEE J. Sel. Top. Appl. Earth Obs. Remote Sens.* **2020**, *13*, 3735–3756. [CrossRef]
38. Lu, X.; Gong, T.; Zheng, X. Multisource compensation network for remote sensing cross-domain scene classification. *IEEE Trans. Geosci. Remote Sens.* **2019**, *58*, 2504–2515. [CrossRef]
39. Cheng, G.; Yang, C.; Yao, X.; Guo, L.; Han, J. When deep learning meets metric learning: Remote sensing image scene classification via learning discriminative CNNs. *IEEE Trans. Geosci. Remote Sens.* **2018**, *56*, 2811–2821. [CrossRef]
40. Zhang, W.; Tang, P.; Zhao, L. Remote sensing image scene classification using CNN-CapsNet. *Remote Sens.* **2019**, *11*, 494. [CrossRef]
41. Sun, H.; Li, S.; Zheng, X.; Lu, X. Remote sensing scene classification by gated bidirectional network. *IEEE Trans. Geosci. Remote Sens.* **2019**, *58*, 82–96. [CrossRef]
42. Pires de Lima, R.; Marfurt, K. Convolutional neural network for remote-sensing scene classification: Transfer learning analysis. *Remote Sens.* **2020**, *12*, 86. [CrossRef]
43. Xie, H.; Chen, Y.; Ghamisi, P. Remote Sensing Image Scene Classification via Label Augmentation and Intra-Class Constraint. *Remote Sens.* **2021**, *13*, 2566. [CrossRef]
44. Shi, C.; Zhao, X.; Wang, L. A Multi-Branch Feature Fusion Strategy Based on an Attention Mechanism for Remote Sensing Image Scene Classification. *Remote Sens.* **2021**, *13*, 1950. [CrossRef]
45. Oreshkin, B.N.; Rodriguez, P.; Lacoste, A. Tadam: Task dependent adaptive metric for improved few-shot learning. *arXiv* **2018**, arXiv:1805.10123.
46. Ren, M.; Triantafillou, E.; Ravi, S.; Snell, J.; Swersky, K.; Tenenbaum, J.B.; Larochelle, H.; Zemel, R.S. Meta-learning for semi-supervised few-shot classification. *arXiv* **2018**, arXiv:1803.00676.
47. Finn, C.; Abbeel, P.; Levine, S. Model-agnostic meta-learning for fast adaptation of deep networks. In Proceedings of the International Conference on Machine Learning, Sydney, Australia, 6–11 August 2017; PMLR: Sydney, Australia, 2017; pp. 1126–1135.
48. Nichol, A.; Achiam, J.; Schulman, J. On first-order meta-learning algorithms. *arXiv* **2018**, arXiv:1803.02999.
49. Sun, Q.; Liu, Y.; Chua, T.S.; Schiele, B. Meta-transfer learning for few-shot learning. In Proceedings of the IEEE Conference on Computer Vision and Pattern Recognition, Long Beach, CA, USA, 16–20 June 2019; pp. 403–412.
50. Jamal, M.A.; Qi, G.J. Task agnostic meta-learning for few-shot learning. In Proceedings of the IEEE Conference on Computer Vision and Pattern Recognition, Long Beach, CA, USA, 16–20 June 2019; pp. 11719–11727.
51. Rusu, A.A.; Rao, D.; Sygnowski, J.; Vinyals, O.; Pascanu, R.; Osindero, S.; Hadsell, R. Meta-learning with latent embedding optimization. *arXiv* **2018**, arXiv:1807.05960.
52. Li, Z.; Zhou, F.; Chen, F.; Li, H. Meta-sgd: Learning to learn quickly for few-shot learning. *arXiv* **2017**, arXiv:1707.09835.
53. Gupta, A.; Thadani, K.; O'Hare, N. Effective few-shot classification with transfer learning. In Proceedings of the 28th International Conference on Computational Linguistics, Barcelona, Spain, 8–13 December 2020; pp. 1061–1066.
54. Dhillon, G.S.; Chaudhari, P.; Ravichandran, A.; Soatto, S. A baseline for few-shot image classification. *arXiv* **2019**, arXiv:1909.02729.
55. Chen, W.Y.; Liu, Y.C.; Kira, Z.; Wang, Y.C.F.; Huang, J.B. A closer look at few-shot classification. In Proceedings of the International Conference on Learning Representations, New Orleans, LA, USA, 6–9 May 2019.
56. Liu, Y.; Lee, J.; Park, M.; Kim, S.; Yang, E.; Hwang, S.J.; Yang, Y. Learning to propagate labels: Transductive propagation network for few-shot learning. *arXiv* **2018**, arXiv:1805.10002.
57. Yang, L.; Li, L.; Zhang, Z.; Zhou, X.; Zhou, E.; Liu, Y. Dpgn: Distribution propagation graph network for few-shot learning. In Proceedings of the IEEE Conference on Computer Vision and Pattern Recognition, Seattle, WA, USA, 13–19 June 2020; pp. 13390–13399.
58. Li, W.; Wang, L.; Xu, J.; Huo, J.; Gao, Y.; Luo, J. Revisiting local descriptor based image-to-class measure for few-shot learning. In Proceedings of the IEEE Conference on Computer Vision and Pattern Recognition, Long Beach, CA, USA, 15–20 June 2019; pp. 7260–7268.
59. Yang, Y.; Newsam, S. Bag-of-visual-words and spatial extensions for land-use classification. In Proceedings of the 18th SIGSPATIAL International Conference on Advances in Geographic Information Systems, San Jose, CA, USA, 2–5 November 2010; pp. 270–279.
60. Sheng, G.; Yang, W.; Xu, T.; Sun, H. High-resolution satellite scene classification using a sparse coding based multiple feature combination. *Int. J. Remote Sens.* **2012**, *33*, 2395–2412. [CrossRef]
61. Cheng, G.; Han, J.; Lu, X. Remote sensing image scene classification: Benchmark and state of the art. *Proc. IEEE* **2017**, *105*, 1865–1883. [CrossRef]
62. Ravi, S.; Larochelle, H. Optimization as a model for few-shot learning. In Proceedings of the ICLR, Toulon, France, 24–26 April 2017.

remote sensing

MDPI

Article

Positive Unlabelled Learning for Satellite Images' Time Series Analysis: An Application to Cereal and Forest Mapping

Johann Desloires [1,2,*], Dino Ienco [1], Antoine Botrel [2] and Nicolas Ranc [2]

1 INRAE, UMR TETIS, University of Montpellier, 34000 Montpellier, France; dino.ienco@inrae.fr
2 Syngenta Seeds, 31790 Saint-Sauveur, France; antoine.botrel@syngenta.com (A.B.); nicolas.ranc@syngenta.com (N.R.)
* Correspondence: johann.desloires@syngenta.com

Abstract: Applications in which researchers aim to extract a single land type from remotely sensed data are quite common in practical scenarios: extract the urban footprint to make connections with socio-economic factors; map the forest extent to subsequently retrieve biophysical variables and detect a particular crop type to successively calibrate and deploy yield prediction models. In this scenario, the (positive) targeted class is well defined, while the negative class is difficult to describe. This one-class classification setting is also referred to as positive unlabelled learning (PUL) in the general field of machine learning. To deal with this challenging setting, when satellite image time series data are available, we propose a new framework named positive and unlabelled learning of satellite image time series (PUL-SITS). PUL-SITS involves two different stages: In the first one, a recurrent neural network autoencoder is trained to reconstruct only positive samples with the aim to higight reliable negative ones. In the second stage, both labelled and unlabelled samples are exploited in a semi-supervised manner to build the final binary classification model. To assess the quality of our approach, experiments were carried out on a real-world benchmark, namely *Haute-Garonne*, located in the southwest area of France. From this study site, we considered two different scenarios: a first one in which the process has the objective to map *Cereals/Oilseeds* cover versus the rest of the land cover classes and a second one in which the class of interest is the *Forest* land cover. The evaluation was carried out by comparing the proposed approach with recent competitors to deal with the considered positive and unlabelled learning scenarios.

Keywords: land cover mapping; positive unlabelled learning; satellite image time series; deep learning

Citation: Desloires, J.; Ienco, D.; Botrel, A.; Ranc, N. Positive Unlabelled Learning for Satellite Images'Time Series Analysis: An Application to Cereal and Forest Mapping. *Remote Sens.* **2022**, *14*, 140. https://doi.org/10.3390/rs14010140

Academic Editor: Parth Sarathi Roy

Received: 25 October 2021
Accepted: 25 December 2021
Published: 29 December 2021

1. Introduction

Currently, modern remote sensing systems provide image acquisitions describing the Earth's surface at a high spatial resolution and revisit time period. A notable example is the Copernicus programme, which, through the Sentinel-2 mission, provides optical multi-spectral images within the visible and near-infrared regions (electromagnetic spectrum) with a spatial resolution between 10 m and 60 m and a revisit time period of approximately 5 days [1]. Such a stream of information can be profitably organised as satellite image time series (SITS) and can support a wide range of application domains, such as ecology [2], agriculture [3], mobility, health, risk assessment [4], land management planning [5], and forest [6], and natural habitat monitoring [7]. Due to the large panel of applications that can be supported by such data, it constitutes a valuable source of information to monitor the dynamics of the Earth's surface.

One of the main tasks related to SITS data analysis is associated with land cover mapping, where a classification model is learned to make the connection between satellite data (i.e., SITS) and the associated land cover classes [5]. SITS data capture the temporal dynamics exhibited by land cover classes, thus resulting in a more effective discrimination among them [8].

Land cover maps can be the final outcome of the analysis or it can constitutes an intermediate step of a more general pipeline [9]. Examples of the latter case are the mapping of a specific culture versus the rest of the land cover classes to successively calibrate and deploy a yield prediction model [9], distinguish between forest areas and the rest of the land cover classes to subsequently retrieve biophysical variables such as canopy height or biomass volume [10], or extract the settlement footprint detecting urban vs. non-urban land covers to make a connection with socio-economic factors [11]. In these particular applications, the objective is to obtain binary land cover maps where one category (the positive one) is well defined since it is associated with the land cover of interest, while the other category (the negative one) is difficult to describe since it gathers together the rest of the possible land cover classes present in a particular study area. This one-class classification setting is also referred to as positive unlabelled learning (PUL) [12] in the general field of machine learning. Different from the standard supervised classification context, in the positive unlabelled learning setting, training data are composed of a set of positive samples and a set of unlabelled ones with the latter involving both positive and negative samples. The objective of a PUL framework is to build a binary classification model that can make predictions on unseen data distinguishing between positive and negative samples.

With the aim to focus on the analysis of SITS data, in this work, we propose a new framework, named PUL-SITS, to deal with positive and unlabelled learning for land cover mapping from satellite image time series. PUL-SITS involves two different stages: In the first one, a recurrent neural network autoencoder is trained only on positive samples. Successively, the same autoencoder model is employed to filter out reliable negative samples from the unlabelled ones. In the second stage, both labelled (positive and reliable negative) and unlabelled samples are exploited in a semi-supervised manner to build the final binary classification model. In this stage, the unlabelled samples are employed to force consistency between the prediction of the classification model and a variant of the same network trained on the SITS data reconstructed by the recurrent neural network autoencoder model.

To assess the effectiveness of PUL-SITS, we considered two different scenarios involving data coming from a particular study area (*Haute-Garonne*) located in the southwest area of France. More precisely, the *Haute-Garonne* study area is heavily characterized by the *Cereals/Oilseeds* and *Forest* land cover classes. For this reason, we designed a first scenario in which the *Cereals/Oilseeds* land cover class has the role of the positive class, while the rest of the land cover classes play the role of the negative class. In the second scenario, the *Forest* land cover class is targeted as the positive class. Experiments were carried out comparing the proposed approach w.r.t. well-suited competitors to deal with the considered positive and unlabelled learning scenarios. The remainder of this paper is structured as follows: First, Section 2 reviews related work in literature at the intersection of positive and unlabelled learning and remote sensing. The proposed framework is described in Section 3. The study site and the associated data are presented in Section 4. The experimental settings, as well as the evaluation are reported and discussed in Section 5. Finally, Section 6 draws the conclusion.

2. Related Work

Regarding the remote sensing domain, Reference [13] investigated the use of positive and unlabelled learning to cope with urban footprint mapping considering high-spatial-resolution aerial photographs. The PUL algorithm trains a classifier on positive and unlabelled data, estimating the probability that a positive training sample has been labelled, and finally, it generates binary predictions for test samples using an adjusted threshold. Reference [14] proposed an ensemble approach to deal with one-class classification under a PUL setting. Here, the class of interest was still the urban land cover, and the approach was evaluated considering different sensors (Landsat 8, WorldView-3, and Gaofen-1). Similar to [13], also in this case, the classification method was deployed on single-date images. In [15], the authors proposed to employ positive and unlabelled learning to characterize landslide susceptibility from several features extracted from multi-source data (elevation, slope angle, curvature, slope aspect, normalized difference vegetation index, soil type, loess erosion intensity, and topographic wetness index). Recently, Reference [16] evaluated PUL-based classifiers on different land cover classes (i.e., trees, bare soil), highlighting that the proposed strategy is well adapted for a plethora of different application contexts.

Unfortunately, despite the fact that the positive unlabelled learning setting naturally fits a wide range of remote sensing applications in which researchers aim to extract a binary land cover map with a focus on a particular class of interest, all the proposed strategies were developed considering mono-date images [13,14,16], while, today, SITS data are ubiquitous in the remote sensing community. According to our literature survey, no research study has yet proposed dealing with SITS data, explicitly with the temporal dynamics carried out by such a rich source of information, under the positive and unlabelled learning setting.

3. Method

In this section, we briefly introduce the positive and unlabelled learning setting, and successively, we describe PUL-SITS, a new framework to deal with land cover mapping from satellite image time series under the PUL setting.

We remind the reader that we considered satellite image time series at the pixel level as the unit of analysis. We refer to a generic SITS (at the pixel level) as $TS = (x_1, \ldots, x_t, \ldots, x_T)$ where T is the length of the time series and a generic element x_t is a multi-variate (or multidimensional) vector with the subscript t, the corresponding timestamp. The PU learning setting considers a scenario in which we dispose of $D = P \cup U$ a set of samples $\{TS_i\}_{i=1}^{|D|}$ composed by a set P of positive samples and a set U of unlabelled ones. The set U contains both positive and negative samples, but their label information is not accessible. The task of learning from both positive and unlabelled data consists of exploiting both labelled P and unlabelled U samples to learn a binary classification model, allowing the assignment of a binary label (positive or negative) to new, previously unseen, satellite image time series.

Regarding PUL-SITS, our framework is constituted of two different stages. The first stage is devoted to selecting a set of reliable negative samples that highly differ from the positive set. This step is addressed via a variational recurrent neural network autoencoder (VAERNN), which is used to model the set of positive samples.

In the second stage, PUL-SITS learns a satellite image time series classifier, based on the recurrent neural network (RNN) model proposed in [17], to cope with the underlying binary classification task. In this stage, a semi-supervised training procedure is designed in order to take advantage of both labelled (positive and reliable negative) and unlabelled samples when the binary classification model is trained.

3.1. Selection of Reliable Negative Samples

The first stage of our framework addresses the identification of a set of reliable negative (RN) samples. A sketch of this procedure is depicted in Figure 1. To this end, we first modelled the positive set of satellite image time series (at the pixel level) via a variational RNN autoencoder (VAERNN), then we used the trained neural network to select reliable negative RN samples from the set of unlabelled satellite image time series data. To this end, the reconstruction error was used as the evaluation metric. Samples with a small reconstruction error are likely to belong to the positive class, while samples on which the reconstruction error is high are probably those representing the negative class. The use of recurrent neural networks (RNNs) to model the data distribution naturally arises due to the fact that such a model is especially tailored to managing multi-variate sequential data [18] and, in particular, compressing sequential information to extract useful multi-variate time series representations [19,20].

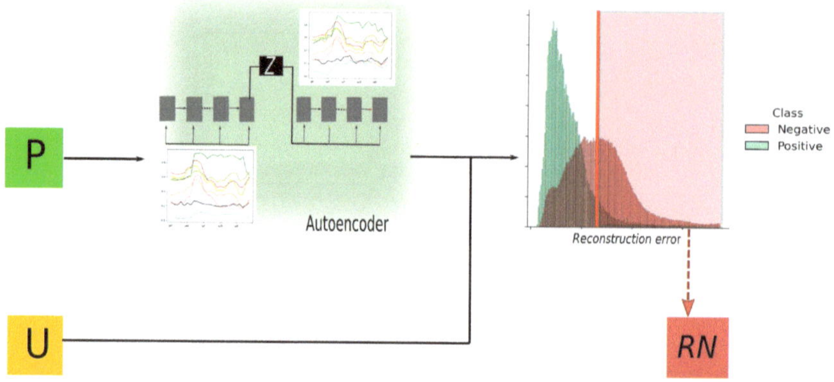

Figure 1. The VAERNN is trained on the set of positively labelled samples P. Then, the model is applied on the set of unlabelled samples U. Successively, the reconstruction error for each $TS \in U$ is derived, and the set of reliable negative samples RN is identified based on the reconstruction error. The vertical red line indicates the average reconstruction error (μ_{rec}) on the set of unlabelled samples U. Samples with a reconstruction error bigger than μ_{rec} are possible candidates for the reliable negative set RN.

In our case, we used a variational autoencoder [21] based on a two-level gated recurrent unit [22]. Gated recurrent units are preferred to long-short term memory [23] since the former model has demonstrated better performance than the latter one in several remote-sensing-related applications [24,25]. The gated recurrent units have fewer parameters than the LSTM cells, and this makes the learned cell more robust and easy to train; finally, GRU cells have also exhibited good performance for reconstruction tasks and, more generally, for unsupervised analysis [20,21].

The gated recurrent unit (GRU) is formally defined as follows:

$$z_t = \sigma(W_{zx}x_t + W_{zh}h_{t-1} + b_z) \tag{1}$$
$$r_t = \sigma(W_{rx}x_t + W_{rh}h_{t-1} + b_r) \tag{2}$$
$$h_t = z_t \odot h_{t-1} + (1 - z_t) \odot \tanh(W_{hx}x_t + W_{hr}(r_t \odot h_{t-1}) + b_h) \tag{3}$$

The $\odot$ symbol indicates an elementwise multiplication, while σ and tanh represent the sigmoid and hyperbolic tangent functions, respectively. x_t is the timestamp input vector, and h_{t-1} is the hidden state of the recurrent unit at time $t-1$. The different weight matrices W_{**} and bias vectors b_* are parameters learned during the training of the model.

This unit follows the general philosophy of modern recurrent neural network models implementing gates and cell states. The *GRU* cell has two gates, update (z_t) and reset (r_t), and one cell state, the hidden state (h_t). Moreover, the two gates combine the current input (x_t) with the information coming from the previous timestamps (h_{t-1}). The update gate effectively controls the tradeoff between how much information from the previous hidden state will carry over to the current hidden state and how much information of the current timestamp needs to be kept. On the other hand, the reset gate monitors how much information of the previous timestamps needs to be integrated with the current information. As each hidden cell has separate reset and update gates, they are able to capture dependencies over different time scales. Cells more prone to capturing short-term dependencies will tend to have a frequently activated reset gate, but those that capture longer-term dependencies will have update gates that remain mostly active [22]. Figure 2 visually depicts the GRU cell.

The encoder involves two GRU cells, while the decoder is symmetric to the encoder. Variational autoencoders, as opposed to standard ones, model the latent space to match a Gaussian distribution [26] (mean and variance). Such a way of compressing the input information can be more appropriate to manage complex data distributions such as in the case of satellite image time series, as recently highlighted in [27].

Furthermore, for an element at timestamps t output by the autoencoder, we performed an additional linear transformation to obtain the reconstructed element $\hat{x_t}$, since, considering regression tasks, the softmax layer is commonly replaced by a fully connected layer with no activation (linear transformation) [28]. Finally, the variational autoencoder model was trained end-to-end by means of a reconstruction loss function on the set of positively labelled satellite image time series P:

$$LOSS_{VAERNN} = \frac{\sum_{TS \in P} HLoss(TS, Linear(VAERNN(TS)))}{|P|} \tag{4}$$

where $(Linear(VAERNN(TS)))$ is the variational recurrent neural network autoencoder followed by a linear transformation and HLoss is the Huber loss [28], defined as follows:

$$HLoss(y, \hat{y}) = \begin{cases} \frac{1}{2}(y - \hat{y})^2 & if\, |y - \hat{y}| < \delta \\ \delta|y - \hat{y}| - \frac{1}{2}\delta^2 & otherwise \end{cases} \tag{5}$$

Basically, the Huber loss behaves as the mean squared error (MSE) loss when the error is lower than a threshold δ, while it mimics the mean absolute error (MAE) loss otherwise. In addition, it is also differentiable at 0. The Huber loss combines the good properties from both the MAE and MSE, and it avoids their limitations. As opposed to the MAE, it bypasses large gradient back propagation when the estimated quantity becomes closer to the real value, and different from the MSE loss, it is more robust to outliers. To manage the balance between the two components of the Huber loss function, we relied on the default parameter setting [29] in which the σ value is set to 1.

Once the VAERNN is trained on the set of positive SITS data, it is successively deployed on the set of unlabelled time series U. For each time series $TS \in U$, a reconstruction error is computed via the Huber loss by comparing the original multi-variate time series with the reconstructed one.

Then, we computed the average reconstruction error μ_{rec} (the mean reconstruction error value on the set of unlabelled time series). After that, we considered multi-variate time series (belonging to the unlabelled set) with a reconstruction error greater than μ_{rec}, and we named such a set $U_{\mu_{rec}}$. Finally, we randomly selected elements from $U_{\mu_{rec}}$ to build the set of reliable negative samples RN with a size equal to the size of the positive set P such that $|P| == |RN|$.

3.2. Semi-Supervised Training Procedure

Once both P and RN sets were available, we trained a classification model to deal with the underlying binary classification task. As a classification backbone for our PUL framework, we considered the RNN architecture recently proposed in [17] and referred to as the fully connected gated recurrent unit (FCGRU). We chose such a model due to the interesting performance it has exhibited w.r.t. other state-of-the-art approaches for land cover mapping from SITS data. The FCGRU cell extends the standard GRU cell with two additional fully connected layers that are applied to the input multi-variate time series, with the aim to enrich the initial representation before deploying the GRU cell, as illustrated in Figure 2. More formally, the FCGRU cell is defined as follows:

$$x_{t'} = \tanh(W_2 \tanh(W_1 x_t + b_1) + b_2) \tag{6}$$
$$z_t = \sigma(W_{zx} x_{t'} + W_{zh} h_{t-1} + b_z) \tag{7}$$
$$r_t = \sigma(W_{rx} x_{t'} + W_{rh} h_{t-1} + b_r) \tag{8}$$
$$h_t = z_t \odot h_{t-1} + (1 - z_t) \odot \tanh(W_{hx} x_{t'} + W_{hr}(r_t \odot h_{t-1}) + b_h) \tag{9}$$

where the $\odot$ symbol indicates an elementwise multiplication, while σ and tanh represent the sigmoid and hyperbolic tangent function, respectively. x_t is the timestamp input vector, and h_{t-1} is the hidden state of the recurrent unit at time $t-1$. The different weight matrices W_{**} and bias vectors b_* are the parameters learned during the training of the model.

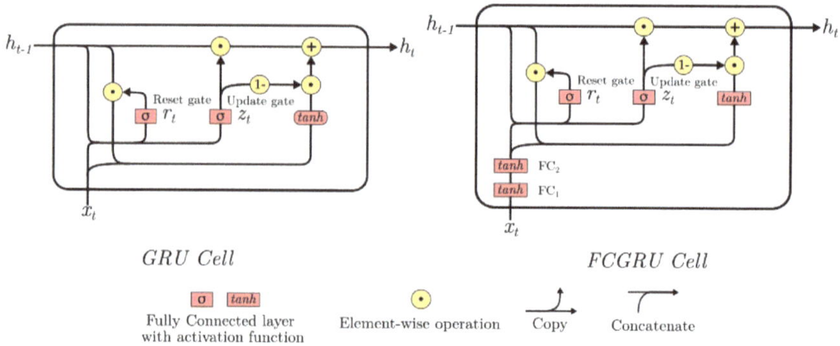

Figure 2. Visual representation of the GRU and FCGRU cells [17].

The classification model is trained via a semi-supervised procedure with the aim to leverage as much as possible the information available considering both labelled and unlabelled samples [30,31]. The procedure is depicted in Figure 3. More precisely, we considered a consistency-regularization-based strategy where the binary classification model is forced to behave similarly to an auxiliary model trained on the perturbed data. Such a consistency regularization strategy was deployed considering a reduced set of unlabelled samples $\tilde{U} = U \setminus U_{\mu_{rec}}$.

We denote by f_θ the binary classification model. When only labelled data P and RN are considered, the model is trained via the standard binary cross-entropy (BCE) loss, defined as:

$$Loss_{BCE}(Y, P \cup RN, f_\theta) = -\frac{1}{|P \cup RN|} \sum_{i=1}^{|P \cup RN|} y_i \log(f_\theta(TS_i)) + (1 - y_i) \log(f_\theta(1 - TS_i)) \tag{10}$$

where $TS_i \in P \cup RN$ is a generic satellite image time series sample with an associated label $y_i \in Y$ and Y is the set of labels.

With the aim to exploit available unlabelled data $\tilde{U}$ in the learning process, we added an extra term to the main loss function as follows:

$$Loss = Loss_{BCE}(Y, P \cup RN, f_\theta) + \lambda \frac{1}{|\tilde{U}|} \sum_{j=1}^{|\tilde{U}|} KL(f_{\hat{\theta}}(\hat{TS_j}) || f_\theta(TS_j)) \tag{11}$$

where $TS_j \in \tilde{U}$ and $f_{\hat{\theta}}$ is an auxiliary classification model that has the same network structure as f_θ, but trained on $\hat{P}$ and $\hat{RN}$, which are the reconstructed (via the VAERNN autoencoder) positive and reliable negative samples, respectively. $KL(X||Z)$ is the standard

Kullback–Leibler divergence, which quantifies how much the distribution Z differs from the distribution X. The second term of the semi-supervised loss function has the goal of leveraging the unlabelled set of SITS data, $\tilde{U}$, regularizing the classification behaviour via consistency assessment [31]. This is achieved by imposing that the output distribution of the f_θ model should behave similarly to the output distribution of a perturbed version of itself, named $f_{\hat{\theta}}$, trained on the set of reconstructed positive and reliable negative samples. Such a strategy implicitly enforces the classifier f_θ to produce similar output distributions considering a specific sample and its perturbed version, thus stressing smoothness on the manifold on which the data are projected by the neural network model [31]. The trade-off hyper-parameter λ is equal to 2.

To implement such a training strategy, at each iteration of an epoch, the classification models f_θ and $f_{\hat{\theta}}$ were trained simultaneously. Firstly, $f_{\hat{\theta}}$ was trained on a batch of samples coming from the perturbed data, and successively, the weight parameters of the f_θ model were updated considering the binary cross-entropy loss on a batch of labelled samples from $P \cup RN$ plus the consistency regularization loss on a batch of unlabelled samples from the set $\tilde{U}$.

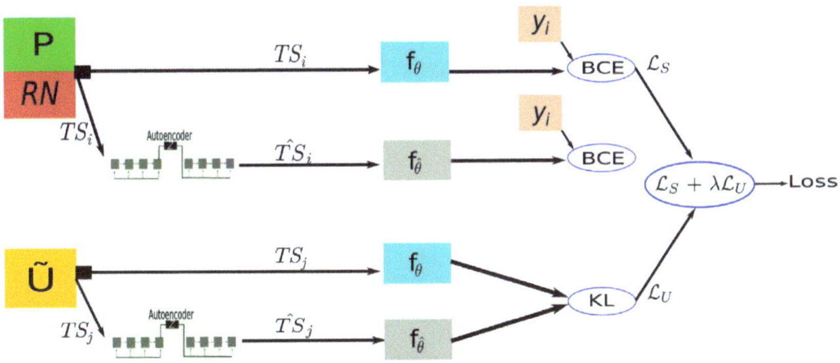

Figure 3. The training strategy involves two models trained simultaneously. The first model $f_{\hat{\theta}}$ is trained on the set of perturbed positive labelled P and reliable negative RN samples by the autoencoder with the supervised BCE loss on true labels y_i. The second f_θ is the final classification model trained on P and RN samples, with a semi-supervised process involving two losses: the binary cross entropy loss on the true labels y_i ($\mathcal{L}_S$) and the KL divergence between the generated soft labels $f_\theta(TS_j)$ and $f_{\hat{\theta}}(\hat{TS}_j)$ to enforce consistency ($\mathcal{L}_U$). The λ hyper-parameter weights the importance of the unsupervised term.

At the end of the training procedure, the classifier f_θ is the model that is employed to perform inference on new unseen samples.

4. Data

The study area we considered in our research study, namely *Haute-Garonne*, is located in the southwest area of France (Figure 4a), and it covers a surface area of 4146.2 km^2. The satellite data describing the study site consist of a Sentinel-2 (S2) time series involving 31 images acquired over the year 2019 from January to December. The temporal distribution of the satellite image time series is depicted in Figure 5.

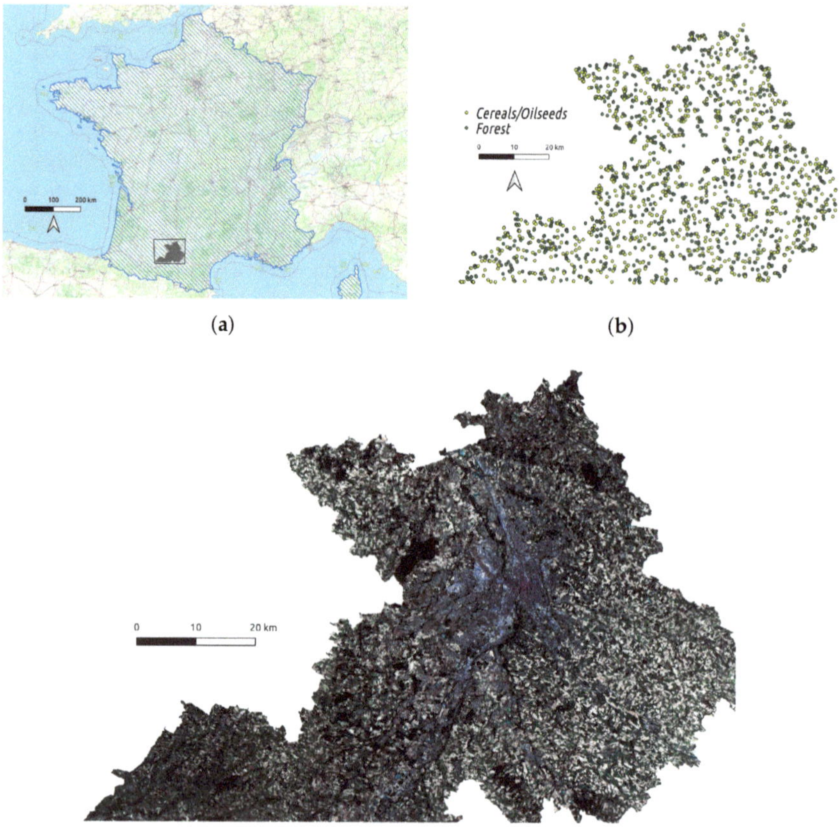

(**a**) (**b**)

(**c**)

Figure 4. *Haute-Garonne* is located in the southwest of France (**a**). The spatial distribution of the land covers (*Cereals/Oilseeds* and *Forest*) considered as positive classes (**b**). The Sentinel-2 RGB composite obtained over the year 2019 (**c**).

Figure 5. Overview of the acquisition dates of Sentinel-2 (S2) images over *Haute-Garonne*. S2 acquisitions are sparsely distributed due to the ubiquitous cloudiness.

The Sentinel-2 images were downloaded from the THEIA pole platform (http://theia.cnes.fr, accessed on 10 November 2020) at level-2A in top-of-canopy reflectance values and provided with cloud masks. Only 10 m spatial resolution bands (blue, green, red, and near-infrared spectrum) were considered in this analysis. Additionally, two common spectral indices were then extracted: the NDVI [32] and the NDWI [33]. A preprocessing step was performed over each band to replace cloudy pixel values as detected by the available cloud masks through a linear multi-temporal interpolation (cf. temporal gap-filling [5]). The entire study site is enclosed in the Sentinel-2 tile T31TCJ.

Table 1. *Haute-Garonne* ground truth characteristics.

Class	# of Objects	# of Pixels
Cereals/Oilseeds	898	336,044
Forest	846	186,870
Other land cover classes	6460	323,925
Total	7358	846,838

Since we are addressing a positive and unlabelled learning setting, we considered two different scenarios where each scenario involves a particular land cover class as the positive class and all the other land cover classes as negative. As positive classes, we considered the *Cereals/Oilseeds* and the *Forest* land covers.

The statistics about the *Cereals/Oilseeds* and *Forest* land cover classes, in the *Haute-Garonne* study site, are reported in Table 1. We also report the statistics for the number of labelled samples belonging to the other land cover classes present in the *Haute-Garonne* study area (i.e., Water, Urban, Orchards, etc.) that are neither *Cereals/Oilseeds* nor *Forest* samples.

The ground truth (GT) was obtained from various sources: (i) the *Registre Parcellaire Graphique* (RPG) reference data for 2019 (https://www.data.gouv.fr/en/datasets, accessed on 10 November 2020), which is part of the European Land Parcel Identification System (LPIS) provided by the French Agency for services and payment ; (ii) the French Topographic Database (BD TOPO) (https://geoservices.ign.fr/blog/2020/10/26/Grand_angle_diffusion_BDTOPO.html, accessed on 10 November 2020), which contains geographical and administrative entities of the national territory; (iii) the French National Forest inventory (BD FORET), which includes information about forest cover at the national scale. The GT was assembled in a Geographic Information System (GIS) vector file, containing a collection of polygons, each attributed to a land cover category (see Table 1). Finally, the polygons were rasterized at the Sentinel-2 spatial resolution (10 m).

5. Experimental Evaluation

In this section, we describe the experimental evaluation we conducted with the aim to assess the behaviour of our framework. To this end, we firstly compared PUL-SITS with different competing approaches on the study site introduced in Section 4. Secondly, we evaluated the different components on which PUL-SITS is based on through an ablation study. Finally, we visually inspected the land cover maps produced by the different competing approaches.

5.1. Competing Methods and Ablation

To assess the behaviour of PUL-SITS, we considered the following competitors:

- A random forest approach based on the positive unlabelled learning framework introduced in [34]. Here, the authors proposed to exploit probability theory to weight unlabelled samples with the aim to bias the learning stage of standard supervised classification methods. We named such a competitor RF_{PUL};
- A one-class support vector machine approach [35]. In this case, the learning approach discards unlabelled samples, and it only leverages positive samples to build the final binary classification model. We named such a competing approach OCSVM;
- An ensemble-based approach for the positive unlabelled learning scenario recently introduced in the field of remote sensing analysis [14]. The ensemble approach combines the prediction of different base learners via a mean-squared-error-based aggregation function. Our ensemble-based competitor involves the following base learners: logistic ridge regression, logistic elastic-net regression, random forest, multilayer perceptron and linear discriminant analysis. We named such competitors $ENSEMBLE_{PUL}$.

In order to disentangle the contributions of the different stages on which PUL-SITS is based, we considered the following ablations of our framework:

- The FCGRU [17] classification model only trained on the positive and reliable negative samples obtained from the first stage of our framework. This ablation did not involve the consistency loss. We named such an ablation PUL-SITS$_{noReg}$;
- The FCGRU classification model trained on the positive and reliable negative samples obtained from the first stage of our framework. Here, the classifier was not trained on the original data, but on the reconstructed data output by the autoencoder model. We named such an ablation PUL-SITS$_{reco}$.

5.2. Experimental Settings

We designed two different positive unlabelled learning experiments according to the data presented in Section 4. We remind the reader that in the first (resp. second) experiment, we considered as the positive class the *Cereals/Oilseeds* (resp. *Forest*) land cover class, while the negative class was constituted by all the samples not belonging to the land cover class representing the positive class.

With the purpose of setting up a fair evaluation protocol to assess the quality of the different competitors, for each experiment, we divided the data (both positive and negative classes) into two sets: training and test. We considered 50% of the positive (resp. negative) class as the test data. The remaining 50% of the data were considered as the training data. Then, the training set was split again into two parts: the positive and the unlabelled set. While the former contained only positive samples, the latter consisted of samples from both the positive and negative classes. Whereas the amount of positive samples can influence the model behaviour, we considered increasing the quantity of positive samples ranging in the set $\{20, 40, 60, 80, 100\}$ in terms of objects. Once the set of positive samples was fixed, the rest of the training data were considered as unlabelled data. The number of objects and pixels for the positive, unlabelled, and negative samples in the training and test sets for the first (*Cereals/Oilseeds* land cover as the positive class) and second (*Forest* land cover as the positive class) experiments is reported in Table 2 (resp. Table 3).

Table 2. Number of positive, unlabelled, and negative objects/pixels in the training and test set when the positive samples belong to the *Cereals/Oilseeds* land cover class.

# of Objects Training (P/U)	# of Pixels Training (P/U)	# of Objects Test (P/N)	# of Pixels Test (P/N)
20/3659	7010/410,712		
40/3639	14,618/403,104		
60/3619	21,001/403,822	451/3228	168,404/260,711
80/3599	28,955/388,767		
100/3579	36,442/381,280		

Table 3. Number of positive, unlabelled, and negative objects/pixels in the training and test set when the positive samples belong to the *Forest* land cover class.

# of Objects Train (P/U)	# of Pixels Train (P/U)	# of Objects Test (P/N)	# of Pixels Test (P/N)
20/3658	4206/413,517		
40/3638	9565/408,157		
60/3618	13,900/403,822	428/3250	95,725/333,390
80/3598	17,865/399,857		
100/3578	22,143/395,579		

We remind the reader that, despite having defined the size of the positive samples in terms of objects, the classification procedure was performed at the pixel level (all the pixels belonging to the object belong exclusively to the training or the test set). We organized the training/test data split procedures in terms of objects with the aim to avoid possible spatial bias in the evaluation procedure [5], since, in this way, pixels belonging to the same ground truth object were assigned exclusively to one of the data partitions. In addition, for time series data, all values were normalized per band in the interval $[0, 1]$ using 2% (resp. 98%), rather than the minimum (resp. maximum) value to rescale the data [8].

Regarding our framework, PUL-SITS, the VAERNN involves an encoder with two levels with 64 and 16 units, respectively, while the decoder is symmetric to the encoder. The FCGRU [17] recurrent cell involved in the binary classification process has a number of units equal to 32. Both models were trained for 50 epochs with a batch size equal to 32 and a learning rate of $10e^{-4}$ through the Adam optimizer. For the latter component, we also employed a dropout rate equal to 0.2. The same learning procedure was employed for the PUL-SITS$_{noReg}$ and PUL-SITS$_{reco}$ ablations. Regarding the competing methods, RF$_{PUL}$, OCSVM, and ENSEMBLE$_{PUL}$, they were employed with the default settings.

The assessment of the quantitative performances was performed considering the *Accuracy*, *F-Measure*, *Sensitivity*, *Specificity*, and *Kappa* metrics [36]. To reduce the bias induced by the training/test data split, all the results were averaged over ten different random splits. Finally, both the average and standard deviation are reported for each evaluation metric.

The experiments were carried out on a workstation with an AMD Ryzen 7 3700X 8-core processor CPU@3.60GHz with 64 GB of RAM and a GeForce RTX 2080 Ti GPU. All the deep-learning methods were implemented using the Python Tensorflow library.

5.3. *Quantitative Results*

As described before, we compared PUL-SITS with respect to the competing methods on two different experiments: the first one involved the *Cereals/Oilseeds* land cover as the class of interest, while the second experiment considered the *Forest* land cover as the positive class. In both cases, we firstly assessed the performances of the competing strategies in terms of the standard evaluation metrics, then we discuss the associate confusion matrices; finally, we provide a qualitative analysis, inspecting some extracts of the produced land cover maps.

5.3.1. Experiments with the *Cereals/Oilseeds* Land Cover as the Positive Class

Tables 4–8 summarise the results in terms of the F-Measure, Accuracy, Sensitivity, Specificity, and Kappa measure, respectively. Generally, we can observe that no matter the size of the training data, our framework always outperformed all the competing approaches considering almost all the evaluation metrics. We can also note that all the approaches exhibited better behaviours as the amount of available positive labelled data increased.

Regarding the competing methods, we can observe that OCSVM achieved the best performance. Such a method, as opposed to RF$_{PUL}$ and ENSEMBLE$_{PUL}$, only exploited the positive samples to build its classification model. This fact underlines that RF$_{PUL}$ and ENSEMBLE$_{PUL}$ are probably impacted by the fact that the negative class (all the land cover classes except *Cereals/Oilseeds*) is characterised by a high intra-heterogeneity, thus influencing the ability of the algorithm to model such a wide range of sample as one class in the original feature space. This fact is illustrated by high Sensitivity (Table 6) and low Specificity (Table 7) values, which highlight that RF$_{PUL}$ and ENSEMBLE$_{PUL}$ simply label the majority of the data points with the negative class. Concerning the ablations associated with our framework, PUL-SITS$_{noReg}$ and PUL-SITS$_{reco}$, we can see that both strategies achieved better performances than the competing methods, with PUL-SITS$_{noReg}$ as a clear winner between them, but the PUL-SITS$_{reco}$ component played an important role in supporting the performance of PUL-SITS, underlying the quality of our two-stage methodology. Figure 6 shows the confusion matrices associated with the three direct competing approaches, as well as PUL-SITS. A closer look at these statistics confirms that PUL-SITS was more precise than the competitors. This consideration emerges from the fact that the corresponding heat map (Figure 6d) has a more visible diagonal structure (the dark green blocks concentrated on the diagonal). This is not the case for the other methods, where the majority of the samples are associated with the negative (zero) class.

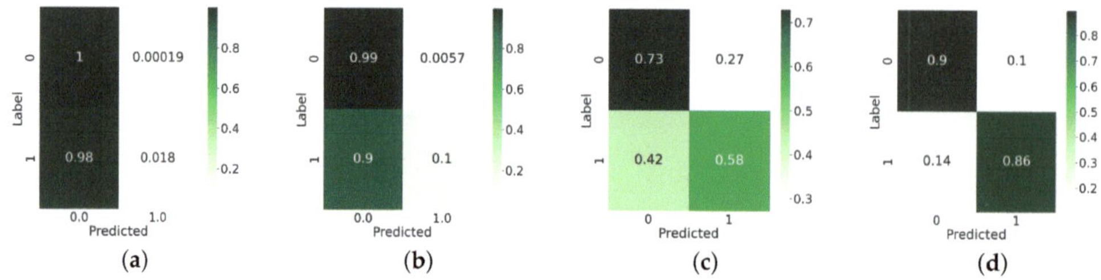

Figure 6. Confusion matrices of (**a**) RF$_{PUL}$, (**b**) ENSEMBLE$_{PUL}$, (**c**) OCSVM, and (**d**) PUL-SITS considering the case in which the land cover class of interest (the positive class) is *Cereals/Oilseeds* with the number of positive training data objects (n) equal to 80.

Table 4. Performances in terms of the F-Measure of the different competing strategies when the land cover class of interest (the positive class) is *Cereals/Oilseeds*.

# of Obj Samples	RF$_{PUL}$	ENSEMBLE$_{PUL}$	OCSVM	PUL-SITS$_{noReg}$	PUL-SITS$_{reco}$	PUL-SITS
20	46.2 ± 1.7	47.3 ± 1.5	61.6 ± 4.0	**80.3** ± 2.3	75.5 ± 3.5	78.9 ± 4.0
40	46.7 ± 1.7	49.3 ± 1.7	63.8 ± 2.7	84.9 ± 1.9	82.0 ± 2.3	**85.3** ± 2.8
60	47.2 ± 1.8	52.4 ± 1.6	64.7 ± 2.0	86.6 ± 2.8	84.2 ± 2.0	**87.3** ± 2.2
80	48.3 ± 2.0	55.7 ± 2.1	64.7 ± 2.3	86.5 ± 2.3	85.0 ± 2.5	**88.0** ± 2.6
100	50.2 ± 1.6	60.1 ± 1.8	64.3 ± 2.0	86.2 ± 3.0	85.5 ± 1.8	**88.6** ± 1.3

Table 5. Performances in terms of the Accuracy of the different competing strategies when the land cover class of interest (the positive class) is *Cereals/Oilseeds*.

# of Obj Samples	RF$_{PUL}$	ENSEMBLE$_{PUL}$	OCSVM	PUL-SITS$_{noReg}$	PUL-SITS$_{reco}$	PUL-SITS
20	60.9 ± 1.4	61.3 ± 1.3	63.9 ± 2.1	**80.9** ± 2.0	76.3 ± 3.1	79.5 ± 3.6
40	61.1 ± 1.4	62.1 ± 1.2	65.1 ± 2.2	85.0 ± 1.9	82.3 ± 2.2	**85.4** ± 2.8
60	61.3 ± 1.4	63.5 ± 1.2	66.0 ± 1.6	86.6 ± 2.8	84.2 ± 2.0	**87.3** ± 2.1
80	61.8 ± 1.5	65.1 ± 1.3	65.4 ± 1.9	86.5 ± 2.3	85.1 ± 2.4	**87.9** ± 2.5
100	62.6 ± 1.3	67.5 ± 1.2	65.2 ± 1.7	86.1 ± 2.9	85.5 ± 1.8	**88.6** ± 1.3

Table 6. Performances in terms of the Sensitivity of the different competing strategies when the land cover class of interest (the positive class) is *Cereals/Oilseeds*.

# of Obj Samples	RF$_{PUL}$	ENSEMBLE$_{PUL}$	OCSVM	PUL-SITS$_{noReg}$	PUL-SITS$_{reco}$	PUL-SITS
20	**100** ± 0.0	99.8 ± 0.1	81.5 ± 5.0	91.7 ± 4.1	88.1 ± 3.5	88.6 ± 8.9
40	**100** ± 0.0	99.6 ± 0.2	79.6 ± 3.9	88.4 ± 2.8	89.0 ± 1.8	90.3 ± 2.0
60	**100** ± 0.0	99.3 ± 0.3	77.3 ± 3.8	86.7 ± 5.8	87.3 ± 3.6	88.3 ± 2.5
80	**100** ± 0.0	99.1 ± 0.2	77.0 ± 3.6	86.6 ± 5.3	88.9 ± 2.5	88.7 ± 2.0
100	**99.9** ± 0.1	98.9 ± 0.2	77.7 ± 3.2	85.1 ± 5.9	87.4 ± 2.8	89.4 ± 2.0

Table 7. Performances in term of the Specificity of the different competing strategies when the land cover class of interest (the positive class) is *Cereals/Oilseeds*.

# of Obj Samples	RF_{PUL}	$ENSEMBLE_{PUL}$	OCSVM	PUL-SITS$_{noReg}$	PUL-SITS$_{reco}$	PUL-SITS
20	0.3 ± 0.3	1.6 ± 0.8	36.9 ± 11.2	65.3 ± 9.8	59.3 ± 10.0	**66.6** ± 13.6
40	0.9 ± 0.6	4.0 ± 2.2	42.7 ± 7.7	**80.0** ± 4.7	72.4 ± 5.2	78.2 ± 5.3
60	1.4 ± 0.7	8.1 ± 2.1	47.4 ± 6.8	85.9 ± 2.9	79.6 ± 5.0	**85.9** ± 5.0
80	2.7 ± 0.9	12.4 ± 3.0	47.7 ± 6.9	86.1 ± 4.2	79.4 ± 5.5	**86.8** ± 4.5
100	5.0 ± 1.2	18.8 ± 2.8	45.9 ± 6.0	**87.6** ± 3.1	82.7 ± 3.7	87.3 ± 4.3

Table 8. Performances in term of the Kappa measure of the different competing strategies when the land cover class of interest (the positive class) is *Cereals/Oilseeds*.

# of Obj Samples	RF_{PUL}	$ENSEMBLE_{PUL}$	OCSVM	PUL-SITS$_{noReg}$	PUL-SITS$_{reco}$	PUL-SITS
20	0.00 ± 0.00	0.02 ± 0.01	0.19 ± 0.07	**0.59** ± 0.05	0.49 ± 0.07	0.56 ± 0.07
40	0.01 ± 0.01	0.04 ± 0.02	0.23 ± 0.05	0.68 ± 0.04	0.62 ± 0.05	**0.69** ± 0.06
60	0.02 ± 0.01	0.09 ± 0.02	0.25 ± 0.04	0.72 ± 0.05	0.67 ± 0.04	**0.74** ± 0.04
80	0.03 ± 0.01	0.14 ± 0.03	0.25 ± 0.04	0.72 ± 0.04	0.69 ± 0.05	**0.75** ± 0.05
100	0.06 ± 0.01	0.21 ± 0.03	0.24 ± 0.04	0.72 ± 0.05	0.70 ± 0.03	**0.76** ± 0.03

5.3.2. Experiments with the *Forest* Land Cover as the Positive Class

Tables 9–13 resume respectively the results in terms of the F-Measure, Accuracy, Sensitivity, Specificity, and Kappa of the different competing methods when the *Forest* land cover was considered as the positive class. In this case also, PUL-SITS outperformed all the other approaches, while RF_{PUL} and $ENSEMBLE_{PUL}$ simply labelled all of the data points as the negative class (as highlighted by the high values of the Sensitivity and the low values of the Sensibility they achieved). OCSVM exhibited the best performance among the state-of-the-art competitors. Concerning the ablations, PUL-SITS$_{noReg}$ did not always achieve better performance over OCSVM, while PUL-SITS$_{reco}$ did. This phenomena clearly underlines the importance of exploiting the available unlabelled data and the appropriateness of the consistency-based procedure in order to regularize the training stage of the final classification model. Figure 7 shows the confusion matrices, which confirm that PUL-SITS achieved a better predictive performance w.r.t. the other approaches, i.e., it has a more visible diagonal structure.

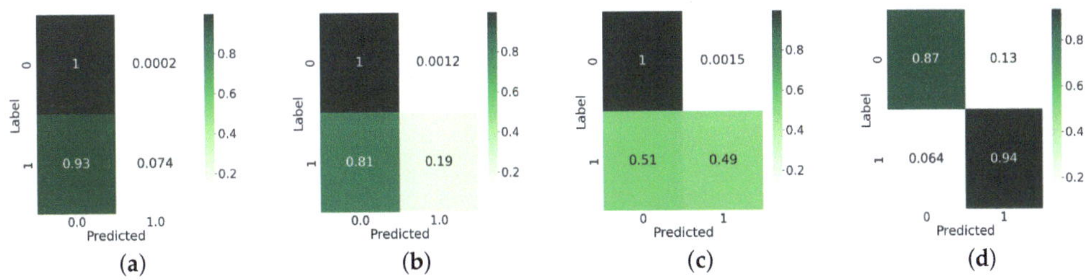

Figure 7. Confusion matrices of (**a**) RF_{PUL}, (**b**) $ENSEMBLE_{PUL}$, (**c**) OCSVM, and (**d**) PUL-SITS considering the case in which the land cover class of interest (the positive class) is *Forest* with the number of positive training objects (n) equal to 80.

Table 9. Performances in terms of the F-Measure of the different competing strategies when the land cover class of interest (the positive class) is *Forest*.

# of Obj Samples	RF_{PUL}	$ENSEMBLE_{PUL}$	OCSVM	PUL-SITS$_{noReg}$	PUL-SITS$_{reco}$	PUL-SITS
20	68.4 ± 2.7	69.8 ± 2.4	82.7 ± 3.0	86.1 ± 1.8	89.7 ± 2.0	**91.4** ± 1.9
40	69.7 ± 2.7	72.8 ± 3.1	83.1 ± 2.6	85.1 ± 3.4	89.8 ± 3.7	**92.9** ± 2.3
60	71.6 ± 2.9	76.4 ± 3.1	84.4 ± 2.5	86.5 ± 2.5	90.1 ± 2.2	**92.6** ± 2.5
80	73.9 ± 3	79.5 ± 3.1	84.7 ± 2.4	86.6 ± 2.9	88.8 ± 2.9	**91.4** ± 3.5
100	76.8 ± 2.7	83.1 ± 2.8	85.2 ± 2.4	85.0 ± 1.7	88.7 ± 2.8	**91.9** ± 1.8

Table 10. Performances in terms of the Accuracy of the different competing strategies when the land cover class of interest (the positive class) is *Forest*.

# of Obj Samples	RF_{PUL}	$ENSEMBLE_{PUL}$	OCSVM	PUL-SITS$_{noReg}$	PUL-SITS$_{reco}$	PUL-SITS
20	77.9 ± 1.9	78.4 ± 1.8	85.0 ± 2.1	83.2 ± 2.3	87.9 ± 2.6	**90.1** ± 2.7
40	78.4 ± 1.9	79.8 ± 2.1	85.8 ± 1.9	81.9 ± 4.6	88.0 ± 4.8	**92.1** ± 2.8
60	79.3 ± 2.0	81.7 ± 2.1	86.7 ± 1.8	83.8 ± 3.8	88.4 ± 2.9	**91.5** ± 3.1
80	80.4 ± 2.0	83.5 ± 2.2	86.9 ± 1.8	83.7 ± 3.9	86.7 ± 3.7	**90.1** ± 4.3
100	81.9 ± 1.9	85.8 ± 2.1	87.2 ± 1.8	81.8 ± 2.3	86.7 ± 3.7	**90.8** ± 2.2

Table 11. Performances in terms of the Sensitivity of the different competing strategies when the land cover class of interest (the positive class) is *Forest*.

# of Obj Samples	RF_{PUL}	$ENSEMBLE_{PUL}$	OCSVM	PUL-SITS$_{noReg}$	PUL-SITS$_{reco}$	PUL-SITS
20	**100** ± 0.0	99.9 ± 0.1	99.7 ± 0.2	81.6 ± 2.5	87.2 ± 3.0	89.7 ± 2.9
40	**100** ± 0.0	99.8 ± 0.1	99.7 ± 0.3	80.1 ± 5.1	87.1 ± 5.5	91.9 ± 3.3
60	**99.9** ± 0.0	99.8 ± 0.2	99.7 ± 0.2	82.0 ± 3.7	87.6 ± 3.4	91.2 ± 3.6
80	**99.9** ± 0.1	99.8 ± 0.1	99.7 ± 0.2	82.2 ± 4.3	85.7 ± 4.3	89.5 ± 5.0
100	**99.8** ± 0.1	99.7 ± 0.1	99.7 ± 0.1	79.9 ± 2.6	85.6 ± 4.3	90.2 ± 2.6

Table 12. Performances in terms of the Specificity of the different competing strategies when the land cover class of interest (the positive class) is *Forest*.

# of Obj Samples	RF_{PUL}	$ENSEMBLE_{PUL}$	OCSVM	PUL-SITS$_{noReg}$	PUL-SITS$_{reco}$	PUL-SITS
20	0.8 ± 0.4	3.7 ± 1.3	36.0 ± 7.8	**97.3** ± 1.01	94.1 ± 1.5	94.0 ± 1.6
40	3.4 ± 2.4	10.1 ± 3.7	37.2 ± 7.2	**97.6** ± 1.26	95.2 ± 1.8	93.6 ± 2.5
60	7.5 ± 3.2	18.7 ± 5.2	41.1 ± 6.9	**97.9** ± 1.0	95.5 ± 1.7	94.3 ± 2.9
80	12.5 ± 4.2	26.8 ± 6.3	42.0 ± 6.5	**97.9** ± 1.0	96.2 ± 1.5	95.2 ± 2.2
100	19.6 ± 4.7	37.3 ± 6.8	43.6 ± 6.5	**98.3** ± 0.7	96.3 ± 1.4	94.9 ± 2.0

Table 13. Performances in terms of the Kappa measure of the different competing strategies when the land cover class of interest (the positive class) is *Forest*.

# of Obj Samples	RF_{PUL}	$ENSEMBLE_{PUL}$	OCSVM	PUL-SITS$_{noReg}$	PUL-SITS$_{reco}$	PUL-SITS
20	0.01 ± 0.01	0.05 ± 0.02	0.46 ± 0.09	0.46 ± 0.05	0.55 ± 0.07	**0.61** ± 0.08
40	0.05 ± 0.04	0.14 ± 0.05	0.47 ± 0.07	0.44 ± 0.08	0.57 ± 0.13	**0.67** ± 0.09
60	0.11 ± 0.05	0.26 ± 0.07	0.51 ± 0.07	0.47 ± 0.08	0.57 ± 0.08	**0.66** ± 0.1
80	0.18 ± 0.06	0.36 ± 0.08	0.52 ± 0.07	0.48 ± 0.08	0.54 ± 0.08	**0.62** ± 0.1
100	0.27 ± 0.06	0.47 ± 0.08	0.54 ± 0.07	0.44 ± 0.05	0.53 ± 0.07	**0.63** ± 0.05

5.4. Visual Inspection of the Binary Classification Maps

Here, we performed a visual inspection of the binary classification maps generated via the different competing approaches regarding the two scenarios we previously described.

Figures 8 and 9 report the results of the mapping process on two different extracts of the study site when the *Cereals/Oilseeds* land cover class was considered as the positive class. The first extract (Figure 8) depicts an agricultural area mainly characterized by cereal and oilseed parcels. We can observe that PUL-SITS clearly recovered the parcel structure, while the other approaches had serious issues in detecting pixels belonging to the *Cereals/Oilseeds* class. While RF_{PUL} and $ENSEMBLE_{PUL}$ exhibited high false-negative rates, the OCSVM method showed reasonable performance on the considered extract, but still (visually) inferior w.r.t. our framework. A similar behaviour was exhibited by all the competing methods on the second extract depicted in Figure 9. In this case also, PUL-SITS provided a more coherent land cover map that clearly discriminated between positive (*Cereals/Oilseeds*) and negative pixels.

Figures 10 and 11 depict the results of the mapping process on two different extracts of the study site when the *Forest* land cover class was considered as the positive class. The first extract (Figure 8) shows a portion of the study site where clear forest cover is present. Here, we can note that PUL-SITS well identified the large forest area in the bottom left part of the extract. All the other methods were not capable of recovering all the forest extent with, also in this case, issues related to false-negative detection. The second extract (Figure 11) describes a zone on which the forest cover is not concentrated, but, conversely, spread over all the area. Similar to what happened in the previous case, PUL-SITS produced a binary land cover map that was consistent with what we can observe in the RGB composite on the left side of the figure. To sum up, we can state that the visual inspection of the binary land cover maps, considering both scenarios, is coherent with the quantitative results we obtained in the previous stage of the evaluation analysis.

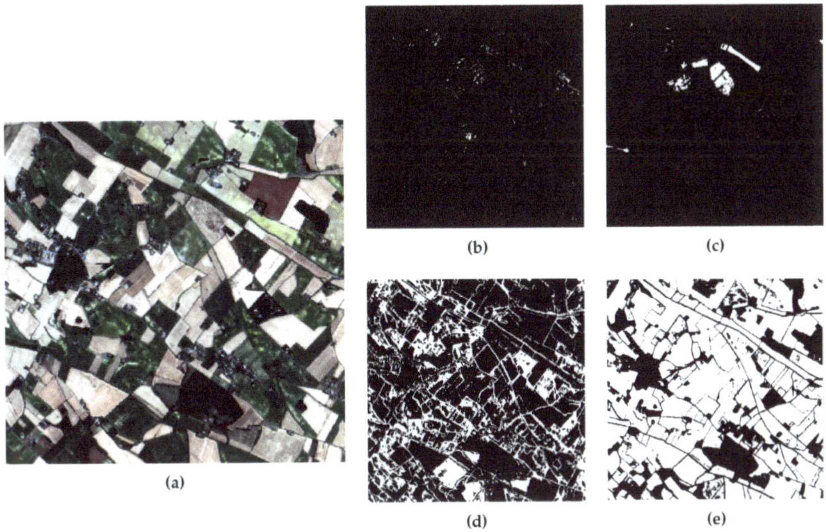

Figure 8. A first visual evaluation zone with (**a**) its RGB composite and the corresponding binary classification maps for (**b**) RF_{PUL}, (**c**) $ENSEMBLE_{PUL}$, (**d**) OCSVM, and (**e**) PUL-SITS for the positive class *Cereals/Oilseeds* with the number of positive training data objects (n) equal to 80.

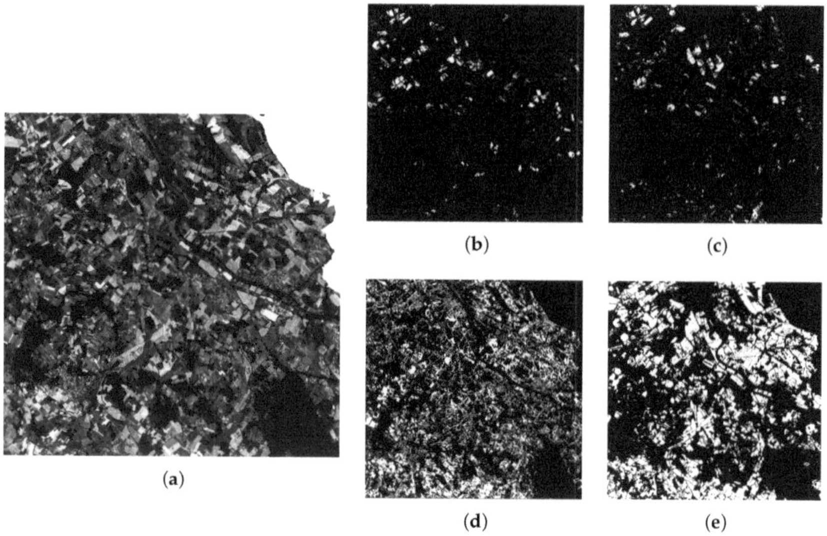

Figure 9. A second visual evaluation zone with (**a**) its RGB composite and the corresponding binary classification maps for (**b**) RF_{PUL}, (**c**) $ENSEMBLE_{PUL}$, (**d**) OCSVM, and (**e**) PUL-SITS for the positive class *Cereals/Oilseeds* with the number of positive training data objects (n) equal to 80.

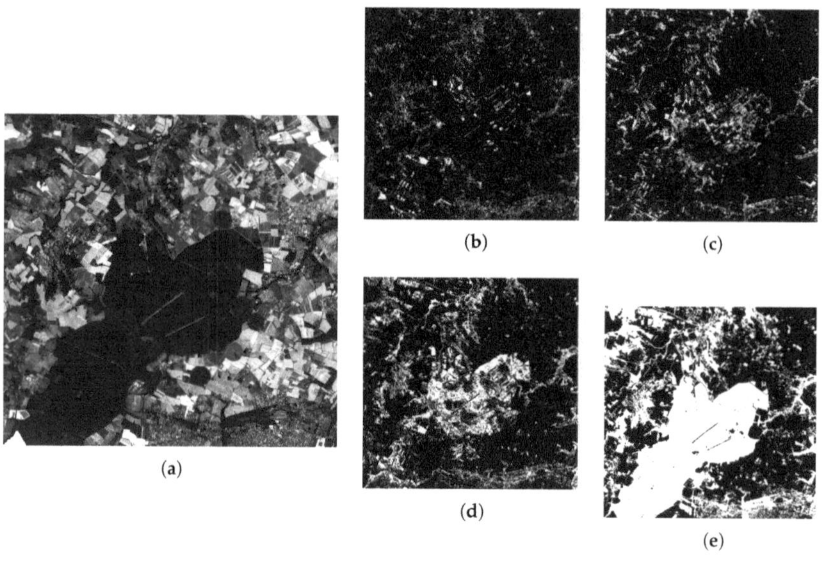

Figure 10. A first visual evaluation zone with (**a**) its RGB composite and the corresponding binary classification maps for (**b**) RF_{PUL}, (**c**) $ENSEMBLE_{PUL}$, (**d**) OCSVM, and (**e**) PUL-SITS for the positive class *Forest* with the number of positive training data objects (n) equal to 80.

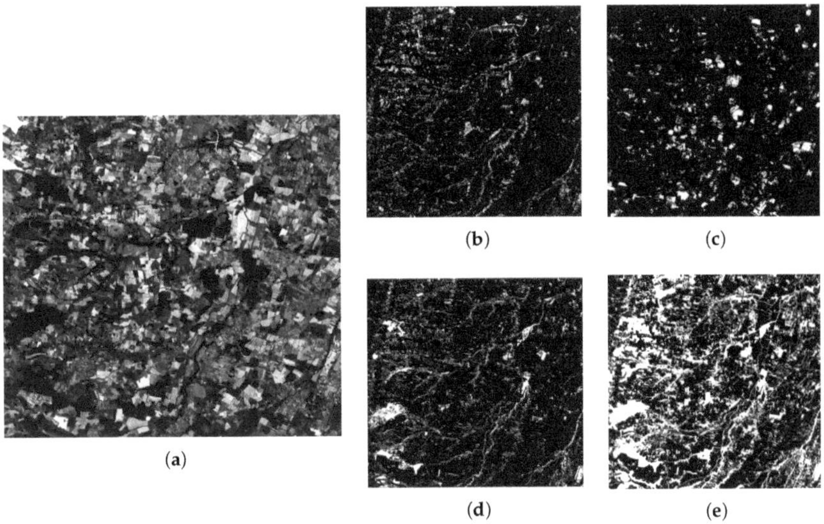

Figure 11. A second visual evaluation zone with (a) its RGB composite and the corresponding binary classification maps for (b) RF_{PUL}, (c) $ENSEMBLE_{PUL}$, (d) OCSVM, and (e) PUL-SITS for the positive class *Forest* with the number of positive training data objects (n) equal to 80.

6. Conclusions

In this research work, we introduced a novel framework, namely PUL-SITS, to deal with the positive unlabelled learning analysis of satellite image time series data for the task of land cover mapping. Our approach was based on a two-step strategy: in the first one, a recurrent neural network autoencoder was trained only on positive samples. Successively, the same autoencoder model was employed to filter out reliable negative samples. In the second stage, both labelled (positive and reliable negative) and unlabelled samples were exploited in a semi-supervised manner to build the final binary classification model. To assess the performances of PUL-SITS, we compared it with state-of-the-art methods for positive unlabelled learning considering two different scenarios: a first one in which the *Cereal/Oilseeds* land cover class has the role of the positive class while the rest of the land cover classes play the role of the negative class and a second scenario where the *Forest* land cover class was targeted as the positive class. Both scenarios underlined the quality of our approach. As future work, we plan to extend our framework towards multi-source analysis and introduce, for instance, satellite image time series of radar sensors (i.e., Sentinel-1) in order to exploit the complementarity between different available sources/modalities.

Author Contributions: D.I. and J.D. conceived and designed the experiments; J.D. performed the experiments; J.D. and D.I. analyzed the results; D.I. and J.D. wrote the article; D.I., A.B. and N.R. revised the paper. All authors have read and agreed to the published version of the manuscript.

Funding: This research was funded by ANRT, grant number CIFRE 2019/1993.

Institutional Review Board Statement: Not applicable.

Informed Consent Statement: Not applicable.

Data Availability Statement: Not applicable.

Acknowledgments: The authors wish also to thank the European Space Agency (ESA) for the Sentinel-2 data and Theia pole for the calibration of the Sentinel-2 images.

Conflicts of Interest: The authors declare no conflict of interest. The funders had no role in the design of the study; in the collection, analyses, or interpretation of data; in the writing of the manuscript, or in the decision to publish the results.

References

1. Berger, M.; Moreno, J.; Johannessen, J.; Levelt, P.; Hanssen, R. ESA's sentinel missions in support of Earth system science Keywords: Sentinel ESA Earth system science Global Monitoring for Environment and Security (GMES). *Remote Sens. Environ.* **2012**, *120*, 84–90. [CrossRef]
2. Chen, L.; Jin, Z.; Michishita, R.; Cai, J.; Yue, T.; Chen, B.; Xu, B. Dynamic monitoring of wetland cover changes using time-series remote sensing imagery. *Ecol. Inform.* **2014**, *24*, 17–26. [CrossRef]
3. Bellón, B.; Bégué, A.; Seen, D.L.; de Almeida, C.A.; Simões, M. A Remote Sensing Approach for Regional-Scale Mapping of Agricultural Land-Use Systems Based on NDVI Time Series. *Remote Sens.* **2017**, *9*, 600. [CrossRef]
4. Olen, S.; Bookhagen, B. Mapping Damage-Affected Areas after Natural Hazard Events Using Sentinel-1 Coherence Time Series. *Remote Sens.* **2018**, *10*, 1272. [CrossRef]
5. Inglada, J.; Vincent, A.; Arias, M.; Tardy, B.; Morin, D.; Rodes, I. Operational High Resolution Land Cover Map Production at the Country Scale Using Satellite Image Time Series. *Remote Sens.* **2017**, *9*, 95. [CrossRef]
6. Wulder, M.A.; Masek, J.G.; Cohen, W.B.; Loveland, T.R.; Woodcock, C.E. Opening the archive: How free data has enabled the science and monitoring promise of Landsat Author links open overlay panel. *Remote Sens. Environ.* **2012**, *122*, 2–10. [CrossRef]
7. Khiali, L.; Ienco, D.; Teisseire, M. Object-oriented satellite image time series analysis using a graph-based representation. *Ecol. Inform.* **2018**, *43*, 52–64. [CrossRef]
8. Pelletier, C.; Webb, G.; Petitjean, F. Temporal Convolutional Neural Network for the Classification of Satellite Image Time Series. *Remote Sens.* **2019**, *11*, 523. [CrossRef]
9. Azzari, G.; Jain, M.; Lobell, D. Towards fine resolution global maps of crop yields: Testing multiple methods and satellites in three countries. *Remote Sens. Environ.* **2017**, *202*, 129–141. [CrossRef]
10. Li, W.; Niu, Z.; Shang, R.; Qin, Y.; Wang, L.; Chen, H. High-resolution mapping of forest canopy height using machine learning by coupling ICESat-2 LiDAR with Sentinel-1, Sentinel-2 and Landsat-8 data. *Int. J. Appl. Earth Obs. Geoinf.* **2020**, *92*, 102163. [CrossRef]
11. Xie, S.M.; Jean, N.; Burke, M.; Lobell, D.B.; Ermon, S. Transfer Learning from Deep Features for Remote Sensing and Poverty Mapping. In Proceedings of the Thirtieth AAAI Conference on Artificial Intelligence, Phoenix, AZ, USA, 12–17 February 2016; pp. 3929–3935.
12. Bekker, J.; Davis, J. Learning from positive and unlabelled data: A survey. *Mach. Learn.* **2020**, *109*, 719–760. [CrossRef]
13. Li, W.; Guo, Q.; Elkan, C. A Positive and Unlabeled Learning Algorithm for One-Class Classification of Remote-Sensing Data. *IEEE Trans. Geosci. Remote Sens.* **2011**, *49*, 717–725. [CrossRef]
14. Ran, L.; Li, W.; Liu, X.; Lu, X.; Li, T.; Guo, Q. An Ensemble of Classifiers Based on Positive and Unlabeled Data in One-Class Remote Sensing Classification. *IEEE J. Sel. Top. Appl. Earth Obs. Remote Sens.* **2018**, *11*, 572–584.
15. Wu, B.; Qiu, W.; Jia, J.; Liu, N. Landslide Susceptibility Modeling Using Bagging-Based Positive-Unlabeled Learning. *IEEE Geosci. Remote Sens. Lett.* **2021**, *18*, 766–770. [CrossRef]
16. Li, W.; Guo, Q.; Elkan, C. One-Class Remote Sensing Classification From Positive and Unlabeled Background Data. *IEEE J. Sel. Top. Appl. Earth Obs. Remote Sens.* **2021**, *14*, 730–746. [CrossRef]
17. Gbodjo, Y.J.E.; Ienco, D.; Leroux, L.; Interdonato, R.; Gaetano, R.; Ndao, B. Object-Based Multi-Temporal and Multi-Source Land Cover Mapping Leveraging Hierarchical Class Relationships. *Remote Sens.* **2020**, *12*, 2814. [CrossRef]
18. Linzen, T.; Dupoux, E.; Goldberg, Y. Assessing the Ability of LSTMs to Learn Syntax-Sensitive Dependencies. *Trans. Assoc. Comput. Linguist.* **2016**, *4*, 521–535. [CrossRef]
19. Kieu, T.; Yang, B.; Guo, C.; Jensen, C.S. Outlier Detection for Time Series with Recurrent Autoencoder Ensembles. In Proceedings of the Twenty-Eighth International Joint Conference on Artificial Intelligence, Macao, China, 10–16 August 2019; pp. 2725–2732.
20. Ienco, D.; Interdonato, R. Deep Multivariate Time Series Embedding Clustering via Attentive-Gated Autoencoder. *Adv. Knowl. Discov. Data Min.* **2020**, *12084*, 318–329.
21. Guo, Y.; Liao, W.; Wang, Q.; Yu, L.; Ji, T.; Li, P. Multidimensional Time Series Anomaly Detection: A GRU-based Gaussian Mixture Variational Autoencoder Approach. In Proceedings of the 10th Asian Conference on Machine Learning, ACML, PMLR, Beijing, China, 14–16 November; Volume 95, pp. 97–112.
22. Cho, K.; van Merrienboer, B.; Gülçehre, Ç.; Bahdanau, D.; Bougares, F.; Schwenk, H.; Bengio, Y. Learning Phrase Representations using RNN Encoder-Decoder for Statistical Machine Translation. *arXiv* **2014**, arXiv:1406.1078.
23. Greff, K.; Srivastava, R.K.; Koutník, J.; Steunebrink, B.R.; Schmidhuber, J. LSTM: A Search Space Odyssey. *IEEE Trans. Neural Netw. Learn. Syst.* **2017**, *28*, 2222–2232. [CrossRef] [PubMed]
24. Ndikumana, E.; Ho Tong Minh, D.; Baghdadi, N.; Courault, D.; Hossard, L. Deep Recurrent Neural Network for Agricultural Classification using multitemporal SAR Sentinel-1 for Camargue, France. *Remote Sens.* **2018**, *10*, 1217. [CrossRef]
25. Bakhti, K.; Larabi, M.E.A. Comparing Deep Recurrent Learning and Convolutional Learning for Multi-Temporal Vegetation Classification. In Proceedings of the IEEE International Geoscience and Remote Sensing Symposium, IGARSS, Brussels, Belgium, 11–16 July 2021, pp. 4392–4395.
26. Doersch, C. Tutorial on Variational Autoencoders. *arXiv* **2016**, arXiv:1606.05908.
27. Sanchez, E.H.; Serrurier, M.; Ortner, M. Learning Disentangled Representations of Satellite Image Time Series. *arXiv* **2019**, arXiv:1903.08863.

28. Lathuilière, S.; Mesejo, P.; Alameda-Pineda, X.; Horaud, R. A Comprehensive Analysis of Deep Regression. *IEEE Trans. Pattern Anal. Mach. Intell.* **2020**, *42*, 2065–2081. [CrossRef]
29. Huber, P.J. Robust Estimation of a Location Parameter. *Ann. Stat.* **1964**, *53*, 73–101. [CrossRef]
30. van Engelen, J.E.; Hoos, H.H. A survey on semi-supervised learning. *Mach. Learn.* **2020**, *109*, 373–440. [CrossRef]
31. Yang, X.; Song, Z.; King, I.; Xu, Z. A Survey on Deep Semi-supervised Learning. *arXiv* **2021**, arXiv:2103.00550.
32. Rouse, J.W.; Hass, R.H.; Schell, J.; Deering, D. Monitoring vegetation systems in the great plains with ERTS. *NASA Spec. Publ.* **1973**, *1*, 309–317.
33. cai Gao, B. NDWI—A normalized difference water index for remote sensing of vegetation liquid water from space. *Remote Sens. Environ.* **1996**, *58*, 257–266.
34. Elkan, C.; Noto, K. Learning Classifiers from Only Positive and Unlabeled Data. In Proceedings of the 14th ACM SIGKDD International Conference on Knowledge Discovery and Data Mining, Las Vegas, NV, USA, 24–27 August 2008; pp. 213–220.
35. Tran, Q.; Li, X.; Duan, H. Efficient performance estimate for one-class support vector machine. *Pattern Recognit. Lett.* **2005**, *26*, 1174–1182. [CrossRef]
36. Tan, P.N.; Steinbach, M.; Kumar, V. *Introduction to Data Mining*; Addison Wesley: Boston, MA, USA, 2005.

MDPI AG
Grosspeteranlage 5
4052 Basel
Switzerland
Tel.: +41 61 683 77 34

MDPI Books Editorial Office
E-mail: books@mdpi.com
www.mdpi.com/books

www.ingramcontent.com/pod-product-compliance
Lightning Source LLC
Chambersburg PA
CBHW041102270926
51850CB00025B/3415
9783725839759